INTRODUCTION TO BIOMEDICAL ENGINEERING

Second Edition

This is a volume in the
ACADEMIC PRESS SERIES IN BIOMEDICAL ENGINEERING

JOSEPH BRONZINO, SERIES EDITOR
Trinity College—Hartford, Connecticut

INTRODUCTION TO BIOMEDICAL ENGINEERING

Second Edition

John D. Enderle
University of Connecticut
Storrs, Connecticut

Susan M. Blanchard
Florida Gulf Coast University
Fort Myers, Florida

Joseph D. Bronzino
Trinity College
Hartford, Connecticut

ELSEVIER
ACADEMIC
PRESS

Amsterdam Boston Heidelberg London New York Oxford
Paris San Diego San Francisco Singapore Sydney Tokyo

Elsevier Academic Press
30 Corporate Drive, Suite 400, Burlington, MA 01803, USA
525 B Street, Suite 1900, San Diego, California 92101-4495, USA
84 Theobald's Road, London WC1X 8RR, UK

This book is printed on acid-free paper. ⊛

Library of Congress Cataloging-in-Publication Data
Introduction to biomedical engineering / edited by John D. Enderle, Joseph
D. Bronzino, and Susan M. Blanchard. —2nd ed.
     p. ;cm.
  Includes biographical references and index.
  ISBN-13: 978-0-12-238662-6   ISBN-10: 0-12-238662-0
    1. Biomedical engineering.
  [DNLM: 1. Biomedical Engineering. QT 36 I615 2005] I. Enderle, John D.
  (John Denis) II. Bronzino, Joseph D., III. Blanchard, Susam M.
  R856.I47 2005
  610′.28—dc22   2004030223

British Library Cataloguing in Publication Data
A catalogue record for this book is available from the British Library

ISBN-13: 978-0-12-238662-6
ISBN-10: 0-12-238662-0

For all information on all Elsevier Academic Press publications visit our Web site at www.books.elsevier.com

Printed in the United States of America
08 09 10 9 8 7 6 5 4

This book is dedicated to our families

CONTENTS

3   ANATOMY AND PHYSIOLOGY   73

4   BIOMECHANICS   127

5   REHABILITATION ENGINEERING AND ASSISTIVE TECHNOLOGY   211

9   BIOMEDICAL SENSORS   505

10   BIOSIGNAL PROCESSING   549

11   BIOELECTRIC PHENOMENA   627

12 PHYSIOLOGICAL MODELING   693

13 GENOMICS AND BIOINFORMATICS   799

14 COMPUTATIONAL CELL BIOLOGY AND COMPLEXITY   833

PREFACE

The purpose of the second edition remains the same as the first edition: that is, to serve as an introduction to and overview of the field of biomedical engineering. Many chapters have undergone major revision from the previous edition with new end-of-chapter problems added. Some chapters were combined and some chapters were eliminated completely, with several new chapters added to reflect changes in the field.

Over the past fifty years, as the discipline of biomedical engineering has evolved, it has become clear that it is a diverse, seemingly all-encompassing field that includes such areas as bioelectric phenomena, bioinformatics, biomaterials, biomechanics, bioinstrumentation, biosensors, biosignal processing, biotechnology, computational biology and complexity, genomics, medical imaging, optics and lasers, radiation imaging, rehabilitation engineering, tissue engineering, and moral and ethical issues. Although it is not possible to cover all of the biomedical engineering domains in this textbook, we have made an effort to focus on most of the major fields of activity in which biomedical engineers are engaged.

The text is written primarily for engineering students who have completed differential equations and a basic course in statics. Students in their sophomore year or junior year should be adequately prepared for this textbook. Students in the biological sciences, including those in the fields of medicine and nursing, can also read and understand this material if they have the appropriate mathematical background.

Although we do attempt to be fairly rigorous with our discussions and proofs, our ultimate aim is to help students grasp the nature of biomedical engineering. Therefore, we have compromised when necessary and have occasionally used less rigorous mathematics in order to be more understandable. A liberal use of illustrative examples amplifies concepts and develops problem-solving skills. Throughout the text, MATLAB® (a matrix equation solver) and SIMULINK® (an extension to MATLAB®

for simulating dynamic systems) are used as computer tools to assist with problem solving. The Appendix provides the necessary background to use MATLAB® and SIMULINK®. MATLAB® and SIMULINK® are available from:

The Mathworks, Inc.
24 Prime Park Way
Natick, Massachusetts 01760
Phone: (508) 647-7000
Email: info@mathworks.com
WWW: http://www.mathworks.com {extend}

Chapters are written to provide some historical perspective of the major developments in a specific biomedical engineering domain as well as the fundamental principles that underlie biomedical engineering design, analysis, and modeling procedures in that domain. In addition, examples of some of the problems encountered, as well as the techniques used to solve them, are provided. Selected problems, ranging from simple to difficult, are presented at the end of each chapter in the same general order as covered in the text.

The material in this textbook has been designed for a one-semester, two-semester, or three-quarter sequence depending on the needs and interests of the instructor. Chapter 1 provides necessary background to understand the history and appreciate the field of biomedical engineering. Chapter 2 presents the vitally important chapter on biomedically-based morals and ethics. Basic anatomy and physiology are provided in Chapter 3. Chapters 4-10 provide the basic core biomedical engineering areas: biomechanics, rehabilitation engineering, biomaterials, tissue engineering, bioinstrumentation, biosensors, and biosignal processing. To assist instructors in planning the sequence of material they may wish to emphasize, it is suggested that the chapters on bioinstrumentation, biosensors, and biosignal processing should be covered together as they are interdependent on each other. The remainder of the textbook presents material on biomedical technology (Chapters 12-17).

A website is available at http://intro-bme-book.bme.uconn.edu/ that provides an errata and extra material.

ACKNOWLEDGEMENTS

Many people have helped us in writing this textbook. Well deserved credit is due to the many contributors who provided chapters and worked under a very tight timeline. Special thanks go to our publisher, Elsevier, especially for the tireless work of the editors, Christine Minihane and Shoshanna Grossman. In addition, we appreciate the work of Karen Forster, the project manager, and Kristin Macek, who supervised the production process.

A great debt of gratitude is extended to Joel Claypool, the editor of the first edition of the book and Diane Grossman from Academic Press. From an initial conversation over coffee in Amsterdam in 1996 to publication in 2000 required a huge effort.

CONTRIBUTORS TO THE FIRST EDITION

Susan M. Blanchard
Florida Gulf Coast University
Fort Myers, Florida

Joseph D. Bronzino
Trinity College
Hartford, Connecticut

Stanley A. Brown
Food and Drug Administration
Gaithersburg, Maryland

Gerard Coté
Texas A&M University
College Station, Texas

Roy B. Davis III
Shriners Hospital for Children
Greenville, South Carolina

John D. Enderle
University of Connecticut
Storrs Connecticut

Robert J. Fisher
University of Massachusetts
Amherst, Massachusetts

Carol Lucas
University of North Carolina
Chapel Hill, North Carolina

Amanda Marley
North Carolina State University
Raleigh, North Carolina

Yitzhak Mendelson
Worcester Polytechnic Institute
Worcester, Massachusetts

Katharine Merritt
Food and Drug Administration
Gaithersburg, Maryland

H. Troy Nagle
North Carolina State University
Raleigh, North Carolina

Joseph Palladino
Trinity College
Hartford, Connecticut

Bernhard Palsson
University of California at San Diego
San Diego, California

Sohi Rastegar
National Science Foundation
Arlington, Virginia

Daniel Schneck
Virginia Polytechnic Institute & State University
Blacksburg, Virginia

Kirk K Shung
Pennsylvania State University
University Park, Pennsylvania

Anne-Marie Stomp
North Carolina State University
Raleigh, North Carolina

Andrew Szeto
San Diego State University
San Diego, California

LiHong Wang
Texas A&M University
College Station, Texas

Steven Wright
Texas A&M University
College Station, Texas

Melanie T. Young
North Carolina State University
Raleigh, North Carolina

CONTRIBUTORS TO THE SECOND EDITION

Susan M. Blanchard
Florida Gulf Coast University
Fort Myers, Florida

Joseph D. Bronzino
Trinity College
Hartford, Connecticut

Stanley A. Brown
Food and Drug Administration
Gaithersburg, Maryland

Gerard Coté
Texas A&M University
College Station, Texas

Charles Coward
Drexel University
Philadelphia, Pennsylvania

Roy B. Davis
Shriners Hospital for Children
Greenville, South Carolina

Robert Dennis
University of North Carolina
Chapel Hill, North Carolina

John D. Enderle
University of Connecticut
Storrs, Connecticut

Monty Escabí
University of Connecticut
Storrs, Connecticut

R.J. Fisher
University of Massachusetts
Amherst, Massachusetts

Liisa Kuhn
University of Connecticut Health Center
Farmington, Connecticut

Carol Lucas
University of North Carolina
Chapel Hill, North Carolina

Jeffrey Mac Donald
University of North Carolina
Chapel Hill, North Carolina

Amanda Marley
North Carolina State University
Raleigh, North Carolina

Randall McClelland
University of North Carolina
Chapel Hill, North Carolina

Yitzhak Mendelson
Worcester Polytechnic Institute
Worcester, Massachusetts

Katharine Merritt
Food and Drug Administration
Gaithersburg, Maryland

Spencer Muse
North Carolina State University
Raleigh, North Carolina

H. Troy Nagle
North Carolina State University
Raleigh, North Carolina

Banu Onaral
Drexel University
Philadelphia, Pennsylvania

Joseph Palladino
Trinity College
Hartford, Connecticut

Bernard Palsson
University of California at San Diego
San Diego, California

Sohi Rastegar
National Science Foundation
Arlington, Virginia

Lola Reid
University of North Carolina
Chapel Hill, North Carolina

Kirk K. Shung
Pennsylvania State University
University Park, Pennsylvania

Anne-Marie Stomp
North Carolina State University
Raleigh, North Carolina

Tom Szabo
Boston University
Boston, Massachusetts

Andrew Szeto
San Diego State University
San Diego, California

LiHong Wang
Texas A&M University
College Station, Texas

Stephen Wright
Texas A&M University
College Station, Texas

Melanie T. Young
North Carolina State University
Raleigh, North Carolina

1 BIOMEDICAL ENGINEERING: A HISTORICAL PERSPECTIVE

Joseph Bronzino PhD, PE

Chapter Contents

At the conclusion of this chapter, students will be able to:

- Identify the major role that advances in medical technology have played in the establishment of the modern health care system.
- Define what is meant by the term biomedical engineering and the roles biomedical engineers play in the health care delivery system.
- Explain why biomedical engineers are professionals.

In the industrialized nations, technological innovation has progressed at such an accelerated pace that it is has permeated almost every facet of our lives. This is especially true in the area of medicine and the delivery of health care services. Although the art of medicine has a long history, the evolution of a technologically based health care system capable of providing a wide range of effective diagnostic and therapeutic treatments is a relatively new phenomenon. Of particular importance in this evolutionary process has been the establishment of the modern hospital as the center of a technologically sophisticated health care system.

Since technology has had such a dramatic impact on medical care, engineering professionals have become intimately involved in many medical ventures. As a result, the discipline of biomedical engineering has emerged as an integrating medium for two dynamic professions, medicine and engineering, and has assisted in the struggle against illness and disease by providing tools (such as biosensors, biomaterials, image processing, and artificial intelligence) that can be utilized for research, diagnosis, and treatment by health care professionals.

Thus, biomedical engineers serve as relatively new members of the health care delivery team that seeks new solutions for the difficult problems confronting modern society. The purpose of this chapter is to provide a broad overview of technology's role in shaping our modern health care system, highlight the basic roles biomedical engineers play, and present a view of the professional status of this dynamic field.

1.1   EVOLUTION OF THE MODERN HEALTH CARE SYSTEM

Primitive humans considered diseases to be "visitations," the whimsical acts of affronted gods or spirits. As a result, medical practice was the domain of the witch doctor and the medicine man and medicine woman. Yet even as magic became an integral part of the healing process, the cult and the art of these early practitioners were never entirely limited to the supernatural. These individuals, by using their natural instincts and learning from experience, developed a primitive science based on empirical laws. For example, through acquisition and coding of certain reliable practices, the arts of herb doctoring, bone setting, surgery, and midwifery were advanced. Just as primitive humans learned from observation that certain plants and grains were good to eat and could be cultivated, so the healers and shamans observed the nature of certain illnesses and then passed on their experiences to other generations.

Evidence indicates that the primitive healer took an active, rather than a simply intuitive interest in the curative arts, acting as a surgeon and a user of tools. For instance, skulls with holes made in them by trephiners have been collected in various parts of Europe, Asia, and South America. These holes were cut out of the bone with flint instruments to gain access to the brain. Although one can only speculate the purpose of these early surgical operations, magic and religious beliefs seem to be the most likely reasons. Perhaps this procedure liberated from the skull the malicious demons that were thought to be the cause of extreme pain (as in the case of migraine) or attacks of falling to the ground (as in epilepsy). That this procedure was carried out on living patients, some of whom actually survived, is evident from the rounded edges on the bone surrounding the hole which indicate that the bone had grown again after the operation. These survivors also achieved a special status of sanctity so that, after their death, pieces of their skull were used as amulets to ward off convulsive attacks. From these beginnings, the practice of medicine has become integral to all human societies and cultures.

It is interesting to note the fate of some of the most successful of these early practitioners. The Egyptians, for example, have held Imhotep, the architect of the first pyramid (3000 BC), in great esteem through the centuries, not as a pyramid builder, but as a doctor. Imhotep's name signified "he who cometh in peace" because he visited the sick to give them "peaceful sleep." This early physician practiced his art so well that he was deified in the Egyptian culture as the god of healing.

Egyptian mythology, like primitive religion, emphasized the interrelationships between the supernatural and one's health. For example, consider the mystic sign Rx, which still adorns all prescriptions today. It has a mythical origin in the legend of the Eye of Horus. It appears that as a child Horus lost his vision after being viciously attacked by Seth, the demon of evil. Then Isis, the mother of Horus, called for assistance to Thoth, the most important god of health, who promptly restored the eye and its powers. Because of this intervention, the Eye of Horus became the Egyptian symbol of godly protection and recovery, and its descendant, Rx, serves as the most visible link between ancient and modern medicine.

The concepts and practices of Imhotep and the medical cult he fostered were duly recorded on papyri and stored in ancient tombs. One scroll (dated c. 1500 BC), acquired by George Elbers in 1873, contains hundreds of remedies for numerous afflictions ranging from crocodile bite to constipation. A second famous papyrus (dated c. 1700 BC), discovered by Edwin Smith in 1862, is considered to be the most important and complete treatise on surgery of all antiquity. These writings outline proper diagnoses, prognoses, and treatment in a series of surgical cases. These two papyri are certainly among the outstanding writings in medical history.

As the influence of ancient Egypt spread, Imhotep was identified by the Greeks with their own god of healing, Aesculapius. According to legend, the god Apollo fathered Aesculapius during one of his many earthly visits. Apparently Apollo was a concerned parent, and, as is the case for many modern parents, he wanted his son to be a physician. He made Chiron, the centaur, tutor Aesculapius in the ways of healing. Chiron's student became so proficient as a healer that he soon surpassed his tutor and kept people so healthy that he began to decrease the population of Hades. Pluto, the god of the underworld, complained so violently about this course of events that Zeus killed Aesculapius with a thunderbolt and in the process promoted Aesculapius to Olympus as a god.

Inevitably, mythology has become entangled with historical facts, and it is not certain whether Aesculapius was in fact an earthly physician like Imhotep, the Egyptian. However, one thing is clear; by 1000 BC, medicine was already a highly respected profession. In Greece, the Aesculapia were temples of the healing cult and may be considered among the first hospitals (Fig. 1.1). In modern terms, these temples were essentially sanatoriums that had strong religious overtones. In them, patients

Figure 1.1   Illustration of a sick child brought into the Temple of Aesculapius (Courtesy of http://www.nouveaunet.com/images/art/84.jpg).

were received and psychologically prepared, through prayer and sacrifice, to appreciate the past achievements of Aesculapius and his physician priests. After the appropriate rituals, they were allowed to enjoy "temple sleep." During the night, "healers" visited their patients, administering medical advice to clients who were awake or interpreting dreams of those who had slept. In this way, patients became convinced that they would be cured by following the prescribed regimen of diet, drugs, or bloodletting. On the other hand, if they remained ill, it would be attributed to their lack of faith. With this approach, patients, not treatments, were at fault if they did not get well. This early use of the power of suggestion was effective then and is still important in medical treatment today. The notion of "healthy mind, healthy body" is still in vogue today.

One of the most celebrated of these "healing" temples was on the island of Cos, the birthplace of Hippocrates, who as a youth became acquainted with the curative arts through his father, also a physician. Hippocrates was not so much an innovative physician as a collector of all the remedies and techniques that existed up to that time. Since he viewed the physician as a scientist instead of a priest, Hippocrates also injected an essential ingredient into medicine: its scientific spirit. For him, diagnostic observation and clinical treatment began to replace superstition. Instead of blaming disease on the gods, Hippocrates taught that disease was a natural process, one that developed in logical steps, and that symptoms were reactions of the body to disease. The body itself, he emphasized, possessed its own means of recovery, and the function of the physician was to aid these natural forces. Hippocrates treated each patient as an original case to be studied and documented. His shrewd descriptions of diseases are models for physicians even today. Hippocrates and the school of Cos trained a number of individuals who then migrated to the corners of the Mediterranean world to practice medicine and spread the philosophies of their preceptor. The work of Hippocrates and the school and tradition that stem from him constitute the first real break from magic and mysticism and the foundation of the rational art of medicine. However, as a practitioner, Hippocrates represented the spirit, not the science, of medicine, embodying the good physician: the friend of the patient and the humane expert.

As the Roman Empire reached its zenith and its influence expanded across half the world, it became heir to the great cultures it absorbed, including their medical advances. Although the Romans themselves did little to advance clinical medicine (the treatment of the individual patient), they did make outstanding contributions to public health. For example, they had a well-organized army medical service, which not only accompanied the legions on their various campaigns to provide "first aid" on the battlefield but also established "base hospitals" for convalescents at strategic points throughout the empire. The construction of sewer systems and aqueducts were truly remarkable Roman accomplishments that provided their empire with the medical and social advantages of sanitary living. Insistence on clean drinking water and unadulterated foods affected the control and prevention of epidemics, and however primitive, made urban existence possible. Unfortunately, without adequate scientific knowledge about diseases, all the preoccupation of the Romans with public health could not avert the periodic medical disasters, particularly the plague, that mercilessly befell its citizens.

Initially, the Roman masters looked upon Greek physicians and their art with disfavor. However, as the years passed, the favorable impression these disciples of Hippocrates made upon the people became widespread. As a reward for their service to the peoples of the Empire, Caesar (46 BC) granted Roman citizenship to all Greek practitioners of medicine in his empire. Their new status became so secure that when Rome suffered from famine that same year, these Greek practitioners were the only foreigners not expelled from the city. On the contrary, they were even offered bonuses to stay!

Ironically, Galen, who is considered the greatest physician in the history of Rome, was himself a Greek. Honored by the emperor for curing his "imperial fever," Galen became the medical celebrity of Rome. He was arrogant and a braggart and, unlike Hippocrates, reported only successful cases. Nevertheless, he was a remarkable physician. For Galen, diagnosis became a fine art; in addition to taking care of his own patients, he responded to requests for medical advice from the far reaches of the empire. He was so industrious that he wrote more than 300 books of anatomical observations, which included selected case histories, the drugs he prescribed, and his boasts. His version of human anatomy, however, was misleading because he objected to human dissection and drew his human analogies solely from the studies of animals. However, because he so dominated the medical scene and was later endorsed by the Roman Catholic Church, Galen actually inhibited medical inquiry. His medical views and writings became both the "bible" and "the law" for the pontiffs and pundits of the ensuing Dark Ages.

With the collapse of the Roman Empire, the Church became the repository of knowledge, particularly of all scholarship that had drifted through the centuries into the Mediterranean. This body of information, including medical knowledge, was literally scattered through the monasteries and dispersed among the many orders of the Church.

The teachings of the early Roman Catholic Church and the belief in divine mercy made inquiry into the causes of death unnecessary and even undesirable. Members of the Church regarded curing patients by rational methods as sinful interference with the will of God. The employment of drugs signified a lack of faith by the doctor and patient, and scientific medicine fell into disrepute. Therefore, for almost a thousand years, medical research stagnated. It was not until the Renaissance in the 1500s that any significant progress in the science of medicine occurred. Hippocrates had once taught that illness was not a punishment sent by the gods but a phenomenon of nature. Now, under the Church and a new God, the older views of the supernatural origins of disease were renewed and promulgated. Since disease implied demonic possession, monks and priests treated the sick through prayer, the laying on of hands, exorcism, penances, and exhibition of holy relics—practices officially sanctioned by the Church.

Although deficient in medical knowledge, the Dark Ages were not entirely lacking in charity toward the sick poor. Christian physicians often treated the rich and poor alike, and the Church assumed responsibility for the sick. Furthermore, the evolution of the modern hospital actually began with the advent of Christianity and is considered one of the major contributions of monastic medicine. With the rise in 335 AD of Constantine I, the first of the Roman emperors to embrace Christianity, all pagan temples of healing were closed, and hospitals were established in every cathedral city. [Note: The word hospital comes from the Latin hospes, meaning, "host" or "guest." The same root has provided hotel and hostel.] These first hospitals were simply houses where weary travelers and the sick could find food, lodging, and nursing care. The Church ran these hospitals, and the attending monks and nuns practiced the art of healing.

As the Christian ethic of faith, humanitarianism, and charity spread throughout Europe and then to the Middle East during the Crusades, so did its hospital system. However, trained "physicians" still practiced their trade primarily in the homes of their patients, and only the weary travelers, the destitute, and those considered hopeless cases found their way to hospitals. Conditions in these early hospitals varied widely. Although a few were well financed and well managed and treated their patients humanely, most were essentially custodial institutions to keep troublesome and infectious people away from the general public. In these establishments, crowding, filth, and high mortality among both patients and attendants were commonplace. Thus, the hospital was viewed as an institution to be feared and shunned.

The Renaissance and Reformation in the fifteenth and sixteenth centuries loosened the Church's stronghold on both the hospital and the conduct of medical practice. During the Renaissance, "true learning"—the desire to pursue the true secrets of nature, including medical knowledge—was again stimulated. The study of human anatomy was advanced and the seeds for further studies were planted by the artists Michelangelo, Raphael, Durer, and, of course, the genius Leonardo da Vinci. They viewed the human body as it really was, not simply as a text passage from Galen. The painters of the Renaissance depicted people in sickness and pain, sketched in great detail, and in the process, demonstrated amazing insight into the workings of the heart, lungs, brain, and muscle structure. They also attempted to portray the individual and to discover emotional as well as physical qualities. In this stimulating era, physicians began to approach their patients and the pursuit of medical knowledge in similar fashion. New medical schools, similar to the most famous of such institutions at Salerno, Bologna, Montpelier, Padua, and Oxford, emerged. These medical training centers once again embraced the Hippocratic doctrine that the patient was human, disease was a natural process, and commonsense therapies were appropriate in assisting the body to conquer its disease.

During the Renaissance, fundamentals received closer examination and the age of measurement began. In 1592, when Galileo visited Padua, Italy, he lectured on mathematics to a large audience of medical students. His famous theories and inventions (the thermoscope and the pendulum, in addition to the telescopic lens) were expounded upon and demonstrated. Using these devices, one of his students, Sanctorius, made comparative studies of the human temperature and pulse. A future graduate of Padua, William Harvey, later applied Galileo's laws of motion and mechanics to the problem of blood circulation. This ability to measure the amount of blood moving through the arteries helped to determine the function of the heart.

Galileo encouraged the use of experimentation and exact measurement as scientific tools that could provide physicians with an effective check against reckless speculation. Quantification meant theories would be verified before being accepted. Individuals involved in medical research incorporated these new methods into their activities. Body temperature and pulse rate became measures that could be related to other symptoms to assist the physician in diagnosing specific illnesses or disease. Concurrently, the development of the microscope amplified human vision, and an unknown world came into focus. Unfortunately, new scientific devices had little effect on the average physician, who continued to blood-let and to disperse noxious ointments. Only in the universities did scientific groups band together to pool their instruments and their various talents.

In England, the medical profession found in Henry VIII a forceful and sympathetic patron. He assisted the doctors in their fight against malpractice and supported the establishment of the College of Physicians, the oldest purely medical institution in Europe. When he suppressed the monastery system in the early sixteenth century, church hospitals were taken over by the cities in which they were located. Consequently, a network of private, nonprofit, voluntary hospitals came into being. Doctors and medical students replaced the nursing sisters and monk physicians. Consequently, the professional nursing class became almost nonexistent in these public institutions.

Only among the religious orders did nursing remain intact, further compounding the poor lot of patients confined within the walls of the public hospitals. These conditions were to continue until Florence Nightingale appeared on the scene years later.

Still another dramatic event occurred. The demands made upon England's hospitals, especially the urban hospitals, became overwhelming as the population of these urban centers continued to expand. It was impossible for the facilities to accommodate the needs of so many. Therefore, during the seventeenth century two of the major urban hospitals in London, St. Bartholomew's and St. Thomas, initiated a policy of admitting and attending to only those patients who could possibly be cured. The incurables were left to meet their destiny in other institutions such as asylums, prisons, or almshouses.

Humanitarian and democratic movements occupied center stage primarily in France and the American colonies during the eighteenth century. The notion of equal rights finally arose, and as urbanization spread, American society concerned itself with the welfare of many of its members. Medical men broadened the scope of their services to include the "unfortunates" of society and helped to ease their suffering by advocating the power of reason and spearheading prison reform, child care, and the hospital movement. Ironically, as the hospital began to take up an active, curative role in medical care in the eighteenth century, the death rate among its patients did not decline but continued to be excessive. In 1788, for example, the death rate among the patients at the Hotel Dru in Paris, thought to be founded in the seventh century and the oldest hospital in existence today, was nearly 25%. These hospitals were lethal not only to patients, but also to the attendants working in them, whose own death rate hovered between 6 and 12% per year.

Essentially, the hospital remained a place to avoid. Under these circumstances, it is not surprising that the first American colonists postponed or delayed building hospitals. For example, the first hospital in America, the Pennsylvania Hospital, was not built until 1751, and the City of Boston took over two hundred years to erect its first hospital, the Massachusetts General, which opened its doors to the public in 1821.

Not until the nineteenth century could hospitals claim to benefit any significant number of patients. This era of progress was due primarily to the improved nursing practices fostered by Florence Nightingale on her return to England from the Crimean War (Fig. 1.2). She demonstrated that hospital deaths were caused more frequently by hospital conditions than by disease. During the latter part of the nineteenth century she was at the height of her influence, and few new hospitals were built anywhere in the world without her advice. During the first half of the nineteenth century Nightingale forced medical attention to focus once more on the care of the patient. Enthusiastically and philosophically, she expressed her views on nursing: "Nursing is putting us in the best possible condition for nature to restore and preserve health.... The art is that of nursing the sick. Please mark, not nursing sickness."

Although these efforts were significant, hospitals remained, until this century, institutions for the sick poor. In the 1870s, for example, when the plans for the projected Johns Hopkins Hospital were reviewed, it was considered quite appropriate to allocate 324 charity and 24 pay beds. Not only did the hospital population before the turn of the century represent a narrow portion of the socioeconomic spectrum,

Figure 1.2   A portrait of Florence Nightingale (Courtesy of http://ginnger.topcities.com/cards/computer/nurses/765x525nightengale.gif).

but it also represented only a limited number of the type of diseases prevalent in the overall population. In 1873, for example, roughly half of America's hospitals did not admit contagious diseases, and many others would not admit incurables. Furthermore, in this period, surgery admissions in general hospitals constituted only 5%, with trauma (injuries incurred by traumatic experience) making up a good portion of these cases.

American hospitals a century ago were rather simple in that their organization required no special provisions for research or technology and demanded only cooking and washing facilities. In addition, since the attending and consulting physicians were normally unsalaried and the nursing costs were quite modest, the great bulk of the hospital's normal operation expenses were for food, drugs, and utilities. Not until the twentieth century did modern medicine come of age in the United States. As we shall see, technology played a significant role in its evolution.

1.2   THE MODERN HEALTH CARE SYSTEM

Modern medical practice actually began at the turn of the twentieth century. Before 1900, medicine had little to offer the average citizen since its resources were mainly physicians, their education, and their little black bags. At this time physicians were in short supply, but for different reasons than exist today. Costs were minimal, demand small, and many of the services provided by the physician also could be obtained from experienced amateurs residing in the community. The individual's dwelling was the major site for treatment and recuperation, and relatives and neighbors constituted an able and willing nursing staff. Midwives delivered babies, and those illnesses not cured by home remedies were left to run their fatal course. Only in the twentieth century did the tremendous explosion in scientific knowledge and technology lead to the development of the American health care system with the hospital as its focal point and the specialist physician and nurse as its most visible operatives.

In the twentieth century, advances in the basic sciences (chemistry, physiology, pharmacology, and so on) began to occur much more rapidly. It was an era of intense interdisciplinary cross-fertilization. Discoveries in the physical sciences enabled medical researchers to take giant strides forward. For example, in 1903 William Einthoven devised the first electrocardiograph and measured the electrical changes that occurred during the beating of the heart. In the process, Einthoven initiated a new age for both cardiovascular medicine and electrical measurement techniques.

Of all the new discoveries that followed one another like intermediates in a chain reaction, the most significant for clinical medicine was the development of x-rays. When W.K. Roentgen described his "new kinds of rays," the human body was opened to medical inspection. Initially these x-rays were used in the diagnosis of bone fractures and dislocations. In the United States, x-ray machines brought this modern technology to most urban hospitals. In the process, separate departments of radiology were established, and the influence of their activities spread, with almost every department of medicine (surgery, gynecology, and so forth) advancing with the aid of this new tool. By the 1930s, x-ray visualization of practically all the organ systems of the body was possible by the use of barium salts and a wide variety of radiopaque materials.

The power this technological innovation gave physicians was enormous. The x-ray permitted them to diagnose a wide variety of diseases and injuries accurately. In addition, being within the hospital, it helped trigger the transformation of the hospital from a passive receptacle for the sick poor to an active curative institution for all citizens of the American society.

The introduction of sulfanilamide in the mid-1930s and penicillin in the early 1940s significantly reduced the main danger of hospitalization: cross infection among patients. With these new drugs in their arsenals, surgeons were able to perform their operations without prohibitive morbidity and mortality due to infection. Also consider that, even though the different blood groups and their incompatibility were discovered in 1900 and sodium citrate was used in 1913 to prevent clotting, the full development of blood banks was not practical until the 1930s when technology provided adequate refrigeration. Until that time, "fresh" donors were bled, and the blood was transfused while it was still warm.

As technology in the United States blossomed so did the prestige of American medicine. From 1900 to 1929 Nobel Prize winners in physiology or medicine came primarily from Europe, with no American among them. In the period 1930 to 1944, just before the end of World War II, seven Americans were honored with this award. During the post-war period of 1945 to 1975, 37 American life scientists earned similar honors, and from 1975–2003, the number was 40. Thus, since 1930 a total of 79 American scientists have performed research significant enough to warrant the distinction of a Nobel Prize. Most of these efforts were made possible by the technology (Fig. 1.3) available to these clinical scientists.

The employment of the available technology assisted in advancing the development of complex surgical procedures (Fig. 1.4). The Drinker respirator was introduced in 1927 and the first heart–lung bypass in 1939. In the 1940s, cardiac catheterization and angiography (the use of a cannula threaded through an arm vein

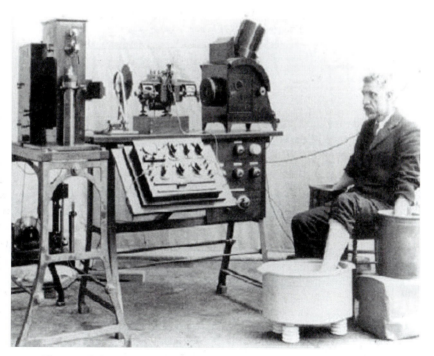

Figure 1.3   Photograph depicting an early electrocardiograph machine.

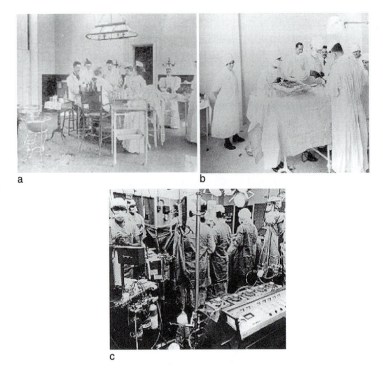

a   b

c

Figure 1.4   Changes in the operating room: (a) the surgical scene at the turn of the century, (b) the surgical scene in the late 1920s and early 1930s, (c) the surgical scene today (from JD Bronzino, Technology for Patient Care, Mosby: St. Louis, 1977; The Biomedical Engineering Handbook, CRC Press: Boca Raton, FL, 1995; 2000; 2005).

and into the heart with the injection of radiopaque dye for the x-ray visualization of lung and heart vessels and valves) were developed. Accurate diagnoses of congenital and acquired heart disease (mainly valve disorders due to rheumatic fever) also became possible, and a new era of cardiac and vascular surgery began.

Another child of this modern technology, the electron microscope, entered the medical scene in the 1950s and provided significant advances in visualizing relatively small cells. Body scanners to detect tumors arose from the same science that brought societies reluctantly into the atomic age. These "tumor detectives" used radioactive material and became commonplace in newly established departments of nuclear medicine in all hospitals.

The impact of these discoveries and many others was profound. The health care system that consisted primarily of the "horse and buggy" physician was gone forever, replaced by the doctor backed by and centered around the hospital, as medicine began to change to accommodate the new technology.

Following World War II, the evolution of comprehensive care greatly accelerated. The advanced technology that had been developed in the pursuit of military objectives now became available for peaceful applications with the medical profession benefiting greatly from this rapid surge of technological finds. For instance, the realm of electronics came into prominence. The techniques for following enemy ships and planes, as well as providing aviators with information concerning altitude, air speed, and the like, were now used extensively in medicine to follow the subtle electrical behavior of the fundamental unit of the central nervous system, the neuron, or to monitor the beating heart of a patient.

Science and technology have leap-frogged past one another throughout recorded history. Anyone seeking a causal relation between the two was just as likely to find technology the cause and science the effect as to find science the cause and technology the effect. As gunnery led to ballistics, and the steam engine to thermodynamics, so powered flight led to aerodynamics. However, with the advent of electronics this causal relation between technology and science changed to a systematic exploitation of scientific research and the pursuit of knowledge that was undertaken with technical uses in mind.

The list becomes endless when one reflects upon the devices produced by the same technology that permitted humans to stand on the moon. What was considered science fiction in the 1930s and the 1940s became reality. Devices continually changed to incorporate the latest innovations, which in many cases became outmoded in a very short period of time. Telemetry devices used to monitor the activity of a patient's heart freed both the physician and the patient from the wires that previously restricted them to the four walls of the hospital room. Computers, similar to those that controlled the flight plans of the Apollo capsules, now completely inundate our society. Since the 1970s, medical researchers have put these electronic brains to work performing complex calculations, keeping records (via artificial intelligence), and even controlling the very instrumentation that sustains life. The development of new medical imaging techniques (Fig. 1.5) such as computerized tomography (CT) and magnetic resonance imaging (MRI) totally depended on a continually advancing computer technology. The citations and technological discoveries are so myriad it is impossible to mention them all.

"Spare parts" surgery is now routine. With the first successful transplantation of a kidney in 1954, the concept of artificial organs gained acceptance and officially came into vogue in the medical arena (Fig. 1.6). Technology to provide prosthetic devices such as artificial heart valves and artificial blood vessels developed. Even an artificial heart program to develop a replacement for a defective or diseased human heart began. Although, to date, the results have not been satisfactory, this program has provided "ventricular assistance" for those who need it. These technological innovations radically altered surgical organization and utilization. The comparison of a hospital in which surgery was a relatively minor activity as it was a century ago to the contemporary hospital in which surgery plays a prominent role dramatically suggests the manner in which this technological effort has revolutionized the health profession and the institution of the hospital.

Through this evolutionary process, the hospital became the central institution that provided medical care. Because of the complex and expensive technology that could be based only in the hospital and the education of doctors oriented both as clinicians and investigators toward highly technological norms, both the patient and the

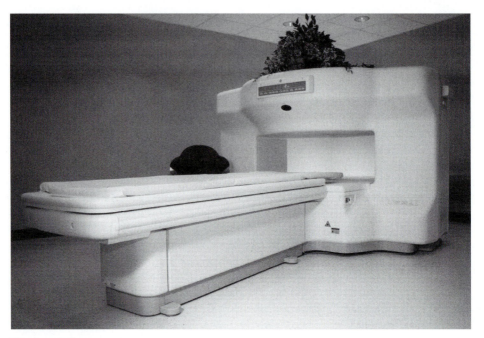

Figure 1.5   Photograph of a modern medical imaging facility (http://137.229.52.100/~physics/p113/hasan/).

physician were pushed even closer to this center of attraction. In addition, the effects of the increasing maldistribution and apparent shortage of physicians during the 1950s and 1960s also forced the patient and the physician to turn increasingly to the ambulatory clinic and the emergency ward of the urban hospital in time of need.

Emergency wards today handle not only an ever-increasing number of accidents (largely related to alcohol and the automobile) and somatic crises such as heart attacks and strokes, but also problems resulting from the social environments that surround the local hospital. Respiratory complaints, cuts, bumps, and minor trauma constitute a significant number of the cases seen in a given day. Added to these individuals are those who live in the neighborhood of the hospital and simply cannot afford their own physician. Often such individuals enter the emergency ward for routine care of colds, hangovers, and even marital problems. Because of these developments, the hospital has evolved as the focal point of the present system of health care delivery. The hospital, as presently organized, specializes in highly technical and complex medical procedures. This evolutionary process became inevitable as technology produced increasingly sophisticated equipment that private practitioners or even large group practices were economically unequipped to acquire and maintain. Only the hospital could provide this type of service. The steady expansion of scientific and technological innovations has not only necessitated specialization for all health professionals (physicians, nurses, and technicians) but has also required the housing of advanced technology within the walls of the modern hospital.

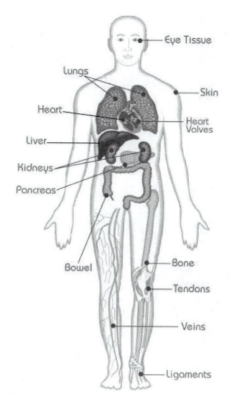

Figure 1.6 Illustration of various transplantation possibilities (http://www.transplant.bc.ca/images/what_organs.GIF).

In recent years, technology has struck medicine like a thunderbolt. The Human Genome Project was perhaps the most prominent scientific and technological effort of the 1990s. Some of the engineering products vital to the effort included automatic sequencers, robotic liquid handling devices, and software for databasing and sequence assembly. As a result, a major transition occurred, moving biomedical engineering to focus on the cellular and molecular level rather than solely on the organ system level. With the success of the genome project, new vistas have been opened (e.g., it is now possible to create individual medications based on one's DNA) (Fig. 1.7). Advances in nanotechnology, tissue engineering, and artificial organs are clear indications that science fiction will continue to become reality. However, the social and economic consequences of this vast outpouring of information and innovation must be fully understood if this technology is to be exploited effectively and efficiently.

As one gazes into the crystal ball, technology offers great potential for affecting health care practices (Fig. 1.8). It can provide health care for individuals in remote rural areas by means of closed-circuit television health clinics with complete communication links to a regional health center. Development of multiphasic screening

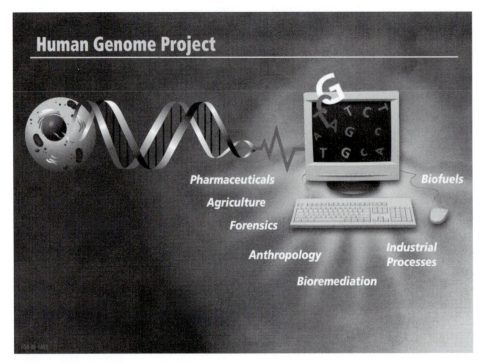

Figure 1.7   The Human Genome Project's potential applications (http://labmed.hallym.ac.kr/genome/genome-photo/98-1453.jpg).

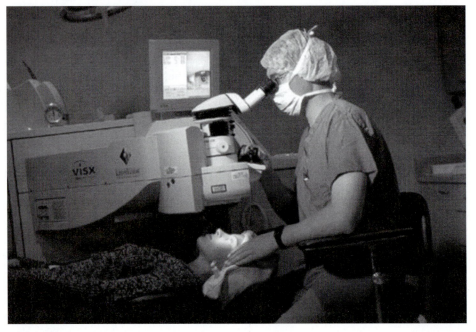

Figure 1.8   Laser surgery, a new tool in the physician's arsenal (http://riggottphoto.com/corporate/lgimg6.html).

systems can provide preventive medicine to the vast majority of the population and restrict admission to the hospital to those needing the diagnostic and treatment facilities housed there. Automation of patient and nursing records can inform physicians of the status of patients during their stay at the hospital and in their homes. With the creation of a central medical records system, anyone who changes residences or becomes ill away from home can have records made available to the attending physician easily and rapidly. Tissue engineering—the application of biological, chemical, and engineering principles towards the repair, restoration, and regeneration of living tissue using biomaterials, cells, and factors alone or in combinations—has gained a great deal of attention and is projected to grow exponentially in the first quarter of the twenty-first century. These are just a few of the possibilities that illustrate the potential of technology in creating the type of medical care system that will indeed be accessible, of high quality, and reasonably priced for all. [Note: for an extensive review of major events in the evolution of biomedical engineering see Nebekar, 2002.]

1.3 WHAT IS BIOMEDICAL ENGINEERING?

Many of the problems confronting health professionals today are of extreme importance to the engineer because they involve the fundamental aspects of device and systems analysis, design, and practical application—all of which lie at the heart of processes that are fundamental to engineering practice. These medically relevant design problems can range from very complex large-scale constructs, such as the design and implementation of automated clinical laboratories, multiphasic screening facilities (i.e., centers that permit many tests to be conducted), and hospital information systems, to the creation of relatively small and simple devices, such as recording electrodes and transducers that may be used to monitor the activity of specific physiological processes in either a research or clinical setting. They encompass the many complexities of remote monitoring and telemetry and include the requirements of emergency vehicles, operating rooms, and intensive care units.

The American health care system, therefore, encompasses many problems that represent challenges to certain members of the engineering profession called biomedical engineers. Since biomedical engineering involves applying the concepts, knowledge, and approaches of virtually all engineering disciplines (e.g., electrical, mechanical, and chemical engineering) to solve specific health care related problems, the opportunities for interaction between engineers and health care professionals are many and varied.

Biomedical engineers may become involved, for example, in the design of a new medical imaging modality or development of new medical prosthetic devices to aid people with disabilities. Although what is included in the field of biomedical engineering is considered by many to be quite clear, many conflicting opinions concerning the field can be traced to disagreements about its definition. For example, consider the terms biomedical engineering, bioengineering, biological engineering, and clinical (or medical) engineer, which are defined in the Bioengineering Education Directory. Although Pacela defined bioengineering as the broad umbrella term used to describe this entire field, bioengineering is usually defined as a basic-research-oriented activity closely related to biotechnology and genetic engineering, that is, the modification of animal or plant cells or parts of cells to improve plants or animals or to develop new microorganisms for beneficial ends. In the food industry, for example, this has meant the improvement of strains of yeast for fermentation. In agriculture, bioengineers may be concerned with the improvement of crop yields by treating plants with organisms to reduce frost damage. It is clear that bioengineers for the future will have tremendous impact on the quality of human life. The full potential of this specialty is difficult to imagine. Typical pursuits include the following:

- Development of improved species of plants and animals for food production
- Invention of new medical diagnostic tests for diseases
- Production of synthetic vaccines from clone cells
- Bioenvironmental engineering to protect human, animal, and plant life from toxicants and pollutants
- Study of protein-surface interactions
- Modeling of the growth kinetics of yeast and hybridoma cells
- Research in immobilized enzyme technology
- Development of therapeutic proteins and monoclonal antibodies

The term biomedical engineering appears to have the most comprehensive meaning. Biomedical engineers apply electrical, chemical, optical, mechanical, and other engineering principles to understand, modify, or control biological (i.e., human and animal) systems. Biomedical engineers working within a hospital or clinic are more properly called clinical engineers, but this theoretical distinction is not always observed in practice, and many professionals working within U.S. hospitals today continue to be called biomedical engineers.

The breadth of activity of biomedical engineers is significant. The field has moved from being concerned primarily with the development of medical devices in the 1950s and 1960s to include a more wide-ranging set of activities. As illustrated in Figure 1.9, the field of biomedical engineering now includes many new career areas.

These areas include

- Application of engineering system analysis (physiologic modeling, simulation, and control to biological problems
- Detection, measurement, and monitoring of physiologic signals (i.e., biosensors and biomedical instrumentation)
- Diagnostic interpretation via signal-processing techniques of bioelectric data
- Therapeutic and rehabilitation procedures and devices (rehabilitation engineering)
- Devices for replacement or augmentation of bodily functions (artificial organs)
- Computer analysis of patient-related data and clinical decision making (i.e., medical informatics and artificial intelligence)
- Medical imaging; that is, the graphical display of anatomic detail or physiologic function

THE WORLD OF BIOMEDICAL ENGINEERING

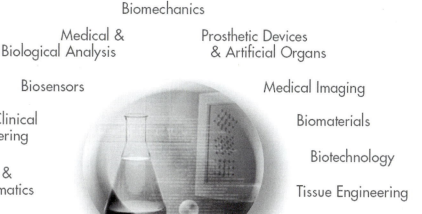

Biomechanics

Medical &
Biological Analysis

Prosthetic Devices
& Artificial Organs

Biosensors

Medical Imaging

Clinical
Engineering

Biomaterials

Biotechnology

Medical &
Bioinformatics

Tissue Engineering

Rehabilitation
Engineering

Neural
Engineering

Physiological
Modeling

Biomedical
Instrumentation

Bionanotechnology

Figure 1.9 The world of biomedical engineering.

- The creation of new biologic products (i.e., biotechnology and tissue engineering)

Typical pursuits of biomedical engineers include

- Research in new materials for implanted artificial organs
- Development of new diagnostic instruments for blood analysis
- Writing software for analysis of medical research data
- Analysis of medical device hazards for safety and efficacy
- Development of new diagnostic imaging systems
- Design of telemetry systems for patient monitoring
- Design of biomedical sensors
- Development of expert systems for diagnosis and treatment of diseases
- Design of closed-loop control systems for drug administration
- Modeling of the physiologic systems of the human body
- Design of instrumentation for sports medicine
- Development of new dental materials
- Design of communication aids for individuals with disabilities
- Study of pulmonary fluid dynamics

■ Study of biomechanics of the human body
■ Development of material to be used as replacement for human skin

The preceding list is not intended to be all-inclusive. Many other applications use the talents and skills of the biomedical engineer. In fact, the list of biomedical engineers' activities depends on the medical environment in which they work. This is especially true for clinical engineers, biomedical engineers employed in hospitals or clinical settings. Clinical engineers are essentially responsible for all the high-technology instruments and systems used in hospitals today; for the training of medical personnel in equipment safety; and for the design, selection, and use of technology to deliver safe and effective health care.

Engineers were first encouraged to enter the clinical scene during the late 1960s in response to concerns about the electrical safety of hospital patients. This safety scare reached its peak when consumer activists, most notably Ralph Nader, claimed that "at the very least, 1,200 Americans are electrocuted annually during routine diagnostic and therapeutic procedures in hospitals." This concern was based primarily on the supposition that catheterized patients with a low-resistance conducting pathway from outside the body into blood vessels near the heart could be electrocuted by voltage differences well below the normal level of sensation. Despite the lack of statistical evidence to substantiate these claims, this outcry served to raise the level of consciousness of health care professionals with respect to the safe use of medical devices.

In response to this concern, a new industry—hospital electrical safety—arose almost overnight. Organizations such as the National Fire Protection Association (NFPA) wrote standards addressing electrical safety in hospitals. Electrical safety analyzer manufacturers and equipment safety consultants became eager to serve the needs of various hospitals that wanted to provide a "safety fix," and some companies developed new products to ensure patient safety, particularly those specializing in power distribution systems (most notably isolation transformers). To alleviate these fears, the Joint Commission on the Accreditation of Healthcare Organizations (then known as the Joint Commission on Accreditation of Hospitals) turned to NFPA codes as the standard for electrical safety and further specified that hospitals must inspect all equipment used on or near a patient for electrical safety at least every six months. To meet this new requirement hospital administrators considered a number of options, including: (1) paying medical device manufacturers to perform these electrical safety inspections, (2) contracting for the services of shared-services organizations, or (3) providing these services with in-house staff. When faced with this decision, most large hospitals opted for in-house service and created whole departments to provide the technological support necessary to address these electrical safety concerns.

As a result, a new engineering discipline—clinical engineering—was born. Many hospitals established centralized clinical engineering departments. Once these departments were in place, however, it soon became obvious that electrical safety failures represented only a small part of the overall problem posed by the presence of medical equipment in the clinical environment. At the time, this equipment was neither totally understood nor properly maintained. Simple visual inspections often revealed broken knobs, frayed wires, and even evidence of liquid spills. Many devices did not perform in accordance with manufacturers' specifications and were not maintained in accordance with manufacturers' recommendations. In short, electrical safety problems were only the tip of the iceberg. By the mid-1970s, complete performance inspections before and after equipment use became the norm and sensible inspection procedures were developed. In the process, these clinical engineering pioneers began to play a more substantial role within the hospital. As new members of the hospital team, they

- Became actively involved in developing cost-effective approaches for using medical technology
- Provided advice to hospital administrators regarding the purchase of medical equipment based on its ability to meet specific technical specifications
- Started utilizing modern scientific methods and working with standards-writing organizations
- Became involved in the training of health care personnel regarding the safe and efficient use of medical equipment

Then, during the 1970s and 1980s, a major expansion of clinical engineering occurred, primarily due to the following events:

- The Veterans' Administration (VA), convinced that clinical engineers were vital to the overall operation of the VA hospital system, divided the country into biomedical engineering districts, with a chief biomedical engineer overseeing all engineering activities in the hospitals in that district.
- Throughout the United States, clinical engineering departments were established in most large medical centers and hospitals and in some smaller clinical facilities with at least 300 beds.
- Health care professionals (i.e., physicians and nurses) needed assistance in utilizing existing technology and incorporating new innovations.
- Certification of clinical engineers became a reality to ensure the continued competence of practicing clinical engineers.

During the 1990s, the evaluation of clinical engineering as a profession continued with the establishment of the American College of Clinical Engineering (ACCE) and the Clinical Engineering Division within the International Federation of Medical and Biological Engineering (IFMBE).

Clinical engineers today provide extensive engineering services for the clinical staff and serve as a significant resource for the entire hospital (Fig. 1.10). Possessing in-depth knowledge regarding available in-house technological capabilities as well as the technical resources available from outside firms, the modern clinical engineer enables the hospital to make effective and efficient use of most if not all of its technological resources.

Biomedical engineering is thus an interdisciplinary branch of engineering heavily based both in engineering and in the life sciences. It ranges from theoretical, nonexperimental undertakings to state-of-the-art applications. It can encompass research, development, implementation, and operation. Accordingly, like medical practice itself, it is unlikely that any single person can acquire expertise that

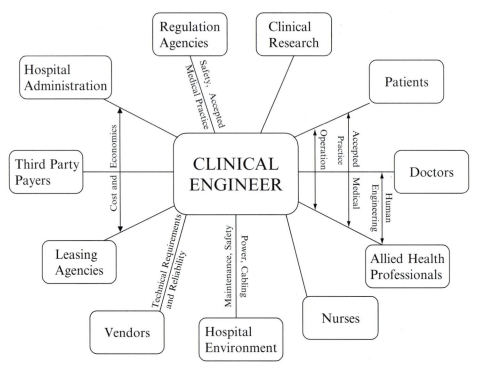

Figure 1.10   The range of interactions with clinical engineers in a hospital setting.

encompasses the entire field. As a result, there has been an explosion of biomedical engineering specialists to cover this broad spectrum of activity. Yet, because of the interdisciplinary nature of this activity, there is considerable interplay and overlapping of interest and effort between them. For example, biomedical engineers engaged in the development of biosensors may interact with those interested in prosthetic devices to develop a means to detect and use the same bioelectric signal to power a prosthetic device. Those engaged in automating the clinical chemistry laboratory may collaborate with those developing expert systems to assist clinicians in making clinical decisions based on specific laboratory data. The possibilities are endless.

Perhaps a greater potential benefit occurring from the utilization of biomedical engineers is the identification of problems and needs of our present health care delivery system that can be solved using existing engineering technology and systems methodology. Consequently, the field of biomedical engineering offers hope in the continuing battle to provide high-quality health care at a reasonable cost. If properly directed towards solving problems related to preventive medical approaches, ambulatory care services, and the like, biomedical engineers can provide the tools and techniques to make our health care system more effective and efficient.

1.4 ROLES PLAYED BY BIOMEDICAL ENGINEERS

In its broadest sense, biomedical engineering involves training essentially three types of individuals: (1) the clinical engineer in health care, (2) the biomedical design engineer for industry, and (3) the research scientist. Currently, one might also distinguish among three specific roles these biomedical engineers can play. Each is different enough to merit a separate description. The first type, the most common, might be called the "problem solver." This biomedical engineer (most likely the clinical engineer or biomedical design engineer) maintains the traditional service relationship with the life scientists who originate a problem that can be solved by applying the specific expertise of the engineer. For this problem-solving process to be efficient and successful, however, some knowledge of each other's language and a ready interchange of information must exist. Biomedical engineers must understand the biological situation to apply their judgment and contribute their knowledge toward the solution of the given problem as well as to defend their methods in terms that the life scientist can understand. If they are unable to do these things, they do not merit the "biomedical" appellation.

The second type, which is more rare, might be called the "technological entrepreneur" (most likely a biomedical design engineer in industry). This individual assumes that the gap between the technological education of the life scientist or physician and present technological capability has become so great that the life scientist cannot pose a problem that will incorporate the application of existing technology. Therefore, technological entrepreneurs examine some portion of the biological or medical front and identify areas in which advanced technology might be advantageous. Thus, they pose their own problem and then proceed to provide the solution, at first conceptually and then in the form of hardware or software. Finally, these individuals must convince the medical community that they can provide a useful tool because, contrary to the situation in which problem solvers find themselves, the entrepreneur's activity is speculative at best and has no ready-made customer for the results. If the venture is successful, however, whether scientifically or commercially, then an advance has been made much earlier than it would have been through the conventional arrangement. Because of the nature of their work, technological entrepreneurs should have a great deal of engineering and medical knowledge as well as experience in numerous medical systems.

The third type of biomedical engineer, the "engineer–scientist" (most likely found in academic institutions and industrial research labs), is primarily interested in applying engineering concepts and techniques to the investigation and exploration of biological processes. The most powerful tool at their disposal is the construction of an appropriate physical or mathematical model of the specific biological system under study. Through simulation techniques and available computing machinery, they can use this model to understand features that are too complex for either analytical computation or intuitive recognition. In addition, this process of simulation facilitates the design of appropriate experiments that can be performed on the actual biological system. The results of these experiments can, in turn, be used to amend the model.

Thus, increased understanding of a biological mechanism results from this iterative process.

This mathematical model can also predict the effect of these changes on a biological system in cases where the actual experiments may be tedious, very difficult, or dangerous. The researchers are thus rewarded with a better understanding of the biological system, and the mathematical description forms a compact, precise language that is easily communicated to others. The activities of the engineer–scientist inevitably involve instrument development because the exploitation of sophisticated measurement techniques is often necessary to perform the biological side of the experimental work. It is essential that engineer–scientists work in a biological environment, particularly when their work may ultimately have a clinical application. It is not enough to emphasize the niceties of mathematical analysis while losing the clinical relevance in the process. This biomedical engineer is a true partner of the biological scientist and has become an integral part of the research teams being formed in many institutes to develop techniques and experiments that will unfold the mysteries of the human organism. Each of these roles envisioned for the biomedical engineer requires a different attitude, as well as a specific degree of knowledge about the biological environment. However, each engineer must be a skilled professional with a significant expertise in engineering technology. Therefore, in preparing new professionals to enter this field at these various levels, biomedical engineering educational programs are continually being challenged to develop curricula that will provide an adequate exposure to and knowledge about the environment, without sacrificing essential engineering skills. As we continue to move into a period characterized by a rapidly growing aging population, rising social and economic expectations, and a need for the development of more adequate techniques for the prevention, diagnosis, and treatment of disease, development and employment of biomedical engineers have become a necessity. This is true not only because they may provide an opportunity to increase our knowledge of living systems, but also because they constitute promising vehicles for expediting the conversion of knowledge to effective action.

The ultimate role of the biomedical engineer, like that of the nurse and physician, is to serve society. This is a profession, not just a skilled technical service. To use this new breed effectively, health care practitioners and administrators should be aware of the needs for these new professionals and the roles for which they are being trained. The great potential, challenge, and promise in this endeavor offer not only significant technological benefits but also humanitarian benefits.

1.5 PROFESSIONAL STATUS OF BIOMEDICAL ENGINEERING

Biomedical engineers are professionals. Professionals have been defined as an aggregate of people finding identity in sharing values and skills absorbed during a common course of intensive training. Whether individuals are professionals is determined by examining whether or not they have internalized certain given professional values. Furthermore, a professional is someone who has internalized professional values and is licensed on the basis of his or her technical competence. Professionals generally accept scientific standards in their work, restrict their work activities to areas in which they are technically competent, avoid emotional involvement, cultivate objectivity in their work, and put their clients' interests before their own.

The concept of a profession that is involved in the design, development, and management of medical technology encompasses three primary occupational models: science, business, and profession. Consider initially the contrast between science and profession. Science is seen as the pursuit of knowledge, its value hinging on providing evidence and communicating with colleagues. Profession, on the other hand, is viewed as providing a service to clients who have problems they cannot handle themselves. Scientists and professionals have in common the exercise of some knowledge, skill, or expertise. However, while scientists practice their skills and report their results to knowledgeable colleagues, professionals such as lawyers, physicians, and engineers serve lay clients. To protect both the professional and the client from the consequences of the layperson's lack of knowledge, the practice of the profession is often regulated through such formal institutions as state licensing. Both professionals and scientists must persuade their clients to accept their findings. Professionals endorse and follow a specific code of ethics to serve society. On the other hand, scientists move their colleagues to accept their findings through persuasion.

Consider, for example, the medical profession. Its members are trained in caring for the sick, with the primary goal of healing them. These professionals not only have a responsibility for the creation, development, and implementation of that tradition, but they are also expected to provide a service to the public, within limits, without regard to self-interest. To ensure proper service, the profession closely monitors the licensing and certification process. Thus, medical professionals themselves may be regarded as a mechanism of social control. However, this does not mean that other facets of society are not involved in exercising oversight and control of physicians in their practice of medicine.

A final attribute of professionals is that of integrity. Physicians tend to be both permissive and supportive in relationships with patients and yet are often confronted with moral dilemmas involving the desires of their patients and social interest. For example, how to honor the wishes of terminally ill patients while not facilitating the patients' deaths is a moral question that health professionals are forced to confront. A detailed discussion of the moral issues posed by medical technology is presented in Chapter 2.

One can determine the status of professionalization by noting the occurrence of six crucial events: (1) the first training school; (2) the first university school; (3) the first local professional association; (4) the first national professional association; (5) the first state license law; and (6) the first formal code of ethics.

The early appearances of the training school and the university affiliation underscore the importance of the cultivation of a knowledge base. The strategic innovative role of the universities and early teachers lies in linking knowledge to practice and creating a rationale for exclusive jurisdiction. Those practitioners pushing for prescribed training then form a professional association. The association defines the tasks of the profession: raising the quality of recruits; redefining their function to permit the use of less technically skilled people to perform the more routine, less involved tasks; and managing internal and external conflicts. In the process, internal conflict may arise between those committed to previously established procedures and newcomers committed to change and innovation. At this stage, some form of professional regulation, such as licensing or certification, surfaces because of a belief that it will ensure minimum standards for the profession, enhance status, and protect the layperson in the process.

The last area of professional development is the establishment of a formal code of ethics, which usually includes rules to exclude unqualified and unscrupulous practitioners, rules to reduce internal competition, and rules to protect clients and emphasize the ideal service to society. A code of ethics usually comes at the end of the professionalization process.

In biomedical engineering, all six of these critical steps have been taken. The field of biomedical engineering, which originated as a professional group interested primarily in medical electronics in the late 1950s, has grown from a few scattered individuals to a very well-established organization. There are approximately 48 international societies throughout the world serving an increasingly expanding community of biomedical engineers. Today, the scope of biomedical engineering is enormously diverse. Over the years, many new disciplines such as tissue engineering and artificial intelligence, which were once considered alien to the field, are now an integral part of the profession.

Professional societies play a major role in bringing together members of this diverse community to share their knowledge and experience in pursuit of new technological applications that will improve the health and quality of life of human beings. Intersocietal cooperation and collaborations, both at national and international levels, are more actively fostered today through professional organizations such as the Biomedical Engineering Society (BMES), the American Institute of Medical and Biological Engineers (AIMBE), and the Engineering in Medicine and Biology Society (EMBS) of the Institute of Electrical and Electronic Engineers (IEEE).

1.6   PROFESSIONAL SOCIETIES

1.6.1   American Institute for Medical and Biological Engineering

The United States has the largest biomedical engineering community in the world. Major professional organizations that address various cross sections of the field and serve biomedical engineering professionals include: (1) the American College of Clinical Engineering, (2) the American Institute of Chemical Engineers, (3) the American Medical Informatics Association, (4) the American Society of Agricultural Engineers, (5) the American Society for Artificial Internal Organs, (6) the American Society of Mechanical Engineers, (7) the Association for the Advancement of Medical Instrumentation, (8) the Biomedical Engineering Society, (9) the IEEE Engineering in Medicine and Biology Society, (10) an interdisciplinary Association for the Advancement of Rehabilitation and Assistive Technologies, and (11) the Society for Biomaterials. In an effort to unify all the disparate components of the biomedical engineering community in the United States as represented by these various societies, the American Institute for Medical and Biological Engineering (AIMBE) was created in 1992. The primary goal of AIMBE is to serve as an umbrella organization in the United States for the purpose of unifying the bioengineering community, addressing public policy issues, and promoting the engineering approach in society's effort to enhance health and quality of life through the judicious use of technology. For information, contact AIMBE, 1901 Pennsylvania Avenue N.W., Suite 401, Washington, D.C. 20006 (http://aimbe.org/; Email: info@aimbe.org).

1.6.2 IEEE Engineering in Medicine and Biology Society

The Institute of Electrical and Electronic Engineers (IEEE) is the largest international professional organization in the world, and it accommodates 37 societies and councils under its umbrella structure. Of these 37, the Engineering in Medicine and Biology Society (EMBS) represents the foremost international organization serving the needs of over 8000 biomedical engineering members around the world. The major interest of the EMBS encompasses the application of concepts and methods from the physical and engineering sciences to biology and medicine. Each year the society sponsors a major international conference while cosponsoring a number of theme-oriented regional conferences throughout the world. Premier publications consist of a monthly journal (*Transactions on Biomedical Engineering*), three quarterly journals (*Transactions on Neural Systems and Rehabilitation Engineering, Transactions on Information Technology in Biomedicine,* and *Transactions on Nanobioscience*), and a bimonthly magazine (*IEEE Engineering in Medicine and Biology Magazine*). Secondary publications, authored in collaboration with other societies, include *Transactions on Medical Imaging, Transactions on Neural Networks, Transactions on Pattern Analysis,* and *Machine Intelligence.* For more information, contact the IEEE EMBS Executive Office, IEEE, 445 Hoes Lane, Piscataway, NJ, 08855–1331 USA (http://www.embs. org/; Email: emb-exec@ieee.org).

1.6.3 Biomedical Engineering Society

Established in 1968, the Biomedical Engineering Society (BMES) was founded to address a need for a society that afforded equal status to representatives of both biomedical and engineering interests. With that in mind, the primary goal of the BMES, as stated in their Articles of Incorporation, is "to promote the increase of biomedical engineering knowledge and its utilization." Regular meetings are scheduled biannually in both the spring and fall. Additionally, special interest meetings are interspersed throughout the year, and are promoted in conjunction with other biomedical engineering societies such as AIMBE and EMBS. The primary publications associated with the BMES include: *Annals of Biomedical Engineering*, a monthly journal presenting original research in several biomedical fields; *BMES Bulletin*, a quarterly newsletter presenting a wider array of subject matter relating both to biomedical engineering and BMES news and events; and the *BMES Membership Directory*, an annual publication listing the contact information of the society's individual constituents. For more information, contact the BMES directly: BMES, 8401 Corporate Drive, Suite 225, Landover, MD 20785–2224, USA (http:// www.bmes.org/default.asp; Email: info@bmes.org).

The activities of these biomedical engineering societies are critical to the continued advancement of the professional status of biomedical engineers. Therefore, all biomedical engineers, including students in the profession, are encouraged to become members of these societies and engage in the activities of true professionals.

EXERCISES

1. Select a specific medical technology from the following list of historical periods. Describe the fundamental principles of operation and discuss their impact on health care delivery: (a) 1900–1939; (b) 1945–1970; (c) 1970–1980; (d) 1980–2003.
2. Provide a review of the effect computer technology has had on health care delivery, citing the computer application and the time frame of its implementation.
3. The term *genetic engineering* implies an engineering function. Is there one? Should this activity be included in the field of biomedical engineering?
4. Discuss in some detail the role the genome project has had and is anticipated to have on the development of new medical technology.
5. Using your crystal ball, what advances in engineering and/or life science do you think will have the greatest effect on clinical care or biomedical research?
6. The organizational structure of a hospital involves three major groups: (1) the board of trustees, (2) administrators, and (3) the medical staff. Specify the major responsibilities of each. In what group should a department of clinical engineering reside? Explain your answer.
7. Based on its definition, what attributes should a clinical engineer have?
8. List at least seven (7) specific activities of clinical engineers.
9. Provide modern examples (i.e., names of individuals and their activities) of the three major roles played by biomedical engineers: (a) The problem solver; (b) The technological entrepreneur; (c) The engineer–scientist.
10. Do the following groups fit the definition of a profession? Discuss how they do or do not: (a) Registered nurses; (b) Biomedical technicians; (c) Respiratory therapists; (d) Hospital administrators.
11. List the areas of knowledge necessary to practice biomedical engineering. Identify where in the normal educational process one can acquire knowledge. How best can administrative skills be acquired?
12. Provide a copy of the home page for a biomedical engineering professional society and a list of the society's major activities for the coming year.
13. What is your view regarding the role biomedical engineers will play in the health care system of tomorrow?

14. Discuss the trade-offs in health care that occur as a result of limited financial resources.

15. Discuss whether medical technology is an economic cost factor, benefit, or both.

REFERENCES AND SUGGESTED READING

Aston, C. (2001). Biological warfare canaries. *IEEE Spectrum* **38:10**, 35–40.

Bankman, I.N. (2000). *Handbook of Medical Imaging*. CRC Press, Boca Raton, FL.

Bronzino, J.D. (2005). *Biomedical Engineering Handbook,* 2nd Ed. CRC Press, Boca Raton, FL.

Bronzino, J.D. (1992). *Management of Medical Technology: A Primer for Clinical Engineering*. Butterworth, Stoneham, MA.

Carson, E. and Cobelli, C. (2001). *Modeling Methodology for Physiology and Medicine*. Academic Press, San Diego, CA.

Laurenchin, C.T. (2003). Repair and restore with tissue engineering. *EMBS Magazine* **22:5**, 16–17.

Nebekar, F. (2002). Golden accomplishments in biomedical engineering. *EMBS Magazine* **21:3**, 17–48.

Pacela, A. (1990). *Bioengineering Education Directory*. Quest Publishing, Brea, CA.

Palsson, B.O. and Bhatia, S.N. (2004). *Tissue Engineering*. Prentice Hall, Englewood, NJ.

The *EMBS Magazine* published by the Institute of Electrical and Electronic Engineers, edited by John Enderle, especially Writing the book on BME, **21:3**, 2002.

2 MORAL AND ETHICAL ISSUES

Joseph Bronzino PhD, PE

Chapter Contents

After completing this chapter, students will be able to:

- Define and distinguish between the terms *morals* and *ethics*.
- Present the rationale underlying two major philosophical schools of thought: utilitarianism and nonconsequentialism.
- Present the codes of ethics for the medical profession, nursing profession, and biomedical engineering.
- Identify the modern moral dilemmas, including redefining death, deciding how to care for the terminally ill, and experimentation on humans, which arise from the two moral norms: beneficence (the provision of benefits) and nonmaleficence (the avoidance of harm).
- Discuss the moral judgments associated with present policies regarding the regulation of the development and use of new medical devices.

The tremendous infusion of technology into the practice of medicine has created a new medical era. Advances in material science have led to the production of artificial limbs, heart valves, and blood vessels, thereby permitting "spare-parts" surgery. Numerous patient disorders are now routinely diagnosed using a wide range of highly sophisticated imaging devices, and the lives of many patients are being extended through significant improvements in resuscitative and supportive devices such as respirators, pacemakers, and artificial kidneys.

These technological advances, however, have not been entirely benign. They have had significant moral consequences. Provided with the ability to develop cardiovascular assist devices, perform organ transplants, and maintain the breathing and heartbeat of terminally ill patients, society has been forced to reexamine the meaning of such terms as *death, quality of life, heroic efforts*, and *acts of mercy*, and consider such moral issues as the right of patients to refuse treatment (living wills) and to participate in experiments (informed consent). As a result, these technological advances have made the moral dimensions of health care more complex, and have posed new and troubling moral dilemmas for medical professionals, biomedical engineers, and society at large.

The purpose of this chapter is to examine some of the moral questions related to the use of new medical technologies. The objective, however, is not to provide solutions or recommendations for these questions. Rather, the intent is to demonstrate that each technological advance has consequences that affect the very core of human values.

Technology and ethics are not foreigners; they are neighbors in the world of human accomplishment. Technology is a human achievement of extraordinary ingenuity and utility and is quite distant from the human accomplishment of ethical values. They face each other, rather than interface. The personal face of ethics looks at the impersonal face of technology to comprehend technology's potential and its limits. The face of technology looks to ethics to be directed to human purposes and benefits.

In the process of making technology and ethics face each other, it is our hope that individuals engaged in the development of new medical devices, as well as those responsible for the care of patients, will be stimulated to examine and evaluate critically "accepted" views and to reach their own conclusions. This chapter, therefore, begins with some definitions related to morality and ethics, followed by a more detailed discussion of some of the moral issues of special importance to biomedical engineers.

2.1 MORALITY AND ETHICS: A DEFINITION OF TERMS

From the very beginning, individuals have raised concerns about the nature of life and its significance. Many of these concerns have been incorporated into the four fundamental questions posed by the German philosopher Immanuel Kant (1724–1804): What can I know? What ought I to do? What can I hope? What is man? Evidence that early societies raised these questions can be found in the generation of rather complex codes of conduct embedded in the customs of the earliest human social organization, the tribe. By 600 BC, the Greeks were successful in reducing many primitive speculations, attitudes, and views on these questions to some type of order or system and integrating them into the general body of wisdom called philosophy. Being seafarers and colonizers, the Greeks had close contact with many different peoples and cultures. In the process, struck by the variety of customs, laws, and institutions that prevailed in the societies that surrounded them, they began to examine and compare all human conduct in these societies. This part of philosophy they called ethics.

The term *ethics* comes from the Greek *ethos*, meaning "custom." On the other hand, the Latin word for custom is *mos*, and its plural, *mores*, is the equivalent of the Greek *ethos* and the root of the words *moral* and *morality*. Although both terms (ethics and morality) are often used interchangeably, there is a distinction between them that should be made.

Philosophers define ethics as a particular kind of study and use morality to refer to its subject matter. For example, customs that result from some abiding principal human interaction are called *morals*. Some examples of morals in our society are telling the truth, paying one's debts, honoring one's parents, and respecting the rights and property of others. Most members of society usually consider such conduct not only customary but also correct or right. Thus, morality encompasses what people believe to be right and good and the reasons they give for it.

Most of us follow these rules of conduct and adjust our lifestyles in accordance with the principles they represent. Many even sacrifice life itself rather than diverge from them, applying them not only to their own conduct, but also to the behavior of others. Individuals who disregard these accepted codes of conduct are considered deviants and, in many cases, are punished for engaging in an activity that society as a whole considers unacceptable. For example, individuals committing "criminal acts" (defined by society) are often "outlawed" and, in many cases, severely punished. These judgments regarding codes of conduct, however, are not inflexible; they must continually be modified to fit changing conditions and thereby avoid the trauma of revolution as the vehicle for change.

Morality represents the codes of conduct of a society, but ethics is the study of right and wrong, of good and evil in human conduct. Ethics is not concerned with providing any judgments or specific rules for human behavior, but rather with providing an objective analysis about what individuals "ought to do." Defined in this way, it represents the philosophical view of morals, and, therefore, is often referred to as moral philosophy.

Consider the following three questions: (1) Should badly deformed infants be kept alive?; (2) Should treatment be stopped to allow a terminally ill patient to die?; (3) Should humans be used in experiments? Are these questions of morality or ethics? In terms of the definitions just provided, all three of these inquiries are questions of moral judgment.

Philosophers argue that all moral judgments are considered to be normative judgments, which can be recognized simply by their characteristic evaluative terms such as good, bad, right, or wrong. Typical normative judgments include

- Stealing is wrong.
- Everyone ought to have access to an education.
- Voluntary euthanasia should not be legalized.

Each of these judgments expresses an evaluation (i.e., conveys a negative or positive attitude toward some state of affairs). Each, therefore, is intended to play an action-guiding function.

Arriving at moral judgments, however, requires knowledge of valid moral standards in our society. How is such knowledge obtained? The efforts to answer this question lie in two competing schools of thought that currently dominate normative ethical theory: utilitarianism, a form of consequentialism, and Kantianism, a form of nonconsequentialism. Consequentialism holds that the morally right action is always the one among the available options that has the best consequences. An important implication of consequentialism is that no specific actions or courses of conduct are automatically ruled out as immoral or ruled in as morally obligatory. The rightness or wrongness of an action is wholly contingent upon its effects.

According to utilitarianism, there are two steps to determining what ought to be done in any situation. First, determine which courses of action are open. Second, determine the consequences of each alternative. When this has been accomplished, the morally right course of action is the one that maximizes pleasure, minimizes pain, or both; the one that does the "greatest good for the greatest number." Because the central motivation driving the design, development, and use of medical devices is improvement of medicine's capacity to protect and restore health, an obvious virtue of utilitarianism is that it assesses medical technology in terms of what many believe makes health valuable: the attainment of well-being and the avoidance of pain.

Utilitarianism, therefore, advocates that the end justifies the means. As long as any form of treatment maximizes good consequences, it should be used. Many people, though, believe that the end does not always justify the means and that individuals have rights that are not to be violated no matter how good the consequences might be.

In opposition to utilitarianism stands the school of normative ethical thought known as nonconsequentialism. Proponents of this school deny that moral evaluation is simply and wholly a matter of determining the consequences of human conduct. They agree that other considerations are relevant to moral assessment and so reject the view that morally right conduct is whatever has the best consequences. Based largely on the views of Immanuel Kant, this ethical school of thought insists that there is something uniquely precious about human beings from the moral point of view. According to Kant's theory, humans have certain rights that do not apply to any other animal. For example, the moral judgments that we should not kill and eat each other for food or hunt each other for sport or experiment on each other for medical science are all based on this view of human rights. Humans are, in short, owed a special kind of respect simply because they are people.

To better understand the Kantian perspective, it may be helpful to recognize that Kant's views are an attempt to capture in secular form a basic tenet of Christian morality. What makes human beings morally special entities deserving a unique type of respect? Christianity answers in terms of the doctrine of ensoulment. This doctrine holds that only human beings are divinely endowed with an eternal soul. According to Christian ethics, the soul makes humans the only beings with intrinsic value. Kant's secular version of the doctrine of ensoulment asserts that human beings are morally unique and deserve special respect because of their autonomy. Autonomy is taken by Kant to be the capacity to make choices based on rational deliberation. The central task of ethics then is to specify what human conduct is required to respect the unique dignity of human beings. For most Kantians, this means determining what limits human beings must observe in the way they treat each other and this, in turn, is taken to be a matter of specifying each individual's fundamental moral rights.

These two ethical schools of thought, therefore, provide some rationale for moral judgments. However, when there is no clear moral judgment, one is faced with a dilemma. In medicine, moral dilemmas arise in those situations that raise fundamental questions about right and wrong in the treatment of sickness and the promotion of health in patients. In many of these situations the health professional faces two alternative choices, neither of which seems to be a satisfactory solution to the problem. For example, is it more important to preserve life or prevent pain? Is it right to withhold treatment when doing so may lead to a shortening of life? Does an individual have the right to refuse treatment when refusing it may lead to death? All these situations seem to have no clear-cut imperative based on our present set of convictions about right and wrong. That is the dilemma raised by Kant: what ought I do?

Case Study: Stem Cell Research

At the moment of conception, that is to say when sperm unites with egg, the process of fertilization occurs (Fig 2.1). The formation of an embryo is initiated. Once the sperm enters the egg, there is an immediate opening of ion channels, which depolarizes the plasma membrane of the cell, and prevents other sperm from fusing with it. DNA replication then begins, and the first cell division occurs approximately 36 hours later. As the process continues, the cell begins to experience cleavage, in which the cells repeatedly divide, cycling between the S (DNA synthesis) and M (mitosis) phases of cell division, essentially skipping the G_1 and G_2 phases, when most cell growth normally occurs. Thus, there is no net growth of the cells, merely subdivision into smaller cells, individually called blastomeres.

(continued)

Case Study: Stem Cell Research (*Continued*)

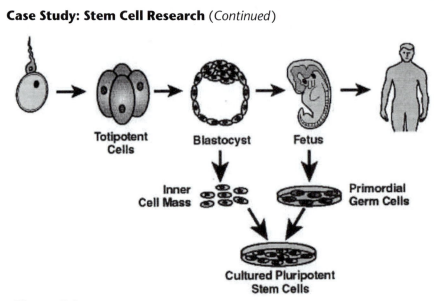

Figure 2.1   An illustration of the use of inner cell mass to form pluripotent stem cells (Courtesy of http://www.nih.gov/news/stemcell/primer.htm).

Five days subsequent to fertilization, the number of cells composing the embryo is in the hundreds, and the cells form tight junctions characteristic of a compact epithelium, which is arranged around a central cavity. This is the embryonic stage known as the blastocyst. Within the cavity exists a mass of cells that protrude inward. These cells are known as the inner cell mass and become the embryo. The exterior cells are the trophoblast and eventually form the placenta. It is the cells from the inner cell mass of the blastocyst that, when isolated and grown in a culture, are identified as embryonic stem cells.

It is important to note that if cell division continues, determination and differentiation happen. Differentiation occurs when a cell begins to exhibit the specific attributes of a predestined specialized cellular role. Determination is related to differentiation, but is somewhat dissimilar. When a cell group that has been determined is transplanted, it will not assimilate with the other cells, but will rather grow into cells that comprised the original organ it was destined to become.

Because the process of obtaining embryonic stem cells destroys the embryo, the following questions arise:

Is the embryo a living human being, entitled to all of the same rights that a human at any other age would be granted? Discuss the answer to this question from a utilitarian and a Kantian point of view.

Should any research that is potentially beneficial to the well-being of mankind be pursued?

Should the federal government support such research?

In the practice of medicine, moral dilemmas are certainly not new. They have been present throughout medical history. As a result, over the years there have been efforts to provide a set of guidelines for those responsible for patient care. These efforts have resulted in the development of specific codes of professional conduct. Let us examine some of these codes or guidelines.

For the medical profession, the World Medical Association adopted a version of the Hippocratic oath entitled the Geneva Convention Code of Medical Ethics in 1949. This declaration contains the following statements:

I solemnly pledge myself to consecrate my life to the services of humanity;
I will give to my teachers the respect and gratitude which is their due;
I will practice my profession with conscience and dignity;
The health of my patient will be my first consideration;
I will respect the secrets which are confided in me;
I will maintain by all the means in my power, the honour and the noble traditions of the medical profession;
My colleagues will be my brothers;
I will not permit considerations of religion, nationality, race, party politics or social standing to intervene between my duty and my patient;
I will maintain the utmost respect for human life from the time of conception; even under threat;
I will not use my medical knowledge contrary to the laws of humanity;
I make these promises solemnly, freely and upon my honour.

In the United States, the American Medical Association (AMA) adopted a set of Principles of Medical Ethics in 1980, and revised them in June, 2001. Following is a comparison of the two sets of principles.

*REVISED PRINCIPLES**

Version adopted by the AMA House of Delegates, June 17, 2001

The medical profession has long subscribed to a body of ethical statements developed primarily for the benefit of the patient. As a member of this profession, a physician must recognize responsibility ~~not only to patients, but also~~ to patients first and foremost, as well as to society, to other health professionals, and to self. The following Principles adopted by the American Medical Association are not laws, but standards of conduct which define the essentials of honorable behavior for the physician.

I. A physician shall be dedicated to providing competent medical care ~~service~~, with compassion and respect for human dignity and rights.
II. A physician shall ~~deal honestly with patients and colleagues~~ uphold the standards of

PREVIOUS PRINCIPLES

As adopted by the AMA's House of Delegates, 1980

The medical profession has long subscribed to a body of ethical statements developed primarily for the benefit of the patient. As a member of this profession, a physician must recognize responsibility not only to patients, but also to society, to other health professionals, and to self. The following Principles adopted by the American Medical Association are not laws, but standards of conduct which define the essentials of honorable behavior for the physician.

I. A physician shall be dedicated to providing competent medical service with compassion and respect for human dignity.
II. A physician shall deal honestly with patients and colleagues, and strive to expose

<u>professionalism, be honest in all professional interactions,</u> and strive to <u>report,</u> ~~expose those~~ physicians deficient in character or competence, or <u>engaging</u> ~~who engage~~ in fraud or deception, <u>to appropriate entities.</u>

III. A physician shall respect the law and also recognize a responsibility to seek changes in those requirements which are contrary to the best interests of the patient.

IV. A physician shall respect the rights of patients, of colleagues, and of other health professionals, and shall safeguard patient confidences <u>and privacy</u> within the constraints of the law.

V. A physician shall continue to study, apply, and advance scientific knowledge, <u>maintain a commitment to medical education,</u> make relevant information available to patients, colleagues, and the public, obtain consultation, and use the talents of other health professionals when indicated.

VI. A physician shall, in the provision of appropriate patient care, except in emergencies, be free to choose whom to serve, with whom to associate, and the environment in which to provide medical <u>care</u> ~~services.~~

VII. A physician shall recognize a responsibility to participate in activities contributing to <u>the improvement of the</u> ~~an improved~~ community <u>and the betterment of public health.</u>

VIII. <u>A physician shall, while caring for a patient, regard responsibility to the patient as paramount.</u>

IX. <u>A physician shall support access to medical care for all people.</u>

those physicians deficient in character or competence, or who engage in fraud or deception.

III. A physician shall respect the law and also recognize a responsibility to seek changes in those requirements which are contrary to the best interests of the patient.

IV. A physician shall respect the rights of patients, of colleagues, and of other health professionals, and shall safeguard patient confidences within the constraints of the law.

V. A physician shall continue to study, apply and advance scientific knowledge, make relevant information available to patients, colleagues, and the public, obtain consultation, and use the talents of other health professionals when indicated.

VI. A physician shall, in the provision of appropriate patient care, except in emergencies, be free to choose whom to serve, with whom to associate, and the environment in which to provide medical services.

VII. A physician shall recognize a responsibility to participate in activities contributing to an improved community.

For the nursing profession, the American Nurses Association formally adopted in 1976 the Code For Nurses, whose statements and interpretations provide guidance for conduct and relationships in carrying out nursing responsibilities.

PREAMBLE: The Code for Nurses is based on belief about the nature of individuals, nursing, health, and society. Recipients and providers of nursing services are viewed as individuals and groups who possess basic rights and responsibilities, and whose values and circumstances command respect at all times. Nursing encompasses the promotion and restoration of health, the prevention of illness, and the alleviation of suffering. The statements of the Code and their interpretation provide guidance for conduct and relationships in carrying out nursing responsibilities consistent with the ethical obligations of the profession and quality in nursing care.

1. The nurse provides services with respect for human dignity and the uniqueness of the client unrestricted by considerations of social or economic status, personal attributes, or the nature of health problems.
2. The nurse safeguards the client's right to privacy by judiciously protecting information of a confidential nature.
3. The nurse acts to safeguard the client and the public when health care and safety are affected by the incompetent, unethical, or illegal practice of any person.
4. The nurse assumes responsibility and accountability for individual nursing judgments and actions.
5. The nurse maintains competence in nursing.
6. The nurse exercises informed judgment and uses individual competence and qualifications as criteria in seeking consultation, accepting responsibilities, and delegating nursing activities to others.
7. The nurse participates in activities that contribute to the ongoing development of the profession's body of knowledge.
8. The nurse participates in the profession's efforts to implement and improve standards of nursing.
9. The nurse participates in the profession's efforts to establish and maintain conditions of employment conducive to high-quality nursing care.
10. The nurse participates in the profession's effort to protect the public from misinformation and misrepresentation and to maintain the integrity of nursing.
11. The nurse collaborates with members of the health professions and other citizens in promoting community and national efforts to meet the health needs of the public.

These codes take as their guiding principle the concepts of service to humankind and respect for human life. When reading these codes of conduct, it is difficult to imagine that anyone could improve on them as summary statements of the primary goals of individuals responsible for the care of patients. However, some believe that such codes fail to provide answers to many of the difficult moral dilemmas confronting health professionals today. For example, in many situations, all the fundamental responsibilities of the nurse cannot be met at the same time. When a patient suffering from a massive insult to the brain is kept alive by artificial means and this equipment is needed elsewhere, it is not clear from these guidelines how "nursing competence is to be maintained to conserve life and promote health." Although it may be argued that the decision to treat or not to treat is a medical and not a nursing decision, both professions are so intimately involved in the care of patients that they are both concerned with the ultimate implications of any such decision.

For biomedical engineers, an increased awareness of the ethical significance of their professional activities has also resulted in the development of codes of professional ethics. Typically consisting of a short list of general rules, these codes express both the minimal standards to which all members of a profession are expected to conform and the ideals for which all members are expected to strive. Such codes provide a practical guide for the ethical conduct of the profession's practitioners. Consider, for example, the code of ethics endorsed by the American College of Clinical Engineers:

As a member of the American College of Clinical Engineering, I subscribe to the established Code of Ethics in that I will:

- Accurately represent my level of responsibility, authority, experience, knowledge, and education.
- Strive to prevent a person from being placed at risk due to the use of technology.
- Reveal conflicts of interest that may affect information provided or received.
- Respect the confidentiality of information.
- Work toward improving the delivery of health care.
- Work toward the containment of costs by the better management and utilization of technology.
- Promote the profession of clinical engineering.

Although these codes can be useful in promoting ethical conduct, such rules obviously cannot provide ethical guidance in every situation. A profession that aims to maximize the ethical conduct of its members must not limit the ethical consciousness of its members to knowledge of their professional code alone. It must also provide them with resources that will enable them to determine what the code requires in a particular concrete situation, and thereby enable them to arrive at ethically sound judgments in situations in which the directives of the code are ambiguous or simply do not apply.

2.2   TWO MORAL NORMS: BENEFICENCE AND NONMALEFICENCE

Two moral norms have remained relatively constant across the various moral codes and oaths that have been formulated for health care providers since the beginnings of Western medicine in classical Greek civilization. They are beneficence, the provision of benefits, and nonmaleficence, the avoidance of doing harm. These norms are traced back to a body of writings from classical antiquity known as the *Hippocratic Corpus*. Although these writings are associated with the name of Hippocrates, the acknowledged founder of Western medicine, medical historians remain uncertain whether any, including the Hippocratic oath, were actually his work. Although portions of the Corpus are believed to have been authored during the sixth century BC, other portions are believed to have been written as late as the beginning of the Christian era. Medical historians agree that many of the specific moral directives of the *Corpus* represent neither the actual practices nor the moral ideals of the majority of physicians of ancient Greece and Rome.

Nonetheless, the general injunction, "As to disease, make a habit of two things: (1) to help or, (2) at least, to do no harm," was accepted as a fundamental medical ethical norm by at least some ancient physicians. With the decline of Hellenistic civilization and the rise of Christianity, beneficence and nonmaleficence became increasingly accepted as the fundamental principles of morally sound medical practice. Although beneficence and nonmaleficence were regarded merely as concomitant to the craft of medicine in classical Greece and Rome, the emphasis upon compassion and the brotherhood of humankind, central to Christianity, increasingly made these norms the only acceptable motives for medical practice. Even today, the provision of benefits and the avoidance of doing harm are stressed just as much in virtually all contemporary Western codes of conduct for health professionals as they were in the oaths and codes that guided the health care providers of past centuries.

Traditionally, the ethics of medical care have given greater prominence to nonmaleficence than to beneficence. This priority was grounded in the fact that, historically, medicine's capacity to do harm far exceeded its capacity to protect and restore health. Providers of health care possessed many treatments that posed clear and genuine risks to patients and that offered little prospect of benefit. Truly effective therapies were all too rare. In this context, it is surely rational to give substantially higher priority to avoiding harm than to providing benefits.

The advent of modern science changed matters dramatically. Knowledge acquired in laboratories, tested in clinics, and verified by statistical methods has increasingly dictated the practice of medicine. This ongoing alliance between medicine and science became a critical source of the plethora of technologies that now pervade medical care. The impressive increases in therapeutic, preventive, and rehabilitative capabilities that these technologies have provided have pushed beneficence to the forefront of medical morality. Some have even gone so far as to hold that the old medical ethic of "Above all, do no harm" should be superseded by the new ethic that "The patient deserves the best." However, the rapid advances in medical technology capabilities have also produced great uncertainty as to what is most beneficial or least harmful for the patient. In other words, along with increases in the ability to be beneficent, medicine's technology has generated much debate about what actually counts as beneficent or nonmaleficent treatment. Having reviewed some of the fundamental concepts of ethics and morality, let us now turn to several specific moral issues posed by the use of medical technology.

2.3 REDEFINING DEATH

Although medicine has long been involved in the observation and certification of death, many of its practitioners have not always expressed philosophical concerns regarding the beginning of life and the onset of death. Since medicine is a clinical and empirical science, it would seem that health professionals had no medical need to consider the concept of death; the fact of death was sufficient. The distinction between life and death was viewed as the comparison of two extreme conditions separated by an infinite chasm. With the advent of technological advances in medicine to assist health professionals to prolong life, this view has changed.

There is no doubt that the use of medical technology has in many instances warded off the coming of the grim reaper. One need only look at the trends in average life expectancy for confirmation. For example, in the United States today, the average life expectancy for males is 74.3 years and for females 76 years, whereas in 1900 the average life expectancy for both sexes was only 47 years. Infant mortality has been significantly reduced in developed nations where technology is an integral part of the culture. Premature births no longer constitute a threat to life because of the artificial environment that medical technology can provide. Today, technology has not only helped individuals avoid early death but has also been effective in delaying the inevitable. Pacemakers, artificial kidneys, and a variety of other medical devices have enabled individuals to add many more productive years to their lives. Technology has been so successful that health professionals responsible for the care of critically ill patients have been able to maintain their "vital signs of life" for extensive periods of time. In the process, however, serious philosophical questions concerning the quality of the life provided to these patients have arisen.

Consider the case of the patient who sustains a serious head injury in an automobile accident. To the attendants in the ambulance who reached the scene of the accident, the patient was unconscious, but still alive with a beating heart. After the victim was rushed to the hospital and into the emergency room, the resident in charge verified the stability of the vital signs of heartbeat and respiration during examination and ordered a computerized tomography (CT) scan to indicate the extent of the head injury. The results of this procedure clearly showed extensive brain damage. When the EEG was obtained from the scalp electrodes placed about the head, it was noted to be significantly abnormal. In this situation, then, the obvious questions arise: What is the status of the patient? Is the patient alive?

Alternatively, consider the events encountered during one open-heart surgery. During this procedure, the patient was placed on the heart bypass machine while the surgeon attempted to correct a malfunctioning valve. As the complex and long operation continued, the EEG monitors that had indicated a normal pattern of electrical activity at the onset of the operation suddenly displayed a relatively straight line indicative of feeble electrical activity. However, since the heart–lung bypass was maintaining the patient's so-called vital signs, what should the surgeon do? Should the medical staff continue on the basis that the patient is alive, or is the patient dead?

The increasing occurrence of these situations has stimulated health professionals to reexamine the definition of death. In essence, advances in medical technology that delay death actually hastened its redefinition. This should not be so surprising because the definition of death has always been closely related to the extent of medical knowledge and available technology. For many centuries, death was defined solely as the absence of breathing. Since it was felt that the spirit of the human being resided in the spiritus (breath), its absence became indicative of death. With the continuing proliferation of scientific information regarding human physiology and the development of techniques to revive a nonbreathing person, attention turned to the pulsating heart as the focal point in determination of death. However, this view was to change through additional medical and technological advances in supportive therapy, resuscitation, cardiovascular assist devices, and organ transplantation.

As understanding of the human organism increased, it became obvious that one of the primary constituents of the blood is oxygen and that any organ deprived of oxygen for a specified period of time will cease to function and die. The higher functions of the brain are particularly vulnerable to this type of insult, and the removal of oxygen from the blood supply even for a short period of time (3 minutes) produces irreversible damage to the brain tissues. Consequently, the evidence of death began to shift from the pulsating heart to the vital, functioning brain. Once medicine was provided with the means to monitor the brain's activity (i.e., the EEG), another factor was introduced in the definition of death. Advocates of the concept of brain death argued that the human brain is truly essential to life. When the brain is irreversibly damaged, so are the functions that are identified with self and our own humanness: memory, feeling, thinking, and knowledge.

As a result, it became widely accepted that in clinical death the spontaneous activity of the lungs, heart, and brain is no longer present. The irreversible cessation of functioning of all three major organs (i.e., heart, lungs, and brain) was required before anyone was pronounced dead. Although damage to any other organ system such as the liver or kidney may ultimately cause the death of the individual through a fatal effect on the essential functions of the heart, lungs, or brain, this aspect was not included in the definition of clinical death.

With the development of modern respirators, however, the medical profession encountered an increasing number of situations in which a patient with irreversible brain damage could be maintained almost indefinitely. Once again, a new technological advance created the need to reexamine the definition of death.

The movement toward redefining death received considerable impetus with the publication of a report sponsored by the Ad Hoc Committee of the Harvard Medical School in 1968, in which the committee offered an alternative definition of death based on the functioning of the brain. The report of this committee was considered a landmark attempt to deal with death in light of technology.

In summary, the criteria for death established by this committee included the following: (1) the patient must be unreceptive and unresponsive, that is, in a state of irreversible coma; (2) the patient must have no movements of breathing when the mechanical respirator is turned off; (3) the patient must not demonstrate any reflexes; and (4) the patient must have a flat EEG for at least 24 hours, indicating no electrical brain activity. When these criteria are satisfied, then death may be declared.

At the time, the committee also strongly recommended that the decision to declare the person dead and then to turn off the respirator should not be made by physicians involved in any later efforts to transplant organs or tissues from the deceased individual. In this way, a prospective donor's death would not be hastened merely for the purpose of transplantation. Thus, complete separation of authority and responsibility for the care of the recipient from the physician or group of physicians responsible for the care of the prospective donor is essential.

The shift to a brain-oriented concept involved deciding that much more than just biological life is necessary to be a human person. The brain death concept was essentially a statement that mere vegetative human life is not personal human life. In other words, an otherwise intact and alive but brain-dead person is not a human person. Many of us have taken for granted the assertion that being truly alive in this world requires an "intact functioning brain." Yet, precisely this issue was at stake in the gradual movement from using heartbeat and respiration as indices of life to using brain-oriented indices instead.

Indeed, total and irreparable loss of brain function, referred to as brainstem death, whole brain death, or, simply, brain death, has been widely accepted as the legal standard for death. By this standard, an individual in a state of brain death is legally indistinguishable from a corpse and may be legally treated as one even though respiratory and circulatory functions may be sustained through the intervention of technology. Many take this legal standard to be the morally appropriate one, noting that once destruction of the brain stem has occurred, the brain cannot function at all, and the body's regulatory mechanisms will fail unless artificially sustained. Thus mechanical sustenance of an individual in a state of brain death is merely postponement of the inevitable and sustains nothing of the personality, character, or consciousness of the individual. It is simply the mechanical intervention that differentiates such an individual from a corpse and a mechanically ventilated corpse is a corpse nonetheless.

Even with a consensus that brainstem death is death, and thus that an individual in such a state is indeed a corpse, difficult cases remain. Consider the case of an individual in a persistent vegetative state, the condition known as neocortical death. Although severe brain injury has been suffered, enough brain function remains to make mechanical sustenance of respiration and circulation unnecessary. In a persistent vegetative state, an individual exhibits no purposeful response to external stimuli and no evidence of self-awareness. The eyes may open periodically and the individual may exhibit sleep–wake cycles. Some patients even yawn, make chewing motions, or swallow spontaneously. Unlike the complete unresponsiveness of individuals in a state of brainstem death, a variety of simple and complex responses can be elicited from an individual in a persistent vegetative state. Nonetheless, the chances that such an individual will regain consciousness are remote. Artificial feeding, kidney dialysis, and the like make it possible to sustain an individual in a state of neocortical death for decades. This sort of condition and the issues it raises are exemplified by the famous case of Karen Ann Quinlan.

In April 1975, this young woman suffered severe brain damage and was reduced to a chronic vegetative state in which she no longer had any cognitive function. Accepting the doctors' judgment that there was no hope of recovery, her parents sought permission from the courts to disconnect the respirator that was keeping her alive in the intensive care unit of a New Jersey hospital.

The trial court, and then the Supreme Court of New Jersey, agreed that Karen's respirator could be removed. So it was disconnected. However, the nurse in charge of her care in the Catholic hospital opposed this decision and, anticipating it, had begun to wean her from the respirator so that by the time it was disconnected she could remain alive without it. So, Karen did not die. She remained alive for ten additional years. In June 1985, she finally died of acute pneumonia. Antibiotics, which would have fought the pneumonia, were not given.

If brainstem death is death, is neocortical death also death? Again, the issue is not a straightforward factual matter. For it, too, is a matter of specifying which features of living individuals distinguish them from corpses and so make treatment of them as corpses morally impermissible. Irreparable cessation of respiration and circulation, the classical criterion for death, would entail that an individual in a persistent vegetative state is not a corpse and so, morally speaking, must not be treated as one. The brainstem death criterion for death would also entail that a person in a state of neocortical death is not yet a corpse. On this criterion, what is crucial is that brain damage be severe enough to cause failure of the regulatory mechanisms of the body.

Is an individual in a state of neocortical death any less in possession of the characteristics that distinguish the living from cadavers than one whose respiration and circulation are mechanically maintained? It is a matter that society must decide. And until society decides, it is not clear what counts as beneficent or nonmaleficent treatment of an individual in a state of neocortical death.

2.4   THE TERMINALLY ILL PATIENT AND EUTHANASIA

Terminally ill patients today often find themselves in a strange world of institutions and technology devoted to assisting them in their fight against death. However, at the same time, this modern technologically oriented medical system may cause patients and their families considerable economic, psychological, and physical pain. In enabling medical science to prolong life, modern technology has in many cases made dying slower and more undignified. As a result of this situation, there is a moral dilemma in medicine. Is it right or wrong for medical professionals to stop treatment or administer a lethal dose to terminally ill patients?

This problem has become a major issue for our society to consider. Although death is all around us in the form of accidents, drug overdose, alcoholism, murder, and suicide, for most of us the end lies in growing older and succumbing to some form of chronic illness. As the aged approach the end of life's journey, they may eventually wish for the day when all troubles can be brought to an end. Such a desire, frequently shared by a compassionate family, is often shattered by therapies provided with only one concern: to prolong life regardless of the situation. As a result, many claim a dignified death is often not compatible with today's standard medical view.

Consider the following hypothetical version of the kind of case that often confronts contemporary patients, their families, health care workers, and society as a whole. Suppose a middle-aged man suffers a brain hemorrhage and loses consciousness as a result of a ruptured aneurysm. Suppose that he never regains consciousness and is hospitalized in a state of neocortical death, a chronic vegetative state. His life is maintained by a surgically implanted gastronomy tube that drips liquid nourishment from a plastic bag directly into his stomach. The care of this individual takes seven and one-half hours of nursing time daily and includes shaving, oral hygiene, grooming, attending to his bowels and bladder, and so forth. Suppose further that his wife undertakes legal action to force his caregivers to end all medical treatment, including nutrition and hydration, so that complete bodily death of her husband will occur. She presents a preponderance of evidence to the court to show that her husband would have wanted just this result in these circumstances.

The central moral issue raised by this sort of case is whether the quality of the individual's life is sufficiently compromised to make intentional termination of that life morally permissible. While alive, he made it clear to both family and friends that he would prefer to be allowed to die rather than be mechanically maintained in a condition of irretrievable loss of consciousness. Deciding whether his judgment in such a case should be allowed requires deciding which capacities and qualities make life worth living, which qualities are sufficient to endow it with value worth sustaining, and whether their absence justifies deliberate termination of a life, at least when this would be the wish of the individual in question. Without this decision, the traditional norms of medical ethics, beneficence and nonmaleficence, provide no guidance. Without this decision, it cannot be determined whether termination of life support is a benefit or harm to the patient.

For many individuals, the fight for life is a correct professional view. They believe that the forces of medicine should always be committed to using innovative ways of prolonging life for the individual. However, this cannot be the only approach to caring for the terminally ill. Certain moral questions regarding the extent to which physicians engage in heroic efforts to prolong life must be addressed if the individual's rights are to be preserved. The goal of those responsible for patient care should not solely be to prolong life as long as possible by the extensive use of drugs, operations, respirators, hemodialyzers, pacemakers, and the like, but rather to provide a reasonable quality of life for each patient. It is out of this new concern that euthanasia has once again become a controversial issue in the practice of medicine.

The term *euthanasia* is derived from two Greek words meaning "good" and "death." Euthanasia was practiced in many primitive societies in varying degrees. For example, on the island of Cos, the ancient Greeks assembled elderly and sick people at an annual banquet to consume a poisonous potion. Even Aristotle advocated euthanasia for gravely deformed children. Other cultures acted in a similar manner toward their aged by abandoning them when they felt these individuals no longer served any useful purpose. However, with the spread of Christianity in the Western world, a new attitude developed toward euthanasia. Because of the Judeo-Christian belief in the biblical statements "Thou shalt not kill" (Exodus 20:13) and "He who kills a man should be put to death" (Leviticus 24:17), the practice of euthanasia decreased. As a result of these moral judgments, killing was considered a sin, and the decision about whether someone should live or die was viewed solely as God's responsibility.

In today's society, euthanasia implies to many "death with dignity," a practice to be followed when life is merely being prolonged by machines and no longer seems to have value. In many instances, it has come to mean a contract for the termination of life to avoid unnecessary suffering at the end of a fatal illness and, therefore, has the connotation of relief from pain.

Discussions of the morality of euthanasia often distinguish active from passive euthanasia, a distinction that rests on the difference between an act of commission and an act of omission. When failure to take steps that could effectively forestall death results in an individual's demise, the resultant death is an act of omission and a case of letting a person die. When a death is the result of doing something to hasten the end of a person's life (for example, giving a lethal injection), that death is caused by an act of commission and is a case of killing a person. The important difference between active and passive euthanasia is that in passive euthanasia, the physician does not do anything to bring about the patient's death. The physician does nothing, and death results due to whatever illness already afflicts the patient. In active euthanasia, however, the physician does something to bring about the patient's death. The physician who gives the patient with cancer a lethal injection has caused the patient's death, whereas if the physician merely ceases treatment, the cancer is the cause of death.

In active euthanasia, someone must do something to bring about the patient's death, and in passive euthanasia, the patient's death is caused by illness rather than by anyone's conduct. Is this notion correct? Suppose a physician deliberately decides not to treat a patient who is terminally ill, and the patient dies. Suppose further that the physician were to attempt self-exoneration by saying, "I did nothing. The patient's death was the result of illness. I was not the cause of death." Under current legal and moral norms, such a response would have no credibility. The physician would be blameworthy for the patient's death as surely as if he or she had actively killed the patient. Thus, the actions taken by a physician to continue treatment to the very end are understood.

Euthanasia may also be classified as involuntary or voluntary. An act of euthanasia is involuntary if it hastens the individual's death for his or her own good, but against the individual's wishes. Involuntary euthanasia, therefore, is no different in any morally relevant way from unjustifiable homicide. However, what happens when the individual is incapable of agreeing or disagreeing? Suppose that a terminally ill person is unconscious and cannot make his or her wishes known. Would hastening that person's death be permissible? It would be if there was substantial evidence that the individual had given prior consent. The individual may have told friends and relatives that, under certain circumstances, efforts to prolong his or her life should not be undertaken or continued and might even have recorded those wishes in the form of a living will or an audiotape or videotape. When this level of substantial evidence of prior consent exists, the decision to hasten death would be morally justified. A case of this sort would be a case of voluntary euthanasia.

For a living will to be valid, the person signing it must be of sound mind at the time the will is made and shown not to have altered his or her opinion in the interim between its signing and the onset of the illness. In addition, the witnesses must not be able to benefit from the individual's death. As the living will itself states, it is not a legally binding document. It is essentially a passive request and depends on moral persuasion. Proponents of the will, however, believe that it is valuable in relieving the burden of guilt often carried by health professionals and the family in making the decision to allow a patient to die.

Those who favor euthanasia point out the importance of individual rights and freedom of choice and look on euthanasia as a kindness ending the misery endured by the patient. The thought of a dignified death is much more attractive than the process of continuous suffering and gradual decay into nothingness. Viewing each person as a rational being possessing a unique mind and personality, proponents argue that terminally ill patients should have the right to control the ending of their own life.

On the other hand, those opposed to euthanasia demand to know who has the right to end the life of another. Some use religious arguments, emphasizing that euthanasia is in direct conflict with the belief that God, and God alone, has the power to decide when a human life ends. Their view is that anyone who practices euthanasia is essentially acting in the place of God, and that no human should ever be considered omnipotent.

Others turn to the established codes, reminding those responsible for the care of patients that they must do whatever is in their power to save a life. Their argument is

A Living Will

TO MY FAMILY, MY PHYSICIAN, MY CLERGYMAN, MY LAWYER:

If the time comes when I can no longer take part in decisions about my own future, let this statement stand as testament of my wishes: If there is no reasonable expectation of my recovery from physical or mental disability, I request that I be allowed to die and not be kept alive by artificial means or heroic measures. Death is as much a reality as birth, growth, maturity, and old age—it is the one certainty. I do not fear death as much as I fear the indignity of deterioration, dependence, and hopeless pain. I ask that drugs be mercifully administered to me for the terminal suffering even if they hasten the moment of death. This request is made after careful consideration. Although this document is not legally binding, you who care for me will, I hope, feel morally bound to follow its mandate. I recognize that it places a heavy burden of responsibility upon you, and it is with the intention of sharing that responsibility and of mitigating any feelings of guilt that this statement is made.

Signed_____

Date_____

Witnessed by_____

that health professionals cannot honor their pledge and still believe that euthanasia is justified. If terminally ill patients are kept alive, there is at least a chance of finding a cure that might be useful to them. Opponents of euthanasia feel that legalizing it would destroy the bonds of trust between doctor and patient. How would sick individuals feel if they could not be sure that their physician and nurse would try everything possible to cure them, knowing that if their condition worsened, they would just lose faith and decide that death would be better? Opponents of euthanasia also question whether it will be truly beneficial to the suffering person or will only be a means to relieve the agony of the family. They believe that destroying life (no matter how minimal) merely to ease the emotional suffering of others is indeed unjust.

Many fear that if euthanasia is legalized, it will be difficult to define and develop clear-cut guidelines that will serve as the basis for carrying out euthanasia. Furthermore, once any form of euthanasia is accepted by society, its detractors fear that many other problems will arise. Even the acceptance of passive euthanasia could, if carried to its logical conclusion, be applied in state hospitals and institutions for the mentally retarded and the elderly. Such places currently house thousands of people who have neither hope nor any prospect of a life that even approaches normality. Legalization of passive euthanasia could prompt an increased number of suits by parents seeking to end the agony of incurably afflicted children or by children seeking to shorten the suffering of aged and terminally ill parents. In Nazi Germany, for example, mercy killing was initially practiced to end the suffering of the terminally ill. Eventually, however, the practice spread, so that even persons with the slightest deviation from the norm (e.g., the mentally ill, minority groups such as Jews and others) were terminated. Clearly, the situation is delicate and thought provoking.

2.5   TAKING CONTROL

Medical care decisions can be tremendously difficult. They often involve unpleasant topics and arise when we are emotionally and physically most vulnerable. Almost always these choices involve new medical information that feels alien and can seem overwhelming. In an attempt to assist individuals to make these decisions, it is often helpful to follow the pathway outlined here:

1. Obtain all the facts (i.e., clarify the medical facts of the situation).
2. Understand all options and their consequences.
3. Place a value on each of the options based on your own set of personal values.

The three-step facts/options/values path concerns the "how" of decisions, but equally important is the "who." Someone must make every single medical decision. Ideally, decisions will be made by the person most intimately involved—the patient. Very often, however, they are made by someone else—spouse, family, physician, or by a group of those people acting on behalf of the patient. It is, therefore, important to recognize the four concentric circles of consent:

- The first, and primary, circle is the patient.
- The second circle is the use of advance directives, that is, choosing in advance through the use of such documents as the living will.
- The third circle is others deciding for the patient (i.e., the move from personal control to surrogate control).
- The fourth and final circle is the courts and bureaucrats. It is the arena of last resort where our society has decreed that we go when the patient may be incapacitated, when there is no clear advance directive, and when it is not clear who should make the decision.

These three steps and four circles are simply attempts to impose some order on the chaos that is medical decision making. They can help individuals take control.

2.6   HUMAN EXPERIMENTATION

Medical research has long held an exalted position in our modern society. It has been acclaimed for its significant achievements that range from the development of the Salk and Sabin vaccines for polio to the development of artificial organs. To determine their effectiveness and value, however, these new drugs and medical devices eventually are used on humans. The issue is, therefore, not only whether humans should be involved in clinical studies designed to benefit themselves or their fellow humans but also clarifying or defining more precisely the conditions under which such studies are to be permitted.

For example, consider the case of a 50-year-old female patient suffering from severe coronary artery disease. What guidelines should be followed in the process of experimenting with new drugs or devices that may or may not help her? Should only those procedures viewed as potentially beneficial to her be tried? Should experimental diagnostic procedures or equipment be tested on this patient to evaluate their effectiveness when compared to more accepted techniques, even though they will not be directly beneficial to the patient?

On the other hand, consider the situation of conducting research on the human fetus. This type of research is possible as a result of the legalization of abortion in the United States and the technological advances that have made fetal studies more practical than in the past. Under what conditions should medical research be conducted on these subjects? Should potentially hazardous drugs be given to women planning to have abortions to determine the effect of these drugs on the fetus? Should the fetus, once aborted, be used in any experimental studies?

Although these questions are difficult to answer, clinical researchers responsible for the well-being of their patients must face the moral issues involved in testing new equipment and procedures and at the same time safeguard the individual rights of their patients.

Case Study: Neonatal Intensive Care Unit (NICU)

Throughout time, low birth weight, oftentimes arising from premature birth, has been a major factor affecting infant survival. Underweight infants, who are typically classified as either low birth weight (LBW) (less than 1500 g) or very low birth weight (VLBW) (less than 1000 g), must be treated with the utmost caution and care to maximize their chances of survival. Advances in premature-infant medical care, such as improved thermoregulation and ventilation techniques, have greatly decreased the mortality rate among LBW and VLBW infants. Included in these advances was the creation of the NICU, where all the necessary equipment needed to sustain the life of the child could be kept conveniently together (Figure 2.2).

One of the most important devices used in the NICU is the incubator. This device, typically molded of see-through plastic, is used to stabilize the body temperature of the infant. In essence, the incubator allows the medical staff to keep the newborn warm without having to wrap it in blankets. The incubator also aids in preventing infection and in stabilizing the humidity of the child's environment. By keeping the temperature and humidity levels of the newborn's environment static, the baby remains well hydrated and water loss is kept to a minimum.

A complication that many preterm infants suffer from is the inability to breathe normally on their own. The child may be completely unable to breathe

Figure 2.2 A neonatal intensive care unit (Courtesy of http://www.hlarch.com/health1.htm).

for itself, or may suffer from a condition known as apnea, where the breathing pattern is either aperiodic or irregular.

In these cases, children susceptible to an apneic event are closely monitored so that if they stop breathing, nurses can rush to the bedside and wake them up. However, it is often minutes before the nurse can arrive at the scene. To facilitate the process of waking the infant experiencing an apneic event, biomedical engineers developed a tactile vibrator that is triggered by such an event to vibrate against the infant's foot and wake him or her. To prove the device is effective and safe, a human experiment must be initiated. In this case, the following questions need to be resolved:

Who is responsible for proposing the conduction of this study?
What should the approval process for such a study include?
What should be the policy on informed consent?
Should changes that were made in the device during the course of the study, which would alter the nature of the initially proposed device, be allowed?

2.7 DEFINITION AND PURPOSE OF EXPERIMENTATION

What exactly constitutes a human experiment? Although experimental protocols may vary, it is generally accepted that human experimentation occurs whenever the clinical situation of the individual is consciously manipulated to gather information regarding the capability of drugs and devices. In the past, experiments involving human subjects have been classified as either therapeutic or nontherapeutic. A therapeutic experiment is one that may have direct benefit for the patient, but the goal of nontherapeutic research is to provide additional knowledge without direct benefit to the person. The central difference is a matter of intent or aim rather than results.

Throughout medical history, there have been numerous examples of therapeutic research projects. The use of nonconventional radiotherapy to inhibit the progress of a malignant cancer, of pacemakers to provide the necessary electrical stimulation for proper heart function, or of artificial kidneys to mimic nature's function and remove poisons from the blood were all, at one time, considered novel approaches that might have some value for the patient. In the process, they were tried and found not only to be beneficial for the individual patient but also for humankind.

Nontherapeutic research has been another important vehicle for medical progress. Experiments designed to study the impact of infection from the hepatitis virus or the malarial parasite or the procedures involved in cardiac catheterization have had significant effects on the advancement of medical science and the ultimate development of appropriate medical procedures for the benefit of all humans.

In the mid-1970s the National Commission for the Protection of Human Subjects of Biomedical and Behavioral Research offered the terms *practice* and *research* to replace the conventional therapeutic and nontherapeutic distinction. Quoting the commission, Alexander Capron in 1986 wrote:

> [T]he term "practice" refers to interventions that are designed solely to enhance the well-being of an individual patient or client and that have a reasonable expectation of success. In the medical sphere, practices usually involve diagnosis, preventive treatment, or therapy; in the social sphere, practices include governmental programs such as transfer payments, education, and the like.
>
> By contrast, the term "research" designates an activity designed to test a hypothesis, to permit conclusions to be drawn, and thereby to develop or contribute to generalizable knowledge (expressed, for example, in theories, principles, statements of relationships). In the polar cases, then, practice uses a proven technique in an attempt to benefit one or more individuals, while research studies a technique in an attempt to increase knowledge.

Although the practice/research dichotomy has the advantage of not implying that therapeutic activities are the only clinical procedures intended to benefit patients, it is also based on intent rather than outcome. Interventions are practices when they are proven techniques intended to benefit the patient, but interventions aimed at increasing generalizable knowledge constitute research. What about those interventions that do not happily fit into either category?

One such intervention is nonvalidated practice, which may encompass prevention as well as diagnosed therapy. The primary purpose of the use of a nonvalidated practice is to benefit the patient while emphasizing that it has not been shown to be safe and efficacious. For humans to be subjected to nonvalidated practice, they must be properly informed and give their consent.

2.8 INFORMED CONSENT

Informed consent has long been considered by many to be the most important moral issue in human experimentation. It is the principal condition that must be satisfied for human experimentation to be considered both lawful and ethical. All adults have the legal capacity to give medical consent (unless specifically denied through some legal process). As a result, issues concerning legal capability are usually limited to minors. Many states, if not all, have some exceptions that allow minors to give consent.

Informed consent is an attempt to preserve the rights of individuals by giving them the opportunity for self-determination, that is, to determine for themselves whether they wish to participate in any experimental effort. In 1964 the World Medical Association (WMA) in Finland endorsed a code of ethics for human experimentation as an attempt to provide some guidelines in this area. In October 2000 the 52nd WMA General Assembly in Edinburgh, Scotland revised these guidelines.

Because it is often essential to use the results obtained in human experiments to further scientific knowledge, the World Medical Association prepared the following recommendations to serve as a guide to physicians all over the world. However, it is important to point out that these guidelines do not relieve physicians, scientists, and engineers from criminal, civil, and ethical responsibilities dictated by the laws of their own countries.

Case Study: The Artificial Heart

In the early 1980s a screening committee had been set up to pick the first candidate for the Jarvik 7, a new (at the time) artificial heart (Figure 2.3). It was decided that the first recipient had to be someone so sick that death was imminent. It was thought unethical to pick someone who might have another year to live because the artificial heart might well kill him or her.

Is this an example of nonvalidated practice?
Is informed consent still required?

A week after the operation, Barney Clark began having seizures from head to toe. During one seizure, Clark's unconscious body quivered for several hours. The seizures and spells of mental confusion continued throughout the next months. As a result, Clark expressed a desire to die. Although he did issue a positive statement during a videotaped interview, Clark was not a happy man, tethered to a huge machine, barely conscious, and in some pain. In March 1983 Barney Clark died of multiple organ collapse.

Discuss in detail the notions of "criteria for success" and quality of life in this case.

(continued)

Case Study: The Artificial Heart (*Continued*)

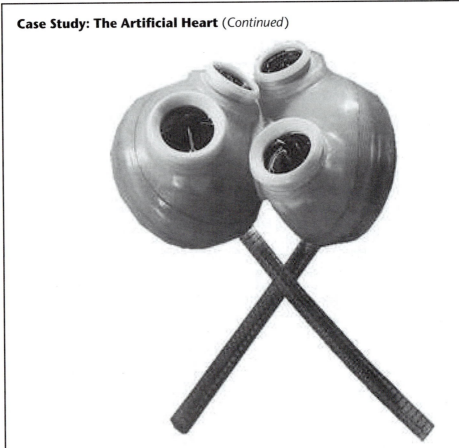

Figure 2.3 Jarvik-7 artificial heart, 1985. In August 1985 at the University Medical Center of the University of Arizona, Dr. Jack G. Copeland implanted this Jarvik-7 artifical heart in Michael Drummond, a patient awaiting a heart transplant. The Jarvik-7 kept Drummond alive until a donor organ became available one week later. Within months of the surgery, the medical center offered this artifact to the Smithsonian, which accepted it as a successful example of "spare parts"—devices or machines designed to function in place of a body part or organ—and an illustration of one of the controversies accompanying advanced medical technology (Courtesy of http://www.smithsonianlegacies.si.edu/object description.cfm?ID=172).

2.8.1 Basic Principles

- Biomedical research involving human subjects must conform to generally accepted scientific principles and should be based on adequately performed laboratory and animal experimentation and on a thorough knowledge of the scientific literature.
- The design and performance of each experimental procedure involving human subjects should be clearly formulated in an experimental protocol, which should

be transmitted to a specially appointed independent committee for consideration, comment, and guidance.

- Biomedical research involving human subjects should be conducted only by scientifically qualified persons and under the supervision of a clinically competent medical person. The responsibility for the human subject must always rest with a medically qualified person and never rest on the subject of the research, even though the subject has given his or her consent.
- Biomedical research involving human subjects cannot legitimately be carried out unless the importance of the objective is in proportion to the inherent risk to the subject.
- Every biomedical research project involving human subjects should be preceded by careful assessment of predictable risks in comparison with foreseeable benefits to the subject or to others. Concern for the interests of the subject must always prevail over the interests of science and society.
- The right of the research subject to safeguard his or her integrity must always be respected. Every precaution should be taken to respect the privacy of the subject and to minimize the impact of the study on the subject's physical and mental integrity and on the personality of the subject.
- Doctors should abstain from engaging in research projects involving human subjects unless they are satisfied that the hazards involved are believed to be predictable. Doctors should cease any investigation if the hazards are found to outweigh the potential benefits.
- In publication of the results of his or her research, the doctor is obliged to preserve the accuracy of the results. Reports of experimentation not in accordance with the principles laid down in this Declaration should not be accepted for publication.
- In any research on human beings, each potential subject must be adequately informed of the aims, methods, anticipated benefits, and potential hazards of the study and the discomfort it may entail. He or she should be informed that he or she is at liberty to abstain from participation in the study and that he or she is free to withdraw his or her consent to participation at any time. The doctor should then obtain the subject's freely given informed consent, preferably in writing.
- When obtaining informed consent for the research project, the doctor should be particularly cautious if the subject is in a dependent relationship to him or her or may consent under duress. In that case, the informed consent should be obtained by a doctor who is not engaged in the investigation and who is completely independent of this official relationship.
- In the case of legal incompetence, informed consent should be obtained from the legal guardian in accordance with national legislation. Where physical or mental incapacity makes it impossible to obtain informed consent, or when the subject is a minor, permission from the responsible relative replaces that of the subject in accordance with national legislation.
- The research protocol should always contain a statement of the ethical considerations involved and should indicate that the principles enunciated in the present Declaration are complied with.

2.8.2　Medical Research Combined with Professional Care (Clinical Research)

- In the treatment of the sick person, the doctor must be free to use a new diagnostic and therapeutic measure, if in his or her judgment it offers hope of saving life, reestablishing health, or alleviating suffering.
- The potential benefits, hazards, and discomfort of a new method should be weighed against the advantages of the best current diagnostic and therapeutic methods.
- In any medical study, every patient—including those of a control group, if any—should be assured of the best-proven diagnostic and therapeutic method.
- The refusal of the patient to participate in a study must never interfere with the doctor–patient relationship.
- If the doctor considers it essential not to obtain informed consent, the specific reasons for this proposal should be stated in the experimental protocol for transmission to the independent committee.
- The doctor can combine medical research with professional care, the objective being the acquisition of new medical knowledge, only to the extent that medical research is justified by its potential diagnostic or therapeutic value for the patient.

2.8.3　Nontherapeutic Biomedical Research Involving Human Subjects (Nonclinical Biomedical Research)

- In the purely scientific application of medical research carried out on a human being, it is the duty of the doctor to remain the protector of the life and health of that person on whom biomedical research is being carried out.
- The subjects should be volunteers (i.e., either healthy persons or patients for whom the experimental design is not related to the patient's illness).
- The investigator or the investigating team should discontinue the research if in his/her or their judgment it may, if continued, be harmful to the individual.
- In research on humans, the interest of science and society should never take precedence over considerations related to the well-being of the subject.

These guidelines generally converge on six basic requirements for ethically sound human experimentation. First, research on humans must be based on prior laboratory research and research on animals, as well as on established scientific fact, so that the point under inquiry is well focused and has been advanced as far as possible by nonhuman means. Second, research on humans should use tests and means of observation that are reasonably believed to be able to provide the information being sought by the research. Methods that are not suited for providing the knowledge sought are pointless and rob the research of its scientific value. Third, research should be conducted only by persons with the relevant scientific expertise. Fourth,

> All foreseeable risks and reasonably probable benefits, to the subject of the investigation and to science, or more broadly to society, must be carefully assessed, and…the com-

parison of those projected risks and benefits must indicate that the latter clearly out-weighs the former. Moreover, the probable benefits must not be obtainable through other less risky means.

Fifth, participation in research should be based on informed and voluntary consent. Sixth, participation by a subject in an experiment should be halted immediately if the subject finds continued participation undesirable or a prudent investigator has cause to believe that the experiment is likely to result in injury, disability, or death to the subject. Conforming to conditions of this sort probably does limit the pace and extent of medical progress, but society's insistence on these conditions is its way of saying that the only medical progress truly worth having must be consistent with a high level of respect for human dignity. Of these conditions, the requirement to obtain informed and voluntary consent from research subjects is widely regarded as one of the most important protections.

A strict interpretation of these criteria for subjects automatically rules out whole classes of individuals from participating in medical research projects. Children, the mentally retarded, and any patient whose capacity to think is affected by illness are excluded on the grounds of their inability to comprehend exactly what is involved in the experiment. In addition, those individuals having a dependent relationship to the clinical investigator, such as the investigator's patients and students, would be eliminated based on this constraint. Since mental capacity also includes the ability of subjects to appreciate the seriousness of the consequences of the proposed procedure, this means that even though some minors have the right to give consent for certain types of treatments, they must be able to understand all the risks involved.

Any research study must clearly define the risks involved. The patient must receive a total disclosure of all known information. In the past, the evaluation of risk and benefit in many situations belonged to the medical professional alone. Once made, it was assumed that this decision would be accepted at face value by the patient. Today, this assumption is not valid. Although the medical staff must still weigh the risks and benefits involved in any procedure they suggest, it is the patient who has the right to make the final determination. The patient cannot, of course, decide whether the procedure is medically correct because that requires more medical expertise than the average individual possesses. However, once the procedure is recommended, the patient then must have enough information to decide whether the hoped-for benefits are sufficient to risk the hazards. Only when this is accomplished can a valid consent be given.

Once informed and voluntary consent has been obtained and recorded, the following protections are in place:

- It represents legal authorization to proceed. The subject cannot later claim assault and battery.
- It usually gives legal authorization to use the data obtained for professional or research purposes. Invasion of privacy cannot later be claimed.
- It eliminates any claims in the event that the subject fails to benefit from the procedure.
- It is defense against any claim of an injury when the risk of the procedure is understood and consented to.

Case Study: Confidentiality, Privacy, and Consent

Integral to the change currently taking place in the United States health care industry is the application of computer technology to the development of a health care information system (Figure 2.4). Although a computerized health care information system is believed to offer opportunities to collect, store, and link data as a whole, implementation of such a system is not without significant challenges and risks.

Figure 2.4   Current technology put to use in monitoring the health care of children (Courtesy of http://www.ustechlab.com/ and http://www.bbc.co.uk/threecounties/do_that/2002/10/black_history_ month.shtml).

In a particular middle-sized city, it had been noted that children from the neighborhood were coming to the emergency room of a local hospital for health care services. A major problem associated with this activity was the absence of any record of treatment when the child showed up at a later date and was treated by another clinician. In an effort to solve this problem, the establishment of a pilot Children's Health Care Network was proposed that would enable clinicians to be aware of the medical treatment record of children coming from a particular school located near the hospital. The system required the creation of a computerized medical record at the school for each child, which could be accessed and updated by the clinicians at the local hospital.

> Discuss at length the degree to which this system should be attentive to the patient's individual rights of confidentiality and privacy.
> Discuss in detail where and how the issue of consent should be handled.

- It protects the investigator against any claim of an injury resulting from the subject's failure to follow safety instructions if the orders were well explained and reasonable.

Nevertheless, can the aims of research ever be reconciled with the traditional moral obligations of physicians? Is the researcher/physician in an untenable position? Informed and voluntary consent once again is the key only if subjects of an experiment agree to participate in the research. What happens to them during and because of the experiment is then a product of their own decision. It is not something that is imposed on them, but rather, in a very real sense, something that they elected to have done to themselves. Because their autonomy is thus respected, they are not made a mere resource for the benefit of others. Although they may suffer harm for the benefit of others, they do so of their own volition, as a result of the exercise of their own autonomy, rather than as a result of having their autonomy limited or diminished.

For consent to be genuine, it must be truly voluntary and not the product of coercion. Not all sources of coercion are as obvious and easy to recognize as physical violence. A subject may be coerced by fear that there is no other recourse for treatment, by the fear that nonconsent will alienate the physician on whom the subject depends for treatment, or even by the fear of disapproval of others. This sort of coercion, if it truly ranks as such, is often difficult to detect and, in turn, to remedy.

Finally, individuals must understand what they are consenting to do. Therefore, they must be given information sufficient to arrive at an intelligent decision concerning whether to participate in the research or not. Although a subject need not be given all the information a researcher has, it is important to determine how much should be provided and what can be omitted without compromising the validity of the subject's consent. Another difficulty lies in knowing whether the subject is competent to understand the information given and to render an intelligent opinion based on it. In any case, efforts must be made to ensure that sufficient relevant information is given and that the subject is sufficiently competent to process it. These are matters of judgment that probably cannot be made with absolute precision and certainty, but rigorous efforts must be made in good faith to prevent research on humans from involving gross violations of human dignity.

2.9 REGULATION OF MEDICAL DEVICE INNOVATION

The Food and Drug Administration (FDA) is the sole federal agency charged by Congress with regulating medical devices in the United States to ensure their safety and effectiveness. Unlike food and drugs, which have been regulated by the FDA since

1906, medical devices first became subject to FDA regulation in 1938. At that time, the FDA's major concern was to ensure that legitimate medical devices were in the marketplace and were truthfully labeled, not misbranded. Over time, the scope of FDA review of medical devices has evolved, as has the technology employed by medical devices. The first substantive legislative attempt to address the premarket review of all medical devices occurred with the Medical Device Amendment of 1976 (Pub. L. No. 94-295, 90 Stat. 539). This statute requires approval from the FDA before new devices are marketed and imposes requirements for the clinical investigation of new medical devices on human subjects. For details related to the FDA process, visit http://www.fda.gov/.

The FDA is organized into five major program centers: the Center for Biologics Evaluation and Research, the Center for Drug Evaluation and Research, the Center for Food Safety and Applied Nutrition, the Center for Veterinary Medicine, and the Center for Devices and Radiological Health (CDRH). Each FDA program center has primary jurisdiction over a different subject area. According to the FDA, the CDRH is responsible for ensuring the safety and effectiveness of medical devices and eliminating unnecessary human exposure to man-made radiation from medical, occupational, and consumer products.

There are six distinct offices located within the CDRH: the Office of Systems and Management, the Office of Compliance, the Office of Science and Technology, the Office of Health and Industry Programs, the Office of Surveillance and Biometrics, and the Office of Device Evaluation (ODE). The ODE has several principal functions, including

- Advising the CDRH Director on all premarket notification 510(k) submissions, premarket approvals (PMAs), device classifications, and investigational device exemptions (IDEs)
- Planning, conducting, and coordinating CDRH actions regarding approval, denial, and withdrawals of 510(k)s, PMAs, and IDEs
- Ongoing review, surveillance, and medical evaluation of the labeling, clinical experience, and required reports submitted by sponsors of approval applications
- Developing and interpreting regulations and guidelines regarding the classification of devices, 510(k)'s, PMAs, and IDEs
- Participating in the development of national and international consensus standards

Everyone who develops or markets a medical device will likely have multiple interactions with ODE before, during, and after the development of a medical device.

In principle, if a manufacturer makes medical claims about a product, it is considered a device, and may be subject to FDA pre- and postmarket regulatory controls (Figure 2.5). The device definition distinguishes a medical device from other FDA-regulated products, such as drugs. According to the FDA, a medical device is

> An instrument, apparatus, machine, contrivance, implant, *in vitro* reagent, or other similar or related article intended for use in the diagnosis of disease or other conditions, or in the cure, mitigation, treatment, or prevention of disease in man or other animals OR

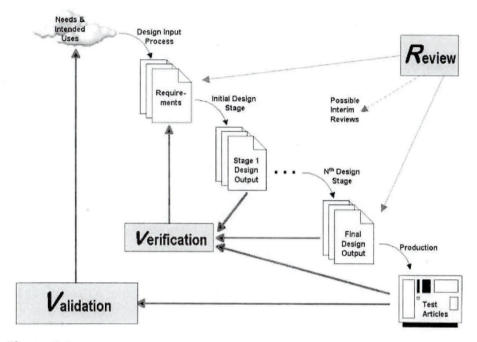

Figure 2.5   The purpose of the regulatory process is to conduct product reviews to: (1) assure device safety and effectiveness, (2) assure quality of design, and (3) provide surveillance to monitor device quality. Therefore, the review process results in verification and validation of the medical device.

intended to affect the structure or any function of the body of man or other animals, and which does not achieve any of its primary intended purposes through chemical action or is not dependent upon being metabolized.

2.10   MARKETING MEDICAL DEVICES

The four principal routes to marketing a medical device in the United States are as follows:

> Premarket Approval (PMA). A marketing approach for high-risk (Class III) medical devices must be accomplished through a PMA unless the device can be marketed through the 510(k) process (see following section). The PMA hinges on the FDA determining that the medical device is safe and effective. The PMA process can be quite costly; the collection of the data required for a PMA may costs hundreds of thousands, if not several million, dollars. Moreover, the timeline for a PMA applicant to collect the requisite data could be several years. However, an approved PMA is akin to a private license granted to the applicant to market a particular medical device, because other firms seeking to market the same type of device for the same use must also have an approved PMA.

Investigational Device Exemption (IDE). The IDE is an approved regulatory mechanism that permits manufacturers to receive an exemption for those devices solely intended for investigational use on human subjects (clinical evaluation). Because an IDE is specifically for clinical testing and not commercial distribution, the FDCA authorizes the FDA to exempt these devices from certain requirements that apply to devices in commercial distribution. The clinical evaluation of all devices may not be cleared for marketing, unless otherwise exempt by resolution, requires an IDE. An IDE may be obtained either by an institutional review board (IRB), or an IRB and the FDA.

Product Development Protocol (PDP). An alternative to the IDE and PMA processes for Class III devices subject to premarket approval. The PDP is a mechanism allowing a sponsor to come to early agreement with the FDA as to what steps are necessary to demonstrate the safety and effectiveness of a new device. In the years immediately subsequent to the enactment of the Medical Device Amendment, the FDA did not focus its energies on the PDP, but worked to effectively implement the major provisions of the Amendment, including device classification systems and the 510(k) and PMA processes.

510(k) Notification. Unless specifically exempted by federal regulation, all manufacturers are required to give the FDA 90 days' notice before they intend to introduce a device to the U.S. market by submitting a 510(k). During that 90-day period, the FDA is charged with determining whether the device is or is not substantially equivalent to a pre-Amendment device. The premarket notification is referred to in the industry as a 510(k) because 510(k) is the relevant section number of the FDCA. The 510(k) is used to demonstrate that the medical device is or is not substantially equivalent to a legally marketed device.

With respect to clinical research on humans, the FDA divides devices into two categories: devices that pose significant risk and those that involve insignificant risk. Examples of the former included orthopedic implants, artificial hearts, and infusion pumps. Examples of the latter include various dental devices and daily-wear contact lenses. Clinical research involving a significant risk device cannot begin until an institutional review board (IRB) has approved both the protocol and the informed consent form, and the FDA itself has given permission. This requirement to submit an IDE application to the FDA is waived in the case of clinical research in which the risk posed is insignificant. In this case, the FDA requires only that approval from an IRB be obtained certifying that the device in question poses only insignificant risk. In deciding whether to approve a proposed clinical investigation of a new device, the IRB and the FDA must determine the following:

1. Risk to subjects is minimized.
2. Risks to subjects are reasonable in relation to anticipated benefit and knowledge to be gained.
3. Subject selection is equitable.
4. Informed consent materials and procedures are adequate.
5. Provisions for monitoring the study and protecting patient information are acceptable.

The FDA allows unapproved medical devices to be used without an IDE in three types of situations: feasibility studies, emergency use, and treatment use.

2.11 ETHICAL ISSUES IN FEASIBILITY STUDIES

In a feasibility study, or limited investigation, human research involving the use of a new device would take place at a single institution and involve no more than 10 human subjects. The sponsor of a limited investigation is required to submit to the FDA a "Notice of Limited Investigation," which includes a description of the device, a summary of the purpose of the investigation, the protocol, a sample of the informed consent form, and a certification of approval by the responsible medical board. In certain circumstances, the FDA could require additional information or require the submission of a full IDE application or suspend the investigation.

Investigations of this kind are limited to: (1) investigations of new uses for existing devices, (2) investigations involving temporary or permanent implants during the early developmental stages, and (3) investigations involving modification of an existing device.

To comprehend adequately the ethical issues posed by clinical use of unapproved medical devices outside the context of an IDE, it is necessary to use the distinctions between practice, nonvalidated practice, and research elaborated upon in the previous pages. How do those definitions apply to feasibility studies?

Clearly, the goal of the feasibility study—generalizable knowledge—makes it an instance of research rather than practice. Manufacturers seek to determine the performance of a device with respect to a particular patient population in an effort to gain information about its efficacy and safety. Such information is important to determine whether further studies (animal or human) need to be conducted, whether the device needs modification before further use, and the like. The main difference between use of an unapproved device in a feasibility study and its use under the terms of an IDE is that the former would be subject to significantly less intensive FDA review than the latter. This, in turn, means that the responsibility for ensuring that the use of the device is ethically sound would fall primarily to the IRB of the institution conducting the study.

The ethical concerns posed here can be best comprehended only with a clear understanding of what justifies research in the first place. Ultimately, no matter how much basic research and animal experimentation has been conducted on a given device, the risks and benefits it poses for humans cannot be adequately determined until it is actually used on humans. The benefit of research on humans lies primarily in the generalizable information that is provided. This information is crucial to medical science's ability to generate new modes of medical treatment that are both efficacious and safe. Therefore, one condition for experimentation to be ethically sound is that it must be scientifically sound.

Although scientific soundness is a necessary condition of ethically sound research on humans, it is not of and by itself sufficient. The human subjects of such research are at risk of being mere research resources, that is, having value only for the ends of the research. Human beings are not valuable wholly or solely for the uses to which they can be put. They are valuable simply by being the kinds of entities they are. To treat them as such is to respect them as people. Treating individuals as people means respecting their autonomy. This requirement is met by ensuring that no competent person is subjected to any clinical intervention without first giving voluntary and informed consent. Furthermore, respect for people means that the physician will not subject a human to unnecessary risks and will minimize the risks to patients in required procedures.

Much of the scrutiny that the FDA imposes upon use of unapproved medical devices in the context of an IDE addresses two conditions of ethically sound research: (1) is the experiment scientifically sound, and (2) does it respect the rights of the human subjects involved? Medical ethicists argue that decreased FDA scrutiny will increase the likelihood that either or both of these conditions will not be met. This possibility exists because many manufacturers of medical devices are, after all, commercial enterprises, companies that are motivated to generate profit and thus to get their devices to market as soon as possible with as little delay and cost as possible. These self-interest motives are likely, at times, to conflict with the requirements of ethically sound research and thus to induce manufacturers to fail to meet these requirements. Profit is not the only motive that might induce manufacturers to contravene the requirements of ethically sound research on humans. A manufacturer may sincerely believe that its product offers great benefit to many people and be prompted to take shortcuts that compromise the quality of the research. Whether the consequences being sought by the research are desired for reasons of self-interest, altruism, or both, the ethical issue is the same. Research subjects may be placed at risk of being treated as mere objects rather than as people.

What about the circumstances under which feasibility studies would take place? Are these not sufficiently different from the "normal" circumstances of research to warrant reduced FDA scrutiny? As noted previously, manufacturers seek to engage in feasibility studies to investigate new uses of existing devices, to investigate temporary or permanent implants during the early developmental stages, and to investigate modifications to an existing device. As was also noted, a feasibility study would take place at only one institution and would involve no more than 10 human subjects. Given these circumstances, is the sort of research that is likely to occur in a feasibility study less likely to be scientifically sound or to fail to respect people than normal research on humans in "normal" circumstances?

Research in feasibility studies would be done on a very small subject pool, and the harm of any ethical lapses would likely affect fewer people than if such lapses occurred under more usual research circumstances. Yet even if the harm done is limited to 10 or fewer subjects in a single feasibility study, the harm is still ethically wrong. To wrong 10 or fewer people is not as bad as to wrong in the same way more than 10 people, but it is to engage in wrongdoing nonetheless.

Are ethical lapses more likely to occur in feasibility studies than in studies that take place within the requirements of an IDE? Although nothing in the preceding discussion provides a definitive answer to this question, it is a question to which the FDA should give high priority. The answer to this question might be quite different when the device at issue is a temporary or permanent implant than when it is an already approved device being put to new uses or modified in some way. Whatever the contemplated use under the feasibility studies mechanism, the FDA would be ethically advised not to allow this kind of exception to IDE use of an unapproved device without a reasonably high level of certainty that research subjects would not be placed in greater jeopardy than in "normal" research circumstances.

2.12   ETHICAL ISSUES IN EMERGENCY USE

What about the mechanism for avoiding the rigors of an IDE for emergency use? The FDA has authorized emergency use in instances where an unapproved device offers the only alternative for saving the life of a dying patient even though an IDE has not yet been approved for the device or its use, or an IDE has been approved but the physician who wishes to use the device is not an investigator under the IDE.

The purpose of emergency use of an unapproved device is to attempt to save a dying patient's life under circumstances where no other alternative is available. This sort of use constitutes practice rather than research. Its aim is primary benefit to the patient rather than provision of new and generalizable information. Because this sort of use occurs before the completion of clinical investigation of the device, it constitutes a nonvalidated practice. What does this mean?

First, it means that although the aim of the use is to save the life of the patient, the nature and likelihood of the potential benefits and risks engendered by use of the device are far more speculative than in the sort of clinical intervention that constitutes validated practice. In validated practice, thorough investigation of a device, including preclinical studies, animals studies, and studies on human subjects, has established its efficacy and safety. The clinician thus has a well-founded basis upon which to judge the benefits and risks such an intervention poses for the patient.

It is precisely this basis that is lacking in the case of a nonvalidated practice. Does this mean that emergency use of an unapproved device should be regarded as immoral? This conclusion would follow only if there were no basis upon which to make an assessment of the risks and benefits of the use of the device. The FDA requires that a physician who engages in emergency use of an unapproved device must have substantial reason to believe that benefits will exist. This means that there should be a body of preclinical and animal tests allowing a prediction of the benefit to a human patient.

Case Study: Medical Expert Systems

Expert systems have been developed in various disciplines, including clinical decision making. These systems have been designed to simulate the decision-making skills of physicians. Their adaptability, however, depends on the presence of an accepted body of knowledge regarding the prescribed path physicians would take given specific input data. These systems have been viewed as advisory systems providing the clinician with suggested/recommended courses of action. The ultimate decision remains with the physician.

(continued)

Case Study: Medical Expert Systems (*Continued*)

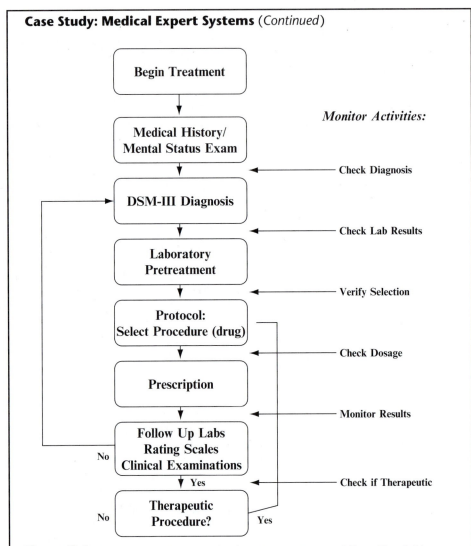

Figure 2.6   Flow diagram illustrating the drug treatment process followed by clinicians.

Consider one such system designed to monitor drug treatment in a psychiatric clinic. This system, designed and implemented by biomedical engineers working with clinicians, begins by the entry of a specific diagnosis and immediately recommends the appropriate drugs to be considered for the treatment of someone with that mental disorder (Figure 2.6). The physician selects one of the recommended drugs and conducts a dose regimen to determine the effectiveness of the drug for the particular patient. During the treatment, blood tests are conducted to ascertain the presence of drug toxicity, and other psychiatric measures obtained to determine if the drug is having the desired effect.

> **Case Study: Medical Expert Systems** (*Continued*)
>
> As these data elements are entered, they are compared with standard expected outcomes, and if the outcomes are outside the expected limits, an alert is sent to the physician indicating further action needs to be taken.
>   In this situation:
>
>   Who is liable for mistreatment: the clinician, the programmer, or the systems administrator?
>   What constitutes mistreatment?
>   What is the role of the designers of such a system (i.e., what constitutes a successful design)?
>   How does the clinic evaluate the performance of a physician using the system, and the system itself?

Thus, although the benefits and risks posed by use of the device are highly speculative, they are not entirely speculative. Although the only way to validate a new technology is to engage in research on humans at some point, not all nonvalidated technologies are equal. Some will be largely uninvestigated, and assessment of their risks and benefits will be wholly or almost wholly speculative. Others will at least have the support of preclinical and animal tests. Although this is not sufficient support for incorporating use of a device into regular clinical practice, it may represent sufficient support to justify use in the desperate circumstances at issue in emergency situations. Desperate circumstances can justify desperate actions, but desperate actions are not the same as reckless actions, hence the ethical soundness of the FDA's requirement that emergency use be supported by solid results from preclinical and animal tests of the unapproved device.

A second requirement that the FDA imposes on emergency use of unapproved devices is the expectation that physicians "exercise reasonable foresight with respect to potential emergencies and...make appropriate arrangements under the IDE procedures." Thus, a physician should not "create" an emergency in order to circumvent IRB review and avoid requesting the sponsor's authorization of the unapproved use of a device. From a Kantian point of view, which is concerned with protecting the dignity of people, this is a particularly important requirement. To create an emergency in order to avoid FDA regulations is to treat the patient as a mere resource whose value is reducible to service to the clinician's goals. Hence, the FDA is quite correct to insist that emergencies are circumstances that reasonable foresight would not anticipate.

Also important here is the nature of the patient's consent. Individuals facing death are especially vulnerable to exploitation and deserve greater measures for their protection than might otherwise be necessary. One such measure would be to ensure that the patient, or the patient's legitimate proxy, knows the highly speculative nature of the intervention being offered. That is, to ensure that it is clearly understood that the clinician's estimation of the intervention's risks and benefits is far less solidly grounded than in the case of validated practices. The patient's consent must be based on an awareness that the device being contemplated has not undergone complete and rigorous testing on humans and that estimations of its potential are based wholly on preclinical and animal studies. Above all, the patient must not be led to believe that the risks and benefits of the intervention are not better understood than they in fact are. Another important point is to ensure that the patient understands all of the options: not simply life or death, but also a life with severely impaired quality. Although desperate circumstances may legitimate desperate actions, the decision to take such actions must rest on the informed and voluntary consent of the patient, certainly for an especially vulnerable patient.

It is important here for a clinician involved in emergency use of an unapproved device to recognize that these activities constitute a form of practice, albeit nonvalidated, and not research. Hence, the primary obligation is to the well-being of the patient. The patient enters into the relationship with the clinician with the same trust that accompanies any normal clinical situation. To treat this sort of intervention as if it were an instance of research, and hence justified by its benefits to science and society, would be an abuse of this trust.

2.13   ETHICAL ISSUES IN TREATMENT USE

The FDA has adopted regulations authorizing the use of investigational new drugs in certain circumstances where a patient has not responded to approved therapies. This treatment use of unapproved new drugs is not limited to life-threatening emergency situations, but also is available to treat serious diseases or conditions. The FDA has not approved treatment use of unapproved medical devices, but it is possible that a manufacturer could obtain such approval by establishing a specific protocol for this kind of use within the context of an IDE.

The criteria for treatment use of unapproved medical devices would be similar to criteria for treatment use of investigational drugs: (1) the device is intended to treat a serious or life-threatening disease or condition, (2) there is no comparable or satisfactory alternative product available to treat that condition, (3) the device is under an IDE, or has received an IDE exemption, or all clinical trials have been completed and the device is awaiting approval, and (4) the sponsor is actively pursuing marketing approval of the investigational device. The treatment use protocol would be submitted as part of the IDE and would describe the intended use of the device, the rationale for use of the device, the available alternatives and why the investigational product is preferable, the criteria for patient selection, the measures to monitor the use of the device and to minimize risk, and technical information that is relevant to the safety and effectiveness of the device for the intended treatment purpose.

Were the FDA to approve treatment use of unapproved medical devices, what ethical issues would be posed? First, because such use is premised on the failure of validated interventions to improve the patient's condition adequately, it is a form of practice rather than research. Second, since the device involved in an instance of treatment use is unapproved, such use would constitute nonvalidated practice. As such, like emergency use, it should be subject to the FDA's requirement that prior preclinical tests and animal studies have been conducted that provide substantial reason to believe that patient benefit will result. As with emergency use, although this does not prevent assessment of the intervention's benefits and risks from being highly speculative, it does prevent assessment from being totally speculative. Here, too, although desperate circumstances can justify desperate action, they do not justify reckless action. Unlike emergency use, the circumstances of treatment use involve serious impairment of health rather than the threat of premature death. Hence, an issue that must be considered is how serious such impairment must be to justify resorting to an intervention with risks and benefits that have not been solidly established.

In cases of emergency use, the FDA requires that physicians not create an exception to an IDE to avoid requirements that would otherwise be in place. As with emergency use of unapproved devices, the patients involved in treatment uses would be particularly vulnerable patients. Although they are not dying, they are facing serious medical conditions and are thereby likely to be less able to avoid exploitation than patients under less desperate circumstances. Consequently, here too it is especially important that patients be informed of the speculative nature of the intervention and of the possibility that treatment may result in little to no benefit to them.

2.14   THE ROLE OF THE BIOMEDICAL ENGINEER IN THE FDA PROCESS

On November 28, 1991, the Safe Medical Devices Act of 1990 (Public Law 101-629) went into effect. This regulation requires a wide range of health care institutions, including hospitals, ambulatory-surgical facilities, nursing homes, and outpatient treatment facilities, to report information that "reasonably suggests" the likelihood that the death, serious injury, or serious illness of a patient at that facility has been caused or contributed to by a medical device. When a death is device-related, a report must be made directly to the FDA and to the manufacturer of the device. When a serious illness or injury is device-related, a report must be made to the manufacturer or to the FDA in cases where the manufacturer is not known. In addition, summaries of previously submitted reports must be submitted to the FDA on a semiannual basis. Prior to this regulation, such reporting was wholly voluntary. This new regulation was designed to enhance the FDA's ability to learn quickly about problems related to medical devices, and it supplements the medical device reporting (MDR) regulations promulgated in 1984. MDR regulations require that manufacturers and importers submit reports of device-related deaths and serious injuries to the FDA. The new law extends this requirement to users of medical devices along with manufacturers and importers. This act gives the FDA authority over device-user facilities.

The FDA regulations are ethically significant because, by attempting to increase the FDA's awareness of medical device-related problems, it attempts to increase that agency's ability to protect the welfare of patients. The main controversy over the FDA's regulation policies is essentially utilitarian in nature. Skeptics of the law are dubious about its ability to provide the FDA with much useful information. They worry that much of the information generated by this new law will simply duplicate information already provided under MDR regulations. If this were the case, little or no benefit to patients would accrue from compliance with the regulation. Furthermore, these regulations, according to the skeptics, are likely to increase lawsuits filed against hospitals and manufacturers and will require device-user facilities to implement formal systems for reporting device-related problems and to provide personnel to operate those systems. This would, of course, add to the costs of health care and thereby exacerbate the problem of access to care, a situation that many believe to be of crisis proportions already. In short, the controversy over FDA policy centers on the worry that its benefits to patients will be marginal and significantly outweighed by its costs.

Biomedical engineers need to be aware of FDA regulations and the process for FDA approval of the use of medical devices and systems. These regulatory policies are, in effect, society's mechanism for controlling the improper use of these devices.

EXERCISES

1. Explain the distinction between the terms *ethics* and *morality*. Provide examples that illustrate this distinction in the medical arena.
2. Provide three examples of medical moral judgments.
3. What do advocates of the utilitarian school of thought believe?
4. What does Kantianism expect in terms of the patient's rights and wishes?
5. Discuss how the code of ethics for clinical engineers provides guidance to practitioners in the field.
6. Discuss what is meant by brainstem death. How is this distinguished from neocortical death?
7. Distinguish between active and passive euthanasia as well as voluntary and involuntary euthanasia. In your view, which, if any, are permissible? Provide your reasoning and any conditions that must be satisfied to meet your approval.
8. If the family of a patient in the intensive care unit submits the individual's living will document, should it be honored immediately or should there be a discussion between physicians and the family? Who should make the decision? Why?
9. What constitutes a human experiment? Under what conditions are they permitted? What safeguards should hospitals have in place?
10. A biomedical engineer has designed a new sleep apnea monitor. Discuss the steps that should be taken before it is used in a clinical setting.
11. Discuss the distinctions between practice, research, and nonvalidated practice. Provide examples of each in the medical arena.
12. What are the two major conditions for ethically sound research?
13. Informed consent is one of the essential factors in permitting humans to participate in medical experiments. What ethical principles are satisfied by informed consent? What should be done to ensure it is truly voluntary? What information should be given to human subjects?
14. What are the distinctions between feasibility studies and emergency use?
15. In the practice of medicine, health care professionals use medical devices to diagnose and treat patients. Therefore, the clinical staff not only needs to

become knowledgeable and skilled in their understanding of human physiology, they must also be competent in using the medical tools at their disposal. This requirement often results in litigation when a device fails. The obvious question is "who is to blame?"

Consider the case of a woman undergoing a surgical procedure that requires the use of a ground plate, i.e., usually an 8-by-11-inch pad that serves as a return path for any electrical current that comes from electrosurgical devices used during the procedure. As a result of the procedure, this woman received a major burn that seriously destroyed tissue at the site of the ground plate.

(a) Discuss the possible individuals and/or organizations that may have been responsible for this injury.

(b) Outside of seeking the appropriate responsible party, are there specific ethical issues here?

SUGGESTED READING

Abrams, N. and Buckner, M.D. (Eds.) (1983). *Medical Ethics*. MIT Press, Cambridge, MA.

Bronzino, J.D., Smith, V.H. and Wade, M.L. (1990). *Medical Technology and Society*. MIT Press, Cambridge, MA.

Bronzino, J.D. (1992). *Management of Medical Technology*. Boston, Butterworth, 1992.

Chapman, A.R. (1997). *Health Care and Information Ethics: Protecting Fundamental Human Rights*. Sheed and Ward, Kansas City, KS.

Dubler, N. and Nimmons, D. (1992). *Ethics on Call*. Harmony, New York.

Jonsen, A.R. (1990). *The New Medicine and the Old Ethics*. Harvard Univ. Press, Cambridge, MA.

Moskop, J.C. and Kopelman, L. (Eds.) (1985). *Ethics and Critical Care Medicine*. Reidel, Boston.

Pence, G.E. (1990). *Classic Cases in Medical Ethics*. McGraw-Hill, New York.

Rachels, J. (1986). *Ethics at the End of Life: Euthanasia and Morality*. Oxford Univ. Press, Oxford.

Reiss, J. (2001). *Bringing Your Medical Device to Market*. FDLI Publishers, Washington DC.

Seebauer, E.G. and Barry, R.L. (2001). *Fundamentals of Ethics for Scientists and Engineers*. Oxford Press, New York.

3 ANATOMY AND PHYSIOLOGY

Susan Blanchard, PhD

Chapter Contents

At the conclusion of this chapter, the reader will be able to:

- Define anatomy and physiology and explain why they are important to biomedical engineering.
- Define important anatomical terms.
- Describe the cell theory.
- List the major types of organic compounds and other elements found in cells.

- Explain how the plasma membrane maintains the volume and internal concentrations of a cell.
- Calculate the internal osmolarity and ionic concentrations of a model cell at equilibrium.
- List and describe the functions of the major organelles found within mammalian cells.
- Describe the similarities, differences, and purposes of replication, transcription, and translation.
- List and describe the major components and functions of five organ systems: Circulatory, respiratory, nervous, skeletal, and muscular.
- Define homeostasis and describe how feedback mechanisms help maintain it.

3.1   INTRODUCTION

Since biomedical engineering is an interdisciplinary field based in both engineering and the life sciences, it is important for biomedical engineers to have knowledge about and be able to communicate in both areas. Biomedical engineers must understand the basic components of the body and how they function well enough to exchange ideas and information with physicians and life scientists. Two of the most basic terms and areas of study in the life sciences are anatomy and physiology. Anatomy refers to the internal and external structures of the body and their physical relationships, whereas physiology refers to the study of the functions of those structures.

Figure 3.1a shows a male body in anatomical position. In this position, the body is erect and facing forward with the arms hanging at the sides and the palms facing outward. This particular view shows the anterior (ventral) side of the body, whereas Figure 3.1c illustrates the posterior (dorsal) view of another male body that is also in anatomical position and Figure 3.1b presents the lateral view of the female body. In clinical practice, directional terms are used to describe the relative positions of various parts of the body. Proximal parts are nearer to the trunk of the body or to the attached end of a limb than are distal parts (Fig. 3.1a). Parts of the body that are located closer to the head than other parts when the body is in anatomical position are said to be superior (Fig. 3.1b), whereas those located closer to the feet than other parts are termed inferior. Medial implies that a part is toward the midline of the body, whereas lateral means away from the midline (Fig. 3.1c). Parts of the body that lie in the direction of the head are said to be in the cranial direction, whereas those parts that lie in the direction of the feet are said to be in the caudal direction (Fig. 3.2).

Anatomical locations can also be described in terms of planes. The plane that divides the body into two symmetric halves along its midline is called the midsaggital plane (Fig. 3.2). Planes that are parallel to the midsaggital plane but do not divide the body into symmetric halves are called sagittal planes. The frontal plane is perpendicular to the midsaggital plane and divides the body into asymmetric anterior and posterior portions. Planes that cut across the body and are perpendicular to the midsaggital and frontal planes are called transverse planes.

Human bodies are divided into two main regions, axial and appendicular. The axial part consists of the head, neck, thorax (chest), abdomen, and pelvis whereas the

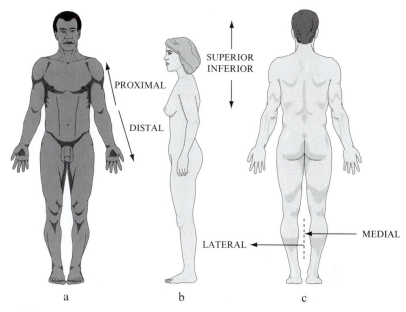

Figure 3.1 (a) Anterior view of male body in anatomical position. (b) Lateral view of female body. (c) Posterior view of male body in anatomical position. Relative directions (proximal and distal, superior and inferior, and medial and lateral) are also shown.

appendicular part consists of the upper and lower extremities. The upper extremities, or limbs, include the shoulders, upper arms, forearms, wrists, and hands whereas the lower extremities include the hips, thighs, lower legs, ankles, and feet. The abdominal region can be further divided into nine regions or four quadrants.

The cavities of the body hold the internal organs. The major cavities are the dorsal and ventral body cavities and the smaller cavities include the nasal, oral, orbital (eye), tympanic (middle ear), and synovial (movable joint) cavities. The dorsal body cavity includes the cranial cavity that holds the brain and the spinal cavity that contains the spinal cord. The ventral body cavity contains the thoracic and abdominopelvic cavities that are separated by the diaphragm. The thoracic cavity contains the lungs and the mediastinum, which contains the heart and its attached blood vessels, the trachea, the esophagus, and all other organs in this region except for the lungs. The abdominopelvic cavity is divided by an imaginary line into the abdominal and pelvic cavities. The former is the largest cavity in the body and holds the stomach, small and large intestines, liver, spleen, pancreas, kidneys, and gall bladder. The latter contains the urinary bladder, the rectum, and the internal portions of the reproductive system.

The anatomical terms described previously are used by physicians, life scientists, and biomedical engineers when discussing the whole human body or its major parts. Correct use of these terms is vital for biomedical engineers to communicate with health care professionals and to understand the medical problem of concern or interest. Although it is important to be able to use the general terms that describe

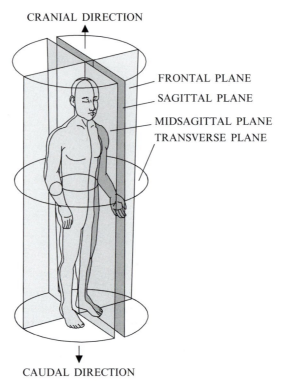

Figure 3.2   The body can be divided into sections by the frontal, sagittal, and transverse planes. The midsagittal plane goes through the midline of the body.

the human body, it is also important for biomedical engineers to have a basic understanding of some of the more detailed aspects of human anatomy and physiology.

3.2   CELLULAR ORGANIZATION

Although there are many smaller units such as enzymes and organelles that perform physiological tasks or have definable structures, the smallest anatomical and physiological unit in the human body that can, under appropriate conditions, live and reproduce on its own is the cell. Cells were first discovered more than 300 years ago shortly after Antony van Leeuwenhoek, a Dutch optician, invented the microscope. With his microscope, van Leeuwenhoek was able to observe "many very small animalcules, the motions of which were very pleasing to behold" in tartar scrapings from his teeth. Following the efforts of van Leeuwenhoek, Robert Hooke, a Curator of Instruments for the Royal Society of England, in the late 1600s further described cells when he used one of the earliest microscopes to look at the plant cell walls that remain in cork. These observations and others led to the cell theory developed by

impermeable to Na^+ and the internal anions? The total osmolarity inside the cell is 250 mOsm (12 mM Na^+, 125 mM K^+, 5 mM Cl^-, 108 mM anions) while the total osmolarity outside the cell is also 250 mOsm (120 mM Na^+, 5 mM K^+, 125 mM Cl^-) so the cell is in osmotic balance (i.e., there will be no net movement of water across the plasma membrane). If the average charge per molecule of the anions inside the cell is considered to be -1.2, then the cell is also approximately in electrical equilibrium ($12 + 125$ positive charges for Na^+ and K^+; $5 + 1.2 \times 108$ negative charges for Cl^- and the other anions). Real cells, however, cannot maintain this equilibrium without expending energy since real cells are slightly permeable to Na^+. In order to maintain equilibrium and keep Na^+ from accumulating intracellularly, mammalian cells must actively pump Na^+ out of the cell against its diffusion and electrical gradients. Since Na^+ is pumped out through specialized protein channels at a rate equivalent to the rate at which it leaks in through other channels, it behaves osmotically as if it cannot cross the plasma membrane. Thus, mammalian cells exist in a steady state, rather than at equilibrium, since energy in the form of ATP must be used to prevent a net movement of ions across the plasma membrane.

Example Problem 3.4

Consider a simple model cell, such as the one in Figure 3.7, which has the following ion concentrations. Is the cell at equilibrium? Explain your answer.

Ion	Intracellular Concentration (mM)	Extracellular Concentration (mM)
K^+	158	4
Na^+	20	163
Cl^-	52	167
A^-	104	–

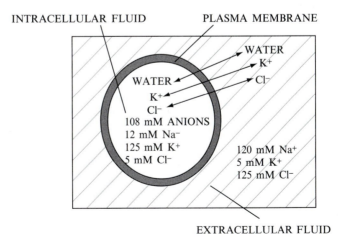

INTRACELLULAR FLUID   PLASMA MEMBRANE

WATER
K^+
Cl^-

WATER
K^+
Cl^-
108 mM ANIONS
12 mM Na^-
125 mM K^+
5 mM Cl^-

120 mM Na^+
5 mM K^+
125 mM Cl^-

EXTRACELLULAR FLUID

Figure 3.7 A model cell with internal and external concentrations similar to those of a typical mammalian cell. The full extent of the extracellular volume is not shown and is much larger than the cell's volume.

Solution

Yes. The cell is both electrically and osmotically at equilibrium because the charges within the inside and outside compartments are equal and the osmolarity inside the cell equals the osmolarity outside of the cell.

	Inside	**Outside**
Positive	$158 + 20 = 178\,\text{m}M$	$4 + 163 = 167\,\text{m}M$
Negative	$52 + 1.2 \times 104 = 177\,\text{m}M$	$167\,\text{m}M$
	$178\,\text{m}M_{\text{pos}} \approx 177\,\text{m}M_{\text{neg}}$	$167\,\text{m}M_{\text{pos}} = 167\,\text{m}M_{\text{neg}}$
Osmolarity	$158 + 20 + 52 + 104 = 334\,\text{m}M$	$4 + 163 + 167 = 334\,\text{m}M$

$$334\,\text{m}M_{\text{inside}} = 334\,\text{m}M_{\text{outside}}$$

∎

One of the consequences of the distribution of charged particles in the intracellular and extracellular fluids is that an electrical potential exists across the plasma membrane. The value of this electrical potential depends on the intracellular and extracellular concentrations of ions that can cross the membrane and will be described more fully in Chapter 11.

In addition to controlling the cell's volume, the plasma membrane also provides a route for moving large molecules and other materials into and out of the cell. Substances can be moved into the cell by means of endocytosis (Fig. 3.8a) and out of the cell by means of exocytosis (Fig. 3.8b). In endocytosis, material (e.g., a bacterium) outside of the cell is engulfed by a portion of the plasma membrane that encircles it to form a vesicle. The vesicle then pinches off from the plasma membrane and moves its contents to the inside of the cell. In exocytosis, material within the cell is surrounded by a membrane to form a vesicle. The vesicle then moves to the edge of the cell where its membrane fuses with the plasma membrane and its contents are released to the exterior of the cell.

3.2.2 Cytoplasm and Organelles

The cytoplasm contains fluid (cytosol) and organelles. Ions (such as Na^+, K^+, and Cl^-) and molecules (such as glucose) are distributed through the cytosol via diffusion. Membrane-bound organelles include the nucleus, rough and smooth endoplasmic reticulum, the Golgi apparatus, lysosomes, and mitochondria. Nonmembranous organelles include nucleoli, ribosomes, centrioles, microvilli, cilia, flagella, and the microtubules, intermediate filaments, and microfilaments of the cytoskeleton.

The nucleus (Fig. 3.4) consists of the nuclear envelope (a double membrane) and the nucleoplasm (a fluid that contains ions, enzymes, nucleotides, proteins, DNA, and small amounts of RNA). Within its DNA, the nucleus contains the instructions for life's processes. Nuclear pores are protein channels that act as connections for ions and RNA, but not proteins or DNA, to leave the nucleus and enter the cytoplasm and for some proteins to enter the nucleoplasm. Most nuclei contain one or more nucleoli.

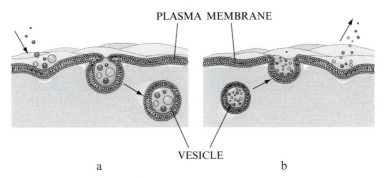

PLASMA MEMBRANE

VESICLE

a   b

Figure 3.8   Substances that are too large to pass through the integral proteins in the plasma membrane can be moved into the cell by means of endocytosis (a) and out of the cell by means of exocytosis (b).

Each nucleolus contains DNA, RNA, and proteins and synthesizes the components of the ribosomes that cells use to make proteins.

The smooth and rough endoplasmic reticulum (ER), Golgi apparatus, and assorted vesicles (Figs. 3.4, 3.9a, and 3.9b) make up the cytomembrane system which delivers proteins and lipids for manufacturing membranes and accumulates and stores proteins and lipids for specific uses. The ER also acts as a storage site for calcium ions. The rough ER differs from the smooth ER in that it has ribosomes attached to its exterior surface. Ribosomes provide the platforms for synthesizing proteins. Those that are synthesized on the rough ER are passed into its interior where nonproteinaceous side chains are attached to them. These modified proteins move to the smooth ER where they are packaged in vesicles. The smooth ER also manufactures and packages lipids into vesicles and is responsible for releasing stored calcium ions. The vesicles leave the smooth ER and become attached to the Golgi apparatus where their contents are released, modified, and repackaged into new vesicles. Some of these vesicles, called lysosomes, contain digestive enzymes which are used to break down materials that move into the cells via endocytosis. Other vesicles contain proteins such as hormones and neurotransmitters that are secreted from the cells by means of exocytosis.

The mitochondria (Figs. 3.9c and 3.10) contain two membranes: an outer membrane that surrounds the organelle and an inner membrane that divides the organelle's interior into two compartments. Approximately 95% of the ATP required by the cell is produced in the mitochondria in a series of oxygen-requiring reactions which produce carbon dioxide as a byproduct. Mitochondria are different from most other organelles in that they contain their own DNA. The majority of the mitochondria in sexually reproducing organisms, such as humans, come from the mother's egg cell because the father's sperm contributes little more than the DNA in a haploid (half) set of chromosomes to the developing offspring.

Microtubules, intermediate filaments, and microfilaments provide structural support and assist with movement. Microtubules are long, hollow, cylindrical structures that radiate from microtubule organizing centers and, during cell division, from centrosomes, a specialized region of the cytoplasm that is located near the nucleus

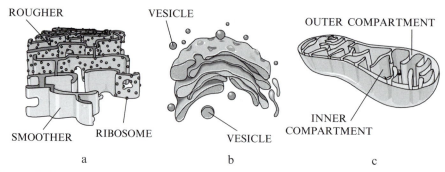

ROUGHER   VESICLE   OUTER COMPARTMENT

SMOOTHER   RIBOSOME   VESICLE   INNER COMPARTMENT

a   b   c

Figure 3.9   Subcellular organelles. The endoplasmic reticulum (a), the Golgi apparatus (b), and vesicles (b) make up the cytomembrane system in the cell. The small circles on the endoplasmic reticulum (ER) represent ribosomes. The area containing ribosomes is called the rough ER and the area that lacks ribosomes is called the smooth ER. The mitochondria (c) have a double membrane system which divides the interior into two compartments that contain different concentrations of enzymes, substrates, and hydrogen ions (H^+). Electrical and chemical gradients between the inner and outer compartments provide the energy needed to generate ATP.

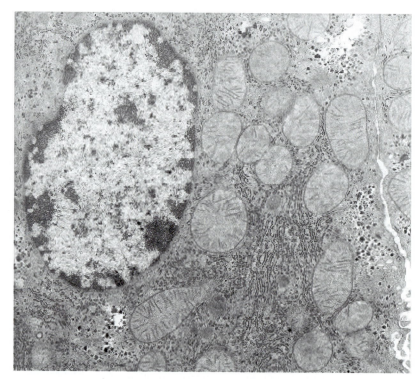

Figure 3.10   Scanning electron micrograph of a normal mouse liver at 8000X magnification. The large round organelle on the left is the nucleus. The smaller round and oblong organelles are mitochondria that have been sliced at different angles. The narrow membranes in parallel rows are endoplasmic reticula. The small black dots on the ERs are ribosomes (Photo courtesy of Valerie Knowlton, Center for Electron Microscopy, North Carolina State University).

and contains two centrioles (Figs. 3.4 and 3.11a) oriented at right angles to each other. Microtubules consist of spiraling subunits of a protein called tubulin, whereas centrioles consist of nine triplet microtubules that radiate from their centers like the spokes of a wheel. Intermediate filaments are hollow and provide structure to the plasma membrane and nuclear envelope. They also aid in cell-to-cell junctions and in maintaining the spatial organization of organelles. Myofilaments are found in most cells and are composed of strings of protein molecules. Cell movement can occur when actin and myosin, protein subunits of myofilaments, interact. Microvilli (Fig. 3.11b) are extensions of the plasma membrane that contain microfilaments. They increase the surface area of a cell to facilitate absorption of extracellular materials.

Cilia (Fig. 3.11c) and flagella are parts of the cytoskeleton that have shafts composed of nine pairs of outer microtubules and two single microtubules in the center. Both types of shafts are anchored by a basal body which has the same structure as a centriole. Flagella function as whiplike tails that propel cells such as sperm. Cilia are generally shorter and more profuse than flagella and can be found on specialized cells such as those that line the respiratory tract. The beating of the cilia helps move mucus-trapped bacteria and particles out of the lungs.

3.2.3 DNA and Gene Expression

DNA (Fig. 3.3) is found in the nucleus and mitochondria of eukaryotic cells. In organisms that reproduce sexually, the DNA in the nucleus contains information from both parents whereas that in the mitochondria comes from the organism's mother. In the nucleus, the DNA is wrapped around protein spools, called nucleosomes, and is organized into pairs of chromosomes. Humans have 22 pairs of autosomal chromosomes and two sex chromosomes, XX for females and XY for males (Fig. 3.12). If the DNA from all 46 chromosomes in a human somatic cell (i.e., any cell

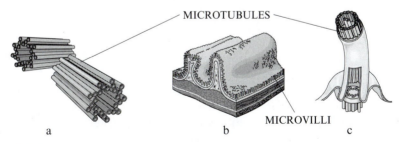

MICROTUBULES

MICROVILLI

a   b   c

Figure 3.11   Centrioles (a) contain microtubules and are located at right angles to each other in the cell's centrosome. These organelles play an important part in cell division by anchoring the microtubules that are used to divide the cell's genetic material. Microvilli (b), which are extensions of the plasma membrane, line the villi (tiny fingerlike protrusions in the mucosa of the small intestine) and help increase the area available for the absorption of nutrients. Cilia (c) line the respiratory tract. The beating of these organelles helps move bacteria and particles trapped in mucus out of the lungs.

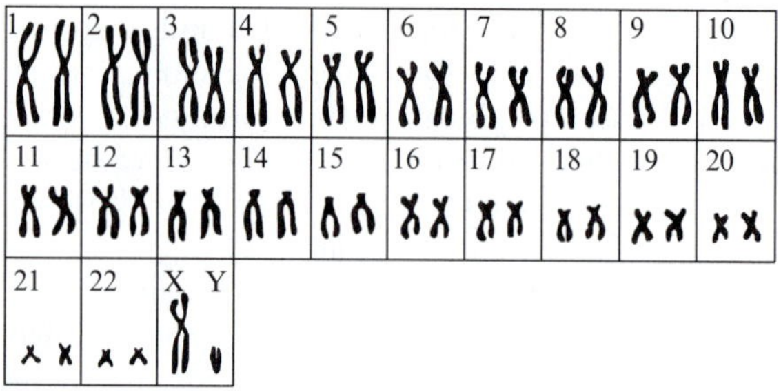

Figure 3.12 This karyotype of a normal human male shows the 22 pairs of autosomal chromosomes in descending order based on size, as well as the X and Y sex chromosomes.

that does not become an egg or sperm cell) was stretched out end to end, it would be about 2 nm wide and 2 m long. Each chromosome contains thousands of individual genes that are the units of information about heritable traits. Each gene has a particular location in a specific chromosome and contains the code for producing one of the three forms of RNA (ribosomal RNA, messenger RNA, and transfer RNA). The Human Genome Project was begun in 1990 and had as its goal to first identify the location of at least 3000 specific human genes and then to determine the sequence of nucleotides (about 3 billion!) in a complete set of haploid human chromosomes (one chromosome from each of the 23 pairs). See Chapter 13 for more information about the Human Genome Project.

DNA replication occurs during cell division (Fig. 3.13). During this semiconservative process, enzymes unzip the double helix, deliver complementary bases to the nucleotides, and bind the delivered nucleotides into the developing complementary strands. Following replication, each strand of DNA is duplicated so that two double helices now exist, each consisting of one strand of the original DNA and one new strand. In this way, each daughter cell gets the same hereditary information that was contained in the original dividing cell. During replication, some enzymes check for accuracy while others repair pairing mistakes so that the error rate is reduced to approximately one per billion.

Since DNA remains in the nucleus where it is protected from the action of the cell's enzymes and proteins are made on ribosomes outside of the nucleus, a method (transcription) exists for transferring information from the DNA to the cytoplasm. During transcription (Fig. 3.14), the sequence of nucleotides in a gene that codes for a protein is transferred to messenger RNA (mRNA) through complementary base pairing of the nucleotide sequence in the gene. For example, a DNA sequence of TACGCTCCGATA would become AUGCGAGGCUAU in the mRNA. The process is somewhat more complicated since the transcript produced directly from the DNA contains sequences of nucleotides, called introns, that are removed before the final mRNA is produced. The mRNA also has a tail, called a poly-A tail, of about 100–200 adenine nucleotides attached to one end. A cap with a nucleotide that has a methyl group and phosphate groups bonded to it is attached at the other end of the mRNA. Transcription differs from replication in that (1) only a certain stretch of DNA acts as

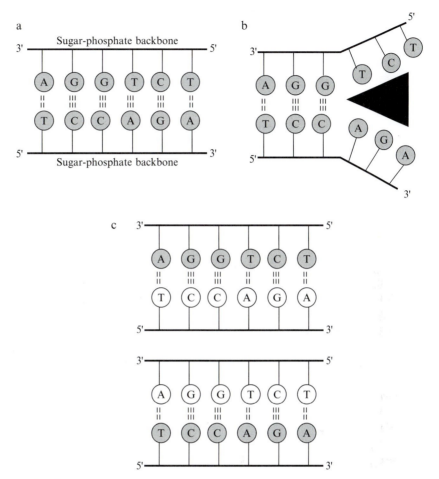

Figure 3.13 During replication, DNA helicase (shown as a black wedge in b) unzips the double helix (a). Another enzyme, DNA polymerase, then copies each side of the unzipped chain in the 5′ to 3′ direction. One side of the chain (5′ to 3′) can be copied continuously while the opposite side (3′ to 5′) is copied in small chunks in the 5′ to 3′ direction that are bound together by another enzyme, DNA ligase. Two identical double strands of DNA are produced as a result of replication.

the template and not the whole strand, (2) different enzymes are used, and (3) only a single strand is produced.

After being transcribed, the mRNA moves out into the cytoplasm through the nuclear pores and binds to specific sites on the surface of the two subunits that make up a ribosome (Fig. 3.15). In addition to the ribosomes, the cytoplasm contains amino acids and another form of RNA, transfer RNA (tRNA). Each tRNA contains a triplet of bases, called an anticodon, and binds at an area away from the triplet to an amino acid that is specific for that particular anticodon. The mRNA that was produced from the gene in the nucleus also contains bases in sets of three. Each triplet in mRNA is called a codon. The four possibilities for nucleotides (A, U, C, G) in each of the three

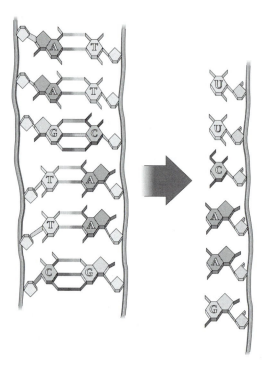

Figure 3.14   During transcription, RNA is formed from genes in the cell's DNA by complementary base pairing to one of the strands. RNA contains uracil (U) rather than thymine (T) so the T in the first two pairs of the DNA become Us in the single stranded RNA.

places give rise to 64 (4^3) possible codons. These 64 codons make up the genetic code. Each codon codes for a specific amino acid, but some amino acids are specified by more than one codon (see Table 3.1). For example, AUG is the only mRNA codon for methionine (the amino acid that always signals the starting place for translation—the process by which the information from a gene is used to produce a protein) whereas UUA, UUG, CUU, CUC, CUA, and CUG are all codons for leucine. The anticodon on the tRNA that delivers the methionine to the ribosome is UAC, whereas tRNAs with anticodons of AAU, AAC, GAA, GAG, GAU, and GAC deliver leucine.

During translation, the mRNA binds to a ribosome and tRNA delivers amino acids to the growing polypeptide chain in accordance with the codons specified by the mRNA. Peptide bonds are formed between each newly delivered amino acid and the previously delivered one. When the amino acid is bound to the growing chain, it is released from the tRNA, and the tRNA moves off into the cytoplasm where it joins with another amino acid that is specified by its anticodon. This process continues until a stop codon (UAA, UAG, or UGA) is reached on the mRNA.

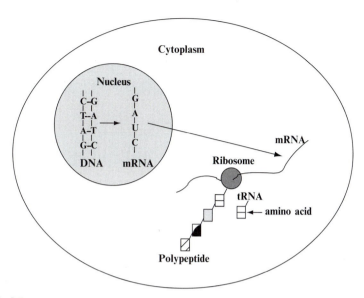

Figure 3.15 Following transcription from DNA and processing in the nucleus, mRNA moves from the nucleus to the cytoplasm. In the cytoplasm, the mRNA joins with a ribosome to begin the process of translation. During translation, tRNA delivers amino acids to the growing polypeptide chain. Which amino acid is delivered depends on the three-base codon specified by the mRNA. Each codon is complementary to the anticodon of a specific tRNA. Each tRNA binds to a particular amino acid at a site that is opposite the location of the anticodon. For example, the codon CUG in mRNA is complementary to the anticodon GAC in the tRNA that carries leucine and will result in adding the amino acid leucine to the polypeptide chain.

The protein is then released into the cytoplasm or into the rough ER for further modifications.

Example Problem 3.5

Consider a protein that contains the amino acids asparagine, phenylalanine, histidine, and serine in sequence. Which nucleotide sequences on DNA (assuming that there were no introns) would result in this series of amino acids? What would be the anticodons for the tRNAs that delivered these amino acids to the ribosomes during translation?

Solution

The genetic code (Table 3.1) provides the sequence for the mRNA codons that specify these amino acids. The mRNA codons can be used to determine the sequence in the original DNA and the anticodons of the tRNA since the mRNA bases must pair with the bases in both DNA and tRNA. Note that DNA contains thymine (T) but no uracil (U) and that both mRNA and tRNA contain U and not T. See Figs. 3.3 and 3.14 for examples of base pairing.

	Asparagine (Asn)	Phenylalanine (Phe)	Histidine (His)	Serine (Ser)
mRNA codon	AAU or AAC	UUU or UUC	CAU or CAC	UC(A, G, U, or C)
DNA	TTA or TTG	AAA or AAG	GTA or GTG	AG(T, C, A, or G)
tRNA anticodon	UUA or UUG	AAA or AAG	GUA or GUG	AG(U, C, A, or G)

■

3.3   TISSUES

Groups of cells and surrounding substances that function together to perform one or more specialized activities are called tissues (Fig. 3.16). There are four primary types of tissue in the human body: epithelial, connective, muscle, and nervous. Epithelial tissues are either composed of cells arranged in sheets that are one or more layers thick or are organized into glands that are adapted for secretion. They are also characterized by having a free surface (e.g., the inside surface of the intestines or the outside of the skin) and a basilar membrane. Typical functions of epithelial tissue include absorption (lining of the small intestine), secretion (glands), transport (kidney

TABLE 3.1   The genetic code

First base	Second base				Third base
	A	**U**	**G**	**C**	
A	Lys	Ile	Arg	Thr	A
	Asn	Ile	Ser	Thr	U
	Lys	Met - Start	Arg	Thr	G
	Asn	Ile	Ser	Thr	C
U	Stop	Leu	Stop	Ser	A
	Tyr	Phe	Cys	Ser	U
	Stop	Leu	Trp	Ser	G
	Tyr	Phe	Cys	Ser	C
G	Glu	Val	Gly	Ala	A
	Asp	Val	Gly	Ala	U
	Glu	Val	Gly	Ala	G
	Asp	Val	Gly	Ala	C
C	Gln	Leu	Arg	Pro	A
	His	Leu	Arg	Pro	U
	Gln	Leu	Arg	Pro	G
	His	Leu	Arg	Pro	C

Amino acid 3-letter and 1-letter codes: Ala (A) = Alanine: Arg (R) = Arginine; Asn (N) = Asparagine; Asp (D) = Aspartic acid; Cys (C) = Cysteine; Glu (E) = Glutamic acid; Gln (Q) = Glutamine; Gly (G) = Glycine; His (H) = Histidine; Ile (I) = Isoleucine; Leu (L) Leucine; Lys (K) = Lysine; Met (M) = Methionine; Phe (F) = Phenylalanine; Pro (P) = Proline; Ser (S) = Serine; Thr (T) = Threonine; Trp (W) = Tryptophan; Tyr (Y) = Tyrosine; Val (V) = Valine.

tubules), excretion (sweat glands), protection (skin, Fig. 3.16a), and sensory reception (taste buds). Connective tissues are the most abundant and widely distributed. Connective tissue proper can be loose (loosely woven fibers found around and between organs), irregularly dense (protective capsules around organs), and regularly dense (ligaments and tendons), whereas specialized connective tissue includes blood (Fig. 3.16b), bone, cartilage, and adipose tissue. Muscle tissue provides movement for the body through its specialized cells that can shorten in response to stimulation and then return to their uncontracted state. Figure 3.16c shows the three types of muscle tissue: skeletal (attached to bones), smooth (found in the walls of blood vessels), and cardiac (found only in the heart). Nervous tissue consists of neurons (Fig. 3.16d) that conduct electrical impulses and glial cells that protect, support, and nourish neurons.

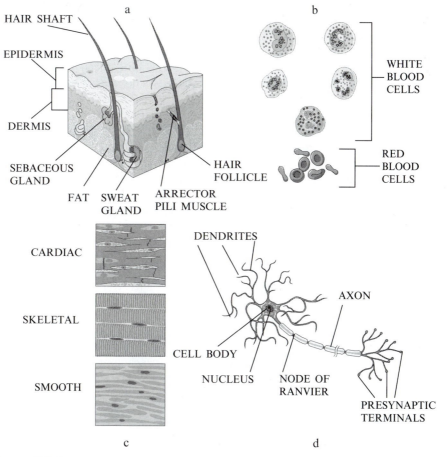

Figure 3.16 Four tissue types. Skin (a) is a type of epithelial tissue that helps protect the body. Blood (b) is a specialized connective tissue. There are three types of muscle tissue (c): cardiac, skeletal, and smooth. Motor neurons (d) are a type of nervous tissue that conduct electrical impulses from the central nervous system to effector organs such as muscles.

3.4   MAJOR ORGAN SYSTEMS

Combinations of tissues that perform complex tasks are called organs, and organs that function together form organ systems. The human body has 11 major organ systems: integumentary, endocrine, lymphatic, digestive, urinary, reproductive, circulatory, respiratory, nervous, skeletal, and muscular. The integumentary system (skin, hair, nails, and various glands) provides protection for the body. The endocrine system (ductless glands such as the thyroid and adrenals) secretes hormones that regulate many chemical actions within cells. The lymphatic system (glands, lymph nodes, lymph, lymphatic vessels) returns excess fluid and protein to the blood and helps defend the body against infection and tissue damage. The digestive system (stomach, intestines, and other structures) ingests food and water, breaks food down into small molecules that can be absorbed and used by cells, and removes solid wastes. The urinary system (kidneys, ureters, urinary bladder, and urethra) maintains the fluid volume of the body, eliminates metabolic wastes, and helps regulate blood pressure and acid–base and water–salt balances. The reproductive system (ovaries, testes, reproductive cells, and accessory glands and ducts) produces eggs or sperm and provides a mechanism for the production and nourishment of offspring. The circulatory system (heart, blood, and blood vessels) serves as a distribution system for the body. The respiratory system (airways and lungs) delivers oxygen to the blood from the air and carries away carbon dioxide. The nervous system (brain, spinal cord, peripheral nerves, and sensory organs) regulates most of the body's activities by detecting and responding to internal and external stimuli. The skeletal system (bones and cartilage) provides protection and support as well as sites for muscle attachments, the production of blood cells, and calcium and phosphorus storage. The muscular system (skeletal muscle) moves the body and its internal parts, maintains posture, and produces heat. Although biomedical engineers have made major contributions to understanding, maintaining, and/or replacing components in each of the eleven major organ systems, only the last five in the preceding list will be examined in greater detail.

3.4.1   Circulatory System

The circulatory system (Fig. 3.17) delivers nutrients and hormones throughout the body, removes waste products from tissues, and provides a mechanism for regulating temperature and removing the heat generated by the metabolic activities of the body's internal organs. Every living cell in the body is no more than 10–100 μm from a capillary (small blood vessels with walls only one cell thick that are 8 μm in diameter, approximately the same size as a red blood cell). This close proximity allows oxygen, carbon dioxide, and most other small solutes to diffuse from the cells into the capillary or from the capillary into the cells with the direction of diffusion determined by concentration and partial pressure gradients.

The heart (Fig. 3.18), the pumping station that moves blood through the blood vessels, consists of two pumps—the right side and the left side. Each side has one

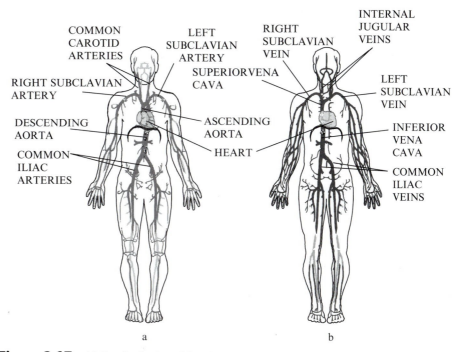

Figure 3.17   (a) The distribution of the main arteries in the body which carry blood away from the heart. (b) The distribution of the main veins in the body which return the blood to the heart.

chamber (the atrium) that receives blood and another chamber (the ventricle) that pumps the blood away from the heart. The right side moves deoxygenated blood that is loaded with carbon dioxide from the body to the lungs, and the left side receives oxygenated blood that has had most of its carbon dioxide removed from the lungs and pumps it to the body. The vessels that lead to and from the lungs make up the pulmonary circulation, and those that lead to and from the rest of the tissues in the body make up the systemic circulation (Fig. 3.19). Blood vessels that carry blood away from the heart are called arteries and those that carry blood toward the heart are called veins. The pulmonary artery is the only artery that carries deoxygenated blood, and the pulmonary vein is the only vein that carries oxygenated blood. The average adult has about 5 L of blood with 80–90% in the systemic circulation at any one time; 75% of the blood is in the systemic circulation in the veins, 20% in the arteries, and 5% in the capillaries. Cardiac output is the product of the heart rate and the volume of blood pumped from the heart with each beat (i.e., the stroke volume). Each time the heart beats, about 80 ml of blood leave the heart. Thus, it takes about 60 beats for the average red blood cell to make one complete cycle of the body.

In the normal heart, the cardiac cycle, which refers to the repeating pattern of contraction (systole) and relaxation (diastole) of the chambers of the heart, begins with a self-generating electrical pulse in the pacemaker cells of the sinoatrial node

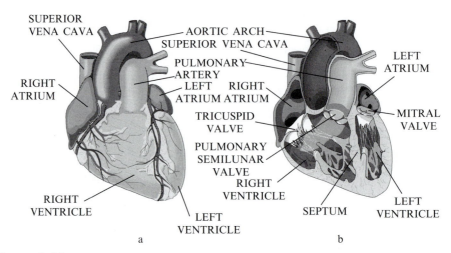

Figure 3.18   (a) The outside of the heart as seen from its anterior side. (b) The same view after the exterior surface of the heart has been removed. The four interior chambers—right and left atria and right and left ventricles—are visible, as are several valves.

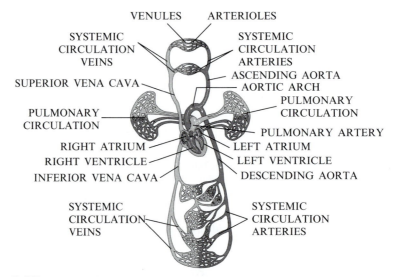

Figure 3.19   Oxygenated blood leaves the heart through the aorta. Some of the blood is sent to the head and upper extremities and torso, whereas the remainder goes to the lower torso and extremities. The blood leaves the aorta and moves into other arteries, then into smaller arterioles, and finally into capillary beds where nutrients, hormones, gases, and waste products are exchanged between the nearby cells and the blood. The blood moves from the capillary beds into venules and then into veins. Blood from the upper part of the body returns to the right atrium of the heart through the superior vena cava, whereas blood from the lower part of the body returns through the inferior vena cava. The blood then moves from the right atrium to the right ventricle and into the pulmonary system through the pulmonary artery. After passing through capillaries in the lungs, the oxygenated blood returns to the left atrium of the heart through the pulmonary vein. It moves from the left atrium to the left ventricle and then out to the systemic circulation through the aorta to begin the same trip over again.

(Fig. 3.20). This rapid electrical change in the cells is the result of the movement of ions across their plasma membranes. The permeability of the plasma membrane to Na^+ changes dramatically and allows these ions to rush into the cell. This change in the electrical potential across the plasma membrane from one in which the interior of the cell is more negative than the extracellular fluid (approximately -90 mV) to one in which the interior of the cell is more positive than the extracellular fluid (approximately 20 mV) is called depolarization. After a very short period of time (<0.3 s), changes in the membrane and activation of the sodium–potassium pumps result in repolarization, the restoration of the original ionic balance in the cells. The entire electrical event in which the polarity of the potential across the plasma membrane rapidly reverses and then becomes reestablished is called an action potential. The cells in the sinoatrial node depolarize on the average of every 0.83 s in a typical adult at rest. This gives a resting heart rate of 72 beats per minute with about $\frac{5}{8}$ of each beat spent in diastole and $\frac{3}{8}$ in systole.

Cardiac cells are linked and tightly coupled so that action potentials spread from one cell to the next. Activation wavefronts move across the atria at a rate of about 1 m/s. When cardiac cells depolarize, they also contract. The contraction process in the atria (atrial systole) moves blood from the right atrium to the right ventricle and from the left atrium to the left ventricle (Fig. 3.21). The activation wavefront then moves to the atrioventricular (AV) node where it slows to a rate of about 0.05 m/s to allow time for the ventricles to completely fill with the blood from the atria. After leaving the AV node, the activation wavefront moves to specialized conduction tissue, the Purkinje system, which spreads the wavefront very rapidly (at about 3 m/s) to many cells in both venricles. The activation wavefront spreads through ventricular tissue at about

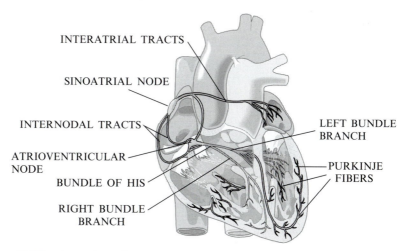

INTERATRIAL TRACTS

SINOATRIAL NODE

INTERNODAL TRACTS

LEFT BUNDLE BRANCH

ATRIOVENTRICULAR NODE

PURKINJE FIBERS

BUNDLE OF HIS

RIGHT BUNDLE BRANCH

Figure 3.20 Pacemaker cells in the sinoatrial node (SA node) depolarize first and send an activation wavefront through the atria. The propagating action potential slows down as it passes through the atrioventricular node (AV node), then moves through the Bundle of His and Purkinje system very rapidly until it reaches the cells of the ventricles.

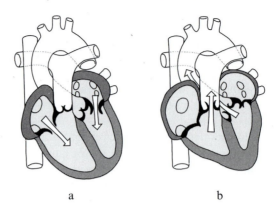

a   b

Figure 3.21   (a) During the first part of the cardiac cycle, the atria contract (atrial systole) and move blood into the ventricles. (b) During the second part of the cardiac cycle, the atria relax (diastole), and the ventricles contract (ventricular systole) and move blood to the lungs (pulmonary circulation) and to the rest of the body (systemic circulation).

0.5 m/s. This results in the simultaneous contraction of both ventricles (ventricular systole) so that blood is forced from the heart into the pulmonary artery from the right ventricle and into the aorta from the left ventricle.

The electrocardiogram (ECG; Fig. 3.22) is an electrical measure of the sum of these ionic changes within the heart. The P wave represents the depolarization of the atria and the QRS represents the depolarization of the ventricles. Ventricular repolarization shows up as the T wave and atrial repolarization is masked by ventricular depolarization. Changes in the amplitude and duration of the different parts of the

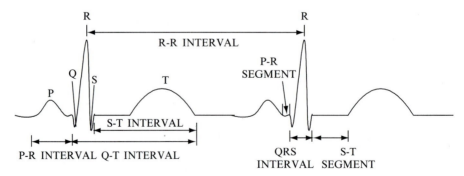

Figure 3.22   Typical Lead II ECG. This electrocardiogram is typical of one that would be recorded from the body's surface by having a positive electrode on the left leg and a negative electrode on the right arm. The vertical direction represents voltage and the horizontal direction represents time. The P, R, and T waves are easily identified and are the result of the movement of ions in cells in different parts of the heart. Different intervals and segments have been identified which provide information about the health of the heart and its conduction system. The R–R interval can be used to determine heart rate.

ECG provide diagnostic information for physicians. Many biomedical engineers have worked on methods for recording and analyzing ECGs.

Example Problem 3.6

What would be the heart rate given by an ECG in which 10 R-waves occurred in 6.4 s?

Solution

A sequence of 10 R-waves represents 9 R–R intervals (see Fig. 3.22) or beats of the heart.

$$\left(\frac{9 \text{ beats}}{6.4 \text{ s}}\right) \left(\frac{60 \text{ s}}{1 \text{ min}}\right) = 84 \text{ bpm} \qquad \blacksquare$$

Example Problem 3.7

What would be the cardiac output of the heart in Example Problem 3.6 if the stroke volume were 75 ml?

Solution

The cardiac output (given in liters per minute) is the product of the heart rate and the stroke volume.

$$CO = 84 \frac{\text{beats}}{\text{min}} \times 75 \frac{\text{ml}}{\text{beat}} = 6300 \frac{\text{ml}}{\text{min}} = 6.3 \frac{\text{liters}}{\text{min}} \qquad \blacksquare$$

During atrial and ventricular systole, special one-way valves (Fig. 3.23a) keep the blood moving in the correct direction. When the atria contract, the atrioventricular valves (tricuspid and mitral) open to allow blood to pass into the ventricles. During ventricular systole, the semilunar valves (aortic and pulmonary) open to allow blood to leave the heart while the atrioventricular valves close and prevent blood from flowing backwards from the ventricles to the atria. The aortic and pulmonary valves prevent blood from flowing back from the pulmonary artery and aorta into the right and left ventricles, respectively. If a valve becomes calcified or diseased or is not properly formed during embryonic development, it can be replaced by an artificial valve (Fig. 3.23b), a device that has been developed by cooperative work between biomedical engineers and physicians.

Blood pressure can be measured directly or indirectly (noninvasively). Direct blood pressure measurements are made by introducing a catheter or needle that is coupled to a pressure transducer into a vein or artery. Indirect methods include sphygmomanometry, in which a cuff is used to apply sufficient pressure to an artery, usually in the arm, to prevent the flow of blood through the artery, and a stethoscope is used to listen to the change in sounds as the cuff is slowly deflated. The first Korotkoff sounds occur when the systolic pressure, the highest pressure reached when the ventricles contract and eject blood, first exceeds the pressure in the cuff so that blood once again flows through the artery beneath the stethoscope. The Korotkoff sounds become muffled and disappear when the pressure in the cuff drops below the diastolic

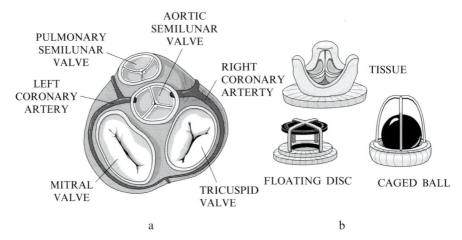

Figure 3.23 (a) The atrioventricular (tricuspid and mitral) and semilunar (pulmonary and aortic) valves. (b) Three types of artificial valves—tissue, floating disc, and caged ball—that can be used to replace diseased or malformed human valves.

pressure, the minimum pressure that occurs at the end of ventricular relaxation. Another indirect measurement is the oscillometric method, which uses a microprocessor to periodically inflate and slowly deflate a cuff. When blood breaks through the occlusion caused by the cuff, the walls of the artery begin to vibrate slightly due to the turbulent nature of the blood flow. The onset of these oscillations in pressure correlates with the systolic pressure. The oscillations decrease in amplitude over time with the diastolic pressure event corresponding to the point at which the rate of amplitude decrease suddenly changes slope. A third indirect measurement, the ultrasonic method, depends on the Doppler shift of sound waves that hit red blood cells that are flowing with the blood.

Blood in the systemic circulation leaves the heart through the aorta with an average internal pressure of about 100 mm Hg (maximum systolic pressure of about 120 mm Hg with a diastolic pressure of about 80 mm Hg in a normal adult) and moves to medium-sized arteries (Fig. 3.17a) and arterioles. Arterioles lead to capillaries (average internal pressure of about 30 mm Hg), which are followed by venules. Venules lead to medium-sized veins, then to large veins, and finally to the venae cavae (average internal pressure of about 10 mm Hg) which return blood to the heart at the right atrium. Blood in the pulmonary circulation (Fig. 3.19) leaves the pulmonary artery and moves to arterioles and then the capillary beds within the lungs. It returns to the heart through the left atrium. Blood flow is highest in the large arteries and veins (30–40 cm/s in the aorta; 5 cm/s in the venae cavae) and slowest in the capillary beds (1 mm/s) where the exchange of nutrients, metabolic wastes, gases, and hormones takes place. Pressures in the pulmonary circulation are lower (25 mm Hg/10 mm Hg) than in the systemic circulation due to the decreased pumping power of the smaller right ventricle as compared to the left and to the lower resistance of blood vessels in the lungs.

Example Problem 3.8

What would be the pulse pressure and the mean arterial pressure for a person with a blood pressure reading of 118 mm Hg/79 mm Hg?

Solution

The pulse pressure is defined as the difference between the systolic (118 mm Hg) and disastolic (79 mm Hg) pressures, which would be 39 mm Hg in this case.

Mean arterial pressure is the average blood pressure in the arteries and is estimated as the diastolic pressure plus one-third of the pulse pressure, which would be 92 mm Hg in this example. ∎

3.4.2 Respiratory System

The respiratory system (Fig. 3.24a) moves air to and from the gas exchange surfaces in the body where diffusion can occur between air and the circulating blood. It includes the conduction zone and the respiratory zone. In the conduction zone (mouth, nose, sinuses, pharynx, trachea, bronchi, and bronchioles), the air that enters the body is warmed, humidified, filtered, and cleaned. Mucus is secreted by cells in the conduction zone and traps small particles ($> 6\,\mu m$) before they can reach the respiratory zone. Epithelial cells that line the trachea and bronchi have cilia that beat in a coordinated fashion to move mucus toward the pharynx where it can be swallowed or expectorated. The respiratory zone, consisting of respiratory bronchioles with outpouchings of alveoli and terminal clusters of alveolar sacs, is where gas exchange

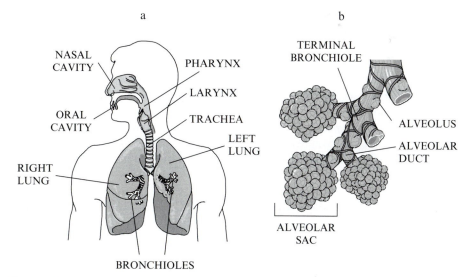

Figure 3.24 (a) The respiratory system consists of the passageways that are used to move air into and out of the body and the lungs. (b) The terminal bronchioles and alveolar sacs within the lungs have alveoli where gas exchange occurs between the lungs and the blood in the surrounding capillaries.

between air and blood occurs (Fig. 3.24b). The respiratory zone comprises most of the mass of the lungs.

Certain physical properties—compliance, elasticity, and surface tension—are characteristic of lungs. Compliance refers to the ease with which lungs can expand under pressure. A normal lung is about 100 times more distensible than a toy balloon. Elasticity refers to the ease with which the lungs and other thoracic structures return to their initial sizes after being distended. This aids in pushing air out of the lungs during expiration. Surface tension is exerted by the thin film of fluid in the alveoli and acts to resist distention. It creates a force that is directed inward and creates pressure in the alveolus which is directly proportional to the surface tension and inversely proportional to the radius of the alveolus (Law of Laplace). Thus, the pressure inside an alveolus with a small radius would be higher than the pressure inside an adjacent alveolus with a larger radius and would result in air flowing from the smaller alveolus into the larger one. This could cause the smaller alveolus to collapse. This does not happen in normal lungs because the fluid inside the alveoli contains a phospholipid that acts as a surfactant. The surfactant lowers the surface tension in the alveoli and allows them to get smaller during expiration without collapsing. Premature babies often suffer from respiratory distress syndrome because their lungs lack sufficient surfactant to prevent their alveoli from collapsing. These babies can be kept alive with mechanical ventilators or surfactant sprays until their lungs mature enough to produce surfactant.

Breathing, or ventilation, is the mechanical process by which air is moved into (inspiration) and out of (expiration) the lungs. A normal adult takes about 15 to 20 breaths per minute. During inspiration, the inspiratory muscles contract and enlarge the thoracic cavity, the portion of the body where the lungs are located. This causes the alveoli to enlarge and the alveolar gas to expand. As the alveolar gas expands, the partial pressure within the respiratory system drops below atmospheric pressure by about 3 mm Hg so that air easily flows in (Boyle's Law). During expiration, the inspiratory muscles relax and return the thoracic cavity to its original volume. Since the volume of the gas inside the respiratory system has decreased, its pressure increases to a value that is about 3 mm Hg above atmospheric pressure. Air now moves out of the lungs and into the atmosphere.

Lung mechanics refers to the study of the mechanical properties of the lung and chest wall, whereas lung statics refers to the mechanical properties of a lung in which the volume is held constant over time. Understanding lung mechanics requires knowledge about the volumes within the lungs. Lung capacities contain two or more volumes. The tidal volume (TV) is the amount of air that moves in and out of the lungs during normal breathing (Fig. 3.25). The total lung capacity (TLC) is the amount of gas contained within the lungs at the end of a maximum inspiration. The vital capacity (VC) is the maximum amount of air that can be exhaled from the lungs after inspiration to TLC. The residual volume (RV) is the amount of gas remaining in the lungs after maximum exhalation. The amount of gas that can be inhaled after inhaling during tidal breathing is called the inspiratory reserve volume (IRV). The amount of gas that can be expelled by a maximal exhalation after exhaling during tidal breathing is called the expiratory reserve volume (ERV). The inspiratory capacity (IC) is the

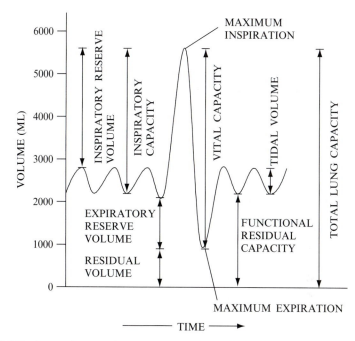

Figure 3.25 Lung volumes and capacities, except for residual volume, functional residual capacity, and total lung capacity, can be measured using spirometry.

maximum amount of gas that can be inspired after a normal exhalation during tidal breathing, and the functional residual capacity (FRC) is the amount of gas that remains in the lungs at this time.

All of the volumes and capacities except those that include the residual volume can be measured with a spirometer. The classic spirometer is an air-filled container that is constructed from two drums of different sizes. One drum contains water and the other air-filled drum is inverted over an air-filled tube and floats in the water. The tube is connected to a mouthpiece used by the patient. When the patient inhales, the level of the floating drum drops. When the patient exhales, the level of the floating drum rises. These changes in floating drum position can be recorded and used to measure lung volumes.

Example Problem 3.9

The total lung capacity of a patient is 5.9 liters. If the patient's inspiratory capacity was found to be 3.3 liters using spirometry, what would be the patient's functional residual capacity? What would you need to measure to determine the patient's residual volume?

Solution

From Figure 3.25, total lung capacity (TLC) is equal to the sum of inspiratory capacity (IC) and functional residual capacity (FRC).

$$TLC = IC + FRC$$

$$5.9 \text{ liters} = 3.3 \text{ liters} + FRC$$

$$FRC = 2.6 \text{ liters}$$

TLC, which cannot be determined by means of spirometry, and vital capacity (VC), which can be measured using spirometry, must be known to determine residual volume (RV) since

$$TLC - VC = RV \qquad \blacksquare$$

Because spirograms record changes in volume over time, flow rates can be determined for different maneuvers. For example, if a patient exhales as forcefully as possible to residual volume following inspiration to TLC, then the forced expiratory volume ($FEV_{1.0}$) is the total volume exhaled at the end of 1 s. The $FEV_{1.0}$ is normally about 80% of the vital capacity. Restrictive diseases, in which inspiration is limited by reduced compliance of the lung or chest wall or by weakness of the inspiratory muscles, result in reduced values for $FEV_{1.0}$ and vital capacity but their ratio remains about the same. In obstructive diseases, such as asthma, the $FEV_{1.0}$ is reduced much more than the vital capacity. In these diseases, the TLC is abnormally large but expiration ends prematurely. Another useful measurement is the forced expiratory flow rate ($FEF_{25-75\%}$), which is the average flow rate measured over the middle half of the expiration (i.e., from 25 to 75% of the vital capacity). Flow-volume loops provide another method for analyzing lung function by relating the rate of inspiration and expiration to the volume of air that is moved during each process.

The TLC can be measured using the gas dilution technique. In this method, patients inspire to TLC from a gas mixture containing a known amount of an inert tracer gas such as helium, and hold their breaths for 10 s. During this time, the inert gas becomes evenly distributed throughout the lungs and airways. Due to conservation of mass, the product of initial tracer gas concentration (which is known) times the amount inhaled (which is measured) equals the product of final tracer gas concentration (which is measured during expiration) times the TLC. Body plethysmography, which provides the most accurate method for measuring lung volumes, uses an airtight chamber in which the patient sits and breathes through a mouthpiece. This method makes use of Boyle's Law, which states that the product of pressure and volume for gas in a chamber is constant under isothermal conditions. Changes in lung volume and pressure at the mouth when the patient pants against a closed shutter can be used to calculate the functional residual capacity. Since the expiratory reserve volume can be measured, the residual volume can be calculated by subtracting it from the functional residual capacity.

Example Problem 3.10

A patient is allowed to breathe a mixture from a 2-liter reservoir that contains 10% of an inert gas (i.e., one that will not cross from the lungs into the circulatory system). At the end of a period that is sufficient for the contents of the reservoir and the lungs to equilibrate, the concentration of the inert gas is measured and is found to be 2.7%. What is the patient's total lung capacity?

Solution

The total amount of inert gas is the same at the beginning and end of the measurement, but its concentration has changed from 10% (C_1) to 2.7% (C_2). At the beginning, it is confined to a 2-liter reservoir (V_1). At the end, it is in both the reservoir and the patient's lungs ($V_2 = V_1 + TLC$).

$$C_1 V_1 = C_2 V_2$$

$$(0.1) \ (2 \text{ liters}) = (0.027) \ (2 \text{ liters} + TLC)$$

$$0.2 \text{ liters} - 0.054 \text{ liters} = 0.027 \text{ TLC}$$

$$5.4 \text{ liters} = TLC \qquad \blacksquare$$

External respiration occurs in the lungs when gases are exchanged between the blood and the alveoli (Fig. 3.26). Each adult lung contains about 3.5×10^8 alveoli, which results in a large surface area ($60 - 70 \text{ m}^2$) for gas exchange to occur. Each alveolus is only one cell layer thick, making the air–blood barrier only two cells thick (an alveolar cell and a capillary endothelial cell) which is about $2 \, \mu\text{m}$. The partial pressure of oxygen in the alveoli is higher than the partial pressure of oxygen in the blood so oxygen moves from the alveoli into the blood. The partial pressure of carbon dioxide in the alveoli is lower than the partial pressure of carbon dioxide in the blood so carbon dioxide moves from the blood into the alveoli. During internal respiration, carbon dioxide and oxygen move between the blood and the extracellular fluid surrounding the body's cells. The direction and rate of movement of a gas depend

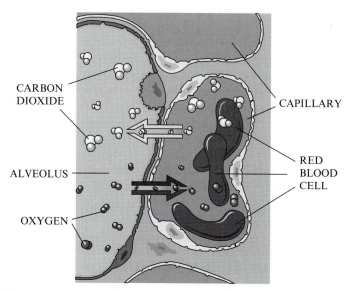

Figure 3.26 During external respiration, oxygen moves from the alveoli to the blood and carbon dioxide moves from the blood to the air within the alveoli.

on the partial pressures of the gas in the blood and the extracellular fluid, the surface area available for diffusion, the thickness of the membrane that the gas must pass through, and a diffusion constant that is related to the solubility and molecular weight of the gas (Fick's Law).

Mechanical ventilators can be used to deliver air or oxygen to a patient. They can be electrically or pneumatically powered and can be controlled by microprocessors. Negative pressure ventilators such as iron lungs surround the thoracic cavity and force air into the lungs by creating a negative pressure around the chest. This type of ventilator greatly limits access to the patient. Positive pressure ventilators apply high-pressure gas at the entrance to the patient's lungs so that air or oxygen flows down a pressure gradient and into the patient. These ventilators can be operated in control mode to breathe for the patient at all times or in assist mode to help with ventilation when the patient initiates the breathing cycle. This type of ventilation changes the pressure within the thoracic cavity to positive during inspiration, which affects venous return to the heart and cardiac output (the amount of blood the heart moves with each beat). High frequency jet ventilators deliver very rapid (60–900 breaths per minute) low-volume bursts of air to the lungs. Oxygen and carbon dioxide are exchanged by molecular diffusion rather than by the mass movement of air. This method causes less interference with cardiac output than does positive pressure ventilation. Extracorporeal membrane oxygenation (ECMO) uses the technology that was developed for cardiopulmonary bypass machines. Blood is removed from the patient and passed through an artificial lung where oxygen and carbon dioxide are exchanged. It is warmed to body temperature before being returned to the patient. This technique allows the patient's lungs to rest and heal themselves and has been used successfully on some cold-water drowning victims and on infants with reversible pulmonary disease.

3.4.3   Nervous System

The nervous system, which is responsible for the integration and control of all the body's functions, has two major divisions: the central nervous system and the peripheral nervous system (Fig. 3.27). The former consists of all nervous tissue enclosed by bone (e.g., the brain and spinal cord), whereas the latter consists of all nervous tissue not enclosed by bone, which enables the body to detect and respond to both internal and external stimuli. The peripheral nervous system consists of the 12 pairs of cranial and 31 pairs of spinal nerves with afferent (sensory) and efferent (motor) neurons.

The nervous system has also been divided into the somatic and autonomic nervous systems. Each of these systems consists of components from both the central and peripheral nervous systems. For example, the somatic peripheral nervous system consists of the sensory neurons, which convey information from receptors for pain, temperature, and mechanical stimuli in the skin, muscles, and joints to the central nervous system, and the motor neurons, which return impulses from the central nervous system to these same areas of the body. The autonomic nervous system is concerned with the involuntary regulation of smooth muscle, cardiac muscle, and glands and consists of the sympathetic and parasympathetic divisions.

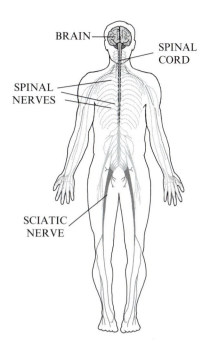

BRAIN

SPINAL
CORD

SPINAL
NERVES

SCIATIC
NERVE

Figure 3.27 The central nervous system (CNS) consists of all nervous tissue that is enclosed by bone (i.e., the brain and spinal cord), whereas the peripheral nervous system (PNS) consists of the nervous tissue that is not encased by bone.

The sympathetic division causes blood vessels in the viscera and skin to constrict, vessels in the skeletal muscles to dilate, and heart rate to increase, whereas the parasympathetic division has the opposite effect on the vessels in the viscera and skin, provides no innervation to the skeletal muscles, and causes heart rate to decrease. Thus, the sympathetic division prepares the body for "fight or flight" and the parasympathetic division returns the body to normal operating conditions.

Specialized cells that conduct electrical impulses (neurons) or protect, support, and nourish neurons (glial cells) make up the different parts of the nervous system. The cell body of the neuron (Fig. 3.16d) gives rise to and nourishes a single axon and multiple, branching dendrites. The dendrites are the main receptor portion of the neuron although the cell body can also receive inputs from other neurons. Dendrites usually receive signals from thousands of contact points (synapses) with other neurons. The axon extends a few millimeters (in the brain) to a meter (from the spinal cord to the foot) and carries nerve signals to other nerve cells in the brain or spinal cord or to glands and muscles in the periphery of the body. Some axons are surrounded by sheaths of myelin that are formed by specialized, nonneural cells called Schwann cells. Each axon has many branches, called presynaptic terminals, at its end. These knoblike protrusions contain synaptic vesicles that hold neurotransmitters. When the neuron is stimulated by receiving a signal at its dendrites, the permeability of the cell's plasma membrane to sodium increases, as occurs in cardiac cells, and an action potential moves from the dendrite to the cell body and then on to the axon. Gaps, called nodes of Ranvier, in the myelin sheaths of some axons allow the action potential to move more rapidly by essentially jumping from one node to the next. The vesicles in the presynaptic terminals release their neurotransmitter into the space between the axon and an adjacent neuron, muscle cell, or gland. The neurotransmitter diffuses across the synapse and causes a response (Fig. 3.28).

Neurons interconnect in several types of circuits. In a divergent circuit, each branch in the axon of the presynaptic neuron connects with the dendrite of a different postsynaptic neuron. In a convergent circuit, axons from several presynaptic neurons meet at the dendrite(s) of a single postsynaptic neuron. In a simple feedback circuit, the axon of a neuron connects with the dendrite of an interneuron that connects back with the dendrites of the first neuron. A two-neuron circuit is one in which a sensory neuron synapses directly with a motor neuron, whereas a three-neuron circuit consists of a sensory neuron, an interneuron in the spinal cord, and a motor neuron. Both of these circuits can be found in reflex arcs (Fig. 3.29). The reflex arc is a special type of neural circuit that begins with a sensory neuron at a receptor (e.g., a pain receptor in the fingertip) and ends with a motor neuron at an effector (e.g., a skeletal muscle). Withdrawal reflexes are elicited primarily by stimuli for pain and heat great enough to be painful and are also known as protective or escape reflexes. They allow the body to respond quickly to dangerous situations without taking additional time to send signals to and from the brain and to process the information.

The brain is a large soft mass of nervous tissue and has three major parts: (1) cerebrum, (2) diencephalon, and (3) brain stem and cerebellum. The cerebrum (Fig. 3.30), which is divided into two hemispheres, is the largest and most obvious portion of the brain and consists of many convoluted ridges (gyri), narrow grooves

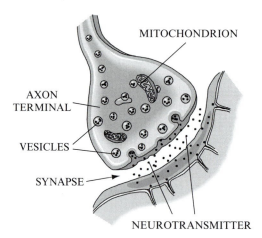

Figure 3.28 Following stimulation, vesicles in the axon terminal move to the synapse by means of exocytosis and release neurotransmitters into the space between the axon and the next cell, which could be the dendrite of another neuron, a muscle fiber, or a gland. The neurotransmitters diffuse across the synapse and elicit a response from the adjacent cell.

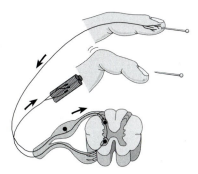

Figure 3.29   This reflex arc begins with a sensory neuron in the finger that senses pain when the fingertip is pricked by the pin. An action potential travels from the sensory neuron to an interneuron and then to a motor neuron that synapses with muscle fibers in the finger. The muscle fibers respond to the stimulus by contracting and removing the fingertip from the pin.

(sulci), and deep fissures which result in a total surface area of about 2.25 m². The outer layer of the cerebrum, the cerebral cortex, is composed of gray matter (neurons with unmyelinated axons) that is 2–4 mm thick and contains over 50 billion neurons and 250 billion glial cells called neuroglia. The thicker inner layer is the white matter that consists of interconnecting groups of myelinated axons that project from the cortex to other cortical areas or from the thalamus (part of the diencephalon) to the cortex. The connection between the two cerebral hemispheres is called the corpus callosum (Fig. 3.30b). The left side of the cortex controls motor and sensory functions from the right side of the body, whereas the right side controls the left side of the body. Association areas that interpret incoming data or coordinate a motor response are connected to the sensory and motor regions of the cortex.

Fissures divide each cerebral hemisphere into a series of lobes that have different functions. The functions of the frontal lobes include initiating voluntary movement of the skeletal muscles, analyzing sensory experiences, providing responses relating to personality, and mediating responses related to memory, emotions, reasoning, judgment, planning, and speaking. The parietal lobes respond to stimuli from cutaneous (skin) and muscle receptors throughout the body. The temporal lobes interpret some sensory experiences, store memories of auditory and visual experiences, and contain auditory centers that receive sensory neurons from the cochlea of the ear. The occipital lobes integrate eye movements by directing and focusing the eye and are responsible for correlating visual images with previous visual experiences and other sensory stimuli. The insula is a deep portion of the cerebrum that lies under the parietal, frontal, and temporal lobes. Little is known about its function, but it seems to be associated with gastrointestinal and other visceral activities.

The diencephalon is the deep part of the brain that connects the midbrain of the brain stem with the cerebral hemispheres. Its main parts are the thalamus, hypothalamus, and epithalamus (Fig. 3.30b). The thalamus is involved with sensory and motor systems, general neural background activity, and the expression of emotion and

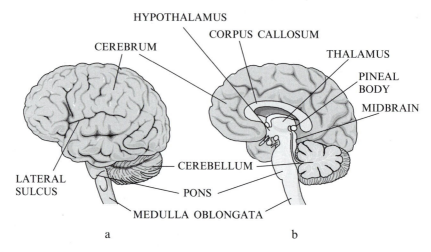

Figure 3.30   (a) The exterior surface of the brain. (b) A midsagittal section through the brain.

uniquely human behaviors. Due to its two-way communication with areas of the cortex, it is linked with thought, creativity, interpretation and understanding of spoken and written words, and identification of objects sensed by touch. The hypothalamus is involved with integration within the autonomic nervous system, temperature regulation, water and electrolyte balance, sleep–wake patterns, food intake, behavioral responses associated with emotion, endocrine control, and sexual responses. The epithalamus contains the pineal body that is thought to have a neuroendocrine function.

The brain stem connects the brain with the spinal cord and automatically controls vital functions such as breathing. Its principal regions include the midbrain, pons, and medulla oblongota (Fig. 3.30b). The midbrain connects the pons and cerebellum with the cerebrum and is located at the upper end of the brain stem. It is involved with visual reflexes, the movement of eyes, focusing of the lenses, and the dilation of the pupils. The pons is a rounded bulge between the midbrain and medulla oblongata which functions with the medulla oblongata to control respiratory functions, acts as a relay station from the medulla oblongata to higher structures in the brain, and is the site of emergence of cranial nerve V. The medulla oblongata is the lowermost portion of the brain stem and connects the pons to the spinal cord. It contains vital centers that regulate heart rate, respiratory rate, constriction and dilation of blood vessels, blood pressure, swallowing, vomiting, sneezing, and coughing. The cerebellum is located behind the pons and is the second largest part of the brain. It processes sensory information that is used by the motor systems and is involved with coordinating skeletal muscle contractions and impulses for voluntary muscular movement that originate in the cerebral cortex. The cerebellum is a processing center that is involved with coordination of balance, body positions, and the precision and timing of movements.

3.4.4 Skeletal System

The average adult skeleton contains 206 bones, but the actual number varies from person to person and decreases with age as some bones become fused. Like the body, the skeletal system is divided into two parts: the axial skeleton and the appendicular skeleton (Fig. 3.31). The axial skeleton contains 80 bones (skull, hyoid bone, vertebral column, and thoracic cage), whereas the appendicular skeleton contains 126 (pectoral and pelvic girdles and upper and lower extremities). The skeletal system protects and supports the body, helps with movement, produces blood cells, and stores important minerals. It is made up of strong, rigid bones that are composed of specialized connective tissue, bear weight, and form the major supporting elements of the body. Some support also comes from cartilage which is a smooth, firm, resilient, nonvascular type of connective tissue. Since the bones of the skeleton are hard, they protect the organs, such as the brain and abdominal organs, that they surround.

There are 8 cranial bones that support, surround, and protect the brain. Fourteen facial bones form the face and serve as attachments for the facial muscles that primarily move skin rather than bone. The facial bones, except for the lower jaw (mandible), are joined with each other and with the cranial bones. There are 6 auditory ossicles, 3 in each ear, that transmit sound waves from the external environment to the inner ear. The hyoid bone, which is near the skull but not part

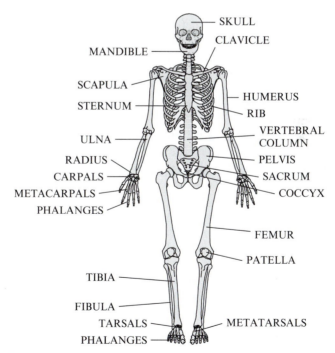

Figure 3.31 The skull, hyoid bone (not shown), vertebral column, and thoracic cage (ribs, cartilage, and sternum) make up the axial skeleton, whereas the pectoral (scapula and clavicle) and pelvic girdles and upper and lower extremities make up the appendicular skeleton.

of it, is a small U-shaped bone that is located in the neck just below the lower jaw. It is attached to the skull and larynx (voice box) by muscles and ligaments and serves as the attachment for several important neck and tongue muscles.

The vertebral column starts out with approximately 34 bones, but only 26 independent ones are left in the average human adult. There are 7 cervical bones, including the axis which acts as a pivot around which the head rotates, and the atlas which sits on the axis and supports the "globe" of the head. These are followed by 5 cervical, 12 thoracic, and 5 lumbar vertebrae and then the sacrum and the coccyx. The last two consist of 5 fused vertebrae. The vertebral column supports the weight of and allows movement of the head and trunk, protects the spinal cord, and provides places for the spinal nerves to exit from the spinal cord. There are 4 major curves (cervical, thoracic, lumbar, and sacral/coccygeal) in the adult vertebral column which allow it to flex and absorb shock. Although movement between any 2 adjacent vertebrae is generally quite limited, the total amount of movement provided by the vertebral column can be extensive. The thoracic cage consists of 12 thoracic vertebrae (which are counted as part of the vertebral column), 12 pairs of ribs and their associated cartilage, and the sternum (breastbone). It protects vital organs and prevents the collapse of the thorax during ventilation.

Bones are classified as long, short, flat, or irregular according to their shape. Long bones, such as the femur and humerus, are longer than they are wide. Short bones, such as those found in the ankle and wrist, are as broad as they are long. Flat bones, such as the sternum and the bones of the skull, have a relatively thin and flattened shape. Irregular bones do not fit into the other categories and include the bones of the vertebral column and the pelvis.

Bones make up about 18% of the mass of the body and have a density of $1.9\,\mathrm{g/cm}^3$. There are two types of bone: spongy and compact (cortical). Spongy bone forms the ends (epiphyses) of the long bones and the interior of other bones and is quite porous. Compact bone forms the shaft (diaphysis) and outer covering of bones and has a tensile strength of $120\,\mathrm{N/mm}^2$, compressive strength of $170\,\mathrm{N/mm}^2$, and Young's modulus of $1.8 \times 104\,\mathrm{N/mm}^2$. The medullary cavity, a hollow space inside the diaphysis, is filled with fatty, yellow marrow or red marrow that contains blood-forming cells.

Bone is a living organ that is constantly being remodeled. Old bone is removed by special cells called osteoclasts, and new bone is deposited by osteoblasts. Bone remodeling occurs during bone growth and to regulate calcium availability. The average skeleton is totally remodeled about three times during a person's lifetime. Osteoporosis is a disorder in which old bone is broken down faster than new bone is produced so that the resulting bones are weak and brittle.

The bones of the skeletal system are attached to each other at fibrous, cartilaginous, or synovial joints (Fig. 3.32). The articulating bones of fibrous joints are bound tightly together by fibrous connective tissue. These joints can be rigid and relatively immovable to slightly movable. This type of joint includes the suture joints in the skull. Cartilage holds together the bones in cartilaginous joints. These joints allow limited motion in response to twisting or compression and include the joints of the vertebral system and the joints that attach the ribs to the vertebral column and to the sternum.

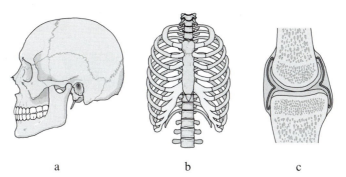

a   b   c

Figure 3.32 Bones of the skeletal system are attached to each other at fibrous (a), cartilaginous (b), or synovial (c) joints.

Synovial joints, such as the knee, are the most complex and varied and have fluid-filled joint cavities, cartilage that covers the articulating bones, and ligaments that help hold the joints together.

Synovial joints are classified into six types based on their structure and the type of motion they permit. Gliding joints (Fig. 3.33) are the simplest type of synovial joint, allow back-and-forth or side-to-side movement, and include the intercarpal articulations in the wrist. Hinge joints such as the elbow permit bending in only one plane and are the most common type of synovial joint. The atlas and axis provide an example of a pivot joint that permits rotation. In condyloid articulations, an oval, convex surface of one bone fits into a concave depression on another bone. Condyloid joints, which include the metacarpophalangeal joints (knuckles) of the fingers, permit flexion–extension and rotation and are considered to be biaxial because rotation is limited to two axes of movement. The saddle joint, represented by the joint at the base of the thumb, is a modified condyloid joint that permits movement in several directions (multiaxial). Ball-and-socket joints allow motion in many directions around a fixed center. In these joints, the ball-shaped head of one bone fits into a cuplike concavity of another bone. This multiaxial joint is the most freely movable of all and includes the shoulder and hip joints. Biomedical engineers have helped develop artificial joints that are routinely used as replacements in diseased or injured hips, shoulders, and knees (Fig. 3.34).

3.4.5 Muscular System

The muscular system (Fig. 3.35) is composed of 600–700 skeletal muscles, depending on whether certain muscles are counted as separate or as pairs, and makes up 40% of the body's mass. The axial musculature makes up about 60% of the skeletal muscles in the body and arises from the axial skeleton (Fig. 3.31). It positions the head and spinal column and moves the rib cage during breathing. The appendicular musculature moves or stabilizes components of the appendicular skeleton.

The skeletal muscles in the muscular system maintain posture, generate heat to maintain the body's temperature, and provide the driving force that is used to move the bones and joints of the body and the skin of the face. Muscles that play a major

GLIDING   PIVOT   SADDLE

BALL-AND-
SOCKET   CONDYLOID   HINGE

Figure 3.33   Synovial joints have fluid-filled cavities and are the most complex and varied types of joints. Each synovial joint is classified into one of six types depending on its structure and type of motion.

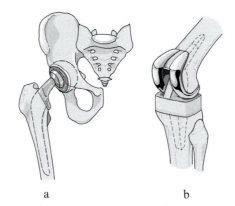

a   b

Figure 3.34   Diseased or damaged hip (a) and knee (b) joints that are nonfunctional or extremely painful can be replaced by prostheses. Artificial joints can be held in place by a special cement [poly-methylmethacrylate (PMMA)] and by bone ingrowth. Special problems occur at the interfaces due to the different elastic moduli of the materials (110 GPa for titanium, 2.2 GPa for PMMA, and 20 GPa for bone).

role in accomplishing a movement are called prime movers, or agonists. Muscles that act in opposition to a prime mover are called antagonists, whereas muscles that assist a prime mover in producing a movement are called synergists. The continual contraction of some skeletal muscles helps maintain the body's posture. If all of these muscles relax, which happens when a person faints, the person collapses.

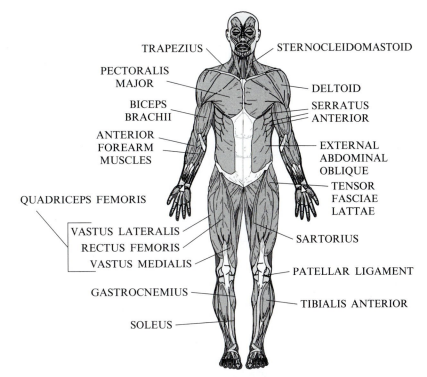

TRAPEZIUS

STERNOCLEIDOMASTOID

PECTORALIS
MAJOR

DELTOID

BICEPS
BRACHII

SERRATUS
ANTERIOR

ANTERIOR
FOREARM
MUSCLES

EXTERNAL
ABDOMINAL
OBLIQUE

QUADRICEPS FEMORIS

TENSOR
FASCIAE
LATTAE

VASTUS LATERALIS
RECTUS FEMORIS
VASTUS MEDIALIS

SARTORIUS

PATELLAR LIGAMENT

GASTROCNEMIUS

TIBIALIS ANTERIOR

SOLEUS

Figure 3.35 Some of the major skeletal muscles on the anterior side of the body are shown.

A system of levers, which consist of rigid lever arms that pivot around fixed points, is used to move skeletal muscle (Fig. 3.36). Two forces act on every lever: the weight to be moved (i.e., the resistance to be overcome) and the pull or effort applied (i.e., the applied force). Bones act as lever arms and joints provide a fulcrum. The resistance to be overcome is the weight of the body part that is moved and the applied force is generated by the contraction of a muscle or muscles at the insertion, the point of attachment of a muscle to the bone it moves. An example of a first-class lever, one in which the fulcrum is between the force and the weight, is the movement of the facial portion of the head when the face is tilted upwards. The fulcrum is formed by the joint between the atlas and the occipital bone of the skull and the vertebral muscles inserted at the back of the head generate the applied force that moves the weight, the facial portion of the head. A second-class lever is one in which the weight is between the force and the fulcrum. This can be found in the body when a person stands on "tip toe." The ball of the foot is the fulcrum and the applied force is generated by the calf muscles on the back of the leg. The weight that is moved is that of the whole body. A third-class lever is one in which the force is between the weight and the fulcrum. When a person has a bent elbow and holds a ball in front of the body, the applied force is generated by the contraction of the biceps brachii muscle. The weight to be moved

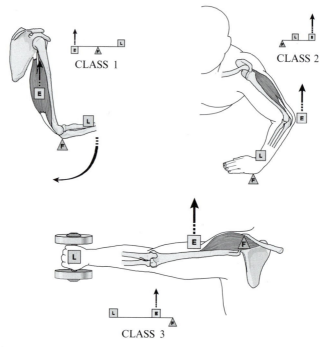

Figure 3.36   Depending on the muscle in use, the location of the load, and the location of the fulcrum, the humerus can act as a class 1 lever, a class 2 lever, or a class 3 lever.

includes the ball and the weight of the forearm and hand, and the elbow acts as the fulcrum.

The three types of muscle tissue—cardiac, skeletal, and smooth—share four important characteristics: (1) contractility, the ability to shorten; (2) excitability, the capacity to receive and respond to a stimulus; (3) extensibility, the ability to be stretched; (4) and elasticity, the ability to return to the original shape after being stretched or contracted. Cardiac muscle tissue is found only in the heart, whereas smooth muscle tissue is found within almost every other organ where it forms sheets, bundles, or sheaths around other tissues. Skeletal muscles are composed of skeletal muscle tissue, connective tissue, blood vessels, and nervous tissue.

Each skeletal muscle is surrounded by a layer of connective tissue (collagen fibers) that separates the muscle from surrounding tissues and organs. These fibers come together at the end of the muscle to form tendons which connect the skeletal muscle to bone, to skin (face), or to the tendons of other muscles (hand). Other connective tissue fibers divide the skeletal muscles into compartments called fascicles that contain bundles of muscle fibers. Within each fascicle, additional connective tissue surrounds each skeletal muscle fiber and ties adjacent ones together. Each skeletal muscle fiber has hundreds of nuclei just beneath the cell membrane. Multiple nuclei provide multiple copies of the genes that direct the production of enzymes and structural proteins needed for normal contraction so that contraction can occur faster.

In muscle fibers, the plasma membrane is called the sarcolemma and the cytoplasm is called the sarcoplasm (Fig. 3.37). Transverse tubules (T tubules) begin at the sarcolemma and extend into the sarcoplasm at right angles to the surface of the sarcolemma. The T tubules, which play a role in coordinating contraction, are filled with extracellular fluid and form passageways through the muscle fiber. They make close contact with expanded chambers, cisternae, of the sarcoplasmic reticulum, a specialized form of the ER. The cisternae contain high concentrations of calcium ions which are needed for contraction to occur.

The sarcoplasm contains cylinders 1 or 2 μm in diameter that are as long as the entire muscle fiber and are called myofibrils. The myofibrils are attached to the sarcolemma at each end of the cell and are responsible for muscle fiber contraction. Myofilaments—protein filaments consisting of thin filaments (primarily actin) and thick filaments (mostly myosin)—are bundled together to make up myofibrils. Repeating functional units of myofilaments are called sarcomeres (Fig. 3.38). The sarcomere is the smallest functional unit of the muscle fiber and has a resting length of about 2.6 μm. The thin filaments are attached to dark bands, called Z lines, which form the ends of each sarcomere. Thick filaments containing double-headed myosin molecules lie between the thin ones. It is this overlap of thin and thick filaments that gives skeletal muscle its banded, striated appearance. The I band is the area in a relaxed muscle fiber that just contains actin filaments, whereas the H zone is the area that just contains myosin filaments. The H zone and the area in which the actin and myosin overlap form the A band.

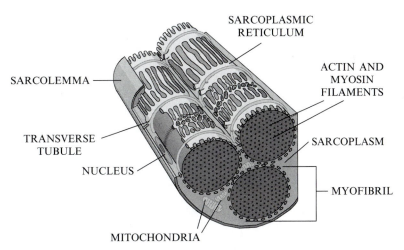

Figure 3.37   Skeletal muscles are composed of muscle fascicles that are composed of muscle fibers such as the one shown here. Muscle fibers have hundreds of nuclei just below the plasma membrane, the sarcolemma. Transverse tubules extend into the sarcoplasm, the cytoplasm of the muscle fiber, and are important in the contraction process because they deliver action potentials that result in the release of stored calcium ions. Calcium ions are needed to create active sites on actin filaments so that cross-bridges can be formed between actin and myosin and the muscle can contract.

When a muscle contracts, myosin molecules in the thick filaments form cross-bridges at active sites in the actin of the thin filaments and pull the thin filaments toward the center of the sarcomere. The cross-bridges are then released and reformed at a different active site further along the thin filament. This results in a motion that is similar to the hand-over-hand motion that is used to pull in a rope. This action, the sliding filament mechanism, is driven by ATP energy and results in shortening of the muscle. Shortening of the muscle components (contraction) results in bringing the muscle's attachments (e.g., bones) closer together (Fig. 3.38).

Muscle fibers have connections with nerves. Sensory nerve endings are sensitive to length, tension, and pain in the muscle and send impulses to the brain via the spinal cord, whereas motor nerve endings receive impulses from the brain and spinal cord that lead to excitation and contraction of the muscle. Each motor axon branches and supplies several muscle fibers. Each of these axon branches loses its myelin sheath and splits up into a number of terminals that make contact with the surface of the muscle. When the nerve is stimulated, vesicles in the axon terminals release a neurotransmitter, acetylcholine, into the synapse between the neuron and the muscle. Acetylcholine diffuses across the synapse and binds to receptors in a special area, the motor end plate, of the sarcolemma. This causes the sodium channels in the sarcolemma to open up, and an action potential is produced in the muscle fiber. The resulting action potential spreads over the entire sarcolemmal surface and travels down all of the T tubules where it triggers a sudden massive release of calcium by the cisternae.

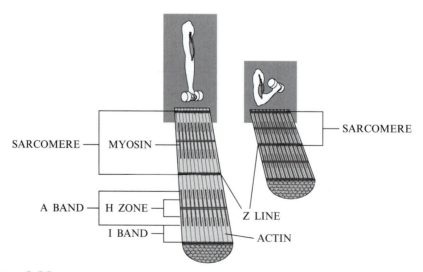

Figure 3.38   The sarcomere is the basic functional unit of skeletal muscles and extends from one Z line to the next. Actin filaments are attached to the Z lines and extend into the A band where they overlap with the thicker myosin filaments. The H zone is the portion of the A band that contains no overlapping actin filaments. When the muscle changes from its extended, relaxed position (left panel) to its contracted state (right panel), the myosin filaments use cross-bridges to slide past the actin filaments and bring the Z lines closer together. This results in shorter sarcomeres and a contracted muscle.

Calcium triggers the production of active sites on the thin filaments so that cross-bridges with myosin can form and contraction occurs. Acetylcholinesterase breaks down the acetylcholine while the contraction process is underway so that the original relatively low permeability of the sarcolemma to sodium is restored.

A motor unit is a complex consisting of one motor neuron and the muscle fibers it innervates. All the muscle fibers in a single motor unit contract at the same time, whereas muscle fibers in the same muscle but belonging to different motor units may contract at different times. When a contracted muscle relaxes, it returns to its original (resting) length if another contracting muscle moves it or if it is acted upon by gravity. During relaxation, ATP is expended to move calcium back to the cisternae. The active sites that were needed for cross-bridge formation become covered so that actin and myosin can no longer interact. When the cross-bridges disappear, the muscle returns to its resting length (i.e., it relaxes).

The human body contains two types of skeletal muscle fibers: fast and slow. Fast fibers can contract in 10 ms or less following stimulation and make up most of the skeletal muscle fibers in the body. They are large in diameter and contain densely packed myofibrils, large glycogen reserves (used to produce ATP), and relatively few mitochondria. These fibers produce powerful contractions that use up massive amounts of ATP and fatigue (can no longer contract in spite of continued neural stimulation) rapidly. Slow fibers take about three times as long to contract as fast fibers. They can continue to contract for extended periods of time because they contain (1) a more extensive network of capillaries so that they can receive more oxygen, (2) a special oxygen-binding molecule called myoglobin, and (3) more mitochondria which can produce more ATP than fast fibers. Muscles contain different amounts of slow and fast fibers. Those that are dominated by fast fibers (e.g., chicken breast muscles) appear white and those that are dominated by slow fibers (e.g., chicken legs) appear red. Most human muscles appear pink because they contain a mixture of both. Genes determine the percentage of fast and slow fibers in each muscle, but the ability of fast muscle fibers to resist fatigue can be increased through athletic training.

3.5 HOMEOSTASIS

Organ systems work together to maintain a constant internal environment within the body. Homeostasis is the process by which physical and chemical conditions within the internal environment of the body are maintained within tolerable ranges even when the external environment changes. Body temperature, blood pressure, and breathing and heart rates are some of the functions that are controlled by homeostatic mechanisms that involve several organ systems working together.

Extracellular fluid—the fluid that surrounds and bathes the body's cells—plays an important role in maintaining homeostasis. It circulates throughout the body and carries materials to and from the cells. It also provides a mechanism for maintaining optimal temperature and pressure levels, the proper balance between acids and bases, and concentrations of oxygen, carbon dioxide, water, nutrients, and many of the chemicals that are found in the blood.

Three components—sensory receptors, integrators, and effectors—interact to maintain homeostasis (Fig. 3.39). Sensory receptors, which may be cells or cell parts, detect stimuli (i.e., changes to their environment) and send information about the stimuli to integrators. Integrators are control points that pull together information from one or more sensory receptors. Integrators then elicit a response from effectors. The brain is an integrator that can send messages to muscles or glands or both. The messages result in some type of response from the effectors. The brain receives information about how parts of the body are operating and can compare this to information about how parts of the body should be operating.

Positive feedback mechanisms are ones in which the initial stimulus is reinforced by the response. There are very few examples of this in the human body since it disrupts homeostasis. Childbirth provides one example. Pressure from the baby's head in the birth canal stimulates receptors in the cervix which send signals to the hypothalamus. The hypothalamus responds to the stimulus by releasing oxytocin which enhances uterine contractions. Uterine contractions increase in intensity and force the baby further into the birth canal which causes additional stretching of the receptors in the cervix. The process continues until the baby is born, the pressure on the cervical stretch receptors ends, and the hypothalamus is no longer stimulated to release oxytocin.

Negative feedback mechanisms result in a response that is opposite in direction to the initiating stimulus. For example, receptors in the skin and elsewhere in the body detect the body's temperature. Temperature information is forwarded to the hypothalamus in the brain which compares the body's current temperature to what the temperature should be (approximately 37°C). If the body's temperature is too low, messages are sent to contract the smooth muscles in blood vessels near the skin (reducing the diameter of the blood vessels and the heat transferred through the skin), to skeletal muscles to start contracting rapidly (shivering), and to the arrector pili muscles (Fig. 3.16a) to erect the hairs and form "goose bumps." The metabolic activity of the muscle contractions generates heat and warms the body. If the body's temperature is too high, messages are sent to relax the smooth muscles in

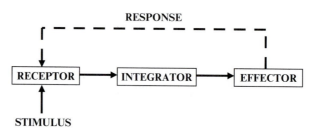

Figure 3.39   Feedback mechanisms are used to help maintain homeostatis. A stimulus is received by a receptor which sends a signal (messenger) to an effector or to an integrator which sends a signal to an effector. The effector responds to the signal. The response feeds back to the receptor and modifies the effect of the stimulus. In negative feedback, the response subtracts from the effect of the stimulus on the receptor. In positive feedback, the response adds to the effect of the stimulus on the receptor.

the blood vessels near the skin (increasing the diameter of the blood vessels and the amount of heat transferred through the skin) and to sweat glands to release moisture and thus increase evaporative cooling of the skin. When the temperature of circulating blood changes enough in the appropriate direction that it reaches the set point of the system, the hypothalamus stops sending signals to the effector muscles and glands.

Another example of a negative feedback mechanism in the body involves the regulation of glucose in the blood stream by clusters of cells, the pancreatic islets (Fig. 3.40). There are between 2×10^5 and 2×10^6 pancreatic islets scattered throughout the adult pancreas. When glucose levels are high, beta cells in the islets produce insulin which facilitates glucose transport across plasma membranes and into cells and enhances the conversion of glucose into glycogen which is stored in the liver. During periods of fasting or whenever the concentration of blood glucose drops below normal (70–110 mg/dl), alpha cells produce glucagon which stimulates the liver to convert glycogen into glucose and the formation of glucose from noncarbohydrate sources such as amino acids and lactic acid. When glucose levels return to normal, the effector cells in the pancreatic islets stop producing their respective hormone (i.e., insulin or glucagon). Some biomedical engineers are working on controlled drug delivery systems that can sense blood glucose levels and emulate the responses of the pancreatic islet cells, whereas other biomedical engineers are trying to develop an artificial pancreas that would effectively maintain appropriate blood glucose levels.

EXERCISES

1. Using as many appropriate anatomical terms as apply, write sentences which describe the positional relationship between your mouth and (1) your left ear, (2) your nose, and (3) the big toe on your right foot.

2. Using as many appropriate anatomical terms as apply, describe the position of the stomach in the body and its position relative to the heart.

3. Search the Internet to find a transverse section of the body that was imaged using computerized tomography (CT) or magnetic resonance imaging (MRI). Print the image and indicate its web address.

4. Search the Internet to find a frontal section of the body that was imaged using CT or MRI. Print the image and indicate its web address.

5. Name and give examples of the four classes of biologically important organic compounds. What are the major functions of each of these groups?

6. What are the molarity and osmolarity of a 1-liter solution that contains half a mole of calcium chloride? How many molecules of chloride would the solution contain?

7. Consider a simple model cell, such as the one in Figure 3.6, that consists of cytoplasm and a plasma membrane. The cell's initial volume is 2 nl and contains $0.2 M$ protein. The cell is placed in a large volume of $0.05 M$ $CaCl_2$. Neither Ca^{++} nor Cl^- can cross the plasma membrane and enter the cell. Is the $0.05 M$ $CaCl_2$ solution hypotonic, isotonic, or hypertonic relative to the

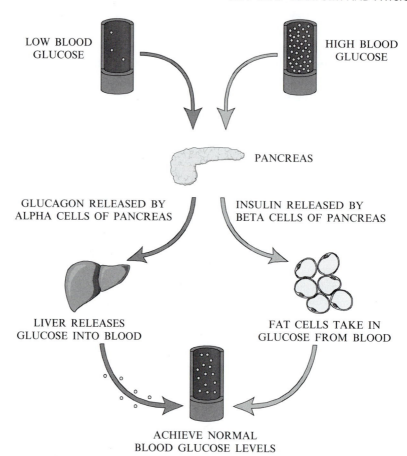

Figure 3.40   Two negative feedback mechanisms help control the level of glucose in the blood.
When blood glucose levels are higher than the body's set point (stimulus), beta cells islets (receptors) in
the pancreatic islets produce insulin (messenger) which facilitates glucose transport across plasma
membranes and enhances the conversion of glucose into glycogen for storage in the liver (effector).
This causes the level of glucose in the blood to drop. When the level equals the body's set point, the beta
cells stop producing insulin. When blood glucose levels are lower than the body's set point (stimulus),
alpha cells (receptors) in the pancreatic islets produce glucagon (messenger) which stimulates the liver
(effector) to convert glycogen into glucose. This causes the level of glucose in the blood to increase.
When the level equals the body's set point, the alpha cells stop producing glucagon.

osmolarity inside the cell? Describe what happens to the cell as it achieves
equilibrium in this new environment. What will be the final osmolarity of
the cell? What will be its final volume?

8. What does the principle of electrical neutrality mean in terms of the
concentration of ions within a cell?

9. Consider the same model cell that was used in Exercise 7, but instead of
being placed in $0.05\,M$ $CaCl_2$, the cell is placed in $0.2\,M$ urea. Unlike Ca^{++}
and Cl^-, urea can cross the plasma membrane and enter the cell. Describe

what happens to the cell as it achieves equilibrium in this environment. What will be the final osmolarity of the cell? What will be its final volume?

10. Briefly describe the path that a protein (e.g., a hormone) which is manufactured on the rough ER would take in order to leave the cell.

11. What major role do mitochondria have in the cell? Why might it be important to have this process contained within an organelle?

12. List and briefly describe three organelles that provide structural support and assist with cell movement.

13. Find a location on the Internet that describes the Human Genome Project. Print its home page and indicate its web address. Find and print an ideogram of a chromosome that shows a gene that causes cystic fibrosis.

14. Briefly describe the major differences between replication and transcription.

15. Describe how the hereditary information contained in genes within the cell's DNA is expressed as proteins which direct the cell's activities.

16. Six different codons code for leucine while only one codes for methionine. Why might this be important for regulating translation and producing proteins?

17. Insulin (Fig. 3.41) was the first protein to be sequenced biochemically. Assuming that there were no introns involved in the process, what are the possible DNA sequences that produced the last four amino acids in the molecule?

18. Copy the title page and abstract of five peer-reviewed journal articles that discuss engineering applications for five different organ systems in the body (one article per organ system). Review articles, conference proceeding papers, copies of keynote addresses and other speeches, book chapters, articles from the popular press and newspapers, and editorials are not acceptable. Good places to look are the *Annals of Biomedical Engineering*, the *IEEE Transactions on Biomedical Engineering*, the *IEEE Engineering in Medicine and Biology Magazine*, and *Medical and Biological Engineering and Computing*. What information in the article indicates that it was peer-reviewed?

19. Trace the path of a single red blood cell from a capillary bed in your right hand to the capillary beds of your right lung and back. What gases are exchanged? Where are they exchanged during this process?

20. Draw and label a block diagram of pulmonary and systemic blood flow that includes the chambers of the heart, valves, major veins and arteries that enter and leave the heart, the lungs, and the capillary bed of the body. Use arrows to indicate the direction of flow through each component.

21. Find on the Internet an example of an ECG representing normal sinus rhythm and use it to demonstrate how heart rate is determined.

22. Why are R waves (Fig. 3.22) used to determine heart rate rather than T waves?

23. How can the stroke volume be determined if a thermal dilution technique is used to determine cardiac output?

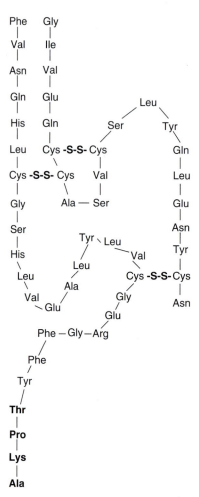

Figure 3.41   Bovine insulin consists of two polypeptide chains that are joined by two disulfide bonds (-S-S-). Hydrogen bonds also exist between the chains and between segments of the same chain. The three-letter names stand for different amino acids (see Table 3.1).

24. What would be the pulse pressure and mean arterial pressure for a hypertensive person with a systolic pressure of 145 mm Hg and a diastolic pressure of 98 mm Hg?

25. The total lung capacity of a patient is 5.5 liters. Find the patient's inspiratory reserve volume if the patient's vital capacity was 4.2 liters, the tidal volume was 500 ml, and the expiratory reserve volume was 1.2 liters.

26. What would you need to know or measure to determine the residual volume of the patient described in Example Problem 3.10?

27. Briefly describe the functions and major components of the central, peripheral, somatic, automatic, sympathetic, and parasympathetic nervous systems. Which ones are subsets of others?

28. Explain how sarcomeres shorten and how that results in muscle contraction.

29. How do the muscular and skeletal systems interact to produce movement?

30. Draw a block diagram to show the negative feedback mechanisms that help regulate glucose levels in the blood. Label the inputs, sensors, integrators, effectors, and outputs.

SUGGESTED READING

Brown, B.H., Smallwood, R.H., Barber, D.C., Lawford, P.V. and Hose, D.R. (1999). *Medical Physics and Biomedical Engineering*. Institute of Physics Publishing, Bristol and Philadelphia.

Cooper, G.M. (2000). *The Cell—A Molecular Approach*, 2nd Ed. ASM Press, Washington, D.C.

Deutsch, S. and Deutsch, A. (1993). *Understanding the Nervous System: An Engineering Perspective*. IEEE Press, New York.

Fox, S.I. (2004). *Human Physiology*, 8th Ed. McGraw-Hill, Boston.

Germann, W.J. and Stanfield, C.L. (2005). *Principles of Human Physiology*, 2nd Ed. Pearson Benjamin Cummings, San Francisco.

Guyton, A.C. (1991). *Basic Neuroscience: Anatomy & Physiology*. W. B. Saunders, Philadelphia.

Guyton, A.C. and Hall, J.E. (2000). *Textbook of Medical Physiology*, 10th Ed. W. B. Saunders, Philadelphia.

Harold, F.M. (2001). *The Way of the Cell: Molecules, Organisms and the Order of Life*. Oxford Univ. Press, New York.

Karp, G. (2002). *Cell and Molecular Biology—Concepts and Experiments*, 3rd Ed. Wiley, New York.

Katz, A.M. (1986). *Physiology of the Heart*. Raven Press, New York.

Keynes, R.D. and Aidley, D.J. (1991). *Nerve & Muscle*, 2nd Ed. Cambridge Univ. Press, Cambridge.

Leff, A.R. and Schumacker, P.T. (1993). *Respiratory Physiology: Basics and Applications*. Saunders, Philadelphia.

Lodish, H., Berk, A., Zipursky, S.L., Matsudaira, P., Baltimore, D. and Darnell, J. (2000). *Molecular Cell Biology*, 4th Ed. W. H. Freeman, New York.

Martini, F.H. (2001). *Fundamentals of Anatomy & Physiology*, 5th Ed. Prentice Hall, Upper Saddle River, NJ.

Matthews, G.G. (1991). *Cellular Physiology of Nerve and Muscle*. Blackwell Scientific, Boston.

Pollack, G.H. (2001). *Cells, Gels and the Engines of Life—A New Unifying Approach to Cell Function*. Ebner & Sons, Seattle, WA.

Rhoades R. and Pflanzer, R. (2003). *Human Physiology*, 4th Ed. Thomson Learning, Pacific Grove, CA.

Silverthorn, D.U. (2004). *Human Physiology—An Integrated Approach*, 3rd Ed. Pearson Benjamin Cummings, San Francisco.

Tözeren A. and Byers, S.W. (2004). *New Biology for Engineers and Computer Scientists*. Pearson Education, Upper Saddle River, NJ.

Van De Graaff, K.M., Fox, S.I. and LaFleur, K.M. (1997). *Synopsis of Human Anatomy & Physiology*. Wm. C. Brown, Dubuque, IA.

West, J.B. (1990). *Respiratory Physiology—The Essentials*, 4th Ed. Williams & Wilkins, Baltimore.

Widmaier, E.P., Raff, H. and Strang, K.T. (2004). *Vander, Sherman, & Luciano's Human Physiology—The Mechanisms of Body Function*. McGraw-Hill, Boston.

4 BIOMECHANICS

Joseph L. Palladino, PhD
Roy B. Davis, PhD

Chapter Contents

At the conclusion of this chapter, students will be able to:

- Understand the application of engineering kinematic relations to biomechanical problems.
- Understand the application of engineering kinetic relations to biomechanical problems.
- Understand the application of engineering mechanics of materials to biological structures.
- Use MATLAB to write and solve biomechanical static and dynamic equations.
- Use Simulink to study viscoelastic properties of biological tissues.
- Understand how kinematic equations of motion are used in clinical analysis of human gait.
- Understand how kinetic equations of motion are used in clinical analysis of human gait.
- Explain how biomechanics applied to human gait is used to quantify pathological conditions, to suggest surgical and clinical treatments, and to quantify their effectiveness.
- Understand basic rheology of biological fluids.
- Understand the development of models that describe blood vessel mechanics.
- Understand basic heart mechanics.
- Explain how biomechanics applied to the cardiovascular system is used to quantify the effectiveness of the heart as a pump, to study heart–vessel interaction, and to develop clinical applications.

4.1   INTRODUCTION

Biomechanics combines engineering and the life sciences by applying principles from classical mechanics to the study of living systems. This relatively new field covers a broad range of topics, including strength of biological materials, biofluid mechanics in the cardiovascular and respiratory systems, material properties and interactions of medical implants and the body, heat and mass transfer into biological tissues (e.g., tumors), biocontrol systems regulating metabolism or voluntary motion, and kinematics and kinetics applied to study human gait. The great breadth of the field of biomechanics arises from the complexities and variety of biological organisms and systems.

The goals of this chapter are twofold—to apply basic engineering principles to biological structures, and then to develop clinical applications. Section 4.2 provides a review of concepts from introductory statics and dynamics. Section 4.3 presents concepts from mechanics of material that are fundamental for engineers and accessible to those with only a statics/dynamics background. Section 4.4 introduces viscoelastic complexities characteristic of biological materials, with the concepts further applied in Section 4.5: Cartilage, Ligament, Tendon, and Muscle. The last two sections bring all of this information together in two "real world" biomechanics applications: human gait analysis and cardiovascular dynamics.

The human body is a complex machine with the skeletal system and ligaments forming the framework and the muscles and tendons serving as the motors and cables. Human gait biomechanics may be viewed as a structure (skeleton) comprised of levers (bones) with pivots (joints) that move as the result of net forces produced by pairs of agonist and antagonist muscles. Consequently, the strength of the structure and the action of muscles will be of fundamental importance. Using a similar functional model, the cardiovascular system may be viewed as a complex pump (heart) pumping a complex fluid (blood) into a complex set of pipes (blood vessels). An extensive suggested reading list for both gait and cardiovascular dynamics permits the reader to go beyond the very introductory nature of this textbook.

The discipline of mechanics has a long history. For lack of more ancient records, the history of mechanics starts with the ancient Greeks and Aristotle (384–322 BC). Hellenic mechanics devised a correct concept of statics, but those of dynamics, fundamental in living systems, did not begin until the end of the Middle Ages and the beginning of the modern era. Starting in the sixteenth century, the field of dynamics advanced rapidly with work by Kepler, Galileo, Descartes, Huygens, and Newton. Dynamic laws were subsequently codified by Euler, LaGrange, and LaPlace (see *A History of Mechanics* by Dugas).

In Galileo's *Two New Sciences* (1638) the subtitle *Attenenti all Mecanica & i Movimenti Locali (Pertaining to Mechanics and Local Motions)* refers to force, motion, and strength of materials. Since then, "mechanics" has been extended to describe the forces and motions of any system, ranging from quanta, atoms, molecules, gases, liquids, solids, structures, stars, and galaxies. The biological world is consequently a natural object for the study of mechanics.

The relatively new field of biomechanics applies mechanical principles to the study of living systems. The eminent professor of biomechanics Dr. Y.C. Fung describes the role of biomechanics in biology, physiology, and medicine:

> Physiology can no more be understood without biomechanics than an airplane can without aerodynamics. For an airplane, mechanics enables us to design its structure and predict its performance. For an organ, biomechanics helps us to understand its normal function, predict changes due to alteration, and propose methods of artificial intervention. Thus diagnosis, surgery and prosthesis are closely associated with biomechanics.[1]

Clearly, biomechanics is essential to assessing and improving human health.

The following is a brief list of biomechanical milestones, especially those related to the topics in this chapter:

Galen of Pergamon (129–199) Published extensively in medicine, including *De Motu Muscularum (On the Movements of Muscles)*. He realized that motion requires muscle contraction.

Leonardo da Vinci (1452–1519) Made the first accurate descriptions of ball-and-socket joints, such as the shoulder and hip, calling the latter the "polo dell'omo"

---

[1]*Biomechanics: Mechanical Properties of Living Tissues*, Y.C. Fung, 1993.

(pole of man). His drawings depicted mechanical force acting along the line of muscle filaments.

Andreas Vesalius (1514–1564) Published *De Humani Corporis Fabrica* (*The Fabric of the Human Body*). Based on human cadaver dissections, his work led to a more accurate anatomical description of human musculature than Galen's and demonstrated that motion results from the contraction of muscles that shorten and thicken.

Galileo Galilei (1564–1642) Studied medicine and physics, integrated measurement and observation in science, and concluded that mathematics is an essential tool of science. His analyses included the biomechanics of jumping and the gait analysis of horses and insects, as well as dimensional analysis of animal bones.

Santorio Santorio (1561–1636) Used Galileo's method of measurement and analysis and found that the human body changes weight with time. This observation led to the study of metabolism and, thereby, ushered in the scientific study of medicine.

William Harvey (1578–1657) Developed an experimental basis for the modern circulation concept of a closed path between arteries and veins. The structural basis, the capillary, was discovered by Malpighi in 1661.

Giovanni Borelli (1608–1679) A mathematician who studied body dynamics, muscle contraction, animal movement, and motion of the heart and intestines. He published *De Motu Animalium* (*On the Motion of Animals*) in 1680.

Jan Swammerdam (1637–1680) Introduced the nerve–muscle preparation, stimulating muscle contraction by pinching the attached nerve in the frog leg. He also showed that muscles contract with little change in volume, refuting the previous belief that muscles contract when "animal spirits" fill them, causing bulging.

Robert Hooke (1635–1703) Devised Hooke's Law, relating the stress and elongation of elastic materials, and used the term *cell* in biology.

Isaac Newton (1642–1727) Not known for biomechanics work, but he developed calculus, the classical laws of motion, and the constitutive equation for viscous fluid, all of which are fundamental to biomechanics.

Nicholas André (1658–1742) Coined the term *orthopaedics* at the age of eighty and believed that muscular imbalances cause skeletal deformities.

Stephen Hales (1677–1761) Was likely the first to measure blood pressure, as described in his book *Statistical Essays: Containing Haemostaticks, or an Account of some Hydraulick and Hydrostatical Experiments made on the Blood and Blood-Vessels of Animals; etc.*, in 1733.

Leonard Euler (1707–1783) Generalized Newton's laws of motion to continuum representations that are used extensively to describe rigid body motion, and studied pulse waves in arteries.

Thomas Young (1773–1829) Studied vibrations and voice, wave theory of light and vision, and devised Young's modulus of elasticity.

Ernst Weber (1795–1878) and Eduard Weber (1806–1871) Published *Die Mechanik der meschlichen Gerwerkzeuge (On the Mechanics of the Human Gait Tools)* in 1836, pioneering the scientific study of human gait.

Hermann von Helmholtz (1821–1894) Studied an immense array of topics, including optics, acoustics, thermodynamics, electrodynamics, physiology, and medicine, including ophthalmoscopy, fluid mechanics, nerve conduction speed, and the heat of muscle contraction.

Etienne Marey (1830–1904) Analyzed the motion of horses, birds, insects, fish, and humans. His inventions included force plates to measure ground reaction forces and the *Chronophotographe a pellicule*, or motion picture camera.

Wilhelm Braune and Otto Fischer (research conducted from 1895–1904) Published *Der Gang des Menschen (The Human Gait)*, containing the mathematical analysis of human gait and introducing methods still in use. They invented "cyclography" (now called interrupted-light photography with active markers), pioneered the use of multiple cameras to reconstruct 3D motion data, and applied Newtonian mechanics to estimate joint forces and limb accelerations.

4.2 BASIC MECHANICS

This section reviews some of the main points from any standard introductory mechanics (statics and dynamics) course. Good references abound, such as *Engineering Mechanics* by Merriam and Kraige (2002). A review of vector mathematics is followed by matrix coordinate transformations, a topic new to some students. Euler's equations of motion (section 4.2.5) may also be new material. For both topics, *Principles of Dynamics* by Greenwood provides a comprehensive reference.

4.2.1 Vector Mathematics

Forces may be written in terms of scalar components and unit vectors (of magnitude equal to one), or in polar form with magnitude and direction. Figure 4.1 shows that

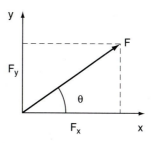

Figure 4.1 2-dimensional representation of vector **F**.

the 2-dimensional vector **F** is comprised of the **i** component, F_x, in the x direction, and the **j** component, F_y, in the y direction, or

$$\mathbf{F} = F_x\mathbf{i} + F_y\mathbf{j} \tag{4.1}$$

as in 20**i**+40**j** lb. In this chapter vectors are set in bold type. This same vector may be written in polar form in terms of the vector's magnitude $|F|$, also called the *norm*, and the vector's angle of orientation, θ:

$$|F| = \sqrt{F_x^2 + F_y^2} \tag{4.2}$$

$$\theta = \arctan\frac{F_y}{F_x} \tag{4.3}$$

yielding $|F| = 44.7$ lb and $\theta = 63.4°$. Vectors are similarly represented in three dimensions in terms of their **i**, **j** and **k** components:

$$\mathbf{F} = F_x\mathbf{i} + F_y\mathbf{j} + F_z\mathbf{k} \tag{4.4}$$

with **k** in the z direction.

Often, a vector's magnitude and two points along its line of action are known. Consider the 3-dimensional vector in Figure 4.2. **F** has magnitude of 10 lb, and its line of action passes from the origin (0,0,0) to the point (2,6,4). **F** is written as the product of the magnitude $|F|$ and a unit vector $\mathbf{e_F}$ that points along its line of action:

$$\mathbf{F} = |F|\mathbf{e_F}$$

$$= 10\,\text{lb}\left(\frac{2\mathbf{i} + 6\mathbf{j} + 4\mathbf{k}}{\sqrt{2^2 + 6^2 + 4^2}}\right)$$

$$\mathbf{F} = 2.67\mathbf{i} + 8.02\mathbf{j} + 5.34\mathbf{k}\ \text{lb}$$

The quantity in parentheses is the unit vector of **F**, or

$$\mathbf{e_F} = \left(\frac{2\mathbf{i} + 6\mathbf{j} + 4\mathbf{k}}{\sqrt{2^2 + 6^2 + 4^2}}\right) = 0.267\mathbf{i} + 0.802\mathbf{j} + 0.534\mathbf{k}$$

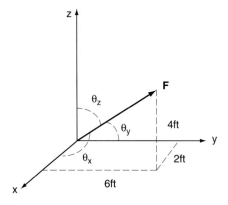

Figure 4.2   3-D vector defined by its magnitude and line of action.

and the magnitude of **F** is

$$|F| = \sqrt{2.67^2 + 8.02^2 + 5.34^2}$$
$$= 10\,\text{lb}$$

The vector **F** in Figure 4.2 may also be defined in 3-D space in terms of the angles between its line of action and each coordinate axis. Consider the angles θ_x, θ_y, and θ_z that are measured from the positive x, y, and z axes, respectively, to **F**. Then

$$\cos\theta_x = \frac{F_x}{|F|} \tag{4.5}$$

$$\cos\theta_y = \frac{F_y}{|F|} \tag{4.6}$$

$$\cos\theta_z = \frac{F_z}{|F|} \tag{4.7}$$

These ratios are termed the *direction cosines* of **F**. The unit vector $\mathbf{e_F}$ is equivalent to

$$\mathbf{e_F} = \cos\theta_x \mathbf{i} + \cos\theta_y \mathbf{j} + \cos\theta_z \mathbf{k} \tag{4.8}$$

or, in general

$$\mathbf{e_F} = \left(\frac{F_x\mathbf{i} + F_y\mathbf{j} + F_z\mathbf{k}}{\sqrt{F_x^2 + F_y^2 + F_z^2}} \right) \tag{4.9}$$

The angles θ_x, θ_y, and θ_z for this example are consequently

$$\theta_x = \arccos\left(\frac{2.67}{10}\right) = 74.5°$$

$$\theta_y = \arccos\left(\frac{8.02}{10}\right) = 36.7°$$

$$\theta_z = \arccos\left(\frac{5.34}{10}\right) = 57.7°$$

Vectors are added by summing their components:

$$\mathbf{A} = A_x\mathbf{i} + A_y\mathbf{j} + A_z\mathbf{k}$$
$$\mathbf{B} = B_x\mathbf{i} + B_y\mathbf{j} + B_z\mathbf{k}$$
$$\mathbf{C} = \mathbf{A} + \mathbf{B} = (A_x + B_x)\mathbf{i} + (A_y + B_y)\mathbf{j} + (A_z + B_z)\mathbf{k}$$

In general, a set of forces may be combined into an equivalent force denoted the resultant **R**, where

$$\mathbf{R} = \sum F_x\mathbf{i} + \sum F_y\mathbf{j} + \sum F_z\mathbf{k} \tag{4.10}$$

as will be illustrated in subsequent sections. Vectors are subtracted similarly by subtracting vector components.

Vector multiplication consists of two distinct operations, the dot and cross products. The dot, or scalar, product of vectors **A** and **B** produces a scalar via

$$\mathbf{A} \cdot \mathbf{B} = AB \cos \theta \tag{4.11}$$

where θ is the angle between the vectors. For an orthogonal coordinate system, where all axes are 90° apart

$$\begin{aligned}
\mathbf{i} \cdot \mathbf{i} &= \mathbf{j} \cdot \mathbf{j} = \mathbf{k} \cdot \mathbf{k} = 1 \\
\mathbf{i} \cdot \mathbf{j} &= \mathbf{j} \cdot \mathbf{k} = \mathbf{k} \cdot \mathbf{i} = \cdots = 0
\end{aligned} \tag{4.12}$$

For example:

$$\mathbf{A} = 3\mathbf{i} + 2\mathbf{j} + \mathbf{k} \text{ ft}$$
$$\mathbf{B} = -2\mathbf{i} + 3\mathbf{j} + 10\mathbf{k} \text{ lb}$$
$$\mathbf{A} \cdot \mathbf{B} = 3(-2) + 2(3) + 1(10) = 10 \text{ ft lb}$$

Note that the dot product is commutative (i.e., $\mathbf{A} \cdot \mathbf{B} \equiv \mathbf{B} \cdot \mathbf{A}$).

The physical interpretation of the dot product $\mathbf{A} \cdot \mathbf{B}$ is the projection of **A** onto **B**, or, equivalently, the projection of **B** onto **A**. For example, *work* is defined as the force that acts in the same direction as the motion of a body. Figure 4.3 (left) shows a force vector **F** dotted with a direction of motion vector **d**. The work W done by **F** is given by $\mathbf{F} \cdot \mathbf{d} \equiv Fd \cos \theta$. Dotting **F** with **d** yields the component of **F** acting in the same direction as **d**.

The *moment* of a force about a point or axis is a measure of its tendency to cause rotation. The cross, or vector, product of two vectors yields a new vector that points along the axis of rotation. For example, Figure 4.3 (right) shows a vector **F** acting in the *x*–*y* plane at a distance from the body's coordinate center O. The vector **r** points from O to the line of action of **F**. The cross product $\mathbf{r} \times \mathbf{F}$ is a vector that points in the *z* direction along the body's axis of rotation. If **F** and **r** are 3-dimensional (**k** components), their cross product will have additional components of rotation about the *x* and *y* axes. The moment **M** resulting from crossing **r** into **F** is written

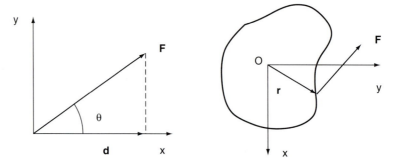

Figure 4.3   (left) The dot, or scalar, product of vectors **F** and **d** is equivalent to the projection of **F** onto **d**. (right) The cross, or vector, product of vectors **r** and **F** is a vector that points along the axis of rotation, the *z* axis coming out of the page.

$$\mathbf{M} = M_x\mathbf{i} + M_y\mathbf{j} + M_z\mathbf{k} \tag{4.13}$$

where M_x, M_y, and M_z cause rotation of the body about the x, y, and z axes, respectively.

Cross products may be taken by crossing each vector component term by term, for example:

$$\mathbf{A} \times \mathbf{B} = 3(-2)\mathbf{i} \times \mathbf{i} + 3(3)\mathbf{i} \times \mathbf{j} + 3(10)\mathbf{i} \times \mathbf{k}$$
$$+ 2(-2)\mathbf{j} \times \mathbf{i} + 2(3)\mathbf{j} \times \mathbf{j} + 2(10)\mathbf{j} \times \mathbf{k}$$
$$+ 1(-2)\mathbf{k} \times \mathbf{i} + 1(3)\mathbf{k} \times \mathbf{j} + 1(10)\mathbf{k} \times \mathbf{k}$$

The magnitude $|\mathbf{A} \times \mathbf{B}| = AB \sin\theta$, where θ is the angle between $\mathbf{A}$ and $\mathbf{B}$. Consequently, for an orthogonal coordinate system the cross products of all like terms equal zero, and $\mathbf{i} \times \mathbf{j} = \mathbf{k}$, $\mathbf{j} \times \mathbf{k} = \mathbf{i}$, $\mathbf{k} \times \mathbf{i} = \mathbf{j}$, $\mathbf{i} \times \mathbf{k} = -\mathbf{j}$, and so on. The previous example yields

$$\mathbf{A} \times \mathbf{B} = 9\mathbf{k} - 30\mathbf{j} + 4\mathbf{k} + 20\mathbf{i} - 2\mathbf{j} - 3\mathbf{i}$$
$$= 17\mathbf{i} - 32\mathbf{j} + 13\mathbf{k} \text{ lb ft}$$

Note that the cross product is not commutative (i.e., $\mathbf{A} \times \mathbf{B} \not\equiv \mathbf{B} \times \mathbf{A}$).

Cross products of vectors are commonly computed using matrices. The previous example $\mathbf{A} \times \mathbf{B}$ is given by the matrix

$$\mathbf{A} \times \mathbf{B} = \begin{vmatrix} \mathbf{i} & \mathbf{j} & \mathbf{k} \\ A_x & A_y & A_z \\ B_x & B_y & B_z \end{vmatrix}$$

$$= \begin{vmatrix} \mathbf{i} & \mathbf{j} & \mathbf{k} \\ 3 & 2 & 1 \\ -2 & 3 & 10 \end{vmatrix} \tag{4.14}$$

$$= \mathbf{i}[(2)(10) - (1)(3)] - \mathbf{j}[(3)(10) - (1)(-2)] + \mathbf{k}[(3)(3) - (2)(-2)]$$
$$= \mathbf{i}(20 - 3) - \mathbf{j}(30 + 2) + \mathbf{k}(9 + 4)$$
$$= 17\mathbf{i} - 32\mathbf{j} + 13\mathbf{k} \text{ lb ft}$$

Example Problem 4.1

The vector $\mathbf{F}$ in Figure 4.4 has a magnitude of 10 kN and points along the dashed line as shown. (a) Write $\mathbf{F}$ as a vector. (b) What is the component of $\mathbf{F}$ in the x–z plane? (c) What moment does $\mathbf{F}$ generate about the origin $(0,0,0)$?

Solution

This example problem is solved using MATLAB. The $\gg$ prompt denotes input and the percent sign, %, precedes comments (ignored by MATLAB). Lines that begin without the $\gg$ prompt are MATLAB output. Some spaces in the following output were omitted to conserve space.

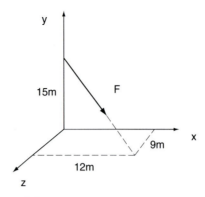

Figure 4.4   Force vector **F** has magnitude of 10 kN.

≫ %(a) First write the direction vector d that points along F
≫ % as a 1D array:
≫ d = [12 −15 9]

d = 12 −15 9
≫ % Now write the unit vector of F, giving its direction:
≫ unit_vector = d/norm (d)

unit_vector = 0.5657   −0.7071   0.4243

≫ % F consists of the magnitude 10 kN times this unit vector
≫ F = 10*unit_vector

F = 5.6569   −7.0711   4.2426

≫ % Or, more directly
≫ F = 10*(d/norm(d))

F = 5.6569   −7.0711   4.2426

≫ % (b) First write the vector r_xz that points in the xz plane:
≫ r_xz = [12 0 9]
r_xz = 12   0   9

≫ % The dot product is given by the sum of all the term by term
≫ % multiplications of elements of vectors F and r_xz
≫ F_dot_r_xz = sum(F.*r_xz)
≫ % or simply, dot(F,r_xz)

F_dot_r_xz = 106.0660

≫ % (c) Cross F with a vector that points from the origin to F.
≫ % The cross product is given by the cross function
≫ r_xz_cross_F = cross(r_xz,F)

r_xz_cross_F = 63.6396   0   −84.8528

≫ % Note that the cross product is not commutative
≫ cross(F,r_xz)
ans = −63.6396   0   84.8528

≫ % Vectors are added and subtracted in MATLAB using the + and −
≫ % operations, respectively.   ∎

4.2.2   Coordinate Transformations

3-D   Direction Cosines

When studying the kinematics of human motion, it is often necessary to transform body or body segment coordinates from one coordinate system to another. For example, coordinates corresponding to a coordinate system determined by markers on the body (a moving coordinate system) must be translated to coordinates with respect to the fixed laboratory (inertial coordinate system). These 3-dimensional transformations use direction cosines that are computed as follows.

Consider the vector **A** measured in terms of the uppercase coordinate system XYZ, shown in Figure 4.5 in terms of the unit vectors **I, J, K**.

$$\mathbf{A} = A_x\mathbf{I} + A_y\mathbf{J} + A_z\mathbf{K} \qquad (4.15)$$

The unit vectors **I, J, K** can be written in terms of **i, j, k** in the xyz system

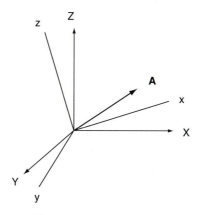

Figure 4.5   Vector **A**, measured with respect to coordinate system XYZ is related to coordinate system xyz via the nine direction cosines of Eq. 4.20.

$$\mathbf{I} = \cos\theta_{xX}\mathbf{i} + \cos\theta_{yX}\mathbf{j} + \cos\theta_{zX}\mathbf{k} \tag{4.16}$$

$$\mathbf{J} = \cos\theta_{xY}\mathbf{i} + \cos\theta_{yY}\mathbf{j} + \cos\theta_{zY}\mathbf{k} \tag{4.17}$$

$$\mathbf{K} = \cos\theta_{xZ}\mathbf{i} + \cos\theta_{yZ}\mathbf{j} + \cos\theta_{zZ}\mathbf{k} \tag{4.18}$$

where θ_{xX} is the angle between $\mathbf{i}$ and $\mathbf{I}$, and similarly for the other angles.

Substituting Eqs. 4.16–4.18 into Eq. 4.15 gives

$$\begin{aligned} \mathbf{A} = &A_x[\cos\theta_{xX}\mathbf{i} + \cos\theta_{yX}\mathbf{j} + \cos\theta_{zX}\mathbf{k}] \\ &+ A_y[\cos\theta_{xY}\mathbf{i} + \cos\theta_{yY}\mathbf{j} + \cos\theta_{zY}\mathbf{k}] \\ &+ A_z[\cos\theta_{xZ}\mathbf{i} + \cos\theta_{yZ}\mathbf{j} + \cos\theta_{zZ}\mathbf{k}] \end{aligned} \tag{4.19}$$

or

$$\begin{aligned} \mathbf{A} = &(A_x\cos\theta_{xX} + A_y\cos\theta_{xY} + A_z\cos\theta_{xZ})\mathbf{i} \\ &+ (A_x\cos\theta_{yX} + A_y\cos\theta_{yY} + A_z\cos\theta_{yZ})\mathbf{j} \\ &+ (A_x\cos\theta_{zX} + A_y\cos\theta_{zY} + A_z\cos\theta_{zZ})\mathbf{k} \end{aligned} \tag{4.20}$$

Consequently, $\mathbf{A}$ may be represented in terms of $\mathbf{I}, \mathbf{J}, \mathbf{K}$ or $\mathbf{i}, \mathbf{j}, \mathbf{k}$.

Euler Angles

The coordinates of a body in one orthogonal coordinate system may be related to another orthogonal coordinate system via Euler angle transformation matrices. For example, one coordinate system might correspond to markers placed on the patient's pelvis and the other coordinate system might correspond to the patient's thigh. The two coordinate systems are related by a series of rotations about each original axis in turn. Figure 4.6 shows the xyz coordinate axes with a y–x–z rotation sequence. First, xyz is rotated about the y axis (top), transforming the $\mathbf{ijk}$ unit vectors into the $\mathbf{i'j'k'}$ unit vectors, via the equations

$$\mathbf{i'} = \cos\theta_y\mathbf{i} - \sin\theta_y\mathbf{k} \tag{4.21}$$

$$\mathbf{j'} = \mathbf{j} \tag{4.22}$$

$$\mathbf{k'} = \sin\theta_y\mathbf{i} + \cos\theta_y\mathbf{k} \tag{4.23}$$

This new primed coordinate system is then rotated about the x axis (Fig. 4.6, middle), giving the double-primed system:

$$\mathbf{i''} = \mathbf{i'} \tag{4.24}$$

$$\mathbf{j''} = \cos\theta_x\mathbf{j'} + \sin\theta_x\mathbf{k'} \tag{4.25}$$

$$\mathbf{k''} = -\sin\theta_x\mathbf{j'} + \cos\theta_x\mathbf{k'} \tag{4.26}$$

Finally, the double-primed system is rotated about the z axis, giving the triple-primed system:

$$\mathbf{i'''} = \cos\theta_z\mathbf{i''} + \sin\theta_z\mathbf{j''} \tag{4.27}$$

$$\mathbf{j'''} = -\sin\theta_z\mathbf{i''} + \cos\theta_z\mathbf{j''} \tag{4.28}$$

$$\mathbf{k'''} = \mathbf{k''} \tag{4.29}$$

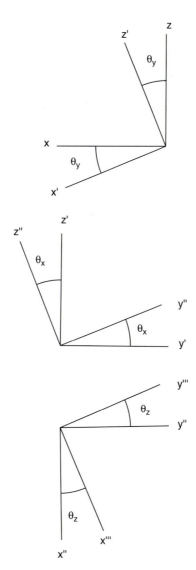

Figure 4.6 The unprimed coordinate system *xyz* undergoes three rotations: about the *y* axis (top), about the *x* axis (middle) and about the *z* axis (bottom), yielding the new triple-primed coordinate system $x'''y'''z'''$.

The three rotations may be written in matrix form to directly translate **ijk** into **i'''j'''k'''**:

$$
\begin{bmatrix} \mathbf{i}''' \\ \mathbf{j}''' \\ \mathbf{k}''' \end{bmatrix} = \begin{bmatrix} \cos\theta_z & \sin\theta_z & 0 \\ -\sin\theta_z & \cos\theta_z & 0 \\ 0 & 0 & 1 \end{bmatrix} \begin{bmatrix} 1 & 0 & 0 \\ 0 & \cos\theta_x & \sin\theta_x \\ 0 & -\sin\theta_x & \cos\theta_x \end{bmatrix} \begin{bmatrix} \cos\theta_y & 0 & -\sin\theta_y \\ 0 & 1 & 0 \\ \sin\theta_y & 0 & \cos\theta_y \end{bmatrix} \begin{bmatrix} \mathbf{i} \\ \mathbf{j} \\ \mathbf{k} \end{bmatrix}
$$

$$
= \begin{bmatrix} \cos\theta_z & \sin\theta_z\cos\theta_x & \sin\theta_z\sin\theta_x \\ -\sin\theta_z & \cos\theta_z\cos\theta_x & \cos\theta_z\sin\theta_x \\ 0 & -\sin\theta_x & \cos\theta_x \end{bmatrix} \begin{bmatrix} \cos\theta_y & 0 & -\sin\theta_y \\ 0 & 1 & 0 \\ \sin\theta_y & 0 & \cos\theta_y \end{bmatrix} \begin{bmatrix} \mathbf{i} \\ \mathbf{j} \\ \mathbf{k} \end{bmatrix}
$$

(4.30)

$$
\begin{bmatrix} \mathbf{i}''' \\ \mathbf{j}''' \\ \mathbf{k}''' \end{bmatrix} = \begin{bmatrix} \cos\theta_z\cos\theta_y + \sin\theta_z\sin\theta_x\sin\theta_y & \sin\theta_z\cos\theta_x & -\cos\theta_z\sin\theta_y + \sin\theta_z\sin\theta_x\cos\theta_y \\ -\sin\theta_z\cos\theta_y + \cos\theta_z\sin\theta_x\sin\theta_y & \cos\theta_z\cos\theta_x & \sin\theta_z\sin\theta_y + \cos\theta_z\sin\theta_x\cos\theta_y \\ \cos\theta_x\sin\theta_y & -\sin\theta_x & \cos\theta_x\cos\theta_y \end{bmatrix} \begin{bmatrix} \mathbf{i} \\ \mathbf{j} \\ \mathbf{k} \end{bmatrix}
$$

(4.31)

If the angles of coordinate system rotation $(\theta_x, \theta_y, \theta_z)$ are known, coordinates in the xyz system can be transformed into the $x'''y'''z'''$ system. Alternatively, if both the unprimed and triple-primed coordinates are known, the angles may be computed as follows

$$
\mathbf{k}''' \cdot \mathbf{j} = -\sin\theta_x
$$
$$
\theta_x = -\arcsin(\mathbf{k}''' \cdot \mathbf{j})
$$

(4.32)

$$
\mathbf{k}''' \cdot \mathbf{i} = \cos\theta_x \sin\theta_y
$$
$$
\theta_y = \arcsin\left[\frac{\mathbf{k}''' \cdot \mathbf{i}}{\cos\theta_x}\right]
$$

(4.33)

$$
\mathbf{i}''' \cdot \mathbf{j} = \sin\theta_z \cos\theta_x
$$
$$
\theta_z = \arcsin\left[\frac{\mathbf{i}''' \cdot \mathbf{j}}{\cos\theta_x}\right]
$$

(4.34)

Example Problem 4.2

Write the Euler angle transformation matrices for the y–x–z rotation sequence using the MATLAB symbolic math toolbox.

Solution

```
% eulerangles.m
%
% Euler angles for y-x-z rotation sequence
% using MATLAB symbolic math toolbox
%
% x, y and z are thetax, thetay and thetaz, respectively
% First define them as symbolic variables
```

syms x y z

% Writing equations 4.21–23 as a matrix A

$$A = \begin{bmatrix} \cos(y), & 0, & -\sin(y); \\ 0, & 1, & 0; \\ \sin(y), & 0, & \cos(y) \end{bmatrix}$$

% equations 4.24–26 as matrix B

$$B = \begin{bmatrix} 1, & 0, & 0; \\ 0, & \cos(x), & \sin(x); \\ 0, & -\sin(x), & \cos(x) \end{bmatrix}$$

% and equations 4.27–29 as matrix C
$$C = \begin{bmatrix} \cos(z), & \sin(z), & 0; \\ -\sin(z), & \cos(z), & 0; \\ 0, & 0, & 1 \end{bmatrix}$$

% The matrix equation 4.30 is created by multiplying matrices C, B
% and A

$$D = C*B*A$$

The resulting transformation matrix from the preceding m-file is
D =
[cos(z)*cos(y)+sin(z)*sin(x)*sin(y), sin(z)*cos(x), −cos(z)*sin(y)+sin(z)*sin(x)*cos(y)]
[−sin(z)*cos(y)+cos(z)*sin(x)*sin(y), cos(z)*cos(x), sin(z)*sin(y)+cos(z)*sin(x)*cos(y)]
[cos(x)*sin(y),   −sin(x),   cos(x)*cos(y)]

Which is the same as Eq. 4.31.   ■

The Euler transformation matrices are used differently depending on the available data. For example, if the body coordinates in both the fixed (unprimed) and body (triple-primed) systems are known, the body angles θ_x, θ_y, and θ_z can be computed (e.g., Eqs. 4.32–4.34 for a y–x–z rotation sequence). Alternatively, the body's initial position and the angles θ_x, θ_y, and θ_z may be used to compute the body's final position.

Example Problem 4.3

An aircraft undergoes 30 degrees of pitch (θ_x), then 20 degrees of roll (θ_y), and finally 10 degrees of yaw (θ_z). Write a MATLAB function that computes the Euler angle transformation matrix for this series of angular rotations.

Solution

Since computers use radians for trigonometric calculations, first write two simple functions to compute cosines and sines in degrees:

function y = cosd(x)
%COSD(X) cosines of the elements of X measured in degrees.
y = cos(pi*x/180);

function y = sind(x)
%SIND(X) sines of the elements of X measured in degrees.
y = sin(pi*x/180);

Next write the *x–y–z* rotation sequence transformation matrix

function D = eulangle (thetax, thetay, thetaz)
%EULANGLE matrix of rotations by Euler's angles.
%   EULANGLE(thetax, thetay, thetaz) yields the matrix of
%   rotation of a system of coordinates by Euler's
%   angles thetax, thetay and thetaz, measured in degrees.

% Now the first rotation is about the *x* axis, so we use eqs. 4.24–26

A = [1   0   0
      0   cosd(thetax)   sind(thetax)
      0   −sind(thetax)   cosd(thetax)];

% Next is the *y* axis rotation (Eqs. 4.21–23)

B = [cosd(thetay)   0   −sind(thetay)
      0   1   0
      sind(thetay)   0   cosd(thetay)];

% Finally, the *z* axis rotation (Eqs. 4.27–29)

C = [cosd(thetaz)   sind(thetaz)   0
       −sind(thetaz)   cosd(thetaz)   0
        0   0   1   ];

% Multiplying rotation matrices C, B and A as in Eq. 4.30 gives the solution:

D=C*B*A;

Now use this function to compute the numerical transformation matrix:

```
≫ eulangle(30,20,10)
ans =
    0.9254    0.3188   -0.2049
   -0.1632    0.8232    0.5438
    0.3420   -0.4698    0.8138
```

This matrix can be used to convert any point in the initial coordinate system (premaneuver) to its position after the roll, pitch, and yaw maneuvers have been executed. ∎

4.2.3 Static Equilibrium

Newton's equations of motion applied to a structure in static equilibrium reduce to the following vector equations

$$\sum \mathbf{F} = 0 \tag{4.35}$$

$$\sum \mathbf{M} = 0 \tag{4.36}$$

These equations are applied to biological systems in the same manner as standard mechanical structures. Analysis begins with a drawing of the free-body diagram of the body segment(s) of interest with all externally applied loads and reaction forces at the supports. Orthopedic joints can be modeled with appropriate ideal joints (e.g., hinge, ball-and-socket, etc.) as discussed in Chapter 3 (Fig. 3.33).

Example Problem 4.4

Figure 4.7 (top) shows a Russell's traction rig used to apply an axial, tensile force to a fractured femur for immobilization. (a) What magnitude weight w must be suspended from the free end of the cable to maintain the leg in static equilibrium? (b) Compute the average tensile force applied to the thigh under these conditions.

Solution

The free body diagram for this system is shown in the lower panel of Figure 4.7. If the pulleys are assumed frictionless and of small radius, the cable tension T is constant throughout. Using Eq. 4.35,

$$\mathbf{F}_1 + \mathbf{F}_2 + \mathbf{F}_3 + \mathbf{F}_{femur} - mg\mathbf{j} = 0$$

Writing each force in vector form,

$$\mathbf{F}_1 = -F_1\mathbf{i} = -T\mathbf{i}$$

$$\mathbf{F}_2 = (-F_2 \cos 30°)\mathbf{i} + (F_2 \sin 30°)\mathbf{j}$$
$$= (-T \cos 30°)\mathbf{i} + (T \sin 30°)\mathbf{j}$$

$$\mathbf{F}_3 = (F_3 \cos 40°)\mathbf{i} + (F_3 \sin 40°)\mathbf{j}$$
$$= (T \cos 40°)\mathbf{i} + (T \sin 40°)\mathbf{j}$$

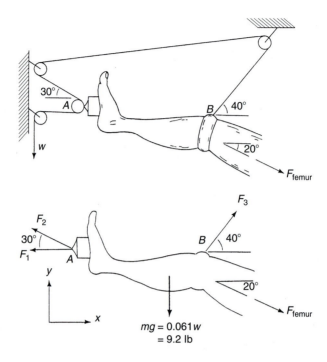

Figure 4.7 (top) Russell's traction mechanism for clinically loading lower extremity limbs. (bottom) Free-body diagram of the leg in traction (adapted from Davis 1986, Figs. 6.14 and 6.15, pp. 206–207).

$$\mathbf{F}_{\text{femur}} = (F_{\text{femur}} \cos 20°)\mathbf{i} - (F_{\text{femur}} \sin 20°)\mathbf{j}$$

Using Table 4.1, and neglecting the weight of the thigh, the weight of the foot and leg is 0.061 multiplied by total body weight, yielding

$$mg\mathbf{j} = (0.061)(150\mathbf{j}\,\text{lb}) = 9.2\mathbf{j}\,\text{lb}$$

Summing the x components gives

$$-T - T \cos 30° + T \cos 40° + F_{\text{femur}} \cos 20° = 0$$

Summing the y components gives

$$T \sin 30° + T \sin 40° - F_{\text{femur}} \sin 20° - mg = 0$$

The last two expressions may be solved simultaneously, giving both T, which is equal to the required externally applied weight, and the axial tensile force, F_{femur}

$$T = 12.4\,\text{lb}$$
$$F_{\text{femur}} = 14.5\,\text{lb}$$

∎

Example Problem 4.5

A 160-lb person is holding a 10-lb weight in his palm with the elbow fixed at 90° flexion (Fig. 4.8, top). (a) What force must the biceps generate to hold the forearm in static equilibrium? (b) What force(s) does the forearm exert on the humerus?

Solution

Figure 4.8 (bottom) shows the free-body diagram of this system. Due to the increased number of unknowns, compared to the previous example, both Eqs. 4.35 and 4.36 will be used. Summing moments about the elbow at point O, the equilibrium equation $\Sigma M = 0$ can be written as

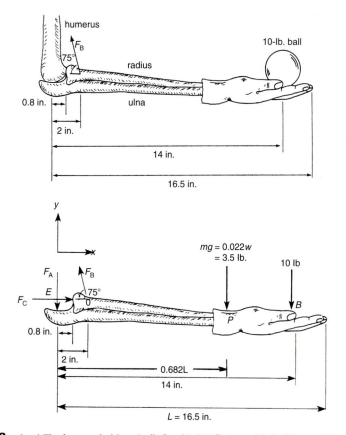

Figure 4.8 (top) The forearm held statically fixed in 90° flexion while holding a 10-lb weight at the hand. (bottom) Free-body diagram of the forearm system (adapted from Davis, 1986, Figs. 6.16 and 6.17, pp. 208–209).

$$-\mathbf{r}_{OE} \times (-\mathbf{F}_A) + \mathbf{r}_{OB} \times (-10 \text{ lb})\mathbf{j} + \mathbf{r}_{OP} \times (-3.5 \text{ lb})\mathbf{j} = 0$$
$$(-2 \text{ in})\mathbf{i} \times (-F_A)\mathbf{j} + (12 \text{ in})\mathbf{i} \times (-10 \text{ lb})\mathbf{j} + (9.25 \text{ in})\mathbf{i} \times (-3.5 \text{ lb})\mathbf{j} = 0$$
$$(2 \text{ in})F_A\mathbf{k} - (120 \text{ lb in})\mathbf{k} - (32.4 \text{ lb in})\mathbf{k} = 0$$

Solving this last expression for the one unknown, F_A, the vertical force at the elbow:

$$F_A = 76.2 \text{ lb}$$

To find the unknown horizontal force at the elbow, F_C, and the unknown force the biceps must generate, F_B, the other equation of equilibrium $\Sigma \mathbf{F} = 0$ is used:

$$F_C\mathbf{i} - F_A\mathbf{j} + (-F_B \cos 75°\mathbf{i} + F_B \sin 75°\mathbf{j}) - 10 \text{ lb } \mathbf{j} - 3.5 \text{ lb } \mathbf{j} = 0$$

Summing the x and y components gives

$$F_C - F_B \cos (75°) = 0$$
$$-F_A + F_B \sin (75°) - 10 \text{ lb} - 3.5 \text{ lb} = 0$$

Solving these last two equations simultaneously and using $F_A = 76.2$ lb gives the force of the biceps muscle, F_B, and the horizontal elbow force, F_C:

$$F_B = 92.9 \text{ lb}$$
$$F_C = 24.1 \text{ lb} \qquad \blacksquare$$

Example Problem 4.6

The force plate depicted in Figure 4.9 has four sensors, one at each corner, that read the vertical forces F_1, F_2, F_3, and F_4. If the plate is square with side of length ℓ and forces $F_1 - F_4$ are known, write two expressions that will give the x and y locations of the resultant force R.

Solution

The resultant magnitude R can be computed from the sum of forces in the z direction:

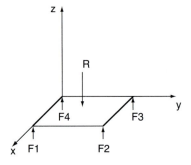

Figure 4.9   A square force plate with sides of length ℓ is loaded with resultant force R and detects the vertical forces at each corner, $F_1 - F_4$.

$$\sum F_z = 0$$

$$F_1 + F_2 + F_3 + F_4 - R = 0$$

$$R = F_1 + F_2 + F_3 + F_4$$

The force plate remains horizontal; hence the sum of the moments about the x and y axes must each be zero. Taking moments about the x axis,

$$\sum M_x = 0$$

$$F_2\ell + F_3\ell - Ry = 0$$

$$y = \frac{(F_2 + F_3)\ell}{R}$$

Similarly, summing moments about the y axis,

$$\sum M_y = 0$$

$$F_1\ell + F_2\ell - Rx = 0$$

$$x = \frac{(F_1 + F_2)\ell}{R}$$

The coordinates x and y locate the resultant R.   ■

4.2.4 **Anthropomorphic Mass Moments of Inertia**

A body's mass resists linear motion; its mass moment of inertia resists rotation. The resistance of a body (or a body segment such as a thigh in gait analysis) to rotation is quantified by the body or body segment's moment of inertia I:

$$I = \int_m r^2 dm \tag{4.37}$$

where m is the body mass and r is the the moment arm to the axis of rotation. The incremental mass dm can be written ρdV. For a body with constant density ρ the moment of inertia can be found by integrating over the body's volume V:

$$I = \rho \int_V r^2 dV \tag{4.38}$$

This general expression can be written in terms of rotation about the x, y, and z axes:

$$I_{xx} = \int_V (y^2 + z^2)\rho dV$$

$$I_{yy} = \int_V (x^2 + z^2)\rho dV \tag{4.39}$$

$$I_{zz} = \int_V (x^2 + y^2)\rho dV$$

The radius of gyration k is the moment arm between the axis of rotation and a single point where all of the body's mass is concentrated. Consequently, a body segment may be treated as a point mass with moment of inertia,

$$I = mk^2 \qquad (4.40)$$

where m is the body segment mass. The moment of inertia with respect to a parallel axis I is related to the moment of inertia with respect to the body's center of mass I_{cm} via the parallel axis theorem:

$$I = I_{cm} + md^2 \qquad (4.41)$$

where d is the perpendicular distance between the two parallel axes. Anthropomorphic data for various body segments are listed in Table 4.1.

Example Problem 4.7

A 150-lb person has a thigh length of 17 in. Find the moment of inertia of this body segment with respect to its center of mass in SI units.

Solution

Thigh length in SI units is

$$\ell_{\text{thigh}} = 17\,\text{in} = 0.432\,\text{m}$$

Table 4.1 lists ratios of segment weight to body weight for different body segments. Starting with body mass,

$$m_{\text{body}} = (150\,\text{lb})(0.454\,\text{kg/lb}) = 68.1\,\text{kg}$$

the thigh segment mass is

$$m_{\text{thigh}} = (0.100)(68.1\,\text{kg}) = 6.81\,\text{kg}$$

Table 4.1 also lists body segment center of mass and radius of gyration as ratios with respect to segment length for each body segment. Table 4.1 gives both proximal and distal segment length ratios. Note that "proximal" for the thigh refers toward the hip and "distal" refers toward the knee. Consequently, the proximal thigh segment length is the distance between the thigh center of mass and the hip, and the distal thigh segment length is the distance between the thigh center of mass and the knee. The moment of inertia of the thigh with respect to the hip is therefore

$$I_{\text{thigh/hip}} = mk^2 = (6.81\,\text{kg})[(0.540)(0.432\,\text{m})]^2 = 0.371\,\text{kg m}^2$$

The thigh's moment of inertia with respect to the hip is related to the thigh's moment of inertia with respect to its center of mass via the parallel axis theorem (Eq. 4.41),

$$I_{\text{thigh/hip}} = I_{\text{thigh/cm}} + md^2$$

TABLE 4.1 Anthropomorphic Data

Segment	Definition	Segment Weight/Body Weight	Center Mass/Segment Length Proximal	Center Mass/Segment Length Distal	Radius Gyration/Segment Length Proximal	Radius Gyration/Segment Length Distal
Hand	Wrist axis/knuckle II middle finger	0.006	0.506	0.494	0.587	0.577
Forearm	Elbow axis/ulnar styloid	0.016	0.430	0.570	0.526	0.647
Upper arm	Glenohumeral axis/elbow axis	0.028	0.436	0.564	0.542	0.645
Forearm and hand	Elbow axis/ulnar styloid	0.022	0.682	0.318	0.827	0.565
Total arm	Glenohumeral joint/ulnar styloid	0.050	0.530	0.470	0.645	0.596
Foot	Lateral malleolus/head metatarsal II	0.0145	0.50	0.50	0.690	0.690
Leg	Femoral condyles/medial malleolus	0.0465	0.433	0.567	0.528	0.643
Thigh	Greater trochanter/femoral condyles	0.100	0.433	0.567	0.540	0.653
Foot and leg	Femoral condyles/medial malleolus	0.061	0.606	0.394	0.735	0.572
Total leg	Greater trochanter/medial malleolus	0.161	0.447	0.553	0.560	0.650
Head and neck	C7-T1 and 1st rib/ear canal	0.081	1.000		1.116	
Shoulder mass	Sternoclavicular joint/glenohumeral axis		0.712	0.288		
Thorax	C7-T1/T12-L1 and diaphragm	0.216	0.82	0.18		
Abdomen	T12-L1/L4-L5	0.139	0.44	0.56		
Pelvis	L4-L5/greater trochanter	0.142	0.105	0.895		
Thorax and abdomen	C7-T1/L4-L5	0.355	0.63	0.37		
Abdomen and pelvis	T12-L1/greater trochanter	0.281	0.27	0.73		
Trunk	Greater trochanter/glenohumeral joint	0.497	0.50	0.50		
Trunk, head, neck	Greater trochanter/glenohumeral joint	0.578	0.66	0.34	0.830	0.607
Head, arm, trunk	Greater trochanter/glenohumeral joint	0.678	0.626	0.374	0.798	0.621

Adapted from Winter, 1990, Table 3.1, pp. 56–57.

so

$$I_{\text{thigh}/\text{cm}} = I_{\text{thigh}/\text{hip}} - md^2$$

In this case, distance d is given by the proximal segment length data:

$$d = (0.432 \, \text{m})(0.433) = 0.187 \, \text{m}$$

and the final result is

$$I_{\text{thigh}/\text{cm}} = 0.371 \, \text{kg} \, \text{m}^2 - (6.81 \, \text{kg})(0.187 \, \text{m})^2 = 0.133 \, \text{kg} \, \text{m}^2 \qquad \blacksquare$$

4.2.5 Equations of Motion

Vector equations of motion are used to describe the translational and rotational kinetics of bodies.

Newton's Equations of Motion

Newton's second law relates the net force $\mathbf{F}$ and the resulting translational motion as

$$\mathbf{F} = m\mathbf{a} \tag{4.42}$$

where $\mathbf{a}$ is the linear acceleration of the body's center of mass for translation. For rotation

$$\mathbf{M} = \mathbf{I}\boldsymbol{\alpha} \tag{4.43}$$

where $\mathbf{I}\boldsymbol{\alpha}$ is the body's angular momentum. Hence, the rate of change of a body's angular momentum is equal to the net moment $\mathbf{M}$ acting on the body. These two vector equations of motion are typically written as a set of six x, y, and z component equations.

Euler's Equations of Motion

Newton's equations of motion describe the motion of the center of mass of a body. More generally, Euler's equations of motion describe the motion of a rigid body with respect to its center of mass. For the special case where the xyz coordinate axes are chosen to coincide with the body's principal axes,

$$\sum M_x = I_{xx}\alpha_x + (I_{zz} - I_{yy})\omega_y\omega_z \tag{4.44}$$

$$\sum M_y = I_{yy}\alpha_y + (I_{xx} - I_{zz})\omega_z\omega_x \tag{4.45}$$

$$\sum M_z = I_{zz}\alpha_z + (I_{yy} - I_{xx})\omega_x\omega_y \tag{4.46}$$

M_i is the net moment, I_{ii} is the body's moment of inertia with respect to the principal axes, and α_i and ω_i are the body's angular acceleration and angular velocity, respectively. Euler's equations require angular measurements in radians. Their derivation is outside the scope of this chapter, but may be found in any intermediate dynamics book. Equations 4.44–4.46 will be used in Section 4.6 to compute intersegmental or joint moments.

4.3 MECHANICS OF MATERIALS

Just as kinematic and kinetic relations may be applied to biological bodies to describe their motion and its associated forces, concepts from mechanics of materials may be used to quantify tissue deformation, to study distributed orthopedic forces, and to predict the performance of orthopedic implants and prostheses and of surgical corrections. Since this topic is very broad, some representative concepts will be illustrated with the following examples.

An orthopedic bone plate is a flat segment of stainless steel used to screw two failed sections of bone together. The bone plate in Figure 4.10 has a rectangular cross-section, A, measuring 4.17 mm by 12 mm and made of 316L stainless steel. An applied axial load, F, of 500 N produces axial stress, σ, (force/area):

$$
\begin{aligned}
\sigma &= \frac{F}{A} \\
&= \frac{500\,\text{N}}{(4.17 \times 10^{-3}\,\text{m})(12 \times 10^{-3}\,\text{m})} = 10\,\text{MPa}
\end{aligned}
\tag{4.47}
$$

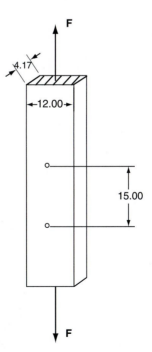

Figure 4.10   Bone plate used to fix bone fractures, with applied axial load. Dimensions are in mm (adapted from Burstein and Wright, 1994, pp. 104–108).

The maximum shear stress, τ_{max}, occurs at a 45° angle to the applied load

$$
\begin{aligned}
\tau_{max} &= \frac{F_{45°}}{A_{45°}} \\
&= \frac{(500\,\text{N})\cos 45°}{\left[\frac{(0.00417\,\text{m})(0.012\,\text{m})}{\cos 45°}\right]} = 5\,\text{MPa}
\end{aligned}
\tag{4.48}
$$

which is 0.5σ, as expected from mechanics of materials principles. Prior to loading, two points were punched 15 mm apart on the long axis of the plate, as shown. After the 500 N load is applied, those marks are an additional 0.00075 mm apart. The plate's strain, ε, relates the change in length, $\Delta\ell$ to the original length, ℓ:

$$
\begin{aligned}
\varepsilon &= \frac{\Delta\ell}{\ell} \\
&= \frac{0.00075\,\text{mm}}{15\,\text{mm}} = 50 \times 10^{-6}
\end{aligned}
\tag{4.49}
$$

often reported as $50\,\mu$ where μ denotes microstrain (10^{-6}).

The elastic modulus, E, relates stress and strain and is a measure of a material's resistance to distortion by a tensile or compressive load. For linearly elastic (Hookean) materials, E is a constant, and a plot of σ as a function of ε is a straight line with slope E:

$$
E = \frac{\sigma}{\varepsilon}
\tag{4.50}
$$

For the bone plate,

$$
E = \frac{10 \times 10^6\,\text{Pa}}{50 \times 10^{-6}} = 200\,\text{GPa}
$$

Materials such as metals and plastics display linearly elastic properties only in limited ranges of applied loads. Biomaterials have even more complex elastic properties. Figure 4.11 shows tensile stress–strain curves measured from longitudinal and transverse sections of bone. Taking the longitudinal curve first, from 0–7000 µ bone behaves as a purely elastic solid with E ≈ 12 GPa. At a tensile stress of approximately 90 MPa, the stress–strain curve becomes nonlinear, yielding into the plastic region of deformation. This sample ultimately fails at 120 MPa. Table 4.2 shows elastic moduli, yield stresses, and ultimate stresses for some common orthopedic materials, both natural and implant.

Figure 4.11 also shows that the elastic properties of bone differ depending on whether the sample is cut in the longitudinal or transverse direction (i.e., bone is anisotropic). Bone is much weaker and less stiff in the transverse compared to the longitudinal direction, as is illustrated by the large differences in the yield and ultimate stresses and the slopes of the stress–strain curves for the two samples.

Figure 4.12 shows that the elastic properties of bone also vary depending on whether the load is being applied or removed, displaying hysteresis. From a thermodynamic view, the energy stored in the bone during loading is not equal to the energy released during unloading. This energy difference becomes greater as the maximum

TABLE 4.2 Tensile Yield and Ultimate Stresses and Elastic Moduli (E) for Some Common Orthopedic Materials

Material	σ_{yield} [MPa]	$\sigma_{ultimate}$ [MPa]	E [GPa]
Stainless steel	700	850	180
Cobalt alloy	490	700	200
Titanium alloy	1100	1250	110
Bone	85	120	18
PMMA (fixative)		35	5
UHMWPE (bearing)	14	27	1
Patellar ligament		58	

Data from Burstein and Wright, 1994, Table 4.2, p. 122.

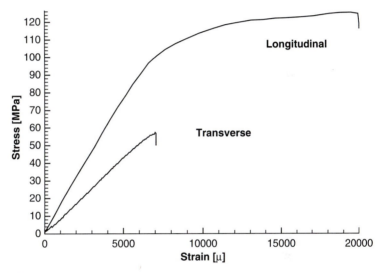

Figure 4.11 Tensile stress–strain curves for longitudinal and transverse sections of bone (adapted from Burstein and Wright, 1994, Fig. 4.12, p. 116).

load increases (curves A to B to C). The "missing" energy is dissipated as heat due to internal friction and damage to the material at high loads.

The anisotropic nature of bone is sufficient that its ultimate stress in compression is 200 MPa while in tension it is only 140 MPa and in torsion 75 MPa. For torsional loading the shear modulus or modulus of rigidity, denoted G, relates the shear stress to the shear strain. The modulus of rigidity is related to the elastic modulus via Poisson's ratio, ν, where

$$\nu = \frac{\varepsilon_{transverse}}{\varepsilon_{longitudinal}} \tag{4.51}$$

Typically, $\nu \approx 0.3$, meaning that longitudinal deformation is three times greater than transverse deformation. For linearly elastic materials, E, G, and ν are related by

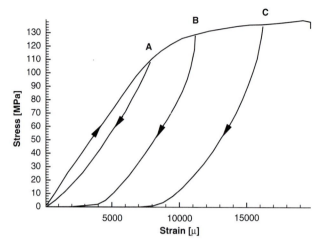

Figure 4.12   Bone shows hysteresis and shifting of stress–strain curves with repeated loading and unloading (adapted from Burstein and Wright, 1994, Fig. 4.15, p. 119).

$$G = \frac{E}{2(1 + v)} \qquad (4.52)$$

One additional complexity of predicting biomaterial failure is the complexity of physiological loading. For example, bone is much stronger in compression than in tension. This property is demonstrated in "boot-top" fractures in skiing. Since the foot is fixed, the skier's forward momentum causes a moment over the ski boot top and produces three-point bending of the tibia. In this bending mode the anterior tibia undergoes compression, while the posterior is in tension and potentially in failure. Contraction of the triceps surae muscle produces high compressive stress at the posterior side, reducing the amount of bone tension. The following example shows how topics from statics and mechanics of materials may be applied to biomechanical problems.

Example Problem 4.8

Figure 4.13 (left) shows an orthopedic nail-plate used to fix an intertrochanteric fracture. The hip applies an external force of 400 N during static standing, as shown. The nail-plate is rectangular stainless steel with cross-sectional dimensions of 10 mm (width) by 5 mm (height), and is well fixed with screws along its vertical axis and friction fit into the trochanteric head (along the x axis). What forces, moments, stresses, and strains will develop in this orthopedic device?

Solution

As for any statics problem, the first task is constructing a free-body diagram, including all applied forces and moments and all reaction forces and moments that develop at

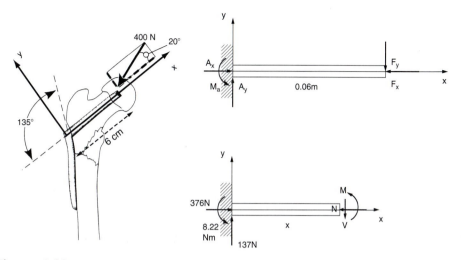

Figure 4.13 (left) An intertrochanteric nail-plate used in bone repair (adapted from Burstein and Wright, 1994, Fig. 5.5, p. 141). To the right is the free-body diagram of the upper section of this device and below it is the free-body diagram of a section of this beam cut at a distance x from the left hand support.

the supports. Because of the instability at the fracture site the nail-plate may be required to carry the entire 400 N load. Consequently, one reasonable model of the nail-plate is a cantilever beam of length 0.06 m with a combined loading, as depicted in Figure 4.13 (right, top). The applied 400 N load consists of both axial and transverse components:

$$F_x = 400\,\text{N}\ \cos 20° = 376\,\text{N}$$
$$F_y = 400\,\text{N}\ \sin 20° = 137\,\text{N}$$

The axial load produces compressive normal stress; from Eq. 4.47,

$$\sigma_x = \frac{F_x}{A}$$
$$= \frac{376\,\text{N}}{(0.005\,\text{m})(0.01\,\text{m})} = 7.52\,\text{MPa}$$

in compression, which is only about 1% of the yield stress for stainless steel (Table 4.2). The maximum shear stress due to the axial load is

$$\tau_{\text{max}} = \frac{\sigma_x}{2} = 3.76\,\text{MPa}$$

and occurs at 45° from the long axis. The axial strain can be computed using the elastic modulus for stainless steel,

$$E = \frac{\sigma}{\varepsilon} = \frac{F/A}{\Delta\ell/\ell}$$

giving an expression for strain:

$$\varepsilon = \frac{F}{EA}$$

$$= \frac{376\,\text{N}}{180 \times 10^9\,\text{Pa}\,(0.005\,\text{m})(0.01\,\text{m})} = 41.8 \times 10^{-6}$$

From this strain the axial deformation can be computed:

$$\Delta\ell_{\text{axial}} = \varepsilon\ell = 2.51 \times 10^{-6}\,\text{m}$$

which is negligible.

The transverse load causes the cantilever section to bend. The equations describing beam bending can be found in any mechanics of materials text (e.g., Roark 1989). Consider the beam in the left panel of Figure 4.14. If this beam is fixed at the left hand side and subjected to a downward load on the right, it will bend with the top of the beam elongating and the bottom shortening. Consequently, the top of the beam is in tension and the bottom in compression. The point of transition, where there is no bending force, is denoted the neutral axis, located at distance c. For a symmetric rectangular beam of height h, c is located at the midline $h/2$. The beam resists bending via its area moment of inertia I. For a rectangular cross section of width b and height h, $I = \frac{1}{12}bh^3$, depicted in the right panel of Figure 4.14.

Beam tip deflection δy is equal to

$$\delta y = \frac{Fx^2}{6EI}(3L - x) \tag{4.53}$$

where x is the axial distance along the beam, L is the total beam length, and I is the beam's cross-sectional area moment of inertia. For this example,

$$I = \frac{1}{12}(10 \times 10^{-3}\,\text{m})(5 \times 10^{-3}\,\text{m})^3 = 10.42 \times 10^{-9}\,\text{m}^4$$

Maximum deflection will occur at $x = L$,

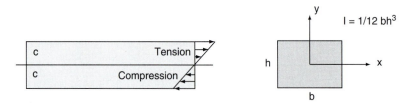

Figure 4.14   (left) A beam fixed on the left and subjected to a downward load on the right undergoes bending, with the top of the beam in tension and the bottom in compression. The position where tension changes to compression is denoted the neutral axis, located at c. (right) A beam of rectangular cross section with width b and height h resists bending via the area moment of inertia $I = \frac{1}{12}bh^3$.

$$\delta y_{max} = \frac{FL^3}{3EI}$$

$$= \frac{137\,\text{N}(0.06\,\text{m})^3}{3(180 \times 10^9\,\text{N/m}^2)(10.42 \times 10^{-9}\,\text{m}^4)} \qquad (4.54)$$

$$= 5.26 \times 10^{-4}\,\text{m} = 0.526\,\text{mm}$$

which is also negligible.

Computation of maximum shear and bending stresses require maximum shear force V and bending moment M. Starting by static analysis of the entire free-body

$$\sum F_x \quad : \quad A_x - 376\,\text{N} = 0$$

$$\sum F_y \quad : \quad A_y - 137\,\text{N} = 0$$

$$\sum M_A \quad : \quad M_a - 137\,\text{N}(0.06\,\text{m}) = 0$$

Solving these equations gives $A_x = 376\,\text{N}$, $A_y = 137\,\text{N}$, and $M_a = 8.22\,\text{N m}$. Taking a cut at any point x to the right of A and isolating the left-hand section gives the free-body in Figure 4.13 (right, bottom). Applying the equations of static equilibrium to this isolated section yields

$$\sum F_x \quad : \quad 376\,\text{N} - N = 0$$

$$N(x) \quad = \quad 376\,\text{N}$$

$$\sum F_y \quad : \quad 137\,\text{N} - V = 0$$

$$V(x) \quad = \quad 137\,\text{N}$$

$$\sum M_A \quad : \quad 8.22\,\text{N m} - (137\,\text{N})(x\,\text{m}) + M = 0$$

$$M(x) \quad = \quad (137\,\text{N m})\,x - 8.22\,\text{N m}$$

These last equations can be plotted easily using MATLAB, giving the axial force, shear force, and bending moment diagrams shown in Figure 4.15.

```
% Use MATLAB to plot axial force, shear force, and bending moment diagrams
% for Example Problem 4.8

X = [0:0.01:0.06];
N = x.*0 + 376;
V = x.*0 + 137;
M = 137.*x - 8.22;

figure
subplot (3,1,1), plot(x,N,x,N, 'x')
xlabel ('x [m]')
ylabel('N [N]')
title ('Axial Force N')
```

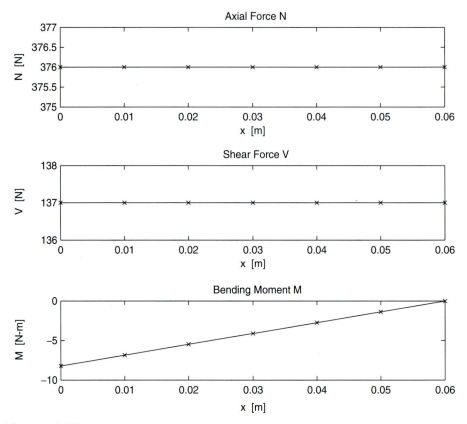

Figure 4.15   Axial force N (top), shear force V (middle), and bending moment M (bottom) computed for the nail-plate in Fig. 4.13 as functions of the distance x along the plate.

subplot(3,1,2), plot(x,V,x,V,'x')
xlabel('x [m]')
ylabel('V [N]')
title('Shear Force V')
subplot(3,1,3), plot(x,M,x,M,'x')
xlabel('x [m]')
ylabel('M [N−m]')
title ('Bending Moment M')

The maximum bending and shear stresses follow as

$$\sigma_{b\max} = \frac{M_{\max}c}{I} \qquad (4.55)$$

where c, the distance to the beam's neutral axis, is $h/2$ for this beam:

$$\sigma_{b\,max} = \frac{-8.22\,\mathrm{Nm}[0.5(5 \times 10^{-3}\ \mathrm{m})]}{10.42 \times 10^{-9}\ \mathrm{m}^4} = -197\,\mathrm{MPa}$$

$$\tau_{b\,max} = \frac{V_{max}h^2}{8I} \tag{4.56}$$

$$= \frac{137\,\mathrm{N}(5 \times 10^{-3}\ \mathrm{m})^2}{8(10.42 \times 10^{-9}\ \mathrm{m}^4)} = 4.11\,\mathrm{MPa}$$

All of these stresses are well below $\sigma_{yield} = 700\,\mathrm{MPa}$ for stainless steel. ∎

4.4 VISCOELASTIC PROPERTIES

The Hookean elastic solid is a valid description of materials only within a narrow loading range. For example, an ideal spring that relates force and elongation by a spring constant k is invalid in nonlinear low-load and high-load regions. Further, if this spring is coupled to a mass and set into motion, the resulting perfect harmonic oscillator will vibrate forever, which experience shows does not occur. Missing is a description of the system's viscous or damping properties. In this case, energy is dissipated as heat in the spring and air friction on the moving system.

Similarly, biomaterials all display viscoelastic properties. Different models of viscoelasticity have been developed to characterize materials with simple constitutive equations. For example, Figure 4.16 shows three such models that consist of a series ideal spring and dashpot (Maxwell), a parallel spring and dashpot (Voight), and a series spring and dashpot with a parallel spring (Kelvin). Each body contains a dashpot, which generates force in proportion to the derivative of its elongation. Consequently, the resulting models exhibit stress and strain properties that vary in time.

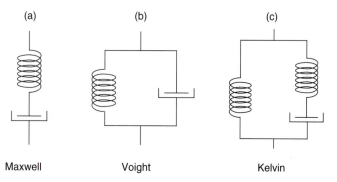

Figure 4.16 Three simple viscoelastic models: (a) the Maxwell model, (b) the Voight model, and (c) the Kelvin body or standard linear solid model.

The dynamic response of each model can be quantified by applying a step change in force F and noting the model's resulting change in length, or position x, denoted the *creep* response. The converse experiment applies a step change in x and measures the resulting change in F, denoted *stress relaxation*. Creep and stress relaxation tests for each dynamic model can be carried out easily using the Simulink program. Figure 4.17 shows a purely elastic material subjected to a step change in applied force F. The material's subsequent position x follows the change in force directly. This material exhibits no creep. Figure 4.18 shows the purely elastic material subjected to a step change in position x. Again, the material responds immediately with a step change in F (i.e., no stress relaxation is observed).

James Clerk Maxwell (1831–1879) used a series combination of ideal spring and dashpot to describe the viscoelastic properties of air. Figure 4.19 shows the Maxwell viscoelastic model subjected to a step change in applied force, and Fig. 4.20 shows the Maxwell model's stress relaxation response. The latter exhibits an initial high stress followed by stress relaxation back to the initial stress level. The creep response, however, shows that this model is not bounded in displacement since an ideal dashpot may be extended forever.

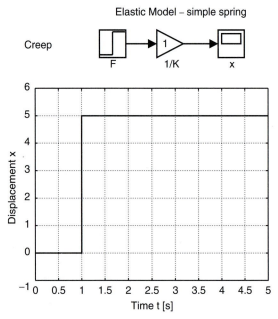

Figure 4.17 Simulink model of the creep test for a purely elastic material (an ideal spring). This model solves the equation $x = F/K$ where x is displacement, F is applied force, and K is the spring constant. Below is the elastic creep response to a step increase in applied force F with $K = 1$ and force changed from 0 to 5 (arbitrary units). The displacement x linearly follows the applied force.

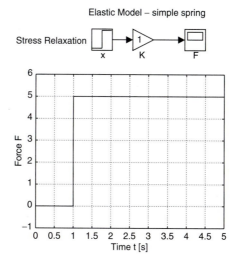

Figure 4.18 Simulink model (top) and stress relaxation test of the purely elastic model which solves $F = Kx$. Applied step displacement $x = 5$ and spring constant $K = 1$. Force linearly follows displacement.

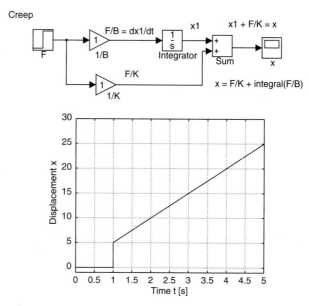

Figure 4.19 Creep of the Maxwell viscoelastic model, a series combination of ideal spring and dashpot (Fig. 4.16a). The spring constant $K = 1$ and dashpot damping coefficient $B = 1$ (arbitrary units). This system is subjected to a step change in force and displacement x arises by solving $x = F/K + \int F/B$. The spring instantly responds, followed by creep of the ideal dashpot, which may extend as long as force is applied.

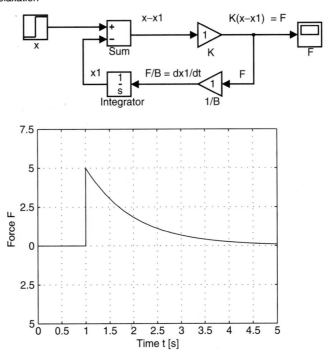

Maxwell Model of Viscoelasticity – series spring and dashpot

Figure 4.20   Stress relaxation of the Maxwell viscoelastic model. This model solves $F = K[x - \int F/B]$, again with $K = B = 1$ and arbitrary units. The ideal spring instantly responds followed by stress relaxation via the dashpot to the steady-state force level.

Woldemar Voight (1850–1919) used the parallel combination of an ideal spring and dashpot in his work with crystallography. Figure 4.21 shows the creep test of the Voight viscoelastic model. Figure 4.22 shows that this model is unbounded in force. That is, when a step change in length is applied, force goes to infinity since the dashpot cannot immediately respond to the length change.

William Thompson (Lord Kelvin, 1824–1907) used the three-element viscoelastic model (Figure 4.16c) to describe the mechanical properties of different solids in the form of a torsional pendulum. Figure 4.23 shows the three-element Kelvin model's creep response. This model has an initial rapid jump in position with subsequent slow creep. Figure 4.24 shows the Kelvin model stress relaxation test. Initially, the material is very stiff with subsequent stress decay to a non zero steady-state level that is due to the extension of the dashpot. The three-element Kelvin model is the simplest lumped viscoelastic model that is bounded both in extension and force.

The three-element viscoelastic model describes the basic features of stress relaxation and creep. Biological materials often exhibit more complex viscoelastic properties. For example, plotting hysteresis as a function of frequency of applied strain gives

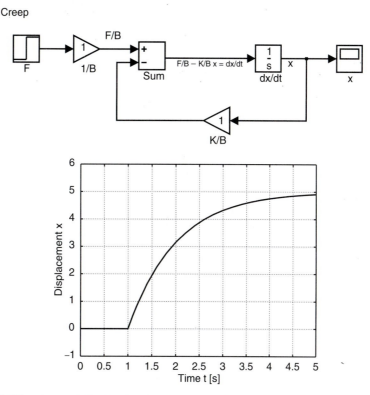

Figure 4.21 Creep of the Voight viscoelastic model, a parallel combination of ideal spring and dashpot (Fig. 4.16b). This model solves the differential equation $dx/dt = 1/B[F - Kx]$ for x. $K = B = 1$ and the step applied force is 5 arbitrary units. Displacement slowly creeps toward its steady-state value.

discrete curves for the lumped viscoelastic models. Biological tissues demonstrate broad, distributed hysteresis properties. One solution is to describe biomaterials with a distributed network of three-element models. A second method is to use the generalized viscoelastic model of Westerhof and Noordergraaf (1990) to describe the viscoelastic wall properties of blood vessels. Making the elastic modulus (mathematically) complex yields a model that includes the frequency dependent elastic modulus, stress relaxation, creep, and hysteresis exhibited by arteries. Further, the Voight and Maxwell models emerge as special (limited) cases of this general approach.

4.5 CARTILAGE, LIGAMENT, TENDON, AND MUSCLE

The articulating surfaces of bones are covered with articular cartilage, a biomaterial composed mainly of collagen. Collagen is the main structural material of hard and

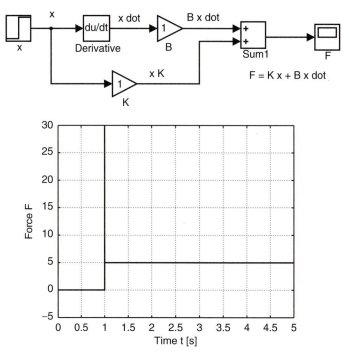

Figure 4.22 Stress relaxation of the Voight viscoelastic model. This model solves the equation $F = Kx + B\,dx/dt$. Since the dashpot is in parallel with the spring, and since it cannot respond immediately to a step change in length, the model force goes to infinity.

soft tissues in animals. Isolated collagen fibers have high tensile strength that is comparable to nylon (50–100 MPa) and an elastic modulus of approximately 1 GPa. Elastin is a protein found in vertebrates and is particularly important in blood vessels and the lungs. Elastin is the most linearly elastic biosolid known, with an elastic modulus of approximately 0.6 MPa. It gives skin and connective tissue their elasticity.

4.5.1 Cartilage

Cartilage serves as the bearing surfaces of joints. It is porous and its complex mechanical properties arise from the motion of fluid in and out of the tissue when subjected to joint loading. Consequently, articular cartilage is strongly viscoelastic with stress relaxation times in compression on the order of 1 to 5 seconds. Cartilage is anisotropic and displays hysteresis during cyclical loading. Ultimate compressive stress of cartilage is on the order of 5 MPa.

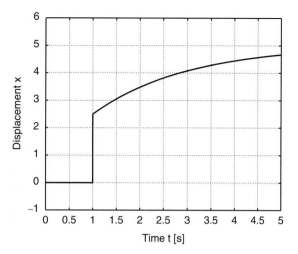

Figure 4.23 Creep of the Kelvin three-element viscoelastic model. This model's equations of motion are left to the reader to derive. After a step change in force, this model has an initial immediate increase in displacement, with a subsequent slow creep to a steady-state level.

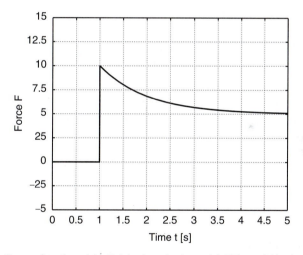

Figure 4.24 Stress relaxation of the Kelvin viscoelastic model. This model has an initial immediate increase in force followed by slower stress relaxation to a steady-state force level.

4.5.2 Ligaments and Tendons

Ligaments join bones together and consequently serve as part of the skeletal framework. Tendons join muscles to bones and transmit forces generated by contracting muscles to cause movement of the jointed limbs. Tendons and ligaments primarily transmit tension; hence they are composed mainly of parallel bundles of collagen fibers and have similar mechanical properties. Human tendon has an ultimate stress of

50–100 MPa and exhibits very nonlinear stress–strain curves. The middle stress–strain range is linear with an elastic modulus of approximately 1–2 GPa. Both tendons and ligaments exhibit hysteresis, viscoelastic creep, and stress relaxation. These materials may also be "preconditioned," whereby initial tensile loading can affect subsequent load-deformation curves. The material properties shift due to changes in the internal tissue structure with repeated loading.

4.5.3   Muscle Mechanics

Chapter 3 introduced muscle as an active, excitable tissue that generates force by forming cross-bridge bonds between the interdigitating actin and myosin myofilaments. The quantitative description of muscle contraction has evolved into two separate foci—lumped descriptions based on A. V. Hill's contractile element, and cross-bridge models based on A.F. Huxley's description of a single sarcomere (Palladino and Noordergraaf, 1998). The earliest quantitative descriptions of muscle are lumped whole muscle models with the simplest mechanical description being a purely elastic spring. Potential energy is stored when the spring is stretched, and shortening occurs when it is released. The idea of muscle elastance can be traced back to Ernst Weber (1846) who considered muscle as an elastic material that changes state during activation via conversion of chemical energy. Subsequently, investigators retained the elastic description but ignored metabolic alteration of muscle stiffness. A purely elastic model of muscle can be refuted on thermodynamic grounds since the potential energy stored during stretching is less than the sum of the energy released during shortening as work and heat. Still, efforts to describe muscle by a combination of traditional springs and dashpots continued. In 1922, Hill coupled the spring with a viscous medium, thereby reintroducing viscoelastic muscle descriptions that can be traced back to the 1840s.

Quick stretch and release experiments show that muscle's viscoelastic properties are strongly time dependent. In general, the faster a change in muscle length occurs, the more severely the contractile force is disturbed. Muscle contraction clearly arises from a more sophisticated mechanism than a damped elastic spring. In 1935, Fenn and Marsh added a series elastic element to Hill's damped elastic model and concluded that "muscle cannot properly be treated as a simple mechanical system." Subsequently, Hill embodied the empirical hyperbolic relation between load and initial velocity of shortening for skeletal muscle as a model building block, denoted the contractile element. Hill's previous viscoelastic model considered muscle to possess a fixed amount of potential energy whose rate of release is controlled by viscosity. Energy is now thought to be controlled by some undefined internal mechanism rather than by friction. This new feature of muscle dynamics varying with load was a step in the right direction; however, subsequent models, including heart studies, built models based essentially on the hyperbolic curve that was measured for tetanized skeletal muscle. This approach can be criticized on two grounds, (1) embodiment of the contractile element by a single force-velocity relation sets a single, fixed relation between muscle energetics and force, and (2) it yields no information on the contractile mechanism behind this relation. Failure of the contractile element to describe a particular loading condition led investigators to add passive springs and dashpots liberally with the number of elements reaching at least nine by the late 1960s. Distributed models of muscle contraction, to date, have been conservative in design and have depended fundamentally on the Hill contractile element. Recent models are limited to tetanized, isometric contractions or to isometric twitch contractions.

A second, independent focus of muscle contraction research works at the ultra-structural level with the sliding filament theory serving as the most widely accepted contraction mechanism. Muscle force generation is viewed as the result of cross-bridge bonds formed between thick and thin filaments at the expense of biochemical energy. The details of bond formation and detachment are under considerable debate, with the mechanism for relaxation particularly uncertain. Prior to actual observation of cross-bridges, A. F. Huxley (1957) devised the cross-bridge model based on structural and energetic assumptions. Bonds between myofilaments are controlled via rate constants f and g that dictate attachment and detachment, respectively. One major shortcoming of this idea was the inability to describe transients resulting from rapid changes in muscle length or load, similar to the creep and stress relaxation tests previously discussed.

Subsequent models adopt increasingly complex bond attachment and detachment rate functions and are often limited in scope to description of a single pair of myofilaments. Each tends to focus on description of a single type of experiment (e.g., quick release). No model has been shown to broadly describe all types of contractile loading conditions. Cross-bridge models have tended to rely on increasingly complex bond attachment and detachment rate functions. This trend has reversed the issue of describing complex muscle dynamics from the underlying (simpler) cross-bridges to adopting complex cross-bridge dynamics to describe a particular experiment.

Alternatively, Palladino and Noordergraaf (1998) proposed a large-scale, distributed muscle model that manifests both contraction and relaxation as the result of fundamental mechanical properties of cross-bridge bonds. As such, muscle's complex contractile properties emerge from the underlying ultrastructure dynamics (i.e., function follows from structure). Bonds between myofilaments, which are biomaterials, are described as viscoelastic material. The initial stimulus for contraction is electrical. Electrical propagation through cardiac muscle occurs at finite speed, implying spatial asynchrony of stimulation. Furthermore, Ca^{++} release from the sarcoplasmic reticulum depends on diffusion for availability at the myosin heads. These effects, as well as nonuniformity of structure, strongly suggest that contraction is asynchronous throughout the muscle. Recognition of muscle's distributed properties by abandoning the assumption of perfect synchrony in contraction and consideration of myofilament mass allow for small movements of thick with respect to thin filaments. Such movements lead to bond detachment and heat production. Gross movement (e.g., muscle shortening) exacerbates this process. Quick transients in muscle length or applied load have particularly strong effects and have been observed experimentally. Muscle relaxation is thereby viewed as a consequence of muscle's distributed properties.

The new distributed muscle model is built from the following main features: sarcomeres consist of overlapping thick and thin filaments connected by cross-bridge

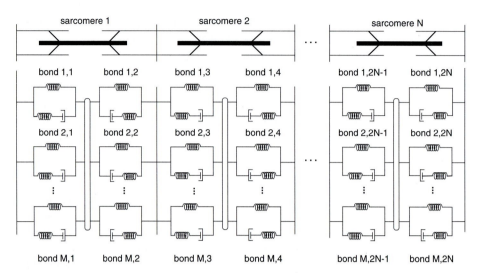

Figure 4.25   Schematic diagram of a muscle fiber built from a distributed network of N sarcomeres. Each sarcomere has M parallel pairs of cross-bridge bonds (adapted from Palladino and Noordergraaf, 1998, Fig. 3.4, p. 44).

bonds which form during activation and detach during relaxation. Figure 4.25 shows a schematic of a muscle fiber (cell) comprised of a string of series sarcomeres. Cross-bridge bonds are each described as three-element viscoelastic solids, and myofilaments as masses. Force is generated due to viscoelastic cross-bridge bonds that form and are stretched between the interdigitating matrix of myofilaments. The number of bonds formed depends on the degree of overlap between thick and thin filaments and is dictated spatially and temporally due to finite electrical and chemical activation rates. Asynchrony in bond formation and unequal numbers of bonds formed in each half sarcomere, as well as mechanical disturbances such as muscle shortening and imposed length transients, cause small movements of the myofilaments. Since myofilament masses are taken into account, these movements take the form of damped vibrations with a spectrum of frequencies due to the distributed system properties. When the stress in a bond goes to zero, the bond detaches. Consequently, myofilament motion and bond stress relaxation lead to bond detachment and produce relaxation without assumption of bond detachment rate functions. In essence, relaxation results from inherent system instability. Although the model is built from linear, time-invariant components (springs, dashpots, and masses), the highly dynamic structure of the model causes its mechanical properties to be highly nonlinear and time-varying, as is found in muscle fibers and strips.

Sensitivity of the model to mechanical disturbances is consistent with experimental evidence from muscle force traces, aequorin measurements of free calcium ion, and high speed X-ray diffraction studies which all suggest enhanced bond detachment. The model is also consistent with sarcomere length feedback studies in which reduced internal motion delays relaxation, and it predicted muscle fiber (cell) dynamics prior to their experimental measurement.

This model proposes a structural mechanism for the origin of muscle's complex mechanical properties and predicts new features of the contractile mechanism (e.g., a mechanism for muscle relaxation and prediction of muscle heat generation). This new approach computes muscle's complex mechanical properties from physical description of muscle anatomical structure, thereby linking subcellular structure to organ-level function.

This chapter describes some of the high points of biological tissues' mechanical properties. More comprehensive references include Fung's *Biomechanics: Mechanical Properties of Living Tissues*, Nigg and Herzog's *Biomechanics of the Musculo-Skeletal System*, and Mow and Hayes' *Basic Orthopaedic Biomechanics*. Muscle contraction research has a long history, as chronicled in the book *Machina Carnis* by Needham. For a more comprehensive history of medicine see Singer and Underwood's (1962) book. The next two sections apply biomechanics concepts introduced in Sections 4.2–4.5 to human gait analysis and to the quantitative study of the cardiovascular system.

4.6   CLINICAL GAIT ANALYSIS

An example of applied dynamics in human movement analysis is clinical gait analysis. Clinical gait analysis involves the measurement of the parameters that characterize a patient's gait pattern, the interpretation of the collected and processed data, and the recommendation of treatment alternatives. It is a highly collaborative process that requires the cooperation of the patient and the expertise of a multidisciplinary team that typically includes a physician, a physical therapist or kinesiologist, and an engineer or technician. The engineer is presented with a number of challenges. The fundamental objective in data collection is to monitor the patient's movements accurately and with sufficient precision for clinical use without altering the patient's typical performance. While measurement devices for clinical gait analysis are established to some degree (i.e., commercially available) the protocols for the use of the equipment continue to develop. The validity of these protocols and associated models and the care with which they are applied ultimately dictate the meaning and quality of the resulting data provided for interpretation. This is one area in which engineers in collaboration with their clinical partners can have a significant impact on the clinical gait analysis process.

Generally, data collection for clinical gait analysis involves the placement of highly reflective markers on the surface of the patient's skin. These external markers then reflect light to an array of video-based motion cameras that surround the measurement volume. The instantaneous location of each of these markers can then be determined stereometrically based on the images obtained simultaneously from two or more cameras. Other aspects of gait can be monitored as well, including ground reactions via force platforms embedded in the walkway and muscle activity via electromyography with either surface or intramuscular fine wire electrodes, depending on the location of the particular muscle.

In keeping with the other material presented in this chapter, the focus of this section will pertain to the biomechanical aspects of clinical gait analysis and includes an outline of the computation of segmental and joint kinematics and joint kinetics and a brief illustration of how the data are interpreted.

4.6.1   The Clinical Gait Model

The gait model is the algorithm that transforms the data collected during walking trials into the information required for clinical interpretation. For example, the gait model uses the data associated with the three-dimensional displacement of the markers on the patient to compute the angles that describe how the patient's body segment and lower extremity joints are moving. The design of the gait model is predicated on a clear understanding of the needs of the clinical interpretation team (e.g., the specific aspects of gait dynamics of interest). To meet these clinical specifications, gait model development is constrained both by the technical limitations of the measurement system and by the broad goal of developing protocols that may be appropriate for a wide range of patient populations that vary in age, gait abnormality, walking ability, etc. An acceptable model must be sufficiently general to be used for many different types of patients (e.g., adults and children with varying physical and cognitive involvement), be sufficiently sophisticated to allow detailed biomechanical questions to be addressed, and be based on repeatable protocols that are feasible in a clinical setting.

4.6.2   Kinematic Data Analysis

Reflective markers placed on the surface of the patient's skin are monitored or tracked in space and time by a system of video-based cameras. These marker trajectories are used to compute coordinate systems that are anatomically aligned and embedded in each body segment under analysis. These anatomical coordinate systems provide the basis for computing the absolute spatial orientation (or attitude) of the body segment or the angular displacement of one segment relative to another (e.g., joint angles). For this analysis, at least three non-colinear markers or points of reference must be placed on or identified for each body segment included in the analysis. These markers form a plane from which a segmentally fixed coordinate system may be derived. Any three markers will allow the segment motion to be monitored, but unless these markers are referenced to the subject's anatomy, such kinematic quantification is of limited clinical value. Markers must either be placed directly over palpable bony landmarks on the segment or at convenient (i.e., visible to the measurement cameras) locations on the segment that are referenced to the underlying bone(s). An examination of the pelvic and thigh segments illustrates these two alternatives.

Pelvic Anatomical Coordinate System

For the pelvis, markers placed over the right and left anterior–superior–iliac–spine (ASIS) and either the right or left posterior–superior–iliac–spine (PSIS) will allow for the computation of an anatomically aligned coordinate system, as described in the following example.

Example Problem 4.9

Given the following three-dimensional locations in meters for a set of pelvic markers expressed relative to an inertially fixed laboratory coordinate system (Figure 4.26),

$$\text{Right ASIS : } \mathbf{RASIS} = -0.850\mathbf{i} - 0.802\mathbf{j} + 0.652\mathbf{k}$$
$$\text{Left ASIS : } \mathbf{LASIS} = -0.831\mathbf{i} - 0.651\mathbf{j} + 0.652\mathbf{k}$$
$$\mathbf{PSIS} = -1.015\mathbf{i} - 0.704\mathbf{j} + 0.686\mathbf{k}$$

compute an anatomical coordinate system for the pelvis.

Solution

These three anatomical markers form a plane. The line between the right ASIS and left ASIS represents one coordinate system axis. Another coordinate axis is perpendicular to the pelvic plane. The third coordinate axis is computed to be orthogonal to the first two:

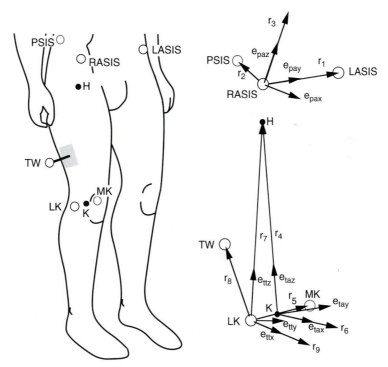

Figure 4.26 Kinematic markers used to define pelvis and thigh coordinate systems. For the pelvis, PSIS denotes posterior–superior–iliac–spine, H is hip center, and RASIS and LASIS denote right and left anterior–superior–iliac–spine markers, respectively. For the thigh, TW is thigh wand, K is knee center, and MK and LK are medial and lateral knee (femoral condyle) markers, respectively.

1. Subtract vector **RASIS** from vector **LASIS**,

$$\textbf{LASIS} - \textbf{RASIS} = (-0.831 - (-0.850))\textbf{i} + (-0.651 - (-0.802))\textbf{j} \\ + (0.652 - 0.652)\textbf{k}$$

to find

$$\textbf{r}_1 = 0.0190\textbf{i} + 0.1510\textbf{j} + 0.0000\textbf{k}$$

and its associated unit vector:

$$e_{r1} = \frac{0.019\textbf{i} + 0.151\textbf{j} + 0.000\textbf{k}}{\sqrt{0.019^2 + 0.151^2 + 0.000^2}}$$
$$e_{r1} = 0.125\textbf{i} + 0.992\textbf{j} + 0.000\textbf{k}$$

Unit vector e_{r1} represents the medial–lateral direction or y axis for the pelvic anatomical coordinate system e_{pay} (Fig. 4.26).

2. A second vector in the pelvic plane is required to compute the coordinate axis that is perpendicular to the plane. Consequently, subtract vector **RASIS** from vector **PSIS** to find

$$\textbf{r}_2 = -0.165\textbf{i} + 0.098\textbf{j} + 0.034\textbf{k}$$

3. Take the vector cross product $e_{pay} \times \textbf{r}_2$ to yield

$$\textbf{r}_3 = \begin{vmatrix} \textbf{i} & \textbf{j} & \textbf{k} \\ 0.125 & 0.992 & 0.000 \\ -0.165 & 0.098 & 0.034 \end{vmatrix}$$
$$= [(0.992)(0.034) - (0.000)(0.098)]\textbf{i}$$
$$+ [(0.000)(-0.165) - (0.125)(0.034)]\textbf{j}$$
$$+ [(0.125)(0.098) - (0.992)(-0.165)]\textbf{k}$$
$$= 0.034\textbf{i} - 0.004\textbf{j} + 0.176\textbf{k}$$

and its associated unit vector:

$$e_{r3} = e_{paz} = 0.188\textbf{i} - 0.024\textbf{j} + 0.982\textbf{k}$$

Unit vector e_{r3} represents the anterior–superior direction or z axis of the pelvic anatomical coordinate system e_{paz} (Fig. 4.26).

4. The third coordinate axis is computed to be orthogonal to the first two. Take the vector cross product $e_{pay} \times e_{paz}$ to compute the fore–aft direction, or x axis, of the pelvic anatomical coordinate system:

$$e_{pax} = 0.974\textbf{i} - 0.123\textbf{j} - 0.190\textbf{k}$$

For this example, the anatomical coordinate system for the pelvis can be expressed as follows:

$$\{e_{pa}\} = \begin{bmatrix} e_{pax} \\ e_{pay} \\ e_{paz} \end{bmatrix} = \begin{bmatrix} 0.974 & -0.123 & -0.190 \\ 0.125 & 0.992 & 0.000 \\ 0.188 & -0.024 & 0.982 \end{bmatrix} \begin{bmatrix} i \\ j \\ k \end{bmatrix}$$

Note that the coefficients associated with these three axes represent the direction cosines that define the orientation of the pelvic coordinate system relative to the laboratory coordinate system. ■

In summary, by monitoring the motion of the three pelvic markers, the instantaneous orientation of an anatomical coordinate system for the pelvis, $\{e_{pa}\}$, comprised of axes e_{pax}, e_{pay}, and e_{paz}, can be determined. The absolute angular displacement of this coordinate system can then be computed via Euler angles as pelvic tilt, obliquity, and rotation using Eqs. 4.32–4.34. An example of these angle computations is presented later in this section.

Thigh Anatomical Coordinate System

The thigh presents a more significant challenge than the pelvis since three bony anatomical landmarks are not readily available as reference points during gait. A model based on markers placed over the medial and lateral femoral condyles and the greater trochanter is appealing but plagued with difficulties. A marker placed over the medial femoral condyle is not always feasible during gait (e.g., with patients whose knees make contact while walking). A marker placed over the greater trochanter is often described in the literature but should not be used as a reference because of its significant movement relative to the underlying greater trochanter during gait (skin motion artifact).

In general, the approach used to quantify thigh motion (and the shank and foot) is to place additional anatomical markers on the segment(s) during a static subject calibration process so that the relationship between these static anatomical markers (that are removed before gait data collection) and the motion markers (that remain on the patient during gait data collection) may be calculated. It is assumed that this mathematical relationship remains constant during gait (i.e., the instrumented body segments are assumed to be rigid). This process is illustrated in the following example.

Example Problem 4.10

Given the following marker coordinate data that have been acquired while the patient stands quietly (also in meters),

lateral femoral condyle marker $\mathbf{LK} = -0.881i - 0.858j + 0.325k$

medial femoral condyle marker $\mathbf{MK} = -0.855i - 0.767j + 0.318k$

compute an anatomical coordinate system for the thigh.

Solution

A thigh plane is formed based on three anatomical markers or points: the hip center, the lateral femoral condyle marker $\mathbf{LK}$, and the medial femoral condyle marker $\mathbf{MK}$.

The knee center location can then be estimated as the midpoint between **LK** and **MK**. With these points, the vector from the knee center to the hip center represents the longitudinal axis of the coordinate system. A second coordinate axis is perpendicular to the thigh plane. The third coordinate axis is computed to be orthogonal to the first two.

The location of the knee center of rotation may be approximated as the midpoint between the medial and lateral femoral condyle markers,

$$\frac{\mathbf{LK} + \mathbf{MK}}{2} = \frac{(-0.881) + (-0.855)}{2}\mathbf{i} + \frac{(-0.858) + (-0.767)}{2}\mathbf{j} + \frac{(0.325) + (0.318)}{2}\mathbf{k}$$

yielding

$$\text{knee center location } \mathbf{K} = -0.868\mathbf{i} - 0.812\mathbf{j} + 0.321\mathbf{k}$$

The location of the center of the head of the femur, referred to as the hip center, is commonly used in this calculation by approximating its location based on patient anthropometry and a statistical model of pelvic geometry that is beyond the scope of this chapter. In this case, it can be located at approximately (Davis et al., 1991)

$$\text{hip center location } \mathbf{H} = -0.906\mathbf{i} - 0.763\mathbf{j} + 0.593\mathbf{k}$$

Now the anatomical coordinate system for the thigh may be computed as follows.

1. Subtract the vector **K** from **H**, giving

$$\mathbf{r}_4 = -0.038\mathbf{i} + 0.049\mathbf{j} + 0.272\mathbf{k}$$

and its associated unit vector

$$\mathbf{e}_{r4} = \mathbf{e}_{taz} = -0.137\mathbf{i} + 0.175\mathbf{j} + 0.975\mathbf{k}$$

Unit vector $\mathbf{e}_{r4}$ represents the longitudinal direction, or z axis, of the thigh anatomical coordinate system $\mathbf{e}_{taz}$.

2. As with the pelvis, a second vector in the thigh plane is required to compute the coordinate axis that is perpendicular to the plane. Consequently, subtract vector **LK** from **MK**:

$$\mathbf{r}_5 = 0.026\mathbf{i} + 0.091\mathbf{j} - 0.007\mathbf{k}$$

3. Form the vector cross product $\mathbf{r}_5 \times \mathbf{e}_{taz}$ to yield

$$\mathbf{r}_6 = 0.090\mathbf{i} - 0.024\mathbf{j} + 0.017\mathbf{k}$$

and its associated unit vector

$$\mathbf{e}_{r6} = \mathbf{e}_{tax} = 0.949\mathbf{i} - 0.258\mathbf{j} + 0.180\mathbf{k}$$

Unit vector $\mathbf{e}_{r6}$ represents the fore–aft direction, or x axis, of the thigh anatomical coordinate system $\mathbf{e}_{tax}$.

4. Again, the third coordinate axis is computed to be orthogonal to the first two. Determine the medial–lateral or y axis of the thigh anatomical coordinate system, e_{tay}, from the cross product $e_{taz} \times e_{tax}$:

$$e_{tay} = 0.284i + 0.950j - 0.131k$$

For this example, the anatomical coordinate system for the thigh can be expressed as

$$\{e_{ta}\} = \begin{bmatrix} e_{tax} \\ e_{tay} \\ e_{taz} \end{bmatrix} = \begin{bmatrix} 0.949 & -0.258 & 0.180 \\ 0.284 & 0.950 & -0.131 \\ -0.137 & 0.175 & 0.975 \end{bmatrix} \begin{bmatrix} i \\ j \\ k \end{bmatrix}$$

This defines an anatomical coordinate system fixed to the thigh, $\{e_{ta}\}$, comprised of axes e_{tax}, e_{tay}, and e_{taz}. Its basis, however, includes an external marker (medial femoral condyle **MK**) that must be removed before the walking trials. Consequently, the location of the knee center cannot be computed as described in the preceeding example. This dilemma is resolved by placing another marker on the surface of the thigh such that it also forms a plane with the hip center and lateral knee marker. These three reference points can then be used to compute a "technical" coordinate system for the thigh to which the knee center location may be mathematically referenced. ∎

Example Problem 4.11

Continuing Example Problem 4.10, and given the coordinates of another marker placed on the thigh but not anatomically aligned,

$$\text{thigh wand marker } \mathbf{TW} = -0.890i - 0.937j + 0.478k$$

compute a technical coordinate system for the thigh.

Solution

A technical coordinate system for the thigh can be computed as follows.

1. Compute the longitudinal direction, or z axis, of the technical thigh coordinate system e_{tt}. Start by subtracting vector **LK** from the hip center **H** to form

$$r_7 = -0.025i + 0.094j + 0.268k$$

and its associated unit vector

$$e_{r7} = e_{ttz} = -0.088i + 0.330j + 0.940k$$

Unit vector e_{r7} represents the z axis of the thigh technical coordinate system, e_{ttz}.

2. To compute the axis that is perpendicular to the plane formed by **LK**, **H** and **TW**, subtract vector **LK** from **TW** to compute

$$r_8 = -0.009i - 0.079j + 0.153k$$

3. Calculate the vector cross product $r_7 \times r_8$ to yield

$$r_9 = 0.036i + 0.001j + 0.003k$$

with its associated unit vector

$$e_{r9} = e_{ttx} = 0.996i + 0.040j + 0.079k$$

Unit vector e_{r9} represents the fore–aft direction, or x axis, of the thigh technical coordinate system e_{ttx}.

4. The third coordinate axis is computed to be orthogonal to the first two axes. Compute the vector cross product $e_{ttz} \times e_{ttx}$ to determine the media–lateral direction, or y axis, of the thigh technical coordinate system:

$$e_{tty} = -0.012i + 0.943j + 0.333k$$

For this example, the technical coordinate system for the thigh can be expressed as

$$\{e_{tt}\} = \begin{bmatrix} e_{ttx} \\ e_{tty} \\ e_{ttz} \end{bmatrix} = \begin{bmatrix} 0.996 & 0.040 & 0.079 \\ -0.012 & 0.943 & 0.333 \\ -0.088 & 0.330 & 0.940 \end{bmatrix} \begin{bmatrix} i \\ j \\ k \end{bmatrix}$$

Note that this thigh technical coordinate system $\{e_{tt}\}$ computed during the standing subject calibration can also be computed after each walking trial. That is, its computation is based on markers (the lateral femoral condyle and thigh wand markers) and an anatomical landmark (the hip center) that are available for both the standing and walking trials. Consequently, the technical coordinate system $\{e_{tt}\}$ becomes the embedded reference coordinate system to which other entities can be related. The thigh anatomical coordinate system $\{e_{ta}\}$ can be related to the thigh technical coordinate system $\{e_{tt}\}$ by using either direction cosines or Euler angles as described in Section 4.2.2. Also, the location of markers that must be removed after the standing subject calibration (e.g., the medial femoral condyle marker MK), or computed anatomical locations (e.g., the knee center), can be transformed into the technical coordinate system $\{e_{tt}\}$ and later retrieved for use in walking trial data reduction. ■

Segment and Joint Angles

Tracking the anatomical coordinate system for each segment allows for the determination of either the absolute angular orientation (or attitude) of each segment in space or the angular position of one segment relative to another. In the preceding example, the three pelvic angles that define the position of the pelvic anatomical coordinate system $\{e_{pa}\}$ relative to the laboratory (inertially fixed) coordinate system can be computed from the Euler angles as described in Section 4.2.2 with Eqs. 4.32–4.34. Note that in these equations the laboratory coordinate system represents the proximal (unprimed) coordinate system and the pelvic anatomical coordinate system $\{e_{pa}\}$ represents the distal (triple primed) coordinate system. Consequently, Eq. 4.32

$$\theta_x = -\arcsin(k''' \cdot j)$$

becomes

$$\begin{aligned}
\theta_x &= -\arcsin(\mathbf{e_{paz}} \cdot \mathbf{j}) \\
&= -\arcsin((0.188\mathbf{i} - 0.024\mathbf{j} + 0.982\mathbf{k}) \cdot \mathbf{j}) \\
&= -\arcsin(-0.024) \\
&= 1° \text{ of pelvic obliquity}
\end{aligned}$$

Similarly, Eq. 4.33

$$\theta_y = \arcsin\left(\frac{(\mathbf{k'''} \cdot \mathbf{i})}{\cos\theta_x}\right)$$

becomes

$$\begin{aligned}
\theta_y &= \arcsin\left(\frac{(\mathbf{e_{paz}} \cdot \mathbf{i})}{\cos\theta_x}\right) \\
&= \arcsin\left(\frac{(0.188\mathbf{i} - 0.024\mathbf{j} + 0.982\mathbf{k}) \cdot \mathbf{i}}{\cos 1°}\right) \\
&= \arcsin\left(\frac{0.188}{\cos 1°}\right) \\
&= 11° \text{ of anterior pelvic tilt}
\end{aligned}$$

and Eq. 4.34

$$\theta_z = \arcsin\left(\frac{(\mathbf{i'''} \cdot \mathbf{j})}{\cos\theta_x}\right)$$

becomes

$$\begin{aligned}
\theta_z &= \arcsin\left(\frac{(\mathbf{e_{pax}} \cdot \mathbf{j})}{\cos\theta_x}\right) \\
&= \arcsin\left(\frac{(0.974\mathbf{i} - 0.123\mathbf{j} - 0.190\mathbf{k}) \cdot \mathbf{j}}{\cos 1°}\right) \\
&= \arcsin\left(\frac{-0.123}{\cos 1°}\right) \\
&= -7° \text{ of pelvic rotation}
\end{aligned}$$

This Euler angle computation may be repeated to solve for the three hip angles that define the position of the thigh anatomical coordinate system $\{\mathbf{e_{ta}}\}$ relative to the pelvic anatomical coordinate system $\{\mathbf{e_{pa}}\}$. For the hip angles, the proximal (un-primed) coordinate system is the pelvis and the distal (triple-primed) coordinate system is the thigh. Substituting the values of $\{\mathbf{e_{pa}}\}$ and $\{\mathbf{e_{ta}}\}$ from Example Problems 4.9 and 4.10 into Eq. 4.32 yields:

$$\begin{aligned}
\theta_x &= -\arcsin(\mathbf{e_{taz}} \cdot \mathbf{e_{pay}}) \\
&= -\arcsin((-0.137\mathbf{i} + 0.175\mathbf{j} + 0.975\mathbf{k}) \cdot (0.125\mathbf{i} + 0.992\mathbf{j} + 0.000\mathbf{k})) \\
&= -\arcsin(0.156) \\
&= -9° \text{ of hip abduction–adduction}
\end{aligned}$$

The negative sign is associated with hip adduction of the left thigh or hip abduction of the right thigh.

Further substitution of values of $\{e_{pa}\}$ and $\{e_{ta}\}$ into Eqs. 4.33 and 4.34 yields

$$\text{hip flexion-extension } \theta_y = 20°$$
$$\text{hip internal-external rotation } \theta_z = -8°$$

For hip internal–external rotation, the negative sign is associated with hip internal rotation of the left thigh or hip external rotation of the right thigh. A negative hip flexion–extension angle corresponds to hip extension, independent of side. This process may be repeated for other body segments such as the shank (or lower leg), foot, trunk, arms, and head with the availability of properly defined anatomical coordinate systems.

4.6.3   Kinetic Data Analysis

The marker displacement or motion data provide an opportunity to appreciate segment and joint kinematics. Kinematic data can be combined with ground reaction data (i.e., forces and torque) and their points of application, referred to as the centers of pressure. Combined with estimates of segment mass and mass moments of inertia, the net joint reactions (i.e., joint forces and moments) may then be computed.

To illustrate the details of this computational process, consider the following determination of the reactions at the ankle (Fig. 4.27) for an individual with mass of 25.2 kg. Data for one instant in the gait cycle are shown in the following table.

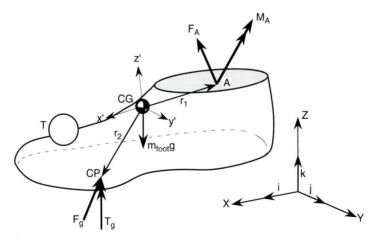

Figure 4.27   Ankle A and toe T marker data are combined with ground reaction force data F_g and segment mass and mass moment of inertia estimates to compute the net joint forces and moments.

	Symbol	Units	x_{lab}	y_{lab}	z_{lab}
ankle center location	A	[m]	0.357	0.823	0.056
toe marker location	T	[m]	0.421	0.819	0.051
center of pressure location	CP	[m]	0.422	0.816	0.000
ground reaction force vector	F_g	[N]	3.94	−15.21	242.36
ground reaction torque vector	T_g	[N-m]	0.000	0.000	0.995
foot anatomical coordinate system	e_{fax}		0.977	−0.0624	−0.202
	e_{fay}		0.0815	0.993	0.0877
	e_{faz}		0.195	−0.102	0.975
foot linear acceleration vector	a_{foot}	[m/s^2]	2.09	−0.357	−0.266
foot angular velocity vector	ω_{foot}	[rad/s]	0.0420	2.22	−0.585
foot angular acceleration vector	α_{foot}	[rad/sec^2]	−0.937	8.85	−5.16
ankle angular velocity vector	ω_{ankle}	[rad/s]	−0.000759	1.47	0.0106

Anthropomorphic relationships presented in Table 4.1 are used to estimate the mass and mass moments of inertia of the foot as well as the location of its center of gravity. The mass of the foot, m_{foot}, may be estimated to be 1.45% of the body mass, or 0.365 kg, and the location of the center of gravity is approximated as 50% of the foot length. The length of the foot ℓ_{foot} may be approximated as the distance between the ankle center and the toe marker, determined as follows:

$$\mathbf{T} - \mathbf{A} = (0.421 - 0.357)\mathbf{i} + (0.819 - 0.823)\mathbf{j} + (0.051 - 0.056)\mathbf{k}$$
$$= 0.064\mathbf{i} - 0.004\mathbf{j} - 0.005\mathbf{k}$$
$$\ell_{foot} = |\mathbf{T} - \mathbf{A}|$$
$$= \sqrt{(0.064)^2 + (-0.004)^2 + (-0.005)^2} = 0.064\,\text{m}$$

Then the location of the center of gravity can be determined relative to the ankle center as

$$\mathbf{A} + \frac{\ell_{foot}}{2}\frac{(\mathbf{T} - \mathbf{A})}{|\mathbf{T} - \mathbf{A}|} = (0.357\mathbf{i} + 0.823\mathbf{j} + 0.056\mathbf{k}) + \left[\frac{0.064}{2}\right]\left[\frac{0.064\mathbf{i} - 0.004\mathbf{j} - 0.005\mathbf{k}}{0.064}\right]$$

giving the location of the center of gravity:

$$\mathbf{CG} = 0.389\mathbf{i} + 0.821\mathbf{j} + 0.054\mathbf{k}$$

which allows computation of position vectors r_1 and r_2 (Fig. 4.27). With a foot length of 0.064 m, a foot mass of 0.365 kg, and a proximal radius of gyration per segment length of 0.690, the mass moment of inertia relative to the ankle center may be estimated with Eq. 4.40 as

$$I_{foot/ankle} = (0.365\,\text{kg})[(0.690)(0.064\,\text{m})]^2$$
$$= 7.12 \times 10^{-4}\,\text{kg m}^2$$

The centroidal mass moment of inertia, located at the foot's center of mass, may then be estimated using the parallel axis theorem (Eq. 4.41):

$$I_{\text{foot/cm}} = I_{\text{foot/ankle}} - m_{\text{foot}}d^2$$

Note that the center of mass is equivalent to the center of gravity in a uniform gravitational field. In this case, d is the distance between the foot's center of mass and the ankle. Table 4.1 shows the ratio of the foot center of mass location relative to its proximal end to be 0.5, so $d = 0.5(\ell_{\text{foot}}) = 0.032$ m. Therefore,

$$I_{\text{foot/cm}} = (7.12 \times 10^{-4} \text{ kg m}^2) - (0.365 \text{ kg})(0.032 \text{ m})^2$$
$$= 3.38 \times 10^{-4} \text{ kg m}^2$$

$I_{\text{foot/cm}}$ represents the centroidal mass moment of inertia about the transverse principal axes of the foot (y' and z' in Fig. 4.27). Consequently,

$$I_{y'y'} = 3.38 \times 10^{-4} \text{ kg m}^2$$
$$I_{z'z'} = 3.38 \times 10^{-4} \text{ kg m}^2$$

The foot is approximated as a cylinder with a length to radius ratio of 6. The ratio of transverse to longitudinal (x') mass moments of inertia can be shown to be approximately 6.5. Then the longitudinal mass moment of inertia (about x' in Fig. 4.27) may be estimated as

$$I_{x'x'} = 5.20 \times 10^{-5} \text{ kg m}^2$$

Having estimated the anthropomorphic values for the foot, the kinetic analysis may now begin. The unknown ankle reaction force, $\mathbf{F_A}$, may be found by using Newton's Second Law, or $\Sigma \mathbf{F} = m\mathbf{a}$:

$$\mathbf{F_g} + \mathbf{F_A} - m_{\text{foot}} g\mathbf{k} = m_{\text{foot}}\mathbf{a}_{\text{foot}}$$
$$\mathbf{F_A} = m_{\text{foot}}\mathbf{a}_{\text{foot}} - \mathbf{F_g} + m_{\text{foot}}g\mathbf{k}$$
$$= (0.365 \text{ kg})[2.09\mathbf{i} - 0.357\mathbf{j} - 0.266\mathbf{k}] \text{ m/s}$$
$$- (3.94\mathbf{i} - 15.21\mathbf{j} + 242.4\mathbf{k}) \text{ N}$$
$$+ (0.365 \text{ kg})(9.81 \text{ m/s}^2)\mathbf{k}$$
$$= -3.18\mathbf{i} + 15.08\mathbf{j} - 238.9\mathbf{k} \text{ N}$$

Euler's equations of motion (Eqs. 4.44–4.46) are then applied to determine the unknown ankle moment reaction $\mathbf{M_A}$. Euler's equations are defined relative to the principal axes fixed to the segment (i.e., x', y', and z' fixed to the foot). It is noted, however, that the data required for the solution presented previously (e.g., ω_{foot} and α_{foot}) are expressed relative to the laboratory coordinate system (x, y, z). Consequently, vectors required for the solution of Euler's equations must first be transformed into the foot coordinate system. In the preceding data set, the foot anatomical coordinate system was given as

$$\mathbf{e}_{\text{fax}} = 0.977\mathbf{i} - 0.0624\mathbf{j} - 0.202\mathbf{k}$$
$$\mathbf{e}_{\text{fay}} = 0.0815\mathbf{i} + 0.993\mathbf{j} + 0.0877\mathbf{k}$$
$$\mathbf{e}_{\text{faz}} = 0.195\mathbf{i} - 0.102\mathbf{j} + 0.975\mathbf{k}$$

where e_{fax}, e_{fay}, and e_{faz} correspond to x', y', and z', or i', j', and k'. Recall from the discussion in Section 4.2.2 that coefficients in the expression for e_{fax} represent the cosines of the angles between x' and x, x' and y, and x' and z, respectively. Similarly, the coefficients in the expression for e_{fay} represent the cosines of the angles between y' and x, y' and y, and y' and z, and the coefficients in the expression for e_{faz} represent the cosines of the angles between z' and x, z' and y, and z' and z. Consequently, these relationships can be transposed as

$$\mathbf{i} = 0.977\mathbf{i}' + 0.0815\mathbf{j}' + 0.195\mathbf{k}'$$
$$\mathbf{j} = -0.0624\mathbf{i}' + 0.993\mathbf{j}' - 0.102\mathbf{k}'$$
$$\mathbf{k} = -0.202\mathbf{i}' + 0.0877\mathbf{j}' + 0.975\mathbf{k}'$$

In this form, these relationships can be used to transform vectors expressed in terms of lab coordinates:

$$\mathbf{A} = A_x\mathbf{i} + A_y\mathbf{j} + A_z\mathbf{k}$$

into foot coordinates:

$$\mathbf{A} = A_x\mathbf{i}' + A_y\mathbf{j}' + A_z\mathbf{k}'$$

To demonstrate this process, consider the foot angular velocity vector

$$\boldsymbol{\omega}_{foot} = 0.042\mathbf{i} + 2.22\mathbf{j} - 0.585\mathbf{k} \text{ rad/s}$$

Substituting the relationships for the lab coordinate system in terms of the foot coordinate system $\boldsymbol{\omega}_{foot}$ becomes,

$$\begin{aligned}
\boldsymbol{\omega}_{foot} &= 0.042(0.977\mathbf{i}' + 0.0815\mathbf{j}' + 0.195\mathbf{k}') \\
&\quad + 2.22(-0.0624\mathbf{i}' + 0.993\mathbf{j}' - 0.102\mathbf{k}') \\
&\quad - 0.585(-0.202\mathbf{i}' + 0.0877\mathbf{j}' + 0.975\mathbf{k}') \\
&= 0.0210\mathbf{i}' + 2.16\mathbf{j}' - 0.789\mathbf{k}' \text{ rad/s}
\end{aligned}$$

In a similar manner, the other vectors required for the computation are transformed into the foot coordinate system:

$$\begin{aligned}
\mathbf{r}_1 &= -0.032\mathbf{i} + 0.002\mathbf{j} + 0.002\mathbf{k} \\
&= -0.032\mathbf{i}' - 0.004\mathbf{k}' \text{ m} \\
\mathbf{r}_2 &= 0.033\mathbf{i} - 0.005\mathbf{j} - 0.054\mathbf{k} \\
&= 0.0435\mathbf{i}' - 0.007\mathbf{j}' - 0.0457\mathbf{k}' \text{m} \\
\mathbf{F}_g &= 3.94\mathbf{i} - 15.21\mathbf{j} + 242.36\mathbf{k} \\
&= -44.16\mathbf{i}' + 6.47\mathbf{j}' + 238.62\mathbf{k}' \text{ N} \\
\mathbf{T}_g &= 0.995\mathbf{k} \\
&= -0.201\mathbf{i}' + 0.0873\mathbf{j}' + 0.970\mathbf{k}' \text{ N m} \\
\mathbf{F}_A &= -3.18\mathbf{i} + 15.1\mathbf{j} - 239\mathbf{k} \\
&= 44.2\mathbf{i}' - 6.23\mathbf{j}' - 235\mathbf{k}'\text{N}
\end{aligned}$$

$$\boldsymbol{\omega}_{\text{foot}} = 0.0420\mathbf{i} + 2.22\mathbf{j} - 0.585\mathbf{k}$$
$$= 0.021\mathbf{i}' + 2.16\mathbf{j}' - 0.789\mathbf{k}' \text{ rad/s}$$
$$\boldsymbol{\alpha}_{\text{foot}} = -0.937\mathbf{i} + 8.85\mathbf{j} - 5.16\mathbf{k}$$
$$= -0.425\mathbf{i}' + 8.26\mathbf{j}' - 6.116\mathbf{k}' \text{ rad/s}^2$$

Expanding Euler's equations of motion (Eqs. 4.44–4.46),

$$M_{Ax'} + (\mathbf{r}_1 \times \mathbf{F}_A)_{x'} + (\mathbf{r}_2 \times \mathbf{F}_g)_{x'} + T_{gx'} = I_{x'x'}\alpha_{x'} + (I_{z'z'} - I_{y'y'})\omega_{y'}\omega_{z'}$$
$$M_{Ay'} + (\mathbf{r}_1 \times \mathbf{F}_A)_{y'} + (\mathbf{r}_2 \times \mathbf{F}_g)_{y'} + T_{gy'} = I_{y'y'}\alpha_{y'} + (I_{x'x'} - I_{z'z'})\omega_{z'}\omega_{x'}$$
$$M_{Az'} + (\mathbf{r}_1 \times \mathbf{F}_A)_{z'} + (\mathbf{r}_2 \times \mathbf{F}_g)_{z'} + T_{gz'} = I_{z'z'}\alpha_{z'} + (I_{y'y'} - I_{x'x'})\omega_{x'}\omega_{y'}$$

where $(\mathbf{r}_1 \times \mathbf{F}_A)_{x'}$ represents the x' component of $\mathbf{r}_1 \times \mathbf{F}_A$, $(\mathbf{r}_2 \times \mathbf{F}_g)_{x'}$ represents the x' component of $\mathbf{r}_2 \times \mathbf{F}_g$, and so forth.

Substitution of the required values and arithmetic reduction yields

$$\mathbf{M}_{A'} = 1.50\mathbf{i}' + 15.9\mathbf{j}' - 1.16\mathbf{k}' \text{ Nm}$$

which can be transformed back into fixed lab coordinates,

$$\mathbf{M}_A = 2.54\mathbf{i} + 15.9\mathbf{j} - 0.037\mathbf{k} \text{ Nm}$$

By combining the ankle moment with the ankle angular velocity, the instantaneous ankle power may be computed as

$$\mathbf{M}_A \cdot \boldsymbol{\omega}_{\text{ankle}} = (2.54\mathbf{i} + 15.9\mathbf{j} - 0.037\mathbf{k} \text{ Nm}) \cdot (-0.000759\mathbf{i} + 1.47\mathbf{j} + 0.0106\mathbf{k} \text{ rad/s})$$
$$= 23.3 \text{ Watts}$$

or

$$\mathbf{M}_{A'} \cdot \boldsymbol{\omega}_{\text{ankle}'} = (1.50\mathbf{i}' + 15.9\mathbf{j}' - 1.16\mathbf{k}' \text{ Nm}) \cdot (-0.0946\mathbf{i}' + 1.46\mathbf{j}' - 0.140\mathbf{k}' \text{ rad/s})$$
$$= 23.3 \text{ Watts}$$

which is thought to represent a quantitative measure of the ankle's contribution to propulsion.

4.6.4 Clinical Gait Interpretation

The information and data provided for treatment decision making in clinical gait analysis include not only the quantitative variables described previously (i.e., 3-D kinematics such as angular displacement of the torso, pelvis, hip, knee, and ankle/foot, and 3-D kinetics such as moments and power of the hip, knee, and ankle) but also

- Clinical examination measures
- Biplanar video recordings of the patient walking
- Stride and temporal gait data such as step length and walking speed
- Electromyographic (EMG) recordings of selected lower extremity muscles

Generally, the interpretation of gait data involves the identification of abnormalities, the determination of the causes of the apparent deviations, and the recommendation of treatment alternatives. As each additional piece of data is incorporated, a coherent picture of the patient's walking ability is developed by correlating corroborating data sets and resolving apparent contradictions in the information. Experience allows the team to distinguish a gait anomaly that presents the difficulty for the patient from a gait compensatory mechanism that aids the patient in circumventing the gait impediment to some degree.

To illustrate aspects of this process, consider the data presented in Figures 4.28–4.30 which were measured from a 9-year-old girl with cerebral palsy spastic diplegia. Cerebral palsy is a nonprogressive neuromuscular disorder that is caused by an injury to the brain during or shortly after birth. The neural motor cortex is most often affected. In the ambulatory patient, this results in reduced control of the muscles required for balance and locomotion, causing overactivity, inappropriately timed activity, and muscle spasticity. Treatment options include physical therapy, bracing (orthoses), spasmolytic medications such as botulinum toxin and Baclofen, and orthopedic surgery and neurosurgery.

The sagittal plane kinematics for the left side of this patient (Fig. 4.28) indicate significant involvement of the hip and knee. Her knee is effectively "locked" in an excessively flexed position throughout stance phase (0–60% of the gait cycle) when her foot is contacting the floor. Knee motion in swing phase (60–100%) is also limited, with the magnitude and timing of peak knee flexion in swing reduced and delayed. The range of motion of her hip during gait is less than normal, failing to reach full extension at the end of stance phase. The motion of her pelvis is significantly greater than normal, tilting anteriorly in early stance coincident with extension of the hip and tilting posteriorly in swing coincident with flexion of the hip.

The deviations noted in these data illustrate neuromuscular problems commonly seen in this patient population. Inappropriate hamstring tightness, observed during the clinical examination, and inappropriate muscle activity during stance, seen in Figure 4.29, prevent the knee from properly extending. This flexed knee position also impedes normal extension of the hip in stance due to hip extensor weakness (also observed during the clinical examination). Hip extension is required in stance to allow the thigh to rotate under the advancing pelvis and upper body. To compensate for her reduced ability to extend the hip, she rotates her pelvis anteriorly in early stance to help move the thigh through its arc of motion. The biphasic pattern of the pelvic curve indicates that this is a bilateral issue to some degree.

The limited knee flexion in swing combines with the plantar flexed ankle position to result in foot clearance problems during swing phase. The inappropriate activity of the rectus femoris muscle (Fig. 4.29) in midswing suggests that spasticity of that muscle, a knee extensor (also observed during clinical examination), impedes knee flexion. Moreover, the inappropriate activity of the ankle plantar flexor, primarily the gastrocnemius muscle, in late swing suggests that it is overpowering the pretibial muscles, primarily the anterior tibialis muscle, resulting in plantar flexion of the ankle or "foot drop."

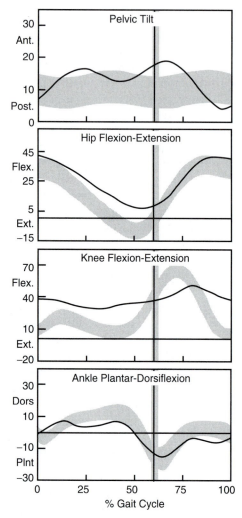

Figure 4.28   Sagittal plane kinematic data for the left side of a 9-year-old patient with cerebral palsy spastic diplegia (solid curves). Shaded bands indicate +/− one standard deviation about the performance of children with normal ambulation. Stance phase is 0–60% of the gait cycle and swing phase is 60–100%, as indicated by the vertical solid lines.

The sagittal joint kinetics for this patient (Fig. 4.30) demonstrate asymmetrical involvement of the right and left sides. Of special note is that her right knee and hip are compensating for some of the dysfunction observed on the left side. Specifically, the progressively increasing right knee flexion beginning at midstance (1st row, center) and continuing into swing aids her contralateral limb in forward advancement during swing (i.e., her pelvis can rock posteriorly along with a flexing hip to advance

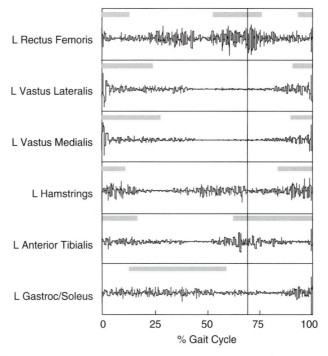

Figure 4.29 Electromyogram (EMG) data for the same cerebral palsy patient as in Fig. 4.28. Plotted are EMG activity signals for each of six left lower extremity muscles, each plotted as functions of percent of gait cycle. Gray bars represent mean normal muscle activation timing.

the thigh). One potentially adverse consequence of this adaptation is the elevated knee extensor moment in late stance that increases patella–femoral loading with indeterminate effects over time. The asymmetrical power production at the hip also illustrates clearly that the right lower extremity, in particular the muscles that cross the hip, provides the propulsion for gait with significant power generation early in stance to pull the body forward and elevate its center of gravity. Moreover, the impressive hip power generation, both with respect to magnitude and timing, at toe-off accelerates the stance limb into swing and facilitates knee flexion in spite of the elevated knee extensor moment magnitude. This is important to appreciate given the bilateral spastic response of the plantar flexor muscles, as evidenced by the premature ankle power generation and the presentation of a spastic stretch reflex in the clinical examination. This girl uses her hip musculature, right more than left, to a much greater degree than her ankle plantar flexors to propel herself forward during gait.

This cursory case examination illustrates the process whereby differences from normal gait are recognized and the associated biomechanical etiology is explored. Some of the effects on gait of neuromuscular pathology in the sagittal plane have been

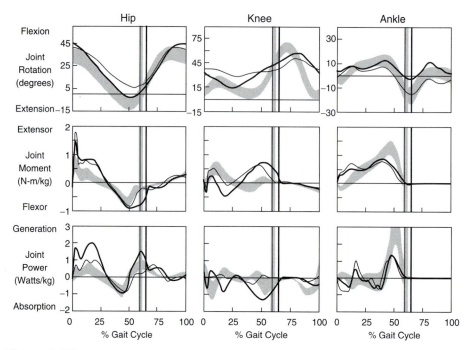

Figure 4.30   Sagittal joint kinetic data for the same cerebral palsy patient of Fig. 4.28. For the hip, knee, and ankle, joint rotation, joint moment, and joint power are plotted as functions of percent of gait cycle. Dark and light solid curves denote right and left sides, respectively. Bands indicate $+/-$ one standard deviation of the normal population.

considered in this discussion. Clinical gait analysis can also document and elucidate gait abnormalities associated with static bony rotational deformities. It also is useful in areas of clinical research by documenting treatment efficacy associated with bracing, surgery, etc. It should be noted, however, that although engineers and applied physicists have been involved in this work for well over a hundred years, there remains significant opportunity for improvement in the biomechanical protocols and analytical tools used in clinical gait analysis—there remains much to learn.

4.7 CARDIOVASCULAR DYNAMICS

One major organ system benefiting from the application of mechanics principles is cardiovascular system dynamics, or *hemodynamics*, the study of the motion of blood. From a functional point of view, the cardiovascular system is comprised of a complex pump, the heart, that generates pressure resulting in the flow of a complex fluid, blood, through a complex network of complex pipes, the blood vessels. Cardiovascular dynamics focuses on the measurement and analysis of blood pressure, volume, and flow within the cardiovascular system. The complexity of this elegant system is such that mechanical models, typically formulated as mathematical equations, are relied on to understand and integrate experimental data, to isolate and identify physiological mechanisms, and to lead ultimately to new clinical measures of heart performance and health and to guide clinical therapies.

As described in Chapter 3, the heart is a four-chambered pump connected to two main collections of blood vessels, the systemic and pulmonary circulations. This pump is electrically triggered and under neural and hormonal control. One-way valves control blood flow. Total human blood volume is approximately 5.2 liters. The left ventricle, the strongest chamber, pumps 5 liters per minute at rest, almost the body's entire blood volume. With each heartbeat, the left ventricle pumps 70 ml, with an average of 72 beats per minute. During exercise, left ventricular output may increase sixfold and heart rate more than doubles. The total length of the circulatory system vessels is estimated at 100,000 km, a distance two and one half times around the earth. The left ventricle generates approximately 1.7 watts of mechanical power at rest, increasing threefold during heavy exercise. One curious constant is the total number of heartbeats in a lifetime, around one billion in mammals (Vogel, 1992). Larger animals have slower heart rates and live longer lives, and vice versa for small animals.

4.7.1 Blood Rheology

Blood is comprised of fluid, called plasma, and suspended cells, including erythrocytes (red blood cells), leukocytes (white cells), and platelets. From a mechanical point of view, a fluid is distinguished from a solid as follows. Figure 4.31 shows a two-dimensional block of solid material (left panel) subjected to two opposite, parallel, transverse external forces, depicted by the solid arrows at the top and bottom surfaces. This applied shear force is resisted by the solid via internally generated reaction forces, depicted by the dashed arrows. When applied to a fluid (right panel), the fluid cannot resist the applied shear but rather flows.

The applied shear forces lead to shear stresses (force per area) and the measure of flow can be quantified by the resulting shear strain rate. In essence, the harder one pushes on a fluid (higher shear stress) the faster the fluid flows (higher shear strain rate). The relationship between shear stress (τ) and shear strain rate ($\dot{\gamma}$) is the fluid's viscosity (μ). Viscosity is often written as η in biomedical applications. As shown in

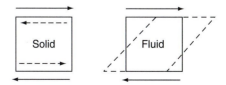

Figure 4.31   (left) A solid material resists applied external shear stress (solid vectors) via internally generated reaction shear stress (dashed vectors). (right) A fluid subjected to applied shear stress is unable to resist and instead flows (dashed lines).

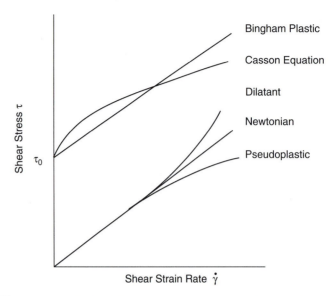

Figure 4.32   Newtonian fluids exhibit a constant viscosity, arising from the linear relation between shear stress and shear strain rate. Non-Newtonian fluids are nonlinear. Blood is often characterized with a Casson equation, but under many conditions may be described as Newtonian.

Figure 4.32, many fluids, including water, are characterized by a constant (linear) viscosity and are called Newtonian. Others possess nonlinear shear stress–strain rate relations, and are non-Newtonian fluids. For example, fluids that behave more viscously as shear strain rate increases (shear thickening) are called dilatant. One example of dilatant behavior is Dow Corning 3179 dilatant compound, a silicone polymer commonly known as "Silly Putty." When pulled slowly, this fluid stretches (plastic deformation); when pulled quickly it behaves as a solid and fractures. Fluids that appear less viscous with higher shear strain rates (shear thinning) are called pseudo-plastic. For example, no-drip latex paint flows when applied with a brush or roller (applied shear stress) but does not flow after application.

Biological fluids are typically non-Newtonian. Blood plasma is Newtonian and is very similar in physical properties to water. Whole blood behaves as a Bingham plastic, whereby a nonzero shear stress (yield stress) is required before this fluid begins to flow. Blood is often characterized by a power law function, of the form

$$\tau = k\dot{\gamma}^n \tag{4.57}$$

where k and n are constants derived from a straight-line fit of $\ln \tau$ plotted as a function of $\ln \dot{\gamma}$, since

$$\ln \tau = \ln k + n \ln \dot{\gamma}$$

Another common description of blood's viscosity is the Casson equation:

$$\tau^{\frac{1}{2}} = \tau_0^{\frac{1}{2}} + k\dot{\gamma}^{\frac{1}{2}} \tag{4.58}$$

From a Casson plot, the yield stress τ_0 can be measured. Rheology, the study of deformation and flow of fluids, focuses on these often complex viscous properties of fluids. Textbooks with rheological data for biofluids include *Biofluid Mechanics* by Mazumdar (1992) and *Basic Transport Phenomena in Biomedical Engineering* by Fournier (1999).

Example Problem 4.12

The following rheological data were measured on a blood sample:

Shear Strain Rate [s^{-1}]	Shear Stress [dyne/cm^2]
1.5	12.5
2.0	16.0
3.2	25.2
6.5	40.0
11.5	62.0
16.0	80.5
25.0	120
50.0	140
100	475

Fit the data to a power law function using a MATLAB m-file.

Solution

```
% Power Law Fit of Blood Data
%
% Store shear strain rate and stress data in arrays
alpha = [1.5,2,3.2,6.5,11.5,16,25,50,100] ;
T = [12.5,16,25.2,40,62,80.5,120,240,475] ;
% Take natural logs of both
x = log(alpha) ;
y = log(T) ;
% Use MATLAB's polyfit function to do linear curve fit
coeff = polyfit (x,y,1)
% Write curve fit coefficients as a new x-y function for plotting
x1=[0;0.01;5]
y1=polyval (coeff,x1)
% Plot the original data as 'o' points
plot(x,y, 'o')
hold on
% Overlay a plot of the curve-fit line
plot(x1,y1)
grid on
```

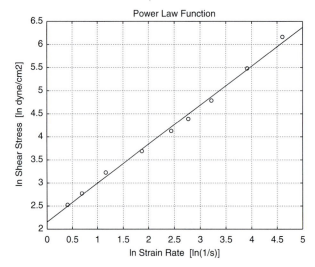

Figure 4.33   The power law curve-fit using MATLAB of the rheological blood data in Example Problem 4.12.

title ('Power Law Function')
xlabel('ln Strain Rate [ln (1/s)]')
ylabel('ln Shear Stress [ln dyne/cm2]')
%
The resulting plot appears in Figure 4.33. ■

When subjected to very low shear rates blood's apparent viscosity is higher than expected. This is due to the aggregation of red blood cells, called rouleaux. Such low shear rates are lower than those typically occurring in major blood vessels or in medical devices. In very small tubes (< 1 mm diameter), blood's apparent viscosity at high shear rates is smaller than in larger tubes, known as the Fahraeus-Lindquist effect, arising from plasma–red cell dynamics. Beyond these two special cases, blood behaves as a Newtonian fluid and is widely accepted as such in the scientific community. We shall see that the assumption of Newtonian fluid greatly simplifies mechanical description of the circulation.

4.7.2 Arterial Vessels

Mechanical description of blood vessels has a long and somewhat complicated history. Much of the advanced mathematics and applied mechanics associated with this work is beyond the scope of this textbook. This section will therefore give an overview of some of the main developments and will present a simplified, reduced arterial system model for use in the following subsection. The reader is referred to the following textbooks for more in-depth coverage: *Circulatory System Dynamics* (1978) by

Noordergraaf, *Hemodynamics* (1989) by Milnor, and *Biofluid Mechanics* (1992) by Mazumdar, and for basic fluid mechanics, *Fluid Mechanics* (2003) by White.

Study of the mechanical properties of the heart as a pump requires the computation of pressures and flows arising from forces and motion of the underlying heart muscle. Consequently, general equations of motion in the cardiovascular system typically arise from the conservation of linear momentum. The Reynold's transport theorem, a conservation equation from fluid mechanics, applied to linear momentum yields the following general equation of motion for any fluid:

$$\rho \mathbf{g} - \nabla_p + \nabla \cdot \tau_{ij} = \rho \frac{d\mathbf{V}}{dt} \qquad (4.59)$$

where ρ is fluid density (mass/volume), p is pressure, τ_{ij} are viscous forces, and $\mathbf{V}$ is velocity. ∇ is the differential operator

$$\nabla = \mathbf{i}\frac{\partial}{\partial x} + \mathbf{j}\frac{\partial}{\partial y} + \mathbf{k}\frac{\partial}{\partial z}$$

The general velocity $\mathbf{V}$ is a vector function of position and time and is written

$$\mathbf{V}(x, \ y, \ z, \ t) = u(x, \ y, \ z, \ t)\mathbf{i} + v(x, \ y, \ z, \ t)\mathbf{j} + w(x, \ y, \ z, \ t)\mathbf{k}$$

where u, v, and w are the local velocities in the x, y, and z directions, respectively.

Equation 4.59 is comprised of four terms: gravitational, pressure, and viscous forces, plus a time-varying term. Note that this is a vector equation and so can be expanded in x, y, and z components as the set of three equations:

$$\rho g_x - \frac{\partial p}{\partial x} + \frac{\partial \tau_{xx}}{\partial x} + \frac{\partial \tau_{yx}}{\partial y} + \frac{\partial \tau_{zx}}{\partial z} = \rho \left(\frac{\partial u}{\partial t} + u\frac{\partial u}{\partial x} + v\frac{\partial u}{\partial y} + w\frac{\partial u}{\partial z} \right) \qquad (4.60)$$

$$\rho g_y - \frac{\partial p}{\partial y} + \frac{\partial \tau_{xy}}{\partial x} + \frac{\partial \tau_{yy}}{\partial y} + \frac{\partial \tau_{zy}}{\partial z} = \rho \left(\frac{\partial v}{\partial t} + u\frac{\partial v}{\partial x} + v\frac{\partial v}{\partial y} + w\frac{\partial v}{\partial z} \right) \qquad (4.61)$$

$$\rho g_z - \frac{\partial p}{\partial z} + \frac{\partial \tau_{xz}}{\partial x} + \frac{\partial \tau_{yz}}{\partial y} + \frac{\partial \tau_{zz}}{\partial z} = \rho \left(\frac{\partial w}{\partial t} + u\frac{\partial w}{\partial x} + v\frac{\partial w}{\partial y} + w\frac{\partial w}{\partial z} \right) \qquad (4.62)$$

This set of nonlinear, partial differential equations is general but not solvable; solution requires making simplifying assumptions. For example, if the fluid's viscous forces are neglected Eq. 4.59 reduces to Euler's equation for inviscid flow. The latter, when integrated along a streamline, yields the famous Bernoulli equation relating pressure and flow. In application, Bernoulli's inviscid, and consequently frictionless, origin is sometimes forgotten.

If flow is steady, the right-hand term of Eq. 4.59 goes to zero. For incompressible fluids, including liquids, density ρ is constant, which greatly simplifies integration of the gravitational and time-varying terms that contain ρ. Similarly, for Newtonian fluids, viscosity μ is constant. In summary, although we can write perfectly general equations of motion, the difficulty of solving these equations requires making reasonable simplifying assumptions.

Two reasonable assumptions for blood flow in major vessels are that of Newtonian and incompressible behavior. These assumptions reduce Eq. 4.59 to the Navier-Stokes equations:

$$\rho g_x - \frac{\partial p}{\partial x} + \mu\left(\frac{\partial^2 u}{\partial x^2} + \frac{\partial^2 u}{\partial y^2} + \frac{\partial^2 u}{\partial z^2}\right) = \rho\frac{du}{dt} \tag{4.63}$$

$$\rho g_y - \frac{\partial p}{\partial y} + \mu\left(\frac{\partial^2 v}{\partial x^2} + \frac{\partial^2 v}{\partial y^2} + \frac{\partial^2 v}{\partial z^2}\right) = \rho\frac{dv}{dt} \tag{4.64}$$

$$\rho g_z - \frac{\partial p}{\partial z} + \mu\left(\frac{\partial^2 w}{\partial x^2} + \frac{\partial^2 w}{\partial y^2} + \frac{\partial^2 w}{\partial z^2}\right) = \rho\frac{dw}{dt} \tag{4.65}$$

Blood vessels are more easily described using a cylindrical coordinate system rather than a rectangular one. Hence the coordinates x, y, and z may be transformed to radius r, angle θ, and longitudinal distance x. If we assume irrotational flow, $\theta = 0$ and two Navier-Stokes equations suffice:

$$-\frac{dP}{dx} = \rho\left(\frac{dw}{dt} + u\frac{dw}{dr} + w\frac{dw}{dx}\right) - \mu\left(\frac{d^2w}{dr^2} + \frac{1}{r}\frac{dw}{dr} + \frac{d^2w}{dx^2}\right) \tag{4.66}$$

$$-\frac{dP}{dr} = \rho\left(\frac{du}{dt} + u\frac{du}{dr} + w\frac{du}{dx}\right) - \mu\left(\frac{d^2u}{dr^2} + \frac{1}{r}\frac{du}{dr} + \frac{d^2u}{dx^2} - \frac{u}{r^2}\right) \tag{4.67}$$

where w is longitudinal velocity dx/dt, and u is radial velocity dr/dt. Most arterial models also make use of the continuity equation, arising from the conservation of mass:

$$\frac{du}{dr} + \frac{u}{r} + \frac{dw}{dx} = 0 \tag{4.68}$$

In essence, the net rate of mass storage in a system is equal to the net rate of mass influx minus the net rate of mass efflux.

Noordergraaf and his colleagues (1969) rewrote the Navier-Stokes Equation 4.66 as

$$-\frac{dP}{dx} = RQ + L\frac{dQ}{dt} \tag{4.69}$$

where P is pressure, Q is the volume rate of flow, R is an equivalent hydraulic resistance, and L is fluid inertance. The Navier-Stokes equations describe fluid mechanics within the blood vessels. Since arterial walls are elastic, equations of motion for the arterial wall are also required. The latter have evolved from linear elastic, linear viscoelastic, to complex viscoelastic (see Noordergraaf, 1978). The most general mechanical description of linear anisotropic arterial wall material requires 21 parameters (see Fung's 1977 text), most of which have never been measured. Noordergraaf et al. divided the arterial system into short segments and combined the fluid mechanical equation (Eq. 4.69) with the continuity equation for each vessel segment. The arterial wall elasticity leads to a time-varying amount of blood stored in the vessel as it bulges with each heartbeat. For a segment of artery, the continuity equation becomes

$$-\frac{dQ}{dx} = GP + C\frac{dP}{dt} \tag{4.70}$$

where G is leakage through the blood vessel wall. This pair of hydraulic equations (Eqs. 4.69 and 4.70) was used to describe each of 125 segments of the arterial system and was the first model sufficiently detailed to explain arterial pressure and flow wave reflection. Arterial branching leads to reflected pressure and flow waves that interact in this pulsatile system. Physical R–L–C circuits were constructed and built into large transmission line networks with measured voltages and currents corresponding to hydraulic pressures and flows, respectively. If distributed arterial properties such as pulse wave reflection are not of interest, the arterial system load seen by the heart can be much reduced, as an electrical network may be reduced to an equivalent circuit.

The most widely used arterial load is the three-element model shown in Figure 4.34. The model appears as an electrical circuit due to its origin prior to the advent of the digital computer. Z_0 is the characteristic impedance of the aorta, in essence the aorta's flow resistance. C_s is transverse arterial compliance, the inverse of elastance, and describes stretch of the arterial system in the radial direction. R_s is the peripheral resistance, describing the systemic arteries' flow resistance downstream of the aorta. This simple network may be used to represent the systemic arterial load seen by the left ventricle. The following ordinary differential equation relates pressure at the left-hand side, $p(t)$, to flow, $Q(t)$:

$$C_s\frac{dp}{dt} + \frac{1}{R_s}p(t) = Q(t)\left(1 + \frac{Z_0}{R_s}\right) + Z_0 C_s\frac{dQ}{dt} \tag{4.71}$$

Example Problem 4.13

Using basic circuit theory, derive the differential equation Eq. 4.71 from Figure 4.34.

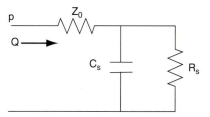

Figure 4.34   Equivalent systemic arterial load. Circuit elements are described in the text.

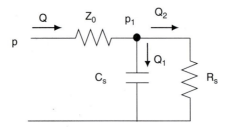

Figure 4.35   Nodal analysis of three-element arterial load.

Solution

Define node 1 as shown in Figure 4.35. By Kirchoff's current law, the flow Q going into node 1 is equal to the sum of the flows Q_1 and Q_2 coming out of the node:

$$Q = Q_1 + Q_2$$

We can write Q_1 and Q_2 as

$$Q_1 = C_s \frac{dp_1}{dt}$$

$$Q_2 = \frac{p_1}{R_s}$$

so

$$Q = C_s \frac{dp_1}{dt} + \frac{p_1}{R_s}$$

From Ohm's law,

$$p - p_1 = Q\, Z_0$$

Solving the last expression for p_1 and substituting back into the flow expression:

$$Q = C_s \frac{d}{dt}[p - Q\, Z_0] + \frac{1}{R_s}[p - Q\, Z_0]$$

$$= C_s \frac{dp}{dt} - Z_0\, C_s \frac{dQ}{dt} + \frac{1}{R_s} p - \frac{Z_0}{R_s} Q$$

Grouping terms for Q on the left and p on the right gives Eq. 4.71.   ∎

4.7.3   Heart Mechanics

Mechanical performance of the heart, more specifically the left ventricle, is typically characterized by estimates of ventricular elastance. The heart is an elastic bag that stiffens and relaxes with each heartbeat. Elastance is a measure of stiffness, classically defined as the differential relation between pressure and volume:

$$E_v = \frac{dp_v}{dV_v} \tag{4.72}$$

Here, p_v and V_v denote ventricular pressure and volume, respectively. For any instant in time, ventricular elastance E_v is the differential change in pressure with respect to volume. Mathematically, this relation is clear. Measurement of E_v is much less clear.

In the 1970s Kennish et al. tried to estimate the differential relation of Eq. 4.72 using the ratio of finite changes in ventricular pressure and volume:

$$E_v = \frac{\Delta p_v}{\Delta V_v} \tag{4.73}$$

This approach leads to physically impossible results. For example, before the aortic valve opens, the left ventricle is generating increasing pressure while there is not yet any change in volume. The ratio in Eq. 4.73 gives an infinite elastance when the denominator is zero. Suga and Sagawa (1974) used the ratio of pressure to volume itself, rather than differential or discrete changes, to estimate elastance:

$$E_v(t) = \frac{p_v(t)}{V_v(t) - V_d} \tag{4.74}$$

In this equation, V_d is a dead volume that remains constant. Now all other terms are allowed to be varying with time. Ventricular elastance measured in this way leads to elastance curves, as depicted in Figure 4.36. These curves show wide variation, as suggested by the large error bars. The distinctive asymmetric shape leads to a major contradiction. A simple experiment involves clamping the aorta, thereby preventing the left ventricle from ejecting blood, denoted an isovolumic beat. Equation 4.74 shows that under isovolumic conditions (V_v is constant) ventricular pressure p_v must have the same shape as elastance $E_v(t)$. However, experiments show that isovolumic

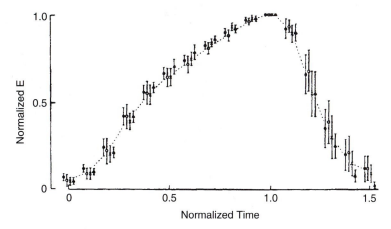

Figure 4.36 Time-varying ventricular elastance curves measured using the definition in Eq. 4.74. Measured elastance curves are distinctive in shape (adapted from Suga and Sagawa, 1974).

pressure curves are symmetric, unlike Figure 4.36. A further complication is the requirement of ejecting beats for measuring $E_v(t)$, which requires not only the heart (a ventricle) but also a circulation (blood vessels). Hence, time-varying elastance curves, such as those in Figure 4.36, are measures of both a particular heart (the source) combined with a particular circulation (its load). Experiments show that elastance curves measured in this way are subject to vascular changes, as well as the desired ventricular properties. As such, this approach cannot uniquely separate out ventricular from vascular properties. Consequently, a new measure of the heart's mechanical properties is required.

The problems just described (i.e., inconsistent isovolumic and ejecting behavior and the combined heart–blood vessel properties) led to the development of a new mechanical description of the left ventricle (Mulier, 1994; Palladino et al., 1997). This model should be simple and versatile, and should have direct physiological significance, in contrast with simulations which merely mimic physiological behavior. The model was developed using isolated canine heart experiments, as depicted in Figure 4.37. The left ventricle was filled with an initial volume of blood, subjected to

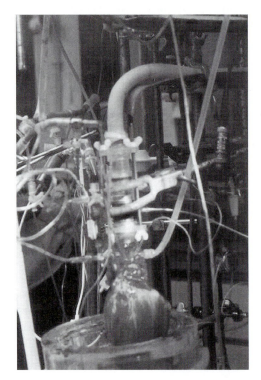

Figure 4.37   Isolated canine left ventricle used to develop a new biomechanical model of the heart (photo courtesy of Dr. Jan Mulier, Leuven, Belgium).

different loading conditions, stimulated, and allowed to beat. Ventricular pressure, and in some experiments ventricular outflow, was then measured and recorded.

Experiments began with measurement of isovolumic ventricular pressure. For each experiment the isolated left ventricle was filled with an initial volume (end-diastolic) and the aorta was clamped to prevent outflow of blood. The ventricle was stimulated and generated ventricular pressure was measured and recorded. The ventricle was then filled to a new end-diastolic volume and the experiment repeated. As in the famous experiments of Otto Frank (c. 1895), isovolumic pressure is directly related to filling. Figure 4.38 shows a set of isovolumic pressure curves measured on a normal canine left ventricle.

These isovolumic pressure curves were then described by the following equation. Ventricular pressure p_v is a function of time t and ventricular volume V_v according to

$$p_v = a(V_v - b)^2 + (cV_v - d)\left[\frac{(1 - e^{-\left(\frac{t}{\tau_c}\right)^\alpha})e^{-\left(\frac{t-t_b}{\tau_r}\right)^\alpha}}{(1 - e^{-\left(\frac{t_p}{\tau_c}\right)^\alpha})e^{-\left(\frac{t_p-t_b}{\tau_r}\right)^\alpha}}\right] \qquad (4.75)$$

or written more compactly,

$$p_v(t,\ V_v) = a(V_v - b)^2 + (cV_v - d)f(t) \qquad (4.76)$$

where $f(t)$ is the activation function in square brackets in Eq. 4.75. The constants a, b, c, d, t_p, τ_c, τ_r, and α were derived from the isolated canine ventricle experiments. Physiologically, Eq. 4.76 says that the ventricle is a time- and volume-dependent

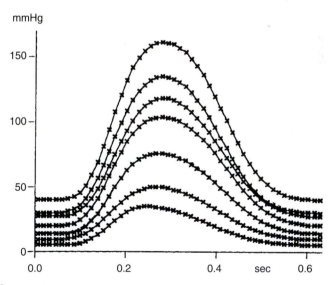

Figure 4.38   Isovolumic ventricular pressure curves. For each curve, the left ventricle is filled with a fixed initial volume, the heart is stimulated, and generated ventricular pressure is measured. Other curves arise from different fixed initial volumes.

pressure generator. The term to the left of the plus sign (including constants a and b) describes the ventricle's passive elastic properties. The term to the right (including c and d) describes its active elastic properties, arising from the active generation of force in the underlying heart muscle. Representative model quantities measured from canine experiments are given in Table 4.3. This model was adapted to describe the human left ventricle using quantities in the right-hand column (Palladino et al., 1997).

Example Problem 4.14

Solve Eq. 4.75 and plot ventricular pressure $p_v(t)$ for one human heartbeat. Use initial ventricular volume of 150 ml and the parameter values in Table 4.3.

Solution

The following MATLAB m-file will perform the required computation and plot the results, shown in Figure 4.39.

```
% ventricle.m
%
% MATLAB m-file to compute isovolumic pressure using Mulier ventricle model
%
% Initial conditions:
%
delt = 0.001; % The interation time step delta t
a = 7e-4;
b = 20.;
c = 2.5;
d = 80.;
tc = 0.264;
tp = 0.371;
```

TABLE 4.3   Ventricle Model Quantities Measured from Animal Experiments and Adapted for the Human Analytical Model

Quantity	Dog (measured)	Human (adapted)
a	0.003 [mmHg/ml^2]	0.0007
b	1.0 [ml]	20.0
c	3.0 [mmHg/ml]	2.5
d	20.0 [mmHg]	80.0
τ_c	0.164 [s]	0.264
t_p	0.271 [s]	0.371
τ_r	0.199 [s]	0.299
t_b	0.233 [s]	0.258
α	2.88	2.88

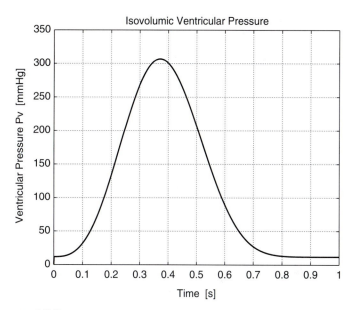

Figure 4.39   Isovolumic ventricular pressure computed for a human heartbeat.

tr = 0.299;
tb = 0.258;
alpha = 2.88;
Vv0 = 150; % Initial (end-diastolic) ventricular volume
%
% Compute an intermediate term denom
% to simplify computations:
%
   denom = ((1. − exp (− (tp/tc)^alpha))) * exp (− ((tp − tb)/tr)^alpha));
%
% Compute for initial time t = 0 (MATLAB does not allow 0 index)
%
  t (1) = 0.;
  Vv (1) = Vv0;
  edp = a * ((Vv0 − b))^2;
  pdp = c * Vv0 − d;
  pp = pdp/denom;
  t1 = 0.; % Time step for first exponential
  t2 = 0.; % Time step for second exponential
  e1 = exp (− (t1/tc)^alpha);
  e2 = exp (− (t2/tr)^alpha);

```
    pv0 = edp + pp * ((1. − e1) * e2);
%
% Main computation loop:
%
    for j=2:1000
  t(j) = t(j − 1) + delt;
  Vv(j) = Vv(j − 1);
%
  edp = a * ((Vv(j) − b))^2;
  pdp = c * Vv(j) − d;
  pp = pdp/denom;
  t1 = t(j);
% Second exponential begins at t > tb
  t2 = t(j) − tb;
  if (t2 < 0.);
    t2 = 0.;
  end
  e1 = exp ( − (t1/tc)^alpha);
  e2 = exp ( − (t2/tr)^alpha);
  pv(j) = edp + pp * ((1. − e1) * e2);
  end
%
plot (t,pv)
grid on
title ('Isovolumic Ventricular Pressure')
xlabel ('Time [s]')
ylabel ('Ventricular Pressure Pv [mmHg]')
```

■

4.7.4 Cardiovascular Modeling

This concise model of the left ventricle was coupled to the reduced arterial load model of Figure 4.34 and allowed to eject blood. Model parameter values for a normal arterial load are given in Table 4.4. Figure 4.40 shows results for a normal canine left

TABLE 4.4 Representative Systemic Arterial Model Element Values

Model Element	Symbol	Control Value
Characteristic aorta impedance	Z_0	0.1 mmHg-s/ml
Systemic arterial compliance	C_s	1.5 ml/mmHg
Peripheral arterial resistance	R_s	1.0 mmHg-s/ml

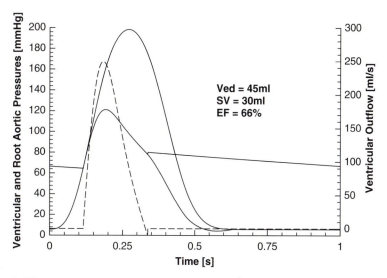

Figure 4.40 Ventricular and root aortic pressures (solid curves, left ordinate) and ventricular outflow (dashed curve, right ordinate) computed using the model of Eq. 4.75 for a normal canine left ventricle pumping into a normal arterial circulation. The topmost solid curve corresponds to a clamped aorta (isovolumic). The ventricle has initial volume of 45 ml and pumps out 30 ml, for an ejection fraction of 66%, about normal.

ventricle ejecting into a normal arterial system. The solid curves (left ordinate) describe ventricular pressure p_v and root aortic pressure as functions of time. Clinically, arterial pressure is reported as two numbers (e.g., 110/60). This corresponds to the maximum and minimum root arterial pulse pressures, in this case about 120/65 mmHg. The dashed curve (right ordinate) shows ventricular outflow. The ventricle was filled with an end-diastolic volume of 45 ml and it ejected 30 ml (stroke volume), giving an ejection fraction of 66%, which is about normal for this size animal.

The same ventricle may be coupled to a pathological arterial system, for example, one with doubled peripheral resistance R_s. This change is equivalent to narrowed blood vessels. As expected, increased peripheral resistance raises arterial blood pressure (to 140/95 mmHg) and impedes the ventricle's ability to eject blood (Fig. 4.41). The ejection fraction decreases to 50% in this experiment. Other experiments, such as altered arterial stiffness, may be performed. The model's flexibility allows description of heart pathology as well as changes in blood vessels. This one ventricular equation with one set of measured parameters is able to describe the wide range of hemodynamics observed experimentally (Palladino et al., 1997).

The previous expressions for ventricular elastance defined in Eqs. 4.73 and 4.74 have the same units as elastance defined classically as Eq. 4.72, but are mathematically not the same. Since ventricular pressure is now defined as an analytical function (Eq. 4.75), ventricular elastance, E_v, defined in the classical sense as $\partial p_v / \partial V_v$, may be calculated as

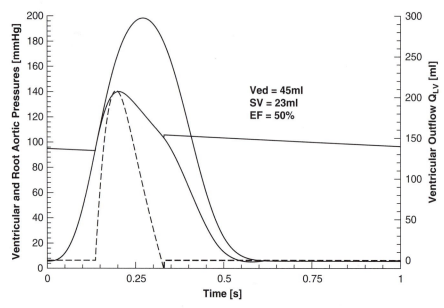

Figure 4.41   The same normal canine ventricle of Fig. 4.40 now pumping into an arterial system with doubled peripheral (flow) resistance. As expected, increased resistance, corresponding to narrowed vessels, leads to increased arterial pulse pressure. Stroke volume is reduced from 66% to 50%.

$$E_v(t, V_v) = 2a(V_v - b) + c \left[\frac{(1 - e^{-\left(\frac{t}{\tau_c}\right)^\alpha}) e^{-\left(\frac{t-t_b}{\tau_r}\right)^\alpha}}{(1 - e^{-\left(\frac{t_p}{\tau_c}\right)^\alpha}) e^{-\left(\frac{t_p-t_b}{\tau_r}\right)^\alpha}} \right] \tag{4.77}$$

or

$$E_v(t, \ V_v) = 2a(V_v - b) + cf(t) \tag{4.78}$$

Figure 4.42 shows ventricular elastance curves computed using the new model's analytical definition of elastance (Eq. 4.77). Elastance was computed for a wide range of ventricular and arterial states, including normal and pathological ventricles, normal and pathological arterial systems, and isovolumic and ejecting beats. These elastance curves are relatively invariant and cluster in two groups—either normal or weakened ventricle contractile state. Consequently, this new measure of elastance may now effectively assess the health of the heart alone, separate from blood vessel pathology.

Experiments showed that when the left ventricle ejects blood, ventricular pressure is somewhat different than expected. As depicted in Figure 4.43, early during ejection the ventricle generates less pressure than expected, denoted *pressure deactivation*. Later in systole the heart generates greater pressure, denoted *hyperactivation*. These two variations with flow have been termed the *ejection effect* (Danielsen et al., 2000). The ejection effect was incorporated into the ventricle model of Eq. 4.76 by adding a flow-dependent term:

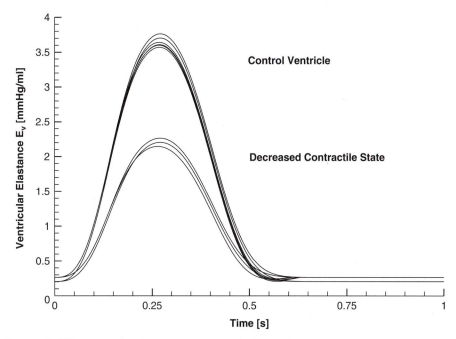

Figure 4.42 Ventricular elastance curves computed using the new analytical function of Eq. 4.77. Elastance curves computed in this way are representative of the ventricle's contractile state–its ability to pump blood.

$$p_v(t, V_v, \; Q_v) = a(V_v - b)^2 + (cV_v - d)F(t, Q_v) \tag{4.79}$$

with F replacing the time function f in Eq. 4.76:

$$F(t, \; Q_v) = f(t) - k_1 Q_v(t) + k_2 Q_v^2(t - \tau), \;\; \tau = \kappa t \tag{4.80}$$

and k_1, k_2, and κ are additional model constants. In summary, the ventricle is a time-volume- and flow-dependent pressure generator. Figure 4.44 shows computed pressures and flows for this formulation at left, compared to the results minus the ejection effect at right for comparison. The ejection effect tends to change the shape of both pressure and flow curves to more closely resemble experimental curves.

Addition of the ejection effect modifies the computed ventricular elastance curves, as depicted in Figure 4.45. All curves are computed using the new function of Eq. 4.77. The dashed curve, minus the ejection effect, is symmetric, as in Figure 4.42. Addition of the ejection effect (solid) makes ventricular elastance asymmetrically skewed to the right, much like the measured curves of Suga and Sagawa depicted in Figure 4.36. As expected, the mechanical process of ejecting blood has a direct effect on the heart's elastance.

In summary, the left ventricle may be described as a time-, volume-, and flow-dependent pressure generator. A small number of experimentally derived parameters is sufficient to describe the wide range of observed cardiovascular dynamics. This

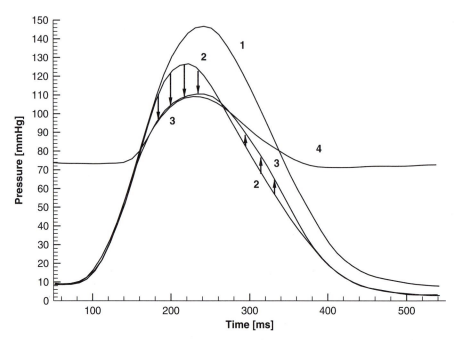

Figure 4.43   The ejection effect, showing that early during blood ejection (systole), the heart generates somewhat less pressure than expected, denoted *deactivation* (down arrows). Later in systole, the heart generates greater pressure, denoted *hyperactivation*. Curves 1 and 2 are ventricular pressures for initial and ejected volumes, respectively. Curve 3 is the measured ejecting pressure curve. Curve 4 is root aortic pressure.

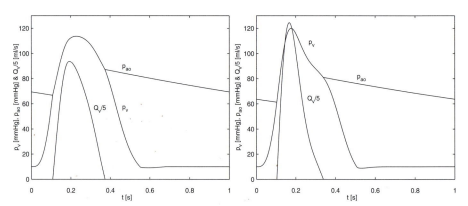

Figure 4.44   The ejection effect incorporated in the ventricle model (left) with the uncorrected ventricular pressure and outflow curves for the same conditions at right for comparison. Ventricular outflow Q_v is normalized by 1/5 to use the same numeric scale as for ventricular pressure.

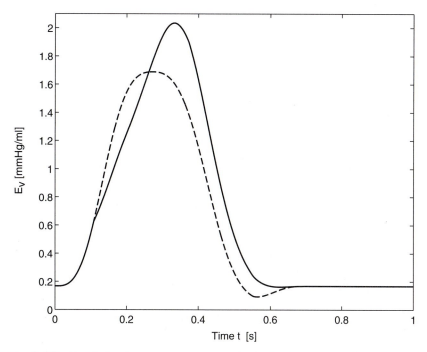

Figure 4.45 Ventricular elastance curves computed using Eq. 4.77 without (dashed) and with the ejection effect (solid).

approach links experiment and theory, leading to new ideas and experiments. Work is currently underway to devise a new measure of cardiovascular health using this model. In essence, it is thought that the magnitude of the observed ejection effect for a particular heart should be directly related to its health.

The left ventricle model of Eq. 4.75 was used to describe each of the four chambers of the human heart, depicted in Figure 4.46 (Palladino et al., 2000). This complete model of the circulatory system displays a remarkable range of cardiovascular physiology with a small set of equations and parameters. For example, plotting ventricular pressure as a function of ventricular volume shows the mechanical work performed by the ventricle, denoted *pressure–volume work loops*. The area within each loop corresponds to external work performed by the heart. Figure 4.47 shows left and right ventricle work loops for the normal heart ejecting into the normal (control) circulatory system, depicted by solid curves. The right ventricle work loop is smaller, as expected, than the left. Figure 4.47 also shows the same two work loops for a weakened left ventricle (dashed curves). As expected, the left ventricle work loop is diminished in size. This work loop also shifts to the right on the volume axis. Since the weaker ventricle ejects less blood, more remains to fill the heart more for the subsequent beat (EDV of 194 instead of 122 ml). This increased filling partially

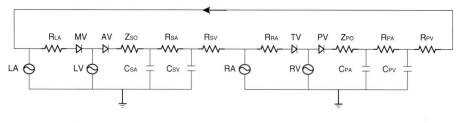

Systemic Circulation

LA = Left Atrium
R_{LA} = Left Atrial Resistance
MV = Mitral Valve
LV = Left Ventricle
AV = Aortic Valve
Z_{SO} = Systemic Characteristic Impedance
C_{SA} = Systemic Arterial Compliance
R_{SA} = Systemic Peripheral Resistance
C_{SV} = Systemic Venous Compliance
R_{SV} = Systemic Venous Resistance

Pulmonary Circulation

RA = Right Atrium
R_{RA} = Right Atrial Resistance
TV = Tricuspid Valve
RV = Right Ventricle
PV = Pulmonic Valve
Z_{PO} = Pulmonic Characteristic Impedance
C_{PA} = Pulmonic Arterial Compliance
R_{PA} = Pulmonic Peripheral Resistance
C_{PV} = Pulmonic Venous Compliance
R_{PV} = Pulmonic Venous Resistance

Figure 4.46 Application of the canine left ventricle model to hemodynamic description of the complete human cardiovascular system. Adapted from Palladino et al., 2000.

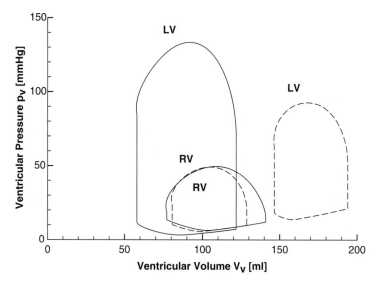

Figure 4.47 Computed work loops for the left and right ventricles under control conditions (solid curves) and for the case of a weakened left ventricle (dashed curves).

compensates for the weakened ventricle via Starling's law (increased pressure for increased filling).

Changing only one model constant, c, in Eq. 4.75 is sufficient to vary chamber contractile state. For example, Table 4.5 shows examples of congestive heart failure, resulting from decreases in c for the left ventricle and for the right ventricle.

TABLE 4.5 Cardiovascular Performance for a Normal Heart, and Weakened Left and Right Ventricles

	SV [ml]		EDV [ml]		EF [%]		P_{AO} [mmHg]	P_{PU} [mmHg]
	LV	RV	LV	RV	LV	RV		
Control	64	64	122	141	53	46	133/70	49/15
Weak LV	48	48	194	129	25	38	93/52	49/21
Weak RV	44	44	102	164	49	24	109/59	29/12

SV denotes stroke volume, EDV denotes end-diastolic volume, EF is ejection fraction and P_{AO} and P_{PU} are root aorta and root pulmonary artery pressures, respectively, for the left (LV) and right (RV) ventricles. Note that SV left and right are equal under all conditions.

Decreasing left ventricular contractile state to one third of the control value (c = 1.0) lowers left ventricular ejection fraction from 53% to 25%, and root aortic pulse pressure decreases from 133/70 to 93/52 mmHg. Left ventricular stroke volume decreases less, from 64 to 48 ml, since it is compensated for by the increased left end-diastolic volume (194 ml) via Starling's law. Decreasing left ventricular contractile state is equivalent to left congestive heart failure. Consequently, pulmonary venous volume increases from 1347 ml to 1981 ml (not shown), indicating pulmonary congestion for this case.

Similar changes are noted when the right ventricle's contractile state is halved (c = 0.5). The right ventricular ejection fraction drops from 46% to 24%, root pulmonary artery pulse pressure decreases from 49/15 to 29/12 mmHg, and right stroke volume decreases from 64 to 44 ml, with an increased end-diastolic volume of 164 ml, from 141 ml. Conversely, c can be increased in any heart chamber to depict administration of an inotropic drug. Although not plotted, pressures, flows, and volumes are available at any circuit site, all as functions of time.

Changes in blood vessel properties may be studied alone or in combination with altered heart properties. Other system parameters such as atrial performance, as well as other experiments, may be examined. The modular form of this model allows its expansion for more detailed studies of particular sites in the circulatory system.

The field of biomechanics applies physical principles to living systems using the language of mathematics. Hemodynamics studies the human cardiovascular system, which is comprised of an interesting pump moving interesting fluid around a complicated network of interesting pipes. In developing hemodynamic principles, experiments and analysis go hand-in-hand, ensuring the validity of principles with experiments and with analysis clarifying, modifying, and often preceding experiments. In this fashion, interpretations of cardiovascular health are further defined.

EXERCISES

1. Write all the vector expressions of Eqs. 4.1 through 4.14 using MATLAB.

2. Repeat Example Problem 4.2 using a z–x–y rotation sequence.
3. Write the free-body diagrams for each of the three orientations of the humerus in Figure 3.36. For a particular load and fixed position, write and solve the equations of static equilibrium.
4. Solve Example Problem 4.5 for forearm orientations angled θ from the horizontal position. Let θ vary from 0–$70°$ (down) from the horizontal in $15°$ increments. Using MATLAB, plot the required biceps muscle force F_B for static equilibrium as a function of θ. By how much does this force vary over this range?
5. Repeat Problem 4, this time plotting forces F_A, F_B, and F_C over the same range of angles θ.
6. Considering the previous problem, explain why Nautilus weight machines at the gym use asymmetric pulleys.
7. The force plate in Figure 4.9 is 70 cm by 70 cm square. At a particular instant of the gait cycle each transducer reads $F_1 = 210$ N, $F_2 = 220$ N, $F_3 = 150$ N, and $F_4 = 180$ N. Compute the resultant force and its location.
8. For your own body, compute the mass moment of inertia of each body segment in Table 4.1 with respect to its center of mass.
9. Repeat Example Problem 4.8 using a titanium rod with circular cross-sectional diameter of 9 mm.
10. Write the Simulink models of the three-element Kelvin viscoelastic description and perform the creep and stress relaxation tests, the results of which appear in Figures 4.23 and 4.24.
11. Use the three-element Kelvin model to describe the stress relaxation of a biomaterial of your choice. Using a stress response curve from the literature, compute the model spring constants K_1 and K_2 and the viscous damping coefficient β.
12. Write and solve the kinematic equations defining an anatomically referenced coordinate system for the pelvis, $\{e_{pa}\}$, using MATLAB.
13. Using the kinematic data of Section 4.6.2, compute the instantaneous ankle power of the 25.2 kg patient using MATLAB.
14. Given the pelvis and thigh anatomical coordinate systems defined in Example Problems 4.9 and 4.10, compute the pelvis tilt, obliquity, and rotation angles using an x–y–z rotation sequence.
15. Repeat Problem 14 (using an x–y–z rotation sequence) solving for hip flexion/extension, hip abduction/adduction, and internal/external hip rotation. Hint: $\{e_{ta}\}$ is the triple-primed coordinate system and $\{e_{pa}\}$ is the unprimed system in this case.
16. Using MATLAB, determine the effect that a 12-mm perturbation in each coordinate direction would have on ankle moment amplitude (Section 4.6.3).
17. Fit the blood rheological data of Example Problem 4.12 to a Casson model and find the yield stress τ_0 for blood.

18. Solve Eq. 4.71 for pressure $p(t)$ when the aortic valve is closed. Using the parameter values in Table 4.4, plot p as a function of time for one heart beat ($t = 0 - 1\,\text{sec}$).

19. Compute isovolumic ventricular pressure $p_v(t)$ for the canine heart with initial volumes $V_v = 30, 40, 50, 60, 70\,\text{ml}$. Overlay these plots as in Figure 4.38.

20. Write a MATLAB m-file to compute ventricular elastance using Eq. 4.77. Compute and plot $E_v(t)$ for the parameter values in Example Problem 4.14.

Suggested Reading

Allard, P., Stokes, I.A.F. and Blanchi, J.P. (Eds.), (1995). *Three-Dimensional Analysis of Human Movement*. Human Kinetics, Champagne, Il.

Burstein, A.H. and Wright, T.M. (1994). *Fundamentals of Orthopaedic Biomechanics*. Williams & Wilkins, Baltimore.

Danielsen, M., Palladino, J.L. and Noordergraaf, A. (2000). The left ventricular ejection effect. In *Mathematical Modelling in Medicine* (J.T. Ottesen and M. Danielsen, Eds.). IOS Press, Amsterdam.

Davis, R.B.D. III. (1986). Musculoskeletal biomechanics: Fundamental measurements and analysis. Chapter 6 in *Biomedical Engineering and Instrumentation* (J.D. Bronzino, Ed.) PWS Engineering, Boston.

Davis R.B, Õunpuu S, Tyburski D.J., Gage J.R. (1991). A gait analysis data collection and reduction technique. *Human Movement Sci.* **10**, 575–587.

Davis R, DeLuca P. (1996). Clinical gait analysis: Current methods and future directions. In *Human Motion Analysis: Current Applications and Future Directions* (G. Harris, and P. Smith, Eds.), IEEE Press, Piscataway, NJ.

Dugas, R. (1988). *A History of Mechanics*, Dover, New York. Reprinted from a 1955 text.

Fournier, R.L. (1999). *Basic Transport Phenomena in Biomedical Engineering*. Taylor & Francis, Philadelphia.

Fung, Y.C. (1977). *A First Course in Continuum Mechanics*, 2nd Ed. Prentice-Hall, Englewood Cliffs, NJ.

Fung, Y.C. (1993). *Biomechanics: Mechanical Properties of Living Tissues*, 2nd Ed. Springer-Verlag, New York.

Gage, J.R. (1991). *Gait Analysis in Cerebral Palsy*. MacKeith, London.

Ganong, W.F. (1997). *Review of Medical Physiology*, 18th Ed. Appleton & Lange, Stamford, CT.

Greenwood, D.T. (1965). *Principles of Dynamics*. Prentice-Hall, Englewood Cliffs, NJ.

Huxley, A.F. (1957). Muscle structure and theories of contraction. *Prog. Biophys.* 7: 255–318.

Kennish, A., Yellin, E. and Frater, R. W. (1995). Dynamic stiffness profiles in the left ventricle. *J. Appl. Physiol.* 39, 565.

Mazumdar, J.N. (1992). *Biofluid Mechanics*. World Scientific, Singapore.

Meriam, J.L. and Kraige, L.G. (2002). *Engineering Mechanics*, 5th Ed., Wiley, New York.

Milnor, W.R. (1989). *Hemodynamics*, 2nd Ed. Williams and Wilkins, Baltimore.

Milnor, W.R. (1990). *Cardiovascular Physiology*. Oxford Univ. Press, New York.

Mow, V.C. and Hayes, W.C. (1999). *Basic Orthopaedic Biomechanics*, 2nd Ed. Lippincott-Raven, Philadelphia.

McCulloch, A.D. (2003). Cardiac biomechanics. In *Biomechanics Principles and Application*. (Schneck, D.J. and Bronzino, J.D., Eds.). CRC Press, Boca Raton, FL.

Mulier, J.P. (1994). Ventricular Pressure as a Function of Volume and Flow. Ph.D. dissertation, Univ. of Leuven, Belgium.

Needham, D.M. (1971). *Machina Carnis: The Biochemistry of Muscular Contraction in its Historical Development*. Cambridge Univ. Press, Cambridge.

Nichols, W.W. and O'Rourke, M.F. (1990). *McDonald's Blood Flow in Arteries: Theoretical, Experimental and Clinical Principles*, 3rd Ed. Edward Arnold, London.

Nigg, B.M. (1994). In *Biomechanics of the Musculo-Skeletal System* (Nig, B.M. and Herzog, W., Eds.). Wiley, New York.

Noordergraaf, A. (1969). Hemodynamics. In *Biological Engineering*. (Schwan, H.P., Ed.). McGraw-Hill, New York.

Noordergraaf, A. (1978). *Circulatory System Dynamics*. Academic, New York.

Palladino, J.L. and Noordergraaf, A. (1998). Muscle contraction mechanics from ultrastructural dynamics. In *Analysis and Assessment of Cardiovascular Function* (Drzewiecki, G.M. and Li, J.K.-J., Eds.). Springer-Verlag, New York.

Palladino, J.L., Mulier, J.P. and Noordergraaf, A. (1999). Closed-loop circulation model based on the Frank mechanism, *Surv. Math. Ind.* **7**, 177–186.

Palladino, J.L., Ribeiro, L.C. and Noordergraaf, A. (2000). Human circulatory system model based on Frank's mechanism. In *Mathematical Modelling in Medicine* (Ottesen, J.T. and Danielsen, M., Eds.). IOS Press, Amsterdam.

Roark, R.J. (1969). *Formulas for Stress and Strain*, 6th Ed. McGraw-Hill, New York.

Singer, C. and Underwood, E.A. (1962). *A Short History of Medicine*, 2nd ed. Oxford Univ. Press, New York.

Suga, H. and Sagawa, K. (1994). Instantaneous pressure–volume relationship under various end-diastolic volume, *Circ. Res.* **35**, 117–126.

Weber, E. (1846). Handwo terbuch der physiologie. Vol. 3B. R. Wagner, ed., Vieweg, Braunschweig.

Westerhof, N. and Noordergraaf, A. (1990). Arterial viscoelasticity: A generalized model, *J. Biomech.* **3**, 357–379.

White, F.M. (2003). *Fluid Mechanics*, 5th Ed. McGraw-Hill, New York.

Winter, D.A. (1990). *Biomechanics and Motor Control of Human Movement*. Wiley, New York.

Vogel, S. (1992). *Vital Circuits: On Pumps, Pipes, and the Workings of Circulatory System*. Oxford Univ. Press, New York.

5 REHABILITATION ENGINEERING AND ASSISTIVE TECHNOLOGY

Andrew Szeto, PhD, PE

Chapter Contents

At the conclusion of this chapter, students will:

- Understand the role played by rehabilitation engineers and assistive technologists in the rehabilitation process.
- Be aware of the major activities in rehabilitation engineering.
- Be familiar with the physical and psychological consequences of disability.

- Know the principles of assistive technology assessment and its objectives and pitfalls.

- Discuss key engineering and ergonomic principles of the field.

- Describe career opportunities and information sources.

5.1   INTRODUCTION

Since the late 1970s, there has been major growth in the application of technology to ameliorate the problems faced by people with disabilities. Various terms have been used to describe this sphere of activity, including prosthetics/orthotics, rehabilitation engineering, assistive technology, assistive device design, rehabilitation technology, and even biomedical engineering applied to disability. With the gradual maturation of this field, several terms have become more widely used, bolstered by their use in some federal legislation.

The two most frequently used terms today are *assistive technology* and *rehabilitation engineering*. Although they are used somewhat interchangeably, they are not identical. In the words of James Reswick (1982), a pioneer in this field, "rehabilitation engineering is the application of science and technology to ameliorate the handicaps of individuals with disabilities." In contrast, assistive technology can be viewed as a product of rehabilitation engineering activities. Such a relationship is analogous to health care being the product of the practice of medicine.

One widely used definition for assistive technology is found in Public Law 100-407. It defines assistive technology as "any item, piece of equipment or product system whether acquired commercially off the shelf, modified, or customized that is used to increase or improve functional capabilities of individuals with disabilities." Notice that this definition views assistive technology as a broad range of devices, strategies, and/or services that help an individual to better carry out a functional activity. Such devices can range from low-technology devices that are inexpensive and simple to make to high-technology devices that are complex and expensive to fabricate. Examples of low-tech devices include dual-handled utensils and mouth sticks for reaching. High-tech examples include computer-based communication devices, reading machines with artificial intelligence, and externally powered artificial arms (Fig. 5.1).

Several other terms often used in this field include rehabilitation technology and orthotics and prosthetics. Rehabilitation technology is that segment of assistive technology that is designed specifically to rehabilitate an individual from his or her present set of limitations due to some disabling condition, permanent or otherwise. In a classical sense, orthotics are devices that augment the function of an extremity, whereas prosthetics replace a body part both structurally and functionally. These two terms now broadly represent all devices that provide some sort of functional replacement. For example, an augmentative communication system is sometimes referred to as a speech prosthesis.

5.1.1   History

A brief discussion of the history of this field will explain how and why so many different yet similar terms have been used to denote the field of assistive technology

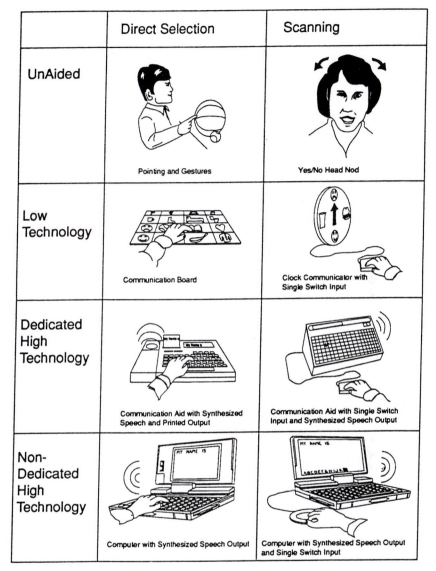

Figure 5.1 Augmentative communication classification system (from Church and Glennen, 1992).

and rehabilitation. Throughout history, people have sought to ameliorate the impact of disabilities by using technology. This effort became more pronounced and concerted in the United States after World War II. The Veterans Administration (VA) realized that something had to be done for the soldiers who returned from war with numerous and serious handicapping conditions. There were too few well-trained artificial limb and brace technicians to meet the needs of the returning soldiers. To train these much-needed providers, the federal government supported the establishment of a number of prosthetic and orthotic schools in the 1950s.

The VA also realized that the state of the art in limbs and braces was primitive and ineffectual. The orthoses and prostheses available in the 1940s were uncomfortable, heavy, and offered limited function. As a result, the federal government established the Veterans Administration Prosthetics Research Board, whose mission was to improve the orthotics and prosthetic appliances that were available. Scientists and engineers formerly engaged in defeating the Axis powers now turned their energies toward helping people, especially veterans with disabilities. As a result of their efforts, artificial limbs, electronic travel guides, and wheelchairs that were more rugged, lighter, cosmetically appealing, and effective were developed.

The field of assistive technology and rehabilitation engineering was nurtured by a two-pronged approach in the federal government. One approach directly funded research and development efforts that would utilize the technological advances created by the war effort toward improving the functioning and independence of injured veterans. The other approach helped to establish centers for the training of prosthetists and orthotists, forerunners of today's assistive technologists.

In the early 1960s, another impetus to rehabilitation engineering came from birth defects in infants born to expectant European women who took thalidomide to combat "morning sickness." The societal need to enable children with severe deformities to lead productive lives broadened the target population of assistive technology and rehabilitation engineering to encompass children as well as adult men. Subsequent medical and technical collaboration in research and development produced externally powered limbs for people of all sizes and genders, automobiles that could be driven by persons with no arms, sensory aids for the blind and deaf, and various assistive devices for controlling a person's environment.

Rehabilitation engineering received formal governmental recognition as an engineering discipline with the landmark passage of the federal Rehabilitation Act of 1973. The act specifically authorized the establishment of several centers of excellence in rehabilitation engineering. The formation and supervision of these centers were put under the jurisdiction of the National Institute for Handicapped Research, which later became the National Institute on Disability and Rehabilitation Research (NIDRR). By 1976, about 15 Rehabilitation Engineering Centers (RECs), each focusing on a different set of problems, were supported by grant funds totaling about $9 million per year. As the key federal agency in the field of rehabilitation, NIDRR also supports rehabilitation engineering and assistive technology through its Rehabilitation Research and Training Centers, Field Initiated Research grants, Research and Demonstration program, and Rehabilitation Fellowships (NIDRR, 1999).

The REC grants initially supported university-based rehabilitation engineering research and provided advanced training for graduate students. Beginning in the mid-1980s, the mandate of the RECs was broadened to include technology transfer and service delivery to persons with disabilities. During this period, the VA also established three of its own RECs to focus on some unique rehabilitation needs of veterans. Areas of investigation by VA and non-VA RECs include prosthetics and orthotics, spinal cord injury, lower and upper limb functional electrical stimulation, sensory aids for the blind and deaf, effects of pressure on tissue, rehabilitation robotics, technology transfer, personal licensed vehicles, accessible telecommunications, applications of wireless technology, and vocational rehabilitation. Another milestone, the formation of the Rehabilitation Engineering Society of North America (RESNA) in 1979, gave greater focus and visibility to rehabilitation engineering. Despite its name, RESNA is an inclusive professional society that welcomes everyone involved with the development, manufacturing, provision, and usage of technology for persons with disabilities. Members of RESNA include occupational and physical therapists, allied health professionals, special educators, and users of assistive technology. RESNA has become an adviser to the government, a developer of standards and credentials, and, via its annual conferences and its journal, a forum for exchange of information and a showcase for state-of-the art rehabilitation technology. In recognition of its expanding role and members who were not engineers, RESNA modified its name in 1995 to the Rehabilitation Engineering and Assistive Technology Society of North America.

Despite the need for and the benefits of providing rehabilitation engineering services, reimbursement for such services by third-party payers (e.g., insurance companies, social service agencies, and government programs) remained very difficult to obtain during much of the 1980s. Reimbursements for rehabilitation engineering services often had to be subsumed under more accepted categories of care such as client assessment, prosthetic/orthotic services, or miscellaneous evaluation. For this reason, the number of practicing rehabilitation engineers remained relatively static despite a steadily growing demand for their services.

The shortage of rehabilitation engineers with suitable training and experience was specifically addressed in the Rehab Act of 1986 and the Technology-Related Assistance Act of 1988. These laws mandated that rehabilitation engineering services had to be available and funded for disabled persons. They also required an individualized work and rehabilitation plan (IWRP) for each vocational rehabilitation client. These two laws were preceded by the original Rehab Act of 1973 which mandated reasonable accommodations in employment and secondary education as defined by a least restrictive environment (LRE). Public Law 95-142 in 1975 extended the reasonable accommodation requirement to children 5–21 years of age and mandated an individual educational plan (IEP) for each eligible child. Table 5.1 summarizes the major United Stated Federal legislation that has affected the field of assistive technology and rehabilitation engineering.

In concert with federal legislation, several federal research programs have attempted to increase the availablity of rehabilitation engineering services for persons with disabilities. The National Science Foundation (NSF), for example, initiated a program called Bioengineering and Research to Aid the Disabled. The program's goals were (1) to provide student-engineered devices or software to disabled individuals that would improve their quality of life and degree of independence, (2) to enhance the education of student engineers through real-world design experiences, and (3) to allow the university an opportunity to serve the local community. The Office of Special Education and Rehabilitation Services in the U.S. Department of Education funded special projects and demonstration programs that addressed identified needs such as model assessment programs in assistive technology, the application of technology for deaf–blind children, interdisciplinary training for students of communicative

TABLE 5.1 Recent Major U.S. Federal Legislation Affecting Assistive Technologies

Legislation	Major Assistive Technology Impact
Rehabilitation Act of 1973, as amended	Mandates reasonable accommodation and least restricted environment in federally funded employment and higher education; requires both assistive technology devices and services be included in state plans and Individualized Written Rehabilitation Plans (IWRP) for each client; Section 508 mandates equal access to electronic office equipment for all federal employees; defines rehabilitation technology as rehabilitation engineering and assistive technology devices and services; mandates rehabilitation technology as primary benefit to be included in IWRP
Individuals with Disabilities Education Act Amendments of 1997	Recognizes the right of every child to a free and appropriate education; includes concept that children with disabilities are to be educated with their peers; extends reasonable accommodation, least restrictive environment (LRE), and assistive technology devices and services to age 3–21 education; mandates Individualized Educational Plan for each child, to include consideration of assistive technologies; also includes mandated services for children from birth to 2 and expanded emphasis on educationally related assistive technologies
Assistive Technology Act of 1998 (replaced Technology Related Assistance for Individuals with Disabilities Act of 1998)	First legislation to specifically address expansion of assistive technology devices and services; mandates consumer-driven assistive technology services, capacity building, advocacy activities, and statewide system change; supports grants to expand and administer alternative financing of assistive technology systems
Developmental Disabilities Assistance and Bill of Rights Act	Provides grants to states for developmental disabilities councils, university-affiliated programs, and protection and advocacy activities for persons with developmental disabilities; provides training and technical assistance to improve access to assistive technology services for individuals with developmental disabilities
Americans with Disabilities Act (ADA) of 1990	Prohibits discrimination on the basis of disability in employment, state and local government, public accommodations, commercial facilities, transportation, and telecommunications, all of which affect the application of assistive technology; use of assistive technology impacts requirement that Title II entities must communicate effectively with people who have hearing, vision, or speech disabilities; addresses telephone and television access for people with hearing and speech disabilities
Medicaid	Income-based ("means-tested") program; eligibility and services differ from state to state; federal government sets general program requirements and provides financial assistance to the states by matching state expenditures; assistive technology benefits differ for adults and children from birth to age 21; assistive technology for adults must be included in state's Medicaid plan or waiver program
Early Periodic Screening, Diagnosis, and Treatment Program	Mandatory service for children from birth through age 21; includes any required or optional service listed in the Medicaid Act; service need not be included in the state's Medicaid plan
Medicare	Major funding source for assistive technology (durable medical equipment); includes individuals 65 or over and those who are permanently and totally disabled; federally administered with consistent rules for all states

From Cook and Hussey (2002).

disorders (speech pathologists), special education, and engineering. In 1993, NIDRR committed $38.6 million to support Rehabilitation Engineering Centers that would focus on the following areas: adaptive computers and information systems, augmentative and alternative communication devices, employability for persons with low back pain, hearing enhancement and assistive devices, prosthetics and orthotics, quantification of physical performance, rehabilitation robotics, technology transfer and evaluation, improving wheelchair mobility, work site modifications and accommodations, geriatric assistive technology, personal licensed vehicles for disabled persons, rehabilitation technology services in vocational rehabilitation, technological aids for blindness and low vision, and technology for children with orthopedic disabilities. In fiscal year 1996, NIDRR funded 16 Rehabilitation Engineering Research Centers at a total cost of $11 million dollars and 45 Rehabilitation Research and Training Centers at a cost of $23 million dollars (NIDRR, 1999).

5.1.2 Sources of Information

Like any other emerging discipline, the knowledge base for rehabilitation engineering was scattered in disparate publications in the early years. Owing to its interdisciplinary nature, rehabilitation engineering research papers appeared in such diverse publications as the *Archives of Physical Medicine & Rehabilitation, Human Factors, Annals of Biomedical Engineering, IEEE Transactions on Biomedical Engineering*, and *Biomechanics*. Some of the papers were very practical and application specific, whereas others were fundamental and philosophical. In the early 1970s, many important papers were published by the Veterans Administration in its *Bulletin of Prosthetic Research*, a highly respected and widely disseminated peer-reviewed periodical. This journal was renamed the *Journal of Rehabilitation R&D* in 1983. In 1989, RESNA began *Assistive Technology*, a quarterly journal that focused on the interests of practitioners engaged in technological service delivery rather than the concerns of engineers engaged in research and development. The IEEE Engineering in Medicine and Biology Society founded the *IEEE Transactions on Rehabilitation Engineering* in 1993 to give scientifically based rehabilitation engineering research papers a much-needed home. This journal, which was renamed *IEEE Transactions on Neural Systems and Rehabilitation Engineering*, is published quarterly and covers the medical aspects of rehabilitation (rehabilitation medicine), its practical design concepts (rehabilitation technology), its scientific aspects (rehabilitation science), and neural systems.

5.1.3 Major Activities in Rehabilitation Engineering

The major activities in this field can be categorized in many ways. Perhaps the simplest way to grasp its breadth and depth is to categorize the main types of assistive technology that rehabilitation engineering has produced (Table 5.2). The development of these technological products required the contributions of mechanical, material, and electrical engineers, orthopedic surgeons, prosthetists and orthotists, allied health professionals, and computer professionals. For example, the use of voice in many assistive devices, as both inputs and outputs, depends on digital signal processing chips, memory chips, and sophisticated software developed by electrical and computer engineers. Figures 5.2 through 5.4 illustrate some of the assistive technologies currently available. As explained in subsequent sections of this chapter, the proper design, development, and application of assistive technology devices

TABLE 5.2 Categories of Assistive Devices

Prosthetics and Orthotics
    Artificial hand, wrist, and arms
    Artificial foot and legs
    Hand splints and upper limb braces
    Functional electrical stimulation orthoses
Assistive Devices for Persons with Severe Visual Impairments
    Devices to aid reading and writing (e.g., closed circuit TV magnifiers, electronic Braille, reading
        machines, talking calculators, auditory and tactile vision substitution systems)
    Devices to aid independent mobility (e.g., Laser cane, Binaural Ultrasonic Eyeglasses, Handheld
        Ultrasonic Torch, electronic enunciators, robotic guide dogs)
Assistive Devices for Persons with Severe Auditory Impairments
    Digital hearing aids
    Telephone aids (e.g., TDD and TTY)
    Lipreading aids
    Speech to text converters
Assistive Devices for Tactile Impairments
    Cushions
    Customized seating
    Sensory substitution
    Pressure relief pumps and alarms
Alternative and Augmentative Communication Devices
    Interface and keyboard emulation
    Specialized switches, sensors, and transducers
    Computer-based communication devices
    Linguistic tools and software
Manipulation and Mobility Aids
    Grabbers, feeders, mounting systems, and page turners
    Environmental controllers
    Robotic aids
    Manual and special-purpose wheelchairs
    Powered wheelchairs, scooters, and recliners
    Adaptive driving aids
    Modified personal licensed vehicles
Recreational Assistive Devices
    Arm-powered cycles
    Sports and racing wheelchairs
    Modified sit-down mono-ski

require the combined efforts of engineers, knowledgeable and competent clinicians, informed end users or consumers, and caregivers.

5.2 THE HUMAN COMPONENT

To knowledgeably apply engineering principles and fabricate devices that will help persons with disabling conditions, it is necessary to have a perspective on the

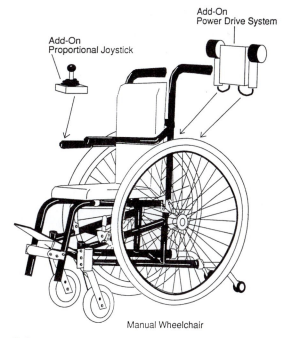

Figure 5.2   Add-on wheelchair system (from Church and Glennen, 1992).

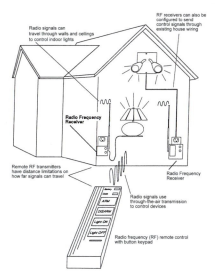

Figure 5.3   Environmental control unit using radio frequency (RF) control (from Church and Glennen, 1992).

human component and the consequence of various impairments. One way to view a human being is as a receptor, processor, and responder of information (Fig. 5.5). The human user of assistive technology perceives the environment via senses and responds or manipulates the environment via effectors. Interposed between the sensors and effectors are central processing functions that include perception, cognition, and movement control. *Perception* is the way in which the human being interprets the incoming sensory data. The mechanism of perception relies on the neural circuitry found in the peripheral nervous system and central psychological factors such as memory of previous sensory experiences. *Cognition* refers to activities that underlie problem solving, decision making, and language formation. *Movement control* utilizes the outcome of the processing functions described previously to form a motor pattern that is executed by the effectors (nerves, muscles, and joints). The impact of the effectors on the environment is then detected by the sensors, thereby providing feedback between the human and the environment. When something goes wrong in the information processing chain, disabilities often result. Table 5.3 lists the prevalence of various disabling conditions in terms of anatomic locations.

Interestingly, rehabilitation engineers have found a modicum of success when trauma or birth defects damage the input (sensory) end of this chain of information processing. When a sensory deficit is present in one of the three primary sensory channels (vision, hearing, and touch), assistive devices can detect important environmental information and present it via one or more of the other remaining senses. For example, sensory aids for severe visual impairments utilize tactile and/or auditory outputs to display important environmental information to the user. Examples of such sensory aids include laser canes, ultrasonic glasses, and robotic guide dogs. Rehabilitation engineers also have been modestly successful at replacing or augmenting some motoric (effector) disabilities (Fig. 5.6). As listed in Table 5.2, these include artificial arms and legs, wheelchairs of all types, environmental controllers, and, in the future, robotic assistants.

However, when dysfunction resides in the "higher information processing centers" of a human being, assistive technology has been much less successful in ameliorating the resultant limitations. For example, rehabilitation engineers and speech pathologists have been unsuccessful in enabling someone to communicate effectively when that person has difficulty formulating a message (aphasia) following a stroke. Despite the variety of modern and sophisticated alternative and augmentative communication devices that are available, none has been able to replace the volitional aspects of the human being. If the user is unable to cognitively formulate a message, an augmentative communication device is often powerless to help.

An awareness of the psychosocial adjustments to chronic disability is desirable because rehabilitation engineering and assistive technology seek to ameliorate the consequences of disabilities. Understanding the emotional and mental states of the person who is or becomes disabled is necessary so that offers of assistance and recommendations of solutions can be appropriate, timely, accepted, and, ultimately, used.

One of the biggest impacts of chronic disability is the minority status and socially devalued position that a disabled person experiences in society. Such loss of social

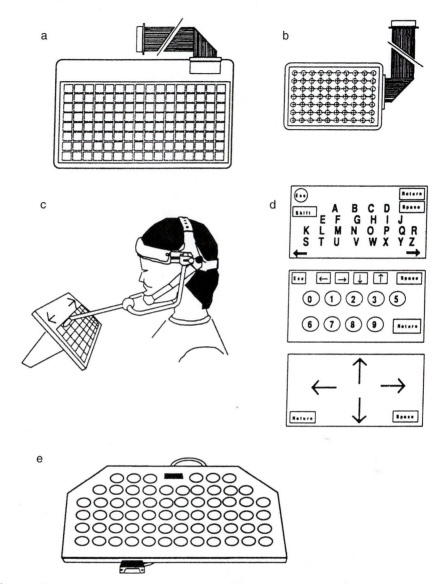

Figure 5.4 Alternative keyboards can replace or operate in addition to the standard keyboard. (a) Expanded keyboards have a matrix of touch-sensitive squares that can be grouped together to form larger squares. (b) Minikeyboards are small keyboards with a matrix of closely spaced touch-sensitive squares. (c) The small size of a minikeyboard ensures that a small range of movement can reach the entire keyboard. (d) Expanded and minikeyboards use standard or customized keyboard overlays. (e) Some alternative keyboards plug directly into the keyboard jack of the computer, needing no special interface or software (from Church and Glennen, 1992).

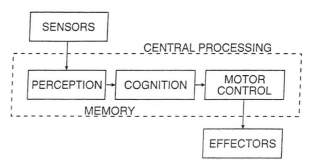

Figure 5.5   An information processing model of the human operator of assistive technologies. Each block represents a group of functions related to the use of technology.

TABLE 5.3   Prevalence of Disabling Conditions in the United States

---

45–50 million persons have disabilities that slightly limit their activities
   32% hearing
   21% sight
   18% back or spine
   16% leg and hip
   5% arm and shoulder
   4% speech
   3% paralysis
   1% limb amputation
7–11 million persons have disabilities that significantly limit their activities
   30% back or spine
   26% leg and hip
   13% paralysis
   9% hearing
   8% sight
   7% arm and shoulder
   4% limb amputation
   3% speech

Data from Stolov and Clowers (1981).

status may result from the direct effects of disability (social isolation) and the indirect effects of disability (economic setbacks). Thus, in addition to the tremendous drop in personal income, a person who is disabled must battle three main psychological consequences of disability: the loss of self-esteem, the tendency to be too dependent on others, and passivity.

For individuals who become disabled through traumatic injuries, the adjustment to disability generally passes through five phases: shock, realization, defensive retreat or denial, acknowledgment, and adaptation or acceptance. During the first days after the onset of disability, the individual is usually in shock, feeling and reacting minimally

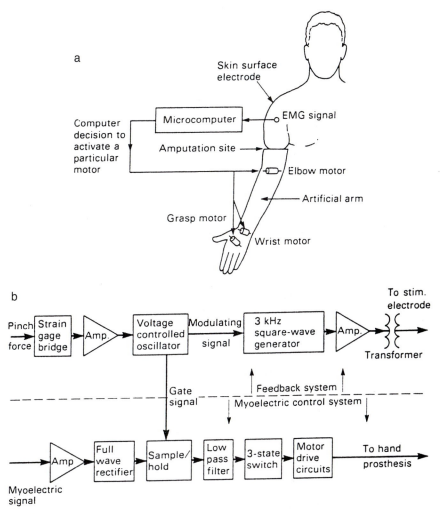

Figure 5.6 (a) This system generates temporal signatures from one set of myoelectric electrodes to control multiple actuators. (b) Electrical stimulaton of the forearm to provide force feedback may be carried out using a system like this one (from Webster et al., 1985).

with the surroundings and showing little awareness of what has happened. Counseling interventions or efforts of rehabilitation technologists are typically not very effective at this time.

After several weeks or months, the individual usually begins to acknowledge the reality and seriousness of the disability. Anxiety, fear, and even panic may be the predominant emotional reactions. Depression and anger may also occasionally appear during this phase. Because of the individual's emotional state, intense or sustained intervention efforts are not likely to be useful during this time.

In the next phase, the individual makes a defensive retreat in order to not be psychologically overwhelmed by anxiety and fear. Predominant among these defenses is denial—claiming that the disability is only temporary and that full recovery will occur. Such denial may persist or reappear occasionally long after the onset of disability.

Acknowledgment of the disability occurs when the individual achieves an accurate understanding of the nature of the disability in terms of its limitations and likely outcome. Persons in this phase may exhibit a thorough understanding of the disability but may not possess a full appreciation of its implications. The gradual recognition of reality is often accompanied by depression and a resultant loss of interest in many activities previously enjoyed.

Adaptation, or the acceptance phase, is the final and ultimate psychological goal of a person's adjustment to disability. An individual in this phase has worked through the major emotional reactions to disability. Such a person is realistic about the likely limitations and is psychologically ready to make the best use of his or her potential. Intervention by rehabilitation engineers or assistive technologists during the acknowledgment and acceptance phases of the psychosocial adjustment to disability is usually appropriate and effective. Involvement of the disabled individual in identifying needs, planning the approach, and choosing among possible alternatives can be very beneficial both psychologically and physically.

5.3   PRINCIPLES OF ASSISTIVE TECHNOLOGY ASSESSMENT

Rehabilitation engineers not only need to know the physical principles that govern their designs, but they also must adhere to some key principles that govern the applications of technology for people with disabilities. To be successful, the needs, preferences, abilities, limitations, and even environment of the individual seeking the assistive technology must be carefully considered. There are at least five major misconceptions that exist in the field of assistive technology:

Misconception #1. *Assistive technology can solve all the problems.* Although assistive devices can making accomplishing tasks easier, technology alone cannot mitigate all the difficulties that accompany a disability.

Misconception #2. *Persons with the same disability need the same assistive devices.* Assistive technology must be individualized because similarly disabled persons can have very different needs, wants, and preferences (Wessels et al., 2003).

Misconception #3. *Assistive technology is necessarily complicated and expensive.* Sometimes low-technology devices are the most appropriate and even preferred for their simplicity, ease of use and maintenance, and low cost.

Misconception #4. *Assistive technology prescriptions are always accurate and optimal.* Experiences clearly demonstrate that the application of technology for persons with disabilities is inexact and will change with time. Changes in the assistive technology user's health, living environment, preferences, and circumstances will require periodic reassessment by the user and those rehabilitation professionals who are giving assistance (Philips and Zhao, 1993).

Misconception #5. *Assistive technology will always be used.* According to data from the 1990 U.S. Census Bureau's National Health Interview Survey, about one-third of the assistive devices not needed for survival are unused or abandoned just 3 months after they were initially acquired.

In addition to avoiding common misconceptions, a rehabilitation engineer and technologist should follow several principles that have proven to be helpful in matching appropriate assistive technology to the person or consumer. Adherence to these principles will increase the likelihood that the resultant assistive technology will be welcomed and fully utilized.

Principle #1. *The user's goals, needs, and tasks must be clearly defined, listed, and incorporated as early as possible in the intervention process.* To avoid overlooking needs and goals, checklists and premade forms should be used. A number of helpful assessment forms can be found in the references given in the suggested reading list at the end of this chapter.

Principle #2. *Involvement of rehabilitation professionals with differing skills and know-how will maximize the probability for a successful outcome.* Depending on the purpose and environment in which the assistive technology device will be used, a number of professionals should participate in the process of matching technology to a person's needs. Table 5.4 lists various technology areas and the responsible professionals.

Principle #3. *The user's preferences, cognitive and physical abilities and limitations, living situation, tolerance for technology, and probable changes in the future must be thoroughly assessed, analyzed, and quantified.* Rehabilitation engineers will find that the highly descriptive vocabulary and qualitative language used by nontechnical professionals needs to be translated into attributes that can be measured and quantified. For example, whether a disabled person can use one or more upper limbs should be quantified in terms of each limb's ability to reach, lift, and grasp.

Principle #4. *Careful and thorough consideration of available technology for meeting the user's needs must be carried out to avoid overlooking potentially useful solutions.* Electronic databases (e.g., assistive technology websites and websites of major technology vendors) can often provide the rehabilitation engineer or assistive technologist with an initial overview of potentially useful devices to prescribe, modify, and deliver to the consumer.

Principle #5. *The user's preferences and choice must be considered in the selection of the assistive technology device.* Surveys indicate that the main reason assistive technology is rejected or poorly utilized is inadequate consideration of the user's

TABLE 5.4   Professional Areas in Assistive Technology

Technology Area	Responsible Professionals*
Academic and vocational skills	Special education
	Vocational rehabilitation
	Psychology
Augmentative communication	Speech–language pathology
	Special education
Computer access	Computer technology
	Vocational rehabilitation
Daily living skills	Occupational therapy
	Rehabilitation technology
Specialized adaptations	Rehabilitation engineering
	Computer technology
	Prosthetics/orthotics
Mobility	Occupational therapy
	Physical therapy
Seating and positioning	Occupational therapy
	Physical therapy
Written communication	Speech–language pathology
	Special education

*Depending on the complexity of technical challenges encountered, an assistive technologist or a rehabilitation engineer can be added to the list of responsible professionals.

needs and preferences. Throughout the process of searching for appropriate technology, the ultimate consumer of that technology should be viewed as a partner and stakeholder rather than as a passive, disinterested recipient of services.

Principle #6. *The assistive technology device must be customized and installed in the location and setting where it primarily will be used.* Often seemingly minor or innocuous situations at the usage site can spell success or failure in the application of assistive technology.

Principle #7. *Not only must the user be trained to use the assistive device, but also the attendants or family members must be made aware of the device's intended purpose, benefits, and limitations.* For example, an augmentative communication device usually will require that the communication partners adopt a different mode of communication and modify their behavior so that the user of this device can communicate a wider array of thoughts and even assume a more active role in the communication paradigm, such as initiating a conversation or changing the conversational topic. Unless the attendants or family members alter their ways of interacting, the newly empowered individual will be dissuaded from utilizing the communication device, regardless of how powerful it may be.

Principle #8. *Follow-up, readjustment, and reassessment of the user's usage patterns and needs are necessary at periodic intervals.* During the first 6 months following the delivery of the assistive technology device, the user and others in that environment learn to accommodate to the new device. As people and the environment change, what worked initially may become inappropriate, and the assistive device may need to be reconfigured or reoptimized. Periodic follow-up and adjustments will lessen technology abandonment and the resultant waste of time and resources.

5.4 PRINCIPLES OF REHABILITATION ENGINEERING

Knowledge and techniques from different disciplines must be utilized to design technological solutions that can alleviate problems caused by various disabling conditions. Since rehabilitation engineering is intrinsically multidisciplinary, identifying universally applicable principles for this emerging field is difficult. Often the most relevant principles depend on the particular problem being examined. For example, principles from the fields of electronic and communication engineering are paramount when designing an environmental control system that is to be integrated with the user's battery-powered wheelchair. However, when the goal is to develop an implanted functional electrical stimulation orthosis for an upper limb impaired by spinal cord injury, principles from neuromuscular physiology, biomechanics, biomaterials, and control systems would be the most applicable.

Whatever the disability to be overcome, however, rehabilitation engineering is inherently design oriented. Rehabilitation engineering design is the creative process of identifying needs and then devising an assistive device to fill those needs. A systematic approach is essential to successfully complete a rehabilitation project. Key elements of the design process involve the following sequential steps: analysis, synthesis, evaluation, decision, and implementation.

Analysis

Inexperienced but enthusiastic rehabilitation engineering students often respond to a plea for help from someone with a disability by immediately thinking about possible solutions. They overlook the important first step of doing a careful analysis of the problem or need. What they discover after much ineffectual effort is that a thorough investigation of the problem is necessary before any meaningful solution can be found. Rehabilitation engineers first must ascertain where, when, and how often the problem arises. What is the environment or the task situation? How have others performed the task? What are the environmental constraints (size, speed, weight, location, physical interface, etc.)? What are the psychosocial constraints (user preferences, support of others, gadget tolerance, cognitive abilities, and limitations)? What are the financial considerations (purchase price, rental fees, trial periods, maintenance and repair arrangements)? Answers to these questions will require diligent investigation and quantitative data such as the weight and size to be lifted, the shape and texture of the object to be manipulated, and the operational features of the desired device. An excellent endpoint of problem analysis would be a list of operational features or performance specifications that the "ideal" solution should possess.

Such a list of performance specifications can serve as a valuable guide for choosing the best solution during later phases of the design process.

Example Problem 5.1

Develop a set of performance specifications for an electromechanical device to raise and lower the lower leg of a wheelchair user (to prevent edema).

Solution

A sample set of performance specifications about the ideal mechanism might be written as follows:

- Be able to raise or lower leg in 5 s
- Independently operable by the wheelchair occupant
- Have an emergency stop switch
- Compatible with existing wheelchair and its leg rests
- Quiet operation
- Entire adaptation weighs no more than five pounds   ∎

Synthesis

A rehabilitation engineer who is able to describe in writing the nature of the problem is likely to have some ideas for solving the problem. Although not strictly sequential, the synthesis of possible solutions usually follows the analysis of the problem. The synthesis of possible solutions is a creative activity that is guided by previously learned engineering principles and supported by handbooks, design magazines, product catalogs, and consultation with other professionals. While making and evaluating the list of possible solutions, a deeper understanding of the problem usually is reached and other, previously not apparent, solutions arise. A recommended endpoint for the synthesis phase of the design process includes sketches and technical descriptions of each trial solution.

Evaluation

Depending on the complexity of the problem and other constraints such as time and money, the two or three most promising solutions should undergo further evaluation, possibly via field trials with mockups, computer simulations, and/or detailed mechanical drawings. Throughout the evaluation process, the end user and other stakeholders in the problem and solution should be consulted. Experimental results from field trials should be carefully recorded, possibly on videotape, for later review. One useful method for evaluating promising solutions is to use a quantitative comparison chart to rate how well each solution meets or exceeds the performance specifications and operational characteristics based on the analysis of the problem.

Decision

The choice of the final solution is often made easier when it is understood that the final solution usually involves a compromise. After comparing the various promising solutions, more than one may appear equally satisfactory. At this point, the final decision may be made based on the preference of the user or some other intangible factor that is difficult to anticipate. Sometimes choosing the final solution may involve consulting with someone else who may have encountered a similar problem. What is most important, however, is careful consideration of the user's preference (principle 5 of assistive technology).

Implementation

To fabricate, fit, and install the final (or best) solution requires additional project planning that, depending on the size of the project, may range from a simple list of tasks to a complex set of scheduled activities involving many people with different skills.

Example Problem 5.2

List the major technical design steps needed to build the automatic battery-powered leg raiser described in Example Problem 5.1.

Solution

The following are some of the key design steps:

- Mechanical design of the linkages to raise the wheelchair's leg rests
- Static determination of the forces needed to raise the occupant's leg
- Determination of the gear ratios and torque needed from the electric motor
- Estimation of the power drain from the wheelchair batteries
- Purchase of the electromechanical components
- Fabrication of custom parts and electronic components
- Assembly, testing, and possible redesign
- Field trials and evaluation of prototype device ∎

5.4.1 Key Engineering Principles

Each discipline and subdiscipline that contributes to rehabilitation engineering has its own set of key principles that should be considered when a design project is begun. For example, a logic family must be selected and a decision whether to use synchronous or asynchronous sequential circuits must be made at the outset in digital design. A few general hardware issues are applicable to a wide variety of design tasks, including worst-case design, computer simulation, temperature effects, reliability, and product safety. In worst-case design, the electronic or mechanical system must continue to operate adequately even when variations in component values degrade performance. Computer simulation and computer-aided design (CAD) software often can be used to predict how well an overall electronic system will perform under different combinations of component values or sizes.

The design also should take into account the effects of temperature and environmental conditions on performance and reliability. For example, temperature extremes can reduce a battery's capacity. Temperature also may affect reliability, so proper venting and use of heat sinks should be employed to prevent excessive temperature increases. For reliability and durability, proper strain relief of wires and connectors should be used in the final design.

Product safety is another very important design principle, especially for rehabilitative or assistive technology. An electromechanical system should always incorporate a panic switch that will quickly halt a device's operation if an emergency arises. Fuses and heavy-duty gauge wiring should be employed throughout for extra margins of safety. Mechanical stops and interlocks should be incorporated to ensure proper interconnections and to prevent dangerous or inappropriate movement.

When the required assistive device must lift or support some part of the body, an analysis of the static and dynamic forces (biomechanics) that are involved should be performed. The simplest analysis is to determine the static forces needed to hold the object or body part in a steady and stable manner. The basic engineering principles needed for static and dynamic analysis usually involve the following steps: (1) Determine the force vectors acting on the object or body part, (2) determine the moment arms, and (3) ascertain the centers of gravity for various components and body segments. Under static conditions, all the forces and moment vectors sum to zero. For dynamic conditions, the governing equation is Newton's second law of motion in which the vector sum of the forces equals mass times an acceleration vector $(F = ma)$.

Example Problem 5.3

Suppose a 125-lb person lies supine on a board resting on knife edges spaced 72 in. apart (Fig. 5.7). Assume that the center of gravity of the lower limb is located through the center line of the limb and 1.5 in. above the knee cap. Estimate the weight of this person's right leg.

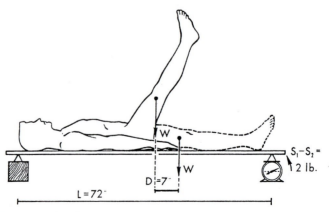

Figure 5.7 Method of weighing body segments with board and scale (from Le Veau, 1976).

Solution

Record the scale reading with both legs resting comfortably on the board and when the right leg is raised almost straight up. Sum the moments about the left knife edge pivot to yield the following static equation:

$$WD = L(S_1 - S_2)$$

where W is the weight of the right limb, L is the length of the board between the supports, S_1 is the scale reading with both legs resting on the board, S_2 is the scale reading with the right leg raised, and D is the horizontal distance through which the limb's center of gravity was moved when the limb was raised. Suppose the two scale readings were 58 lbs for S_1 and 56 lbs for S_2 and $D = 7$ in. Substituting these values into the equations would yield an estimate of 20.6 lbs as the weight of the right leg. ∎

Example Problem 5.4

A patient is exercising his shoulder extensor muscles with wall pulleys (Fig. 5.8). Weights of 20, 10, and 5 lbs are loaded on the weight pan, which weighs 4 lbs. The patient is able to exert 45 lbs on the pulley. What is the resultant force of the entire system? What are the magnitude and direction of acceleration of the weights?

Solution

All the weights and the pan act straight down, whereas the 45 lbs of tension on the pulley's cable exerts an upward force. The net force (F) is 6 lbs upward. Using Newton's second law of motion, $F = ma$, where m is the mass of the weights and the pan and a is the acceleration of the weights and pan. The mass, m, is found by dividing the weight of 39 lbs by the acceleration of gravity ($32.2 \, \text{ft/s}^2$) to yield $m = 1.21$ slugs. Substituting these values into $a = F/m$ yields an acceleration of $4.96 \, \text{ft/s}^2$ in the upward direction.

5.4.2 Key Ergonomic Principles

Ergonomics or human factors is another indispensable part of rehabilitation engineering and assistive technology design. Applying information about human behavior, abilities, limitations, and other characteristics to the design of tools, adaptations, electronic devices, tasks, and interfaces is especially important when designing assistive technology because persons with disabilities generally will be less able to accommodate poorly designed or ill-fitted assistive devices. Several ergonomic principles that are especially germane to rehabilitation engineering are discussed in the following sections.

Principle of Proper Positioning

Without proper positioning or support, an individual who has lost the ability to maintain a stable posture against gravity may appear to have greater deformities and functional limitations than truly exist. For example, the lack of proper arm support may make the operation of even an enlarged keyboard unnecessarily slow

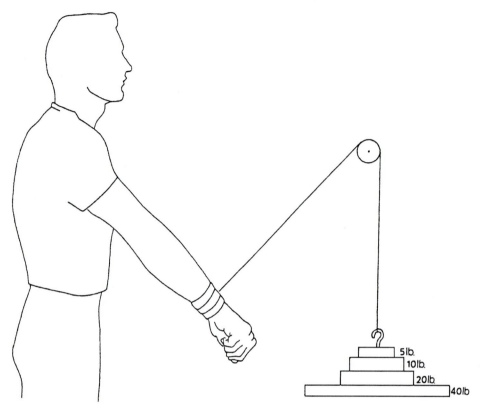

Figure 5.8 Patient exercising his shoulder extensor muscles with wall pulleys (from Le Veau, 1976).

or mistake prone. Also, the lack of proper upper trunk stability may unduly limit the use of an individual's arms because the person is relying on them for support.

During all phases of the design process, the rehabilitation engineer must ensure that whatever adaptation or assistive technology is being planned, the person's trunk, lower back, legs, and arms will have the necessary stability and support at all times (Fig. 5.9). Consultation with a physical therapist or occupational therapist familiar with the focus individual during the initial design phases should be considered if postural support appears to be a concern. Common conditions that require consider-ations of seating and positioning are listed in Table 5.5.

Principle of the Anatomical Control Site

Since assistive devices receive command signals from the users, users must be able to reliably indicate their intent by using overt, volitional actions. Given the variety of switches and sensors that are available, any part of the body over which the user has reliable control in terms of speed and dependability can serve as the anatomical control site. Once the best site has been chosen, an appropriate interface for that

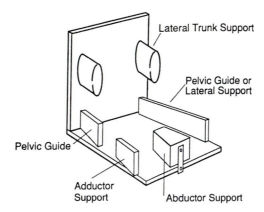

Figure 5.9 Chair adaptations for proper positioning (from Church and Glennen, 1992).

TABLE 5.5 Conditions That Require Consideration of Seating and Positioning

Condition	Description and Characteristics	Seating Considerations
Cerebral palsy	Nonprogressive neuromuscular	
Increased tone (high tone)	Fixed deformity, decreased movements, abnormal patterns	Correct deformities, improve alignment, decrease tone
Decreased tone (low tone)	Subluxations, decreased active movement, hypermobility	Provide support for upright positioning, promote development of muscular control
Athetoid (mixed tone)	Excessive active movement, decreased stability	Provide stability, but allow controlled mobility for function
Muscular dystrophies	Degenerative neuromuscular	
Duchenne	Loss of muscular control proximal to distal	Provide stable seating base, allow person to find balance point
Multiple sclerosis	Series of exacerbations and remissions	Prepare for flexibility of system to follow needs
Spina bifida	Congenital anomaly consisting of a deficit in one or more of the vertebral arches, decreased or absent sensation	Reduce high risk for pressure concerns, allow for typically good upper extremity and head control
Spinal cord injury	Insult to spinal cord, partial or complete loss of function below level of injury, nonprogressive once stabilized, decreased or absent sensation, possible scoliosis/kyphosis	Reduce high risk for pressure concerns, allow for trunk movements used for function
Osteogenesis imperfecta	Connective tissue disorder, brittle bone disease, limited functional range, multiple fractures	Provide protection
Orthopedic impairments	Fixed or flexible	If fixed, support, if flexible, correct

(continued)

TABLE 5.5   Conditions That Require Consideration of Seating and Positioning (Continued)

Condition	Description and Characteristics	Seating Considerations
Traumatic brain injury	Severity dependent on extent of central nervous system damage, may have cognitive component, nonprogressive once stabilized	Allow for functional improvement as rehabilitation progresses, establish a system that is flexible to changing needs
Elderly		
Typical aged	Often, fixed kyphosis, decreased bone mass, and decreased strength, incontinence	Provide comfort and visual orientation, moisture-proof, accommodate kyphosis
Aged secondary to primary disability	Example—older patients with cerebral palsy may have fixed deformities	Provide comfort, support deformities

Adapted with permission from *Evaluating, Selecting, and Using Appropriate Assistive Technology*, J. C. Galvin, M. J. Scherer, p. 66, © 1996 Aspen Publishers, Inc.

site can be designed by using various transducers, switches, joysticks, and keyboards. In addition to the obvious control sites such as the finger, elbow, shoulder, and knee, subtle movements such as raising an eyebrow or tensing a particular muscle can also be employed as the control signal for an assistive device. Often, the potential control sites can and should be analyzed and quantitatively compared for their relative speed, reliability, distinctiveness, and repeatability of control actions. Field trials using mockups, stopwatches, measuring tapes, and a video camera can be very helpful for collecting such performance data.

When an individual's physical abilities do not permit direct selection from among a set of possible choices, single switch activation by the anatomical control site in combination with automated row-column scanning of a matrix is often used. In row-column scanning, each row of a matrix lights up sequentially from the top to the bottom. When the row containing the desired item is highlighted, the user selects it using a switch. Then each item in that row is scanned (from left to right) until the desired item is chosen by a second switch activation. The speed with which a two-dimensional array can be used to compose messages depends on the placement of the letters in that array. Two popular arrangements of alphanumeric symbols—the alphabetic arrangement and the frequency of occurrence arrangement of the alphabet—are shown in Example Problem 5.5.

Example Problem 5.5

Assume that a communication device has either an alphabetical arrangement of letters or a frequency arrangement and does row-column scanning as follows: (1) Two switch activations are needed to select a particular item in the array; (2) The dwell time for each row (starting at the top) is 1.5 s; (3) The dwell time along a selected row (starting from the left) is 1.5 s; and (4) The scan begins at the top row after a successful selection.

For both arrangements, calculate the predicted time needed to generate the phrase "I WANT TO GO TO SEA WORLD." Assume zero errors or missed opportunities.

Alphabetical Arrangement of Letters

SPACE	A	B	C	D	E	F
G	H	I	J	K	L	M
N	O	P	Q	R	S	T
U	V	W	X	Y	Z	TH
IN	ER	RE	AN	HE	.	,

Frequency Arrangement of Letters

SPACE	E	A	I	L	HE	Y
T	O	S	D	P	AN	ER
N	R	C	F	IN	ES	Q
H	TH	M	B	V	X	Z
U	W	G	K	J	.	,

Solution

The time needed to compose the target sentence is equal to the number of steps needed to select each letter and space in that sentence. For the alphabetically arranged array, 5 dwell steps (2nd row plus 3rd column) at 1.5 s per step are needed to reach the letter *I*. For the frequency of occurrence array, 5 dwell steps (1st row plus 4th column) also are needed to reach the letter *I*. To insert a space, both arrays require 2 dwell steps (1st row plus 1st column). For the letter *W*, the same number of dwell steps (7) are needed in both arrays. For the letter *T*, however, 10 dwell steps are needed in the alphabetical array but just 3 dwell steps are needed in the frequency of occurrence array. Each time the letter *T* is used, 7 dwell steps (or 10.5 s) are saved with the frequency of occurrence array. Thus, the time needed to produce the sample sentence, assuming no errors, is 213 s when using the alphabetical array and 180 s when using the frequency array. Notice that even for a 7-word sentence, over half a minute can be saved with the faster frequency arrangement array and that additional time was saved by using the double letter combination *AN* rather than selecting the single letters *A* and *N* separately. ■

Principle of Simplicity and Intuitive Operation

The universal goal of equipment design is to achieve intuitively simple operation, and this is especially true for electronic and computer-based assistive devices. The key to intuitively simple operation lies in the proper choice of compatible and optimal controls and displays. *Compatibility* refers to the degree to which relationships between the control actions and indicator movements are consistent, respectively, with expectations of the equipment's response and behavior. When compatibility relationships are incorporated into an assistive device, learning is faster, reaction time is shorter, fewer errors occur, and the user's satisfaction is higher. Although people can and do learn to use adaptations that do not conform to their expectations, they do so at a price (producing more errors, working more slowly, and/or requiring more attention). Hence, the rehabilitation engineer needs to be aware of and follow some common compatibility relationships and basic ergonomic guidelines, such as:

- The display and corresponding control should bear a physical resemblance to each other.
- The display and corresponding control should have similar physical arrangements and/or be aided by guides or markers.
- The display and corresponding control should move in the same direction and within the same spatial plane (e.g., rotary dials matched with rotary displays, linear vertical sliders matched with vertical displays).
- The relative movement between a switch or dial should be mindful of population stereotypic expectations (e.g., an upward activation to turn something on, a clockwise rotation to increase something, and scale numbers that increase from left to right).

Additional guidelines for choosing among various types of visual displays are given in Table 5.6.

Principle of Display Suitability

In selecting or designing displays for transmission of information, the selection of the sensory modality is sometimes a foregone conclusion, such as when designing a warning signal for a visually impaired person. When there is an option, however, the rehabilitation engineer must take advantage of the intrinsic advantages of one sensory modality over another for the type of message or information to be conveyed. For example, audition tends to have an advantage over vision in vigilance types of warnings because of its attention-getting qualities. A more extensive comparison of auditory and visual forms of message presentation is presented in Table 5.7.

Principle of Allowance for Recovery from Errors

Both rehabilitation engineering and human factors or ergonomics seek to design assistive technology that will expand an individual's capabilities while minimizing errors. However, human error is unavoidable no matter how well something is designed. Hence, the assistive device must provide some sort of allowance for errors without seriously compromising system performance or safety. Errors can be classified as errors of omission, errors of commission, sequencing errors, and timing errors.

A well-designed computer-based electronic assistive device will incorporate one or more of the following attributes:

- The design makes it inherently impossible to commit the error (e.g., using jacks and plugs that can fit together only one way or the device automatically rejects inappropriate responses while giving a warning).
- The design makes it less likely, but not impossible to commit the error (e.g., using color-coded wires accompanied by easily understood wiring diagrams).
- The design reduces the damaging consequences of errors without necessarily reducing the likelihood of errors (e.g., using fuses and mechanical stops that limit excessive electrical current, mechanical movement, or speed).

TABLE 5.6 General Guide to Visual Display Selection

To Display	Select	Because	Example
Go, no go, start, stop, on, off	Light	Normally easy to tell if it is on or off.	
Identification	Light	Easy to see (may be coded by spacing, color, location, or flashing rate; may also have label for panel applications).	
Warning or caution	Light	Attracts attention and can be seen at great distance if bright enough (may flash intermittently to increase conspicuity).	
Verbal instruction (operating sequence)	Enunciator light	Simple "action instruction" reduces time required for decision making.	RELEASE EJECT
Exact quantity	Digital counter	Only one number can be seen, thus reducing chance of reading error.	
Approximate quantity	Moving pointer against fixed scale	General position of pointer gives rapid clue to the quantity plus relative rate of change.	
Set-in quantity	Moving pointer against fixed scale	Natural relationship between control and display motions.	
Tracking	Single pointer or cross pointers against fixed index	Provides error information for easy correction.	
Vehicle attitude	Either mechanical or electronic display of position of vehicle against established reference (may be graphic or pictorial)	Provides direct comparison of own position against known reference or base line.	

Abstracted from Human Factors in Engineering and Design, 7th Ed., by Sanders and McCormick, 1993.

TABLE 5.7   Choosing Between Auditory and Visual Forms of Presentation

Use Auditory Presentation if	Use Visual Presentation if
The message is simple.	The message is complex.
The message is short.	The message is long.
The message will not be referred to later.	The message will be referred to later.
The message deals with events in time.	The message deals with location in space.
The message calls for immediate action.	The message does not call for immediate action.
The visual system of the person is overburdened.	The auditory system of the person is overburdened.
The message is to be perceived by persons not in the area.	The message is to be perceived by someone very close by.
Use artificially generated speech if the listener cannot read.	Use visual display if the message contains graphical elements.

Adapted and modified from Saunders and McCormick (1993, p. 53, Table 3-1).

- The design incorporates an "undo," "escape," or "go-back" command in devices that involve the selection of options within menus.

Principle of Adaptability and Flexibility

One fundamental assumption in ergonomics is that devices should be designed to accommodate the user and not vice versa. As circumstances change and/or as the user gains greater skill and facility in the operation of an assistive device, its operational characteristics must adapt accordingly. In the case of an augmentative electronic communication device, its vocabulary set should be changed easily as the user's needs, skills, or communication environment change. The method of selection and feedback also should be flexible, perhaps offering direct selection of the vocabulary choices in one situation while reverting to a simpler row-column scanning in another setting. The user should also be given the choice of having auditory, visual, or a combination of both as feedback indicators.

Principle of Mental and Chronological Age Appropriateness

When working with someone who has had lifelong and significant disabilities, the rehabilitation engineer cannot presume that the mental and behavioral age of the individual with disabilities will correspond closely with that person's chronological age. In general, people with congenital disabilities tend to have more limited variety, diversity, and quantity of life experiences. Consequently, their reactions and behavioral tendencies often mimic those of someone much younger. Thus, during assessment and problem definition, the rehabilitation engineer should ascertain the functional age of the individual to be helped. Behavioral and biographical information can be gathered by direct observation and by interviewing family members, teachers, and social workers.

Special human factor considerations also need to be employed when designing assistive technology for very young children and elderly individuals. When designing adaptations for such individuals, the rehabilitation engineer must consider that they may have a reduced ability to process and retain information. For example, generally more time is required for very young children and older people to retrieve information from long-term memory, to choose among response alternatives, and to execute correct responses. Studies have shown that elderly persons are much slower in searching for material in long-term memory, in shifting attention from one task to another, and in coping with conceptual, spatial, and movement incongruities.

The preceding findings suggest that the following design guidelines be incorporated into any assistive device intended for an elderly person:

- Strengthen the displayed signals by making them louder, brighter, larger, etc.
- Simplify the controls and displays to reduce irrelevant details that could act as sources of confusion.
- Maintain a high level of conceptual, spatial, and movement congruity, i.e., compatibility between the controls, display, and device's response.
- Reduce the requirements for monitoring and responding to multiple tasks.
- Provide more time between the execution of a response and the need for the next response. Where possible, let the user set the pace of the task.
- Allow more time and practice for learning the material or task to be performed.

5.5 PRACTICE OF REHABILITATION ENGINEERING AND ASSISTIVE TECHNOLOGY

5.5.1 Career Opportunities

As efforts to constrain health care costs intensify, it is reasonable to wonder whether career opportunities will exist for rehabilitation engineers and assistive technologists. Given an aging population, the rising number of children born with cognitive and physical developmental disorders, the impact of recent legislative mandates (Table 5.1), and the proven cost benefits of successful rehabilitation, the demand for assistive technology (new and existent) will likely increase rather than decrease. Correspondingly, employment opportunities for technically oriented persons interested in the development and delivery of assistive technology should steadily increase as well.

In the early 1980s, the value of rehabilitation engineers and assistive technologists was unappreciated and thus required significant educational efforts. Although the battle for proper recognition may not be entirely over, much progress has been made during the last two decades. For example, Medi-Cal, the California version of the federally funded medical assistance program, now funds the purchase and customization of augmentative communication devices. Many states routinely fund technology devices that enable people with impairments to function more independently or to achieve gainful employment.

Career opportunities for rehabilitation engineers and assistive technologists currently can be found in hospital-based rehabilitation centers, public schools, vocational rehabilitation agencies, manufacturers, and community-based rehabilitation technology suppliers; opportunities also exist as independent contractors. For example, a job announcement for a rehabilitation engineer contained the following job description (Department of Rehabilitative Services, Commonwealth of Virginia, 1997):

Provide rehabilitation engineering services and technical assistance to persons with disabilities, staff, community agencies, and employers in the area of employment and reasonable accommodations. Manage and design modifications and manufacture of adaptive equipment.... Requires working knowledge of the design, manufacturing techniques, and appropriate engineering problem-solving techniques for persons with disabilities. Skill in the operation of equipment and tools and the ability to direct others involved in the manufacturing of assistive devices. Ability to develop and effectively present educational programs related to rehabilitation engineering. Formal training in engineering with a concentration in rehabilitation engineering, mechanical engineering, or biomedical engineering or demonstrated equivalent experience a requirement.

The salary and benefits of the job in this announcement were competitive with other types of engineering employment opportunities. Similar announcements regularly appear in trade magazines such as *Rehab Management* and *TeamRehab* and in newsletters of RESNA.

An example of employment opportunities in a hospital-based rehabilitation center can be seen in the Bryn Mawr Rehabilitation Center in Malvern, Pennsylvania. The Center is part of the Jefferson Health System, a nonprofit network of hospitals and long-term, home care, and nursing agencies. Bryn Mawr's assistive technology center provides rehabilitation engineering and assistive technology services. Its geriatric rehabilitation clinic brings together several of the facility's departments to work at keeping senior citizens in their own homes longer. This clinic charges Medicare for assessments and the technology needed for independent living. Support for this program stems from the potential cost savings related to keeping older people well and in their own homes.

Rehabilitation engineers and assistive technologists also can work for school districts that need to comply with the Individuals with Disabilities Education Act. A rehabilitation engineer working in such an environment would perform assessments, make equipment modifications, customize assistive devices, assist special education professionals in classroom adaptations, and advocate to funding agencies for needed educationally related technologies. An ability to work well with nontechnical people such as teachers, parents, students, and school administrators is a must.

One promising employment opportunity for rehabilitation engineers and assistive technologists is in community-based service providers such as the local United Cerebral Palsy Association or the local chapter of the National Easter Seals Society. Through the combination of fees for service, donations, and insurance payments, shared rehabilitation engineering services in a community service center can be financially viable. The center would employ assistive technology professionals to provide information, assessments, customized adaptations, and training.

Rehabilitation engineers also can work as independent contractors or as employees of companies that manufacture assistive technology. Because rehabilitation engineers understand technology and the nature of many disabling conditions, they can serve as a liaison between the manufacturer and its potential consumers. In this capacity, they could help identify and evaluate new product opportunities. Rehabilitation engineers, as independent consultants, also could offer knowledgeable and trusted advice to consumers, funding agencies, and worker compensation insurance companies. Such consultation work often involves providing information about relevant assistive technologies, performing client evaluations, and assessing the appropriateness of assistive devices. It is important that a rehabilitation engineer who wishes to work as an independent consultant be properly licensed as a Professional Engineer (PE) and be certified through RESNA as described in the next section. The usual first step in attaining the Professional Engineer's license is to pass the Fundamentals of Engineering Examination given by each state's licensing board.

5.5.2 Rehabilitation Engineering Outlook

Rehabilitation engineering has reached adolescence as a separate discipline. It has a clearly defined application. For example, rehabilitation engineering research and development has been responsible for the application of new materials in the design of wheelchairs and orthotic and prosthetic limbs, the development of assistive technology that provides a better and more independent quality of life and better employment outcomes for people with disabilities, the removal of barriers to telecommunications and information technology through the application of universal design principles, the development of hearing aids and communication devices that exploit digital technology and advanced signal processing techniques, and the commercialization of neural prostheses that aid hand function, respiration, standing, and even limited walking.

Beginning with the Rehabilitation Act of 1973 and its subsequent amendments in 1992 and 1998, rehabilitation engineering in the United States has been recognized as an activity that is worthy of support by many governments, and many universities offer formal graduate programs in this field. Fees for such services have been reimbursed by public and private insurance policies. Job advertisements for rehabilitation engineers appear regularly in newsletters and employment notices. In 1990, the Americans with Disabilities Act granted civil rights to persons with disabilities and made reasonable accommodations mandatory for all companies having more than 25 employees. Archival journals publish research papers that deal with all facets of rehabilitation engineering. Student interest in this field is rising. What is next?

Based on some recent developments, several trends will likely dominate the practice of rehabilitation engineering and its research and development activities during the next decade.

- Certification of rehabilitation engineers will be fully established in the United States. Certification is the process by which a nongovernmental agency or professional association validates an individual's qualifications and knowledge in a defined functional or clinical area. RESNA is leading such a credentialing effort for providers of assistive technology. RESNA will certify someone as a Professional Rehabilitation Engineer if that person is a registered Professional Engineer (a legally recognized title), possesses the requisite relevant work experience in rehabilitation technology, and passes an examination that contains 200 multiple-choice questions. For nonengineers, certification as an Assistive Technology Practitioner (ATP) or Assistant Technology Supplier (ATS) is available. Sample questions from RESNA's credentialing examination are provided at the end of the chapter.

■ Education and training of rehabilitation technologists and engineers will expand worldwide. International exchange of information has been occurring informally. Initiatives by government entities and professional associations such as RESNA have given impetus to this trend. For example, the U.S. Department of Education supports a consortium of several American and European universities in the training of rehabilitation engineers. One indirect goal of this initiative is to foster formal exchanges of information, students, and investigators.

■ Universal access and universal design of consumer items will become commonplace. Technological advances in the consumer field have greatly benefited people with disabilities. Voice-recognition systems have enabled people with limited movement to use their computers as an interface to their homes and the world. Telecommuting permits gainful employment without requiring a disabled person to be physically at a specified location. Ironically, benefits are beginning to flow in the opposite direction. Consumer items that once were earmarked for the disabled population (e.g., larger knobs, easy-to-use door and cabinet handles, curb cuts, closed-caption television programming, larger visual displays) have become popular with everyone. In the future, the trend toward universal access and products that can be used easily by everyone will expand as the citizenry ages and the number of people with limitations increases. Universal design—which includes interchangeability, component modularity, and user friendliness—will be expected and widespread.

■ Ergonomic issues will play a more visible role in rehabilitation engineering. When designing for people with limitations, ergonomics and human factors play crucial roles, often determining the success of a product. In recognition of this, *IEEE Transactions on Rehabilitation Engineering* published a special issue on "Rehabilitation Ergonomics and Human Factors" in September 1994. The Human Factors and Ergonomics Society has a special interest group on "Medical Systems and Rehabilitation." In the next decade, more and more rehabilitation engineering training programs will offer required courses in ergonomics and human factors. The understanding and appreciation of human factors by rehabilitation engineers will be commonplace. The integration of good human factors designed into specialized products for people with disabilities will be expected.

■ Cost-benefit analysis regarding the impact of rehabilitation engineering services will become imperative. This trend parallels the medical field in that cost containment and improved efficiency have become everyone's concern. Econometric models and socioeconomic analysis of intervention efforts by rehabilitation engineers and assistive technologists will soon be mandated by the federal government. It is inevitable that health maintenance organizations and managed care groups will not continue to accept anecdotal reports as sufficient justification for supporting rehabilitation engineering and assistive technology (Gelderbom & de Witte, 2002; Andrich, 2002). Longitudinal and quantitative studies in rehabilitation, performed by unbiased investigators, will likely be the next major initiative from funding agencies.

- Quality assurance and performance standards for categories of assistive devices will be established. As expenditures for rehabilitation engineering services and assistive devices increase, there will undoubtedly be demand for some objective assurance of quality and skill level. One example of this trend is the ongoing work of the Wheelchair Standards Committee jointly formed by RESNA and the American National Standards Institute. Another example of this trend is the drive for certifying assistive technology providers and assistive technology suppliers.

- Applications of wireless technology will greatly increase the independence and capabilities of persons with disabilities. For example, navigational aids that utilize the Global Positioning System, Internet maps, cellular base station tri-angulation, and ubiquitous radio frequency identification tags will enable the blind to find their way indoors and outdoors as easily as their sighted counter-parts. Wireless technology also will assist people with cognitive limitations in their performance of daily activities. Reminders, cueing devices, trackers and wandering devices, and portable personal data assistants will enable them to remember appointments and medications, locate themselves positionally, follow common instructions, and obtain assistance.

- Technology will become a powerful equalizer as it reduces the limitations of manipulation, distance, location, mobility, and communication that are the common consequences of disabilities. Sometime in the next 20 years, rehabilitation engineers will utilize technologies that will enable disabled individuals to manipulate data and information and to alter system behavior remotely through their voice-controlled, Internet-based, wireless computer workstation embedded in their nuclear-powered wheelchairs. Rather than commuting daily to work, persons with disabilities will or can work at home in an environment uniquely suited to their needs. They will possess assistive technology that will expand their abilities. Their dysarthric speech will be automatically recognized and converted into intelligible speech in real time by a powerful voice-recognition system. Given the breathtaking speed at which technological advances occur, these futuristic devices are not mere dreams but realistic extrapolations of the current rate of progress.

Students interested in rehabilitation engineering and assistive technology R & D will be able to contribute toward making such dreams a reality shortly after they complete their formal training. The overall role of future practicing rehabilitation engineers, however, will not change. They still will need to assess someone's needs and limitations, apply many of the principles outlined in this chapter, and design, prescribe, modify, or build assistive devices.

EXERCISES

Like the engineering design process described earlier in this chapter, answers to the following study questions may require searching beyond this textbook for the necessary information. A good place to begin is the Suggested Reading section. You also may try looking for the desired information using the Internet.

1. The fields of rehabilitation engineering and assistive technology have been strongly influenced by the federal government. Describe the impact federal legislation has had on the prevalence of rehabilitation engineers and the market for their work in assistive technology. Explain and provide examples.

2. As a school-based rehabilitation engineer, you received a request from a teacher to design and build a gadget that would enable an 8-year-old, second-grade student to signal her desire to respond to questions or make a request in class. This young student uses a powered wheelchair, has multiple disabilities, cannot move her upper arms very much, and is unable to produce understandable speech. Prepare a list of quantitative and qualitative questions that will guide your detailed analysis of this problem. Produce a hypothetical set of performance specifications for such a signaling device.

3. Write a sample set of performance specifications for a voice-output oscilloscope to be used by a visually impaired electrical engineering student for a laboratory exercise having to do with operational amplifiers. What features would be needed in the proposed oscilloscope?

4. Write a sample set of performance specifications for a foldable lap tray that will mount on a manual wheelchair. Hints: What should its maximum and minimum dimensions be? How much weight must it bear? Will your add-on lap tray user make the wheelchair user more or less independent? What type of materials should be used?

5. Sketch how the leg raiser described in Example Problem 5.1 might fit onto a battery-powered wheelchair. Draw a side view and rear view of the leg-raiser-equipped wheelchair.

6. Do a careful search of commercially available electronic communication devices that meet the following performance specifications: speech output, icon-based membrane keyboard, portable, weigh less than 7.5 lbs, no more than 2.5 in. thick, no larger than a standard three-ring binder, and able to be customized by the user to quickly produce frequently used phrases. Hints: Consult "The Closing the Gap Product Directory" and the "Cooperative Electronic Library on Disability." The latter is available from the Trace Research and Development Center at the University of Wisconsin, Madison. Also try visiting the applicable websites.

7. A person's disabilities and abilities often depend on his or her medical condition.
   a) A person is known to have spinal cord injury (SCI) at the C5–C6 level. What does this mean in terms of this person's probable motoric and sensory abilities and limitations?
   b) Repeat part (a) for a person with multiple sclerosis. Include the prognosis of the second individual in contrast to the person with SCI.

8. To be portable, an electronic assistive device must be battery powered. Based on your study of technical manuals and battery handbooks, list the pros and cons of using disposable alkaline batteries versus lithium-hydride rechargeable batteries. Include in your comparison an analysis of the technical issues (e.g., battery capacity, weight, and charging circuitry), cost issues, and

practicality issues (e.g., user preferences, potential for misplacement or improper usage of charger, and user convenience).

9. A young person with paraplegia wishes to resume skiing, canoeing, sailing, and golfing. For each of these sports, list four or five adaptations or equipment modifications that are likely to be needed. Sketch and briefly describe these adaptations.

10. A 21-year-old female who has muscular dystrophy requested assistance with computer access, particularly for writing, using spreadsheets, and playing computer games. She lacks movement in all four extremities except for some wrist and finger movements. With her left hand, she is able to reach about 6 in. past her midline. With her right hand, she is able to reach only 2 in. past her midline. Both her hands can reach out about 8 in. from the body. If given wrist support, she has good control of both index fingers. Based on this description, sketch the work area that she appears able to reach with her two hands. Describe the adaptations to a standard or contracted keyboard that she would need to access her home computer. For additional information, consult the "Closing the Gap Product Directory," the "Cooperative Electronic Library on Disability," and the suggested reading materials listed at the end of this chapter.

11. The two main computer user interfaces are the command line interface (CLI), as exemplified by UNIX commands, and the graphical user interface (GUI), as exemplified by the Windows XP or Apple's OS X operating systems. For someone with limited motoric abilities, each type of interface has its advantages and disadvantages. List and compare the advantages and disadvantages of CLI and GUI. Under what circumstances and for what kinds of disabilities would the CLI be superior to or be preferred over the GUI?

12. One of the major categories of assistive devices is alternative and augmentative communication devices. Describe the electronic data processing steps needed for text-to-speech conversion. How have the technological advances in personal computing made this conversion faster and the speech output more lifelike?

13. What would the second scale reading (S_2) be if the person in Example Problem 5.3 raised both of his legs straight up and D was known to be 14 in.?

14. How much tension would be exerted on the pulley in Example Problem 5.4 if the weights were observed to be falling at 1.5 ft/s²?

15. How much contraction force must the flexor muscles generate in order for a person to hold a 25-lb weight in his hand, 14 in. from the elbow joint? Assume that the flexor muscle inserts at 90° to the forearm 2 in. from the elbow joint and that his forearm weighs 4.4 lbs. Use the equilibrium equation, $\Sigma F_X = \Sigma F_y = \Sigma M = 0$, and Figure 5.10 to aid your analysis.

16. How much force will the head of the femur experience when a 200-lb person stands on one foot? Hint: Apply the equilibrium equation, $\Sigma F_X = \Sigma F_y = \Sigma M = 0$, to the skeletal force diagram in Figure 5.11 in your analysis.

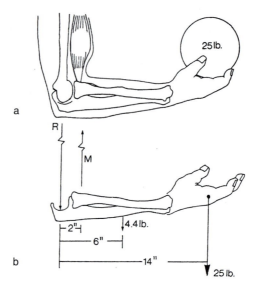

Figure 5.10   Static forces about the elbow joint during an elbow flexor exercise (from Le Veau, 1976).

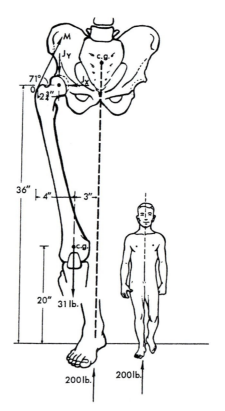

Figure 5.11   Determination of the compression force on the supporting femoral head in unilateral weight bearing (from Le Veau, 1976).

17. Under static or constant velocity conditions, the wheelchair will tip backwards if the vertical projection of the combined center of gravity (*CG*) of the wheelchair and occupant falls behind the point of contact between the rear wheels and the ramp surfaces. As shown in Figure 5.12, the rearward tipover angle (θ_r) is determined by the horizontal distance (d_1) and the vertical distance (d_2) between *CG* and the wheelchair's rear axles.

   a) Using static analysis, derive the equation relating θ_r, d_1, and d_2.
   b) Using the platform approach depicted in Figure 5.7, suggest a method for determining d_1.
   c) Assuming that d_1 and d_2 averaged 13 cm and 24 cm, respectively, for able-bodied individuals, what would θ_r be?
   d) How would d_1 and d_2 change if the wheelchair occupant leaned forward instead of sitting back against the chair? How would θ_r be affected by this postural shift?

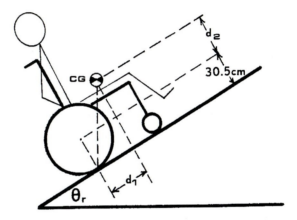

d_1 = horizontal distance in cm (relative to the incline) between the CG and the rear axles.

d_2 = vertical distance in cm (relative to the incline) between the CG and the rear axles.

θ_r = static rearward tipover angle.

Figure 5.12 Conditions under which the occupied wheelchair will begin to tip backwards. The tipover threshold occurs when the vertical projection of the combined *CG* falls behind the rearwheel's contact point with the inclined surface (Szeto, A.Y.J. and R.N. White, "Evaluation of a Curb-Climbing Aid for Manual Wheelchairs: Considerations of Stability, Effort, and Safety," *Journal of Rehabilitation Research & Development*, BPR 10–38, Vol. 20(1), pp. 45–56, July 1983).

18. Perform a static analysis of the situation depicted in Figure 5.13 and derive the equation for the probable forward tipover angle (θ_r) using the data and dimensions shown. Assuming that d_1 and d_2 were the same as given in problem 17 and d_4 and d_5 averaged 27 cm and 49 cm, respectively, what would θ_r be?

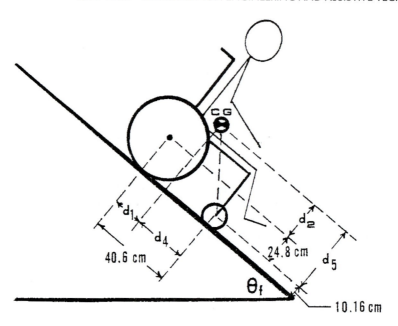

d_1 = horizontal distance in cm (relative to the incline) between the CG and the rear axles.

d_2 = vertical distance in cm (relative to the incline) between the CG and the rear axles.

d_4 = horizontal distance in cm (relative to the incline) between the CG and the casters.

d_5 = vertical distance in cm (relative to the incline) between the CG and the rear axles.

θ_f = forward tipover angle.

Figure 5.13 Conditions under which the occupied wheelchair will begin to tip forward. The tipover threshold occurs when the vertical projection of the combined CG falls behind the caster wheel's contact point with the inclined surface (Szeto, A.Y.J. and R.N. White, "Evaluation of a Curb-Climbing Aid for Manual Wheelchairs: Considerations of Stability, Effort, and Safety," *Journal of Rehabilitation Research & Development*, BPR 10–38, Vol. 20(1), pp. 45–56, July 1983).

19. For persons with good head control and little else, the Head Master (by Prentke Romich Co., Wooster, OH) has been used to emulate the mouse input signals for a computer. The Head Master consists of a headset connected to the computer by a cable. The headset includes a sensor that detects head movements and translates such movements into a signal interpreted as 2-dimensional movements of the mouse. A puff-and-sip pneumatic switch is also attached to headset and substitutes for clicking of the mouse. Based on this brief description of the Head Master, draw a block diagram of how this device might work and the basic components that might be needed in the Head Master. Include in your block diagram the ultrasonic signal source, detectors, timers, and signal processors that would be needed.

20. Based on the frequency of use data shown in Tables 5.8, 5.9, and 5.10, design an optimized general purpose communication array using row-column scanning. Recall that row-column scanning is a technique whereby a vocabulary element is first highlighted row by row from the top to bottom of the array. When the row containing the desired element is highlighted, the user activates a switch to select it. Following the switch activation, the scanning proceeds within the selected row from left to right. When the desired vocabulary element is highlighted again, a second switch activation is made. In row-column scanning, the first press of a switch selects the row and the second press selects the column. Hint: Arrange the most frequently

TABLE 5.8 Simple English Letter Frequency from 10,000 Letters of English Literary Text

E = 1231	L = 403	B = 162
T = 959	D = 365	G = 161
A = 805	C = 320	V = 93
O = 794	U = 310	K = 52
N = 719	P = 229	Q = 20
I = 718	F = 228	X = 20
S = 659	M = 225	J = 10
R = 603	W = 203	Z = 9
H = 514	Y = 188	

Data from Webster et al. (1985).

TABLE 5.9 Frequency of English Two- and Three-Letter Combinations from 25,000 Letters of English Literary Text

Two-letter Combinations

TH = 1582	HE = 542	ON = 420	NT = 337	RA = 275
IN = 784	EN = 511	OU = 361	HI = 330	RO = 275
ER = 667	TI = 510	IT = 356	VE = 321	LI = 273
RE = 625	TE = 492	ES = 343	CO = 296	IO = 270
AN = 542	AT = 440	OR = 339	DE = 275	

Three-letter Combinations

THE = 1182	ERE = 173	HAT = 138	NCE = 113	MAN = 01
ING = 356	HER = 170	ERS = 135	ALL = 111	RED = 101
AND = 284	ATE = 165	HIS = 130	EVE = 111	THI = 100
ION = 252	VER = 159	RES = 125	ITH = 111	IVE = 96
ENT = 246	TER = 157	ILL = 118	TED = 110	
FOR = 246	THA = 155	ARE = 117	AIN = 108	
TIO = 188	ATI = 148	CON = 114	EST = 106	

Data from Webster et al. (1985).

TABLE 5.10   Frequency of English Words from 242,432 Words of English Literary Text

THE = 15,568	FOR = 1869	HAVE = 1344	THIS = 1021
OF = 9757	AS = 1853	YOU = 1336	MY = 963
AND = 7638	WITH = 1849	WHICH = 1291	THEY = 959
TO = 5739	WAS = 1761	ARE = 1222	ALL = 881
A = 5074	HIS = 1732	ON = 1155	THEIR = 824
IN = 4312	HE = 1721	OR = 1101	AN = 789
THAT = 3017	BE = 1535	HER = 1093	SHE = 775
IS = 2509	NOT = 1496	HAD = 1062	HAS = 753
I = 2292	BY = 1392	AT = 1053	WHERE = 753
IT = 2255	BUT = 1379	FROM = 1039	ME = 752

Data from Webster et al. (1985).

used vocabulary elements earliest in the scanning order. See Example Problem 5.5.

21. An electronic guide dog has been proposed as an electronic travel aid for a blind person. List some of the specific tasks that such a device must perform and the information processing steps involved in performing these tasks. List as many items and give as many details as possible. Hints: Consider the problems of obstacle detection, information display, propulsion system, inertial guidance, route recall, power supply, etc.

22. The ability of the user to visually scan an array of options and make appropriate choices is fundamental to many assistive devices. Analyze the difference between visual pursuit tracking and visual scanning in terms of the oculomotor mechanisms that underlie these two activities.

23. Based on Table 5.7, what type of speech synthesis technology would be the most appropriate for the following situations: (a) an augmentative communication system capable of unlimited vocabulary for someone who can spell? (b) a voice output system for a blind person that reads the entire screen of a computer display? (c) an augmentative communication system for a young girl who needs a limited vocabulary set? (d) voice feedback for an environmental control system that echoes back simple one-word commands, such as "on," "off," "lights," "bed," "TV," and "drapes." Explain or justify your answer.

24. Safe and independent mobility by persons with severe visual impairments remains a challenge. To relieve such persons of their dependence on guide dogs or a sighted human guide, various portable navigational aids using a Global Positioning System (GPS) receiver have been marketed.

  a) Conduct an Internet investigation of GPS as the basis for a portable navigational aid for the blind. Address the following issues: How does GPS work? Can GPS signals be reliably received at every location? How accurate are GPS signals in terms of resolution? Is this level of resolution sufficient for finding the entrance to a building? Can dead reckoning and inertial guidance help when GPS signals are lost?

b) Describe the various operational requirements of an ideal portable navigational aid for the blind. Consider such ergonomic issues as the user interface, input and output requirements, and target retail price. List some of the human factor design issues involved.

Sample Multiple-Choice Questions from RESNA's Credentialing Examination in Assistive Technology

1. Which of the following abilities is necessary for development of skilled upper-extremity movements?
   a. Equilibrium reactions in the standing position
   b. Ability to cross midline
   c. Good postural control of the trunk and head
   d. Pincer grasp

2. A 12-year-old male with Duchenne's muscular dystrophy is being evaluated for a mobility system. The therapist notes that he has lateral bending of the trunk and leans to the left. The most appropriate next step is assessment for
   a. Kyphosis
   b. Lordosis
   c. Left-sided weakness
   d. Scoliosis

3. The most appropriate location for training and instruction in functional use of an assistive technology device is
   a. A quiet area with few distractions
   b. The individual's home environment
   c. The environment in which the device will be used
   d. A training center where several therapists are available

4. An architect with C4–C5 quadriplegia would like to use a computer-assisted design (autoCAD) system when he returns to work. The most appropriate first step is assessment of the client's ability to use
   a. Mouthstick
   b. Eye-blink switch
   c. Alternate mouse input
   d. Sip-and-puff switch

5. Under the Individuals with Disabilities Education Act, assistive technology is defined as a device that
   a. Increases functional capability
   b. Improves mobility or communication
   c. Compensates for physical or sensory impairment
   d. Is considered durable medical equipment

6. In addition to the diagnosis, which information must be included in a physician's letter of medical necessity?
   a. Cost of assistive technology requested
   b. Client's prognosis
   c. Client's range of motion
   d. Client's muscle tone

7. Plastic is an ideal seat base for the person with incontinence because it is
   a. Light weight
   b. Less costly than wood
   c. Nonabsorbent
   d. Detachable from wheelchair
8. When considering structural modification of a newly purchased commercial device, which of the following is the *most* important concern?
   a. Future use by other individuals
   b. Voidance of warranty
   c. Resale value
   d. Product appearance
9. A client is interested in using a voice-recognition system to access the computer. Which of the following factors is *least* critical to success with this method?
   a. Hand function
   b. Voice clarity
   c. Voice-recognition system training
   d. Type of computer system used
10. A 9-year-old is no longer able to drive her power-base wheelchair. Training was provided following delivery of the wheelchair 2 years ago. Which of the following is the first step in evaluation?
    a. Interview the parents and child
    b. Perform a cognitive evaluation
    c. Reevaluate access in the wheelchair
    d. Contact the wheelchair manufacturer

SUGGESTED READING

Adams, R.C., Daniel, A.N., McCubbin, J.A. and Rullman, L. (1982). *Games, Sports, and Exercises for the Physically Handicapped,* 3rd Ed. Lea & Febiger, Philadelphia.

American Academy of Orthopaedic Surgeons. (1975). *Atlas of Orthotics: Biomechanical Principles and Application.* Mosby, St. Louis.

Andrich, R. (2002). The SCAI instrument: Measuring costs of individual assistive technology. *Technol. & Disabil.* **14(3):** 95–99.

Bailey, R.W. (1989). *Human Performance Engineering,* 2nd Ed. Prentice Hall, Englewood Cliffs, NJ.

Blackstone, S. (Ed.). (1986). *Augmentative Communication: An Introduction.* American Speech-Language-Hearing Association, Rockville, MD.

Childress, D.S. (1993). Medical/technical collaboration in prosthetics research and development. *J. Rehab. R&D* **30(2):** vii–viii.

Church, G. and Glennen, S. (1992). *The Handbook of Assistive Technology.* Singular, San Diego, CA.

Cook, A.M. and Hussey, S.M. (2002). *Assistive Technology: Principles and Practice,* 2nd Ed. Mosby, St. Louis.

Cooper, R.A., Robertson, R.N., VanSickel, D.P., Stewart, K.J. and Albright, S.J. (1994). Wheelchair impact response to ISO test pendulum and ISO standard curb. *IEEE Trans. Rehab. Eng.* **2(4):** 240–246.

Daus, C. (1996, April/May). Credentialing rehabilitation engineers. *REHAB Management,* 115–116.

Deatherage, B. (1972). Auditory and other sensory forms of information presentation. In H. Van Cott and R. Kincade (Eds.), *Human Engineering Guide to Equipment Design.* Government Printing Office, Washington, DC.

Department of Rehabilitative Services, Commonwealth of Virginia. (1997, Spring). Job Announcement for a Rehabilitation Engineer.

Galvin, J.C. and Scherer, M.J. (1996). *Evaluating, Selecting, and Using Appropriate Assistive Technology.* Aspen, Gaithersburg, MD.

Gaines, H.F. (1940). *Elementary Cryptanalysis.* Chapman and Hall, London.

Gelderblom, G.J. and de Witte, L.P. (2002). The assessment of assistive technology outcomes, effects and costs. *Technol. & Disabil.* **14(3):** 91–94.

Goldenson, R.M., Dunham, J.R. and Dunham, C.S. (1978). *Disability and Rehabilitation Handbook.* McGraw Hill, New York.

Guthrie, M. (1997, July). Where will the jobs be? *TeamRehab Rep.,* 14–23.

Hammer, G. (1993, September). The next wave of rehab technology. *TeamRehab Rep.,* 41–44.

Jones, M., Sanford, J. and Bell, R.B. (1997, October). Disability demographics: How are they changing? *TeamRehab Rep.,* 36–44.

Lange, M. (1997, September). What's new and different in environmental control systems. *TeamRehab Rep.,* 19–23.

Le Veau, B. (1976). *Biomechanics of Human Motion,* 2nd Ed. Saunders, Philadelphia.

McLaurin, C.A. (1991, Aug/Sep). A history of rehabilitation engineering. *REHAB Manag.,* 70–77.

Mercier, J., Mollica, B.M. and Peischl, D. (1997, August). Plain talk: A guide to sorting out AAC cevices. *TeamRehab Rep.,* 19–23.

National Institute on Disability and Rehabilitation Research (NIDRR). (1999, December). Long Range Plan 1999–2003, Washington, DC.

Philips, B. and Zhao, H. (1993). Predictors of assistive technology abandonment. *Assistive Technol.* **5(1):** 36–45.

Reswick, J. (1982). What is a rehabilitation engineer? *Ann. Rev. Rehabilitation* **2.**

Sanders, M.S. and McCormick, E.J. (1993). *Human Factors in Engineering and Design,* 7th Ed. McGraw-Hill, New York.

Scherer, M. (1993). *Living in the State of Stuck: How Technology Impacts the Lives of People with Disabilities.* Brookline Books, Cambridge, MA.

Smith, R.V. and Leslie, J.H., Jr. (Eds.). (1990). *Rehabilitation Engineering.* CRC, Boca Raton, FL.

Stolov, W.C. and Clowers, M.R. (Eds.). (1981). *Handbook of Severe Disability.* U.S. Department of Education, Washington, DC.

Szeto, A.Y.J., Allen, E.J. and Rumelhart, M.A. (1987). "Employability Enhancement through Technical Communication Devices," *American Rehabilitation,* Vol. 13(2):8–11 & 26–29, April-June.

Szeto, A.Y.J. and Riso, R.R. (1990). "Sensory Feedback using Electrical Stimulation of the Tactile Sense," in *Rehabilitation Engineering,* R.V. Smith and J.H. Leslie, Jr. (Eds.), CRC Press, pp. 29–78.

Szeto, A.Y.J. and R.N. White, "Evaluation of a Curb-Climbing Aid for Manual Wheelchairs: Considerations of Stability, Effort, and Safety," *Journal of Rehabilitation Research & Development,* BPR 10–38, Vol. 20(1), pp. 45–56, July 1983

Szeto, A.Y.J., Valerio, N. and Novak, R. (1991). "Audible Pedestrian Traffic Signals, Part 1. Prevalence and Impact, Part 2. Analysis of Sounds Emitted, Part 3. Detectability," *Journal of Rehabilitation R & D,* 28(2):57–78.

Szeto, A.Y.J., Allen, E.J. and Littrell, M.C. (1993) "Comparison of Speed and Accuracy for Selected Electronic Communication Devices and Input Methods," *Augmentative and Alternative Communication*, Vol. 9, Vol. 4, December, pp. 229–242.

U.S. Census Bureau. (1990). *National Health Interview Survey on Assistive Devices (NHIS-AD).* Washington, DC.

Webster, J.G., Cook, A.M., Tompkins, W.J. and Vanderheiden, G.C. (Eds.). (1985). *Electronic Devices for Rehabilitation*. Wiley Medical, New York.

Wessels, R., Dijcks, B., Soede, M., Gelderblom, G.J. and De Witte, L. (2003). Non-use of provided assistive technology devices: A literature overview. *Technol. & Disabil.,* Vol. 15, 2003, 231–238.

Wilcox, A.D. (1990). *Engineering Design for Electrical Engineers*. Prentice Hall, Englewood Cliffs, NJ.

6   BIOMATERIALS

Liisa T. Kuhn, PhD*

Chapter Contents

*With contributions from Katharine Merritt and Stanley A. Brown.

255

At the conclusion of this chapter, the reader will be able to:

- Understand the complexity of natural tissue construction.

- Describe several different types of biological responses to implanted materials.

- Understand the benefits and differences between the various classes of materials used in medicine.

- Design bio-inspired medical device features.

- Describe how biomaterials can be modified to enhance or modify cellular interactions.

- Understand the various methods to prepare scaffolds for tissue engineering.

- Know where to find the appropriate established testing protocols to demonstrate medical product safety.

6.1   MATERIALS IN MEDICINE: FROM PROSTHETICS TO REGENERATION

Throughout the ages, materials used in medicine (biomaterials) have made an enormous impact on the treatment of injury and disease of the human body. Biomaterials use increased rapidly in the late 1800s, particularly after the advent of aseptic surgical technique by Dr. Joseph Lister in the 1860s. The first metal devices to fix bone fractures were used as early as the late eighteenth to nineteenth century; the first total hip replacement prosthesis was implanted in 1938; and in the 1950s and 1960s, polymers were introduced for cornea replacements and as blood vessel replacements. Today, biomaterials are used throughout the body (Fig. 6.1). Estimates of the numbers of biomedical devices incorporating biomaterials used in the United States in 2002 include

- Total hip joint replacements: 448,000
- Knee joint replacements: 452,000
- Shoulder joint replacements: 24,000
- Dental implants: 854,000

Impact of Biomaterials

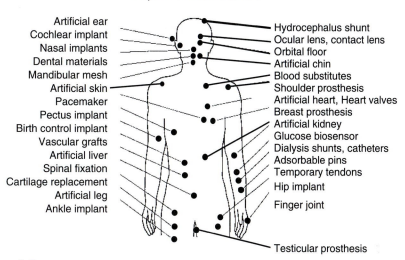

Artificial ear — Hydrocephalus shunt
Cochlear implant — Ocular lens, contact lens
Nasal implants — Orbital floor
Dental materials — Artificial chin
Mandibular mesh — Blood substitutes
Artificial skin — Shoulder prosthesis
Pacemaker — Artificial heart, Heart valves
Pectus implant — Breast prosthesis
Birth control implant — Artificial kidney
Vascular grafts — Glucose biosensor
Artificial liver — Dialysis shunts, catheters
Spinal fixation — Adsorbable pins
Cartilage replacement — Temporary tendons
Artificial leg — Hip implant
Ankle implant — Finger joint
Testicular prosthesis

Figure 6.1   Biomaterials have made an enormous impact on the treatment of injury and disease and are used throughout the body.

- Coronary stents: 1,204,000
- Coronary catheters: 1,328,000

Millions of lives have been saved due to biomaterials and the quality of life for millions more is improved every year due to biomaterials. The field remains a rich area for research and invention because no one material is suitable for all biomaterial applications and new applications are continually being developed as medicine advances. In addition, there are still many unanswered questions regarding the biological response to biomaterials and the optimal role of biomaterials in tissue regeneration that continue to motivate biomaterials research and new product development.

Over most of history, minimal understanding of the biological mechanisms of tissues meant that the biomedical engineering approach was to completely replace the tissue with lost function with a simple biomaterial. As our understanding of tissues, disease, and trauma improved, the concept of attempting to repair damaged tissues emerged. More recently, with the advent of stem cell research, medicine believes it will be possible to regenerate damaged or diseased tissues by cell-based tissue engineering approaches (see Chapter 7). The notion of a biomaterial has evolved over time in step with changing medical concepts. Williams in 1987 defined a biomaterial as "a nonviable material used in a medical device, intended to interact with biological systems." This definition still holds true today and encompasses the earliest use of biomaterials replacing form (e.g., wooden leg, glass eye) as well as the current use of biomaterials in regenerative medical devices such as a biodegradable scaffold used to deliver cells for tissue engineering. While the definition has remained the same, there have been dramatic changes in understanding of the level of interaction of biomaterials with the biological system (in this case, the human body). The expectations for biomaterial function have advanced from remaining relatively inert in the body to being "bioactive" and assisting with regeneration. Bioactive materials have the capability to initiate a biological response after implantation such as cell adhesion, proliferation, or more excitingly, the differentiation of a stem cell leading to regeneration of a damaged tissue or whole organ.

Due to the complexity of cell and tissue reactions to biomaterials, it has proven advantageous to look to nature for guidance on biomaterials design, selection, synthesis, and fabrication. This approach is known as biomimetics. Within the discipline of biomaterials, biomimetics involves imitating aspects of natural materials or living tissues such as their chemistry, microstructure, or fabrication method. This does not always lead to the desired outcome since many of the functionalities of natural tissues are as yet unknown. Furthermore, the desirable or optimal properties of a biomaterial vary enormously depending on the biomedical application. Therefore, in addition to presenting general strategies for guiding tissue repair by varying the chemistry, structure, and properties of biomaterials, this chapter includes application-specific biomaterials solutions for several of the major organ systems in the body and for drug delivery applications. This chapter also includes a section on the standards and regulatory agencies that play an essential role in establishing and ensuring the safety and efficacy of medical products.

6.2 BIOMATERIALS: PROPERTIES, TYPES, AND APPLICATIONS

6.2.1 Mechanical Properties and Mechanical Testing

Some basic terminology regarding the mechanical properties of materials is necessary for a discussion of materials and their interactions with biological tissues. The most common way to determine mechanical properties is to pull a specimen apart and measure the force and deformation. Materials are also tested by crushing them in compression or by bending them. The terminology is essentially the same in either case—only the mathematics are different. Standardized test protocols have been developed to facilitate comparison of data generated from different laboratories. The vast majority of those used in the biomaterials field are from the American Society for Testing and Materials (ASTM). For example, tensile testing of metals can be done according to ASTM E8, ASTM D412 is for rubber materials, and ASTM D638 is for tensile testing of rigid plastics. These methods describe specimen shapes and dimensions, conditions for testing, and methods for calculating and reporting the results.

Tensile testing according to ASTM E8 is done with a "dog bone" shaped specimen that has its large ends held in some sort of a grip while its narrow midsection is the "test" section. The midportion is marked as the "gage length" where deformation is measured. A mechanical test machine uses rotating screws or hydraulics to stretch the specimen. Force is measured in Newtons (N), and how much the specimen stretches—deformation—is measured in millimeters. Since specimens of different dimensions can be tested, measurements must be normalized to be independent of size. Stress, σ (N/m² or Pascals), is calculated as force divided by the original cross-sectional area, and strain, ε (%), is calculated as change in length divided by the original length.

$$\sigma(N/m^2) = force/cross\text{-}sectional\ area \qquad (6.1)$$
$$\varepsilon(\%) = [(deformed\ length - original\ length)/original\ length] * 100\% \qquad (6.2)$$

A stress–strain curve can be generated from these data (Fig. 6.2), and there are a number of material properties that can be calculated. Region A is known as the elastic portion of the curve. If a small stress is applied to a metal, such as up to point (1), it will deform elastically.

This means that, like a rubber elastic band, it will return to its original length when the stress is removed. The slope of the elastic portion of the stress–strain curve is a measure of the stiffness of the material and is called the elastic modulus (E) or Young's modulus.

$$E = \sigma/\varepsilon\ initial\ slope = stress/strain \qquad (6.3)$$

As the applied stress is increased, a point is reached at which the metal begins to deform permanently, the yield point (YS). If at point (2) the stress is now released, the

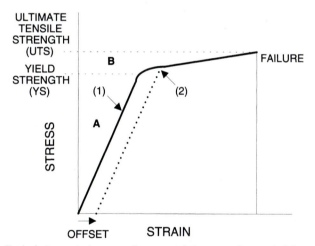

Figure 6.2 Typical stress–strain curve for a metal that stretches and deforms (yields) before breaking. Stress is measured in N/m² (Pa) while strain is measured as a percentage of the original length. The minimum stress that results in permanent deformation of the material is called the yield strength (YS). The ultimate strength (UTS) is the maximum stress that is tolerated by the material before rupturing. The stress at which failure occurs is called the failure strength (FS). Region A represents the elastic region since the strain increases in direct proportion to the applied stress. If a small stress is applied (e.g., to point 1), the material will return to its original length when the stress is removed. Region B represents the plastic region in which changes in strain are no longer proportional to changes in stress. Stresses in this region result in permanent deformation of the material. If a stress is applied that results in the strain at point (2), the material will follow the dotted line back to the baseline when the stress is removed and will be permanently deformed by the amount indicated by the offset.

stress–strain recording will come down the dotted line parallel to the elastic region. The permanent amount of deformation is now shown as the offset yield. Since it may be difficult to determine the yield point for a material, an offset yield point often is used in place of the original yield point. For metals, yield is typically defined as 0.2% while a 2% offset is often used for plastics. If the metal is loaded again, the recording will follow the dotted line starting at the offset yield, reaching the upper curve, and continuing to show a gradual increase in stress with increasing strain. This is known as the plastic region of the curve. The peak stress that is attained is called the tensile or ultimate tensile strength (UTS). Eventually the metal will break at the failure or fracture strength (FS), and the percentage of elongation or compression to failure can be determined.

Example Problem 6.1

A 7-mm cube of bone was subjected to a compression loading test in which it was compressed in increments of approximately 0.05 mm. The force required to produce each amount of deformation was measured, and a table of values was generated. Plot a stress–strain curve for this test. Determine the elastic modulus and the ultimate tensile strength of the bone.

Deformation (mm)	Force (N)
0.00	0
0.10	67.9
0.15	267.6
0.20	640.2
0.26	990.2
0.31	1265.1
0.36	1259.9
0.41	1190.9
0.46	1080.8
0.51	968.6
0.56	814.2

Solution

First, determine the cross-sectional area of the cube in meters (0.007 m × 0.007 m = 49×10^{-6} m). Use this value to determine the stress at each measuring point where stress (σ) equals force divided by cross-sectional area. For example, the stress when the cube was compressed by 0.10 mm was 67.9 N/0.000049 m^2 = 1.39 MPa (1 N/m^2 = 1 Pa). Next, determine the strain at each measuring point. The deformed length when the cube was compressed by 0.10 mm was 7 mm − 0.10 mm equals 6.9 mm, so the strain was [(6.9 mm − 7.0 mm)/7.0 mm]* 100% equals 1.43%. This is the same value that would be obtained by dividing the amount of deformation by the original length. The minus sign is ignored because it merely indicates that the sample was subjected to compression rather than to tension.

Strain (%)	Stress (MPa)
0.00	0
1.43	1.39
2.14	5.46
2.86	13.07
3.71	20.21
4.43	25.82
5.14	25.71
5.86	24.30
6.57	22.06
7.29	19.77
8.00	16.61

The resulting stress–strain curve is shown in Figure 6.3. Linear regression was used to determine the line shown in Figure 6.3. The elastic modulus (i.e., the slope of the line) was 8.4, and the ultimate tensile strength was 25.82 MPa. ∎

Several other terms are applied to the test results. The slope of the elastic portion, the elastic modulus, is often called stiffness. If the metal stretches a great deal before failure it is said to be ductile (Fig. 6.4). If the material does not deform or yield much before failure, it is said to be brittle. The area under the curve has units of energy and

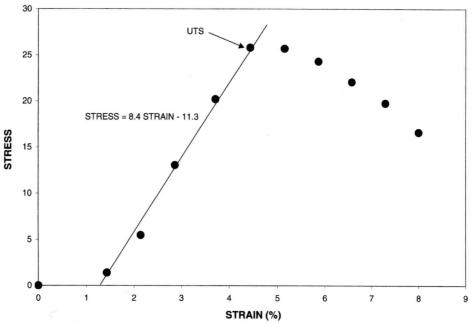

Figure 6.3 The stress–strain curve for the bone data from Example Problem 6.1. Linear regression analysis was used to find the line that best fit the data for strains of 1.43% to 4.43% (i.e., the linear portion of the curve). The slope of the line, 8.4, represents the elastic modulus of the bone. The ultimate tensile strength of the material (UTS) was 25.82 MPa.

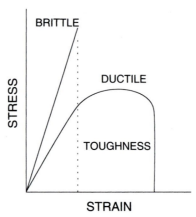

Figure 6.4   Brittle materials reach failure with only a small amount of deformation (strain) while ductile materials stretch or compress a great deal before failure. The area under the stress–strain curve is called toughness and is equal to the integral from ε_0 to $\varepsilon_f \sigma d\varepsilon$.

is called toughness. Although not directly available from a stress–strain curve, the strength of a material can be related to its hardness. Stronger materials are typically harder. Hardness is tested by measuring the indentation caused by a sharp object that is dropped onto the surface with a known force. Hardness is perhaps the most important property when considering a material's wear resistance.

An additional property that is not depicted in the figure is the fatigue strength or endurance limit of a material. If the material were tested as in Figure 6.2, but was loaded to point (2) and unloaded, it would become permanently deformed. If this were repeated several times, like bending a paper clip back and forth, the material would eventually break. If, however, the metal were loaded to point (1) and then unloaded, it would not be deformed or broken. If it were loaded again to point (1) and unloaded, it still would not break. For some metals, there is a stress level below which the part can theoretically be loaded and unloaded an infinite number of times without failure. In reality, a fatigue limit is defined at a specified number of cycles, such as 10^6 or 10^7. Clearly, fatigue strength is a critical property in the design of load-bearing devices such as total hips which are loaded on average a million times a year or heart valves which are loaded 40 million times a year.

6.2.2   Metals

Metals used as biomaterials have high strength and resistance to fracture and are designed to resist corrosion. Examples of metals used in medical devices and their mechanical properties are shown in Tables 6.1 and 6.2. Many orthopedic devices are made of metal, such as hip and knee joint replacements (Figs. 6.5 and 6.6). The implants provide relief from pain and restore function to joints in which the natural cartilage has been worn down or damaged. Plates and screws that hold fractured bone

TABLE 6.1 Materials and Their Medical Uses

Class of Material	Current Uses
Metal	
Stainless steel	Joint replacements, bone fracture fixation, heart valves, electrodes
Titanium and titanium alloys	Joint replacements, dental bridges and dental implants, coronary stents
Cobalt-chrome alloys	Joint replacements, bone fracture fixation
Gold	Dental fillings and crowns, electrodes
Silver	Pacemaker wires, suture materials, dental amalgams
Platinum	Electrodes, neural stimulation devices
Ceramics	
Aluminum oxides	Hip implants, dental implants, cochlear replacement
Zirconia	Hip implants
Calcium phosphate	Bone graft substitutes, surface coatings on total joint replacements, cell scaffolds
Calcium sulfate	Bone graft substitutes
Carbon	Heart valve coatings, orthopedic implants
Glass	Bone graft substitutes, fillers for dental materials
Polymers	
Nylon	Surgical sutures, gastrointestinal segments, tracheal tubes
Silicone rubber	Finger joints, artificial skin, breast implants, intraocular lenses, catheters
Polyester	Resorbable sutures, fracture fixation, cell scaffolds, skin wound coverings, drug delivery devices
Polyethylene (PE)	Hip and knee implants, artificial tendons and ligaments, synthetic vascular grafts, dentures, and facial implants
Polymethylmethacrylate (PMMA)	Bone cement, intraocular lenses
Polyvinylchloride (PVC)	Tubing, facial prostheses
Natural Materials	
Collagen and gelatin	Cosmetic surgery, wound dressings, tissue engineering, cell scaffold
Cellulose	Drug delivery
Chitin	Wound dressings, cell scaffold, drug delivery
Ceramics or demineralized ceramics	Bone graft substitute
Alginate	Drug delivery, cell encapsulation
Hyaluronic acid	Postoperative adhesion prevention, ophthalmic and orthopedic lubricant, drug delivery, cell scaffold

together during healing also are made of metal and are shown in Figure 6.7. Sometimes the metallic plates and screws are retrieved after successful healing, but in other cases they are left in place. Metallic devices are also used to fuse segments of the spine together when the disk has degenerated (Fig. 6.8) and as dental root prosthetic implants (Fig. 6.9).

Materials selection for a medical device is complicated. The selection depends on a number of factors, including the mechanical loading requirements, chemical and

TABLE 6.2   Mechanical Properties of Materials with Literature Values or Minimum Values from Standards

	Yield	UTS	Deform	Modulus
	MPa	MPa	%	GPa
METALS				
High-strength carbon steel	1600	2000	7	206
F138[1], annealed	170	480	40	200
F138, cold worked	690	860	12	200
F138, wire	-	1035	15	200
F75[2], cast	450	655	8	200
F799[3], forged	827	1172	12	200
F136[4] Ti64	795	860	10	105
Gold		2-300	30	97
Aluminum, 2024-T4	303	414	35	73
POLYMERS				
PEEK		93	50	3.6
PMMA Cast		45-75	1.3	2-3
Acetal (POM)		65	40	3.1
UHMWPE		30	200	0.5
Silicone rubber		7	800	0.03
CERAMICS				
Alumina		400	0.1	380
Zirconia, Mg partially stabilized		634		200
Zirconia, Yttria stabilized		900		200
CARBONS AND COMPOSITES				
LTI pyrolytic carbon + 5–12% Si		600	2.0	30
PAN AS4 fiber		3980	1.65	240
PEEK, 61% C fiber, long		2130	1.4	125
PEEK, 61% C fiber, +-45		300	17.2	47
PEEK-30% C fiber, chopped		208	1.3	17
BIOLOGIC TISSUES				
Hydroxyapatite (HA) mineral		100	0.001	114–130
Bone (cortical)		80–150	1.5	18–20
Collagen		50		1.2

[1]F138, wrought stainless steel: 17–19 Cr, 13–15.5 Ni, 2–3 Mo, <2 Mn, <0.08 or <0.03 C
[2]F75, cast cobalt-chromium-molybdenum alloy: 27–30 Cr, <1.0 Ni, 5–7 Mo, <1 Mn
[3]F799, wrought Co-Cr-Mo alloy: 26–30 Cr, <1.0 Ni, 5–7 Mo, <1.0 Mn, <1.5 Fe, <1.5 C
[4]F136 Titanium 6Al-4V alloy: 5.5–6.5 Al, 3.5–4.5 V, <0.015 N, < 0.13 O, <0.08 C

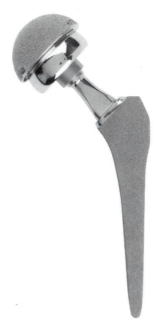

Figure 6.5   A typical total hip joint replacement is made primarily of metal. The ball of the femoral hip stem fits into a pelvic acetabular cup that is lined with ultra high molecular weight polyethylene (UHMWPE) for friction-free motion. (Photograph of the PROFEMUR ® Z minimally invasive hip stem with modular necks courtesy of Wright Medical Technology, Inc.)

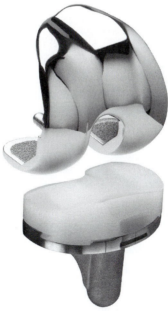

Figure 6.6   A metallic artificial knee joint with an ultra high molecular weight polyethylene bearing surface. (Photograph of the ADVANCE ® medial-pivot knee system courtesy of Wright Medical Technology, Inc.)

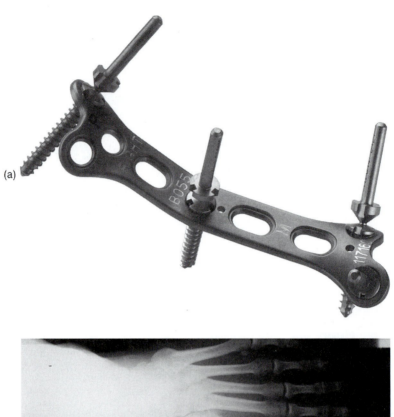

(a)

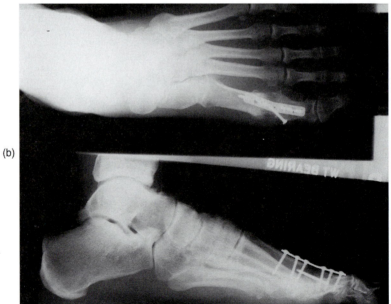

(b)

Figure 6.7   (a) Metal plates and screws are used to hold fractured bone segments together during healing. Depending on the extent of injury, the plates and screws or rods may be removed when the bone is fully repaired. (Photograph of the HALLU®-FIX MTP Fusion System (registered mark of NEW-DEAL) is courtesy of Wright Medical Technology, Inc.) (b) Through the use of x-rays an implanted metal plate with screws can be visualized in this patient's foot and hand. (X-ray courtesy of Wright Medical Technology, Inc.)

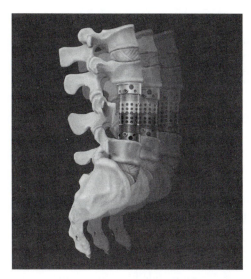

Figure 6.8   Metallic devices are used to fuse segments of the spine together when vertebral bones are fractured due to osteoporosis or back injury. The metal cage can accommodate the patient's own bone particles to assist with new bone formation which will eventually span and fuse the adjacent vertebral bones. (Photograph of the VERTESPAN® spinal fusion cage courtesy of Medtronic Sofamor Danek.)

Figure 6.9   As an alternative to dentures, patients can have metallic dental root prosthetics implanted to replace each missing tooth. The implant is then topped with a porcelain crown. One advantage of dental implants over dentures is that the implant transmits mechanical forces into the jaw bone and stimulates it, resulting in less bone recession over time. (Photograph courtesy of Dr. Martin Freilich of the University of Connecticut Health Center.)

structural properties of the material itself, and the biological requirements. The longstanding use of metals for knee and hip joints, bone plates, and spinal fusion devices is due to the high mechanical strength requirements of these applications and proven biocompatibility in these settings. The advantages of metals over other materials such as ceramics and polymers are that they are strong, tough, and ductile (or deformable, particularly as compared to ceramics). Disadvantages include susceptibility to corrosion due to the nature of the metallic bond (free electrons). In fact, the steels that were used in the early 1900s for hip implants corroded rapidly in the body and caused adverse effects on the healing process. This has led to the preferred selection of alloys of titanium or cobalt-chrome for hip, knee, and dental implants. Other typical properties of metallic materials include a high density and much greater stiffness than most natural materials they replace, which lead to undesirable stress shielding. Stress shielding has been observed after implantation of metal joint replacements and leads to loss of adjacent bone because the bone is not exposed to normal levels of mechanical loading. Certain metals known as shape memory alloys (e.g., nitinol) can be bent or deformed and still return to their original shape when the stress is released. These metals have found application in eye glasses and coronary artery stents that can be inserted through a catheter while collapsed and then spring into a cylindrical shape once they are pushed beyond the confines of the catheter.

Metallic devices are typically made by investment casting, computer-aided design and machining (CAD/CAM), grinding, or powder metallurgy techniques. The specific steps involved in the fabrication of a medical device will depend on factors such as final geometry of the implant, the forming and machining properties of the metal, and the costs of alternative fabrication methods.

Example Problem 6.2

A stent is a helical, woven device that is implanted into an occluded artery to permit increased blood flow. A permanent yet flexible device is needed for use as a vascular stent. What material meets that need? In addition to information contained in this chapter, search the web for information on current materials selections using keywords such as "stents" and "metals." The National Institutes of Health PUBMED website catalogs scientific publications within the biological, biomedical, and medical sciences (http://www.ncbi.nlm.nih.gov/entrez). Corporate web pages can provide additional information. Guidant and Boston Scientific are two companies that currently produce coronary stents. The United States patent office provides another very useful web page for researching uses of materials in surgical and medical devices (www.uspto.gov).

Solution

The preferred material for a stent is a metal such as platinum or titanium. Shape memory alloys such as nitinol also have been used. Stents made of nitinol are self-expanding and "remember" their manufactured shape when they are deployed in the body. They are particularly good for curved or tapered vessels. Metal materials have the strength required for this application and have been shown to be biocompatible after implantation. ∎

6.2.3 Ceramics and Glasses

The advantages of the class of materials known as ceramics are that they are very biocompatible (particularly with bone), are inert, have low wear rates, are resistant to microbial attack, and are strong in compression. Some disadvantages include brittleness, the potential to fail catastrophically, and being difficult to machine. These properties arise from the atomic structure of ceramics: unlike metal, in which atoms are loosely bound and able to move, ceramics are composed of atoms that are ionocovalently bound into compound forms. This atomic immobility means that ceramics do not conduct heat or electricity. Two very obvious properties that are different from metals are melting point and brittleness. Ceramics have very high melting points, generally above 1000°C, and are brittle. Examples of ceramics used in medical devices are shown in Table 6.1. A photograph of a ceramic femoral head of a hip implant is shown in Figure 6.10 and an example of a pelletized calcium sulfate bone graft substitute is shown in Figure 6.11.

Certain compositions of ceramics, glasses, glass-ceramics, and composites have been shown to stimulate direct bone bonding, which is important in securing orthopedic medical devices such as replacement hips and knees and spinal fusion devices. These types of materials are known as bioactive ceramics. Studies on retrieved implants have shown that a biologically active calcium phosphate forms on the biomaterial surface upon implantation in the body. Since the calcium phosphate that forms is much like that found in our bones, bone cells are able to form an intimate attachment to the biomaterial surface after this bone mineral-like layer has formed. The same results are attained by implanting a material that already has a bonelike calcium phosphate surface.

There are several different atomic structures (or phases) of calcium phosphate that have been used in medical applications, including hydroxyapatite, carbonated apatite, di-calcium phosphate or brushite, beta-tricalcium phosphate or tetracalcium phosphate, and amorphous calcium phosphate. The stability of a given calcium phosphate medical device depends on the crystal phase, the crystal size and perfection, the temperature used during processing, the density, and the in-use environment. At physiological temperature and pH, hydroxyapatite is the stable phase, and it generally takes a long time to resorb via physiochemical dissolution. However, bone cells and other cells called macrophages can initiate cell-mediated resorption by changing the local pH to acidic. Resorbable ceramics are typically nonhydroxyapatite phases of calcium phosphate or other calcium-based biomaterials such as calcium carbonate or calcium sulfate.

Due to the high melting point of most ceramics, which prevents them from being cast or extruded, ceramic components are typically made from powdered stock. The powders are formed by wet synthesis methods or by pulverizing raw materials. The

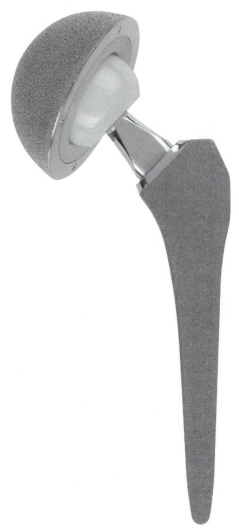

Figure 6.10   In this artificial hip joint, the polymer bearing surface and some of the metallic components have been replaced by ceramics to improve the durability of the joint replacement. This design features a ceramic femoral head and acetabular cup. (Photograph of the LINEAGE ® ceramic–ceramic acetabular cup system is courtesy of Wright Medical Technology, Inc.)

ceramic powder is either added to a liquid with binders to form a slurry that is cast in a mold or dry pressed to form "green ware." The green ware must be finally sintered or fired to densify the powders and remove the porosity between the powder particles. In weight-bearing applications, the porosity must be nearly totally removed or the residual porosity acts as microcracks within the material and weakens it. In other applications such as bone graft substitutes it is desirable to have large pores like those in trabecular or cancellous bone so that cells can infiltrate the material and grow new

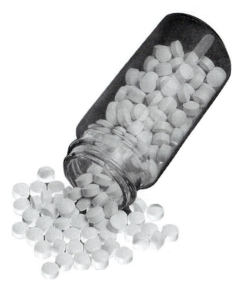

Figure 6.11 If there is an insufficient amount of the patient's own bone or donor bone available to fill a bone defect, synthetic bone graft substitutes made of calcium phosphate or calcium sulfate may be used. (Photograph of OSTEOSET® surgical grade calcium sulfate resorbable beads is courtesy of Wright Medical Technology, Inc.)

vital tissue. In this case, pores are typically created by using second phases, such as polymer beads, that maintain pore space during the early processing steps and are then burned out during the final sintering stage. More detailed descriptions of how porous scaffolds are formed are included later in this chapter. Glasses are silica based. Silica is a network-forming oxide that can be heated to its melting point and, unlike most ceramics, is more easily manufactured.

Example Problem 6.3

What material is preferred for the acetabular cup of a hip implant? What design parameters are utilized during the selection process? Use scientific, corporate, and patent websites to locate information on this topic using keywords such as "ceramic" and "hip replacement."

Solution

Acetabular cups are currently made with a metal support structure and a polyethylene cup; however, problems with wear debris from the soft polyethylene have led to new products with ceramic acetabular cups and femoral balls (alumina or zirconia). The cup must resist wear and deformation and be a low-friction surface because it is in contact with the ball component of the artificial joint. The ceramic materials generate less wear debris during use than does the traditional metal on plastic design. Thus, in theory, the risk that ceramic total hips will fail is low compared to the traditional metal on plastic design. Since the ceramic components are fragile relative to the metallic components and not well tolerated by osteoporotic bone, both types of hip replacements are utilized today.   ■

6.2.4   Polymers

Polymers are well suited for biomedical applications because of their diverse properties. For example, polymers can be flexible or rigid, can be low strength or high strength, are resistant to protein attachment or can be modified to encourage protein attachment, can be biodegradable or permanent, and can be fabricated into complex shapes by many methods. Some disadvantages of polymers are that they tend to have lower strengths than metals or ceramics, deform with time, may deteriorate during sterilization, and may degrade in the body catastrophically or by release of toxic by-products. Examples of polymers used in medical devices and their mechanical properties are listed in Tables 6.1 and 6.2.

The large macromolecules of commercially useful polymers are synthesized by combining smaller molecules (mers) in a process termed *polymerization*. Polymerization may proceed by addition (or chain reaction) polymerization, in which monomer units are attached one at a time and then terminated, or by condensation (or step reaction) polymerization, in which several monomer chains are combined and a by-product of the reaction, such as water, is generated. Additives such as fillers, plasticizers, stabilizers, and colorants typically are used in polymer synthesis to enhance the mechanical, chemical, and physical properties.

Polymers can be classified as thermoplastic or thermosetting. A thermoplastic polymer has a linear or branched structure. As a solid it is like a bowl of spaghetti in that the chains can slide over one another. With heating, the chains can slide more easily, and the polymer melts or flows. Thus, thermoplastic polymers can be heated, melted, molded, and recycled. Differences in properties can be achieved with the addition of different ligands. PVC is more rigid than PE because the chlorine atoms are larger and tend to prevent the sliding of one molecule over another. PMMA, as shown in Table 6.2, is stronger, stiffer, and much more brittle than UHMWPE. In this case, 2 of the 4 hydrogen atoms are replaced, one with a methyl group (CH_3) and the other with an acrylic group ($COOCH_3$). These large side-groups make sliding much more difficult, hence the increase in strength and modulus. They also make it difficult for the molecules to orient in an orderly, crystalline pattern. As a result of this amorphous structure, PMMA (Plexiglas® or Lucite®) is optically transparent.

In contrast, a thermosetting polymer is composed of chains that are cross linked. They do not melt with heating but degrade. The term *thermoset* implies that there is a chemical reaction, often involving heat, which results in setting a three-dimensional cross-linked structure. A common example is "5-minute epoxy." When the two parts are mixed, the catalyst causes setting and cross linking of the epoxy. Once set, it cannot be heated and reused. The amount of cross linking affects the mechanical properties. Few cross links are used to make rubber gloves. Adding more sulfur and cross linking produces a car tire. Even more cross links are added to make the hard casing of a car battery.

Hydrogels are water-swollen, cross-linked polymeric structures that have received significant attention because of their many applications in biomedical applications. Hydrogels are prepared by cross linking polymer chains by irradiation or chemical methods and then expanding the network of chains by swelling the structure with water. The most widely used hydrogel is cross-linked polyhydroxyethylmethacrylate (PHEMA). The PHEMA structure has a water content similar to living tissue, has resistance to degradation, is not absorbed by the body, withstands heat sterilization without damage, and can be prepared in a variety of shapes and forms. Applications of hydrogels include contact lenses, drug delivery vehicles, wound healing adhesives, sexual organ reconstruction materials, artificial kidney membranes, and vocal chord replacement materials.

Quite a variety of techniques are employed in forming polymer medical devices. The technique depends on several factors such as whether the material is thermosetting or thermoplastic, and if thermoplastic, the temperature at which it softens. Thermosetting polymers must be prepared as a liquid linear polymer and then cured in a mold. They cannot be molded after this step. Thermoplastic polymers can be molded repeatedly (by compression, injection, extrusion, etc.), cast, and formed into fibers or films by extrusion followed by drawing or rolling to improve properties such as strength. Precipitation after being dissolved in a solvent by introduction of a nonsolvent is a way to form porous polymer scaffolds for tissue engineering. Large pores suitable for cellular infiltration can be created by adding particulates with the desired pore size to the polymer/solvent mixtures. Pores are created when the particulates are washed out after solvent removal. Additional methods for creating porous polymer scaffolds are described later in this chapter.

Example Problem 6.4

What material is preferred to produce a blood bag? A dialysis bag? What design parameters are involved?

Solution

PVC has been used for blood bags since the 1950s. Since PVC is naturally brittle, phthalate plasticizers are used to make it flexible. These leach out over time from the plastic bag and into the liquid which they contain. When fed in large quantities to rats, the plasticizers can cause cancer; therefore, other plastics are being investigated. Dialysis bags are made of low-density polyethylene (LDPE). In these examples, materials selection has been governed by the fact that the material must be flexible, chemically stable, and relatively inert. ■

6.2.5 Natural Materials

Natural materials are synthesized by an organism or plant and are typically more chemically and structurally complicated than synthetic materials. Examples of natural biomaterials currently used in medical devices are listed in Table 6.1. Proteins and polysaccharides are nature's form of polymers and are used in medical devices. The directional bonds within proteins give rise to the very high mechanical properties of natural polymers. For example, the ultimate tensile strength of silk is higher than that of drawn nylon, one of the strongest synthetic polymers. Furthermore, the elastic modulus of silk is nearly thirteen times that of the elastic modulus of nylon. There are also natural ceramic materials. Natural ceramics are typically calcium based, such as calcium phosphate bone crystals or calcium carbonate coral or sea shells. Natural ceramics are typically much tougher (resistant to fracture) than synthetic ceramics due to their highly organized microstructure which prevents crack propagation. Small ceramic crystals are precisely arranged and aligned and are separated by thin sheets of organic matrix material. A crack in the material is forced to follow this tortuous organic matrix path. The category of natural materials also encompasses donor tissue such as bone or skin which may be patient derived (autograft), from another human (allograft), or from a different species such as bovine or porcine (xenograft).

Natural materials exhibit a lower incidence of toxicity and inflammation as compared to synthetic materials; however, it is often expensive to produce or isolate natural materials. There is also variability between lots of natural materials, which makes it difficult to maintain consistency and sometimes prevents widespread commercial use. The isolation or purification steps typically involve the use of solvents to extract the desired component from the rest of the tissue or the use of solvents to remove the undesired components from the tissue and leave the desired natural material intact. Collagen can be prepared by either method. If it is labeled as soluble collagen, it has been removed by pepsin enzymatic treatment from natural tissues such as cock's comb. Fibrillar collagen is prepared from natural tissue such as tendon by salt and lipid and acid extraction steps to remove the noncollagenous proteins and molecules, leaving the collagen fibers intact.

Biopolymers also may be produced by bacteria. Production of polyhydroxybutyrate (PHB) is carried out through a fermentation procedure. The bacteria produce the polymer in granules within their cytoplasm when they are fed a precise combination of glucose and propionic acid. The cells are then disrupted, and the granules are washed and collected by centrifugation and then dried. This polymer has properties similar to polypropylene and polyethylene yet it degrades into natural components found in the body. Because of the desirable environmental characteristics of biopolymers, they are rapidly finding use in several nonmedical niche markets such as biodegradable monofilament fishing nets. The more rapid and widespread introduction of biopolymers has been hindered by their high price (up to 10 times) relative to that of petroleum-based polymers.

Biopolymers also can be produced by chemically polymerizing naturally occurring monomers. Although these polymers are not produced by biological systems, the fact that they are derived from basic biological building blocks makes them biocompatible, nontoxic, and biodegradable. Lactic acid–based polymers have been used widely for many years for medical devices ranging from biodegradable sutures to tissue engineering scaffolds. Lactic acid is found in blood and muscle tissue and is produced commercially by microbial fermentation of sugars such as glucose or hexose. Polylactides are frequently used in combination with polyglycolic acid.

Example Problem 6.5

What materials are preferred by surgeons for repairing large surgical or traumatic defects in bone? What factors influence this decision? Search the Internet for companies producing bone repair products. Go to the website of the American Academy of Orthopaedic Surgeons and the American Society of Plastic Surgeons.

Solution

Autograft bone is the first choice for surgeons for the repair of bony defects. The tissues are vital and contain living cells and growth factors that are required for bone regeneration. Allograft bone or demineralized allograft bone matrix is the second choice. Demineralized bone has advantages over as-harvested allograft bone because it is flexible and can conform to the defect site, resorbs more rapidly (within months as compared to years for nondemineralized allograft bone), and releases the bone inductive proteins known as bone morphogenetic growth factors originally discovered by Dr. M.R. Urist in the 1950s. ∎

6.2.6 Composites

Composite materials consist of two or more distinct parts. Although a pure material may have distinct structural subunits such as grains or molecules, the term *composite* is reserved for materials consisting of two of more chemically distinct constituents that are separated by a distinct interface. Examples of composites used in biomedical applications include carbon fiber reinforced polyethylene and hydroxyapatite particle reinforced polylactic acid polymers for bone healing applications. The discontinuous phase is typically harder and stronger than the continuous phase and is called the reinforcement.

Composites are made by mixing two components and molding, compacting, or chemically reacting them together. The fibers are typically coated or impregnated with the polymer phase so that the composite can be heated and pressurized to densify the assembly. A chemical reaction may be utilized to form composites in which a second phase precipitates or forms upon reaction. There may be a filament winding process if high-strength hollow cylinders are being formed. Composites are well suited for devices that require a combination of properties such as total joint replacements, dental fillings, and bone plates. The advantages of composites are that the properties can be tailored to fit nearly any application; however, it is difficult to make a composite with an ideal structure. There are typically problems with dispersion of the second phase or weak interfacial bonds between the two phases which leads to less than ideal mechanical properties and, hence, poor product performance. However, in many cases the actual performance is still much better than any single component biomaterial and so composites are becoming more widely used in biomedical applications.

Example Problem 6.6

What materials are preferred for reconstructive dental applications? What are some advantages of the composite structure over a monolithic structure?

Solution

Mercury amalgams made of mercury, silver, and tin are the most commonly used filling materials. Aesthetically pleasing tooth-colored filling materials also can be used and are made of filled resins [e.g., large molecule bifunctional methyacrylates (BisGMA) filled with micro- and nanoparticulate silica]. PMMA is the predominant material used for complete and partial dentures. Chrome-based alloys are used for the framework of removable dentures. Crowns and bridges are made of a cast metal frame veneered with tooth-colored porcelain. All-ceramic systems are available as well. Recently, composites made of light curing resins reinforced with glass fibers have been developed for dental bridges. The metallic post typically used to provide structural support for crowns is now being replaced with this type of glass fiber reinforced composite, primarily for better aesthetic results—the polymer post does not show through the porcelain crown like the metallic post does.   ■

6.3   LESSONS FROM NATURE ON BIOMATERIAL DESIGN AND SELECTION

6.3.1   An Overview of Natural Tissue Construction

Biomedical engineers are asked to design medical devices or systems that repair, monitor, or assist the functions of the human body. Approaches that mimic or replicate nature's techniques, known as biomimetics, are often at the heart of a successful medical device or therapy. There is an incredible complexity to the genesis of natural tissues and organs which is still far beyond the capacity of scientists to replicate. Furthermore, the precise function of every aspect of the tissues or organs is not known. For these reasons, it is very difficult to theoretically design medical devices, and the field has progressed through a fair amount of trial and error. Nonetheless, there are several general concepts that have emerged from the study of structural biology that provide design strategies and guidance for a biomaterials scientist involved in tissue/organ regeneration.

- Cells are programmed by their genetic code to build the tissues and organs of our bodies.
- Cells produce proteins, polysaccharides, glycoproteins, and lipids that self-assemble into composite extracellular matrices that have multiple diverse forms and serve to support tissue growth.
- Cells communicate via growth factors and their recruitment, and even cellular fate is determined by protein signals.
- Blood vessels play a crucial role in tissue growth by providing nutrients, a means for waste removal, and a supply of additional cells to support further growth.

- The nervous system is responsible for the integration and control of all the body's functions.
- Skeletal tissues are made hard and stiff by the protein-controlled nucleation and growth of small, discrete, nanometer-sized mineral crystals within a collagen matrix microenvironment.

Overall, natural tissue design is hierarchical: that is, a structure within a structure, like a nested set of eggs or the branches of a tree. The same structural motif is repeated at multiple length scales to endow the tissue with strength and efficiency of function. Biomimetic paradigms that have been derived from these basic structural and developmental biology concepts provide a rational starting point for the design and fabrication of biomaterials, especially for regenerative tissue-engineered medical devices.

6.3.2 Cells Build Natural Tissues

There are over 200 different cell types in the human body and approximately 100 trillion cells per person. Bone tissue alone has more than ten types of cells:

- Osteoblasts which make bone
- Osteoclasts which degrade bone
- Osteocytes which maintain bone
- Cells in blood vessels that run through bone [erythrocytes (red blood cells) and immune cells (lymphocytes, monocytes, macrophages, neutrophils, and eosinophils)]
- Lipid cells and mesenchymal stem cells that are not yet differentiated into a specific type of cell, which are both found in bone marrow

The existence of a cell population that has the potential to differentiate into a number of tissue cell types is a key design strategy that our bodies use to repair damaged tissues. Even as adults, we retain a small capacity to generate new tissues through stem cells. Stem cells are discussed further in the chapter on tissue engineering (Chapter 7). Stem cells differentiate into specific cell types in response to contact from neighboring cells and their chemical and physical environment and by mechanical forces transmitted by their support matrix.

Cells not only manufacture the tissue constituents, they also maintain the tissue and adapt the tissue structure to the changed environments, including mechanical load environments. Key elements in the assembly of tissues are:

- *Cell–cell signaling:* Cellular interactions are controlled by complex molecular communication between cells. Cell signaling molecules are typically proteins, known as cytokines and chemokines. Cytokines (also called growth factors) cause proliferation (cell replication) and differentiation. Chemokines induce cell migration. Cells attach to each other by various types of junctions. These cell-to-cell linkages provide additional mechanical strength to the tissue and provide semipermeable barrier layers that regulate cell, fluid, ion, and protein transport.
- *Apoptosis:* As part of the coordinated function of tissue growth and morphogenesis (pattern or shape formation), certain cells are programmed to die in a

process known as apoptosis. This is another design strategy that can be utilized by the biomaterials engineer: initiation of a biological event is controlled, and there is a mechanism for stopping it. In a very tidy manner, the unneeded cells shrink, condense, and then fragment into membrane-bound granules that can be cleaned up by macrophages.

- *Necrosis:* In contrast to apoptosis, another form of cell death due to trauma or toxin is known as necrosis, in which the cell swells and bursts, releasing its intracellular contents. Since cells contain signaling proteins, necrotic cell death leads to unwanted, uncontrolled cell recruitment and activity. Therefore, apoptotic cell death is the typical mode of cell death during normal tissue functioning.

- *Angiogenesis:* Cells cannot survive without a supply of oxygen and nutrients and a way to remove waste products. Blood vessels provide these functions, and the lymphatics assist with additional waste removal. The formation of new blood vessels, or angiogenesis, must occur rapidly during new tissue growth or the newly formed tissue will die or necrose. Angiogenic factors are released from nearby macrophages, platelets, or the extracellular matrix.

- *Neurogenesis:* The nervous system consists of several cell types, neurons that conduct electrical impulses; glial cells that protect, support, and nourish neurons; and Schwann cells that build the fatty insulating material called the myelin sheath. As described in Chapter 3, this system of cells is responsible for the integration and control of all of the body's functions. Newly regenerated tissue must establish connections with the nervous system in order to function.

Example Problem 6.7

Is angiogenesis always necessary and desirable?

Solution

In some situations, such as growth of new tissues, angiogenesis is necessary and desirable. In other situations, blood vessels are not desired because they may bring an oversupply of immune cells, causing autoimmune disorders, or may enhance tumor growth by supplying cancer cells with nutrients. This has led to the development of anti-angiogenic factors by the pharmaceutical industry and their delivery by biomaterials. ■

6.3.3   The Extracellular Matrix: Nature's Biomaterial Scaffold

The extracellular matrix (ECM) provides a physical and chemical support structure for cells. From a biomaterials perspective, it is a complex woven polymeric material (fibrous proteins) with reinforcing struts (other fibrous proteins) and gels to resist compaction (water-swollen proteoglycans). A full description of the components of the ECM and their functions can be found in Chapter 7. Basically, the ECM provides a microenvironment for the cells. It provides a surface for cells to adhere to and to migrate across, attachment sites for proteoglycans, and a means of controlling the release of the cell signaling proteins. It is a scaffold that interconnects all the cells physically and mechanically. Cells attach to the ECM via specific cell surface receptors that govern proliferation, differentiation, and protein expression. Interactions of the cells with the ECM thus play a crucial role during tissue regeneration.

The ECM is made by spontaneous folding of proteins produced by cells, typically different collagenous proteins (there are more than 10) and large glycoproteins (e.g., fibronectin, laminin, ostepontin). It is dynamic and is constantly being modified by the cells. The spontaneous folding of peptides is known as self-assembly. The collagen molecule self-assembles from three alpha chains into a triple helix. Within the triple helix, glycine must be present as every third amino acid, and proline and hydroxyproline are required to form and stabilize the triple helix. The collagen molecules then assemble together, a few molecules at a time, with a quarter overlap to form a staggered linear array. The linear aggregates then laterally associate into bundles (Fig. 6.12). Hole zones are left open between the collagen molecule terminal groups for subsequent mineralization (as described further in the example on bone biomineralization later in this chapter). The molecular assembly of DNA is somewhat similar to that of collagen: two preexisting complementary DNA chains combine to form a double helix and then the helices assemble. Self-assembly is one of the key design strategies of tissue formation and must be involved in producing highly ordered structures at the molecular level. Biomaterials scientists are now utilizing this principle to fabricate complicated microstructures at the nanometer length scale.

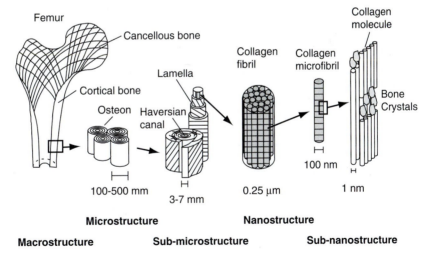

Figure 6.12 The hierarchidal structure of bone. There are at least 5 levels of microarchitecture in bone: (a) the femur, (b) the layers of lamellar bone shown within a cylindrical osteon, that are the building blocks of the cortical bone of (c) fibrils composed of the microfibrils, which form (d) microfibrils with periodic gaps between the ends of the molecules that provide nucleation sites for bone mineral, (e) collagen molecules.

6.3.4 Hierarchical Design

Similar to a nested set of eggs or Russian dolls, in which you open up one just to find another smaller one inside, natural tissues have nested structures. This is known as hierarchical design. The same structural motif is repeated at multiple length scales, and this endows the tissue with efficiency of assembly, properties, and function. During cell-mediated tissue construction, the smallest units self-assemble first, then these units self-assemble to form larger units, and finally the larger units self-assemble. This is how a functional tissue or organ is built. Cells typically become embedded or trapped within the layers of extracellular matrix as the tissue is built. The cells survive and act to maintain the tissue around them. Lamellar bone has similar layered fibrillar structures at the nano, micro, and macro levels. At the smallest level there are the alpha chains that have assembled in a triple helix to form the Type I collagen molecule, the collagen molecules then assemble into microfibrils, microfibrils assemble into larger fibers that assemble by alignment into sheets, and the sheets (lamellae) are layered like plywood in a criss-crossed orientation around blood vessels to form osteons, which have a tubular or large fiber appearance (Fig. 6.12). Skeletal muscle also has a hierarchical structure. The actin and myosin filaments organize into linear constructs known as sarcomeres, which are bundled into myofibrils, which are bundled together to form a muscle fiber (which is the muscle cell with hundreds of nuclei), and multiple muscle fibers form the muscle fascicles we call muscle. A diagram of this can be found in Chapter 3 of this book.

The biological structure of chromosomes is also governed by hierarchical design. Nature efficiently compacts a 7-cm long strand of DNA until it is 10,000 times smaller by twisting and coiling at multiple length scales. At the nanometer level, DNA is a double helix of single-stranded DNA. Subsections of the twisted DNA strand (146 base pairs each) are then subcoiled around protein (histone) cores to form chromatosomes which cause the strand to resemble a beaded necklace. The linked chromatosomes are then coiled to generate a shorter, thicker fiber. The thicker fiber is then coiled upon itself to further shorten its length to form the chromatid, two of which are linked together to form a chromosome.

6.3.5 Biomineralized Tissue Example

Bones, antlers, teeth, coral, eggshells, and seashells are examples of biomineralized tissues. Although they serve different functions and have considerably different external shapes, they are all composed of numerous small, isolated calcium phosphate or calcium carbonate crystals that are held together by a protein matrix. The nucleation and growth of the mineral crystals are regulated by the organic protein component that is secreted by cells and self-assembles to provide a template for the mineral growth and nucleation. Three general biological processing principles governing the composition, architecture, and methods of assembly of a variety of mineralized tissues have been identified that have significant implications for material scientists and engineers:

Biomineralization occurs within specific subunit compartments of microenvironments, which implies stimulation of crystal production at certain functional sites and inhibition or prevention of the process at other sites.

A specific mineral phase is produced with a defined crystal size (frequently in the nanometer range), shape, and orientation.

Macroscopic shape forming is accomplished by packaging many incremental units together, which results in unique net-shaped composites with layered microarchitectures that impart exceptional material properties.

Another powerful feature of biomineralization is that, in most systems, remodeling of the original mineral structure occurs as needed to optimize strength, accommodate organism growth, maintain mineral ion equilibrium, and effect repairs.

In bone, controlled mineral nucleation and growth are accomplished within the microcompartments formed by the collagen matrix. The Type I collagen molecules secreted by osteoblasts self-assemble into microfibrils with a specific tertiary structure having a 67-nm periodicity and 40-nm gaps or holes between the ends of the molecules (Fig. 6.12). The holes localize a microenvironment containing free mineral ions and bound side chain groups from phosphoproteins attached to the collagen. The molecular periodicity of the functional groups serves to nucleate the mineral phase heterogeneously. The nucleation of the thin, platelike apatite crystals of bone occurs within the discrete spaces within the collagen fibrils, thereby limiting the possible primary growth of the mineral crystals and forcing them to be discrete and discontinuous. Only one phase of calcium phosphate is nucleated during normal, nonpathological mineralization processes (carbonated apatite), and the minerals grow with a specific crystalline orientation. In this example on bone tissue formation, the key elements of natural tissue fabrication—cells, an ECM defining a microenvironment for both cells and mineral, protein signaling, and hierarchical design—are all present.

6.4 TISSUE–BIOMATERIAL INTERACTIONS

6.4.1 Interactions with Blood and Proteins

The implantation of a biomaterial often creates a wound, and bleeding generally ensues. Blood thus typically makes first contact with the implanted biomaterial. Blood is a mixture of water, various kinds of cells and cell fragments (platelets), salts, and proteins (plasma). Proteins, the primary group of molecules responsible for making life possible, are built of long chains of only 20 amino acids that are strung together by peptide bonds. Proteins have a myriad of functions in the human body. They can function as enzymes that catalyze thousands of important chemical reactions essential to life. Cell signaling molecules responsible for cell migration and proliferation are made of proteins. Proteins are the building blocks of the supporting extracellular matrix of many tissues. Changes in the levels of proteins or the structure of proteins lead to altered function and are responsible for many diseases. This has led to blood screening for certain proteins that may indicate a diseased state or cancer.

Proteins play an important role in determining the final nature of the tissue–implant interface. Biomaterials can promote cell/tissue attachment and activity by allowing selective protein adsorption or can inhibit tissue interactions by repelling protein. Importantly, changes in microenvironment which can occur after biomaterial implantation, such as pH and ionic strength, can alter the conformation of a nearby protein and hence its function. Proteins also can experience structural alterations during interaction with the solid surfaces of biomaterials and lose some of their biological activity. Albumin is the most common protein in blood, followed by the protective immune system proteins known as immunoglobulins. However, because exchange between absorbed proteins occurs, the final layer of absorbed protein may be fibrinogen, which although less abundant, may have a greater affinity for the biomaterial surface.

Blood coagulation is directed by attachment of the protein clotting factor XII, which is found in blood, to the foreign biomaterial surface. After attachment of this factor, platelets from the blood can and will adhere to the biomaterial, which leads to fibrin clot formation. A cascading chain of cellular reactions that is governed by the initial protein attachment begins. Blood contact provides the cells and cytokines that participate in the biological interaction with the biomaterial. Therefore, every biomaterial that contacts blood will elicit biological responses from the body.

6.4.2 The Wound Healing Response after Biomaterial Implantation

The implantation of a biomaterial creates a disruption of the anatomic continuity of tissue and, as such, creates a wound. The body has a highly developed wound healing response that is immediately triggered by the biomaterial implantation. Much is known about the normal cellular events that transpire after the initiation of a wound (see Chapter 7), and this knowledge provides a foundation to understand and anticipate tissue–biomaterial interactions. From the perspective of tissue–biomaterial interactions, there are four overlapping phases that will always occur (Fig. 6.13):

Hemostasis: Platelet cells control bleeding through coagulation by adhering to the proteins attached to the biomaterial surface and by releasing clot-forming proteins. The clot that is formed acts as a provisional matrix for the initiation of repair tissue and fills the gaps around the implanted biomaterial.

Inflammation: Clot formation induces the production of cell signaling molecules (cytokines) that induce the recruitment of inflammatory cells from a nearby bloodstream. These cells (neutrophils, monocytes, lymphocytes, and macrophages) arrive and attempt to digest tissue debris and the biomaterial by a process known as phagocytosis. The growth factors released at the wound site by the inflammatory cells initiate mitosis (cell replication) of sedentary connective tissue cells at the wound margin.

Proliferation/initial repair: As a result of all the growth factor signaling by the inflammatory cells, there is a proliferation and population of the biomaterial with cells that can recreate the lost or damaged tissue. A nondegrading biomater-

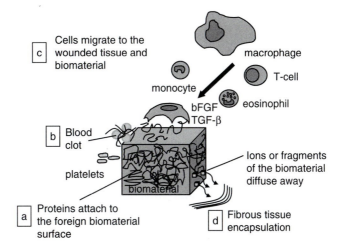

Figure 6.13 Normal tissue–biomaterial interactions involve the four overlapping and interdependent phases of wound healing: hemostasis, inflammation, proliferation/repair, and tissue remodeling. (a) Protein attachment to the biomaterial surface guides cellular interactions. (b) Hemostasis is accomplished by clot formation. (c) Cells found in blood and other inflammatory cells attempt to process the foreign biomaterial and repair adjacent material. (d) The host protects itself from the foreign biomaterial through encapsulation with fibrous tissue.

ial located in the center of the wound typically becomes encapsulated with tight fibrous tissue. The fibrous capsule isolates the material from the biological environment and protects the host. The extent of the inflammatory foreign body response governs the thickness of the fibrous capsule. The chemical characteristics, the shape and physical properties of the biomaterial implant, and the rate of release, accumulation, and bioactivity of released chemicals and corrosion products from implanted materials all also affect the thickness of the fibrous capsule. If the implant is permanent and does not biodegrade, then a small capsule remains throughout the life of the implant, except in bone where there is direct bone apposition on calcium phosphate surfaces without an intervening fibrous tissue layer.

Remodeling: The rapidly formed neotissue will be remodeled by cells into functional tissue more similar to the original tissue, although typically a scar remains.

6.4.3 Metallic Corrosion

There are a number of mechanisms by which metals can corrode, and corrosion resistance is one of the most important properties of metals used for implants. The mechanisms of most significance to implant applications in aqueous saline solutions are galvanic (or mixed metal) corrosion, crevice corrosion, and fretting corrosion.

Galvanic (mixed metal) corrosion results when two dissimilar metals in electrical contact are immersed in an electrolyte. There are four essential components that must exist for a galvanic reaction to occur: an anode, a cathode, an electrolyte, and an external electrical conductor. The *in vivo* environment contains electrolytes. A patient with two total hip replacements (THRs) made of different alloys is not subject to mixed metal corrosion since there is no electrical connection. However, a THR made of two alloys or a fracture plate of one metal fixed with screws of another metal may be susceptible to mixed metal corrosion.

When two dissimilar metals are connected in an electrochemical cell, one will act as an anode while the other will be the cathode. Metal oxidation will occur at the anode, as shown in Equation 6.4. The reaction at the cathode will depend on the pH of the environment (Equations 6.5 and 6.6). The direction of the reaction can be determined by examining the electromotive force (EMF) series, a short listing of which is shown in Table 6.3. These potentials represent half-cell potentials of metals in equilibrium with 1 molar solutions of their ionic species. The potential for hydrogen is defined as zero. As shown, the standard potential for iron is -0.44 V. If iron is connected to copper with an EMF of $+0.34$ V, the potential difference is 0.78 V. Since iron is the anode, iron oxidation will occur according to the reaction shown in Eq. 6.4. The reaction at the copper cathode will depend on the pH of the solution, as shown in Eq. 6.5 and 6.6.

$$\text{anodic reaction } Fe \rightarrow Fe^{+2} + 2e- \tag{6.4}$$

$$\text{cathodic reaction (acidic solution) } 2H^+ + 2e^- \rightarrow H_2 \tag{6.5}$$

$$\text{cathodic reaction (neutral or basic) } O_2 + 2H_2O + 4e^- \rightarrow 4OH^- \tag{6.6}$$

Because the free energy per mole of any dissolved species depends on its concentration, the free energy change and electrode potential of any cell depends on the composition of the electrolyte. Thus, the direction and rate of the reactions also depends on the concentration of the solution. Increasing the concentration of Fe^{+2}

TABLE 6.3   Electromotive Force Series: Standard Reduction Potentials (E^0 V) in Aqueous Solution at 25°C

K $= K^+ + e^-$	-2.93 Active (more anodic)
Na $= Na^+ + e^-$	-2.71
Al $= Al^{+3} + 3e^-$	-1.66
Ti $= Ti^{+2} + 2e^-$	-1.63
Zn $= Zn^{+2} + 2e^-$	-0.76
Cr $= Cr^{+3} + 3e^-$	-0.74
Fe $= Fe^{+2} + 2e^-$	-0.44
Co $= Co^{+2} + 2e^-$	-0.28
Ni $= Ni^{+2} + 2e^-$	-0.25
Sn $= Sn^{+2} + 2e^-$	-0.14
$H_2 = 2H^+ + 2e^-$	0.000
Cu $= Cu^{+2} + 2e^-$	$+0.34$
Ag $= Ag^+ + e^-$	$+0.80$
Pt $= Pt^{+2} + 2e^-$	$+1.20$
Au $= Au^{+3} + 3e^-$	$+1.50$ Noble (more cathodic)

in the environment will shift the potential in the positive or noble direction. As the Fe^{+2} concentration increases, the potential difference between the iron and copper will become less as the iron becomes more cathodic. Similarly, the concentration of oxygen at the cathode will affect the EMF of the cell. Increasing O_2 will make it more noble while decreasing O_2 will make it more anodic. In fact, crevice corrosion is initiated by changes in oxygen concentration as is discussed in a following paragraph.

Galvanic cells occur not only with different alloys but also with differences within an alloy. Carbides, grain boundaries, and different phases within an alloy also present differences in EMF and thus the possibility for localized galvanic cells. Cold working also increases the free energy of metal and thus its susceptibility to corrosion. Bending a plate or pounding on a nail head causes localized cold working and makes that area anodic to the rest of the piece.

Galvanic corrosion can also be utilized to prevent corrosion by cathodically polarizing the part to be protected. Steel ships are protected from rusting by the attachment of blocks of zinc. The zinc blocks ("zincs") serve as a sacrificial anode and protect the steel hull. Metal pumps and other metallic components on ships are also protected with zincs. A power supply can be attached to a part, such as in a steel underground pipeline, to make the pipe cathodic to a replaceable anode. This protects the pipeline.

Electrode size also has an effect on galvanic reaction rates. The classic example is the difference between galvanized and tin-plated iron. As Table 6.3 shows, zinc is anodic to iron. Thus, galvanization results in coating the iron with an anodic material. When the zinc is scratched and the iron is exposed, the small size of the iron cathode limits the reaction, and there is minimal corrosion. In contrast, tin is cathodic to iron. When a tin plate is scratched, the small iron anode is coupled with a large cathode. Anodic corrosion and removal of iron from the scratch results in an increased area of the exposed iron and thus an increase in corrosion rate. This self-accelerating corrosion can be prevented by coating the tin cathode with a nonconductive material such as paint or varnish.

Crevice corrosion can occur in a confined space that is exposed to a chloride solution. The space can be in the form of a gasket-type connection between a metal and a nonmetal or between two pieces of metal bolted or clamped together. Crevice corrosion involves a number of steps that lead to the development of a concentration cell, and it may take 6 months to 2 years to develop. Crevice corrosion has been observed in some implanted devices where metals were in contact, such as in some total hip replacement devices, screws and plates used in fracture fixation, and some orthodontic appliances.

The initial stage is uniform corrosion within the crevice and on the surfaces outside the crevice. Anodic and cathodic reactions occur everywhere, with metal oxidation at the anode and reduction of oxygen and OH-production at the cathode. After a time, the oxygen within the crevice becomes depleted because of the restricted convection of the large oxygen molecule. The cathodic oxygen reduction reaction ceases within the crevice, but the oxidation of the metal within the crevice continues. Metal oxidation within the crevice releases electrons which are conducted through the metal and consumed by the reduction reaction on the free surfaces outside the crevice. This creates an excess positive charge within the crevice that is balanced by an influx of negatively charged chloride ions. Metal chlorides hydrolyze in water and dissociate into an insoluble metal hydroxide and a free acid (Eq. 6.7). This results in an ever-increasing acid concentration in the crevice and a self-accelerating reaction.

$$M^+Cl^- + H_2O \rightarrow MOH + H^+Cl^- \tag{6.7}$$

The surgical alloys in use today all owe their corrosion resistance to the formation of stable, passive oxide films, a process called passivation. Titanium, which appears as an active metal on the EMF series in Table 6.3, forms a tenacious oxide that prevents further corrosion. Stainless steels and cobalt alloys form chromium oxide films. As indicated in Table 6.3, they are active, but in the environment where this oxide film is formed, they become passive or noble.

To be self-passivating, stainless steels must contain at least 12% chromium. However, carbon has a strong affinity for chromium, and chromium carbides form with the average stoichiometry of $Cr_{23}C_6$. The formation of a carbide results from the migration of chromium atoms from the bulk stainless steel alloy into the carbide. The result is that the carbide has high chromium content while the alloy surrounding the carbide is depleted in chromium. If the chromium content is depleted and drops below 12% Cr, then there is insufficient Cr for effective repassivation, and the stainless steel becomes susceptible to corrosion. As a safety factor, surgical stainless contains 17–19% chromium, and the carbon content in surgical alloys is kept low at <0.08% or <0.03%.

The problem of carbide formation is especially important with welded stainless steel parts. If steel is heated to the "sensitizing range" of 425°C to 870°C, the chromium can diffuse in the solid and form carbides. At temperatures above 870°C, the carbon is soluble in the atomic lattice. Below 425°C, the mobility is too low for carbide formation. If the peak temperature in the metal away from the weld is in the sensitizing range, carbides can form. This is known as weld decay or corrosion of the sensitized metal on each side of the weld. By heat treating after welding, the carbides can be redissolved, and the metal quickly quenched to avoid reformation.

With the oxide film intact, surgical alloys are passive and noble. If the film is damaged, as with scratching or fretting, the exposed metal is active. Reformation or repassivation results in restoration of the passive condition. Fretting corrosion involves continuous disruption of the film and the associated oxidation of exposed metal. Devices that undergo crevice corrosion are also examples in which fretting corrosion has accelerated crevice corrosion.

Example Problem 6.8

Old cars in the northern United States are often rusted on the bottom of their doors and on their trunk lids, and the tailgates of old pickup trucks are also often rusted. Name and discuss five reasons for this. Would bolting on a zinc block help?

Solution

(1) The edges and bottoms of doors and lids are formed by bending the metal back on itself. This causes cold working at the bend, which makes it anodic to the rest of the

metal. (2) The crimps are then spot welded closed. This creates an area of different microstructure, which leads to a galvanic situation. (3) The roads are salted in the winter, and the saltwater spray gets caught in the crimp. That is the electrolyte. Sand may also get in the crimped space and help maintain a moist environment. (4) Car manufacturers put a decorative strip of chromium-plated steel along the bottom of the lid. The chromium is cathodic to the steel and provides another source of galvanic corrosion. (5) There are potholes in the roads. This causes bouncing of the lid against the frame. This can chip the paint and expose the unprotected metal, or it may be a cause of fretting corrosion. ■

Bolting on a zinc block would not help except for corrosion of metal in the same electrolyte pool. A piece of zinc in the crevice would help slow the corrosion of the crevice. If the car fell into the ocean, then the zinc block would protect the whole car.

Example Problem 6.9

Your grandmother has a stainless steel total hip. Now she needs the other hip replaced, and the doctor wants to use one made of a cobalt chromium alloy. Is that a problem for corrosion?

Solution

No. One of the four essential elements for galvanic corrosion is missing. There is an anode, the stainless steel. There is a cathode, the cobalt alloy. There is an electrolyte, the saltwater of the body. However, there is no electrical connection, so there is no problem. If she were to fall and fracture her pelvis, and the break was repaired with an external fixator, then there might be an electrical connection and a problem. However, these alloys are so corrosion resistant and similar electrochemically that there is probably no need to worry. ■

6.4.4 Biomaterial Degradation and Resorption

Biomaterials may be permanent or degradable. The degradation process may be chemically driven or accomplished by cells. Bioresorbable implants are designed to degrade gradually over time in the biological environment and be replaced with natural tissues. The goal is to meet the requirements of strength and cell support while the regeneration of tissues is occurring. Small changes in biomaterial chemistry and structure may greatly alter the resorption rate, allowing for materials to be tailored for various applications or leading to unexpected product failure. Collagen and the lactic acid and/or glycolic acid polymers (PLLA and PGA or copolymer PLGA) are the most commonly used for resorbable applications. PLLA and PGA degrade through a process of hydrolytic degradation of the polyester bond. At low molecular weights, the implant can disintegrate and produce small fragments that elicit an immune response from macrophages. PLLA and PGA degrade in a time period of 6 months to several years depending on initial molecular weight and crystallinity.

Copolymers of the two typically degrade into fragments in a few months. The lactic acid and glycolic acid fragments are eventually metabolized into carbon dioxide and water. Tricalcium phosphate ceramics degrade through a surface dissolution process into calcium and phosphate salts, which are also present naturally in the body.

Biomaterial degradation may lead to chronic nonhealing wounds that are arrested at one of the normal phases of wound healing. This may happen if a biomaterial degrades too quickly and releases particulate matter that extends the inflammation stage. Persistent inflammation leads to the formation of giant multinucleated cells that continue to attempt to remove the offending material. They are the trademark of a foreign body response and may necessitate surgical removal of the implanted device. If the healing passes through to the fibrous capsule formation stage, there may still be complications. For example, a drug delivery implant may eventually no longer function due to impaired drug release by the fibrous encapsulation in response to the degrading drug delivery implant.

Example Problem 6.10

Is it possible to successfully pass through all four wound healing phases only to have the biomaterial degrade and lead to wound healing reversal? Explain your answer.

Solution

Yes, this has happened in some patients with total joint replacements. Total joint replacements such as artificial hips typically consist of two metallic components that meet at a polymeric bearing surface (typically UHMWPE). During bending of the joint, wear debris is produced as the metal surface rubs against the softer polymeric surface. This leads to the recruitment of macrophages that identify the particles as foreign and attempt to remove them. Since the synthetic particles cannot be degraded by cell enzymes, inflammation continues indefinitely. The excessive production of inflammatory cytokines leads to resorption of the newly healed adjacent bone that supports the implant, resulting in implant loosening. Fortunately, there have been improvements in the processing of polyethylene so that wear debris is no longer generated at the high rate observed in some of the earliest used hip replacements. Ceramic-on-ceramic hip joints also have been developed, which have better wear properties and are not susceptible to corrosion; however, ceramic hip replacements are badly tolerated by elderly osteoporotic bone because the material is very hard. ∎

6.4.5   Immunogenicity

Immunogenicity is the tendency for an object to stimulate the immune response. Examples of immunogens are bacteria, pollen from grass or trees, small or absorbable biomaterials, and proteins in food that lead to allergies or inflammation. Basically, our immune system protects us through a combination of physical barriers such as skin, chemical barriers such as enzymes and antibodies, and cellular barriers such as targeted cytotoxic T lymphocytes (T cells). When a biomaterial is implanted in the body, the immune system associated proteins immediately attach to the surface, thereby directing subsequent cell behavior toward the biomaterial. Once again it must be emphasized that the surface chemistry and structure of the biomaterial play important roles in determining the extent and type of protein attachment and, hence, the tissue–biomaterial interaction. Proteins of all types will be competing for attachment sites on the biomaterial surface. Depending on the conformation of the attached proteins, a variety of messages may be sent to the nearby cells. Methods for modifying the biomaterial surface to control tissue–biomaterial interactions are discussed later in this chapter.

When allogenic (human) graft biomaterials are implanted in another human, acute rejection can occur if the major histocompatibility complex (MHC) groups on the cells in the graft are of different types than the donor's MHCs. MHCs are a class of cell-surface molecules that provide information as to what has been identified as foreign in the past to the cytotoxic T cells. With nonmatching MHC groups, the T cells receive two sets of instructions as to what is foreign, and this causes an extremely vigorous immune response. Tissue typing can reduce this type of rejection, although the patient usually still requires long-term medication to suppress some of the activity of the immune system. Rejection can also occur against cell–biomaterial scaffolds in tissue engineering applications. The implanted cells may be recognized as foreign and be damaged directly by the attacking immune cells such as macrophages or be starved to death by the lack of nutrients passing through a thick fibrous capsule created through the inflammatory process to protect the host. Therefore, in some tissue engineering applications, the implanted cells are protected from the immune system by enclosing the cells in selectively permeable biomaterials (e.g., islet cells that produce insulin in alginate hydrogels).

Corrosion of metallic implants releases metal ions that can cause metal sensitivity or allergic reactions in some individuals. Allergic reactions can lead to slow or inadequate bone fusion or skin dermatitis. Both of these conditions usually require removal of the implant. Once again, this demonstrates that biomaterials are not inert.

Biomaterials are sometimes deliberately designed to enhance the immune system's response. For example, vaccines are typically given with a particulate biomaterial known as adjuvant for enhanced and longer lasting immunity. Vaccine adjuvants have their own immunogenic properties, resulting in a stronger local stimulus to the immune system. Adjuvants can be simple particles that adsorb the weak immunogen, increasing the effective size of the weak immunogen and enhancing phagocytosis of the particle by macrophages. Adjuvants also work like controlled-release vehicles by prolonging the local retention of a weak immunogen and increasing the chance of a local immune response. Vaccine adjuvant selection represents a compromise between a requirement for adjuvanicity and an acceptable low level of side effects. The FDA has approved only three materials for human use (all of which are mineral salts): aluminum phosphate, aluminum hydroxide, and calcium phosphate. Aluminum compounds are often incorrectly identified in the scientific literature as alum. Alum is potassium aluminum sulfate, which is used as the starting solution to precipitate antigens with either aluminum phosphate or aluminum hydroxide. Other biomaterial adjuvants used in research include oil emulsions, lipopolysaccharide products from bacteria (LPS), and their synthetic derivatives (liposomes).

6.5   GUIDING TISSUE REPAIR WITH BIO-INSPIRED BIOMATERIALS

6.5.1   Surface Chemistry Modifications (1-D)

The interaction of cells and tissues with biomaterial surfaces is critically important to promote new tissue deposition and for healthy integration with the surrounding extracellular matrix. When a biomaterial surface is placed in the body, many different chemical and molecular level events occur in the biomaterial that can affect the cellular response (Fig. 6.14). For example, at the surface of a traditional metal implant, metal and oxide ions diffuse away from the implant surface, while biological ions (Ca, P, Na, Cl) are incorporated. Concurrently, at the implant surface, there is adsorption and desorption of native biological molecules. As discussed in Section 6.4, the adsorption of certain proteins onto the implant surface will initiate a cascade of events such as blood clotting and inflammatory cell recruitment or cell differentiation. In nature, the surface chemistry of every cell and extracellular matrix is carefully controlled to obtain the desired response. Cell–extracellular matrix interactions are dominated by the interaction of cell-adhesion proteins (ligands) bound to the ECM and cell surface receptors (integrins). In medical device applications, the biomaterial replaces the ECM and sends signals to the cells interacting with it through similar mechanisms (Fig. 6.14). Thus, the biomaterial surface plays a very important role in determining tissue–biomaterial interactions, and this concept has become very important to the biomaterials scientist.

One biomimetic approach to controlling cell–surface reactions is to preadsorb proteins on the implant surface that mimics those most involved with cell adhesion. The three-amino-acid sequence of Arg-Gly-Asp (RGD) found in fibronectin and bone sialoprotein is now well known for mediating adhesion of cells to surfaces; therefore, RGD-containing peptides are now being deposited on surfaces to promote cell attachment. Heparin and heparin-sulfate binding peptides that mimic proteoglycan activity also have been found to enhance cell adhesion. Adsorption of other biological molecules such as growth factors to the surfaces of implants can control the tissue–biomaterial interaction and lead to enhanced cell activity and more differentiation than will the cell adhesion molecules alone.

There are many chemical reactions that can be used to attach a biomimetic peptide sequence to a biomaterial. For example, a protein can be immobilized on a surface through a technique known as organosilane chemistry (Fig. 6.15a). The details of the chemical coupling and derivitization processes are beyond the scope of this text. Basically, there are coupling agents such as salines used to create a covalent bond between the biomaterial surface and the protein to be attached. Well-ordered protein attachment results. A wide variety of solid surface modification techniques are available to create the reactive coupling groups, such as photochemical grafting, chemical

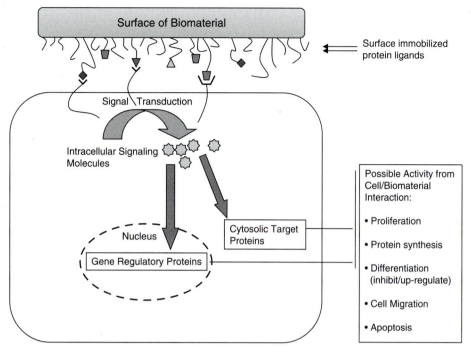

Figure 6.14   Like the extracellular matrix, a protein-covered biomaterial sends signals to the cell interacting with it through ligand-receptor mechanisms. Primary cell signal transduction is facilitated through multiple pathways, leading to the synthesis of various intracellular signaling molecules. Acting both on the genetic regulatory proteins in the nucleus and other cytosolic target proteins, the signaling molecules can induce various phenotypic expressions. Ideally, the biomaterials scientist can engineer the proper surface treatment to elicit the desired cellular activity.

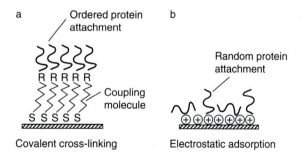

Figure 6.15   Two types of chemical reactions used to modify a biomaterial surface: (a) covalent coupling techniques and (b) physical adsorption methods utilizing electrostatic interactions.

derivation, and plasma gas discharge. Physical adsorption methods utilizing other types of bonding, such as van der Waals and electrostatic binding, can also be used to immobilize proteins (Fig. 6.15b). Physical and electrostatic adsorption is the easiest technique; however, it is the least specific and tends to readily release the adsorbed molecule. Lipid groups and dye molecules can also be used to immobilize proteins on surfaces.

A critical component of surface modification is the resulting ligand density. If protein adsorption is too low, the addition of more functional groups to a relatively inert polymer can be accomplished by plasma glow discharge treatment. The greater reactivity of the surfaces with higher surface energy after plasma treatment generally leads to increased tissue adhesion.

Surface modification also can be used to produce protein-resistant surfaces that are needed in blood-contacting applications such as vascular grafts. For example, polyethylene oxide has been attached to surfaces to reduce protein adsorption. Cell adhesion was significantly reduced on these treated surfaces. Anticoagulants can also be attached to biomaterial surfaces to decrease unwanted cell attachment. Various hydrophilic biomaterials have been shown to reduce platelet adhesion and thrombus formation. Hydrophilic materials have also been shown to hinder bone healing, so what is appropriate for one biomaterial application does not necessarily apply to another.

Hyaluronan is a biomolecule found in cartilage extracellular matrix. It is responsible for tethering a proteoglycan (aggrecan) to the collagen matrix. Studies have shown that it can guide the differentiation of mesenchymal stem cells to cartilage chondrocytes. In those experiments, hyaluronan was chemically bound to tissue culture dishes, and undifferentiated cells were added. It was found that a specific molecular size (200,000–400,000 daltons) was optimal to initiate cartilage formation. Biomolecules such as enzymes, antibodies, antigens, lipids, cell-surface receptors, nucleic acids, DNA, antibiotics, and anticancer agents can all be immobilized on or within polymeric, ceramic, or metal surfaces.

Surface deposition of calcium ions or the use of calcium-containing biomaterials strongly influences the attachment of bone cells. Hydroxyapatite-coated hip implants show decreased fibrous tissue formation and increased direct bone bonding. Better bone attachment has also been found for hydroxyapatite-coated dental implants and spinal fusion cages. Recent studies have shown that hydroxyapatite ceramics are selective in cell recruitment from the bone marrow. This may be due to an intermediate step of selective protein adsorption.

6.5.2   Surface Topography (2-D)

Surface topography of a biomaterial has been shown to provide cues to cells that elicit a large range of cellular responses, including control of adhesion, cell morphology, apoptosis, and gene regulation. Modification of a biomaterial surface can therefore have dramatic effects on guiding tissue growth. Early total joint replacements had smooth surfaces; however, rough, porous coatings or grooved surfaces are now being used on hip implants to achieve bony ingrowth. The hip and knee implants in Figures 6.5 and 6.6 have rough, porous coatings. The porous coatings on orthopedic implants are achieved by partial fusion of small metallic spheres to the implant surface. The interparticle spaces are the pores. The pores, as well as grooves, seem to encourage bone cells to migrate into or along them. It has been observed that the stretching out and aligning of the cells along a surface feature (greater than 5 µm) causes them to initiate bone deposition. This leads to a mechanical interlocking between the bone tissue and the implant that increases the bond strength. The directed activity of cells is known as contact guidance, and it has also been used to increase migration of other cell types into tissue engineering scaffolds. In addition to guiding cell migration downwards into a pore, surface topography variations can be used to restrict cell spreading and force it back upon itself. For example, by restricting cell growth to a long narrow path, endothelial cells were induced to grow upwards and form a three-dimensional capillary tube.

Photolithographic techniques can be utilized to micropattern a surface with proteins, molecules, or functional groups (Fig. 6.16). This technique involves using a photoresist layer in which patterns can be created by selectively exposing certain areas to light. The light degrades the exposed portions of photoresist, leaving a bare biomaterial surface. Proteins or molecules can then be selectively attached to these exposed areas. The remaining photoresist is then removed to obtain a biomaterial surface that has protein or molecular patterning. Cell culture studies have demonstrated preferred cell attachment to the chemically modified areas. Other methods for modifying surface topography include surface roughening by laser ablation or wet etching with a corrosive solvent. It has been observed that the roughness must be on a biologically relevant scale (1–10 µm) to affect cell growth and attachment.

Smooth surfaces such as pyrolitic carbon resist protein and cell attachment and are ideal for heart valve applications. Bioprostheses made of bovine or porcine heart valves are even more superior for valve applications and reduce coagulation and embolism by a combination of an ideal surface topography and surface chemistry. However, they typically fail due to calcification, which is enhanced on the natural collagen surface as compared to the smooth pyrolitic carbon.

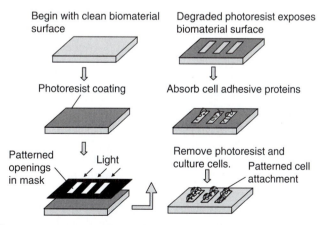

Figure 6.16 Photolithography techniques for micropatterning a biomaterial surface with immobilized proteins or functional molecules.

6.5.3   Scaffolds (3-D)

Tissue repair of large defects is best accomplished by filling the defect space with a scaffold material that can simulate the microenvironment provided by the embryonic extracellular matrix. There are many functions the scaffold must serve: it must provide sites for growth factor attachment, cell migration and attachment, and new tissue deposition. Natural tissues are the ideal choice; however, due to the limited supply of natural donor tissue, there is a large market for synthetic tissue analogs. While metallic or polymeric plates and screws may be useful for temporarily holding the healing tissue together, they do not themselves recreate new tissue. Either natural tissue or analogs to the extracellular matrices of natural tissue and its cells and growth factors are needed to reconstitute a functioning vital tissue. The goal of a scaffold is to recreate important aspects of the cell microenvironment that will allow cell proliferation, differentiation, and synthesis of extracellular matrix. Synthetically produced regenerative materials now available commercially include biomaterial scaffolds (ceramic and polymeric) with or without a tissue stimulating biological molecule and with or without cells. One of the most critical elements of the scaffold biomaterial is that it mimic the ECM scaffold that normally serves to maintain space, support cells, and organize cells into tissues. The section on surface chemistry (Section 6.5.1) gave an example of how mimicking components of the adhesive proteins of the ECM can enhance cell attachment and differentiation. This section focuses on the structural and physical characteristics of the ECM scaffold that appear to be critical to imitate in synthetic scaffolds in order to stimulate cells and lead to the functional regeneration of tissues.

Pore size is a very important parameter of biomaterial scaffolds used for tissue regeneration. Through trial and error, optimal ranges of pore sizes have been determined for different tissues and for different biomaterials. There are now some rules of thumb, such as that the pores must be at least $5–10 \, \mu m$ for a cell to fit through. Successful bone scaffolds typically have pores that traverse the full thickness of the scaffold and are $100–250 \, \mu m$ in size. New blood vessel formation, or neovascularization, has been shown to require pores within polymer scaffolds that are between $0.8–8 \, \mu m$ and to not be possible within polymers with pores less than $0.02 \, \mu m$. Typically, the acceptable pore size in polymers is smaller than in ceramics or metals, perhaps due to pore size expansion which can occur in the body due to degradation or swelling of the polymer.

The pore size determines many aspects of the scaffold, such as mechanical strength and permeability to gases, fluids, and nutrients, in addition to cell ingrowth. Interconnected porosity is essential for tissue engineering applications requiring nutrient diffusion and tissue ingrowth. A highly porous material degrades more quickly than a solid block of material. The biodegradation rate of a scaffold biomaterial is usually found through prolonged immersion of the biomaterial in a fluid that simulates body fluid.

There are several techniques for producing porous biomaterials. Some methods are better suited for drug delivery or for creating a low-density, stiff reinforcing structure than for cell scaffold tissue engineering applications. One approach to creating pores within a material is to dissolve the polymer in a solvent and mix in particulate materials that are stable in the solvent but can be dissolved later (porogens). The solvent is allowed to evaporate, leaving a film of polymer containing porogens (like a chocolate bar with nuts). The porogens are then dissolved by washing in a different solvent, such as water, to form pores. This is known as solvent casting/particulate leaching. Related to this technique is coaservation, in which the polymer is dissolved in a solvent, the porogen is added, and then a nonsolvent for the polymer is added. A polymer precipitate forms, entrapping the porogen particulates. The precipitate can be collected and then compacted prior to removal of the porogen. After washing to remove the porogen, a porous structure is revealed, as shown in the composite bone crystal/polyhydroxybutyrate hydroxyvalerate (PHBHV) polymer of Figure 6.17. Organic solvents are used with these procedures, which precludes the possibility of adding pharmaceutical agents to the scaffold during fabrication. Also, the porogen leaching step significantly increases the scaffold preparation time. This has led to the development of other techniques such as gas foaming. In one variation of gas foaming, solid disks of polymer are formed and then exposed to high pressure CO_2 for three days. Pores are created when these gas-containing disks are suddenly returned to atmospheric conditions. The gas forms bubbles of up to 100 μm, and porosities of up to 93% can be obtained. Other methods for gas foaming include adding a foaming agent that chemically produces a gas upon heating. The pores formed by the gas foaming technique are often not interconnected, making cell seeding and cell migration within the foam difficult.

Phase separation/emulsification can also be used to fabricate porous polymer scaffolds. In this technique, a polymer is dissolved in a solvent and then an immiscible

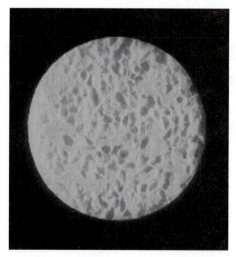

Figure 6.17 A composite biomaterial scaffold made of bone crystals and the resorbable polyhydroxybutyrate hydroxyvalerate (PHBHV) copolymer. The coaservation technique was used to precipitate the polymer from solution to produce this scaffold. Dissolvable porogens were used to create the large pores needed for bone repair applications.

liquid (such as water) is added and mixed to form an emulsion. The polymer/water mixture is cast into a mold, rapidly frozen and then freeze-dried, which is known as lyophilization. The space that was water becomes a pore. Scaffolds with high porosity (up to 95%) have been formed by this method, but the small pore sizes (13–35 μm) are a drawback. Fiber bonding methods in which preformed fibers are layered or woven and then hot melted or glued together by solvent exposure is another technique for forming porous materials. The advantage is that the pore sizes are controllable and interconnected. The drawback is that the pore channels are rectangular and regular, unlike natural extracellular matrix structure. Furthermore, the use of high temperatures and solvents prevents the incorporation of bioactive molecules during processing.

Solid freeform fabrication is also used to make biomaterial scaffolds (in addition to being used for rapid prototyping of automotive parts). Stereolithography utilizes a focused laser that follows a pattern dictated by computer-assisted design drawing to selectively cure only certain areas within a thin layer of liquid polymer. The depth of the liquid is raised around the part being fabricated, and the laser is again sent on a computer-assisted path to form the next layer of the component. This is repeated over and over again to produce a complex three-dimensional shape. Similarly, a technique known as 3-D printing can be used to create porous scaffolds or complex shapes layer by layer, this time using a print head to deposit "glue" over computer-specified areas of a powder bed. After all of the layers have been "printed," the final part is picked up, and the unbonded particles fall away, revealing the three-dimensional component. Pieces of replacement bone have been made using these techniques with computer assisted tomography (CAT) scans of x-ray images. The features of the scaffolds are limited to 10–1000 μm with these techniques which have been used to prepare polymer and ceramic scaffolds.

6.6 SAFETY TESTING AND REGULATION OF BIOMATERIALS

6.6.1 Product Characterization

Ensuring product purity and identity is one of the first steps in developing a safe product. There are extensive data documenting the safety of various biomaterials; however, since processing methods may include additives and the final sterilization step may alter the biomaterial, it is crucial to always verify the end product purity and identity. The American Society for Testing and Materials (ASTM, www.astm.org) has developed many standards that manufacturers of medical device products can use as guidelines to evaluate product purity and identity, as well as safety. Under ASTM specifications for metals used in medical devices, there are restrictions on the composition, microstructure, phase and grain size, inclusion size, defect size, and macro- and microporosity to help ensure safety. There are ASTM standards for ceramic materials that specify chemical composition, phase determination, grain size, and impurities such as sintering aids, which may decrease fatigue resistance. New standards are being written by ASTM to help ensure quality and reliability of tissue engineering scaffold materials. In addition to specifying the correct tests and techniques to determine the chemical identity of the tissue engineering scaffold materials, standardized tests for measuring porosity and permeability have also been developed. ASTM also has standards for dissolution testing, degradation testing, and stability testing. ASTM standards are written through a consensus process and represent the best available knowledge from a wide cross section of manufacturers, users, and general interest groups.

There are many techniques for evaluating the composition and structure of biomaterials. Unfortunately, no single technique is capable of providing all of the needed information. Thorough characterization requires the use of multiple analytical methods. For example, if the material is crystalline, x-ray diffraction can be used for bulk product identification (typically, powdered samples are analyzed). The x-ray beam is diffracted as it goes through the rows of atoms, and a detector measures the reflected intensities as a function of angle from the surface. The intensities at various angles provide a unique signature of the material structure and can be compared to existing data files for product identification purposes.

Fourier transform infrared spectroscopy is a method that complements the structural information gained by x-ray diffraction by providing information about the chemical groups found within the structure. The material does not need to be crystalline and can be gas, liquid, or solid. The sample is exposed to infrared radiation, and the molecular vibrations induced by the radiation are observed. Radiation at frequencies matching the fundamental modes of vibration is absorbed, causing oscillating dipoles perpendicular to the surface. These are detected as a function of wavelength and provide a chemical "fingerprint" that can be compared to existing databases to identify the material. Most characterization methods capable of identifying an unknown substance involve bombarding the material with some type of energy, quantitating the interaction with the material, and then searching a database for similar results. Other techniques utilizing this basic procedure include secondary ion mass spectroscopy (SIMS) and x-ray photoelectron spectroscopy (XPS), both of which are well suited for identifying the surface chemistry of a biomaterial.

Scanning electron microscopy (SEM) is very useful for characterizing the two-dimensional surface topography of a biomaterial. In SEM, a beam of high-energy electrons is scanned across the sample, causing the material to emit secondary electrons. The intensity of the secondary electrons primarily depends on the topography of the surface. An image can be recreated by recording the intensity of the current generated from the secondary electrons. The resolution of an SEM allows magnifications of up to 100,000×. When greater resolution of a material surface is needed, atomic force microscopy (AFM) can be used. In AFM, an atomically sharp tip attached to a cantilever is dragged across the surface of a material but actually does not touch the material. The interactions of the atoms of the material being analyzed with the tip cause either repulsion or attraction. The height adjustments or changes in interatomic forces are recorded and used to construct images of the surface topography. Under proper conditions, images showing individual atoms can be obtained.

Images in three dimensions can be obtained using computer-aided x-ray tomography (micro-CT) or nuclear magnetic resonance (NMR) imaging. An image obtained

Figure 6.18   The scaffold of Figure 6.17 imaged by computer-aided x-ray tomography (micro-CT). (Courtesy of Douglas J. Adams, Ph.D., Micro-CT Facility, University of Connecticut Health Center.)

by micro-CT of a biodegradable porous scaffold is shown in Figure 6.18. In both techniques, the samples are scanned in all directions, and then the image is created mathematically by a merging of all the directional information. Solid or porosity volume or volume fraction can be measured nondestructively. Direct measurement of solid or pore characteristic dimensions (width, diameter, thickness) and spacing (or period) of repeating structure can be made. In micro-CT, image contrast is achieved via attenuation of X-radiation; thus, polymers (or other scaffold substrates) with relatively higher attenuation coefficients will provide higher-contrast images and improved computational segmentation of scaffolds versus the background. Commercial desktop micro-CT instruments are available with a spatial resolution of approximately 5 μm.

6.6.2   Methods for Testing and Evaluating Safety and Biocompatibility

It has been emphasized repeatedly in this chapter that the body can affect the biomaterial and the biomaterial can affect the body. Therefore, both of these aspects must be investigated prior to biomaterial implantation in the human body. This is known as biocompatibility and safety testing. Before initiating biocompatibility and safety testing, the structure and chemistry of the material should be fully characterized by a combination of techniques as described in Section 6.6.1. This is necessary to confirm the purity and identity of the biomaterial and to ensure that no unintended foreign substances have been introduced during the synthesis, manufacturing, and sterilization procedures.

Biocompatibility and safety tests include *in vitro* assays (using cells and tissues), *in vivo* models (in animals), and, finally, human clinical trials. Several guidelines and procedures have been developed by the standards organizations of the world [ASTM and the International Standards Organization (ISO)] and federal regulatory agencies (e.g., Food and Drug Administration). ASTM Standard F-748 and ISO 10993 provide detailed methods for completing adequate safety and biocompatibility testing and are followed by all implant and medical device manufacturers. The tests are separated into various categories based on intended use. For example, there is a matrix of tests appropriate for surface devices, external communicating devices, or implanted devices. The recommended tests are further categorized based on contact duration (short term, prolonged contact, permanent). The basic tests include cytotoxicity, sensitization, irritation or intracutaneous reactivity, acute systemic toxicity, subacute systemic toxicity, genotoxicity, implantation, hemocompatibility, chronic toxicity, and carcinogenicity. The preferred test sample is the intact medical device that has been processed and sterilized in the same manner as the medical device that will be used in humans. However, it is not always practical to use the intact medical device due to the constraints of the biological tests. Therefore, an extract of the leachable components or the degradation components of the implant are often tested first in the *in vitro* assays (cytotoxicity) and also in the preliminary *in vivo* tests (sensitization and both systemic toxicity tests). Completion of the preclinical tests described in F-748 and ISO 10993 typically takes up to 2 years, even if a qualified and experienced facility is conducting the testing.

It is difficult to correlate *in vitro* testing to *in vivo* testing because the *in vivo* system is much more complex and involves many more variables. Typically, cell culture assays are more sensitive than *in vivo* tests; however, demonstration of cytotoxicity *in vitro* may not necessarily mean that the material cannot be used *in vivo*. Both false negatives and false positives can be obtained by cell culture testing; therefore, animal testing is a required step in understanding safety and biocompatibility, and also for an initial evaluation of the product's performance and effectiveness. There are, however, variations in response to biomaterials and drugs among species of animals. The guinea pig has been found to be the most sensitive animal for assessing delayed immune hypersensitivity (the sensitization test). The rabbit has been found to be the most sensitive animal model for detecting pyrogens *in vivo*. Although animal testing does provide a useful screen for restricting the implantation of most toxic components in humans, the final and ultimate biocompatibility and safety testing occurs during human clinical trials. In some cases, products that demonstrate efficacy in a mouse or dog model may not always perform as well in humans, particularly in the case of a new drug. The effective dose sometimes varies greatly between species, as well as between two humans. Therefore dose escalation schemes are incorporated into biocompatibility and safety and efficacy testing, and large numbers of patients must be used in clinical trials.

6.6.3 The Regulatory Process

Regulatory approval by the Food and Drug Administration (FDA) is required in the United States prior to administering a new drug or biologic or implanting a new medical device in a human and also prior to marketing the new product. The FDA is currently divided into six individual centers that regulate devices and radiological health (CDRH), drugs (CDER), biologics (CBER), food and cosmetics, veterinary medicine, and toxicology. An assessment must be made as to which mode of action—drug, device, or biological—contributes the most to the therapeutic benefits of the overall product. Based on this criterion, the FDA decides which center will take the lead on the regulatory review. Each FDA center has different procedures and requirements that must be completed and met to gain FDA approval. Some products are a combination of a biologic and a device or a drug and a device, and, therefore, FDA requirements for two or more centers must be met before the product is granted approval. Tissue-engineered medical products are examples of combination products. Combination products are typically the most difficult to regulate and take the longest to reach the market. Biomaterial scaffolds alone without cells or growth factors are typically regulated by the FDA as a device and are under CDRH jurisdiction.

The regulatory procedures are sufficiently complicated that typically a specialist in this area is hired to manage this aspect of new product development. However, it is important to have some idea of product regulation early on, since it, along with good research and development, is needed to bring a medical product to clinical success. Toward that purpose, the procedures utilized by the CDRH branch of the FDA will now be described.

The CDRH branch of the FDA utilizes classification of medical devices to assist with determining the requirements for approval and the extent of regulatory control (http://www.fda.gov/cdrh/devadvice/). Medical devices are placed into one of three classes. Class I devices are those that have limited body contact and essentially pose no significant risk. Class II devices require special controls and must usually meet some performance standards to provide some assurance of safety. A device is placed in Class III if there is insufficient information to determine that general or special controls are sufficient to provide reasonable assurance of its safety and effectiveness. A new device that is not substantially equivalent to a device on the market will automatically be placed in Class III. Examples of Class I medical devices include dental floss, a tongue depressor, a surgeon's glove, and a clinical chemistry test system such as a pregnancy test. Examples of Class II medical devices include a blood pressure cuff, an oxygen mask, dental impression material, and an electrocardiograph. Examples of Class III medical devices include a heart valve, an automated blood warming device, and a silicone inflatable or gel-filled breast prosthesis. Classification is an important step of the FDA approval process since it will determine the extent of the testing required prior to use in humans and when the device can be sold.

At least 90 days prior to commercially distributing a new or substantially modified device, a manufacturer must submit a premarket notification to the FDA. More than 99% of the applications received by CDRH are cleared for marketing through the 510(k) Premarket Notification process. The goal of this process is to demonstrate to the FDA that the new device is substantially equivalent to an already approved predicate device. This is accomplished through careful characterization by several complementary methods that confirm the identity and purity of the substances involved, followed by completion of an abbreviated form of the ASTM F-748 and/or

ISO 10993 test protocols and adherence to quality system regulations. Quality system regulations include compiling a Device Master Record and Design History File. Together these two files contain documentation of the procurement process, the manufacturing details, all testing results—including assay verification tests—and the details of the design rationale and design verification testing.

If a device does not qualify for 510(k) approval, then a full premarket application (PMA) must be submitted containing all the required information on the safety and the effectiveness of the device as determined through preclinical and clinical testing. There is a decision tree in ISO 10993 that helps define which biocompatibility and safety testing is necessary based on length of contact with the body and/or blood. Clinical trials are highly regulated so that human subjects are not exposed to significant risks without their knowledge. Carefully documented and successfully completed *in vitro* and *in vivo* animal safety testing is required at the time of application to begin a clinical trial. Human clinical trials typically are divided into four phases: safety testing, efficacy testing, blinded efficacy compared to a clinically acceptable alternative (these three take approximately 5 years to complete), and finally, post-market surveillance (gathered from product use by the general public after FDA approval). Prior to initiating a clinical trial, a manufacturer must obtain an investigational device exemption (IDE) that includes all manufacturing and quality control procedures, the plan for the clinical study, and the lists of the review boards that have reviewed the proposed plan (see Chapter 2). The FDA has 30 days to approve or disapprove the IDE. If the PMA application is considered complete, the FDA has 180 days to approve or disapprove the application. If approved, the product can be marketed for human use for the purposes declared in the application, which are to be described in the product labeling. Another FDA submission for review is required prior to legally marketing the product for a new use (off-label use). For those interested in further information on regulation by CDRH or any of the other FDA centers, FDA guidance documents that cover all aspects of regulatory approval are readily available online (www.fda.gov).

6.7 APPLICATION-SPECIFIC STRATEGIES FOR THE DESIGN AND SELECTION OF BIOMATERIALS

6.7.1 Musculoskeletal Repair

The design and selection of the biomaterials components for an implant should be based on restoring the biological function of the damaged or diseased tissue. The principal function of musculoskeletal tissues is to provide a framework to support the organs and to provide a means of locomotion. Bone, cartilage, tendons, ligaments, and muscles are all part of the group of musculoskeletal tissues; however, they have different functions and different biological properties. Each one must be considered individually in terms of implant design and biomaterials selection. Bone is the only tissue capable of undergoing spontaneous regeneration. It is constantly in a state of remodeling, always optimizing its structure to best meet the needs of the body. This ongoing cellular activity is why astronauts rapidly lose bone mass during zero gravity conditions. Unlike bone, cartilage is acellular and has a very limited capacity for repair. Therefore, damage to cartilage is often permanent and often progressive. Cartilage provides an articulating surface enabling low-friction movements between opposing bone surfaces. Ligaments are not simply passive joint restraints; they also provide electromechanical signals for joint stabilizing muscle contractions.

Replacement of damaged or diseased tissues or organs is best accomplished by autograft or allograft donor tissue, but they have limited availability, and a biomimetic synthetic substitute is the next best alternative. Biomimetic calcium phosphate materials (hydroxyapatite) have been shown clearly to enhance bone cell activity and are either used alone or in combination with collagen or other polymers as bone graft substitutes. Hydroxyapatite not only influences bone cell attachment, but it also appears to control the differentiation of stem cells to bone forming cells. This is particularly important for tissue engineering approaches that aim to not only restore function, but also to restore the actual biological tissue. In tissue engineering product development, materials selection is more complicated than it is in traditional approaches; typically, a biomaterial must have sufficient mechanical strength and also have a surface chemistry conducive to cell attachment and proliferation, must also perhaps serve as a drug delivery vehicle and release growth factors, and must also resorb or biodegrade once the new tissue has been formed.

Autografting osteochondral tissue from a healthy portion of a joint to a cleared defect is a current strategy for articular cartilage repair. Another approach involves filling the bulk of the defect with cells that can facilitate the growth of appropriate cartilage and/or bone tissue. This approach recently has been approved by the FDA and utilizes chondrocytes from the patients themselves. The chondrocytes are harvested, are expanded *in vitro*, and then injected into a surgically created compartment over the defect site. The cartilaginous tissue created by this process has been found to degrade faster than cartilage formed during fetal development. Perhaps a vital cell signaling message has been left out during this artificially stimulated biological process.

Further research is being conducted to improve tissue-engineered cartilage reconstruction. Chondrocytes or stem cells are being seeded on soft tissue scaffolds such as collagen, fibrin, and polylactic acid and precultured in bioreactors containing exogenous growth factors to form neocartilage. This approach may offer an improvement over the cell-only approaches, mainly due to the use of a biomaterial scaffold. As an example, a mixture of mesenchymal stem cells suspended in hyaluronan is being developed for direct injection to a damaged knee meniscus. Similar to how hydroxyapatite stimulates differentiation of stem cells to bone cells and is well suited for bony applications, hyaluronan (found in embryonic extracellular matrix) has a chondroinductive and antiangiogenic potential and shows promise as a biomaterial scaffold for cartilaginous tissues.

Biomimetic scaffold materials are clearly more than an inert structural support for cells. When properly selected for the given application they provide a receptive framework and lead to induction of a cell down a specific differentiation pathway (Fig. 6.14). Since mechanical forces do play a role in most tissue function, it may also be that certain scaffolds are superior due to their biomimetic mechanical properties.

Traditionally, the load-bearing requirements of long bone repair applications have been accomplished by metallic biomaterials. Eventually, scientists and engineers will learn how to engineer total joints and all of the tissues they comprise, but for now, the best option is a total prosthesis that has femoral, tibial, and/or patellar components made of traditional metals and polymers or ceramics (Figs. 6.5, 6.6, and 6.10). Based on the trends of successful current research, most musculoskeletal injuries or defects in the future will probably be treated at an earlier, smaller stage. Resorbable composites delivering cells and growth factors will probably become the material of choice for not only bone, but also ligament and cartilage. The role of the traditional metallic load-bearing material may be reduced in the future to serving as a cast or brace that provides protection from biomechanical loading until the new tissue has fully regenerated internally.

6.7.2 Skin Regeneration

As the largest organ in human physiology, the skin plays a vital role in maintaining homeostasis, providing immunity, and supporting sensory feedback. Not surprisingly, wounding skin compromises its ability to maintain these critical functions and makes dehydration and infection much more of a threat, burns being an extreme example of the latter. Traditionally, medicine has used allogenic skin grafts as a primary means of treatment for most skin injuries. Because of additional wounding that results from skin grafts, considerable research has been devoted to the development of a skin equivalent. In this context, biomaterials improve the capabilities of skin substitutes to truly act as replacements for the original.

Macroscopically, skin is a dual-layer organ consisting of the dermis and epidermis, the latter of which has its own secondary hierarchy (Fig. 6.19). The dermis, owing to its collagen component, is mostly responsible for the structural integrity of the skin. In addition, dermal cells, specifically fibroblasts, produce chemical factors that are essential for the proper proliferation and differentiation of the epidermis. Between the dermal and epidermal layers lies a well-defined basal lamina. A proteinaceous extracellular layer, the basal lamina serves as the substrate upon which basal cells of the epidermis adhere and proliferate. Indeed, one common way to evaluate artificial skin is to assay for the presence of the basal lamina, as this would indicate dermal–epidermal integration. Basal keratinocyte cells form the first layer adjacent to the basal lamina. These cells undergo a program of proliferation and differentiation to form the full epidermal layer. The differentiated forms of epidermal cells from inner- to outermost include basal, spinous, and granular keratinocytes. The stratum corneum, a highly cross-linked layer of packed keratinocytes, serves as the outermost barrier and provides the ultimate protection against desiccation.

In the biomimetic tradition, it has been thought that effective skin substitutes should mimic natural skin structure. In addition, well-designed skin substitutes should possess the proper physical integrity to withstand the environment and, at the same time, support the growth of cells. A schematic of artificial skin is shown in

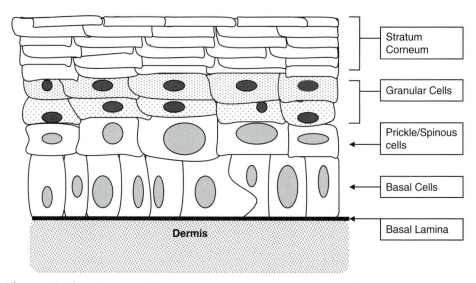

Figure 6.19   Skin is a dual-layer organ consisting of the dermis and epidermis. The main cellular components of the dermis are fibroblasts, and the main cellular components of the epidermis are the keratinocytes that are at various stages of differentiation. (Illustration drawn by Venkatasu Seetharaman.)

Figure 6.20. Biomaterials, in addition to providing the proper substrates for cellular growth, are excellent sources of mechanical strength. Collagen, a cross-linked polymer, is found as the primary dermal component of most skin substitutes. When cross-linked, collagen provides a mechanically sound matrix upon which to grow an epidermal layer and can even be impregnated with fibroblasts to provide chemical factors for proper epidermal differentiation and proliferation.

Type I bovine-derived collagen, typically used for most skin substitutes, is optimized for physiological conditions by altering pH values and moisture content. Once

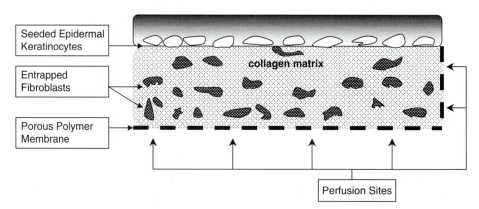

Figure 6.20   A schematic of artificial skin illustrating the similarities to natural skin. (Illustration drawn by Venkatasu Seetharaman.)

prepared, the liquid solution is poured into a specimen plate and allowed to polymerize, thereby taking on a more gel-like consistency. Polymerization shrinks the collagen by way of cross-linking. Depending on how the collagen is prepared and the length of polymerization, it is possible to control the gel consistency. It is important to note that the collagen gel or sponge is quite permeable to most molecules and permits diffusion of essential chemical factors. These can be provided to the epidermis by seeding the collagen with dermal fibroblasts, essentially trapping them within the cross-linked collagen structure. To complete the dual-layer skin substitute, epidermal keratinocytes are seeded on the dermis and allowed to differentiate into the proper epithelial cellular arrangement. Other skin substitutes actually use a thin layer of silicone as a functional epithelial barrier while the dermal layer integrates into the wound site. A subsequent procedure replaces the silicone with an epidermal matrix, thus completing the skin. Both collagen and silicone are examples of how biomaterials can be used to recreate the functionality of normal skin.

6.7.3 Cardiovascular Devices

The primary requirements for biomaterials used for blood-contacting circulatory applications such as heart valves and blood vessel replacements or stents are resistance to platelet and thrombus deposition, biomechanical strength and durability, and biocompatibility and nontoxicity. These key requirements were identified by studying the primary mode of action of the tissue to be replaced. For example, arteries consist of three layers that perform various biological functions: the intima, media, and adventitia. The intima is on the blood vessel interior and has a nonthrombogenic surface—it prevents blood contact with the thrombogenic media tissue. The cells of the intima produce a myriad of biomolecules, including growth factors, vasoactive molecules, and adhesion molecules. The media or parenchymal tissue is the middle muscular layer that provides the required strength while remaining viscoelastic. The media layer is made up of multiple layers of aligned smooth muscle cells. The outer adventitia layer acts as a stiff sheath that protects the smooth muscle media layer from biomechanical overload or overdistention.

As with most tissues, autograft is the preferred biomaterial for vascular tissue replacement. For example, one of the patient's own veins can be harvested and used to replace a clogged artery. Vein grafts have a failure rate as high as 20% in one year. Since vein grafts from the patient are unavailable and unsuitable in approximately 30% of all patients, synthetic graft materials have been developed. The observation that the intact lining of blood vessels (intima) does not induce coagulation has led scientists and engineers to produce more blood-compatible biomaterials by mimicking certain properties of the endothelium. For example, very smooth materials, surfaces with negative charges, and hydrophilic biomaterials are now used with limited success in blood-contacting applications. Knitted Dacron® (polyethylene terephthalate) and Gortex [polytetrafluoroethylene (PTFE)] vascular grafts are commonly used. The use of synthetic grafts has resulted in reasonable degrees of success (approximately 40% experience thrombosis at 6 months when synthetic grafts are used to bypass arteries that are smaller than 6 mm in diameter). Improved vascular products have incorporated anticoagulants such as heparin on the blood-contacting surfaces.

As discussed in the section on wound healing (Section 6.4.2), most biomaterials are recognized by the body as foreign and lead to platelet deposition and thrombus or coagulation (blood clot formation). This limits their use in blood-contacting applications. Therefore the tissue engineering approach of growing and implanting a living multilayered cell construct holds great promise for cardiovascular applications. It is only recently that the cell culture conditions and scaffold material selection that will promote the smooth muscle cell alignment and tight endothelial cell packing of blood vessels have been identified. A great deal of additional research must be completed prior to commercial availability of a functional tissue-engineered artery. For example, the mechanical burst strength of the highly cellular tissue-engineered arteries is not yet sufficient to withstand what the heart can generate, although these artificial arteries are nonthrombogenic.

Diseased human heart valves can currently be replaced with mechanical prostheses (synthetic biomaterials) or bioprostheses (made of biological tissue). The two mechanical heart valves shown in Figure 6.21 have four essential components: an occluder such as a disc or ball; a seating ring against which the occluder sits when the valve is closed; a capture mechanism, such as a cage, that constrains the occluder when the valve is open; and a sewing ring that permits attachment of the valve to the heart. The occluder bounces back and forth from the seating ring to the capture mechanism with each heartbeat. The more promptly it moves, the more efficient the opening and closing. Thus, weight, or mass, of the occluder and wear resistance of the occluder material are critical features. Early materials that were used included light-weight plastics or hollow metal balls. Silicone rubber proved very effective as a ball. The advent of low-profile disc valves for the mitral position set a new material constraint, namely stiffness. Silicone rubber was too soft for the disc designs. Poly-oxymethylene (POM) or poly acetal is stronger and stiffer than silicone rubber, and therefore was used in early disc designs. However, it had a problem with wear. The discs were supposed to be free to spin in the cage, so they could distribute wear evenly around the edge of the disc. However, as the disc moved up and down in the cage, wear tracks developed on the edge, preventing spinning and leading to the development of deep wear tracks and valve failure. Due to its high fatigue strength and wear resistance, pyrolytic carbon was selected as one of the prime materials for the occluder.

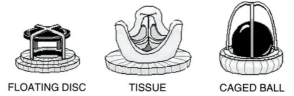

FLOATING DISC   TISSUE   CAGED BALL

Figure 6.21   Three types of artificial valves that are used to replace diseased or malformed human valves. The tissue valve generally contains valve leaflets from pigs; the other two types are composed entirely of manufactured materials.

The capture mechanism and seating ring required strength and stiffness to maintain their shape. Furthermore, they had to be made of a material that could be sprung open for insertion of the occluders. Metal has typically been used, either as machined parts or as separate parts that are welded together. Early designs used the cobalt alloys due to their strength and corrosion resistance. More recently, titanium alloys have been used, due in part to their being lighter than cobalt alloys. Concerns of allergic reactions to cobalt alloy cages have also been expressed over the years.

Both the occluder and containment system are in direct contact with blood. Contact with foreign materials can cause blood to clot and can lead to platelet attachment and clotting. If the design results in eddy flow or stagnation, the clotting factors can accumulate and lead to thrombus formation. Therefore, both the material selection and device design must consider problems of blood contact and blood flow. Since these problems have not been solved, patients with mechanical heart valves receive medication to reduce their tendency to form blood clots.

Biological heart valves resist thrombo-embolism much better than synthetic materials do; however, they are less durable. The valves are harvested from 7- to 12-month-old pigs and preserved with gluteraldehyde fixation. The gluteraldehyde fixation slowly leads to unwanted calcification of the biological valves and eventual failure. Tissue engineering approaches are now being developed to synthesize living heart valves, but durability remains a large concern. The muscle of the heart has never withstood grafting; however, stem cells might be able to rebuild the damaged tissue *in situ*. Recent results from experiments in which mesenchymal stem cells have been injected directly into the ventricle wall have shown that the cells incorporate in the heart muscle and initiate regeneration of damaged heart muscle. Who knows what this will do to the life expectancy of future generations?

6.7.4 Drug Delivery

Biomaterials play an important role as delivery vehicles for pharmaceuticals and biomolecules. The pharmaceutical industry has long made use of powdered biomaterials such as talc and calcium carbonate to form pills and tablets containing a drug. The goal of drug delivery research is to prepare formulations that will result in sustained active drug levels in the body, leading to improved drug efficacy. Controlled-release formulations accomplish this by various techniques that involve conjugation of the drug to a biomaterial. For example, by delivering basic fibroblast growth factor bound to heparin, the blood circulation time (as measured by a half-life) is increased by a factor of three. Conjugation to polyethylene glycol (PEG) is a well-established approach for *in vivo* protein stabilization.

Liposomal drug formulations also exhibit extended circulating half-lives after intravenous injection. Liposomes are made from phospholipids that form hydrophobic and hydrophilic compartments within an aqueous environment. A unilamellar liposome has spherical lipid bilayers that surround an aqueous core. Water soluble drugs can be entrapped in the core and lipid soluble drugs can be dissolved in the bilayer. The drug concentration in plasma over time is elevated three to ten times when incorporated in liposomes.

Polymers are widely used in drug delivery systems. Nondegradable hydrophobic polymers (silicone elastomers) have been used most extensively as semipermeable membranes around drug reservoirs. Alternative formulations involve mixtures of drug and resorbable polymer that release drug when the polymer degrades (Fig. 6.22). Increasing the loading of the protein or drug within the matrix increases the release rate of the compound. As a rule of thumb, as the average molecular weight of the polymer in the matrix increases, the rate of protein/drug release decreases. Polymer microspheres can be formed around drug solutions, thereby encapsulating the drug as another means of preparing a controlled-release formulation. Polymers that undergo bulk erosion tend to have burst release rates as compared to polymers that undergo surface erosion. The targeting of a drug to a particular cell type can be accomplished by using cell-specific ligands. Microparticles that have been modified with ligands do not diffuse uniformly throughout the body. Instead, the drug is delivered directly to only certain cell types through a receptor–ligand interaction. For example, liposomes have been prepared with folic acid ligands that are preferentially taken up by cancer cells because cancer cells express more folate receptors than do other cell types.

Direct injection of the drug into the targeted tissue is another means of obtaining high local drug concentrations. Controlled release of the drug at the site is accomplished through conjugation of the drug to a biomaterial drug delivery vehicle. This is particularly useful when the drug has toxic side effects that can be minimized by exposing only the tissue of interest to high drug levels. With local drug delivery, the toxic peak blood levels of drug are reduced, and there are sustained levels of active drug over a longer time period, leading to better efficacy (Fig. 6.23). As an alternative to direct injection, chemotherapy-carrying magnetic particles (magnetoliposomes)

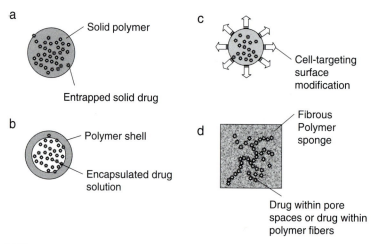

Figure 6.22   Polymers provide a versatile matrix for controlled release of drugs and biomolecules. Pharmaceutical agents can be contained within microparticles, microcapsules, or porous polymer blocks or conjugated to single chains of polymer. Drug targeting can be accomplished through the use of cell-specific ligands.

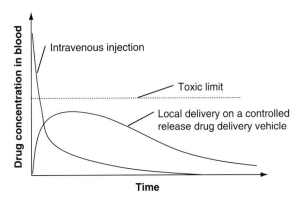

Figure 6.23   The primary advantage of local delivery is a reduction of the systemic blood levels of the drug to below toxic levels. High local levels at the site of injection or implantation allow for increased drug efficacy and reduced side effects. This technology is particularly applicable for chemotherapy drugs and therapeutic hormones.

also have been developed. After systemic injection, the drug-loaded particles are then localized to the cancer site by guidance with magnets.

Some drugs do not need to be solubilized to initiate biological activity. For example, nerve growth factor is effective when immobilized on a surface. Even higher levels of activity may be obtained after immobilization of a protein onto a surface due to conformational changes that occur upon immobilization. The active portions of the protein may be better exposed after immobilization. Alternatively, all biological activity may be lost after immobilization, so it is always necessary to conduct separate tests to confirm drug activity after immobilization or conjugation to a biomaterial surface.

Collagen is commonly used for the delivery of bone growth factors. The protein growth factors are typically not covalently immobilized on the collagen surface, leading to rapid release. Hydroxyapatite materials also are used for delivery of bone growth factors. The drug or protein often attaches quite strongly to the hydroxyapatite surface and release is greatly delayed as compared to a collagen drug delivery vehicle. Depending on the drug and its mode of action, longer release times may or may not be desired.

Rather than attempting to design a controlled-release drug delivery vehicle with the perfect release profile, drug delivery chips that can be programmed to open up drug compartments by an external signal are being developed. The ultimate smart drug delivery vehicle is the cell. Unlike a passive drug delivery device that acts independently, cells produce cytokines, growth factors, and extracellular matrix materials based on the signals from the *in vivo* environment. Attempts are being made to exploit this with the implantation of encapsulated xenograft pancreatic islet cells. Pancreatic islet cells produce insulin in response to the circulating blood levels. The xenograft cells need to be encapsulated within a biomaterial (typically alginate) to evade immune surveillance activity that can be toxic to the cells. Current research in this area is focused on varying the properties of the alginate to maintain sufficient permeability to keep the cells vital, yet protected from immune cell toxins, while still allowing diffusion of insulin out of the device—a tall order for one material. Materials selection for medical devices will always involve this type of balancing act between properties.

EXERCISES

1. List and briefly describe the five basic categories of biomaterials described in this chapter.

2. Describe which type of biomaterial you would select for the construction of the following implantable devices. Explain which properties will be important and why. More than one material can be used in the same device.
   a. Skin substitute
   b. Guidance tube for nerve regeneration
   c. Hip replacement stem
   d. Dental braces
   e. Urinary catheter
   f. Tissue-engineered bone

3. List three biomaterials commonly used for the following applications: (a) sutures, (b) heart valves, (c) endosseous dental root implants, (d) contact lenses, and (e) hip prosthesis. Provide the specific chemical name.

4. When selecting a biomaterial to be used as an orthopedic implant, what are some of the properties or characteristics of the material that should be considered?

5. What would happen to the mechanical functionality of bone if the carbonated apatite crystals were not discontinuous or discrete and instead were long fibers similar to a fiber composite? The modulus of bone mineral is 114–130 GPa. The modulus of cortical (normal) bone is 19–20 GPa.

6. Discuss three advantages and three disadvantages of natural biomaterials for medical devices. What is the most commonly used natural biomaterial? List three medical applications of the most commonly used natural material.

7. The type of implant–tissue response that occurs at the site of implantation is a major predictor for the success and stability of the device. List four types of implant–tissue responses that can occur, beginning with what happens if the material is toxic.

8. What is the purpose of a suture? What are some of the important properties that must be considered when selecting a biomaterial for use as a suture? List four biomaterials that are commonly used for sutures.

9. Define calcification. What type of application is at the most risk for failure due to biomaterial calcification?

10. What is a bioactive material? What is a biomimetic material?

11. In the past, an implanted biomaterial was considered biocompatible if it became encapsulated with fibrous tissue and did not elicit a further response from the host. Why has this definition of biocompatibility changed?

12. If a protein is attached to a biomaterial surface, specifically a growth factor in order to attract cells once implanted, why would it be important to know the structure of the protein? Think about how a growth factor would work to attract the cell to the surface and how these proteins could be anchored to the surface so that this functionality is not compromised.

13. Many cell types can be found adjacent to a biomaterial implant. What type of cells would you expect to find depositing bone adjacent to an orthopedic implant? What type of cells would you expect to find clearing the site of biomaterial debris?

14. There is a well-defined wound healing response following the implantation of a biomaterial. Briefly describe the four phases of wound healing.

15. You have recently designed a new implantable biomaterial and have conducted an *in vivo* implantation study. Subsequent extraction of the sample reveals a thin, fibrous capsule surrounding your material. What does this experiment reveal regarding the biocompatibility of this device, specifically regarding the inflammation response?

16. Is the growth of a fibrous tissue layer around an implanted material a positive aspect in all applications? Why or why not? In what applications is a fibrous tissue layer not desired?

17. Describe two methods for making a biomaterial porous.

18. What biomimetic polymer appears to be particularly well suited for the differentiation of cartilage cells and why?

19. What biomaterial leads to direct bone bonding without an intervening fibrous tissue layer?

20. Describe a way in which a drug-carrying microparticle or nanoparticle can be modified to make it targeted to a specific cell type.

21. Classify the following biomedical devices according to the FDA definitions (www.fda.gov) for Class I, II, and III devices: (a) intraocular lens, (b) heart valve, (c) preformed tooth crown, (d) oxygen mask, (e) stethoscope, (f) dental amalgam.

SUGGESTED READING

Annual Book of ASTM Standards, Vol 13.01. (2004). Medical and Surgical Materials and Devices. American Society of Testing and Materials (ASTM) International, West Conshohocken, Pennsylvania.

Atala, A., Mooney, D., Vacanti, J.P. and Langer R. (1997). *Synthetic Biodegradable Polymer Scaffolds*. Birkhauser, Boston.

Dee, K.C., Puleo, D.A. and Bizios, R. (2002). *An Introduction to Tissue Biomaterial Interactions*. Wiley-Liss, Hoboken, New Jersey.

DeFrances, C.J. and Hall, M.J. (2004). *2002 National Hospital Discharge Survey: Advance Data from Vital and Health Statistics*, No. 342. National Center for Health Statistics, Hyattsville, Maryland.

Dillow, A.K. and Lowman, A.M. (2002). *Biomimetics Materials and Design*. Marcel Dekker, New York.

Helmus, M.N. (2003). *Biomaterials in the Design and Reliability of Medical Devices*. Landes Bioscience, Texas and Kluwer Academic, New York.

Hench, L.L. and Wilson, J. (1993). *An Introduction to Bioceramics*. (Advanced Series in Bioceramics, Vol.1.) World Scientific, River Edge, New Jersey.

Guilak, F., Butler, D.L., Goldstein, S.A. and Mooney D.J. (2003). *Functional Tissue Engineering*. Springer, New York.

Lanza, R.P., Langer, R. and Vacanti, J.P. (2000). *Principles of Tissue Engineering*, 2nd ed. Academic Press, San Diego, California.

Mann S. (1996). *Biomimetic Materials Chemistry*. VCH Publishers, New York.

Palsson, B.O. and Bhatia, S.N. (2004). *Tissue Engineering*. Pearson Prentice Hall, Upper Saddle River, New Jersey.

Park, J.B. and Bronzino, J.D. (2003). *Biomaterials Principles and Applications*. CRC, Boca Raton, Florida.

Saltzman, W.M. (2001). *Drug Delivery: Engineering Principles for Drug Therapy*. Oxford Univ. Press, New York.

Shalaby, S.W. and Burg, K.J.L. (2004). *Absorbable and Biodegradable Polymers*. CRC, Boca Raton, Florida.

Silver, F.H. and Christiansen, D.L. (1999). *Biomaterials Science and Biocompatibility*. Springer, New York.

Ratner, B.D., Hoffman, A.S., Schoen, F.J. and Lemons, J.E. (2004). *Biomaterials Science: An introduction to Materials in Medicine*. 2nd edition, Academic Press. San Diego, California.

Rho, J-Y., Kuhn-Spearing, L. and Zioupos, P. (1998). Mechanical properties and the hierarchical structure of bone. *Med. Engineering & Phys*. **20**, 92–102.

Williams, D.F. Ed. (1987). Definitions in biomaterials. In *Progress in Biomedical Engineering*. **4**, 67. Edited by D.F. Williams. Elsevier, Amsterdam.

7 TISSUE ENGINEERING

Randall McClelland, PhD
Robert Dennis, PhD
Lola M. Reid, PhD
Bernard Palsson, PhD
Jeffery M. Macdonald, PhD

Chapter Contents

At the conclusion of this chapter, the reader will be able to:

- Understand the growing area of cellular therapies.

- Describe the three general categories of extracorporeal bioreactors.

- Understand the cellular dynamics underlying tissue function.

- Qualitatively describe the importance of stem cells in tissue function.

- Quantitatively describe cellular fate processes.

- Analytically describe mass transfer in three different configurations.

- Understand the parameters that characterize the tissue environment and how to approach mimicking them *in vitro*.

- Describe the issues fundamental to scale-up.

- Define functional tissue engineering and how the "-omics" sciences is driving this newly created research area.

- Discuss many of the issues encountered when implementing cellular therapies or bioartificial organs to patients.

7.1 WHAT IS TISSUE ENGINEERING?

Tissue engineering is a biomedical engineering discipline integrating biology with engineering to create tissues or cellular products outside the body (*ex vivo*) or to make use of gained knowledge to better manage the repair of tissues within the body (*in vivo*). This discipline requires understanding of diverse biological fields, including cell and molecular biology, physiology and systems integration, stem cell proliferation and differentiation with lineage attributes, extracellular matrix chemistry and compounds, and endocrinology. It also requires knowledge of many engineering fields, including biochemical and mechanical engineering, polymer sciences, bioreactor design and application, mass transfer analysis of gas and liquid metabolites, and biomaterials. Translation of tissue engineering constructs to clinical applications will involve other scientific disciplines so that novel engineered tissues will be easily accepted and used by clinicians. The combination of these sciences has spawned the field of regenerative medicine which has, at present, two strategic clinical goals: (1) cell therapies for the repair of damaged tissues, involving injection or engraftment of cells or cellular suspensions—sometimes in combination with scaffolding material, or (2) establishment of tissue *ex vivo* for use as grafts or extracorporeal organs to assist or supplement ailing *in vivo* organs. Clinical trials with cell therapies or extracorporeal organs are underway for cartilage, bone, skin, neural, and liver tissues—and are projected as repair mechanisms for many other tissues within the coming years. Both scientific and economic issues will define the success of these future therapeutic modalities.

7.1.1 Challenges Facing the Tissue Engineer

Some of the fundamental scaling challenges that face the tissue engineer in the implementation of cell therapies or creation of grafts and bioartificial organs are shown in Figure 7.1. In these bioartificial structures, the tissue compartments must be scaled to mimic nutrient transport existent within natural capillary beds found in the body. Generally, these *in vivo* beds are composed of 100 µm thick tissue slabs sandwiched by vasculature. Thus, the microenvironment of *ex vivo* bioartificial devices must also maintain these limitations in order to provide appropriate concentrations of nutrients to tissue cultures. Once this *ex vivo* capillary scaling is achieved, then more advanced engineered devices must be designed in order to support larger cell masses. One such design is a bioreactor that maintains appropriate radial scaling parameters while longitudinally extending the cellular space. In Figure 7.1, this bioreactor is illustrated as a 10 cm longitudinal design. Additionally, for automated cell cultures and manufactured devices, the scaling up (or down for individual cell analysis) must maintain correct proportions of nutrient and biomass interactions; therefore, based upon need, distinctive scaling parameters are necessary for unique applications. Some of these applications are disussed with a few examples below:

1. The reconstitution of physical (mass transfer) and biological (soluble and insoluble signals) microenvironments for the development of tissue function
2. To overcome scale-up problems in order to generate cellular microenvironments that are clinically meaningful
3. The system automation to perform on clinically meaningful scales
4. The implementation of devices in clinical settings, with cell handling and preservation procedures that are required for cell therapies

The primary focus of this chapter concentrates on items 1 and 2, although some of the challenges faced with items 3 and 4 are discussed in Section 7.4. For an overall understanding, items 1 and 2 are further illustrated in Figure 7.2 from the viewpoints of "every cell in the body."

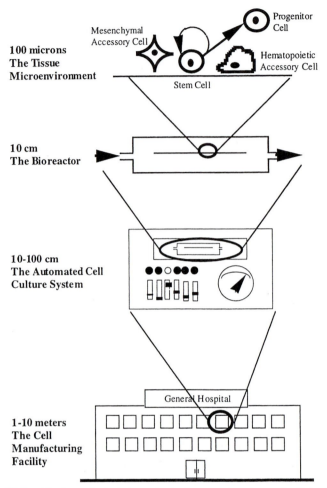

Figure 7.1   The four principal size scales in tissue engineering and cellular therapies.

At the center of Figure 7.2 is a cell. The diagram represents the environment that influences every cell in the body. First, there are the chemical components of the microenvironment: the extracellular matrix, hormones, cyto/chemokines, nutrients, and gases. Physically, it is characterized by its geometry, the dynamics of respiration, and the removal of metabolic by-products. From these characterized observations, expanded details of each component will be discussed such that biological understandings precede the physical considerations. Finally, both biological and physical environments are combined to help integrate research and clinical activities for the future development of tissue engineered products.

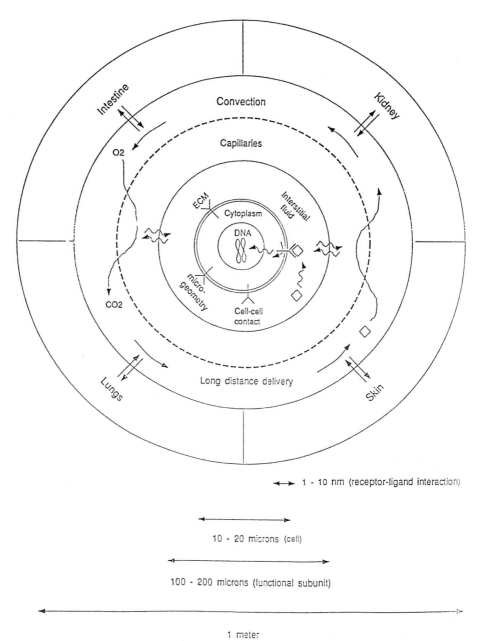

Figure 7.2   A cell and its communication with other body parts (modified from Lightfoot, 1974).

7.1.2   Cellular Therapies, Grafts, and Extracorporeal Bioartificial Organs

The development of cell therapies initially arises from advancing knowledge within the cell and molecular biology science domains. Transferring these developments into the clinical arena becomes a design challenge that requires organized culture control and exploitation of cell metabolites. In this way, many scientific fields, such as bioengineering, biochemical engineering, and biomaterial sciences, are all needed for the implementation of cell therapies. However, current challenges revolve around the design of tissues and tissue-like conditions so that sufficient numbers of cells can be isolated and grown for clinical and commercial programs. For an historical comparison, the discovery of penicillin alone was not enough to affect the delivery of health care. Mass production methods of clinical grade material had to be developed. The development of large-scale production of antibiotics arguably represents the most significant contribution of engineering to the delivery of health care. In a similar fashion, the development of industrial scale methods for isolation, expansion, and cryopreservation of human cells will enable routine uses for cell therapies. To meet these clinical needs, tissue engineering is focused on the manipulation and design of human tissues.

Several basic issues of cell biology need to be understood and quantitatively characterized. These include the key cellular processes of cell differentiation, hyperplastic and hypertrophic growth, migration (motion), and death (necrosis or apoptosis) that are aspects of tissue function. Basic information about stem cell and maturational lineage biology, and the role of determined stem cells in organ function, genesis, and repair will be presented. Tissues are comprised of multiple cell types that are interacting dynamically with each other. Therefore, tissue-specific functions are observed only with co-cultures of those multiple cell types or with cultures of a particular cell type in combination with the signals from the others. Those signals include insoluble factors in the extracellular matrix; signals from direct cell–cell contact; and soluble signals from autocrine, paracrine, and endocrine interactions.

Many bioengineering challenges must be met. For example, bioengineering considerations in *cell therapies* include injection needle design and procedure protocols. For this application, needles must be optimized to reduce shear stress on cell membranes. Nutrient mass transfer must be analyzed to determine adequate cell aggregate sizes for sustained viable tissues. Engraftment techniques and seed site selections must be established, so that cells will prosper and assist in system homeostasis. Detrimental events such as the formation of emboli need to be prevented. For device consideration—such as *bioreactors*—the function, choice, manufacturing, and treatment of biomaterials for cell growth and device construction are important. Fluid mechanics and mass transfer play important roles in normal tissue function and therefore become critical issues in *ex vivo* cellular device designs. Systems analysis of metabolism, cell–cell communication, and other cellular processes play key roles in replicating normalcy and defining bioartificial organ specifications. A properly designed *ex vivo* culture system must balance appropriately the rates of biological and physico-chemical attributes for suitable tissue functions. By mathematically integrating this balancing effect, dimensionless computation groups can be formulated that describe characteristic ratios of time constants. In this way, new dimensionless values will evolve to include ratios of "physical times" with "biological times."

The implementation of cell therapies and grafts in the clinic requires the recognition and resolution of several difficult issues. These include tissue harvest, cell processing and isolation, safety testing, cell activation/differentiation, assay and medium development, storage and stability, and quality assurance and quality control issues. These challenges will be described, but not analyzed in detail. Manufacturing of cell therapy products are and will continue to be fundamentally challenging.

7.1.3 Human Cells and Grafts as Therapeutic Agents

Cell therapies use human cells as therapeutic agents to alleviate a pathological condition. It is important to note that cell therapies are not new. One existing type of cell therapy is blood transfusion, which has been practiced for decades with great therapeutic success and benefit. This therapy uses red blood cells (RBCs) as the transplant product into anemic patients, such that RBCs help restore adequate oxygen transport. Similarly, platelets have been transfused successfully into patients who have blood clotting problems. Bone marrow transplantation (BMT) has been practiced for almost two decades with tens of thousands of cancer patients undergoing high-dose chemo- and radiotherapies followed by a BMT. More recently, transplantation of hematopoietic stem cells is occurring with increasing frequency to correct hematological disorders. These are all applications of cell therapies associated with blood cells and blood cell generation, or hematopoiesis (Greek *hemato*, meaning "blood", and *poiesis*, meaning "generation of"). In this way, numerous patients are currently receiving therapeutic benefit from cellular therapies. Transplants can be xenogeneic (donor and recipient are members of different species), allogeneic (donor and recipient are members of the same species but are not genetically identical), or syngeneic (donor and recipient are genetically identical, e.g., identical twins). Syngeneic transplants include autologous transplants (cells from a patient being isolated and given back to the same person). Historically, the issues relevant to allogeneic transplants are well known and apply to almost all forms of organ transplantation. However, with the advent of cell culture and advancing cell manipulation procedures, autologous transplantation is becoming more common. There are, in addition, attempts underway in several laboratories to make "universal donor" cell sources and cell lines. Such cell sources should have minimal or no immunogenicity problems (analogy: type O, Rh$^-$ blood).

The ability to reconstitute tissues *ex vivo*—a form of tissue engineering—and produce cells in clinically meaningful numbers has a wide spectrum of applications. Table 7.1 summarizes the supply-and-demand incidence of organ and tissue deficiencies versus the number of procedures performed annually in the United States. Although the number of procedures is limited, the overall cost of these procedures was still estimated at a staggering $400 billion per year. The potential socioeconomic impact of cellular therapies is therefore substantial. By effectively decreasing these costs, extensive research to improve patient treatments will become monetarily

TABLE 7.1   Incidence of Organ and Tissue Deficiencies, or the Number of Surgical Procedures Related to These Deficiencies in the United States[a]

Indicator	Procedure or Patients per Year
Skin	
Burns[b]	2,150,000
Pressure sores	150,000
Venous stasis ulcers	500,000
Diabetic ulcers	600,000
Neuromuscular disorders	200,000
Spinal cord and nerves	40,000
Bone	
Joint replacement	558,200
Bone graft	275,000
Internal fixation	480,000
Facial reconstruction	30,000
Cartilage	
Patella resurfacing	216,000
Chondromalacia patellae	103,400
Meniscal repair	250,000
Arthritis (knee)	149,900
Arthritis (hip)	219,300
Fingers and small joints	179,000
Osteochondritis dissecans	14,500
Tendon repair	33,000
Ligament repair	90,000
Blood Vessels	
Heart	754,000
Large and small vessels	606,000
Liver	
Metabolic disorders	5,000
Liver cirrhosis	175,000
Liver cancer	25,000
Pancreas (diabetes)	728,000
Intestine	100,000
Kidney	600,000
Bladder	57,200
Ureter	30,000
Urethra	51,900
Hernia	290,000
Breast	261,000
Blood Transfusions	18,000,000
Dental	10,000,000

[a]From Langer and Vacanti (1993).
[b]Approximately 150,000 of these individuals are hospitalized and 10,000 die annually.

obtainable. In this way, new medical products will greatly improve the quality of life for ailing individuals and reestablish their productivity as citizens within society.

    The concept of directly engineering tissues was pioneered by Y.C. Fung in 1985. The first symposium on the subject was organized by Richard Skalak and Fred Fox in

1988, and since then the field of tissue engineering has grown significantly. Numerous articles have appeared, and the journal, *Tissue Engineering*, is now being published. Cutting-edge cell therapies include various forms of immunotherapies, chondrocytes for cartilage repair, liver and kidney cells for extracorporeal support devices, β-islet cells for diabetes, skin cells for patients with ulcers or burns, and genetically modified myocytes for treatment of muscular dystrophy. In addition, cell therapies include tissue engineered grafts to replace damaged or defective tissues such as blood vessels, ureters and urethras, and other tissues. The challenges faced with each tissue are different. A few examples are provided in the following sections for illustrative purposes.

Bone Marrow Transplantation (BMT)

Bone marrow is the body's most prolific organ. It produces on the order of 400 billion myeloid cells daily, all of which originate from a small number of pluripotent stem cells (Fig. 7.3). The bone marrow is comprised of 500 to 1000 billion cells and regenerates itself every 2 to 3 days, which represents normal hematopoietic function. Individuals under hematopoietic stress, such as systemic infection or sickle-cell anemia, will have blood cell production rates that exceed the basal level. The prolific nature of bone marrow cells makes it especially susceptible to damage from radio- and chemotherapies. Bone marrow damage limits the extent of these therapies, and some regimens are fully myoablative. Without any hematopoietic support, patients who receive myoablative dose regimens will die due to hematopoietic failure.

BMT was developed to overcome this problem. In an autologous setting, the bone marrow is harvested from the patient prior to radio- and chemotherapies. It is cryopreserved during the time period that the patient undergoes treatment. After chemotherapeutic drug application and a few half-lives have passed, the bone marrow is rapidly thawed and returned to the patient. The bone marrow cells are simply put into circulation, and the bone marrow stem cells "home" to the marrow cavity and reconstitute bone marrow function. In other words, the hematopoietic tissue is rebuilt *in vivo* by these cells. This process takes several weeks to complete, during which time the patient is immuno-compromised.

Autologous BMT as a form of cellular therapy simply involves removing the cells from the patient and storing them temporarily outside the patient's body. There are several advantages to growing the harvested cells as newer therapies and treatments are being developed based on *ex vivo* culture of hematopoietic cells. Indeed, newer methods to harvest bone marrow stem cells have been developed. These methods rely on using cytokines or cytotoxic agents to "mobilize" the stem cells into circulation. The hematopoietic stem and progenitor cells are then collected from the circulation using leukopheresis.

Myoablative regimens are used in allogeneic settings. In the case of leukemia, this not only removes the bone marrow but also, it is hoped, the disease. The donor's cells are "home to" the marrow and repopulate the bone cavity, just as in the autologous setting. The primary difficulty with allogeneic transplants is high mortality (10 to 15%) primarily due to graft-versus-host disease, in which the transplanted cells

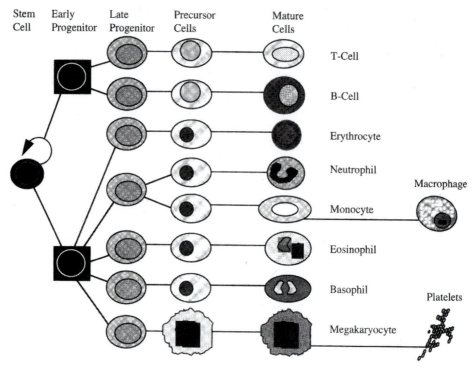

Figure 7.3   Hematopoietic cell production. The production fluxes through the lineages can be estimated based on the known steady-state concentration of cells in circulation, the total volume of blood, and the half-lives of the cells. Note that the 400 billion cells produced per day arise from a small number of stem cells (from Koller and Palsson, 1993).

recognize the recipient's tissues as foreign. Overcoming this rejection problem would significantly advance the use of allogeneic transplantation.

BMT is a well-developed and accepted cellular therapy for a number of indications, including allogeneic transplants for diseases such as leukemia, and autologous transplants for diseases such as lymphoma and breast and testicular cancer. Significant growth has occurred in the use of BMT since the mid-1980s, and approximately 20,000 patients were treated worldwide in 1995 (about evenly split between autologous and allogeneic transplants).

Skin and Vascular Grafts

Skin is the body's third most prolific tissue. It basically consists of two layers: (1) the dermis, the main cellular components of which are stroma or fibroblasts, and (2) the epidermis, the main cellular components of which are epidermal cells that are at various stages of differentiation into keratinocytes (see Fig. 7.4). Both cell types grow very well in culture, and *ex vivo* cultivation is not the limiting factor with this tissue. Interestingly, transplanted dermal fibroblasts have proven to be surprisingly nonimmunogenic.

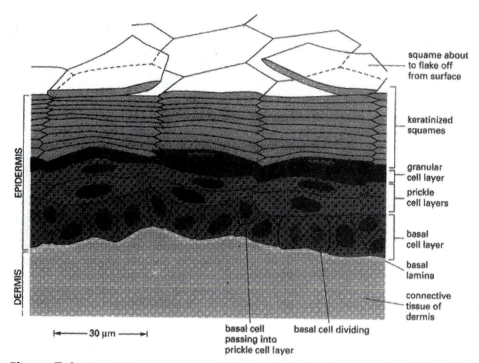

squame about to flake off from surface

keratinized squames

granular cell layer

prickle cell layers

basal cell layer

basal lamina

connective tissue of dermis

EPIDERMIS

DERMIS

|← 30 μm →|

basal cell passing into prickle cell layer

basal cell dividing

Figure 7.4 The cellular arrangement and differentiation in skin. The cross section of skin and the cellular arrangement in the epidermis and the differentiation stages that the cells undergo (from Alberts et al., 1994).

This transplant technology is easily applied to victims of burns and patients with diabetic ulcers who have severe problems with skin healing. To treat these problems, skin can be cultured *ex vivo* and applied to the affected areas. Technologically, this cell therapy is relatively well developed. However, several commercial concerns exist, such as how to "make available" and "utilize" human skin equivalents for clinical purposes. Figure 7.5 shows an expansion bioreactor for creating skin grafts from human foreskins developed by Advanced Tissue Sciences. The bioreactor is constructed with one mechanical hinge and two ports for constant "bleed-feed" flow (left panel). The graft products—which are similar to autologous skin grafts without hair follicles—are easily removed from the bioreactor by the clinician (shown in the right panel). A main limitation in the use of these products is cost. Due to U.S. Food and Drug Agency (FDA) requirements, the cost per single foreskin was exorbitant, and large surface areas ($\sim 4000\,\text{m}^2$) are required for expansion. Additionally, the cost for FDA-approved media components are costly and require lengthy recordkeeping for each utilized component. Therefore, the profit margin is meager and a single contamination of a foreskin culture would result in a million dollar loss. Even so, vascular grafts have been generated using mesenchymal stem cells grown on biodegradable tubes. Initial basic research studies are now awaiting the clinic.

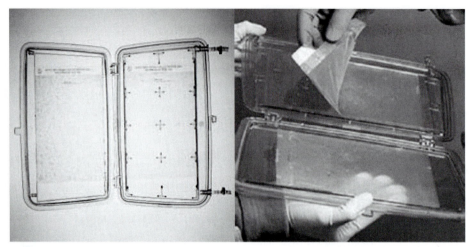

Figure 7.5 Advanced Tissue Sciences bioreactor for culture of their skin product, Trancyte®, derived from human foreskins.

Pancreatic β-Islet Cells

In insulin-dependent diabetics, the pancreas has lost its ability to produce and secrete insulin. Pancreases from cadaveric donors can be used as sources of islets of Langerhans, which contain the insulin-secreting β-islet cells. These cells can be injected into the portal vein leading to the liver. The islets then lodge in the liver and secrete life-sustaining insulin. Unfortunately, at present it is not possible to grow β-islet cells in culture without them losing their essential properties, and thus this procedure is constrained by the severely limited supply of tissue. This form of cellular therapy is an allogeneic transplantation procedure, and the duration of the graft can be prolonged by immunosuppressing the recipient.

The immune rejection problem is an important concern in cellular therapies and is treated in a separate section below. However, ways to overcome rejection include the physical separation of the donor's cells from the immune system by a method that allows exchange between the graft and the host across a semipermeable membrane (Fig. 7.6).

Cartilage and Chondrocytes

Cartilage is an unusual tissue. It is avascular, alymphatic, and aneural. It consists mostly of extracellular matrix in which chondrocytes are dispersed at low densities— on the order of (1 million)/cc. This low value is compared against cell densities of a (few hundred million cells)/cc found in most tissues. Chondrocytes can be cultured *ex vivo* to increase their cell numbers by approximately 10-fold. Using this technique, deep cartilage defects in the knee can be treated by autologous cell transplantation. In this case, a biopsy is collected from the patient's knee outside the affected area (Fig. 7.7). The chondrocytes are liberated from the matrix by an enzymatic treatment and allowed to grow in a two-stage cell culture process. The cells are then harvested

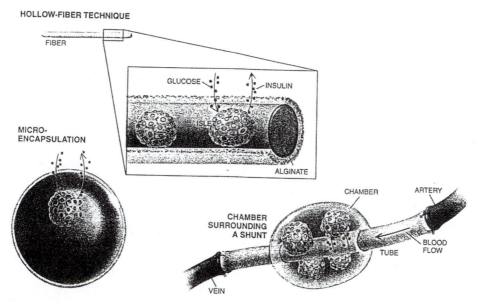

Figure 7.6   Encapsulation of islets in semiporous plastic is one promising way to protect them from attack by the immune system (from Lacey, 1995).

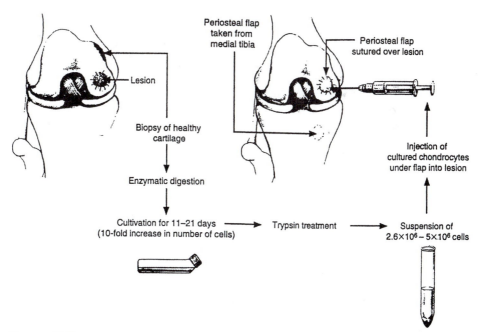

Figure 7.7   Diagram of chondrocyte transplantation in the right femoral condyle (from Brittberg, 1994).

from the culture and introduced into the affected area in the knee joint. The transplanted chondrocytes are able to heal the defect. This procedure is relatively simple in principle, but still has challenges in clinical implementation. Numerous patients are afflicted with knee problems that effectively leave them immobile. Over 200,000 patients are candidates for this type of cellular therapy annually in the United States alone.

Clinical Trials for Liver, Neuronal, and Cardiovascular Stem Cell Therapy

Other forms of cell therapies that are still under development include liver stem cell therapies, neuronal cell therapies, and cardiovascular cell therapies, for which many are in clinical trials and are on the verge of completion. However, implantation efficiency is still a problem. If enough cells implant, they will grow and function to some extent. Various stem cell populations are being used, from adult stem cells derived from cadaveric livers (Vesta Inc.) to human neuronal stem cell populations (Stem Cell Inc.). Still other companies use transformed cell lines and even embryonic stem cells. It is clear that bioengineering new implantation hardware or designing mass transfer computational models for predictive analysis would be of great help for clinical study success. But at present, this is not a conventional procedure required by FDA. Typically, existing technologies are being used and little modeling of mass transfer has been performed. Questions such as, "What maximum cell aggregate diameters are necessary for adequate tissue function and sustained cell viability?" or "What optimum infusion rates are necessary for cell therapy to be effective in patients?" must be answered.

7.1.4   Mechanisms Governing Tissues

Normal tissue functions define the engineering specifications and, therefore, the laws of nature are used as a guide for management of cells to generate tissues. Two dynamic sets of mechanisms govern tissues: (1) the relationship between epithelium and mesenchymal cells and (2) stem cell and maturational lineage biology.

Once the mechanisms underlying tissue processes are understood, one can focus on issues of relevance to engineering. The smallest physiological unit defines the mass transfer dimensions, whereas the number of cells required to replace the physiological functions defines the overall dimension of an engineered product. These numbers help set overall goals of the clinical device or engineered tissue.

The Epithelial–Mesenchymal Relationship

The most fundamental paradigm defining tissues is the epithelial–mesenchymal relationship consisting of a layer of epithelia bound onto a layer of mesenchymal cells (stroma, endothelia, smooth muscle cells) (Fig. 7.8). Normal epithelial cells require a constant relationship with an appropriate mesenchymal partner or with matrix and soluble signals mimicking that relationship. Signaling between and within the two cell layers coordinates local cell activities and is affected by soluble signals (autocrine, paracine, and endocrine factors) working synergistically with extracellular matrix

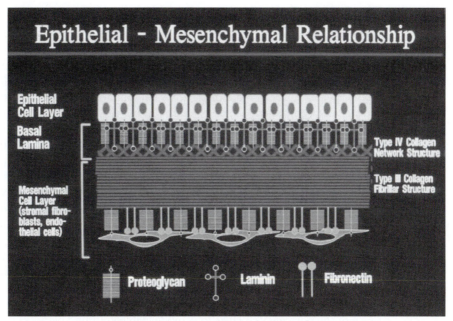

Figure 7.8   Epithelial–mesenchymal relationship.

(ECM), where the ECM is an insoluble complex of proteins and carbohydrates found outside of the cells.

The Soluble Signals

There are three categories of soluble signals:

Autocrine factors: Factors produced by cells and active on the same cell
Paracrine factors: Factors produced by cells and active on their neighbors
Endocrine or systemic factors: Factors produced by cells at a distance from the target cells and carried through the blood or lymphatic fluid to the target cells

There are multiple, large families of factors that operate as autocrine or paracrine signals. All of them can act as mitogens (i.e., elicit growth responses from cells) or can drive differentiation (i.e., induce expression of specialized tissue-specific functions) depending on the chemistry of the extracellular matrix associated with the cells. Some of the most well studied of these factors include the insulin-like growth factors (IGFs), epidermal growth factors (EGFs), fibroblast growth factors (FGFs), colony stimulating factors (CSFs), platelet derived growth factors (PDGFs), tumor growth factors (TGFs), and cytokines such as interleukins. Do not be confused by the fact that most are called "growth factors;" this is due to the fact that all were identified initially by assays in which the scientists were looking for a growth response in some type of cell.

The Extracellular Matrix

The ECM is found on the lateral and basal surfaces of cells. The lateral extracellular matrix is found on the lateral borders of coupled homotypic cells (e.g., epithelia to epithelia). The basal extracellular matrix is present between heterotypic cell types (e.g., epithelial cells to mesenchymal cells) and is referred to also as the basement membrane. The extracellular matrix may contain up to 1/2 of the proteins in the body. Indeed, one component of the extracellular matrix—the collagens—consists of more than 25 different families of molecules and accounts for 1/3 of the body's proteins. For many years, the extracellular matrix was thought to play an entirely mechanical role, binding together cells in specific arrays. Over the last 20 years, it has been found that the extracellular matrix is a major signaling system, one that is in a solid state form and that confers chronic or persistent signaling mechanisms. Two of the primary components of the lateral extracellular matrix are: (1) cell adhesion molecules (CAMs) that are age- and tissue-specific, and (2) proteoglycans—molecules containing a protein core to which are attached polymers of sulfated (negatively charged) sugars called glycosaminoglycans (e.g., heparan sulfates, heparins, chrondroitin sulfates, or dermatan sulfates). The basal ECM consists of basal adhesion molecules (e.g., laminins or fibronectins) that bind the cells via matrix receptors (integrins) to one or more types of collagen scaffoldings. The collagens of one cell layer are cross-linked to those of the adjacent cell layer to provide stable coupling between the layers of cells. In addition, proteoglycans are bound to the basal adhesion molecules, to the collagens, and/or to the basal cell surface.

The lateral and basal ECM components provide direct signaling to cells in the form of chronic or persistent signaling. Indirectly, the components also facilitate signaling by:

1. Stabilizing cells in appropriate configurations of ion channels, receptors, antigens, etc.
2. Influencing intracellular pathways
3. Inducing appropriate cell shapes (flattened or three-dimensional)

These reactions enable the cells to respond rapidly to soluble signals which can derive from local or distance sources. The soluble factors act by binding to high-affinity molecules called receptors, which can be on the cell surface or be intracellular. When the signal binds to its receptor, a signal transduction response is triggered.

Cell Numbers *In Vivo* and Orders of Magnitude

The cell densities in human tissues are on the order of 1–3 billion cells/ml. The volume of a 70 kg human is about 70,000 ml. Therefore, the human body consists of about 100 trillion (trillion $= 10^{12}$) cells. A typical organ is about 100 to 500 mls in size and, therefore an organ contains about 100 to 1,500 billion (10^9) cells. Organs are comprised of functional subunits, as illustrated in Section 7.4. Their typical linear dimensions are approximately 100 μm. The cell number in a cube that is 100 μm on each side is estimated to be about 500 to 1000. These cell numbers are summarized in Table 7.2.

TABLE 7.2 Cell Numbers in Tissue Biology and Tissue Engineering: Orders of Magnitude

Cell numbers *in vivo*	
Whole body	10^{14}
Human organ	$10^9 - 10^{11}$
Functional subunit	$10^2 - 10^3$
Cell production *in vivo*	
Theoretical maximum from a single cell (Hayflick limit)	$2^{30-50} < 10^{15}$
Myeloid blood cells produced over a lifetime	10^{16}
Small intestine epithelial cells produced over a lifetime	5×10^{14}
Cell production *ex vivo*	
Requirements for a typical cellular therapy	$10^7 - 10^9$
Expansion potential[a] of human tissues	
Hematopoietic cells	
Mononuclear cells	10-fold
CD34 enriched	100-fold
Two or three antigen enrichment	10^6- to 10^7-fold
T cells	10^3- to 10^4-fold
Chondrocytes	10- to 20-fold
Muscle, dermal fibroblasts	$> 10^6$-fold

[a]Expansion potential refers to the number of cells that can be generated from a single cell in culture.

Therefore, a typical organ will have a few hundred million functional subunits. This number will be decided by the capability of each subunit and the overall physiological need for its particular function. For example, the number of nephrons in the kidney is determined by the maximal clearance need of toxic by-products and the clearance capability that each nephron possesses.

Several important conclusions can be derived from these numbers. The fundamental functional subunit of most tissues contains only a few hundred cells, and further, this is a mixed cell population. As outlined in Section 7.3, most organs have accessory cells that can be as much as 30% of the total cell number. Further, as illustrated below, the tissue-type-specific cells may be present at many stages of differentiation.

The nature of tissue microenvironments—along with cellular dynamics, communication, and metabolic processes—must be understood in order to reconstitute tissue function accurately. Thus, generating a therapeutic dose of cells requires a large number of microenvironments. These microenvironments must be relatively similar to have all the functional subunits perform in comparable fashions. Therefore, the design of cell culture devices must produce uniformity in supporting factors such as nutrient, oxygen, and growth factor/hormone concentrations. These input applications must be reasonably homogeneous down to 100 μm distances. Below this size scale, nonuniformity would be expected and in fact needed for proper functioning of tissue function subunits.

These considerations influence the approach to cell transplantation. At present, it is unlikely that fully functioning organs will be produced *ex vivo* for transplantation purposes. Integration of such *ex vivo* grown tissues into the recipient may prove problematic. The alternative is to produce stem and progenitor cells that sufficiently proliferate *in vivo* to regenerate organ function. BMT, for instance, relies on this approach.

7.1.5   Clinical Considerations

What Are Clinically Meaningful Numbers of Cells?

Currently, the cells required for clinical practice and experimental cell therapy protocols fall into the range of "a few tens of millions" to "a few billion" (Table 7.2). Since tissuelike cultures require densities above 10 million cells per milliliter, the size of the cell culture devices for cell therapies appears to fall into the range of 10 to a few hundred milliliters in volume.

What Are the Fundamental Limitations to the Production of Normal Cells?

The number of divisions a cell can undergo is dependent on its maturational lineage stage with key stages being the (1) stem cells (diploid, pluripotent), (2) diploid somatic cell subpopulations (unipotent), and (3) polyploid cell subpopulations. The stem cells can self-replicate and undergo unlimited numbers of divisions (stem cells are the potential source of immortalized cell lines and of tumor cells). Diploid cells are subject to the so-called Hayflick limit in the number of divisions for which they are capable. Normal, somatic, diploid human cells can undergo about 30 to 50 doublings in culture. Therefore, a single diploid cell can produce, theoretically, 10^{10} to 10^{15} cells in culture. Given the requirements for cell therapies, the Hayflick limit does not present a stifling limitation for cell therapies based on diploid cells—either stem cells or the somatic cells. Furthermore, the expansion potential is minimal or negligible for the subpopulations of polyploid cells found in all tissues but in very high numbers in various quiescent tissues (e.g., liver and heart). In this way, regenerative stimuli cause the polyploid cells to undergo DNA synthesis with limited or negligible capability to undergo cytokinesis. This results in an increase in their level of ploidy, an increase in cell volume, and a phenomenon called hypertrophy. Thus, the regeneration of tissues is a combination of hyperplasia (the diploid subpopulations) and of hypertrophy (the polyploid subpopulations) activities.

How Rapidly Do Normal Cells Grow in Culture?

Normal cells vary greatly in their growth rates in culture. Hematopoietic progenitors have been individually clocked at 11- or 12-hr doubling times, which represents the minimum cycle time known for adult human cells. Dermal foreskin fibroblasts grow with doubling times of 15 hrs, a fairly rapid rate which may be partially attributable to the fact that they are isolated from neonatal tissue containing a high proportion of stem/progenitor cells. Adult chondrocytes grow slowly in culture with doubling times of about 24 to 48 hrs.

How Are These Cells Currently Produced?

Expansion of cells is done in a variety of culture dishes (T flasks, roller bottles), in bags (T cells), in suspension cultures (certain hemopoietic cells or cell lines). Novel scaffolding material is permitting *ex vivo* expansion of most cell types. The ability to achieve the maximum expansion potential for specific cell types requires precise culture conditions. These conditions may comprise specific forms of extracellular matrix, defined mixtures of hormones and growth factors, nutritional supplements, and basal media containing specific concentrations of calcium, trace elements, and gases. Reviews of these conditions for many cell types are available in recently published textbooks (see Suggested Reading).

7.2 BIOLOGICAL CONSIDERATIONS

7.2.1 Stem Cells

All tissues consist of a stem cell compartment that produces cellular offspring that mature into all tissue phenotypes. The maturational process includes: (a) *commitment*, in which the pluripotent stem cells produce daughter cells with restricted genetic potentials appropriate for single cell activities (unipotent), and (b) *differentiation*, in which sets of genes are activated and/or altered in their levels of expression. The following sections include information about these major tissue subdivisions to explain "how they work together to generate or repair tissues" in order to provide tissue functions.

Stem Cells and the "Niche" Hypothesis

Pluripotent stem cells are cells capable of producing daughter cells with more than one fate, are able to self-replicate, and have the ability to produce daughter cells identical to the parent. *Totipotent stem cells* are cells that can generate all the cell types of the organism. *Determined stem cells* are cells in which the genetic potential is restricted to a subset of the possible fates; they can produce some, but not all, of the cell types in the organism. For example, determined stem cells of the skin can produce all the cell types within the skin but not those of the heart; the determined stem cells of the liver can produce all liver cell types but not brain cells. The lay press refers to determined stem cells as "adult stem cells," a misnomer because determined stem cells are present in fetal and adult tissue. The determined stem cells give rise to unipotent progenitors, also called committed progenitors, with genetic potential restricted to only one fate. These unipotent progenitors rapidly proliferate into large numbers of cells that then differentiate into mature cells. The stem cells and the unipotent progenitors are the normal counterparts to tumor cells and to immortalized cell lines. Determined stem cells identified to date are small (typical diameters of $6-10\,\mu m$), have a high nucleus to cytoplasmic ratio (blastlike cells), and express certain early genes (e.g., alpha-fetoprotein) and antigens (e.g., CD34, CD117). They have chromatin that binds particular dyes at levels lower than that of the chromatin in mature cells, enabling them to be isolated as "side-pocket" cells using flow cytometric technologies. Stem cells express an enzyme, telomerase, that maintains the telomeres of their chromosomes at constant length, a factor in their ability to divide indefinitely *in vivo* and *ex vivo*. Multiple parameters must be used to permit isolation and purification of any determined stem cell type, since there is no one parameter (e.g., antigen, size, cell density) sufficient to define any determined stem cells. Furthermore, they appear to grow very slowly *in vivo* and may commit to growth and differentiation stochastically. A first-order rate constant for hemopoietic stem cells is about 1/(6 weeks), and their cycling times have been measured by means of time-lapse videography. They commit to differentiation in culture. The first and second doubling take about 60 hrs and then the cycling rate speeds up to about a 24-hr cycling time. By the fifth and sixth doubling they are dividing at a maximal rate of a 12- to 14-hr doubling time.

What Evidence Is There That Stem Cells Exist?

Lethally irradiated mice, which otherwise die from complete hematopoietic failure, can be rescued with as few as 20 selected stem cells. These animals reconstitute the multiple lineages of hematopoiesis as predicted by the stem cell model. Genetically marked mesenchymal stem cells, found in bone marrow, will "in sublethally irradiated animals" give rise to cells in multiple organs over a long time period. These investigations and many others have established conclusively the presence of stem cells, their multilineage potential, and that they persist over long periods of time *in vivo*.

Stem Cell Niches

The field of stem cell niches is a relatively new one in which the localized microenvironment of the stem cell compartment is being defined. It is such a young field that it is likely to change dramatically over the coming years. Still, there are already generalizations that have emerged and are proving to be excellent guides for defining *ex vivo* expansion conditions for the cells.

- Stem cells do not have the enzymatic machinery to generate all their lipid derivatives from single lipid sources and so require complex mixtures of lipids for survival and functioning.
- Calcium concentrations are quite critical in defining whether the cells will expand or undergo differentiation. The mechanisms underlying the phenomenology are poorly understood.
- Specific trace elements such as copper can cause more rapid differentiation of some determined stem cell types. It is unknown whether this applies to all stem cells, and the mechanism(s) is (are) not known.
- Specific mixtures of hormones and growth factors are required, the most common requirements being insulin and transferrin/fe. Addition of other factors can result in expansion of committed progenitors and/or lineage restriction of the stem cells towards specific fates.

The matrix chemistry of known stem cell compartments consists of age-specific and cell-type specific cell adhesion molecules, laminins, embryonic collagens (e.g., type III and IV collagen), hyaluronans, and certain embryonic/fetal proteoglycans. With maturation of the stem cells toward specific cell fates, the matrix chemistry changes in a gradient fashion toward one typical for the mature cells. Although the matrix chemistry of the mature cells is unique for each cell type, a general pattern is that it contains adult-specific cell adhesion molecules, various fibrillar collagens (e.g., type I and II collagen), fibronectins, and adult-specific proteoglycans. A major variant is that in foreskin skin and neuronal cells, in which the mature cells lose expression of collagens, fibronectins, and laminin, the matrix chemistry of the mature cells of these cell types is dominated by the CAMs and the proteoglycans.

Stem cells are dependent upon signals from age- and tissue-specific stroma. The signals from the stroma are only partially defined but include signals such as leukemia inhibitory factor (LIF), various fibroblast growth factors (FGFs), and various interleukins (e.g., IL 8, IL 11).

The most poorly understood of all the signals defining the stem cell niche are those from feedback loops, which are initiated from mature cells, where these signals inhibit stem cell proliferations. Implicit evidence for feedback loops is that cell expansion *ex vivo* requires separation between cells capable of cell division (the diploid subpopulations) and the mature nonproliferating cells.

Which Tissues Have Stem Cells?

For decades it was assumed that stem cell compartments existed only in the rapidly proliferating tissues such as skin, bone marrow, and intestine. Now, there is increasing evidence that essentially all tissues have stem cell compartments, even the central nervous system (Tramontin et al., 2003).

Roles of Stem Cells

The stem cell compartment of a tissue is the ultimate source of cells for turnover and regenerative processes. Stem cell commitment initiates cell replacement and genesis of the tissue and, therefore, tissue repair and maintenance of tissue functions. Stem cell depletion, due to disease or toxic influences (e.g., drugs) eventually leads to partial or complete loss of organ function. Mutational events affecting the stem cells can result in tumors for which both altered and normal stem cells are actively present. Thus, tumors are now considered transformed stem cells, an idea now confirmed by current stem cell biologists (Galli et al., 2004).

7.2.2 Maturational Lineage Biology

All stem cells are pluripotent, giving rise to multiple, distinct lineages of daughter cells that differentiate, stepwise, into all of the mature cells of the tissue. A general model for the production of mature cells from tissue-specific stem cells is shown in Figure 7.9. Determined stem cells (pluripotent) replicate slowly *in vivo* with rates influenced by various systemic signals. Their immediate descendents are committed progenitors (unipotent) capable of rapid proliferation and shown in some tissues (e.g., skin) to be the acute responders to mild to moderate regenerative stimuli. The unipotent progenitors mature, stepwise, through intermediate stages into fully mature cells. Characteristically, various tissue-specific functions are expressed in cells throughout the maturational lineage and in a lineage-dependent fashion. One can

	Stem Cell	Early Progenitor	Late Progenitor	Precursor Cell	Mature Cell
Cell Number	Potential 2^{30}-2^{50} per cell				Need About 10^{16} total over lifetime
Cell Cycling	Very slow ($t_d \sim 1/6$wks.)	Slow ($t_d \sim 60$-100hrs.)	Very rapid ($t_d \sim 12$ hrs.)	Slow	Zero (can be activated in special cases)
Apoptosis	Inactive	Inactive	Very Active (1:5000 survives)	Slow	Inactive (can be induced)
Motility	Zero (except during homing)	Zero	Low	Higher	Function of Physiological State
Regulation	Cell-Cell Contact	Cell-Cell Contact	Soluble Growth Factors	Soluble Growth Factors	Soluble Growth Factors

Figure 7.9   Model for cell production in prolific tissues. This model was derived from decades long research in hematology. The columns represent increasingly differentiated cells while the rows indicate the cellular fate processes (Fig. 17.4) and other events that cells undergo at different states of differentiation (t_d denotes doubling time.)

generalize about these gradual phenotypic changes by categorizing the functions as "early," "intermediate," and "late" taskings. In some cases, a specific gene is expressed uniquely only at a specific stage; in some, there are isoforms of genes that are expressed in a pattern along the maturational lineage; and in others, there are changes in the levels of expression of the gene. Finally, the cells progress to senescence, also called apoptosis, with a phenomenon that typify aging cells. The maturation of the cells is dictated in part by mechanisms inherent in the cells (e.g., changes in the chromatin) and in others by matrix and/or soluble signals in their microenvironment. The microenvironment can include signaling affecting growth, differentiation, or apoptosis. For example, some growth factors are survival factors with anti-apoptotic effects.

Examples of Stem Cell–Fed Maturational Lineages

The stem cell models best characterized are the hematopoietic stem cells (those that generate blood cells) the skin stem cells, and the intestinal stem cells.

Bone Marrow and Blood Cell Formation

Hematopoiesis was the first tissue function for which a stem cell model was established (see Figure 7.3). The reconstitution of the multiple lineages of hematopoietic cells following a stem cell transplant in mice has been demonstrated to occur with a surprisingly small number of cells. Once the cells have reached certain maturational stages, they leave the bone marrow and enter the circulation, where they perform their mature cell functions. The mature cells eventually die and must be replaced. The rate of death of mature cells (by apoptosis or necrosis) sets the need for the cell production rate in the tissue. Ultimately, this death rate determines the number of stem cell commitments that are required. The specific hematopoietic lineage cell production is given in Figure 7.3.

The Villi in the Small Intestine

The lining of the small intestine is comprised of villi that absorb nutrients, as shown schematically in Figure 7.10. The intestinal epithelial cell layer is highly dynamic. Its cellular content turns over approximately every 5 days, and it is the body's second most prolific tissue. In between the villi are tube-shaped epithelial infoldings. An infolding is know as a cryptus. All intestinal epithelial cell production takes place in specific positions within the crypt. Once the cells are mature, they migrate to the outer edge of the crypt. Then they move over a period of about 5 days, from the base of villus to the top, where they die and slough off. During this passage the cells carry out their organ-specific function as mature parenchymal cells. They function in the absorption and digestion of nutrients that come from the lumen of the gut. Towards the bottom of the crypt is a ring of slowly dividing determined stem cells. The number of stem cells per crypt is about 20. After division, the daughter cell moves up the crypt where it becomes a rapidly cycling progenitor cell, with a cycling time on the order of 12 hrs. The cells that are produced move up the crypt and differentiate. Once they leave the crypt, they are mature and enter the base of the villi.

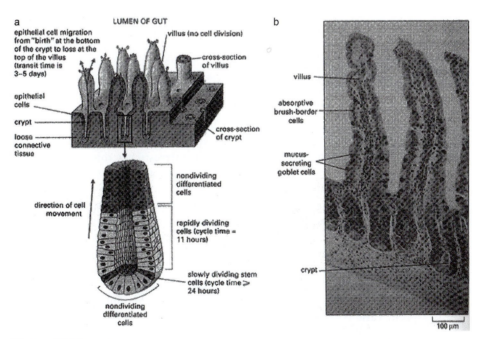

Figure 7.10 Villi in the small intestine. (a) A schematic showing the villi and the crypt indicating the mitotic state of the cells in various loactions. (b) Rows of villi of epithelial intestinal cells (the diameter of a villi is about 80 μm) (from Alberts et al., 1994).

Skin

Human skin has two principal cell layers, an epidermis and dermis, separated by a form of extracellular matrix called a basal lamina (Fig. 7.4). In this extracellular matrix, collagen VII is an important component. The two layers undulate with respect to each other, the undulations producing deep pockets (distant from the skin surface) and other pockets that are more shallow. The relationship pattern between layers is unique in each type of skin (e.g., that on the trunk of the body vs. that in the palms of the hand vs. that on the face). The dermis is a connective tissue layer under the basal lamina and comprised primarily of fibroblasts (also called stroma). The epidermis consists of multiple layers of epithelia comprised of differentiating keratinocytes with the least differentiated cells located at the basal lamina. Thin skin has a squamous columnar (each column about $30\,\mu m$ in diameter) organization. A stem cell compartment has been identified in the deep pockets within the epidermis/dermis undulations and also in bulges near hair follicles. The stem cells produce committed progenitors, also called transit amplifying cells, which migrate into the shallow pockets and from there into cells that line the entire basal lamina. Only the epidermal cells adherent to the basal lamina are cycling, whereas cells that lose their attachment to the basal lamina move upward and differentiate into succeeding stages of cells. Ultimately, they turn into granular cells, then into keratinized squames that eventually flake off. Keratins are proteins defining the differentiated cells of the skin and are evident in myriad forms: from those that provide skin mechanical protection to those that are present in body hair (or in feathers in birds or scales in other vertebrates). The net proliferative rate of skin depends on the region of the body. In particular, the turnover of skin is on the order of a few weeks, making it the body's third most prolific tissue. It is estimated that about half of dust found in houses originates from human skin.

Mesenchymal Stem Cells (MSC)

The existence of a cell population that has the potential to differentiate into a number of connective tissue cell types has been demonstrated. The developmental hierarchy is illustrated in Figure 7.11. Interestingly, this cell population is present in the bone marrow. The existence of a multilineage potential MSC is not fully proven yet, as culture experiments from single cells have not shown the ability to produce multiple mature cell types. The mature cell types that can be derived from a purified population of putative MSC include osteocytes (bone), chondrocytes (cartilage), myoblasts (muscle), fibroblasts (tendon), and possibly adipocytes. The creation of a transgenic mouse carrying an assayable collagen gene has been used to show that multiple tissue types can be obtained from a subpopulation that resides in bone marrow. The marrow from the transgenic mouse was transplanted into a sublethally radiated mouse and the presence of the collagen mini-gene was monitored over time in various tissues. The results show that there is a subpopulation present in bone marrow that has the capability to produce a number of mature cell types in different organs. These observations support, but do not prove, the existence of a multipotent MSC.

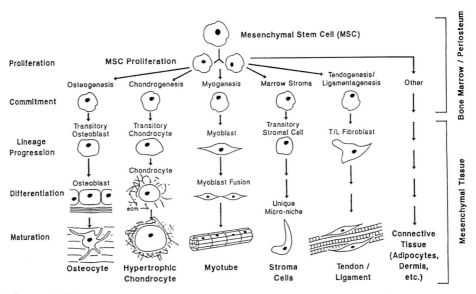

Figure 7.11   The differentiation of mesenchymal stem cells into different mesenchymal cell types (from Bruder et al., 1994).

The Liver

The liver is being recognized increasingly as a maturational lineage system, including a stem cell compartment, analogous to those in the bone marrow, skin, and gut. The liver's lineage is organized physically within the acinus, the structural and functional unit of the liver. In cross section, the acinus is organized like a wheel around two distinct vascular beds; six sets of portal triads, each with a portal venule, hepatic arteriole, and a bile duct, form the periphery, and the central vein forms the hub. The parenchyma, effectively the "spokes" of the wheel, consist of plates of cells lined on both sides by the fenestrated sinusoidal endothelium. By convention, the liver is demarcated into three zones: zone 1 is periportal, zone 2 is midacinar, and zone 3 is pericentral. Blood flows from the portal venules and hepatic arterioles at the portal triads, through sinusoids which line plates of parenchyma, to the terminal hepatic venules, and into the central vein. The stem cell compartment is present around the portal triads (zone 1), and identified in anatomical entities called Canals of Hering. The stem cells are bipotent and produce daughter cells that become either biliary cells (bile duct epithelia) or hepatocytes. Hepatocytes display marked morphologic, biochemical, and functional heterogeneity based on their zonal location (it is assumed that the same occurs for the biliary cells, but they have not been as well characterized as hepatocytes). The size of hepatocytes increases from zone 1 to zone 3, and one can observe distinctive zonal variations in morphological features of the cells such as mitochondria, endoplasmic reticulum, and glycogen granules. Hepatocytes show dramatic differences in DNA content from zone 1 to zone 3 with periportal cells being diploid and with a gradual shift to polyploid cells midacinar to pericentrally.

Octaploid cells in the pericentral zone show evidence of apoptosis. Adult rodent livers (rats and mice) are 90–95% polyploid; adult human livers are 40–50% polyploid, whereas fetal and neonatal liver cells are entirely diploid. The transition to the adult ploidy pattern is observed by 3 to 4 weeks of age in rats and mice and by late teenage years in humans. With age, the liver becomes increasingly polyploid in all mammalian species surveyed. This may help to explain the reduction in regenerative capacity of the liver with age.

Representative of the zonation of liver functions is the cell division potential of parenchymal cells *in vitro* and *in vivo* in which the maximum is observed in the diploid periportal cells and negligible cell division is observed pericentrally in the polyploid cells (Table 7.3). Only the diploid parenchymal cell subpopulations are capable of undergoing complete cell division, where these comprise the subpopulations of stem cells and unipotent progenitors ($< 15\,\mu m$ in diameter) and the diploid adult hepatocytes (or "small hepatocytes") having an average diameter of 18–$22\,\mu m$. Moreover, there remains a difference in cell division potential between these two diploid subpopulations: a single small hepatocyte will yield 120 daughter cells in a 20-day time period, whereas a single stem/progenitor cell in the same time period and under the same conditions will yield 4000–5000 daughter cells. Mature, polyploid cells can undergo DNA synthesis but have limited, if any, cytokinesis under even the most optimal expansion conditions in culture due to downregulation of factors regulating cytokinesis. The only subpopulations that undergo complete cell division are those that are diploid; the polyploid cells can be triggered by various stimuli to undergo DNA synthesis but with limited, if any, cytokinesis. Thus, observations of nuclei by standard "proliferation" assays (e.g., data such as BUDR, Ki67, H^3-thymidine) can be evidence for either hyperplastic (complete cell division) or hypertrophic (DNA synthesis but no cytokinesis) growth. Reviews of this system have been presented by a number of investigators in recent years.

TABLE 7.3 Cell and Culture Parameters with Respect to Species and Acinar Zonal Locations

Parameter	Species	Zone 1	Zone 2	Zone 3
Ploidy[**]	Rats	2N	4N	4N and 8N
	Mice	2N and 4N	4N and 8 N	Up to 32 N
	Humans	2N	2N	2N and 4N
Growth		Maximal	Intermediate	Negligible
Extracellular Matrix Composition		Type IV and III collagen, laminin, hyaluronans, HS-PG	Gradient from zone 1 to 3 ➝	Type I and III collagen, fibronectins, HP-PGs
Gene Expression		Early	Intermediate	Late

In all mammalian species evaluated, the diameters of the adult (mature) hepatocytes depend on ploidy. The cells that are 2N are 18–$22\,\mu m$; those that are 4N are 22–$35\,\mu m$; those that are 8N and above can reach up to $75\,\mu m$ in diameter.

7.2.3 Models for Stem Cell Proliferative Behavior

It should be clear by now that the replication functions of stem cells are critical to tissue function and tissue engineering. How do stem cells divide and what happens when they divide? There are three models that describe the dynamic behavior of the stem cell population.

The Clonal Succession Concept

Cellular systems are maintained by a reservoir of cells that either grow very slowly or may be in a dormant state. In these cases, the reservoir of cells are available throughout the tissues lifespan and can be challenged to enter a complex process of cell proliferation and differentiation. Once triggered, such a stem cell would give rise to a large clone of mature cells. Any one of these clones would have a limited life-span as feedback signals dissipate the need to maintain cell production fluxes. After time, such a clone will burn out and, if needed, a new stem cell clone would take over the cell production role.

Deterministic Self-Maintenance and Self-Renewal

This model relies on an assumption that stem cells can self replicate. Following a stem cell division, there is a 50% probability that one of the daughter cells maintains the stem cell characteristics, while the other undergoes differentiation. The probability of self-renewal is regulated and may not be exactly 50% depending on the dynamic state of the tissue.

Stochastic Models

This model considers that the progeny of a stem cell division can generate zero, one, or two stem cells as daughter cells (notice that the clonal succession model assumes zero, and the deterministic model one). The assumption is that each of the three outcomes has a particular probability.

7.2.4 Stem Cells and Tissue Engineering

Stem cells build tissues. Stem cells are the source for cells in tissues. Thus, stem cells build and maintain tissues *in vivo* and can be used to generate tissue *ex vivo*, as in bioartificial organs.

Ex vivo growth and manipulation of stem cells. Being able to grow stem cells *ex vivo*, that is to have them divide without differentiation, would theoretically provide an inexhaustible source of stem cells. However, *ex vivo* conditions for truly self-replicative expansion have not been fully achieved for any stem cell family, though there have been considerable improvements towards this goal.

Isolation of stem cells for scientific and clinical purposes. Methods have been developed to isolate stem cells or enrich the stem cell content of a cell population. Since at present, no stem cell type can be purified with a single parameter, the most effective protocols make use of multiparametric isolation strategies. These strategies may comprise immunoselection for cells with specific antigenic profiles that are in combination with cell selection methods such as diameter, cell density, and levels of "granularity" (the extent of cytoplasmic particles such as mitochondria). The published protocols include flow cytometry and/or immunoselection with magnetic columns, affinity columns, and counterflow elutriation.

Proof that one has identified stem cells can be done either *ex vivo* or *in vivo*. The *ex vivo* assays are done with clonogenic expansion assays in which a single cell is expanded in cultures under precise conditions and then observed to give rise to daughter cells of more than one fate. The *in vivo* assays are ones in which very small numbers of candidate stem cells are assessed for their ability to reconstitute damaged tissues.

Stem cells are rare. Hematopoietic stem cells are believed to be 1 in 100,000 cells in human bone marrow and 1 in 10,000 in mouse bone marrow.

7.2.5   Stem Cell Aging

Telomerases, DNA Stability, and Natural Cell Senescence

When linear DNA is replicated, the lagging strand is synthesized discontinuously through the formation of the so-called Okazaki fragments. The last fragment cannot be initiated, and therefore the lagging strand will be shorter than the leading strand. Linear chromosomes have noncoding repeating sequences on their ends, which are called telomeres. These telomeres can be rebuilt using an enzyme called telomerase. Telomerase is a ribonucleoprotein DNA polymerase that elongates telomeres. When expressed, telomerase maintains the telomere length in growing cells. The telomere hypothesis implicates short telomere length and telomerase activation as critical players in cellular immortalization. This enzyme is active in microorganisms such as yeast, in stem cells, and in the transformed derivatives of the stem cell (i.e., tumor cells). Commitment of stem cells toward their unipotent descendents results in the loss of telomerase activity. Thus, normal somatic cells lack this activity, and the telomeres are shortened by about 50 to 200 bp per replication. This shortening gives rise to the so-called mitotic clock. The length of the telomeres is about 9 to 11 kbp and when it reaches about 5 to 7 kbp the chromosomes become unstable and replication ceases (Figure 7.12). This mechanism is believed to underlie the Hayflick limit.

Telomerase activity is found in somatic hematopoietic cells, but at a low activity level. There is evidence that telomeres in immature hematopoietic cells do shorten with ontogeny and with increased cell doublings *in vitro*. The rate of telomere shortening in stem cells is finite, but may be slower than in other somatic cells. Numerical evaluation of the consequences of stem cell aging strongly suggest that there has to be some form of self-renewal of stem cells in adults (see Exercise 1).

7.2.6   Tissue Dynamics

Tissues are comprised of many cell types of various developmental origins (Figure 7.13). The dynamic behavior of cells and their interactions determine overall

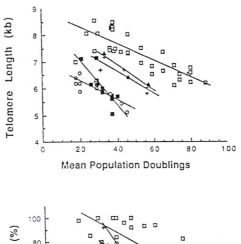

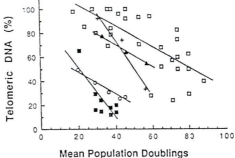

Figure 7.12   Primary experimental data showing the shortening of telomere length with increasing cellular doubling in cell culture (from Harley et al., 1990).

tissue formation, state, and function. The activities of individual cells are often substantial. However, the time scales that relate cellular activities with tissue function are relatively long, and therefore their importance tends to be overlooked. The processes at a cellular level that underlie dynamic states of tissue function fall into the following categories:

1. *Hyperplasia:* Cell replication or proliferation resulting in an increase in cell number due to complete cell division (both DNA synthesis and cytokinesis)
2. *Hypertrophy:* DNA synthesis not accompanied by cytokinesis and resulting in an increase in the cell volume
3. *Cell maturation and differentiation:* Changes in gene expression, and the acquisition of particular functions with the changes occurring in a maturationally lineage-dependent pattern
4. *Cell apoptosis:* Aging cells undergoing senescence, a process distinguishable from necrosis

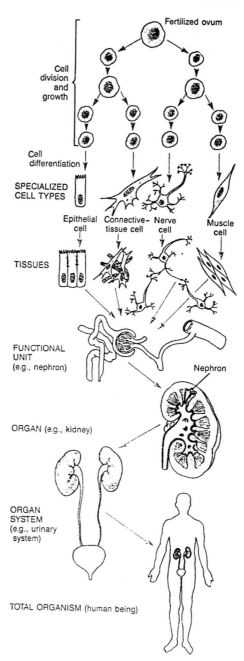

Figure 7.13   Levels of cellular organization in tissues and the diverse developmental origins of cells found in tissues (from Vander et al., 1994).

5. *Cell adhesion:* The physical binding of a cell to its immediate environment, which may be a neighboring cell, extracellular matrix, or an artificial surface
6. *Cell motility:* The motion of a cell into a particular niche, or location

These processes are illustrated in Figure 7.14. What is known about each one of these processes will be briefly described in the following sections, with particular emphasis on quantitative and dynamic descriptions. The processes contribute to three dynamic states at a tissue level:

1. *Histogenesis:* The maturational lineages of cells are derived from a tissue's stem cell compartment. A tissue's overall functions are the net sum of contributions from all the cells within the maturational lineages of a tissue.
2. *Tissue formation:* The formation of tissue has been characterized by studies comprising the field of developmental biology. Tissues vary in their proportion of stem cells, diploid cells, and polyploid cells depending on age. The tissue, when isolated from young donors (e.g., infants) will have tissues with maturational lineages skewed toward the young cells in the lineage (stem cells, diploid somatic cells), whereas those from geriatric donors will have tissues skewed towards the later stages of the lineage (polyploid cells). This phenomenon is the explanation for why tissues procured from young donors have greater expansion potential *ex vivo* and are probably going to have greater potential for most forms of cell therapy programs.
3. *Tissue repair:* Repair of damaged tissues involves production of cells from the stem cell compartment, proliferation of the cells, and their differentiation into the fully mature cells. The repair process can also involve migration of cells, if need be, to a site of damage.

The dynamic processes just described involve interplay among many different cell types. The cells communicate and coordinate their efforts through the principal

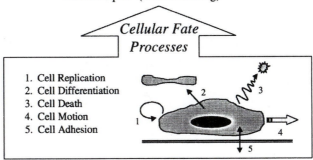

Figure 7.14 Tissue dynamics. The three dynamic states of tissues and the underlying cellular fate processes.

cellular processes (Fig. 7.14). The biology and dynamics of these processes are discussed in detail in subsequent sections.

Tissue Histogenesis

All tissues are dynamic. For instance, tissue dynamics can be illustrated by comparing the cell numbers that some organs produce over a lifetime to the total number of cells in the human body. As stated in Section 7.1, the human bone marrow produces about 400 billion myeloid cells daily in a homeostatic state. Over a 70-year lifetime, the cell production from bone marrow accumulates to a staggering 10^{16} cells. This cell number is several hundred times greater than the total number of cells that are in the body at any given time. Similarly, the intestinal epithelium, the body's second most prolific tissue produces about 5×10^{14} cells over a lifetime—ten times the number of total number of cells in the human body.

Tissues have their own characteristic turnover rates (Table 7.4). The lining of the intestine and myelopoiesis have ranges of 1–2 days, followed by erythropoiesis and hematopoiesis at 2.5 days, and then the epidermis at 7 days. The turnover rate in quiescent tissues is on the order of months to even years. For example, the turnover of the liver of rodents is estimated to be about one year; the turnover rate in the livers of humans is not known. The tissue of the central nervous system is hypothesized to be even slower, but is indeed occurring, with its stem cell compartment now identified to be lining the ventricles of the brain. (Gage, 1994).

Tissue Genesis

The preceding overview is of steady-state tissues focused on stem cell compartments and maturing lineages of cells. That steady state is achieved gradually during embryonic development and the process of organogenesis, which is quite complicated as exemplified by hematopoiesis or the formation of blood cells. During vertebrate

TABLE 7.4 Cell Renewal Rates in Tissues

Tissue	Species	Turnover Time (days)
Erythropoiesis	Rat	2.5
Myelopoiesis	Rat	1.4
Hematopoiesis	Human	2.5
Small intestinal epithelium	Human	4–6
	Rat	1–2
Epidermis	Human	7–100
Corneal epithelium	Human	7
Lymphatic cells	Rat (thymus)	7
	Rat (spleen)	15
Epithelial cells	Rat (vagina)	3.9
	Human (cervix)	5.7
Spermatogonia	Human	74
Renal interstitial cells	Mouse	165
Hepatic cells	Rat	400–500

ontogeny, hematopoiesis sequentially occupies the yolk sac, fetal liver, spleen, and bone marrow. Variations in this pattern exist among vertebrate species. The earliest identification of hematopoietic cells is their assignment to the progeny of the C4 blastomere in the 32-cell embryo. The blastula grows to about 1000 cells (ten doublings) and assumes a spherical shape. Then, the blastula undergoes gastrulation—not unlike pushing a finger into an inflated balloon. The point of invagination is the endoderm (the vegetal pole) that eventually forms the gut. After gastrulation, the ectoderm is brought into a position with the endoderm and a third germ layer is formed between the two: the mesoderm. This middle layer is called the marginal zone and is formed via cell–cell interactions and soluble growth factor action. Several determined stem cells originate from the mesoderm, including hematopoietic tissue, mesenchymal tissue, muscle, kidney, and notochord. Blood cells originate from the ventral mesoderm. Some hematopoietic cells migrate into the yolk sac to form blood islands—mostly erythroid cells ("primitive" hematopoiesis). Intra-embryonic hematopoiesis originates from the aortic region in the embryo and leads to "definitive" hematopoiesis. It appears that the embryonic origin of hematopoietic cells is from bipotent cells that give rise to both the vasculature (the endothelium) and hematopoietic cells. Hematopoietic stem cells are then found in the liver in the fetus. Around birth, the hematopoietic stem cells migrate from the liver into the bone marrow where they reside during postnatal life. Interestingly, the umbilical cord blood contains hematopoietic stem cells capable of engrafting pedriatic, juvenile, and small adult patients. This developmental process illustrates the asymmetric nature of stem cell division during development and increasing restriction in developmental potential. Furthermore, the migration of stem cells during development is important. Understanding the regulatory and dynamic characteristics of the stem cell fate processes is very important to tissue engineering.

Tissue Repair

When tissue is injured, a healing response is induced. The wound healing process is comprised of a coordinated series of cellular events. These events vary with ontological age. Fetal wound healing proceeds rapidly and leads to the restoration of scarless tissue. In contrast, postnatal healing is slower and often leads to scarring, which generally permits satisfactory tissue restoration, though not always fully restoring normal tissue structure. Some pathological states resemble wound healing. A variety of fibrotic diseases involve processes similar to tissue repair and subsequent scarring. The increasing appreciation of stem cell compartments and their descendent maturational lineages and the changes that occur with them with age are likely to provide improved understanding of wound healing phenomena.

The Sequence of Events That Underlie Wound Healing

Immediately following injury, control of bleeding starts with the rapid adhesion of circulating platelets to the site of damage. Within seconds, the platelets are activated, secrete contents from their storage granules, spread, and recruit more platelets to the thrombus that has started to develop. Within minutes of injury, the extent of hemorrhaging is contained through the constriction of surrounding blood vessels.

The next phase of the wound healing process involves the release of agents from the platelets at the injured site that cause vasodilatation and increased permeability of neighboring blood vessels. The clotting cascade is initiated and results in the cleavage of fibrinogen by thrombin to form a fibrin plug. The fibrin plug, along with fibronectin, holds the tissue together and forms a provisional matrix. This matrix plays a role in the early recruitment of inflammatory cells and later in the migration of fibroblasts and other accessory cells.

Inflammatory cells now migrate into the injured site. Neutrophils migrate from circulating blood and arrive early on the scene. As the neutrophils degranulate and die, the abundance of macrophages at the site increases. All tissues have resident macrophages, and their number at the injury site is enhanced by macrophages migrating from circulation. They act in concert with the neutrophils to phagocytose cellular debris, combat any invading microorganisms, and provide the source of chemoattractants and mitogens. These factors induce the migration of endothelial cells and fibroblasts to the wound site and stimulate their subsequent proliferation. If the infiltration of macrophages into the wound site is prevented, the healing process is severely impaired.

At this time, so-called granulation tissue has formed. It is comprised of a dense population of fibroblasts, macrophages, and developing vasculature that is embedded in a matrix that is comprised mainly of fibronectin, collagen, and hyaluronic acid. The invading fibroblasts begin to produce collagen, mostly type I and III. The collagen increases the tensile strength of the wound. Myofibroblasts contract at this time, shrinking the size of the wound by pulling the wound margins together.

The matrix undergoes remodeling which involves the coordinated synthesis and degradation of connective tissue protein. Remodeling leads to a change in the composition of the matrix with time. For instance, collagen type III is abundant early on, but gives way to collagen type I with time. The balance of these processes determines scar formation. Although the wound appears healed at this time, chemical and structural changes continue to occur within the wound site. The final step of the wound healing process is the resolution of the scar. The formation and degradation of matrix components returns to its normal state in a process that may take many months. Finally, the composition of the matrix and the spatial location of the cells have returned to close to the original state.

Understanding the wound healing process is important to the tissue engineer since the placement of disaggregated tissues in *ex vivo* culture induces responses reminiscent of the wound healing process.

7.2.7   Cell Differentiation

The coordinated activity of the cellular fate processes determines the dynamic state of tissue function (Fig. 7.14). There is growing information available about these processes in genetic, biochemical, and kinetic terms. The dynamics considerations that arise from the interplay of the major cellular fate processes are introduced at the end of the chapter, and the associated bioengineering challenges are described.

Describing Cell Differentiation Biologically

Differentiation is a process by which a cell undergoes phenotypic changes to an overtly specialized cell type. The specialized cell type then carries out the physiological function for which it is designed. This process begins with a lineage and differentiation commitment and is followed by a coordinated series of gene expression events.

The term differentiation is derived from differential gene expression. Differentiation involves a change in the set of genes that are expressed in the cell, and this change is usually an irreversible change towards a particular functional state. This process involves a carefully orchestrated switching "off" and "on" of gene families. The final genes expressed are those that pertain to the function of the mature cell. In animals, the process of differentiation is irreversible.

Example Problem 7.1

Production of red blood cells: How long does it take to produce a red blood cell?

Solution

Erythropoiesis replaces decaying mature red blood cells. About 200 billion need to be produced daily in a human adult. Many of the changes that a cell undergoes during the process of erythropoiesis are well known (Fig. 7.15). The cell size, the rates of RNA and DNA synthesis, and protein content all change in a progressive and coordinated fashion. The differentiation from a pro-normoblast (earlier precursor stage) to a fully mature enucleated erythrocyte takes about 180 hrs, or about one week. The replication activity is the highest at the preprogenitor and progenitor stage, but once the precursor stage is reached, replication activity ceases sharply. ■

Experimental Observations of Differentiation

The process of differentiation can be observed directly using fluorescent surface markers and/or light microscopy for morphological observation. Ultrastructural changes are also used to define the stage of differentiation.

A flow cytometer is often used to monitor the process of cellular differentiation. The basis for this approach is the fact that characteristic surface proteins are found on cells at different stages of differentiation. These surface markers can be used as binding sites for fluorescently conjugated monoclonal antibodies. The flow cytometer can be used to trace the expression of several surface markers.

Erythropoiesis can be traced based on expression of the transferrin receptor (CD71) and glycophorin A. The latter is an erythroid-specific surface protein that is highly negatively charged and serves to prevent red cell aggregation in dense red cell suspensions. The transferrin receptor plays a critical role during the stages in which iron is sequestered in hemoglobin. The measurement of this process is shown in Figure 7.16.

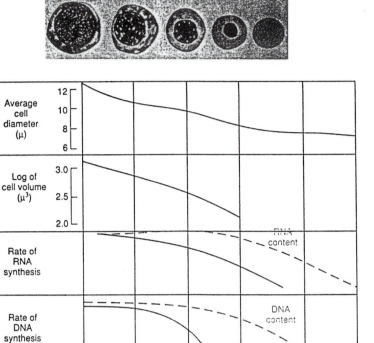

Figure 7.15   Erythroid maturation sequence (from Granick and Levere, 1964).

Describing the Kinetics of Cell Differentiation

The process of differentiation is a slow one, often taking days or weeks to complete. The kinetics of this complex process can be described mathematically using two different approaches.

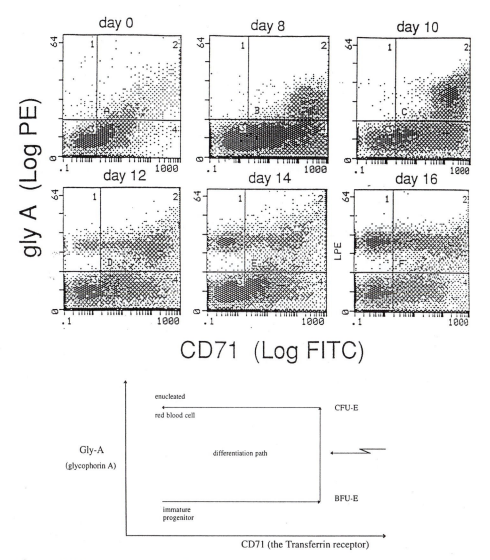

Figure 7.16 Two-parameter definition of erythropoietic differentiation. Glycophorin A is found on erythroid cells post the blast-forming unit-erythroid (BFU-E) stage, whereas transferrin (CD71) is expressed at the progenitor stage [BFU-E and colony-forming unit-erythroid (CFU-E)]. By measuring the two simultaneously using a flow cytometer, this differentiation process can be traced as a U-shaped path on a bivariate dot plot (from Rogers et al., 1995).

Compartmental Models

The traditional approach to describing cell growth and differentiation is to use compartmental models. The differentiation process involves a series of changes in cell phenotype and morphology, typically becoming more pronounced at the latter stages of the process:

$$X_0 \rightarrow X_1 \rightarrow X_2 \rightarrow \ldots\ldots\ldots X_i \rightarrow \ldots\ldots X_n \rightarrow turnover \qquad (7.1)$$

where n can be as high as 16 to 18.

In the use of compartmental models in the past, the transition from one stage to the next was assumed to represent cell division. Thus, these models couple differentiation with cell replication. Mathematically, this model is described by a set of ordinary differential equations, as

$$\frac{dX_i}{dt} = 2k_{i-1}X_{i-1} - k_iX_i \qquad (7.2)$$

The transition rate from one stage to the next is proportional to the number of cells present at that stage. This transition rate can clearly be a function of growth factor concentration and a number of other variables.

Differentiation as a Continuous Process

An alternative view is to consider the differentiation process to be a continuous process. Once the commitment to differentiation has been made, the differentiation process proceeds at a fixed rate. This viewpoint leads to a mathematical description in the form of first-order partial differential equations:

$$\frac{dX}{dt} + \delta\frac{dX}{da} = [\mu(a) - \alpha(a)\,]X \qquad (7.3)$$

where δ is the rate of differentiation and a is a parameter that measures the differentiation state of the cell. μ and α are the growth and death rates, respectively, and vary between zero and unity. Both μ and α can be a function of a.

Cell Motion

The Biological Roles of Cell Migration

Cell migration plays an important role in all physiological functions of tissues and also some pathological processes. Cell migration is important during organogenesis and embryonic development. It plays a role in the tissue repair response in both wound healing and angiogenesis. The immune system relies on cell migration, and pathological situations, such as cancer metastasis, are characterized by cell motility. Cell migration represents an integrated molecular process.

Animal cells exhibit dynamic surface extensions when they migrate or change shape. Such extensions, called lamellipodia and filopodia, are capable of dynamic formation and retraction. Local actin polymerization at the plasma membrane is a significant process in the generation of these structures. These extensions in turn rely on a complex underlying process of multiprotein interactions. In neurites, filopodia are believed to play a role in the progression of elongation by aiding the assembly of microtubules which are a significant component of these cells. The filopodia in neurites extend from the lamellipodial region and act as radial sensors. Filopodia on neurites have been found to be crucial to growth cone navigation. The filopodia on neurite growth cones also have been found to carry receptors for certain cell adhesion molecules. Mature leukocytes can extend cytoplasmic extensions. Recently, structures termed uropods have been found on T lymphocytes. These cytoplasmic projections form during lymphocyte–endothelial interaction. There is a redistribution of adhesion molecules, including ICAM-1, ICAM-3, CD43, and CD44, to this structure. T cells have been found to use the uropods to contact and communicate directly with other T cells. Uropod development was promoted by physiologic factors such as chemokines. Cytoplasmic extensions, therefore, can perform a spectrum of functions in different cells that are related to migration and communication.

Describing Cell Motion Kinetically

Whole populations. The motion of whole, nonreplicating cell populations is described with

$$\frac{dX}{dx} = J \tag{7.4}$$

where J is the flux vector of the system boundary (cells/distance/time in a two-dimensional system), X is the cell number, and x is the flux dimension. The cellular fluxes are then related to cellular concentration and chemokine concentrations with

$$J = \text{random motility} + \text{chemokinesis} + \text{chemotaxis}$$

$$= \sigma \frac{dX}{dx} - \left(\frac{X}{2}\right)\left(\frac{d\sigma}{da}\right)\left(\frac{da}{dx}\right) + \chi \frac{Xda}{dx} \tag{7.5}$$

where σ is the random motility coefficient, χ is the chemotactic coefficient, and a is the concentration of a chemoattractant. The first term is similar to a diffusion term in mass transfer and is a measure of the dispersion of the cell population. The chemokinesis term is a measure of the changes in cell speed with concentration and is normally negligible.

If $\chi = 0$ and chemokinesis is negligible, then the motion of the cells is like random walk and is formally analogous to the process of molecular diffusion. The biased movement of the cells directly attributable to concentration gradients is given by the chemotaxis term, which is akin to convective mass transfer.

As just described, motile cells appear to undergo a random walk process that mathematically can be described in a fashion similar to the diffusion process. Unlike the diffusion process, the motion of each moving entity can be directly determined (Fig. 7.17). Thus, the motility characteristics of migrating cells can be measured on an individual cell basis.

Such a description is given in terms of the cell speed (s), the persistence time (p), which is the length of time that the cell moves without changing its direction, and the orientation bias (θ), which is due to action of chemoattractants. The random motility coefficient (σ) is related to these parameters as:

$$\sigma = s^2 p \tag{7.6}$$

Typical values for these parameters are given in Table 7.5.

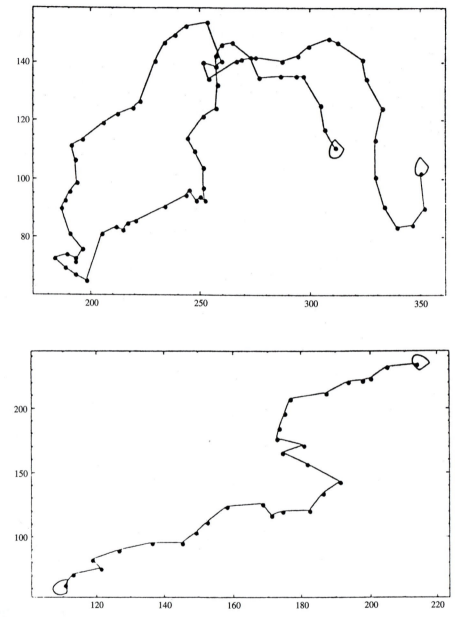

Figure 7.17   Experimental paths of individual neutrophil leukocytes undergoing random motility in uniform environments (from Lauffenburger and Linderman, 1993).

TABLE 7.5 Random Motion—Measured Cell Speeds and Persistence Times

Cell Type	Speed	Persistence Time
Rabbit neutrophils	$20\,\mu m/min$	4 min
Rat alveolar macrophages	$2\,\mu m/min$	30 min
Mouse fibroblasts	$30\,\mu m/h$	1 h
Human microvessel endothelial cells	25–$30\,\mu m/h$	4–5 h

Cell Replication

The Cell Cycle

The process of cell division is increasingly well known and understood in terms of molecular mechanisms (Fig. 7.18). The cycle is driven by a series of regulatory proteins known as cyclin-dependent kinases. The cyclins exist in phosphorylated and dephosphorylated states, basically a binary system. The network goes through a well-orchestrated series of switches, during which the cyclins are turned off and on. When in an "on" state, they serve to drive the biochemical processes that are needed during that part of the cell cycle. The sequence of events that describes the mammalian cell cycle has been the subject of many reviews. It should be noted that there are several important decisions associated with moving "in and out of cycle," namely to move a cell from a quiescent state (the so-called G_0) to a cycling state and vice versa. The cell cycle is such that the time duration of the S, G_2, and M phases is relatively fixed. Once a cell determines it needs to and can divide, it initiates DNA synthesis and subsequent cell division. This process has the overall characteristics of a zero-order

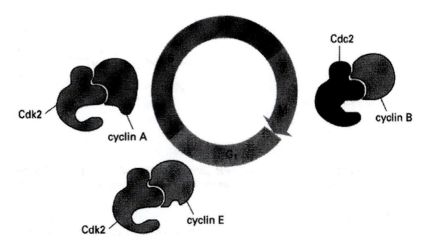

Figure 7.18 Schematic representation of the eukarotic cell cycle (G_1–S–G_2–M) and the presence of cyclin-dependent kinases.

kinetic process. Once initiated, it proceeds at a certain rate. The minimal cycling time of human cells is about 12 hrs. Progenitor cells can cycle at this rate.

Describing the Cell Cycle Dynamics

The dynamics of growth can be described in different ways.

1. *Exponential growth:* If growth is unconstrained, the rate of formation of new cells is simply proportional to the number of cells present:

$$\frac{dX}{dt} = \mu X \quad X(t = 0) = X_0 \Rightarrow X(t) = X_0 \exp(\mu t) \qquad (7.7)$$

   and exponential growth results. The growth rate μ is equal to $\ln(2)/t_d$, where t_d is the doubling time.

2. *Age–time structured descriptions:* If the phases of the cell cycle are to be described, then the cell cycle status of the cell needs to be incorporated in the dynamic description. This leads to first-order partial differential equations in time and cell cycle status:

$$\frac{dX}{dt} + v\frac{dX}{da} = \alpha(a)x \qquad (7.8)$$

   where v is the rate at which the cell moves through the cell cycle, a is a variable that describes the cell cycle status ($a = 0$ newborn cell, $a = 1$ cell completing mitosis), and α is the death rate of the cell that can be cell cycle dependent. This population balance equation can be solved under the appropriate initial and boundary conditions.

3. *Molecular mechanisms:* The cascade of cyclin-dependent kinases that constitute the molecular mechanism for cell cycling have been unraveled. Based on this knowledge, it is possible to describe the cell cycle in terms of the underlying molecular determinants. Such models have many components and need to be solved using a computer.

Interacting Cellular Fate Processes Determine Overall Tissue Dynamics

The differentiation process involves a series of changes in cell phenotype and morphology that typically become more pronounced at the latter stages of the process. The key event is milieu-dependent differentiation (or differential gene expression), which basically is the organogenic process to yield mature cells of a certain type and functionality. Similarly, embryonic induction can be described as a series of such events. Branching in this process will be described in the following paragraphs. This process is schematically presented in Equation 7.1.

This progression of changes is typically coupled to fundamental "driving forces" such as cell cycling (mitosis) and cell death (apoptosis). Thus, the basic cellular processes can be accounted for in a population balance on each stage of differentiation:

$$\Delta \text{s in cell} = \text{entry by input} - \text{differentiation} - \text{exit by apoptosis} + \text{entry by cell division}$$

or in mathematical terms:

$$\frac{dX}{dt} = I - \delta X - \alpha X + \mu X$$
$$= I - (\delta + \alpha - \mu)X \tag{7.9}$$

where δ, α, and μ are the rates of differentiation, apoptosis, and replication, respectively. This equation can be rewritten as:

$$\frac{dX}{dt} + kX = I \tag{7.10}$$

where $k = \delta + \alpha - \mu$ and $t = 1/k$. The parameter k is the reciprocal of the time constant t that characterizes the dynamics of changes in the number of cells of type X. It is therefore evident that the ratios of μ/δ and α/μ are the key dimensionless groups that determine the overall cell production in tissue.

7.2.8 Cellular Communications

How Do Cells Communicate?

Cells in tissues communicate with one another in three principal ways (Fig. 7.19).

1. They secrete soluble signals, known as cyto- and chemokines.
2. They touch each other and communicate via direct cell–cell contact.
3. They make proteins that alter the chemical microenvironment (ECM).

All these means of cellular communication differ in terms of their characteristic time and length scales and in terms of their specificity. Consequently, each is suitable to convey a particular type of a message.

Soluble Growth Factors

Growth factors are small proteins that are on the order of 15 to 20 kDa in size (one Da is the weight of the hydrogen atom). They are relatively chemically stable and have long half-lives unless they are specifically degraded. Initially, growth factors were discovered as active factors that originated in biological fluids. For instance, erythropoietin was first isolated from urine and the colony stimulating factors from conditioned medium from particular cell lines. The protein could be subsequently purified and characterized. With the advent of DNA technology, growth factors can be cloned directly as ligands for known receptors. This procedure was used to isolate thrombopoietin as the *c-mpl* ligand and the stem cell factor as the *c-kit* ligand. Growth factors are produced by a signaling cell and secreted to reach a target cell.

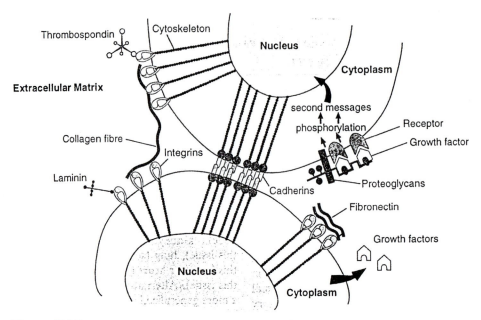

Figure 7.19   Cell and extracellular matrix (ECM) protein interactions (from Mutsaers et al., 1997).

Example Problem 7.2

What is the maximal secretion rate of protein?

Solution

The maximal secretion rate of a protein from a single gene in a cell can be estimated based on the maximal rates of transcription and translation. Such estimates have been carried out for the production of immunoglobulins (MW = 150 kDa) whose genes are under the control of strong promoters. The estimate shows that the maximal secretion rate is on the order of 2000 to 8000 protein molecules per cell per second, which corresponds to about 1 pg per cell per hour. This estimate compares favorably with measured maximal secretion rates. Since growth factors tend to be about one-tenth the size of immunoglobulins, a correspondingly higher secretion rate in terms of molecules per cell per time would be expected, although the total mass would stay the same. The secretion rates of protein from cells are expected to be some fraction of this maximum rate since the cell is making a large number of proteins at any given time. ■

Growth factors bind to their receptors, which are found in cellular membranes, with high affinities. Their binding constants are as low as 10 to 100 pM. The binding of a growth factor to a receptor is described as

$$G + R \Leftrightarrow G\!:\!R, \quad \text{and} \quad \frac{[G][R]}{[G\!:\!R]} = K_a \tag{7.11}$$

where $[G]$, $[R]$, $[G\!:\!R]$ are the concentrations of the growth factor receptor and the bound complex, respectively, and K_a is the binding constant. Since the total number of receptors (R_{tot}) in the system is constant, the mass conservation quantity is given by:

$$R_{tot} = [R] + [G\!:\!R] \tag{7.12}$$

and therefore equation (7.12) can be written as

$$R_{tot} = [G\!:\!R]\frac{K_a}{[G]} + [G\!:\!R] = [G\!:\!R](1 + \frac{K_a}{[G]}) \tag{7.13}$$

or

$$\left(\frac{[G\!:\!R]}{R_{tot}}\right) = \frac{[G]}{([G] + K_a)} \tag{7.14}$$

Receptor occupancy rates ($[G\!:\!R]/R_{tot}$) need to be on the order of 0.25 to 0.5 to reach a significant stimulation (Fig. 7.20), and therefore growth factor concentrations as low as 10 pM are sufficient to generate cellular response.

In many cases, the receptor:ligand complex is internalized, and a typical time constant for internalization is 15 to 30 min. The absolute values for growth factor uptake rates have been measured. For instance, interleukin-3 (IL-3) and the stem cell factor (SCF) are consumed by immature hematopoietic cells at rates of about 10 and 100 ng per million cells per day. Further, it has been shown that 10,000 to 70,000 growth factor molecules need to be consumed to stimulate cell division in complex cell cultures.

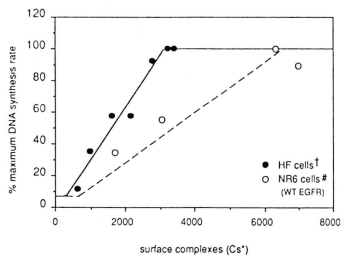

Figure 7.20 Relationship between steady-state surface complexes and mitogenic response to EGF, for NR6 cells (○) and human fibroblasts (●) (data from Knauer et al., 1984).

Example Problem 7.3

How far can soluble signals propagate and how long does it take?

Solution

The maximum signaling distance can be estimated from a simple diffusion model that describes the secretion from a sphere. Under steady-state conditions, it can be shown that the concentration of a secreted molecule (c) as a function of the distance from the cell is described as

$$\frac{c}{K_a} = \alpha \frac{R}{r} \quad \text{where } \alpha = \frac{(R^2/D)}{(K_a R/F)} \tag{7.15}$$

where R is the radius of the cell, F is the secretion rate, and D is the diffusion coefficient of the growth factor. The distance that a signal reaches when $c = K_a$ is estimated by

$$r_{\text{critical}}/R = \alpha \tag{7.16}$$

Thus, the maximal secretion rate (α) is given by the ratio of two time constants, the time constant for diffusion away from the cell (R^2/D) and the secretion time constant ($K_a R/F$). Since it takes an infinite amount of time to reach a steady state, a more reasonable estimate of the signal propagation distance is the time at which the signal reaches one-half of its ultimate value. This leads to a time constant estimate for the signaling process of about 20 minutes and a maximal distance of about 200 µm. This distance is shortened proportionally to the secretion rate (F) and inversely proportionally to the affinity (K_a), both of which are limited.   ■

The binding of a growth factor to its receptor triggers a complex signal transduction process (Fig. 7.21). Typically the receptor complex changes in such a way that its intracellular components take on catalytic activities. These activities are in many cases due to the Janus kinases (JAKs). These kinases then operate on the signal transducers and activators of transcription (STATs) that transmit the signal into the nucleus. The kinetics of this process are complex, and detailed analyses are becoming available for the epidermal growth factor signal transduction process.

Direct Cell–Cell Contact—Insoluble Factors 1

Cells are equipped with proteins on their surface that are called the cell adhesion molecules (CAMs). These include the cadhedrins (adhesion belts, desmosomes) and connexins (gap junctions). These molecules are involved in direct cell-to-cell contact.

Some of these are known as the cell junction molecules since junctions are formed between the adjacent cells allowing for direct cytoplasmic communication. Such junctions are typically on the order of 1.5 nm in diameter and allow molecules below about 1 kDa to pass between cells.

A growing body of literature is showing how fluid mechanical shear forces influence cell and tissue function. Tissue function has a significant mechanical dimension. At the cellular level, the mechanical role of the cytoskeleton is becoming better

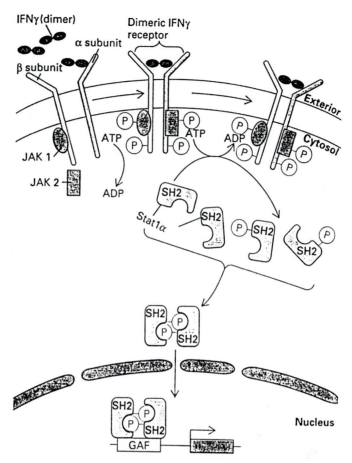

Figure 7.21 A schematic representation of the interferon gamma signal transduction pathway (from Lodish et al., 1995).

understood. Signals may be delivered to the nucleus by cellular stretching in a way that is similar to the method in which growth factor binding to a receptor delivers signals. The integrins thus can perform as "mechanical transducers" of important signals. Further, cells do know their location within a tissue and long-distance information must be transmitted between cells via weak mechanical interactions. The mechanical characteristics of the cellular microenvironment are thus of importance, as are the mechanical properties of the cells themselves.

Cell–Matrix Interactions—Insoluble Factors 2

The extracellular matrix (ECM) comprises complex weaves, glues, struts, and gels that interconnect all the cells in a tissue and their cytoskeletal elements. The ECM is multifunctional and provides tissue with mechanical support and cells with a substrate on which to migrate, as well as a place to locate signals for communications. The ECM thus has structural and informational functions. It is dynamic and is constantly being modified. For instance, ECM components are degraded by metalloproteases. In cardiac muscle, about 3% of the matrix is turned over daily.

The ECM is composed of a large number of components that have varying structural and regulatory functions. On the cell surface, there are a number of adhesion and ECM receptor molecules that facilitate cell–ECM communications. These signals contain instructions for migration, replication, differentiation, and apoptosis. The nature of these signals is governed by the composition of the ECM, which in turn can be altered by the cells found in the tissue. Thus, all the cellular fate processes can be directed by the ECM, and it provides a means for cells to communicate. The signals in the ECM are more stable and can be made more specific and stronger than those delivered by diffusible growth factors.

The components of the ECM and their functions are summarized in Table 7.6.

Example Problem 7.4

How many receptor sites are needed for various cellular functions?

Solution

The RGD (arginine–glycine–aspartic acid) tripeptide binding sequence has been immobilized on a cell growth surface at varying densities. Cell attachment, spreading, and growth were examined as a function of the surface density of RGD binding sites for fibroblasts. The results showed that an average receptor spacing of 440 nm was sufficient for cell attachment and spreading and 160 nm for focal point adhesion formation. ∎

Many tissue engineering efforts to date have been aimed at artificially constructing an ECM. These matrices have taken the form of polymeric materials, which in some cases are bioresorbable to allow the cells to replace this artificial ECM as they establish themselves and reconstruct tissue function. The properties of this matrix are hard to design since most of the functionalities are unknown and since it carries two-way communication between cells. The full spectrum of ECM functionalities can be provided only by the cells themselves (Table 7.7).

7.3   PHYSICAL CONSIDERATIONS

7.3.1   Organization of Tissues into Functional Subunits

The body has eleven major organ systems (Table 7.8), with the muscular and skeletal systems often considered together as the musculoskeletal system. These organ systems carry out major physiological functions such as respiration, digestion, and mechanical motion. Each one of these organs systems in turn is comprised of organs. The major

TABLE 7.6 Components of the Extracellular Matrix[a]

Component	Function	Location
Collagens	Tissue architecture, tensile strength Cell–matrix interactions Matrix–matrix interactions	Ubiquitously distributed
Elastin	Tissue architecture and elasticity	Tissues requiring elasticity (e.g., lung, blood vessels, heart, skin)
Proteoglycans	Cell–matrix interactions Matrix–matrix interactions Cell proliferation Binding and storage of growth factors	Ubiquitously distributed
Hyaluronan	Cell–matrix interactions Matrix–matrix interactions Cell proliferation Cell migration	Ubiquitously distributed
Laminin	Basement membrane component Cell migration	Basement membranes
Epiligrin	Basement membrane component (epithelium)	Basement membranes
Entactin (nidogen)	Basement membrane component	Basement membranes
Fibronectin	Tissue architecture Cell–matrix interactions Matrix–matrix interactions Cell proliferation Cell migration Opsonin	Ubiquitously distributed
Vitronectin	Cell–matrix interactions Matrix–matrix interactions Hemostasis	Blood Sites of wound formation
Fibrinogen	Cell proliferation Cell migration Hemostasis	Blood Sites of wound formation
Fibrillin	Microfibrillar component of elastic fibers	Tissues requiring elasticity (e.g., lung, blood vessels, heart, skin)
Tenascin	Modulates cell–matrix interaction Antiadhesive Antiproliferative	Transiently expressed associated with remodeling matrix
SPARC[b] (osteonectin)	Modulates cell–matrix interaction Antiadhesive Antiproliferative	Transiently expressed associated with remodeling matrix
Thrombospodin	Modulates cell–matrix interaction	Platelet α granules
Adhesion molecules	Cell surface proteins mediating cell adhesion to matrix or adjacent cells Mediators of transmembrane signals	Ubiquitously distributed
von Willebrand factor	Mediates platelet adhesion Carrier for procoagulant factor VIII	Plasma protein Subendothelium

[a]Mutsaers et al., 1997.
[b]SPARC, secreted protein acidic and rich in cysteine.

TABLE 7.7 Adhesion Molecules with the Potential to Regulate Cell–ECM Interactions

Adhesion Molecule	Ligand
Integrins	
$\alpha_1\beta_1$	Collagen (I, IV, VI), laminin
$\alpha_2\beta_1$	Collagen (I–IV, VI), laminin
$\alpha_3\beta_1$	Collagen (I), laminin, fibronectin, entactin, epiligrin
$\alpha_4\beta_1$	Fibronectin$_{ALT}$, VCAM-1, thrombospondin
$\alpha_5\beta_1$	Fibronectin, thrombospondin
$\alpha_6\beta_1$	Laminin
$\alpha_V\beta_1$	Fibronectin
$\alpha_L\beta_2$	ICAM-1, ICAM-2, ICAM-3
$\alpha_M\beta_2$	ICAM-1, iC3b, fibrinogen, factor X, denatured protein
$\alpha_x\beta_2$	Fibrinogen, iC3b, denatured protein
$\alpha_V\beta_3$	Vitronectin, fibrinogen, fibronectin, thrombospondin
$\alpha_V\beta_5$	Vitronectin
$\alpha_V\beta_6$	Fibronectin
$\alpha_0\beta_4$	Laminin
$\alpha_4\beta_7$	Fibronectin$_{ALT}$ VCAM-1, MAdCAM-1
$\alpha_{1th}\beta_3$	Fibrinogen, fibronectin, vitronectin, vWF
LRI[h]	Fibrinogen, fibronectin, vitronectin, vWF, collagen (IV), entactin

organs that participate in digestion are shown in Figure 7.22. Each organ system and organ have homeostatic functions that can be defined based on their physiological requirements. These can be thought of as "spec sheets." An example is given in Table 7.9.

There are a significant number of important conclusions that can be arrived at using simple order of magnitude analysis of the information found in such spec sheets. Insightful and judicious order of magnitude analysis will be the primary mode of analysis of tissue function for some time to come. However, detailed analysis and calculations cannot substitute for focused and well-justified experimental work.

Organs function at a basal rate but have the ability to respond to stress. The total circulation rate and organ distribution under strenuous exercise differ significantly from that at rest. Similarly, under hematopoietic stress, such as infection or in patients with sickle cell anemia, the basal blood cell production rate can significantly exceed the basal rate given in Figure 7.3.

Organs, in turn, are made up of functional subunits. These subunits include the alveoli in the lung and the nephron in the kidney (Fig. 7.23). These functional units are composed of a mixture of different kinds of cells that together constitute tissue function. Separating the functional subunits into their individual cell cohorts leads to the loss of tissue-specific function, but specific cell properties can be studied with such purified preparations.

TABLE 7.8 Major Organ Systems of the Body

Circulatory	Heart, blood vessels, blood (some classifications also include lymphatic vessels and lymph in this system)	Transport of blood throughout the body's tissues
Respiratory	Nose, pharynx, larynx, trachea, bronchi, lungs	Exchange of carbon dioxide and oxygen; regulation of hydrogen–ion concentration
Digestive	Mouth, pharynx, esophagus, stomach, intestines, salivary glands, pancreas, liver, gallbladder	Digestion and absorption of organic nutrients, salts, and water
Urinary	Kidneys, ureters, bladder, urethra	Regulation of plasma composition through controlled excretion of salts, water, and organic wastes
Musculoskeletal	Cartilage, bone, ligaments, tendons, joints, skeletal muscle	Support, protection, and movement of the body; production of blood cells
Immune	Spleen, thymus, and other lymphoid tissues	Defense against foreign invaders; return of extracellular fluid to blood; formation of white blood cells
Nervous	Brain, spinal cord, peripheral nerves and ganglia, special sense organs	Regulation and coordination of many activities in the body; detection of changes in the internal and external environments; states of consciousness; learning; cognition
Endocrine	All glands secreting hormones: Pancreas, testes, ovaries, hypothalamus, kidneys, pituitary, thyroid, parathyroid, adrenal, intestinal, thymus, heart, pineal	Regulation and coordination of many activities in the body
Reproductive	Male: Testes, penis, and associated ducts and glands	Production of sperm; transfer of sperm to female
	Female: Ovaries, uterine tubes, uterus, vagina, mammary glands	Production of eggs; provision of a nutritive environment for the developing embryo and fetus; nutrition of the infant
Integumentary	Skin	Protection against injury and dehydration; defense against foreign invaders; regulation of temperature

A fundamental statement follows from this observation: tissue function is a property of cell–cell interactions. The size of the functional subunits is on the order of $100\,\mu m$, whereas the size scale of a cell is $10\,\mu m$. Each organ is then composed of tens to hundreds of millions of functional subunits. The sizing of organs represents an evolutionary challenge that is also faced by tissue engineering in scaling up the function of reconstituted tissues *ex vivo*.

The microenvironment is thus very complex. To achieve proper reconstitution, these dynamic, chemical, and geometric variables must be accurately replicated. This task is difficult. A significant fraction of this chapter will be devoted to developing quantitative methods to describe the microenvironment. These methods can then be

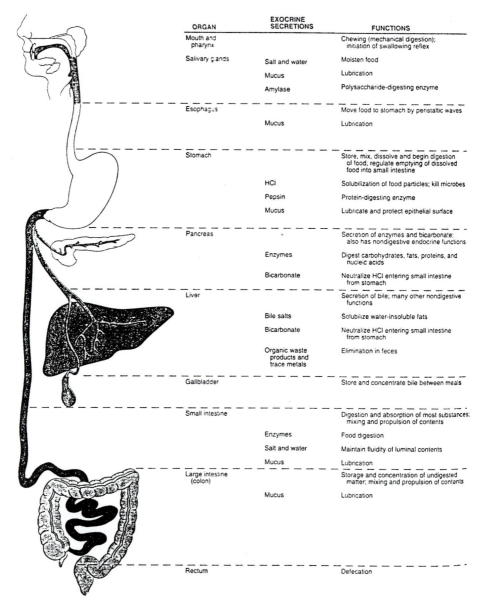

ORGAN	EXOCRINE SECRETIONS	FUNCTIONS
Mouth and pharynx		Chewing (mechanical digestion); initiation of swallowing reflex
Salivary glands	Salt and water	Moisten food
	Mucus	Lubrication
	Amylase	Polysaccharide-digesting enzyme
Esophagus		Move food to stomach by peristaltic waves
	Mucus	Lubrication
Stomach		Store, mix, dissolve and begin digestion of food; regulate emptying of dissolved food into small intestine
	HCl	Solubilization of food particles; kill microbes
	Pepsin	Protein-digesting enzyme
	Mucus	Lubricate and protect epithelial surface
Pancreas		Secretion of enzymes and bicarbonate; also has nondigestive endocrine functions
	Enzymes	Digest carbohydrates, fats, proteins, and nucleic acids
	Bicarbonate	Neutralize HCl entering small intestine from stomach
Liver		Secretion of bile; many other nondigestive functions
	Bile salts	Solubilize water-insoluble fats
	Bicarbonate	Neutralize HCl entering small intestine from stomach
	Organic waste products and trace metals	Elimination in feces
Gallbladder		Store and concentrate bile between meals
Small intestine		Digestion and absorption of most substances; mixing and propulsion of contents
	Enzymes	Food digestion
	Salt and water	Maintain fluidity of luminal contents
	Mucus	Lubrication
Large intestine (colon)		Storage and concentration of undigested matter; mixing and propulsion of contents
	Mucus	Lubrication
Rectum		Defecation

Figure 7.22   Functions and organization of the gastrointestinal organs (from Vander et al., 1994).

used to develop an understanding of key problems, formulation of solution strategy, and analysis for its experimental implementation.

The microcirculation connects all the microenvironments in every tissue to their larger whole-body environment. With few exceptions, essentially all metabolically active cells in the body are located within a few hundred μm from a capillary. The

TABLE 7.9 Standard American Male[a]

Age	30 years
Height	5 ft 8 in or 1.86 m
Weight	150 lb or 68 kg
External surface	19.5 ft^2 or 1.8 m^2
Normal body temperature	37.0°C
Normal mean skin temperature	34°C
Heat capacity	0.86 cal/(g) (°C)
Capacities	
Body fat	10.2 kg or 15%
Subcutaneous fat layer	5 mm
Body fluids	ca.51 liters or 75%
Blood volume	5.0 liters (includes formed elements, primarily red cells, as well as plasma)
	Hematrocrit = 0.43
Lungs	
Total lung capacity	6.0 liters
Vital capacity	4.2 liters
Tidal volume	500 ml
Dead space	150 ml
Mass transfer area	90 m^2
Mass and energy balances at rest	
Energy conversion rate	72 kcal/h or 1730 kcal/day [40 kcal/(m^2)(h)]
O_2 consumption	250 ml/min (respiratory quotient = 0.8)
CO_2 production	200 ml/min
Heart rate	65/min
Cardiac output	5.01/min (rest)
	3.0 + 8 M in general ($M = O_2$ consumption in liters per minute)
Systemic blood pressure	120/80 mmHg

[a]From Lightfoot (1974)

capillaries provide a perfusion environment that connects every cell (the cell at the center of the diagram) to a source of oxygen and sink for carbon dioxide (the lungs), a source of nutrients (the small intestine), the clearance of waste products (the kidney), and so forth. The engineering of these functions *ex vivo* is the domain of bioreactor design. Such culture devices have to appropriately simulate and provide respiratory, gastrointestinal, and renal functions. Further, these cell culture devices have to respect the need for the formation of microenvironments and thus have to have perfusion characteristics that allow for uniformity down to the 100 μm length scale. These are stringent design requirements.

7.3.2 Estimating Tissue Function from Spec Sheets

Most of the useful analysis in tissue engineering is performed with approximate calculations that are based on physiological and cell biological data—a tissue "spec sheet" (e.g., Table 7.9). These calculations are useful to interpret organ physiology

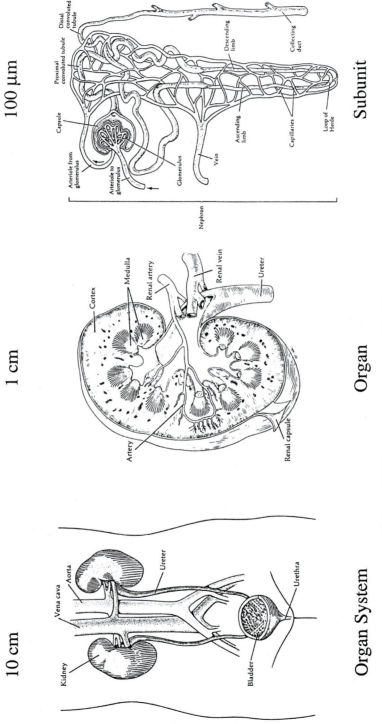

Figure 7.23   An organ system, an organ, and a functional subunit.

and to provide a starting point for an experimental program. One example will be provided in each category.

The Respiratory Functions of Blood

Remarkably insightful calculations leading to interpretation of the physiological respiratory function of blood have been carried out. The basic functionalities and biological design challenges can be directly derived from tissue spec sheets.

Blood needs to deliver about 10 mM of O_2 per min to the body. The gross circulation rate is about 5 liters per minute. Therefore, blood has to deliver to tissues about 2 mM oxygen per liter during each pass through the circulation. The pO_2 of blood leaving the lungs is about 90 to 100 mmHg, while pO_2 in venous blood at rest is about 35 to 40 mmHg. During strenuous exercise, the venous pO_2 drops to about 27 mmHg. These facts state the basic requirements that circulating blood must meet to deliver adequate oxygen to tissues.

The solubility of oxygen in aqueous media is low. Its solubility is given by

$$[O_2] = \alpha_{O2} pO_2 \tag{7.17}$$

where the Henry's law coefficient is about 0.0013 mM per mmHg. The oxygen that can be delivered with a partial pressure change of $95 - 40 = 55$ mmHg is thus about 0.07 mM, far below the required 2 mM (by a factor of about 30-fold). Therefore the solubility or oxygen content of blood must be substantially increased and the concentration dependency of the partial pressure must be such that the 2 mM are given up when the partial pressure changes from 95 to 40 mmHg. Furthermore, during strenuous exercise, the oxygen demand doubles and 4 mM must be liberated for a partial pressure change from 95 to 27 mmHg.

The evolutionary solution is to put an oxygen binding protein into circulating blood to increase the oxygen content of blood. To stay within the vascular bed, such a protein would have to be 50 to 100 kD in size. With a single binding site, the required protein concentration for 10 mM oxygen is 500 to 1000 g/l, which is too concentrated from an osmolarity standpoint, and the viscosity of such a solution may be tenfold that of circulating blood, which is clearly impractical. Further, circulating proteases would lead to a short plasma half-life of such a protein.

Four sites per oxygen-carrying molecule would reduce the protein concentration to 2.3 mM and confining it to a red cell would solve both the viscosity and proteolysis problem. These indeed are the chief characteristics of hemoglobin. A more elaborate kinetic study of the binding characteristics of hemoglobin shows that positive cooperativity will give the desired oxygen transfer capabilities both at rest and under strenuous exercise (Fig. 7.24).

Perfusion Rates in Human Bone Marrow Cultures

The question of how often the medium should be replenished is important in designing cell culture conditions. Normally, the *in vivo* situation provides a good starting point for experimental optimization. Thus, a dynamic similarity analysis is in order.

The blood perfusion through bone marrow is about 0.08 ml per cc per minute. Cellularity in marrow is about 500 million cells per cc. Therefore, the cell-specific

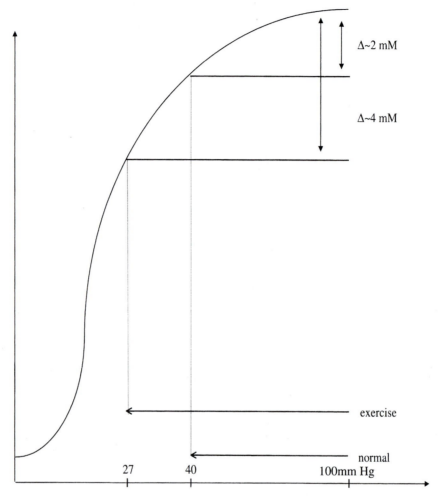

Figure 7.24   A schematic of hemoglobin–oxygen binding curves and oxygen delivery. The change in oxygen concentration during passage through tissues (i.e., oxygen delivery) is shown as a function of the concentration of oxygen in blood (modified from Garby and Meldon, 1977).

perfusion rate is about 2.3 ml/10 million cells/day. Cultures of mononuclear cell populations from murine bone marrow were developed in the mid to late 1970s. These cultures had long-term viability. Initial attempts to use the same culture protocols for human bone marrow cultures in the early 1980s were largely unsuccessful. The culture protocol called for medium exchange about once per week.

To perform a dynamic similarity analysis of perfusion rates, or medium exchange rates, between *in vivo* and *in vitro*, the per cell medium exchange rate in culture is calibrated to that calculated for the preceding *in vivo* situation. Cell cultures are typically started with cell densities on the order of a million cells per ml. Therefore, 10 million cells would be placed in 10 mls of culture medium which contains about 20% serum (vol/vol). A full daily medium exchange would hence correspond to replacing the serum at 2 ml/10 million cells/day, which is similar to the number calculated previously.

Experiments using this perfusion rate and the cell densities were performed in the late 1980s and led to the development of prolific cell cultures of human bone marrow. These cultures were subsequently scaled up to produce a clinically meaningful number of cells and are currently undergoing clinical trials. Thus, a simple similarity analysis of the *in vivo* and *in vitro* tissue dynamics led to the development of culture protocols that are of clinical significance. Such conclusions can be derived from tissue spec sheets.

These examples serve to illustrate the type of approximate calculations that will govern initial analysis in tissue engineering. Well organized fact, or spec sheets provide the basic data. Then, characteristic time constants, length constants, fluxes, rates, concentrations, etc., can be estimated. The relative magnitudes of such characteristics serve as a basis for order of magnitude judgments.

7.3.3 Mass Transfer in 3-D Configurations

Determination of mass transfer in biological systems is essential in tissue engineering because mass transfer rates must be known to design and deliver cell therapy products or extracorporeal organs. Although spec sheets give a global target for mass transfer, the details of the capillary bed of the smallest physiological units of the tissue are required. Mass transfer depends on the diffusion and convection of nutrients and waste to and from tissue and the consumption of nutrients and production of waste by the tissue. Typical convection of blood, as shown in Table 7.10, in the vasculature range from 140 cm/sec in the aorta to 10^{-3} cm/sec in the capillary beds. Convection is driven by pressure differences and dominates in the vasculature, whereas diffusion is driven by concentration gradients and dominates in the tissues. Diffusion can be described by Fick's law and one can modify this equation to describe mass transfer in three configurations: rectangle, cylinders, and slabs (Figure 7.25).

The key to a successful bioreactor design for extracorporeal support or expansion of stem cells is maintaining adequate mass transfer while at the same time providing a local environment conducive to the differentiated state. The dominant mechanism of transport of low molecular weight nutrients (e.g., oxygen and glucose) within tissue or

TABLE 7.10 Peak Convection Rates of Blood in the Vasculature

Compartment	cm/sec
Aorta	140 ± 40
Common carotid	100 ± 20
Vertebral[a]	36 ± 9
Superficial femoral	90 ± 13
Liver sinusoid[b]	$10^{-2} - 10^{-3}$

Data from DeWitt and Wechsler (1988). [a]Data from Jager et al. (1985). [b]Data from McCuskey (1984).

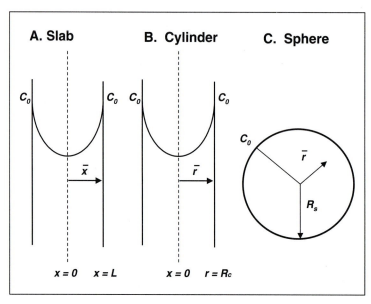

Figure 7.25   The three possible configurations considered by the mass transfer equations. (Reproduced from Macdonald et al., 1999.)

cell aggregates is diffusion. The length scale for diffusive transport (i.e., the distance over which oxygen penetrates into the tissue from the nutrient stream before it is completely consumed by the cells) depends on the volumetric concentration of metabolite in the nutrient stream, C_0, the rate at which the cells consume the nutrient, Q_i, the diffusion coefficient for the metabolite in the tissue, D_t, and the system geometry. The metabolic consumption rate Q_i is generally a function of the nutrient concentration in the cell mass, C. The most common rate expression is the Michaelis–Menton type: $Q_i = (V_{max} \times C)/(K_m + C)$, which reduces to zero-order kinetics for high concentrations of nutrient (i.e., $C \gg K_m$) and to first-order kinetics for low nutrient concentrations (i.e., $C \ll K_m$). It is reasonable to consider the zero-order limit, because the condition $C \gg K_m$ is met for many important nutrients under normal physiological conditions.

An estimate of the length scale for nutrient diffusion in a 3-D cell mass can be obtained by mathematical modeling. The general equation governing the balance between steady-state diffusion and metabolic consumption is

$$D_i \blacktriangledown^2 C = Q_i \qquad\qquad (7.18)$$

where C is the concentration of the nutrient within the cell mass. Three simple geometries amenable to analytical solutions are shown in Figure 7.25: a slab, a cylinder, and a sphere of cells bathed in medium containing the nutrient at concentration C_0. Expanding the gradient operator for each of the these geometries, Eq. 7.18 can be written as

$$\text{slab: } D_t \frac{d^2C}{dx^2} = Q_i \tag{7.19a}$$

$$\text{cylinder: } D_t \frac{1}{r}\frac{d}{dr}\left(\frac{rdC}{dr}\right) = Q_i \tag{7.19b}$$

$$\text{sphere: } D_t \frac{1}{r^2}\frac{d}{dr}\left(\frac{r^2dC}{dr}\right) = Q_i \tag{7.19c}$$

The manner in which the distance variables are defined in each case is shown in Figure 7.25. A standard technique in the analysis of differential equations describing transport phenomena is to scale the variables so that they are dimensionless and range in value 0–1. This allows the relative magnitude of the various terms of the equations to be evaluated easily and allows the solutions to be plotted in a set of graphs as a function of variables that are universally applicable. For all geometries, C is scaled by its maximum value, C_0 (i.e., $\bar{C} = C/C_o$). Distance is scaled with the diffusion path length, so that for a slab, $\bar{x} = x/L$, for a cylinder, $\bar{r} = r/R_c$, and for a sphere, $\bar{r} = r/R_s$. With these definitions, the boundary conditions for all three geometries are no flux of nutrient at the center ($d\bar{c}/d\bar{x}$, $d\bar{c}/d\bar{r} = 0$ at $\bar{x}$, $\bar{r}$, $= 0$) and that $\bar{C} = 1$ at the surface ($\bar{x}$, $\bar{r}$, $= 1$).

The use of scaling allows the solutions for all three geometries to collapse to a common form:

$$\bar{C} = 1 - \frac{\phi^2}{2}(1 - \bar{x}^2) \tag{7.20}$$

All of the system parameters are lumped together in the dimensionless parameter ϕ^2, which is often called the Thiele modulus. The Thiele modulus represents the relative rates of reaction and diffusion and is defined slightly differently for each geometry:

$$\phi^2_{slab} = \frac{QL^2}{C_oD_t}, \qquad \phi^2_{cylinder} = \frac{QR_c^2}{2C_oD_t}, \qquad \phi^2_{sphere} = \frac{QR_s^2}{3C_oD}$$

7.3.4 The Tissue Microenvironment: Cell Therapy and Bioreactor Design Principles

Up to this point, the overall states of tissue function and the fact that this behavior is based on interacting cellular fate processes have been discussed. How these cellular fate processes are quantitatively described and the fact that there are complex genetic circuits that drive these cellular fate processes have been shown.

Synthesis is the next question. How can tissue function be built, reconstructed, and modified? This task is one for tissue engineering. To set the stage, some fundamentals can be stated axiomatically:

- The developmental program and the wound healing response require the systematic and regulated unfolding of the information on the DNA through coordinated execution of "genetic subroutines and programs." Participating cells require detailed information about the activities of their neighbors. Proper cellular communications are of key concern and concepts of the stem cell niche hypothesis are required.

- Upon completion of organogenesis or wound healing, the function of fully formed organs is strongly dependent on the coordinated function of multiple cell types. Tissues function based on multicellular aggregates, so-called functional subunits of tissues.

- The microenvironment has a major effect on the function of an individual cell. The characteristic length scale of the microenvironment is 100 μm.

- The microenvironment is characterized by: (1) neighboring cells: cell–cell contact, soluble growth factors, etc.; (2) the chemical environment: the extracellular matrix, the dynamics of the nutritional environment; and (3) the local geometry.

Cellular Function *In Vivo*: Its Tissue Microenvironment and Communication with Other Organs

An important requirement for successful tissue function is a physiologically acceptable environment in which the cells will express the desired tissue function. Most likely, but not necessarily, this environment should recreate or mimic the physiological *in vivo* situation. How can a physiologically acceptable tissue microenvironment be understood and recreated?

The communication of every cell with its immediate environment and other tissues can be illustrated using a topological representation of the organization of the body. The DNA, sitting in the center of the diagram, contains the information that the tissue engineer wishes to express and manage (Fig. 7.2). This cell is in a microenvironment that has important spatio-temporal characteristics. Examples of tissue microenvironments are shown in Figure 7.26, and some are discussed in Section 7.2.8.

Signals to the nucleus are delivered at the cell membrane and transmitted through the cytoplasm by a variety of signal transduction mechanisms. Some signals are delivered by soluble growth factors after binding to the appropriate receptors. These growth factors may originate from the circulating blood or from neighboring cells. Nutrients, metabolic waste products, and respiratory gases traffic in and out of the microenvironment in a highly dynamic manner. The microenvironment is also characterized by its cellular composition, the ECM, and local geometry, and these components also provide the cell with important signals. The size scale of the microenvironment is on the order of 100 μm. Within this domain, cells function in a coordinated manner. If this arrangement is disrupted, cells that are unable to provide tissue function are obtained.

Example Problem 7.5

Describe the interactions and surrounding microenvironments of hepatic stellate cells (HSC) within normal and abnormal liver tissues.

Liver

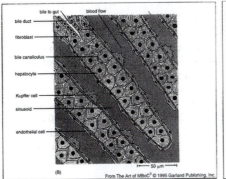

Intestinal Epithelium

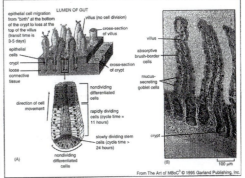

Skin

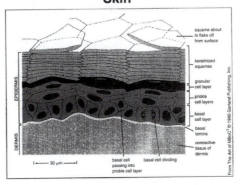

Figure 7.26 *In vivo* tissue microenvironments. (from Alberts et al.)

Solution

The hepatic stellate cell is a mesenchymal cell that is located in the space of Disse which is found between the sheets of hepatocytes and the hepatic sinusoidal endothelium (Fig. 7.27). In the normal histogenic state of liver, hepatic stellate cells are found in close proximity to a basement membrane-like matrix. This matrix consists of collagen type IV, laminin, and heparan sulphate proteoglycans. If the liver is injured, hepatic stellate cells are activated, and they begin to make matrix protein, mostly collagen type I and III. As a result of hepatic stellate cell activation, the phenotype of both the hepatocytes (which lose their brush border) and endothelium (which lose their fenestration) changes. If this condition persists, liver fibrosis results. Thus, a lack of coordination in cellular function in the tissue microenvironment can lead to loss of tissue function. ∎

Tissues are perfused by the microcirculation and they are homogeneous from this standpoint down to a length scale of about 100 μm. The microcirculation then interfaces with the long-distance convective transport system that connects all the

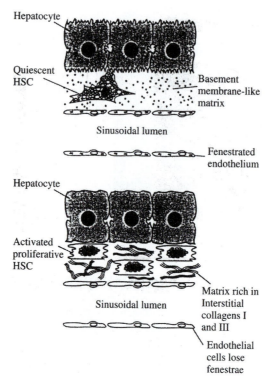

Figure 7.27   Schematic of hepatic stellate cell activation and the process of liver fibrosis (from Iredale, 1997).

tissues in the body and, thus, all cells to a nutritional supply, exchange of respiratory gases, removal of toxic products, etc.

All tissues are composed of functional subunits. This subunit is the irreducible cellular arrangement that generates tissue function. Separating the cells in this microenvironment leads to cellular function, but all of them together form tissue function. These microenvironments have cross-species similarities. The number of microenvironments per tissue is then determined by the overall requirements on the particular organ. The kidney in an elephant must be much larger than the kidney in a mouse, although the fundamental unit, the nephron, may be very similar.

From a tissue engineering standpoint, these features lead to problem decomposition (i.e., designing the macro- and the microenvironments). The challenges that face the tissue engineer can be broken down into two basic categories:

1. The microenvironment (its chemical, geometric, cell architectural, and diffusional characteristics)
2. Interactions with other tissues (source of nutrients, exchange of respiratory gases, removal of waste products, and delivery of soluble protein such as growth factors)

A device that mimics the tissue function must provide the function for the two outermost layers and locally produce an acceptable environment, represented by the remaining layers. Although much experience was gained during the 1980s in bioreactor design for cultures of pure continuous cell lines for protein production, tissue bioreactors clearly represent a series of new challenges. This topic will be addressed in the following section. In this section, the discussion will focus on the variables that influence the microenvironment: its cellularity, its chemistry, its dynamics, and its geometry.

Cellularity

The packing density of cells is on the order of a billion cells per cc. Tissues are typically operating at 1/3 to 1/2 of packing density, leaving typical cell densities in tissues on the order of 100 to 500 million cells per cc. Since the master length scale is about 100 μm, the order of magnitude of the number of cells found in a tissue microenvironment can be estimated. A 100 μm cube at 500 million cells per cc contains about 500 cells. Simple multicellular organisms such as *C. elegans*, a much studied small worm, have about 1000 cells, providing an interesting comparison.

The cellularity of the tissue microenvironment varies among tissues. At the low end there is cartilage. The function of chondrocytes in cartilage is to maintain the extracellular matrix. Cartilage is avascular, alymphatic, and aneural. Thus, many of the cell types found in other tissues are not in cartilage. The cellularity of cartilage is about a million cells per cc or about one cell per cubic 100 μm. Thus, the microenvironment here is simply one cell that maintains its surrounding ECM.

Many cell types are found in tissue microenvironments (Table 7.11). In addition to the parenchymal cells (the tissue-type cells such as hepatocytes in liver), a variety of accessory cells are found in all tissues. These are in the following categories:

TABLE 7.11 Cells That Contribute to the Tissue Microenvironment

Stromal cells: derivates of a common precursor cell
  Mesenchyme
  Fibroblasts
  Myofibroblasts
  Osteogenic/chondrogenic cells
  Adipocytes
Stromal-associated cells: histogenically distinct from stromal cells, permanent residents of a tissue
  Endothelial cells
  Macrophages
Transient cells: cells that migrate into a tissue for host defense either prior to or following an inflammatory stimulus
  B lymphocytes/plasma cells
  Cytotoxic T cells and natural killer (NK) cells
  Granulocytes
Parenchymal cells: cells that occupy most of the tissue volume, express functions that are definitive for the tissue, and interact with all other cell types to facilitate the expression of differentiated function

- Mesenchymal cells (such as fibroblasts and smooth muscle cells) are present in all tissues. These cells are of connective tissue type.
- Monocytes are present in all tissues taking on very different morphologies (Fig. 7.28). Monocytes can differentiate into macrophages that, once activated, produce a variety of cytokines and chemokines that influence the behavior of neighboring cells.
- Endothelial cells are associated with the vasculature found in all tissues. These cells play a major role in the trafficking of cells in and out of tissue and may play a major role in determining tissue metabolism.
- Lymphocytes and neutrophils have a transient presence in tissues, typically as part of a host defense response or other clean-up functions.

These accessory cells are typically about 30% of the cellularity of tissue while the parenchymal cells make up the balance. An example is provided in Table 7.12 that lists the cellularity of the liver.

Example Problem 7.6

What happens if these accessory cells are removed?

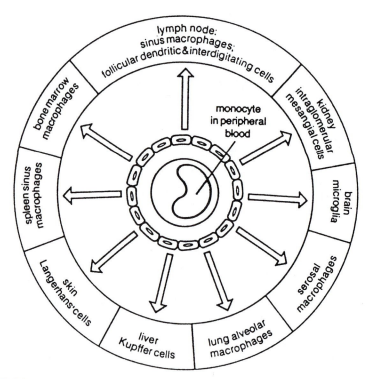

Figure 7.28   Distribution of macrophages and their presence in different tissues (from Hoffbrand and Pettit, 1988).

TABLE 7.12 Cells Contributing to the Hepatic Microenvironment

Cell Type	Size (μm)	Relative Percentage of Total Cells
Stroma		
Kupffer cells	12–16	8
Vascular endothelia	11–12	9
Biliary endothelia	10–12	5
Fat-storing cells	14–18	3
Fibroblasts	11–14	7
Pit cells	11–15	1–2 (variable)
Parenchymal cells		
Mononuclear (type I)	17–22	35
Binuclear (type II)	20–27	27
Acidophilic (type III)	25–32	5

Solution

The role of accessory cell (called stroma) function in bone marrow cultures has been systematically studied. Since the immature cells can be isolated based on known antigens, the relative composition of the key parenchymal cells and the accessory cells can be varied. Such an experiment amounts to a titration of the accessory cell activity. The results from such an experiment are shown in Figure 7.29. All cell production indices (total cells, progenitor cells, and preprogenitor cells) decline sharply as the accessory cells are removed. Supplying preformed irradiated stroma restores the production of total and progenitor cells but not the preprogenitors. This result is consistent with the expectation that specific accessory cell–parenchymal cell interactions are important for immature cells. ■

Dynamics

The microenvironment is highly dynamic and displays a multitude of time constants.

Oxygenation

Generally, mammalian cells do not consume oxygen rapidly compared to microorganisms, but their uptake rate is large compared to the amount of oxygen available at any given time in blood or in culture media (Table 7.13). At 37°C, air-saturated aqueous media contain only 0.21 mM of 210 μmol oxygen per liter. Mammalian cells consume oxygen at a rate in the range of 0.05–0.5 μmol/10^6cells/h. With tissue cellularities of 500 million cells per cc, these oxygen uptake rates call for volumetric oxygen delivery rates of 25 to 250 μmol oxygen/cc/hour. This rate needs to be balanced with the volumetric perfusion rates of tissues and the oxygen concentration in blood.

Metabolically active tissues and cell cultures, even at relatively low cell densities, quickly deplete the available oxygen. For instance, at cell densities of 10^6cells/ml, oxygen is depleted in about 0.4 to 4 hours. Oxygen thus must be continually supplied. To date, numbers of primary cell types (e.g., hepatocytes, keratinocytes, chondrocytes, and hematopoietic cells) have been grown *ex vivo* for the purpose of

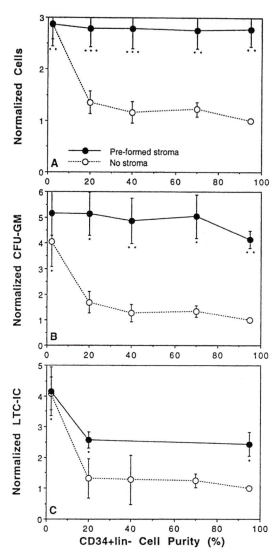

Figure 7.29 Effects of CD34$^+$ lin$^-$ cell (a population of primitive hematopoietic cells) purity on culture output. With increasing purity, the performance on a per cell basis drops due to loss of accessory cell function. CFU-GM = colony-forming units granulocyte/macrophage; LTC-IC = long-term culture-initiating cells (from Koller et al., 1995a).

cell therapy. The effects of oxygen on hepatocytes has been systematically investigated. The reported specific oxygen uptake rate (OUR) for hepatocytes is around $1.0 \, \mu mol/10^6 cells/hr$, which is relatively high for mammalian cells. Conversely, a much lower oxygen consumption rate of about $0.02 \, \mu mol/10^6 cells/hr$ has been reported for rat bone marrow cells.

TABLE 7.13 Measured Oxygen-Demand Rates of Human Cells in Culture

Human	μmol $O_2/10^6$ cells/h
HeLa	0.1–0.0047
HLM (liver)	0.37
LIR (liver)	0.30
AM-57 (amnion)	0.045–0.13
Skin fibroblast	0.064
Detroit 6 (bone marrow)	0.43
Conjunctiva	0.28
Leukemia MCN	0.22
Lymphoblastoid (namalioa)	0.053
Lung	0.24
Intestine	0.40
Diploid embryo WI-38	0.15
MAF-E	0.38
FS-4	0.05

In addition to the supply requirements, the concentration of oxygen close to the cells must be within a specific range. Oxygenation affects a variety of physiological processes, ranging from cell attachment and spreading to growth and differentiation. An insufficient concentration retards growth while an excess concentration may be inhibitory, even toxic. For instance, several studies have shown that the formation of hematopoietic cell colonies in colony assays is significantly enhanced by using oxygen concentrations that are 5% of saturation relative to air, and an optimal oxygen concentration for bioreactor bone marrow cell culture has been shown to exist. The oxygen uptake rate of cells is thus an important parameter in the design of primary cell cultures.

Metabolism and Cell Signaling

Typically, there is not a transport limitation for major nutrients, although cells can respond to their local concentrations. The reason is that their concentrations can be much higher than that of oxygen, especially for the nutrients consumed at high rates. Typical uptake rates of glucose are on the order of 0.2 µmols/million cells/h whereas the consumption rates of amino acids are in the range of 0.1–0.5 µmols per billion cells per hour. The transport and uptake rates of growth factors face more serious transport limitations. The expected diffusional response times are given in Figure 7.30.

Perfusion

The circulatory system provides blood flow to organs that is then distributed into the microenvironments. Overall, the perfusion rates in a human are about 5 liters/min/70 kg, or about 0.07 ml/cc/min. With 500 million cells per cc, this is equivalent to 0.14 µl/million cells/min. These numbers represent a whole-body average. There are differences in the perfusion rates of different organs that typically correlate with their metabolic activity.

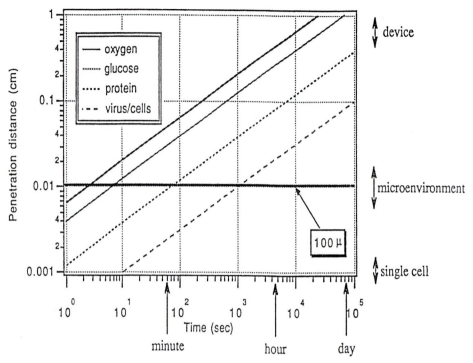

Figure 7.30 The diffusional penetration lengths as a function of time for several classes of biomolecules.

Cell Motion

As described in an earlier section, cells are motile and move at different rates. Neutrophils can move several cell diameters in a minute whereas adherent cell types such as keratinocytes, fibroblasts, and endothelial cells move a cell diameter per hour. These motilities represent rapid processes compared to cell replication and differentiation. Neutrophils have to be able to respond rapidly to invading micro-organisms while the adherent cell types mentioned move in response to dynamic tissue needs.

Matrix Turnover

The ECM is constantly being turned over. In the heart, about 3% of the ECM is turned over on a daily basis.

Geometry and Size

Geometry

The geometric shapes of microenvironments vary (Fig 7.26) and so do their dimensionalities. Many microenvironments are effectively curved 2-D surfaces. The cellular arrangement in bone marrow has been shown to have a fractal geometry with effective dimensionality of 2.7, whereas the brain is a 3-D structure.

Size

What determines the size of the microenvironment? The answer to this question is not clear. At the lower limit is the size of a single cell, about 10 μm. A cell aggregate must be some multiple of this distance. The factors determining the upper limit are not clear; however, estimates of effective growth factor signal propagation distances and experimental evidence of oxygen penetration distances suggest that the dynamics of cell communication and cell metabolic rates are important in determining the size scale of the microenvironment. These distances are determined by the process of diffusion. In both cases, the estimated length scale is about 100 to 200 μm. The stability issues associated with coordinating cellular functions grow with increased numbers of cells and may represent a limitation in practice.

7.3.5 Biomaterials

Biomaterials for tissue engineering present several challenging problems. There appear to be three length scales of primary interest. The shortest is at the biochemical level, where concerns include the specific chemical features of the ECM and interactions with cellular receptors (see Section 7.2.8). Intact ECM components can be used to coat support material to ensure appropriate interactions between the cells and their immediate environment. More sophisticated treatments involve the synthesis of specific binding sequences from ECM protein and presenting them in various ways to the cells. Particular cellular arrangements can be obtained by micropatterning such materials. A combination of material manufacturing, biochemistry, and genetics is required to address these issues.

The next size scale of interest is the 100 μm size scale—the size of a typical organ microenvironment. Many organs have highly specify local geometries that may have to be engineered in an *ex vivo* system. Hence, a particular microgeometry with particular mechanical properties may have to be produced. Clearly, challenging material manufacturing issues arise. Further, the support matrix may have to be biodegradable after transplantation and the degradation products non-toxic. Lactic and glycolic acid-based polymers are promising materials in this regard. If little restructuring of implants occurs following grafting, then the geometry of the support matrix over larger size scales may be important.

The largest size scale is that of the bioreactor itself. Bioreactors in tissue engineering are likely to be small, with dimensions on the order of about 10 cm. The materials issues that arise here are primarily those of biocompatibility. Although manufacturing technology exists for tissue culture plastic, it is likely that additional issues will arise. The tissue culture plastic that is commercially available is designed to promote adhesion, binding, and spreading of continuous cell lines. Although such features may be desirable for continuous cell lines, they may not be so for various primary cells.

7.4   SCALING UP

7.4.1   Fundamental Concept

As discussed previously, tissue dynamics are comprised of intricate interplay between the cellular fate processes of cell replication, differentiation, and apoptosis. They are properly balanced under *in vivo* conditions. The dynamics of the *in vivo* conditions are a balance of these biological dynamics and the constraining physico-chemical processes. The basic concept of design in tissue reconstruction is to engineer a proper balance between the biological and physico-chemical rates so that normal tissue function can occur.

7.4.2   Key Design Challenges

Within this philosophical framework many of the engineering issues associated with successful reconstitution of tissues can be delineated. This short survey is not intended to be a complete enumeration of the important issues of tissue engineering, nor is it intended to be a comprehensive representation of published literature. Its main goal is to provide an engineering perspective of tissue engineering and help define the productive and critical role that engineering needs to play in the *ex vivo* reconstruction of human tissues.

Important design challenges have arisen around the following issues:

- Oxygenation (i.e., providing adequate flux of oxygen at physiological concentrations)
- Provision and removal of cyto- and chemokines
- Physiological perfusion rates and uniformity in distribution
- Biomaterials—functional, structural, toxicity, and manufacturing characteristics

These four issues will be discussed briefly. Other issues associated with the clinical implementation of cellular therapies include the design of disposable devices, optimization of medium composition, initial cell distribution, meeting FDA requirements, and operation in a clinical setting. These cannot be detailed herein, but some of these issues are addressed in the following section.

7.4.3   Time Scales of Mass Transfer

The importance of mass transfer in tissue and cellular function is often overlooked. The limitations imposed by molecular diffusion become clear if the average displacement distance with time is plotted for diffusion coefficients that are typical for biological entities of interest in tissue function (Fig. 7.30). The diffusional penetration lengths over physiological time scales are surprisingly short and constrain the *in vivo* design and architecture of organs. The same constraints are faced in the construction of an *ex vivo* device, and high mass transfer rates into cell beds at physiological cell densities may be difficult to achieve.

The biochemical characteristics of the microenvironment are critical to obtaining proper tissue function. Much information exists about the biochemical requirements for the growth of continuous cell lines. For continuous cell lines, these issues revolve around the provision of nutrients and the removal of waste products. In cultures of primary cells, the nutrients may have other roles and directly influence the physiological performance of the culture. For instance, recently it has been shown that proline and oxygen levels play an important role in hepatocyte cultures.

In most cases, oxygen delivery is likely to be an important consideration. Too much oxygen will be inhibitory or toxic, whereas too little may alter metabolism. Some tissues, such as liver, kidney, and brain, have high oxygen requirements, whereas others require less. Controlling oxygen at the desired concentration levels, at the desired uniformity, and at the fluxes needed at high cell densities is likely to prove challenging. Further, the oxygen and nutritional requirements may vary among cell types in a reconstituted tissue that is comprised of multiple cell types. These requirements further complicate nutrient delivery. Thus, defining, designing, and controlling the biochemical characteristics of the microenvironment may prove difficult, especially given the constraints imposed by diffusion and any requirements for a particular microgeometry.

Example Problem 7.7

How are specific oxygen uptake rates measured?

Solution

Most data on specific oxygen uptake rates are obtained with cells in suspension using standard respirometers. However, to obtain accurate and representative data for primary cells, they need to be adherent. This challenge has led to the design of a novel *in situ* respirometer, shown in Figure 7.31. This respirometer has been used to measure the oxygen uptake rates of hepatocytes in culture (OUR of 1.0 μmols/million cells/h). A similar device has been used to measure the OUR in bone marrow cultures, giving results of 0.03 to 0.04 μmols/million cells/h which are similar to the *in vivo* uptake rates. ∎

Example Problem 7.8

How can oxygen be delivered?

Solution

Oxygen can be delivered *in situ* over an oxygenation membrane as illustrated in Figure 7.32. At slow flow rates compared to the lateral diffusion rate of oxygen, the oxygen in the incoming stream is quickly depleted, and the bulk of the cell bed is oxygenated via diffusion from the gas phase. This leads to oxygen delivery that can be controlled independently of all other operating variables and that is uniform. The gas phase composition can be used to control the oxygen delivery. ∎

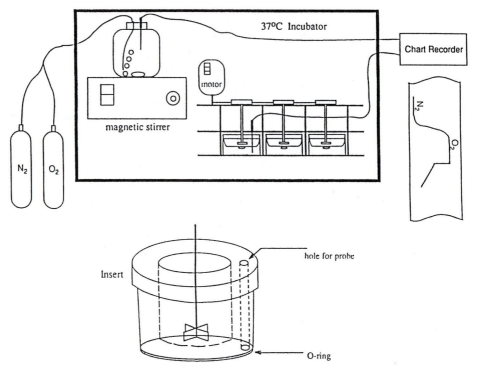

Figure 7.31   Schematic diagram of the apparatus for measuring oxygen uptake rate (from Peng and Palsson, 1996a).

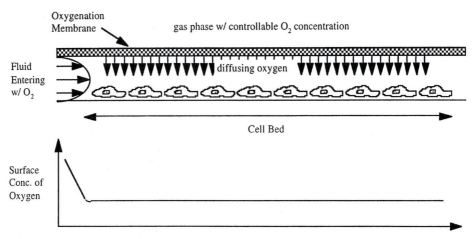

Figure 7.32   Oxygen delivery across a membrane that is placed on top of a fluid that is flowing across a cell bed. If the fluid transit time is much slower than the diffusional time for oxygen, then oxygen is delivered primarily via diffusion. This leads to a small entrance effect in which the oxygen in the incoming stream is consumed while the oxygen concentration over the rest of the cell bed is relatively constant.

7.4.4 Fluid Flow and Uniformity

The size scale of the microenvironment is 100 μm. A device that carries tens of millions of microenvironments must thus be uniform in delivery and removal of gases, nutrients, and growth factors. Achieving such uniformity is difficult. This difficulty arises from the fact that fluid has a no-slip condition as it flows past a solid surface. Thus, there will always be slower flowing regions close to any side walls in a bioreactor. These slow flowing regions under conditions of axial Graetz numbers of unity lead to mass transfer boundary layers that extend beyond the hydrodynamic boundary layer.

The origin of this problem is illustrated in Figure 7.33. Fluid is flowing down a thin slit, representing an on-end view of the chamber shown in Figure 7.34. Thin slits with high aspect ratios will satisfy the Hele–Shaw flow approximations, in which the flow is essentially the same over the entire width of the slit except close to the edges. The width of the slow flowing regions is on the order of the depth of the slit (R). Thus, if the aspect ratio is 10, fluid will flow the same way over about 90% of the width of the slit. In the remaining 10%, the slow flow close to the wall can create microenvironments with a different property than in the rest of the cell bed.

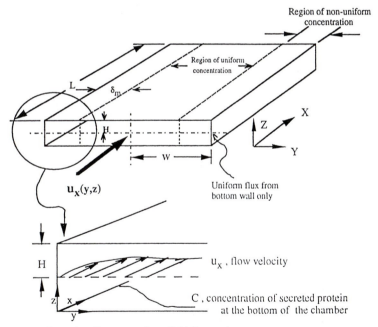

Figure 7.33 Coordinate system for a rectangular chamber with production of biological factors secreted by cells lodged on the bottom wall. The fluid is slowed down close to the side wall, creating a different concentration than that found in the middle of the slit. This leads to a very different microenvironment for cell growth and development of tissue function at the wall than elsewhere in the chamber (from Peng and Palsson, 1996b).

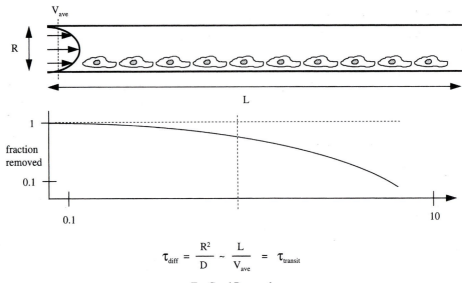

$$\tau_{diff} = \frac{R^2}{D} \sim \frac{L}{V_{ave}} = \tau_{transit}$$

For Good Removal

Figure 7.34   A schematic of the concentration of growth factors in a liquid that is flowing across a bed of cells that consume it. (Bottom) The fractional removal of the growth factor as a function of the relative time constants of lateral diffusion and transit across the cell bed. The ratio of the two is the Graetz number (Gz). If the diffusional response time is slower than that of transit (Gz > 1), then there is insufficient time for the diffusing growth factor molecule to make good contact with the cell bed. Most of the growth factor leaves the system in the exit stream. Conversely, if the diffusion time is much shorter than that of transit, there will be ample time for the growth factor to make it to the cell bed. If it is rapidly consumed there, then negligible amounts will leave the system.

Such nonuniformity can lead to differences in cell growth rates and to migration of cells toward the wall. Such nonuniformity in growth close to walls has been reported. This problem can be overcome by using radial flow.

The example shown is for one type of a bioreactor for cell culturing. Other configurations will have similar difficulties in achieving acceptable uniformity in conditions, and careful mass and momentum transfer analyses need to be performed to guide detailed design.

7.5   IMPLEMENTATION OF TISSUE ENGINEERED PRODUCTS

7.5.1   Delivering Cellular Therapies in a Clinical Setting

In the previous section, the design challenges facing the tissue engineer when it comes to scaling up microenvironments to produce cell numbers that are of clinical significance were surveyed. These problems are only a part of the challenges that must be met in implementing cellular therapies. In this section, some of the other problems that need to be solved will be discussed.

Donor-to-Donor Variability

The genetic variability in the human population is substantial. The outcome from an *ex vivo* growth procedure can vary significantly between donors. The standard deviation can exceed 50% of the mean value for key performance indices. Thus, even if the cell growth process, the production of materials, and the formulation of medium is essentially identical, the fact remains that a large variation in outcomes will be observed. A large number of experiments (10 to 12) need to be performed to obtain satisfactory answers.

This variability is due to intrinsic biological factors. Some regularization in performance can be achieved by using a full complement of accessory cells. Although interdonor variability is considerable, the behavior of the same tissue source is internally consistent. An example is shown in Figure 7.35, in which the relative uptake rates of growth factors are shown for a number of donor samples. The uptake rates of growth factors can be highly correlated within many donor samples. The

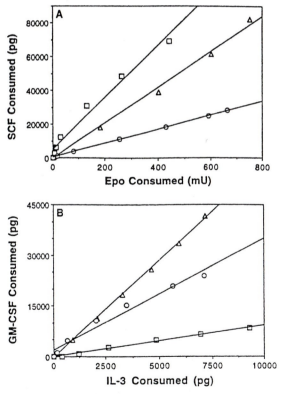

Figure 7.35   The uptake of growth factors, i.e., stem cell factor (SCF) and erythropoietin (Epo), is highly correlated in cultures of human bone marrow. The correlation is strong for a single tissue sample. The slopes of the curves vary significantly among donors (from Koller et al., 1995b).

quantitative nature of the correlation changes from donor to donor, making it difficult to develop a correlation that would represent a large donor population.

Strongly Interacting Variables

Primary cell cultures are sensitive to many of the variables that can be controlled. Many, if not most, of these variables interact strongly. Thus, any change in a single cell culture variable will change the optimal value for all the others. Thus, a statistical experimental design procedure that allows the fastest way to search over many experimental variables should be employed. An example of such a two-dimensional search is shown in Figure 7.36, where the optimal progenitor production performance of human cell cultures is shown as a function of the inoculum density and the medium flow rate.

Figure 7.36 shows the optimal performance of the cultures as measured by two different objectives. The top panel shows the optimal number of progenitors produced per unit cell growth area. Optimizing this objective would lead to the smallest cell culture device possible for a specified total number of cells that is needed. The bottom panel shows the optimal expansion of progenitor cells (i.e., the output number relative to the input). This objective would be used in situations where the starting material is limited and the maximum number of additional progenitors is desired. Note that the two objectives are found under different conditions. Thus, it is critical to clearly delineate the objective of the optimization condition from the outset. Sometimes this choice is difficult.

Immune Rejection

Allogeneic transplants face immune rejection problems. A variety of situations are encountered. Dermal fibroblasts, used for skin ulcers, for unknown reasons are effectively nonimmunogenic. This fact makes it possible to make a large number of grafts from a single source and transplant into many patients. Conversely, β-islets face certain rejection (see Fig. 7.6). Finally, in an allogeneic bone marrow transplantation, the graft may reject the immuno-compromised host. Graft-versus-host disease is the main cause for the mortality resulting from allogeneic bone marrow transplants.

The cellular and molecular basis for the immune response is becoming better understood. The rejection problem is a dynamic process that relies on the interaction between subsets of CD4$^+$ cells (CD4$^+$ is a surface antigen on certain T cells), Th1, and Th2 that differ in their cytokine secretion characteristics and effector functions. The components of the underlying regulatory network are becoming known. Quantitative and systemic analysis of this system is likely to lead to rational strategies for manipulating immune responses for prophylaxis and therapy.

Tissue Procurement

The source of the starting material for a growth process is of critical importance. The source of dermal fibroblasts used for skin replacement is human foreskin obtained from circumcision. This source is prolific and can be used to generate a large number of grafts. Since the source is the same, the biological variability (see Section 7.5.1) in

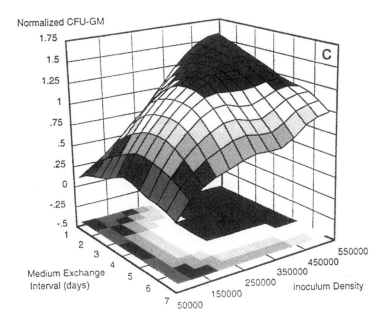

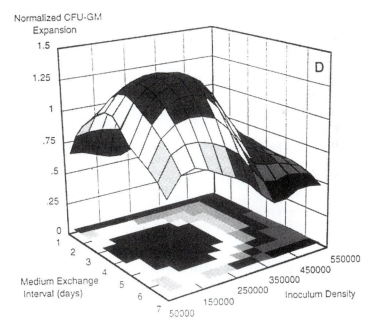

Figure 7.36 The performance of a bone marrow culture system over a wide range of inoculum densities and medium exchange intervals. Note that the variables interact. The performance (on the z axis) in the top panel is the total number of progenitor cells (CFU-GM) per unit area, whereas in the bottom panel the expansion ratio is depicted (number out divided by number in). Note the two measures are optimized under vastly different conditions (from Koller et al., 1995c).

the growth process is greatly diminished. The costs associated with tissue procurement as reflected in the final product are minimal. Conversely, an adult autologous source may be expensive to obtain and will display highly variable performance. Ultimately, the most desirable starting material for cellular therapies is a tissue-specific stem cell whose rate of self-renewal and commitment to differentiation can be controlled. Further, if such a source could be made nonimmunogenic, most of the problems associated with tissue procurement would be solved. The variability in the tissue manufacturing process would be reduced and any quality assurance (QA) and quality control (QC) procedures would be made easier.

Cryopreservation

The scientific basis for cryopreservation is rooted in basic biophysics, chemistry, and engineering. Current cryopreservation procedures are clinically accepted for a number of tissues, including bone marrow, blood cells, cornea, germ cells, and vascular tissue. Recent experience has shown that, in general, the same procedures cannot be applied to human cells that have been grown *ex vivo*. New procedures need to be invented, developed, and implemented. Any cryopreservation used in existing cellular therapies that rely on *ex vivo* manipulation calls for freezing the primary tissue prior to the desired manipulation.

7.6   FUTURE DIRECTIONS: FUNCTIONAL TISSUE ENGINEERING AND THE "-OMICS" SCIENCES

7.6.1   Cellular Aspects

Some cell populations that are to be transplanted may contain subpopulations of unwanted cells. The primary example of this is the contamination of autologously harvested hematopoietic cell populations with the patient's tumor cells. Ideally, any such contamination should be removed prior to transplantation. Similarly, many biopsies are contaminated with accessory cells that may grow faster than the desired parenchymal cells. Fibroblasts are a difficult contaminant to eliminate in many biopsies and they often show superior growth characteristics in culture. For this reason, it is difficult to develop primary cell lines from many tumor types.

7.6.2   Functional Tissue Engineering

Tissue engineering is distinguished from cell biology by the focus on the emergent function that arises from the organization of large numbers of cells into higher-order structures, variously called tissues or organs, depending on the level of anatomical complexity and structural integration. The re-engineering of complex human anatomical structures such as limbs or organ systems is by definition a systems engineering problem. Though it can be argued that all tissue functions arise from fundamental cellular mechanisms, the system-level organization of tissues and organs confers function that is not possible to achieve with individual cells or masses of unorganized cells in a scaffold. A pile of bricks does not provide the functionality of a house, nor does a crate full of car parts function like an automobile. Analogously, engineered tissues must be viewed at the systems level, and the success or failure of the engineering effort ultimately rests upon a quantitative assessment of the organ-level function of the engineered tissue or organ. The use of molecular biological techniques to verify the presence of one or more critical subcellular constituents is simply not an adequate demonstration of end organ function. Thus it is important to develop the necessary facilities to actually quantify the organ-level function of tissue engineered constructs. Of course, different tissues and organ systems have different functions. When designing a tissue or organ, it is therefore essential to develop a design specification for the engineered tissue, with well-defined, quantitative, functional assessments— also called figures of merit (FoM)—as well as a defined method by which to assess these values.

Presumably, the tissues or organs will be cultured for some time prior to their use to permit growth and development. Many tissues have a measurable function that changes during development, so it is most desirable to identify one or more quantities that may be measured nondestructively during the course of the development of the tissues in culture. A specific example is instructive: the contractility of mammalian skeletal muscle changes throughout the early stages of development into adulthood. Muscle phenotype is defined largely by the myosin heavy chain content of individual muscle fibers, but these can be quantitatively inferred by nondestructive measurements of the isometric and dynamic contractility of the muscle tissue. The same is true for tendon. The tangent modulus, tensile strength, and fracture toughness increase during development, whereas the size of the "toe region" of the stress–strain curve tends to decrease, presumably due to the increasingly well-ordered collagen structure during pre- and early postnatal development. It is not possible to nondestructively test the tensile strength and fracture toughness of a cultured tendon specimen, but the tangent modulus and the characteristics of the toe region can be readily measured with minimal disruption to the tendon tissue in culture. With musculoskeletal tissues, it often happens to be the case that the electromechanical signals that are required for nondestructive quantitative assessment of the tissues in culture are essentially the same as those that would be applied chronically to the tissues in culture to guide and promote development. For example, electrically elicited contractions of skeletal and cardiac muscle are currently in use in an attempt to promote development, and the application of mechanical strain has been used since the 1980s on many musculoskeletal (muscle, bone, tendon, cartilage, ligament) and cardiovascular tissues to promote development in culture.

The future challenge is to develop bioreactor systems that permit the application of the stimulus signals while simultaneously allowing the functional properties of the tissues to be nondestructively measured and recorded. If the functional properties of the developing tissue are measured in real time in a bioreactor system, it then becomes possible to assess the current developmental status of each tissue specimen and to use this information to modify the stimulus parameters accordingly. This permits stimulus feedback control of the tissue during development and represents a significant increase in the level of sophistication and effectiveness of functional tissue engineering technology. This constitutes an important aspect of current research in the areas of both musculoskeletal and cardiovascular functional tissue engineering.

7.6.3 Bioartificial Liver Specifics

The development of bioartificial liver specific (BAL) devices arose from society's desire to cure the ill and extend the lives of the human species when confronted with liver failure. This need quickly arises because "backup" systems to replace deficient liver functions are nonexistent, as compared to other tissue duplications such as dual lungs, two kidneys, and fibular crutches. Consequently, this "lone" organ entity that provides multiple functions to assist with body homeostasis is in need of replacement systems when confronted with organ failures such as fulminant hepatic failure (FHF).

Reports of liver treatment can be dated from the 1950s when low protein diets were recommended to improve mental impairment and hepatic encephalopathy and the 1960s' novel concepts of liver assist devices, such that treatment options to replace specific liver functions began its evolution process. Some of these precedent artificial assist systems (Artif-S) currently remain on the research bench or have entered into preliminary FDA trials due to their intrinsic capabilities of treating patients suffering from FHF or other liver specific malfunctions. A few of these systems include charcoal filters for ammonia detoxification, mechanical dialysis permitting toxin transfers, and plasmapheresis for removal of diseased circulating substances. Investigations have shown that many Artif-S are successful in their focused purpose, but they are not complete solutions to replace organ responsiveness because few functional tasks are represented in their applications. Although improvements of Artif-S are frequently updated to improve market potential—such as making use of advanced design parameters found in kidney dialysis machines—the multitude of tasks performed by the healthy *in vivo* liver are insufficiently replicated through mechanical interpretations of liver cell functions.

One successful hurdle in the treatment of patients with FHF is the process of tissue transplantation. This technique exchanges a nonfunctioning liver with a healthy organ capable of performing all metabolic reactions. In this way, successful replacement surgeries alleviate the burden of using mechanical devices in concert with cellular activity. The drawback, although minimal in terms of alternatives, is that patients must remain on immunosuppressants to lessen tissue rejection responses. Even though transplantation options are successful, the limited supply of donor liver organs along with tissue matching requirements illustrate that the demand is $\sim 300\%$ greater than the supply. Ultimately, an alternative to bridge or dissolve the "waiting gap" for patients expiring while on liver donor lists must be resolved.

So again, liver research is recycled from original tissue sources and back into engineered substitutions. This recycled process is a means of evolving technologies for present and future successful liver treatment applications. Thus, the realization that FHF treatment continues to be limited by (1) inadequate artificial systems incapable of replicating the entire organ's function and (2) the supply-and-demand imbalances for tissue transplantation suggests that a combined alternative of embedding living tissues within Artif-S materials may be one avenue in which to develop future medical treatment options.

7.7 CONCLUSIONS

Many new cellular therapies are being developed that create challenges for engineering tissue function. To successfully understand tissue function, it must be possible to quantitatively describe the underlying cellular fate processes and manipulate them. Design can be accomplished by designing the physico-chemical rate processes so that they match the requirements of the cellular processes that underlie tissue function. Tissue engineering is an effort that is still in an embryonic stage, but the use of order-of-magnitude and dimensional analysis is proving to be valuable in designing and reconstituting tissue function. The ability to engineer tissues will undoubtedly markedly improve over the coming decades.

7.8 GLOSSARY

Allogeneic: Transplanation of cells or tissues from a donor into a recipient of the same species but a different strain.

Apoptosis: A cellular process of aging that leads to cell death. This process is initiated by the cell itself.

Autologous transplant: Cells or tissue removed from and then given back to the same donor.

BMT: Bone marrow transplantation.

Cellular therapies: The use of grafted or transfused primary human cells into a patient to affect a pathological condition.

Chondrocytes: Cells found in cartilage.

Colony-forming assay: Assay carried out in semisolid medium under growth factor stimulation. Progenitor cells divide, and progeny are held in place so that a microscopically identifiable colony results after 2 weeks.

Cytokinesis: The process occurring after DNA synthesis and resulting in completion of cell division from one cell into two.

Differentiation: The irreversible progression of a cell or cell population to a more mature state. Distinct sets of genes are expressed and at varying levels in the cells during the differentiation process.

Engraftment: The attainment of a safe number of circulating mature blood cells after a BMT.

Extracellular matrix (ECM): An insoluble complex of proteins and carbohydrates found between cells serving to physically connect cell populations and to persistently signal them to behave in specific ways.

Flow cytometry: Technique for cell analysis using fluorescently conjugated monoclonal antibodies that identify certain cell types. More sophisticated instruments are capable of sorting cells into different populations as they are analyzed.

Functional subunits: The irreducible unit in organs that gives tissue function (i.e., alveoli in lung and nephron in kidney).

Graft-versus-host disease: The immunologic response of transplanted cells against the tissue of their new host. This response is often a severe consequence of allogeneic BMT and can lead to death (acute GVHD) or long-term disability (chronic GVHD).

Hematopoiesis: The regulated production of mature blood cells through a scheme of multilineage proliferation and differentiation.

Hyperplasia: Growth process involving complete cell division, both DNA synthesis and cytokinesis.

Hypertrophy: Growth process involving DNA synthesis but not cytokinesis and resulting in cells of higher ploidy which, secondarily, causes cells to become larger. The late stages of many, if not most, tissue lineages have cells that undergo hypertrophy in response to regenerative stimuli.

Lineage: Refers to cells at all stages of differentiation leading to a particular mature cell type.

Long-term culture-initiating cell: Cell that is measured by a 7–12 week *in vitro* assay. LTC-IC are thought to be very primitive, and the population contains stem cells. However, the population is heterogeneous, so not every LTC-IC is a stem cell.

Mesenchymal cells: Cells of connective type tissue, such as fibroblasts, osteoblasts (bone), chondrocytes (cartilage), adipocytes (fat), etc.

Microenvironment: Refers to the environment surrounding a given cell *in vivo*.

Mitosis: The cellular process that leads to cell division.

Mononuclear cell: Refers to the cell population obtained after density centrifugation of whole bone marrow. This population excludes cells without a nucleus (erythrocytes) and polymorphonuclear cells (granulocytes).

Myoablation: The death of all myeloid (red, white, and platelets) cells, as occurs in a patient undergoing high-dose chemotherapy.

Parenchymal cells: The essential and distinctive cells of a particular organ (i.e., hepatocytes in the liver, or mytocytes in muscle).

Progenitor cells: Unipotent cells that derive from stem cells and will differentiate into mature cells.

Self-renewal: Generation of a daughter cell with identical characteristics to the parent cells. Most often used to refer to stem cell division, which results in the formation of new stem cells.

Stem cells: Pluripotent cells that are capable of self-replication (and therefore unlimited proliferative potential). Malignant tumor cells are aberrant forms of stem cells.

Stromal cells: Mesenchymal cells that partner with epithelial cells. They are age- and tissue-specific. Their roles in regulating the expansion and/or differentiation of epithelia have long been known. For example, heterogeneous mixture of support or accessory cells of the BM, also referred to as the adherent layer, is requisite for BM cultures.

Syngeneic: Transplantation of cells or tissues from a donor into a genetically identical recipient.

Xenogeneic: Transplantation of cells or tissues from a donor of one species into a recipient of a different species.

EXERCISES

1. Given the following data, assess whether human hematopoietic stem cells can truly self-renew *in vivo*.
   - About 400 billion mature hematopoietic cells are produced daily.
   - The best estimate of the Hayflick limit is 44 to 50.
   - About 1:5000 progenitors do NOT apoptose.
   - About 50 to 1000 mature cells are made per progenitor (6 to 10 doublings).
   - The entire differentiation pathway may be 17 to 20 doublings (soft fact).
   - About 10 to 30 million cells at most can be made from a highly purified single hematopoietic stem/preprogenitor cell *in vitro*.

   Present order of magnitude calculations in constructing your decision. Also perform parameter sensitivity analysis of each of the parameters that govern the hematopoietic process. How important are the parameters that govern telemorase activity in determining the total number of mature progeny produced over a person's lifetime?

2. At 1 pM concentration, how many molecules are found in a volume of liquid that is equal to the volume of one cell (use a radius of 5 μm)?

3. Use the continuum approach (Eq. 7.3) to show that in a steady state the number of cells produced during a differentiation process that involves replication but no apoptosis ($\alpha = 0$) is

$$\frac{X_{\text{out}}}{X_{\text{in}}} = e^{\mu/\delta}$$

   and is thus primarily a function of the ratio α/μ. a $= 0$ is the completely undifferentiated state and a $= 1$ is the completely differentiated state. What is X_{out} if μ and α are the same orders of magnitude and if α is 10 times slower? Which scenario is a more reasonable possibility in a physiological situation? Note: If the rates are comparable, only two mature progeny will be produced. On the other hand, if the differentiation rate is ten times slower than the replication rate (probably close to many physiologic situations, i.e., 20 h doubling time, and 200 h = 8 days differentiation time) then about one million cells will be produced. Thus the overall dynamic state tissue is strongly dependent on the relative rates of the cellular fate processes.

4. Kinetics of differentiation/continuous model.
   a. Derive the first-order PDE that describes the population balance.
   b. Make time dimensionless relative to the rate of differentiation.
   c. Describe the two resulting dimensionless groups (call the dimensionless group for apoptosis A, and the one for the cell cycle B).
   d. Solve the equation in steady state for $A = 0$.
   e. Solve the equation(s) where A is nonzero for a portion of the differentiation process (i.e., between a_1 and a_2).
   f. Solve the transient equation for $A = 0$.

5. Consider two cells on a flat surface. One cell secretes a chemokine to which the other responds. Show that the steady-state concentration profile of chemokine emanating from the first cell is:

$$C(r) = \frac{R^2 F}{D} \bullet \frac{1}{r}$$

where R is the radius of the cell, F is the secretion rate (molecules/area time), and D is the diffusion coefficient of the chemokine. The distance from the cell surface is r. Use the cell flux equation to calculate the time it would take for the responding cell to migrate to the signaling cell if there is no random motion ($\mu = 0$) given the following values:

$\chi(C) = 20 \, \text{cm}^2/\text{sec-M}$
$D = 10^{-6} \, \text{cm}^2/\text{sec}$
$R = 5 \, \mu\text{m}$
Production rate $= 5000$ molecules/cell/sec

Hint: Show that the constitutive equation for J reduces to $v = \chi dC/dr$ where v is the velocity of the cell.

6. The flux (F) of a molecule present at concentration C through a circular hole of diameter d on a surface that is adjacent to a fluid that it is diffusing in is given by

$$F = 4DdC$$

The total that can be transferred is the per pore capacity times the number of pores formed.

Calculate the flux allowed though each pore if the diffusion coefficient is $10^{-6} \, \text{cm}^2/\text{sec}$, the concentration is 1 mM, and the pore diameter is 1.5 nm. Discuss your results and try to estimate how many pores are needed to reach meaningful cell-to-cell communications. With the per pore flux estimated above, derive the time constant for transfer of a metabolite from a particular cell to a neighboring cell. Assume that these are two epithelial cells whose geometry can be approximated as a box and that the two adjacent boxes are connected with transfer occurring through n pores.

7. If the cellularity in cartilage is about one million cells per cc, estimate the average distance between the cells. Discuss the characteristics of this micro-environment.

8. Use a one-dimensional analysis of the diffusion of oxygen into a layer of adherent cells to show that the maximum oxygen delivery per unit area (Noxmax) in Example Problem 7.7 is given by

$$N_{ox}^{max} = DC^*/R$$

where C^* is the saturation concentration of oxygen and R is the thickness of the liquid layer.

9. Consider a neuron growth cone that is being influenced by a chemoattractant produced by a target cell. The geometry of the model system is shown in the figure. The target cell secretes a chemoattractant at a rate, P_r, which diffuses into a 3-dimensional volume, with a diffusivity D. The governing equation for mass transport for a spherical source is

$$\frac{\partial C}{\partial t} = D\frac{1}{r^2}\frac{\partial}{\partial r}\left(r^2\frac{\partial C}{\partial r}\right)$$

a. What are the boundary conditions for the system?

b. Considering a steady state, derive the concentration profile as a function of r.

It has been found that the growth cone senses a target cell when the concentration difference across the growth cone is higher than 2%. It is believed that growth cones develop filopodia that extend radially out of the cone as a means to enhance their chemosensing ability. Let β_1 be the angle that a filopodia makes with the center radius line, R, and β_2 be the angle made

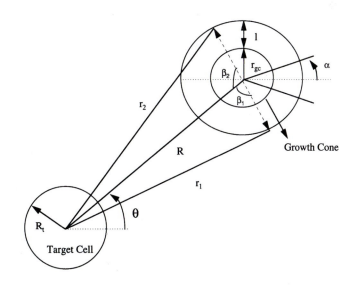

by a filopodia extending in the diametrically opposite direction. Filopodia can extend radially from the growth cone surface except where the cone connects to the axon as defined by α.

c. What are the appropriate limits of β_1 and β_2?

d. What are r_1 and r_2 as a function of β_1?

e. What is the percentage change in concentration, $\Delta C = \frac{|C_1 - C_2|}{C_0}$, across the effective growth cone radius (e.g., $R_{gc} + l$, as a function of r_1 and r_2 and consequently β_1?

f. What is β_1^{max}, for which ΔC is maximum. Plot the gradient change for the entire range of β_1 for filopodial lengths of 1, 5, 8 μm compared to no filopodia.

g. The limit for chemosensing ability is a concentration difference of 2%. For a target cell 500 μm away, what is the effect of filopodial length on enhancing chemosensing ability? Calculate for β_1^{max}.

$$C_0 = \frac{PrR_t^2}{Dr_o}, \quad Pr = \frac{S_r}{4\pi R_t^2}$$

$D = 10^{-6} \, cm^2/sec$ and $S_r = 5000 \, molecules/cell/sec$. $R_0 = R - R_{gc}$

$R_t = 20 \, \mu m, \quad R_{gc} = 2.5 \, \mu m, \quad \alpha = 30°$

10. A coaxial bioreactor has been developed to mimic the liver ascinus.

a. Knowing the dimensions of the ascinus (shown in the following figure), what should be the distance between the two fibers?

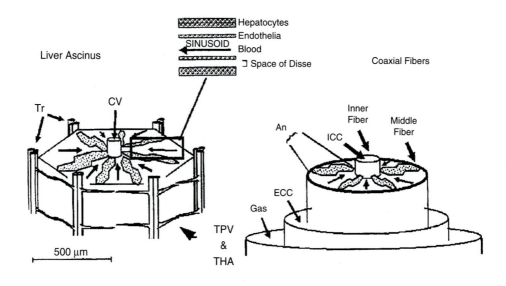

b. From what you've learned about biological aspects what type of cells and soluble and insoluble factors, would you choose to induce an environment of cell growth? What type of cells and what soluble and insoluble factors would you choose to induce an environment of cell differentiation?

11. Use a compartmental model to calculate the number of mature cells produced from a single cell in a particular compartment. Use a doubling time of 24 hours $(\ln(2)/\mu)$ and a mature cell half-life of 8 hours $(\ln(2)/k_d)$. Assume that self-renewal can take place only in the first compartment with a probability of 0.5. Use a total of 10 compartments and calculate the number of cells as a function of time with the initial conditions:

$$X_i(0) = 0 \text{ except } X_j(0) = 1$$

and vary j between 1 and 8. Plot all curves on the same plot. Discuss the implications of your results for transplantation.

12. Kinetics of differentiation/feedback control in a compartmental model.

a. Consider a 6-stage differentiation process ($N = 6$), in which the last population, X_6, produces a cytokine, G, at a per cell rate of q_G. This cytokine has a half-life $t_{0.5}$ ($= 2\,\text{hrs}$) and influences the growth rate of the stem cells (i.e., $u_1 = fn([G])$, where $[G]$ is the concentration of the growth factor G. Extend the base set of differential equations to describe the dynamics of $[G]$.

b. Incorporate into the equations

$$u_1([G]) = u/(1 + K[G])$$

where K is the binding constant for the growth factor. What does the function $f([G])$ describe physiologically

c. Make the equations dimensionless using the growth rate as the scaling factor for time and K for the cell concentration.

d. Describe the meaning of the dimensionless groups and estimate their numerical values.

e. Obtain the numerical values for a simulation starting from a single stem cell. Examine the effect of varying the numerical values of the parameters.

f. Obtain the numerical values for a simulation starting from the steady-state solution and perturb the value of X_3 by 20%. Discuss your results.

SUGGESTED READING

Alberts, B., Bray, D., Lewis, J., Raff, M., Roberts, K. and Watson, J.D. (1994). *Molecular Biology of the Cell,* 3rd ed. Garland, New York.

Brittberg, M. (1994). Treatment of deep cartilage defects in the knee with autologous chondrocyte transplantation. *N. Engl. J. Med.* **331**, 8999.

Bruder, S.P., Fink, D.J. and Caplan, A. (1994). Mesenchymal stem cells in bone development, bone repair, and skeletal regeneration therapy. *Cell. Biochem.* **56**, 283–294.

DeWitt, L. and Wechsler, L. (1988). Transcranial Doppler. *Stroke* **19**(7), 915.

Fawcett, D.W. (1986). *A Textbook of Histology*, 11th ed. Saunders, Philadelphia.

Gage, F. (1994). Neuronal stem cells: Their characterization and utilization. *Neurobiology of Aging* **15**(2). S191.

Galli, R., Binda, E., Orfanelli, U., Cipelletti, B., Gritti, A., De Vitis, S., Fiocco, R., Foroni, C., Dimeco, F. and Vescovi, A. (2004). Isolation and characterization of tumorigenic, stem-like neural precursors from human glioblastoma. *Cancer Res.* **64**(19), 7011–7021.

Garby, L. and Meldon, J. (1977). *The Repiratory Functions of Blood*. Plenum, New York.

Granick, S. and Levere, R. (1964). Heme synthesis in erythroidal cells. In *Progress in Hematology* (C. Moore and E. Brown, Eds.). Grune & Stratton, Orlando, FL.

Gerlyng, P., Abyholm, A., Grotmol, T., Erikstein, B., Huitfeldt, H., Stokke, T. and Seglan, P. (1993). Binucleation and polyploidization patterns in developmental and regenerative rat liver growth. *Cell Proliferation* **26**, 557–565.

Harley, C.B., Futcher, A.B. and Greider, C.W. (1990). Telomeres shorten during aging of human fibroblasts. *Nature* **345**, 458–460.

Hoffbrand, A. and Pettit, J. (1988). Clinical hematology. In *Clinical Hematology* (A. Hoffbrand and J. Pettit, Eds.). Gower Medical, London.

Iredale, J.P. (1997). Tissue inhibitors of metalloproteinases in liver fibrosis. *Int. J. Biochem. & Cell Biol.* **29**, 43–54.

Jager, Ricketts, and Strandness. (1985). *Duplex scanning for the evaluation of lower limb arterial disease. Non invasive diagnostic techniques in vascular disease*. Mosby, St. Louis.

Knauer, D.J., Wiley, H.S. and Cunningham, D.D. (1984). Relationship between epidermal growth factor receptor occupancy and mitogenic response: Quantative analysis using a steady state model system. *J. Biol. Chem.* **259**(9): 5623–5631.

Koller, M.R. and Palsson, B.O. (1993). Tissue engineering: Reconstitution of human hematopoiesis ex vivo. *Biotechnol. & Bioeng.* **42**, 909–930.

Koller, M.R., Bradley, M.S. and Palsson, B.O. (1995a). Growth factor consumption and production in perfusion cultures of human bone marrow correlates with specific cell production. *Exper. Hematol.* **23**, 1275–1283.

Koller, M.R., Manchel, I., Palsson, M.A., Maher, R.J. and Palsson, B.O. (1995b). Different measures of ex vivo hematopoiesis culture performance is optimized under vastly different conditions. *Biotechnol. Bioeng.* **50**, 505–513.

Koller, M.R., Palsson, M.A., Manchel, I. and Palsson, B.O. (1995c). Long-term culture-initiating cell expansion is dependent on frequent medium exchange combined with stromal and other accessory cell effects. *Blood* **86**, 1784–1793.

Krstic, R. (1985). English edition, General histology of the mammal. Springer Verlag, NY, page 83, plate 40. Translated for the original German edition, Krstic, R. (1978). *Die Gewebes des Menschen und der Saugertiere*. Springer Verlag, Berlin.

Lacey, P.E., (1995, July). Treating diabetes with transplanted cells. *Sci. Am.* 50–58.

Langer, R. and Vacanti, J.P. (1993). Tissue engineering. *Science* **260**, 920–926.

Lanza, R., Langer, R., and Vacanti, J. (2000). *Principles of Tissue Engineering*. Academic Press, Inc., San Diego, CA.

Lanza, R., Jackson, R., Sullivan, A., Ringeling, J., McGrath, C., Kuhtreiber, W. and Chick, W. (1999). Xenotransplantation of cells using biodegradable microcapsules. *Transplantation* **67**(8), 1105–1111.

Lauffenburger, D.A. and Linderman, J.J. (1993). *Receptors: Models for Binding, Trafficking, and Signaling*. Oxford Univ. Press, New York.

Lightfoot, E.N. (1974). *Transport Phenoment and Living Systems*. Wiley & Sons, New York.

Lodish, H., Baltimore, D., Berk, A., Zipursky, S.L., Matsudaira, P. and Darnell, J. (1995). *Molecular Cell Biology,* 3rd ed. Scientific American Books, New York.

MacDonald, J., Griffin, J., Kubota, H., Griffith, L., Fair, J. and Reid, L. (1999). *Bioartificial Livers*. Birkhauser, Boston.

McCuskey, R. (1994). *The Hepatic Microvasculature System. The Liver: Biology and Pathology*. Raven Press Ltd., New York.

Mutsaers, S.E., Bishop, J.E., McGrouther, G. and Laurent, G.J. (1997). Mechanisms of tissue repair: From wound healing to fibrosis. *Int. J. Biochem. & Cell Biol.* **29**(1): 5–17.

Naughton, B. (1995). The importance of stromal cells. In *CRC Handbook on Biomedical Engineering* (J.D. Bronzino, Ed.). CRC, Boca Raton, FL.

Niklason, L. and Langer, R. (2002). Morphologic and mechanical characteristics of engineered bovine arteries. *J. Vascular Surg.* **33**(3): 628–638.

Orlowski, T., Godlewska, E., Moscicka, M. and Sitarek, E. (2003). The influence of intraperitoneal transplantation of free and encapsulated Langerhans islets on the second set phenomenon. *Artificial Organs* **27**(12), 1062–1067.

Peng, C. and Palsson, B.O. (1996a). Cell growth and differentiation on feeder layers is predicted to be influenced by bioreactor geometry. *Biotechnol. & Bioeng.* **50**, 479–492.

Peng, C.A. and Palsson, B.O. (1996b). Determination of specific oxygen uptake rates in human hematopoietic cultures and implications for bioreactor design. *Ann. Biomed. Eng.* **24**, 373–381.

Rhodin, J.A.G. (1974). *Histology: Text and Atlas*. Oxford University Press, Oxford.

Rogers, C.E., Bradley, M.S., Palsson, B.O. and Koller, M.R. (1995). Flow cytometric analysis of human bone marrow perfusion cultures: Erythroid development and relationship with burst-forming ynits-erythroid (BFU-E). *Exp. Hematol.* **24**, 597–604.

Sigal, S. and Brill, S. (1992). Invited review: The liver as a stem cell and lineage system. *Am. J. Physiol.* **263**, 139–148.

Schnitzler, C., Mesquita, J. (1998). Bone marrow composition and bone microarchitecture and turnover in blacks and whites. *J. of Bone and Mineral Res.* **13**(8), 1300–1307.

Tramontin, A.D., Garcia-Verdugo, J.M., Lim, D.A. and Alverex-Buylla, A. (2003). Postnatal development of radial glia and the ventricular zone (VZ): A continuum of the neural stem cell compartment. *Cereb Cortex.* **13** (6), 580–587.

Vander, A.J., Sherman, J.H. and Luciano, D.S. (1994). *Human Physiology: The Mechanisms of Body Function,* 6th ed. McGraw Hill, New York.

8 BIOINSTRUMENTATION

John Enderle, PhD*

Chapter Contents

*With contributions by Susan Blanchard, Amanda Marley, and H. Troy Nagle.

After completing this chapter, the student will be able to:

- Describe the components of a basic instrumentation system.

- Analyze linear circuits using the node-voltage method.

- Simplify complex circuits using Thévenin's and Norton's equivalent circuits.

- Solve circuits involving resistors, capacitors, and inductors of any order.

- Analyze circuits that use operational amplifiers.

- Determine the steady-state response to sinusoidal inputs and work in the phasor domain.

- Understand the basic concepts of analog filter design and design basic filters.

- Design a low-pass, high-pass, and band-pass filter.

- Explain the different types of noise in a biomedical instrument system.

8.1   INTRODUCTION

This chapter provides basic information about bioinstrumentation and electric circuit theory used in other chapters. As described in Chapter 9, many biomedical instruments use a transducer or sensor to convert a signal created by the body into an electric signal. Our goal in this chapter is to develop expertise in electric circuit theory applied to bioinstrumentation. We begin with a description of variables used in circuit theory, charge, current, voltage, power, and energy. Next, Kirchhoff's current and voltage laws are introduced, followed by resistance, simplifications of resistive circuits, and voltage and current calculations. Circuit analysis techniques are then presented, followed by inductance and capacitance, and solutions of circuits using the differential equation method. Finally, the operational amplifier and time varying signals are introduced.

Before 1900, medicine had little to offer the typical citizen because its resources were mainly the education and little black bag of the physician. The origins of the changes that occurred within medical science are found in several developments that took place in the applied sciences. During the early nineteenth century, diagnosis was based on physical examination, and treatment was designed to heal the structural abnormality. By the late nineteenth century, diagnosis was based on laboratory tests, and treatment was designed to remove the cause of the disorder. The trend toward the use of technology accelerated throughout the twentieth century. During this period, hospitals became institutions of research and technology. Professionals in the areas of chemistry, physics, mechanical engineering, and electrical engineering began to work in conjunction with the medical field, and biomedical engineering became a recognized profession. As a result, medical technology advanced more in the twentieth century than it had in the rest of history combined (Fig. 8.1).

During this period the area of electronics had a significant impact on the development of new medical technology. Men such as Richard Caton and Augustus Desire proved that the human brain and heart depended on bioelectric events. In 1903, William Einthoven expanded on these ideas after he created the first string galvanometer. Einthoven placed two skin sensors on a man and attached them to the ends of a silvered wire that was suspended through holes drilled in both ends of a large permanent magnet. The suspended silvered wire moved rhythmically with the subject's heartbeat. By projecting a tiny light beam across the silvered wire, Einthoven was able to record the movement of the wire as waves on a scroll of moving photographic paper. Thus, the invention of the string galvanometer led to the creation of the electrocardiogram (ECG), which is routinely used today to measure and record the electrical activity of abnormal hearts and to compare those signals to normal ones.

In 1929, Hans Berger created the first electroencephalogram (EEG), which is used to measure and record electrical activity of the brain. In 1935, electrical amplifiers were used to prove that the electrical activity of the cortex has a specific rhythm, and, in 1960, electrical amplifiers were used in devices such as the first implantable pacemaker which was created by William Chardack and Wilson Greatbatch. These are just a small sample of the many examples in which the field of electronics has been used to significantly advance medical technology.

Many other advancements that were made in medical technology originated from research in basic and applied physics. In 1895, the x-ray machine, one of the most important technological inventions in the medical field, was created when W. K. Roentgen found that x-rays could be used to give pictures of the internal structures of the body. Thus, the x-ray machine was the first imaging device to be created. (Radiation imaging is discussed in detail in Chapter 15.)

Another important addition to medical technology was provided by the invention of the computer, which allowed much faster and more complicated analyses and functions to be performed. One of the first computer-based instruments in the field of medicine, the sequential multiple analyzer plus computer, was used to store a vast amount of data pertaining to clinical laboratory information. The invention of the computer made it possible for laboratory tests to be performed and analyzed more quickly and accurately.

The first large-scale computer-based medical instrument was created in 1972 when the computerized axial tomography (CAT) machine was invented. The CAT machine created an image that showed all of the internal structures that lie in a single plane of the body. This new type of image made it possible to have more accurate and easier

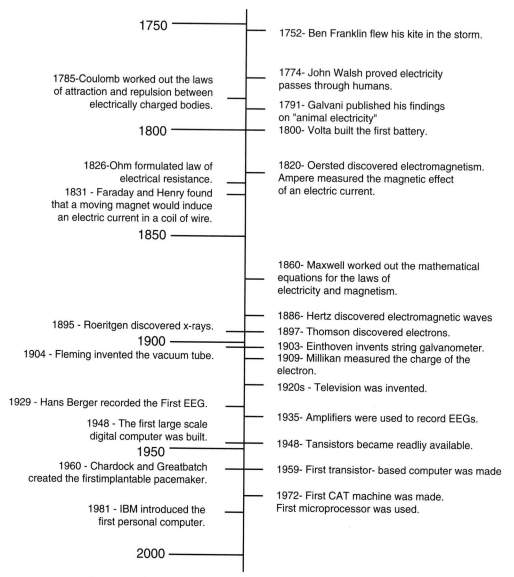

Figure 8.1   Timeline for major inventions and discoveries that led to modern medical instrumentation.

diagnosis of tumors, hemorrhages, and other internal damage from information that was obtained noninvasively (for details, see Chapter 15).

Telemedicine, which uses computer technology to transmit information from one medical site to another, is being explored to permit access to health care for patients in remote locations. Telemedicine can be used to let a specialist in a major hospital receive information on a patient in a rural area and send back a plan of treatment specifically for that patient.

Today, there is a wide variety of medical devices and instrumentation systems. Some are used to monitor patient conditions or acquire information for diagnostic purposes (e.g. ECG and EEG machines) whereas others are used to control physiological functions (e.g., pacemakers and ventilators). Some devices, such as pacemakers, are implantable whereas many others are used noninvasively. This chapter will focus on those features that are common to devices that are used to acquire and process physiological data.

8.2 BASIC BIOINSTRUMENTATION SYSTEM

The quantity, property, or condition that is measured by an instrumentation system is called the measurand (Fig. 8.2). This can be a bioelectric signal, such as those generated by muscles or the brain, or a chemical or mechanical signal that is converted to an electrical signal. As explained in Chapter 9, sensors are used to convert physical measurands into electric outputs. The outputs from these biosensors are analog signals, i.e., continuous signals, which are sent to the analog processing and digital conversion block. There, the signals are amplified, filtered, conditioned, and converted to digital form. Methods for modifying analog signals, such as amplifying and filtering an ECG signal, are discussed later in this chapter. Once the analog signals have been digitized and converted to a form that can be stored and processed by digital computers, many more methods of signal conditioning can be applied (for details see Chapter 10).

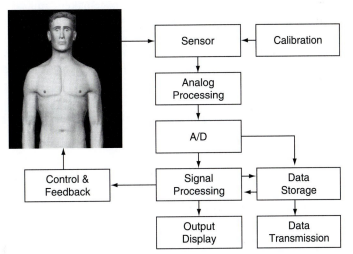

Figure 8.2 Basic instrumentation systems using sensors to measure a signal with data acquisition, storage and display capabilities, and control and feedback.

Basic instrumentation systems also include output display devices that enable human operators to view the signal in a format that is easy to understand. These displays may be numerical or graphical, discrete or continuous, and permanent or temporary. Most output display devices are intended to be observed visually, but some also provide audible output (e.g., a beeping sound with each heartbeat).

In addition to displaying data, many instrumentation systems have the capability of storing data. In some devices, the signal is stored briefly so that further processing can take place or so that an operator can examine the data. In other cases, the signals are stored permanently so that different signal processing schemes can be applied at a later time. Holter monitors, for example, acquire 24 hours of ECG data that is later processed to determine arrhythmic activity and other important diagnostic characteristics.

With the invention of the telephone and now with the Internet, signals can be acquired with a device in one location, perhaps in a patient's home, and transmitted to another device for processing and/or storage. This has made it possible, for example, to provide quick diagnostic feedback if a patient has an unusual heart rhythm while at home. It has also allowed medical facilities in rural areas to transmit diagnostic images to tertiary care hospitals so that specialized physicians can help general practitioners arrive at more accurate diagnoses.

Two other components play important roles in instrumentation systems. The first is the calibration signal. A signal with known amplitude and frequency content is applied to the instrumentation system at the sensor's input. The calibration signal allows the components of the system to be adjusted so that the output and input have a known, measured relationship. Without this information, it is impossible to convert the output of an instrument system into a meaningful representation of the measurand.

Another important component, a feedback element, is not a part of all instrumentation systems. These devices include pacemakers and ventilators that stimulate the heart or the lungs. Some feedback devices collect physiological data and stimulate a response (e.g., a heartbeat or breath) when needed, or are part of biofeedback systems in which the patient is made aware of a physiological measurement (e.g., blood pressure) and uses conscious control to change the physiological response.

8.3   CHARGE, CURRENT, VOLTAGE, POWER, AND ENERGY

8.3.1   Charge

Two kinds of charge, positive and negative, are carried by protons and electrons, respectively. The negative charge carried by an electron, q_e, is the smallest amount of charge that exists and is measured in units called coulombs (C).

$$q_e = -1.602 \cdot 10^{-19} \quad \text{C}$$

The symbol $q(t)$ is used to represent a charge that changes with time, and Q is for constant charge. The charge carried by a proton is the opposite of the electron.

8.3.2 Current

Electric current, $i(t)$, is defined as the change in the amount of charge that passes through a given point or area in a specified time period. Current is measured in amperes (A). By definition, one ampere equals one coulomb/second (C/s).

$$i(t) = \frac{dq}{dt} \tag{8.1}$$

and

$$q(t) = \int_{t_0}^{t} i(\lambda)d\lambda + q(t_0) \tag{8.2}$$

Current, defined by Eq. 8.1, also depends on the direction of flow, as illustrated in the circuit in Figure 8.3.

Current is defined as positive if

1. A positive charge is moving in the direction of the arrow
2. A negative charge is moving in the opposite direction of the arrow

Since these two possibilities produce the same outcome, there is no need to be concerned as to which is responsible for the current. In electric circuits, current is carried by electrons in metallic conductors.

Current is typically a function of time, as given by Eq. 8.1. Consider Figure 8.4, with the current entering terminal 1 in the circuit on the right. In the time interval 0 to 1.5 s, current is positive and enters terminal 1. In the time interval 1.5 to 3 s, the current is negative and enters terminal 2 with a positive value. We typically refer to a constant current as a DC current, and denote it with a capital letter such as I indicating it doesn't change with time. We denote a time-varying current with a lower-case letter such as $i(t)$, or just i.

Kirchhoff's Current Law

Current can flow only in a closed circuit, as shown in Figure 8.1. No current is lost as it flows around the circuit because net charge cannot accumulate within a circuit element and charge must be conserved. Whatever current enters one terminal must

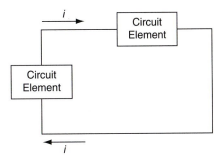

Figure 8.3   A simple electric circuit illustrating current flowing around a closed loop.

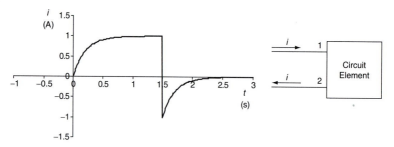

Figure 8.4   (Left) A sample current waveform. (Right) A circuit element with current entering terminal 1 and leaving terminal 2. Passive circuit elements have two terminals with a known voltage–current relationship. Examples of passive circuit elements include resistors, capacitors, and inductors.

leave at the other terminal. Since charge cannot be created and must be conserved, the sum of the currents at any node—that is, a point at which two or more circuit elements have a common connection—must equal zero so no net charge accumulates. This principle is known as Kirchhoff's current law (KCL), given as

$$\sum_{n=1}^{N} i_n(t) = 0 \tag{8.3}$$

where there are N currents leaving the node. Consider the circuit in Figure 8.5. Using Eq. 8.3 and applying KCL for the currents leaving the node gives

$$-i_1 - i_2 + i_4 + i_3 = 0$$

The previous equation is equivalently written for the currents entering the node, as

$$i_1 + i_2 - i_4 - i_3 = 0$$

It should be clear that the application of KCL is for all currents, whether they are all leaving or all entering the node.

In describing a circuit, we define its characteristics with the terms *node*, *branch*, *path*, *closed path*, and *mesh* as follows.

- *Node:* A point at which two or more circuit elements have a common connection.
- *Branch:* A circuit element or connected group of circuit elements; a connected group of circuit elements usually connect nodes together.

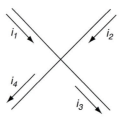

Figure 8.5   A node with 4 currents.

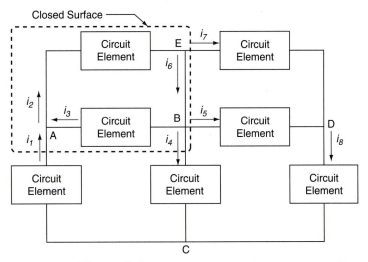

Figure 8.6 A circuit with a closed surface.

- *Path:* A connected group of circuit elements in which none is repeated.
- *Closed path:* A path that starts and ends at the same node.
- *Mesh:* A closed path that does not contain any other closed paths within it.
- *Essential node:* A point at which three or more circuit elements have a common connection.
- *Essential branch:* A branch that connects two essential nodes.

In Figure 8.6, there are five nodes, A, B, C, D, and E, which are all essential nodes. Kirchhoff's current law is applied to each of the nodes as follows.

$$\text{Node A: } -i_1 + i_2 - i_3 = 0$$
$$\text{Node B: } i_3 + i_4 + i_5 - i_6 = 0$$
$$\text{Node C: } i_1 - i_4 - i_8 = 0$$
$$\text{Node D: } -i_7 - i_5 + i_8 = 0$$
$$\text{Node E: } -i_2 + i_6 + i_7 = 0$$

Kirchhoff's current law is also applicable to any closed surface surrounding a part of the circuit. It is understood that the closed surface does not intersect any of the circuit elements. Consider the closed surface drawn with dashed lines in Figure 8.6. Kirchhoff's current law applied to the closed surface gives

$$-i_1 + i_4 + i_5 + i_7 = 0$$

8.3.3 Voltage

Voltage represents the work per unit charge associated with moving a charge between two points (A and B in Fig. 8.7) and given as

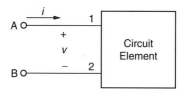

Figure 8.7   Voltage and current convention.

$$v = \frac{dw}{dq} \tag{8.4}$$

The unit of measurement for voltage is the volt (V). A constant (DC) voltage source is denoted by V while a time-varying voltage is denoted by $v(t)$, or just v. In Figure 8.7, the voltage, v, between two points (A and B) is the amount of energy required to move a charge from point A to point B.

Kirchhoff's Voltage Law

Kirchhoff's voltage law (KVL) states the sum of all voltages in a closed path is zero, or

$$\sum_{n=1}^{N} v_n(t) = 0 \tag{8.5}$$

where there are N voltage drops assigned around the closed path, with $v_n(t)$ denoting the individual voltage drops. The sign for each voltage drop in Eq. 8.5 is the first sign encountered while moving around the closed path.

Consider the circuit in Figure 8.8, with each circuit element assigned a voltage, v_n, with a given polarity, and three closed paths, CP1, CP2, and CP3. Kirchhoff's voltage V_n law for each closed path is given as

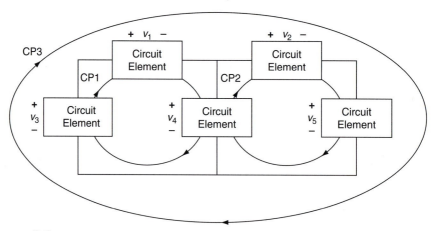

Figure 8.8   Circuit illustrating Kirchhoff's voltage law. Closed paths are identified as CP1, CP2, and CP3.

$$CP1: -v_3 + v_1 + v_4 = 0$$
$$CP2: -v_4 + v_2 + v_5 = 0$$
$$CP3: -v_3 + v_1 + v_2 + v_5 = 0$$

Kirchhoff's laws are applied in electric circuit analysis to determine unknown voltages and currents. Each unknown variable has its distinct equation. To solve for the unknowns using MATLAB, we create a matrix representation of the set of equations and solve using the techniques described in the appendix. This method is demonstrated in many examples in this chapter.

8.3.4 Power and Energy

Power is the rate of energy expenditure given as

$$p = \frac{dw}{dt} = \frac{dw}{dq}\frac{dq}{dt} = vi \tag{8.6}$$

where p is power measured in watts (W), and w is energy measured in joules (J). Power is usually determined by the product of voltage across a circuit element and the current through it. By convention, we assume that a positive value for power indicates that power is being delivered (or absorbed or consumed) by the circuit element. A negative value for power indicates that power is being extracted or generated by the circuit element (i.e., a battery).

Figure 8.9 illustrates the four possible cases for a circuit element's current and voltage configuration. According to convention, if both i and v are positive, with the arrow and polarity shown in Figure 8.9A, energy is absorbed (either lost by heat or

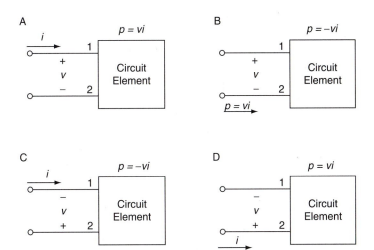

Figure 8.9 Polarity references for four cases of current and voltage. Cases A and D result in positive power being consumed by the circuit element. Cases B and C result in negative power being extracted from the circuit element.

stored). If either the current arrow or voltage polarity is reversed as in B and C, energy is supplied to the circuit. Note that if both the current direction and voltage polarity are reversed together, as in D, energy is absorbed.

A passive circuit element is defined as an element whose power is always positive or zero, which may be dissipated as heat (resistance) stored in an electric field (capacitor) or stored in a magnetic field (inductor). We define an active circuit element as one whose power is negative and capable of generating energy.

Energy is given by

$$w(t) = \int_{-\infty}^{t} p\,dt \tag{8.7}$$

8.3.5   Sources

Sources are two terminal devices that provide energy to a circuit. There is no direct voltage–current relationship for a source; when one of the two variables is given, the other cannot be determined without knowledge of the rest of the circuit. Independent sources are devices for which the voltage or current is given and the device maintains its value regardless of the rest of the circuit. A device that generates a prescribed voltage at its terminals, regardless of the current flow, is called an ideal voltage source. Figures 8.10a and b show the general symbols for an ideal voltage source. Figure 8.10c shows an ideal current source that delivers a prescribed current to the attached circuit. The voltage generated by an ideal current source depends on the elements in the rest of the circuit.

Shown in Figure 8.11 are a dependent voltage and current source. A dependent source takes on a value equaling a known function of some other voltage or current value in the circuit. We use a diamond-shaped symbol to represent a dependent source. Often, a dependent source is called a controlled source. The current generated

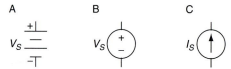

Figure 8.10   Basic symbols used for independent sources: (A) battery, (B) ideal voltage source—V_s can be a constant DC source (battery) or a time-varying source, and (C) ideal current source, I_s.

Figure 8.11   Basic symbols used for dependent or controlled sources: (left) controlled voltage source—the voltage v_s is a known function of some other voltage or current in the circuit; (right) controlled current source—the current i_s is a known function of some other voltage or current in the circuit.

for a dependent voltage source and the voltage for a dependent current source depend on circuit elements in the rest of the circuit. Dependent sources are very important in electronics. Later in this chapter, we will see that the operational amplifier uses a controlled voltage source for its operation.

8.4 RESISTANCE

8.4.1 Resistors

A resistor is a circuit element that limits the flow of current through it and is denoted with the symbol ⩗⩗. Resistors are made of different materials and their ability to impede current is given with a value of resistance, denoted R. Resistance is measured in ohms (Ω), where $1\ \Omega = 1$ V/A. A theoretical bare wire that connects circuit elements together has a resistance of zero. A gap between circuit elements has a resistance of infinity. An ideal resistor follows Ohm's law, which describes a linear relationship between voltage and current, with a slope equal to the resistance.

There are two ways to write Ohm's law, depending on the current direction and voltage polarity. Ohm's law is written for Fig. 8.12A as

$$v = iR \qquad (8.8)$$

and for Fig. 8.12B as

$$v = -iR \qquad (8.9)$$

In this book, we will use the convention shown in Figure 8.12a to write the voltage drop across a resistor. As described, the voltage across a resistor is equal to the product of the current flowing through the element and its resistance, R. This linear relationship does not apply at very high voltages and currents. Some electrically conducting materials have a very small range of currents and voltages in which they exhibit linear behavior. This is true of many physiological models as well: linearity is observed only within a range of values. Outside this range, the model is nonlinear. We define a short circuit as shown in Figure 8.13A with $R = 0$, and having a 0 V voltage drop. We define an open circuit as shown in Figure 8.13B with $R = \infty$, and having 0 A current pass through it.

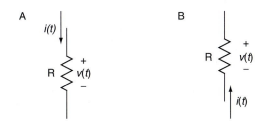

Figure 8.12 An ideal resistor with resistance R in ohms (Ω).

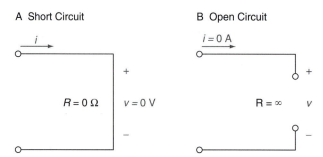

Figure 8.13   Short and open circuits.

Each material has a property called resistivity (ρ) that indicates the resistance of the material. Conductivity (σ) is the inverse of resistivity, and conductance (G) is the inverse of resistance. Conductance is measured in units called siemens (S) and has units of A/V. In terms of conductance, Ohm's Law is written as

$$i = Gv \tag{8.10}$$

Example Problem 8.1

From the following circuit, find I_2, I_3, and V_1.

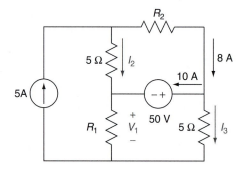

Solution

Find I_2 first by applying KCL at the node in the upper left of the circuit.

$$-5 + I_2 + 8 = 0$$

and

$$I_2 = -3\,\text{A}$$

Current I_3 is determined by applying KCL at the node on the right of the circuit.

$$10 + I_3 - 8 = 0$$

and

$$I_3 = -2\,A$$

Voltage V_1 is determined by applying KVL around the lower right closed path and using Ohm's law.

$$-V_1 - 50 + 5I_3 = 0$$
$$V_1 = -50 + 5 \times (-2) = -60\,V$$ ∎

8.4.2 Power

The power consumed by a resistor is given by the combination of Eq. 8.6 and either Eq. 8.8 or 8.9 as

$$p = vi = \frac{v^2}{R} = i^2 R \tag{8.11}$$

and given off as heat. Eq. 8.11 demonstrates that regardless of the voltage polarity and current direction, power is consumed by a resistor. Power is always positive for a resistor, which is true for any passive element.

Example Problem 8.2

Electric safety is of paramount importance in a hospital or clinical environment. If sufficient current is allowed to flow through the body, significant damage can occur, as illustrated in the following figure.

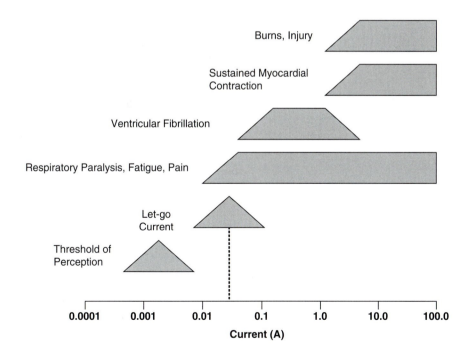

For instance, a current of magnitude 50 mA (dashed line) is enough to cause ventricular fibrillation, as well as other conditions. The figure on the left shows the current distribution from a macroshock from one arm to another. A crude electric circuit model of the body consisting of two arms (each with resistance R_A), two legs (each with resistance R_L), body trunk (with resistance R_T), and head (with resistance R_H) is shown in the following figure on the right.

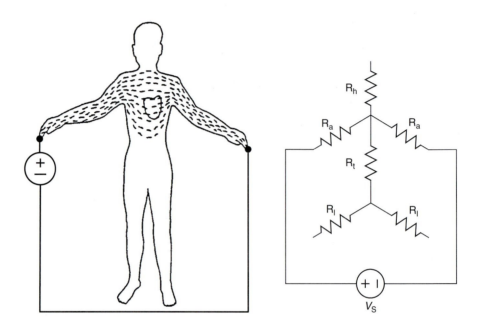

Since the only elements that form a closed path that current can flow is given by the source in series with the two arms, we reduce the body electric circuits to

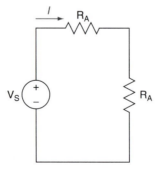

If $R_A = 400\ \Omega$ and $V_s = 120\ \text{V}$, then find I.

Solution

Using Ohm's law, gives

$$I = \frac{V_s}{R_A + R_A} = \frac{120}{800} = 0.15\,\text{A}$$

The current I is the current passing through the heart, and at this level it would cause ventricular fibrillation. ∎

8.4.3 Equivalent Resistance

It is sometimes possible to reduce complex circuits into simpler, equivalent circuits. Consider two circuits equivalent if they cannot be distinguished from each other by voltage and current measurements—that is, the two circuits behave identically. Consider the two circuits A and B in Figure 8.14, consisting of combinations of resistors, each stimulated by a DC voltage V_s. These two circuits are equivalent if $I_A = I_B$. We represent the resistance of either circuit using Ohm's law as

$$R_{EQ} = \frac{V_s}{I_A} = \frac{V_s}{I_B} \tag{8.12}$$

Thus it follows that any circuit consisting of resistances can be replaced by an equivalent circuit as shown in Figure 8.15. In another section on a Thévenin equivalent circuit, we will expand this remark to include any combination of sources and resistances.

8.4.4 Series and Parallel Combinations of Resistance

Resistors in Series

If the same current flows from one resistor to another, the two are said to be in series. If these two resistors are connected to a third and the same current flows through all of

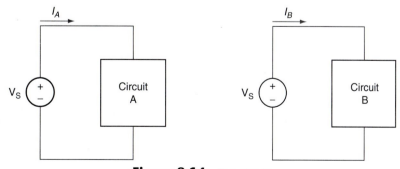

Figure 8.14 Two circuits.

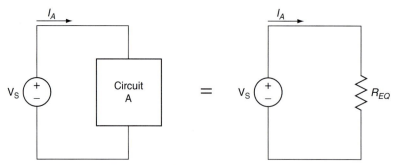

Figure 8.15   Equivalent circuits.

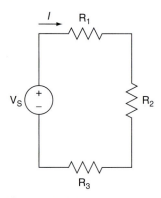

Figure 8.16   A series circuit.

them, then the three resistors are in series. In general, if the same current flows through N resistors, then the N resistors are in series. Consider Figure 8.16 with three resistors in series. An equivalent circuit can be derived through KVL as

$$-V_s + IR_1 + IR_2 + IR_3 = 0$$

or rewritten in terms of an equivalent resistance R_{EQ} as

$$R_{EQ} = \frac{V_s}{I} = R_1 + R_2 + R_3$$

In general, if we have N resistors in series,

$$R_{EQ} = \sum_{i=1}^{N} R_i \qquad (8.13)$$

Resistors in Parallel

Two or more elements are said to be in parallel if the same voltage is across each of the resistors. Consider the three parallel resistors as shown in Figure 8.17. We use a short-

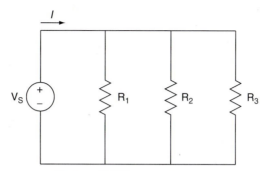

Figure 8.17 A parallel circuit.

hand notation to represent resistors in parallel using the ‖ symbol. Thus in Figure 8.16, $R_{EQ} = R_1 \parallel R_2 \parallel R_3$. An equivalent circuit for Figure 8.16 is derived through KCL as

$$-I + \frac{V_s}{R_1} + \frac{V_s}{R_2} + \frac{V_s}{R_3} = 0$$

or rewritten in terms of an equivalent resistance R_{EQ} as

$$R_{EQ} = \frac{V_s}{I} = \frac{1}{\frac{1}{R_1} + \frac{1}{R_2} + \frac{1}{R_3}}$$

In general, if we have N resistors in parallel,

$$R_{EQ} = \frac{1}{\frac{1}{R_1} + \frac{1}{R_2} + \cdots + \frac{1}{R_N}} \tag{8.14}$$

For just two resistors in parallel, Eq. 8.14 is written as

$$R_{EQ} = R_1 \parallel R_2 = \frac{R_1 R_2}{R_1 + R_2} \tag{8.15}$$

Example Problem 8.3

Find R_{EQ} and the power supplied by the source for the following circuit.

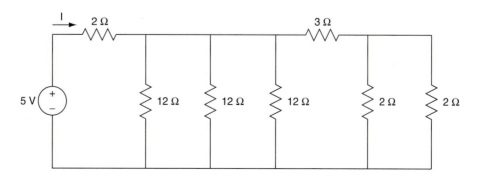

Solution

To solve for R_{EQ}, apply from right to left the parallel and series combinations. First, we have two 2-Ω resistors in parallel that are in series with the 3-Ω resistor. Next, this group is in parallel with the three 12-Ω resistors. Finally, this group is in series with the 2-Ω resistor. These combinations are shown in the following figure and calculation:

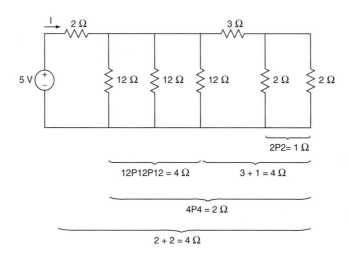

$$R_{EQ} = 2\ \Omega + ((12\ \Omega \parallel 12\ \Omega \parallel 12\ \Omega) \parallel (3\ \Omega + (2\ \Omega \parallel 2\ \Omega)))$$

$$= 2 + \left(\left(\frac{1}{\frac{1}{12} + \frac{1}{12} + \frac{1}{12}}\right) \parallel \left(3 + \frac{1}{\frac{1}{2} + \frac{1}{2}}\right)\right)$$

$$= 2 + ((4) \parallel (3 + 1)) = 2 + 2 = 4\ \Omega$$

Accordingly,

$$I = \frac{5}{R_{EQ}} = \frac{5}{4} = 1.25\ A$$

and

$$p = 5 \times I = 6.25\ W \qquad \blacksquare$$

8.4.5 Voltage and Current Divider Rules

Let us now extend the concept of equivalent resistance, $R_{EQ} = \frac{V}{I}$, to allow us to quickly calculate voltages in series resistor circuits and currents in parallel resistor circuits without digressing to the fundamentals.

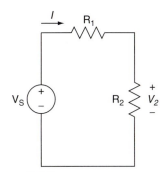

Figure 8.18 Voltage divider rule circuit.

Voltage Divider Rule

The voltage divider rule allows us to easily calculate the voltage across a given resistor in a series circuit. Consider finding V_2 in the series circuit shown in Figure 8.18, where $R_{EQ} = R_1 + R_2$. Accordingly,

$$I = \frac{V_s}{R_{EQ}} = \frac{V_s}{R_1 + R_2}$$

and therefore

$$V_2 = IR_2 = V_s \frac{R_2}{R_1 + R_2}$$

This same analysis can be used to find V_1 as

$$V_1 = V_s \frac{R_1}{R_1 + R_2}$$

In general, if a circuit contains N resistors in series, the voltage divider rule gives the voltage across any one of the resistors, R_i, as

$$V_i = V_s \frac{R_i}{R_1 + R_2 + \cdots R_N} \tag{8.16}$$

Current Divider Rule

The current divider rule allows us to easily calculate the current through any resistor in parallel resistor circuits. Consider finding I_2 in the parallel circuit shown in Figure 8.19, where $R_{EQ} = \frac{R_1 R_2}{R_1 + R_2}$. Accordingly,

$$I_2 = \frac{V_s}{R_2}$$

and

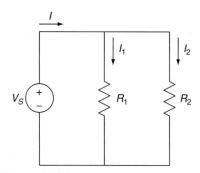

Figure 8.19   Current divider rule circuit.

$$V_s = I\frac{R_1 R_2}{R_1 + R_2}$$

yielding after substituting V_s

$$I_2 = I\frac{\frac{1}{R_2}}{\frac{1}{R_1} + \frac{1}{R_2}}$$

In general, if a circuit contains N resistors in parallel, the current divider rule gives the current through any one of the resistors, R_i, as

$$I_i = I\frac{\frac{1}{R_i}}{\frac{1}{R_1} + \frac{1}{R_2}\cdots + \frac{1}{R_N}} \tag{8.17}$$

Example Problem 8.4

For the following circuit, find I_1.

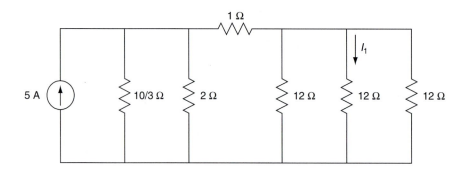

Solution

This circuit problem is solved in two parts, as is evident from the redrawn circuit that follows, by first finding I_2 and then I_1.

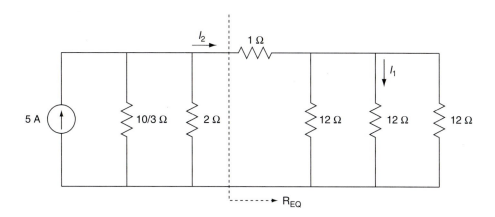

To begin, first find R_{EQ}, which, when placed into the circuit, reduces to three parallel resistors from which I_2 is calculated. The equivalent resistance is found as

$$R_{EQ} = 1 + (12 \parallel 12 \parallel 12) = 1 + \frac{1}{\frac{1}{12} + \frac{1}{12} + \frac{1}{12}} = 5 \ \Omega$$

Applying the current divider rule on the three parallel resistors, $\frac{10}{3} \parallel 2 \parallel R_{EQ}$, we have

$$I_2 = 5 \left(\frac{\frac{1}{5}}{\frac{3}{10} + \frac{1}{2} + \frac{1}{5}} \right) = 1\mathrm{A}$$

I_2 flows through the 1 Ω resistor, and then divides into three equal currents of 1/3 A through each 12 Ω resistor. The current I_1 can also be found by applying the current divider rule as

$$I_1 = I_2 \left(\frac{\frac{1}{12}}{\frac{1}{12} + \frac{1}{12} + \frac{1}{12}} \right) = \frac{\frac{1}{12}}{\frac{1}{12} + \frac{1}{12} + \frac{1}{12}} = \frac{1}{3}\mathrm{A} \qquad ■$$

8.5   LINEAR NETWORK ANALYSIS

Our methods for solving circuit problems up to this point have included applying Ohm's law and Kirchhoff's laws, resistive circuit simplification, and the voltage and current divider rules. This approach works for all circuit problems, but as the circuit complexity increases, it becomes more difficult to solve problems. In this section, we introduce the node-voltage method to provide a systematic and easy solution of circuit problems. The application of the node-voltage method involves expressing the branch currents in terms of one or more node voltages and applying KCL at each of the nodes. This method provides a systematic approach that leads to a solution that is efficient and robust, resulting in a minimum number of simultaneous equations, which saves time and effort.

The use of node equations provides a systematic method for solving circuit analysis problems by the application of KCL at each essential node. The node-voltage method involves the following two steps:

1. Assign each node a voltage with respect to a reference node (ground). The reference node is usually the one with the most branches connected to it and is denoted with the symbol $\perp$. All voltages are written with respect to the reference node.
2. Except for the reference node, we write KCL at each of the N-1 nodes.

The current through a resistor is written using Ohm's law, with the voltage expressed as the difference between the potential on either end of the resistor with respect to the reference node as shown in Figure 8.20. We express node-voltage equations as the currents leaving the node. Two adjacent nodes give rise to the current moving to the right (like Fig. 8.20a) for one node, and the current moving to the left (like Fig. 8.20b) for the other node. The current is written for Figure 8.20a as $I_A = \frac{V}{R} = \frac{V_1 - V_2}{R}$ and for 8.20b as $I_B = \frac{V}{R} = \frac{V_2 - V_1}{R}$. It is easy to verify in 8.20a that $V = V_1 - V_2$ by applying KVL.

If one of the branches located between an essential node and the reference node contains an independent or dependent voltage source, we do not write a node equation for this node, because the node voltage is known. This reduces the number of independent node equations by one and the amount of work in solving for the node voltages. In writing the node equations for the other nodes, we write the value of the independent voltage source in those equations. Consider Figure 8.20a and assume the voltage V_2 results from an independent voltage source of 5 V. Since the node voltage is known, we do not write a node voltage equation for node 2 in this case. When writing the node voltage equation

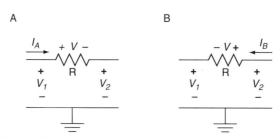

Figure 8.20   Ohm's law written in terms of node voltages.

for node 1, the current I_A is written as $I_A = \frac{V_1 - 5}{R}$. Example Problem 8.5 further illustrates this case.

Example Problem 8.5

Find V_1 using the node-voltage method.

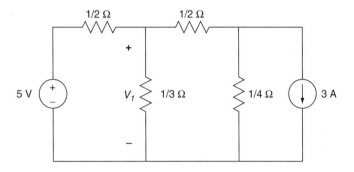

Solution

This circuit has two essential nodes, labeled 1 and 2 in the redrawn circuit that follows, with the reference node and two node voltages, V_1 and V_2, indicated. The node involving the 5 V source has a known node voltage and therefore we do not write a node equation for it.

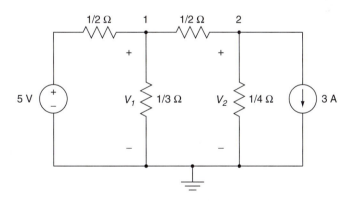

Summing the currents leaving node 1 gives

$$2(V_1 - 5) + 3V_1 + 2(V_1 - V_2) = 0$$

which simplifies to

$$7V_1 - 2V_2 = 10$$

Summing the currents leaving node 2 gives

$$2(V_2 - V_1) + 4V_2 + 3 = 0$$

which simplifies to

$$-2V_1 + 6V_2 = -3$$

The two node equations are written in matrix format as

$$\begin{bmatrix} 7 & -2 \\ -2 & 6 \end{bmatrix}\begin{bmatrix} V_1 \\ V_2 \end{bmatrix} = \begin{bmatrix} 10 \\ -3 \end{bmatrix}$$

and solved with MATLAB as follows:

```
≫ A = [7 -2; -2 6];
≫ F = [10; -3];
≫ V = A\F
V =
1.4211
-0.0263
```

Thus, $V_1 = 1.4211$ V.   ∎

Example Problem 8.6

For the following circuit, find V_3 using the node-voltage method.

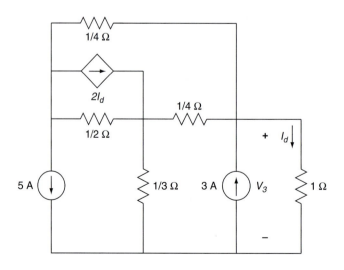

Solution

Notice that this circuit has three essential nodes and a dependent current source. We label the essential nodes 1, 2, and 3 in the redrawn circuit, with the reference node at the bottom of the circuit and three node voltages V_1, V_2, and V_3 as indicated.

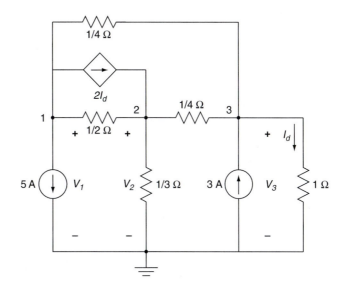

Note that $I_d = V_3$ according to Ohm's law. Summing the currents leaving node 1 gives

$$5 + 2(V_1 - V_2) + 2I_d + 4(V_1 - V_3) = 0$$

which reduces to

$$6V_1 - 2V_2 - 2V_3 = -5$$

Summing the currents leaving node 2 gives

$$-2I_d + 2(V_2 - V_1) + 3V_2 + 4(V_2 - V_3) = 0$$

which simplifies to

$$-2V_1 + 9V_2 - 6V_3 = 0$$

Summing the currents leaving node 3 gives

$$4(V_3 - V_2) - 3 + V_3 + 4(V_3 - V_1) = 0$$

reducing to

$$-4V_1 - 4V_2 + 9V_3 = 3$$

The three node equations are written in matrix format as

$$\begin{bmatrix} 6 & -2 & -2 \\ -2 & 9 & -6 \\ -4 & -4 & 9 \end{bmatrix} \begin{bmatrix} V_1 \\ V_2 \\ V_3 \end{bmatrix} = \begin{bmatrix} -5 \\ 0 \\ 3 \end{bmatrix}$$

Notice that the system matrix is no longer symmetrical because of the dependent current source, and two of the three nodes have a current source giving rise to a nonzero term on the right-hand side of the matrix equation.

Solving with MATLAB gives

≫ A = [6 −2 −2; −2 9 −6; −4 −4 9];
≫ F = [−5; 0; 3];
≫ V = A\F
V =
−1.1471
−0.5294
−0.4118

Thus $V_3 = -0.4118$ V.   ∎

If one of the branches has an independent or controlled voltage source located between two essential nodes as shown in Figure 8.21, the current through the source is not easily expressed in terms of node voltages. In this situation, we form a *supernode* by combining the two nodes. The supernode technique requires only one node equation in which the current, I_A, is passed through the source and written in terms of currents leaving node 2. Specifically, we replace I_A with $I_B + I_C + I_D$ in terms of node voltages. Because we have two unknowns and one supernode equation, we write a second equation by applying KVL for the two node voltages 1 and 2 and the source as

$$-V_1 - V_\Delta + V_2 = 0$$

or

$$V_\Delta = V_1 - V_2$$

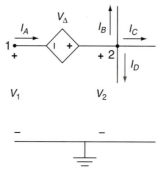

Figure 8.21   A dependent voltage source is located between nodes 1 and 2.

Example Problem 8.7

For the following circuit, find V_3.

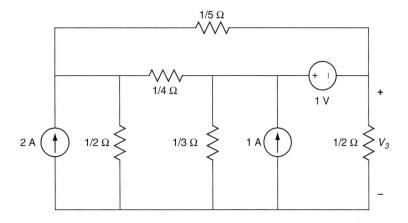

Solution

The circuit has three essential nodes, two of which are connected to an independent voltage source and form a supernode. We label the essential nodes as 1, 2, and 3 in the redrawn circuit, with the reference node at the bottom of the circuit and three node voltages, V_1, V_2, and V_3 as indicated.

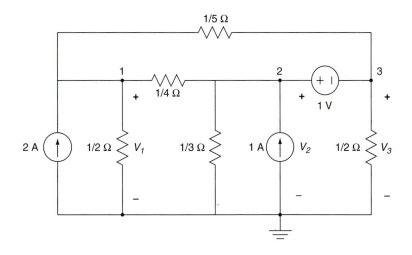

Summing the currents leaving node 1 gives

$$-2 + 2V_1 + 5(V_1 - V_3) + 4(V_1 - V_2) = 0$$

Simplifying gives

$$11V_1 - 4V_2 - 5V_3 = 2$$

Nodes 2 and 3 are connected by an independent voltage source, so we form a supernode $2 + 3$. Summing the currents leaving the supernode $2 + 3$ gives

$$4(V_2 - V_1) + 3V_2 - 1 + 2V_3 + 5(V_3 - V_1) = 0$$

Simplifying yields

$$-9V_1 + 7V_2 + 7V_3 = 1$$

The second supernode equation is KVL through the node voltages and the independent source, giving

$$-V_2 + 1 + V_3 = 0$$

or

$$-V_2 + V_3 = -1$$

The two node and KVL equations are written in matrix format as

$$\begin{bmatrix} 11 & -4 & -5 \\ -9 & 7 & 7 \\ 0 & -1 & 1 \end{bmatrix} \begin{bmatrix} V_1 \\ V_2 \\ V_3 \end{bmatrix} = \begin{bmatrix} 2 \\ 1 \\ -1 \end{bmatrix}$$

Solving with MATLAB gives

```
≫ A = [11 −4 −5; −9 7 7;0 −11];
≫ F = [2; 1; −1];
≫ V = A\F
V =
0.4110
0.8356
−0.1644
```

Thus $V_3 = -0.1644$.   ■

8.6 LINEARITY AND SUPERPOSITION

If a linear system is excited by two or more independent sources, then the total response is the sum of the separate individual responses to each input. This property is called the principle of superposition. Specifically for circuits, the response to several independent sources is the sum of responses to each independent source with the other independent sources dead, where

- A dead voltage source is a short circuit
- A dead current source is an open circuit

In linear circuits with multiple independent sources, the total response is the sum of each independent source taken one at a time. This analysis is carried out by removing all of the sources except one, and assuming the other sources are dead. After the circuit is analyzed with the first source, it is set equal to a dead source and the next source is applied with the remaining sources dead. When each of the sources have been analyzed, the total response is obtained by summing the individual responses. Note carefully that this principle holds true solely for independent sources. Dependent sources must remain in the circuit when applying this technique, and they must be analyzed based on the current or voltage for which they are defined. It should be apparent that voltages and currents in one circuit differ among circuits, and that we cannot mix and match voltages and currents from one circuit with another.

Generally, superposition provides a simpler solution than is obtained by evaluating the total response with all of the applied sources. This property is especially valuable when dealing with an input consisting of a pulse or delays. These are considered in future sections.

Example Problem 8.8

Using superposition, find V_0 as shown in the following figure.

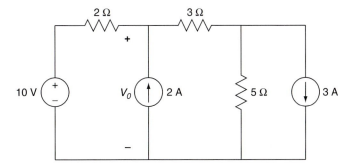

Solution

Start by analyzing the circuit with just the 10 V source active and the two current sources dead, as shown in the following figure.

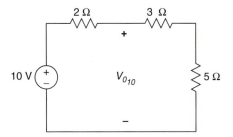

The voltage divider rule easily gives the response, $V_{0_{10}}$, due to the 10 V source

$$V_{0_{10}} = 10\left(\frac{8}{2+8}\right) = 8\text{ V}$$

Next consider the 2 A source active, and the other two sources dead, as shown in the following circuit.

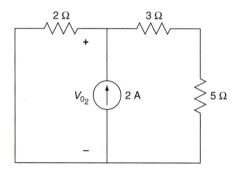

Combining the resistors in an equivalent resistance, $R_{EQ} = 2\|(3+5) = \frac{2\times8}{2+8} = 1.6\Omega$, and then applying Ohm's law gives $V_{0_2} = 2 \times 1.6 = 3.2\text{ V}$.

Finally, consider the response, V_{0_3}, to the 3 A source as shown in the following figure.

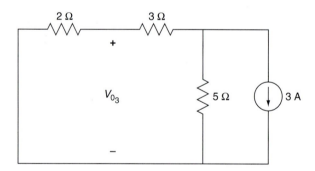

To find V_{0_3}, note that the 3 A current splits into 1.5 A through each branch $(2+3\Omega$ and $5\Omega)$, and $V_{0_3} = -1.5 \times 2 = -3\text{ V}$.

The total response is given by the sum of the individual responses as

$$V_0 = V_{0_{10}} + V_{0_2} + V_{0_3} = 8 + 3.2 - 3 = 8.2\text{ V}$$

This is the same result we would have found if we analyzed the original circuit directly using the node-voltage method.   ∎

Example Problem 8.9

Find the voltage across the 5 A current source, V_5, in the following figure using superposition.

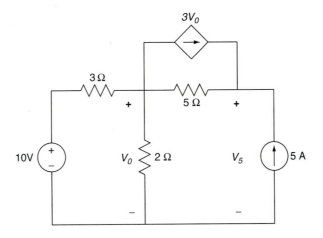

Solution

First consider finding the response, $V_{0_{10}}$ due to the 10 V source only with the 5 A source dead as shown in the following figure. As required during the analysis, the dependent current source is kept in the modified circuit and should not be set dead.

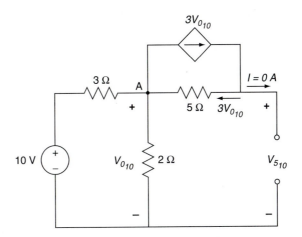

Notice that no current flows through the open circuit created by the dead current source, and that the current flowing through the 5Ω resistor is $3V_0$. Therefore, applying KCL at node A gives

$$\frac{V_{0_{10}} - 10}{3} + \frac{V_{0_{10}}}{2} + 3V_{0_{10}} - 3V_{0_{10}} = 0$$

which gives $V_{0_{10}} = 4$ V. KVL gives $-V_{0_{10}} - 5 \cdot 3V_{0_{10}} + V_{5_{10}} = 0$, and therefore $V_{5_{10}} = 64$ V.

Next consider finding the response, V_{0_5}, due to the 5 A source, with the 10 V source dead.

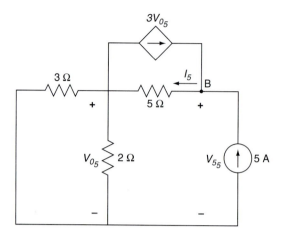

First combine the two resistors in parallel $(3\Omega || 2\Omega)$, giving $1.2~\Omega V_{0_5}$, is easily calculated by Ohm's Law as $V_{0_5} = 5 \cdot 1.2 = 6$ V. KCL is then applied at node B to find I_5, giving

$$-3V_{0_5} + I_5 - 5 = 0$$

With $V_{0_5} = 6$ V, $I_5 = 3.6 + 5 = 23$ A. Finally, apply KVL around the closed path

$$-V_{0_5} - 5I_5 + V_{5_5} = 0$$

or $V_{5_3} = V_{0_5} + 5I_5 = 6 + 5 \times 23 = 121$ V. The total response is given by the sum of the individual responses as

$$V_5 = V_{5_{10}} + V_{5_5} = 64 + 121 = 185 \text{ V}$$ ∎

8.7 THÉVENIN'S THEOREM

Any combination of resistances, controlled sources, and independent sources with two external terminals (A and B, denoted A,B) can be replaced by a single resistance and an independent source, as shown in Figure 8.22. A Thévenin equivalent circuit reduces the original circuit into a voltage source in series with a resistor (Fig. 8.22). This theorem helps reduce complex circuits into simpler circuits. We refer to the circuit elements connected across the terminals A,B (that are not shown) as the load. The Thévenin equivalent circuit is equivalent to the original circuit in that the same voltage and current are observed across any load. Usually the load is not included in the simplification because it is important for other analysis, such as maximum power expended by the load. Although we focus here on sources and resistors, this theorem can be extended to any circuit composed of linear elements with two terminals.

Thévenin's theorem states that an equivalent circuit consisting of an ideal voltage source, V_{OC}, in series with an equivalent resistance, R_{EQ}, can be used to replace any

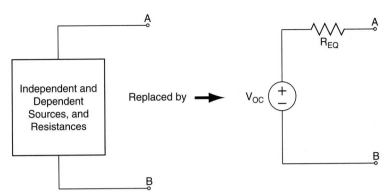

Figure 8.22 A general circuit consisting of independent and dependent sources can be replaced by a voltage source (V_{OC}) in series with a resistor (R_{EQ}).

circuit that consists of independent and dependent voltage and current sources and resistors. V_{OC} is equal to the open circuit voltage across terminals A,B as shown in Figure 8.23, and calculated using standard techniques such as the node-voltage method.

The resistor R_{EQ} is the resistance seen across the terminals A,B when all sources are dead. Recall that a dead voltage source is a short circuit and a dead current source is an open circuit.

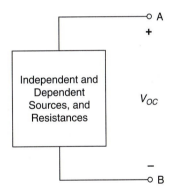

Figure 8.23 The open circuit voltage V_{OC} is calculated across the terminals A,B using standard techniques such as node-voltage or mesh-current methods.

Example Problem 8.10

Find the Thévenin equivalent circuit with respect to terminals A,B for the following circuit.

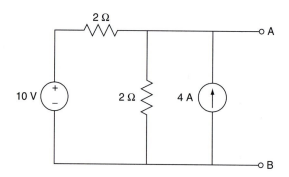

Solution

The solution to finding the Thévenin equivalent circuit is done in two parts, first finding V_{OC} and then solving for R_{EQ}. The open circuit voltage, V_{OC}, is easily found using the node-voltage method as shown in the following circuit.

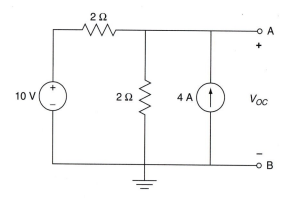

The sum of currents leaving the node is

$$\frac{V_{OC} - 10}{2} + \frac{V_{OC}}{2} - 4 = 0$$

and $V_{OC} = 9\,\text{V}$.

Next, R_{EQ} is found by first setting all sources dead (the current source is an open circuit and the voltage source is a short circuit), and then finding the resistance seen from the terminals A,B as shown in the following figure.

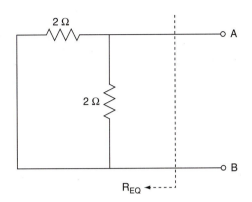

From the previous circuit, it is clear that R_{EQ} is equal to 1 Ω (that is, 2 Ω||2 Ω). Thus the Thévenin equivalent circuit is

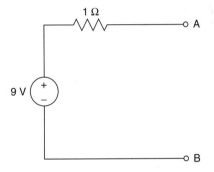

It is important to note that the circuit used in finding V_{OC} is not to be used in finding R_{EQ} as not all voltages and currents are relevant in the other circuit and one cannot simply mix and match. ∎

If the terminals A,B are shorted as shown in Figure 8.24, the current that flows is denoted I_{SC}, and the following relationship holds:

$$R_{EQ} = \frac{V_{OC}}{I_{SC}} \tag{8.18}$$

8.8 INDUCTORS

In the previous sections of this chapter, we considered circuits involving sources and resistors that are described with algebraic equations. Any changes in the source are instantaneously observed in the response. In this section we examine the inductor, a passive element that relates the voltage–current relationship with a differential equation. Circuits that contain inductors are written in terms of derivatives

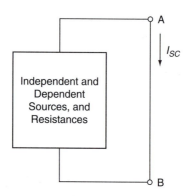

Figure 8.24

and integrals. Any changes in the source with circuits that contain inductors (i.e., a step input) have a response that is not instantaneous, but has a natural response that changes exponentially and a forced response that is the same form as the source.

An inductor is a passive element that is able to store energy in a magnetic field and is made by winding a coil of wire around a core that is an insulator or a ferromagnetic material. A magnetic field is established when current flows through the coil. We use the symbol ⌒⌒⌒⌒ to represent the inductor in a circuit; the unit of measure for inductance is the henry or henries (H), where $1\text{H} = 1\,\text{V} - \text{s/A}$. The relationship between voltage and current for an inductor is given by

$$v = L\frac{di}{dt} \qquad (8.19)$$

The convention for writing the voltage drop across an inductor is similar to that of a resistor, as shown in Figure 8.25.

Physically, current cannot change instantaneously through an inductor since an infinite voltage is required according to Eq. 8.19 (i.e., the derivative of current at the time of the instantaneous change is infinity). Mathematically, a step change in current through an inductor is possible by applying a voltage that is a Dirac delta function. For convenience, when a circuit has just DC currents (or voltages), the inductors can be replaced by short circuits since the voltage drop across the inductors is zero.

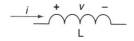

Figure 8.25   An inductor.

Example Problem 8.11

Find v in the following circuit.

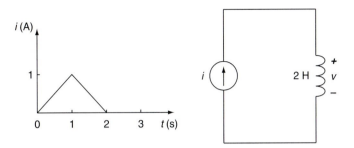

Solution

The solution to this problem is best approached by breaking it up into time intervals consistent with the changes in input current. Clearly for $t < 0$ and $t > 2$, the current is zero and therefore $v = 0$. We use Eq. 8.19 to determine the voltage in the other two intervals as follows.

For $0 < t < 1$

In this interval, the input is $i = t$, and

$$v = L\frac{di}{dt} = 2\frac{d(t)}{dt} = 2\,V$$

For $1 \leq t \leq 2$

In this interval, the input is $i = -(t - 2)$, and

$$v = L\frac{di}{dt} = 2\frac{d(-(t-2))}{dt} = -2\,V$$

Eq. 8.19 defines the voltage across an inductor for a given current. Suppose one is given a voltage across an inductor and asked to find the current. We start from Eq. 8.19 by multiplying both sides by dt, giving

$$v(t)dt = Ldi$$

Integrating both sides yields

$$\int_{t_0}^{t} v(\lambda)d\lambda = L\int_{i(t_0)}^{i(t)} d\alpha$$

or

$$i(t) = \frac{1}{L}\int_{t_0}^{t} v(\lambda)d\lambda + i(t_0) \tag{8.20}$$

For $t_0 = 0$, as is often the case in solving circuit problems, Eq. 8.20 reduces to

$$i(t) = \frac{1}{L}\int_{0}^{t} v(\lambda)d\lambda + i(0) \tag{8.21}$$

and for $t_0 = -\infty$, the initial current is by definition equal to zero, and therefore Eq. 8.20 reduces to

$$i(t) = \frac{1}{L} \int_{-\infty}^{t} v(\lambda)d\lambda \qquad (8.22)$$

The initial current in Eq. 8.20, $i(t_0)$, is usually defined in the same direction as i, which means $i(t_0)$ is a positive quantity. If the direction of $i(t_0)$ is in the opposite direction of i (as will happen when we write node equations), then $i(t_0)$ is negative. ∎

Example Problem 8.12

Find i for $t \geq 0$ if $i(0) = 2\,A$ and $v(t) = 4e^{-3t}u(t)$ in the following circuit.

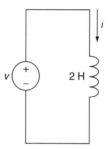

Solution

From Eq. 8.20, we have

$$i(t) = \frac{1}{L} \int_{t_0}^{t} v d\lambda + i(t_0) = \frac{1}{2} \int_{0}^{t} 4e^{-3\lambda} d\lambda + i(0) = \frac{1}{2} \int_{0}^{t} 4e^{-3\lambda} d\lambda + 2$$

$$= 2\frac{e^{-3\lambda}}{-3} \big|_{\lambda=0}^{t} + 2$$

$$= \frac{2}{3}\left(4 - e^{-3t}\right)u(t)V \qquad\qquad ∎$$

8.9   CAPACITORS

A capacitor is a device that stores energy in an electric field by charge separation when appropriately polarized by a voltage. Simple capacitors consist of parallel plates of conducting material that are separated by a gap filled with a dielectric material. Dielectric materials—that is, air, mica, or Teflon—contain a large number of electric dipoles that become polarized in the presence of an electric field. The charge separation caused by the polarization of the dielectric is proportional to the external voltage and given by

$$q(t) = Cv(t) \tag{8.23}$$

where C represents the capacitance of the element. The unit of measure for capacitance is the farad or farads (F), where $1\,\text{F} = 1\,\text{C/V}$. We use the symbol $\dashv\vdash_C$ to denote a capacitor; most capacitors are measured in terms of microfarads ($1\,\mu\text{F} = 10^{-6}\,\text{F}$) or picofarads ($1\,\text{pF} = 10^{-12}\,\text{F}$). Figure 8.26 illustrates a capacitor in a circuit.

Using the relationship between current and charge, Eq. 8.23 is written in a more useful form for circuit analysis problems as

$$i = \frac{dq}{dt} = C\frac{dv}{dt} \tag{8.24}$$

The capacitance of a capacitor is determined by the permittivity of the dielectric ($\varepsilon = 8.854 \times 10^{-12}\frac{\text{F}}{m}$ for air) that fills the gap between the parallel plates, the size of the gap between the plates, d, and the cross-sectional area of the plates, A, as

$$C = \frac{\varepsilon A}{d} \tag{8.25}$$

As described, the capacitor physically consists of two conducting surfaces that store charge, separated by a thin insulating material that has a very large resistance. In actuality, current does not flow through the capacitor plates. Rather, as James Clerk Maxwell hypothesized when he described the unified electromagnetic theory, a displacement current flows internally between capacitor plates and this current equals the current flowing into the capacitor and out of the capacitor. Thus KCL is maintained. It should be clear from Eq. 8.23 that dielectric materials do not conduct DC currents; capacitors act as open circuits when DC currents are present.

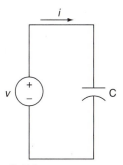

Figure 8.26 Circuit with a capacitor.

Example Problem 8.13

Find i for the following circuit.

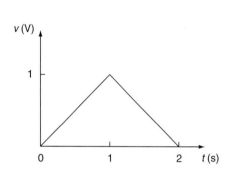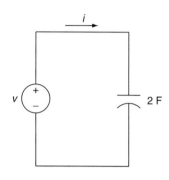

Solution

For $t < 0$ and $t > 2$, $v = 0$V, and therefore $i = 0$ in this interval. For nonzero values, the voltage waveform is described with two different functions, $v = t$ V for $0 \le t \le 1$, and $v = -(t - 2)$V for $1 < t \le 2$. Eq. 8.24 is used to determine the current for each interval as follows.

For $0 < t < 1$

$$i = C\frac{dv}{dt} = 2 \times \frac{d}{dt}(t) = 2\,\text{A}$$

For $1 \le t \le 2$

$$i = C\frac{dv}{dt} = 2 \times \frac{d}{dt}(-(t - 2)) = -2\,\text{A} \qquad \blacksquare$$

Voltage cannot change instantaneously across a capacitor. To have a step change in voltage across a capacitor requires that an infinite current flow through the capacitor, which is not physically possible. Of course, this is mathematically possible using a Dirac delta function.

Eq. 8.24 defines the current through a capacitor for a given voltage. Suppose one is given a current through a capacitor and asked to find the voltage. To find the voltage, we start from Eq. 8.24 by multiplying both sides by dt, giving

$$i(t)dt = Cdv$$

Integrating both sides yields

$$\int_{t_o}^{t} i(\lambda)d\lambda = C \int_{v(t_o)}^{v(t)} dv$$

or

$$v(t) = \frac{1}{C} \int_{t_o}^{t} idt + v(t_o) \tag{8.26}$$

For $t_0 = 0$, Eq. 8.26 reduces to

$$v(t) = \frac{1}{C} \int_{0}^{t} idt + v(0) \tag{8.27}$$

and for $t_0 = -\infty$ Eq. 8.27 reduces to

$$v(t) = \frac{1}{C} \int_{-\infty}^{t} i(\lambda)d\lambda \tag{8.28}$$

The initial voltage in Eq. 8.26, $v(t_0)$, is usually defined with the same polarity as v, which means $v(t_0)$ is a positive quantity. If the polarity of $v(t_0)$ is in the opposite direction of v, then $v(t_0)$ is negative.

Example Problem 8.14

Find v for the circuit that follows.

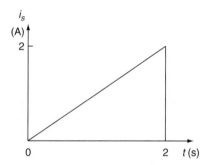 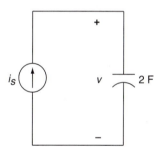

Solution

The current waveform is described with three different functions: for the interval $t \leq 0$, for the interval $0 < t \leq 2$, and for $t > 2$. To find the voltage, we apply Eq. 8.28 for each interval as follows.

For $t < 0$

$$v(t) = \frac{1}{C} \int_{-\infty}^{t} idt = \frac{1}{2} \int_{-\infty}^{0} 0dt = 0 \text{ V}$$

For $0 \leq t \leq 2$

$$v(t) = \frac{1}{C} \int_{0}^{t} idt + v(0)$$

and with $v(0) = 0$, we have

$$v(t) = \frac{1}{2}\int_0^t \lambda d\lambda = \frac{1}{2}\left(\frac{\lambda^2}{2}\right)\Big|_0^t = \frac{t^2}{4}\,\text{V}$$

The voltage at $t = 2$ needed for the initial condition in the next part is

$$v(2) = \frac{t^2}{4}\Big|_{t=2} = 1\,\text{V}$$

For $t > 2$

$$v(t) = \frac{1}{C}\int_2^t i\,dt + v(2) = \frac{1}{2}\int_2^t 0\,dt + v(2) = 1\,\text{V} \qquad \blacksquare$$

8.10   A GENERAL APPROACH TO SOLVING CIRCUITS INVOLVING RESISTORS, CAPACITORS, AND INDUCTORS

Sometimes a circuit consisting of resistors, inductors, and capacitors cannot be simplified by bringing together like elements in series and parallel combinations. Consider the cirucuit shown in Figure 8.27. In this case, the absence of parallel or series combinations of resistors, inductors, or capacitors prevents us from simplifying the circuit for ease in solution. In this section, the node-voltage method is applied to write equations involving integrals and differentials using element relationships for resistors, inductors, and capacitors. From these equations, any unknown currents and voltages of interest can be solved using the standard differential equation approach.

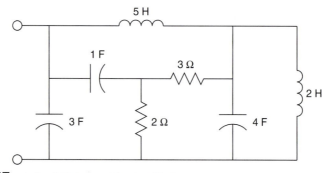

Figure 8.27   A circuit that cannot be simplified.

Example Problem 8.15

Write the node equations for the following circuit for $t \geq 0$ if the initial conditions are zero.

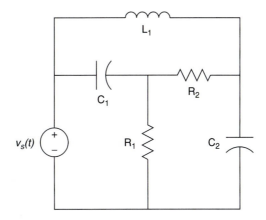

Solution

With the reference node at the bottom of the circuit, we have two essential nodes, as shown in the following redrawn circuit. Recall that the node involving the voltage source is a known voltage and that we do not write a node equation for it. When writing the node-voltage equations, the current through a capacitor is $i_c = C\Delta\dot{v}$, where $\Delta\dot{v}$ is the derivative of the voltage across the capacitor, and the current through an inductor is $i_L = \frac{1}{L}\int_0^t \Delta v\,d\lambda + i_L(0)$, where Δv is the voltage across the inductor. Since the initial conditions are zero, the term $i_L(0) = 0$.

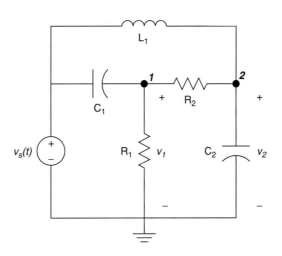

Summing the currents leaving node 1 gives

$$C_1(\dot{v}_1 - \dot{v}_s) + \frac{v_1}{R_1} + \frac{v_1 - v_2}{R_2} = 0$$

which simplifies to

$$C_1 \dot{v}_1 + \left(\frac{1}{R_1} + \frac{1}{R_2} \right) v_1 - \frac{1}{R_2} v_2 = C_1 \dot{v}_s$$

Summing the currents leaving node 2 gives

$$\frac{v_2 - v_1}{R_2} + C_2 \dot{v}_2 + \frac{1}{L_1} \int_0^t (v_2 - v_s) d\lambda = 0$$

Typically, we eliminate integrals in the node equations by differentiating. When applied to the previous expression, this gives

$$\frac{1}{R_2} \dot{v}_2 - \frac{1}{R_2} \dot{v}_1 + C_2 \ddot{v}_2 + \frac{1}{L_1} v_2 - \frac{1}{L_1} v_s = 0$$

and after rearranging yields

$$\ddot{v}_2 + \frac{1}{C_2 R_2} \dot{v}_2 + \frac{1}{C_2 L_1} v_2 - \frac{1}{C_2 R_2} \dot{v}_1 = \frac{1}{C_2 L_1} v_s \qquad \blacksquare$$

When applying the node-voltage method, we generate one equation for each essential node. To write a single differential equation involving just one node voltage and the inputs, we use the other node equations and substitute into the node equation of the desired node voltage. Sometimes this involves differentiation as well as substitution. The easiest case involves a node equation containing an undesired node voltage without its derivatives. Another method for creating a single differential equation is to use the D operator.

Consider the node equations for Example Problem 8.15, and assume that we are interested in obtaining a single differential equation involving node voltage v_1 and its derivatives, and the input. For ease in analysis, let us assume that the values for the circuit elements are

$R_1 = R_2 = 1\,\Omega$, $C_1 = C_2 = 1\,\mathrm{F}$, and $L_1 = 1\,\mathrm{H}$, giving us

$$\dot{v}_1 + 2v_1 - v_2 = \dot{v}_s$$

and

$$\ddot{v}_2 + \dot{v}_2 + v_2 - \dot{v}_1 = v_s$$

Using the first equation, we solve for v_2, calculate $\dot{v}_2$ and $\ddot{v}_2$, and then substitute into the second equation as follows.

$$v_2 = \dot{v}_1 + 2v_1 - \dot{v}_s$$
$$\dot{v}_2 = \ddot{v}_1 + 2\dot{v}_1 - \ddot{v}_s$$
$$\ddot{v}_2 = \dddot{v}_1 + 2\ddot{v}_1 - \dddot{v}_s$$

After substituting into the second node equation, we have

$$\dddot{v}_1 + 2\ddot{v}_1 - \dddot{v}_s + \ddot{v}_1 + 2\dot{v}_1 - \ddot{v}_s + \dot{v}_1 + 2v_1 - \dot{v}_s - \dot{v}_1 = v_s$$

and, after simplifying,

$$\dddot{v}_1 + 3\ddot{v}_1 + 2\dot{v}_1 + 2v_1 = \dddot{v}_s + \ddot{v}_s + \dot{v}_s - v_s$$

In general, the order of the differential equation relating a single output variable and the inputs is equal to the number of energy-storing elements in the circuit (capacitors and inductors). In some circuits, the order of the differential equation is less than the number of capacitors and inductors in the circuit. This occurs when capacitor voltages and inductor currents are not independent—that is, there is an algebraic relationship between the capacitor, specifically voltages and the inputs, or the inductor currents and the inputs. This occurs when capacitors are connected directly to a voltage source or when inductors are connected directly to a current source.

The previous example involved a circuit with zero initial conditions. When circuits involve nonzero initial conditions, our approach remains the same as before except that the initial inductor currents are included when writing the node-voltage equations.

Example Problem 8.16

Write the node equations for the following circuit for $t \geq 0$ assuming the initial conditions are $i_{L_1}(0) = 8\text{A}$ and $i_{L_2}(0) = -4\text{A}$.

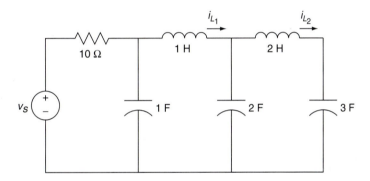

Solution

With the reference node at the bottom of the circuit, there are three essential nodes as shown in the redrawn circuit that follows.

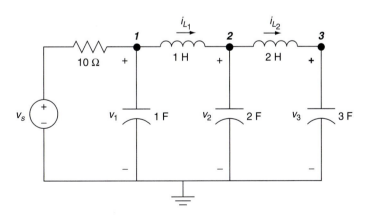

Summing the currents leaving node 1 gives

$$\frac{(v_1 - v_s)}{10} + \dot{v}_1 + \int_0^t (v_1 - v_2)d\lambda + 8 = 0$$

where $i_{L_1}(0) = 8A$.

Summing the currents leaving node 2 gives

$$\int_0^t (v_2 - v_1)d\lambda - 8 + 2\dot{v}_2 + \frac{1}{2}\int_0^t (v_2 - v_3)d\lambda - 4 = 0$$

where $i_{L_2}(0) = -4A$. Notice that the sign for the initial inductor current is negative because the direction is from right to left and the current is defined on the circuit diagram in the opposite direction for the node-two equation.

Summing the currents leaving node 3 gives

$$\frac{1}{2}\int_0^t (v_3 - v_2)d\lambda + 4 + 3\dot{v}_3 = 0$$

In this example, the node equations were not simplified by differentiating to remove the integral, which would have eliminated the initial inductor currents from the node equations. If we were to write a single differential equation involving just one node voltage and the input, a fifth-order differential equation would result because there are five energy-storing elements in the circuit. To solve the differential equation, we would need five initial conditions, the initial node voltage for the variable selected, and the first through fourth derivatives at time zero. ∎

8.10.1 Discontinuities and Initial Conditions in a Circuit

Discontinuities in voltage and current occur when an input such as a unit step is applied or a switch is thrown in a circuit. As we have seen, when solving an nth order differential equation one must know n initial conditions, typically the output variable and its (n-1) derivatives at the time the input is applied or the switch is thrown. As we will see, if the inputs to a circuit are known for all time, we can solve for initial conditions directly based on energy considerations and not depend on being provided with them in the problem statement. Almost all of our problems involve the input applied at time zero, so our discussion here is focused on time zero, but may be easily extended to any time an input is applied.

Energy cannot change instantaneously for elements that store energy. Thus, there are no discontinuities allowed in current through an inductor or voltage across a capacitor at any time—specifically, the value of the variable remains the same at $t = 0^-$ and $t = 0^+$. In the previous problem, when we were given initial conditions for the inductors and capacitors, this implied $i_{L_1}(0^-) = i_{L_1}(0^+)$ and $i_{L_2}(0^-) = i_{L_2}(0^+)$, and $v_1(0^-) = v_1(0^+), v_2(0^-) = v_2(0^+)$, and $v_3(0^-) = v_3(0^+)$. With the exception of variables associated with current through an inductor and voltage across a capacitor, other variables can have discontinuities, especially at a time when a unit step is applied or when a switch is thrown; however, these variables must obey KVL and KCL.

Although it may not seem obvious at first, a discontinuity is allowed for the derivative of the current through an inductor and voltage across a capacitor at $t = 0^-$ and $t = 0^+$ since

$$\frac{di_L(0+)}{dt} = \frac{v_L(0+)}{L} \text{ and } \frac{dv_C(0+)}{dt} = \frac{i_C(0+)}{L}$$

as discontinuities are allowed in $v_L(0+)$ and $i_C(0+)$. Keep in mind that the derivatives in the previous expression are evaluated at zero after differentiation, that is

$$\frac{di_L(0+)}{dt} = \frac{di_L(t)}{dt}\bigg|_{t=0^+} \text{ and } \frac{dv_C(0+)}{dt} = \frac{dv_C(t)}{dt}\bigg|_{t=0^+}$$

In calculations to determine the derivatives of variables not associated with current through an inductor and voltage across a capacitor, the derivative of a unit step input may be needed. Here we assume the derivative of a unit step input is zero at $t = 0^+$.

The initial conditions for variables not associated with current through an inductor and voltage across a capacitor at times of a discontinuity are determined only from the initial conditions from variables associated with current through an inductor and voltage across a capacitor and any applicable sources. The analysis is done in two steps involving KCL and KVL or using the node-voltage method.

1. First, we analyze the circuit at $t = 0^-$. Recall that when a circuit is at steady state, an inductor acts as a short circuit and a capacitor acts as an open circuit. Thus at steady state at $t = 0^-$ we replace all inductors by short circuits and capacitors by open circuits in the circuit. We then solve for the appropriate currents and voltages in the circuit to find the currents through the inductors (actually the shorts connecting the sources and resistors) and voltages across the capacitors (actually the open circuits among the sources and resistors).

2. Second, we analyze the circuit at $t = 0^+$. Since the inductor current cannot change in going from $t = 0^-$ to $t = 0^+$, we replace the inductors with current sources whose values are the currents at $t = 0^-$. Moreover, since the capacitor voltage cannot change in going from $t = 0^-$ to $t = 0^+$, we replace the capacitors with voltage sources whose values are the voltages at $t = 0^-$. From this circuit we solve for all desired initial conditions necessary to solve the differential equation.

Example Problem 8.17

Use the node-voltage method to find v_c for the following circuit for $t \geq 0$.

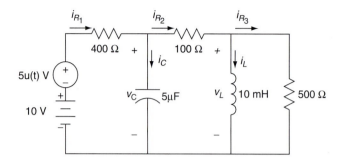

Solution

For $t \geq 0$, the circuit is redrawn for analysis in the following figure.

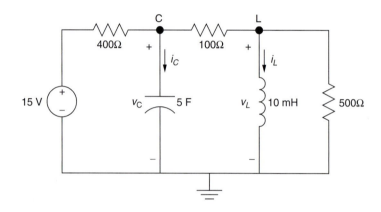

Summing the currents leaving node C gives

$$\frac{v_C - 15}{400} + 5 \times 10^{-6}\dot{v}_C + \frac{v_C - v_L}{100} = 0$$

which simplifies to

$$\dot{v}_C + 2500v_C - 2000v_L = 7500$$

Summing the currents leaving node L gives

$$\frac{v_L - v_C}{100} + \frac{1}{10 \times 10^{-3}} \int_0^t v_L d\lambda + i_L(0^+) + \frac{v_L}{500} = 0$$

which, after multiplying by 500 and differentiating, simplifies to

$$6\dot{v}_L + 50 \times 10^3 v_L - 5\dot{v}_C = 0$$

Using the D operator method, the two differential equations are written as

$$Dv_C + 2500v_C - 2000v_L = 7500 \text{ or } (D + 2500)v_C - 2000v_L = 7500$$
$$6Dv_L + 50 \times 10^3 v_L - 5Dv_C = 0 \text{ or } (6D + 50 \times 10^3)v_L - 5Dv_C = 0$$

We then solve for v_L from the first equation:

$$v_L = (0.5 \times 10^{-3}D + 1.25)v_C - 3.75$$

and then substitute v_L into the second equation, giving

$$(6D + 50 \times 10^3)v_L - 5Dv_C = (6D + 50 \times 10^3)((0.5 \times 10^{-3}D + 1.25)v_C - 3.750)$$
$$- 5Dv_C = 0$$

Reducing this expression yields

$$D^2 v_C + 10.417 \times 10^3 Dv_C + 20.83 \times 10^6 v_C = 62.5 \times 10^6$$

Returning to the time domain gives

$$\ddot{v}_C + 10.417 \times 10^3 \dot{v}_C + 20.83 \times 10^6 v_C = 62.5 \times 10^6$$

The characteristic equation for the previous differential equation is

$$s^2 + 10.417 \times 10^3 s + 20.833 \times 10^6 = 0$$

with roots -7.718×10^3 and -2.7×10^3 and the natural solution

$$v_{C_n}(t) = K_1 e^{-7.718 \times 10^3 t} + K_2 e^{-2.7 \times 10^3 t} \text{V}$$

Next we solve for the forced response, assuming that $v_{C_f}(t) = K_3$. After substituting into the differential equation, this gives

$$20.833 \times 10^6 K_3 = 62.5 \times 10^6$$

or $K_3 = 3$. Thus our solution is now

$$v_C(t) = v_{C_n}(t) + v_{C_f}(t) = K_1 e^{-7.718 \times 10^3 t} + K_2 e^{-2.7 \times 10^3 t} + 3 \text{V}$$

Initial conditions for $v_C(0^+)$ and $\dot{v}_C(0^+)$ are necessary to solve for K_1 and K_2. For $t = 0^-$, the capacitor is replaced by an open circuit and the inductor by a short circuit as shown in the following circuit.

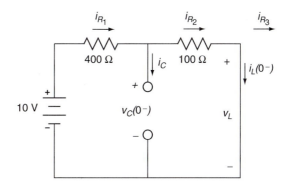

Notice that $v_L(0^-) = 0V$ because the inductor is a short circuit. Also note that the 500-Ω resistor is not shown in the circuit since it is shorted out by the inductor, and so $i_{R_3}(0^-) = 0A$. Using the voltage divider rule, we have

$$v_C(0^-) = 10 \times \frac{100}{400 + 100} = 2V$$

and by Ohm's law

$$i_L(0^-) = \frac{10}{100 + 400} = 0.02A$$

It follows that $i_{R_1}(0^-) = i_{R_2}(0^-) = i_L(0^-) = 0.02A$. Because voltage across a capacitor and current through an inductor are not allowed to change from $t = 0^-$ to $t = 0^+$ we have $v_C(0^+) = v_C(0^-) = 2V$ and $i_L(0^-) = i_L(0^+) = 0.02A$.

The circuit for $t = 0^+$ is drawn by replacing the inductors in the original circuit with current sources whose values equal the inductor currents at $t = 0^-$ and the capacitors with voltage sources whose values equal the capacitor voltages at $t = 0^-$ as shown in the following figure with nodes C and L and reference. Note also that the input is now $10 + 5u(t) = 15V$.

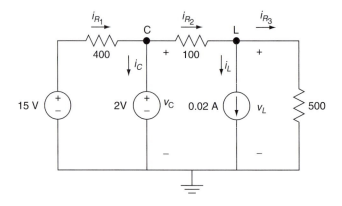

To find $v_L(0^+)$, we sum the currents leaving node L, yielding

$$\frac{v_L - 2}{100} + 0.02 + \frac{v_L}{500} = 0$$

which gives $v_L(0^+) = 0V$. Now

$$i_{R_3}(0^+) = \frac{v_L(0^+)}{500} = 0A, \ i_{R_2}(0^+) = 0.02 + i_{R_3}(0^+) = 0.02A, \ \text{and}$$

$$i_{R_1}(0^+) = \frac{15 - 2}{400} = 0.0325A.$$

To find $i_c(0^+)$, we write KCL at node C, giving

$$-i_{R_1}(0^+) + i_C(0^+) + i_{R_2}(0^+) = 0$$

or

$$i_C(0^+) = i_{R_1}(0^+) - i_{R_2}(0^+) = 0.0325 - 0.02 = 0.125 \text{A}$$

To find $\dot{v}_C(0^+)$, note that $i_C(0^+) = C\dot{v}_C(0^+)$ or

$$\dot{v}_C(0^+) = \frac{i_C(0^+)}{C} = \frac{0.0125}{5 \times 10^{-6}} = 2.5 \times 10^3 \frac{\text{V}}{\text{s}}.$$

With the initial conditions, the constants K_1 and K_2 are solved as

$$v_C(0) = 2 = K_1 + K_2 + 3$$

Next

$$\dot{v}_C(t) = -7.718 \times 10^3 K_1 e^{-7.718 \times 10^3 t} - 2.7 \times 10^3 K_2 e^{-2.7 \times 10^3 t}$$

and at $t = 0$,

$$\dot{v}_C(0) = 2.5 \times 10^3 = -7.718 \times 10^3 K_1 - 2.7 \times 10^3 K_2$$

Solving gives $K_1 = 0.04$ and $K_2 = -1.04$. Substituting these values into the solution gives

$$v_C(t) = 0.04e^{-7.718 \times 10^3 t} - 1.04e^{-2.7 \times 10^3 t} + 3\text{V}$$

for $t \geq 0$. ∎

8.11 OPERATIONAL AMPLIFIERS

In Section 8.3, controlled voltage and current sources dependent on a voltage or current elsewhere in a circuit was introduced. These devices were modeled as a two-terminal device. Presented in this section is the operational amplifier, also known as an op amp, which is a multiterminal device. An operational amplifier is an electronic device that consists of large numbers of transistors, resistors, and capacitors. To fully understand its operation requires knowledge of diodes and transistors, topics not covered in this book. However, to appreciate how an operational amplifier operates in a circuit involves a topic already covered: the controlled voltage source. Circuits involving operational amplifiers form the cornerstone for any bioinstrumentation, from amplifiers to filters. Amplifiers used in biomedical applications have very high input impedance to keep the current drawn from the system being measured low. Most body signals have very small magnitudes. For example, an ECG has a magnitude in the millivolts and the EEG has a magnitude in the microvolts. Analog filters are often used to remove noise from a signal, typically through use of frequency-domain analysis to design the filter. Analog filters are described in Section 8.13. The theory behind filters is covered in Chapter 10.

As the name implies, the operation amplifier is an amplifier. But as we will see, when it is combined with other circuit elements it also integrates, differentiates, sums, and subtracts. One of the first operational amplifiers appeared as an eight-lead dual-in-line package (DIP), as shown in Figure 8.28.

Differing from previous circuit elements, this device has two input terminals and one output terminal. Rather than draw the operational amplifier using Figure 8.28, the operational amplifier is drawn with the symbols in Figure 8.29. The input terminals are labeled the noninverting input (+) and the inverting input (−). The power supply terminals are labeled V+ and V−, which are frequently omitted since

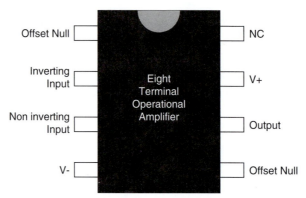

Figure 8.28 An 8-terminal operational amplifier. The terminal NC is not connected, and the two terminal offset nulls are used to correct imperfections (typically not connected). V+ and V− are terminal power to provide energy to the circuit. Keep in mind that a ground exists for both V+ and V−, a ground that is shared by other elements in the circuit. Modern operational amplifiers have 10 or more terminals.

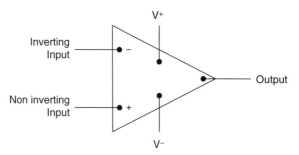

Figure 8.29 Circuit element symbol for the operational amplifier.

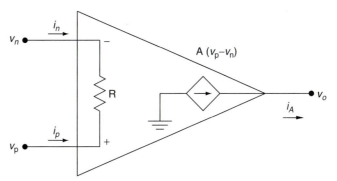

Figure 8.30 An internal model of the op amp. The internal resistance between the input terminals, R, is very large—exceeding $1\,M\Omega$. The gain of the amplifier, A, is also large—exceeding 10^4. Power supply terminals are omitted for simplicity.

they do not affect the circuit behavior except in saturation conditions, as will be described. Most people shorten the name of the operational amplifier to the *op amp*.

Illustrated in Figure 8.30 is a model of the op amp focusing on the internal behavior of the input and output terminals. The input–output relationship is

$$v_o = A(v_p - v_n) \tag{8.29}$$

Because the internal resistance is very large, we will replace it with an open circuit to simplify analysis, leaving us with the op amp model shown in Figure 8.31.

With the replacement of the internal resistance with an open circuit, the currents $i_n = i_p = 0$ A. In addition, current i_A, the current flowing out of the op amp, is not zero. Because i_A is unknown, seldom is KCL applied at the output junction. In solving op amp problems, KCL is almost always applied at input terminals.

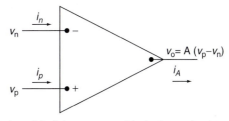

Figure 8.31 Idealized model of the op amp with the internal resistance R replaced by an open circuit.

Example Problem 8.18

Find v_o for the following circuit.

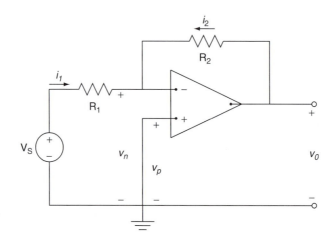

Solution

Using the op amp model of Figure 8.31, we apply KCL at the inverting terminal giving

$$-i_1 - i_2 = 0$$

because no current flows into the op amp's input terminals. Replacing the current using Ohm's law gives

$$\frac{v_s - v_1}{R_1} + \frac{v_o - v_1}{R_2} = 0$$

Multiplying by $R_1 R_2$ and collecting like terms, we have

$$R_2 v_s = (R_1 + R_2)v_1 - R_1 v_o$$

Now $v_o = A(v_p - v_n)$ and since the noninverting terminal is connected to ground, $v_p = 0$ so

$$v_o = -A v_n$$

or

$$v_n = -\frac{v_o}{A}$$

Substituting v_n into the KCL inverting input equation gives

$$R_s v_s = (R_1 + R_2)\left(-\frac{v_o}{A}\right) - R_1 v_o$$

$$= \left(\frac{R_1 + R_2}{A} + R_1\right) v_o$$

or

$$v_o = \frac{-R_2 v_s}{\left(R_1 + \frac{R_1 + R_2}{A}\right)}$$

As A goes to infinity, the previous equation goes to

$$v_o = -\frac{R_2}{R_1} v_s$$

Interestingly, with A going to infinity, v_0 remains finite due to the resistor R_2. This happens because a negative feedback path exists between the output and the inverting input terminal through R_2. This circuit is called an inverting amplifier with an overall gain of $-\frac{R_2}{R_1}$. ∎

An operational amplifier with a gain of infinity is known as an ideal op amp. Because of the infinite gain, there must be a feedback path between the output and input and we cannot connect a voltage source directly between the inverting and noninverting input terminals. When analyzing an ideal op amp circuit, we simplify the analysis by letting

$$v_n = v_p$$

Consider the previous example. Because $v_p = 0$, $v_n = 0$. Applying KCL at the inverting input gives

$$-\frac{v_s}{R_1} + \frac{-v_o}{R_2} = 0$$

or

$$v_o = -\frac{R_2}{R_1} v_s$$

Notice how simple the analysis becomes when we assume $v_n = v_p$. Keep in mind that this approximation is valid as long as A is very large (infinity) and a feedback is included.

Example Problem 8.19

Find the overall gain for the following circuit.

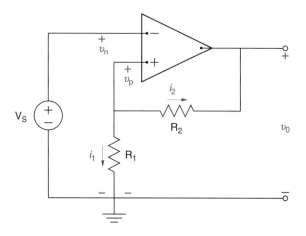

Solution

Assuming the op amp is ideal, we start with $v_n = v_p$. Then since the op amp's noninverting terminal is connected to the source, $v_n = v_p = v_s$. Because no current flows into the op amp, by KCL we have

$$i_1 + i_2 = 0$$

and

$$\frac{v_s}{R_1} + \frac{v_s - v_o}{R_2} = 0$$

or

$$v_o = \left(\frac{R_1 + R_2}{R_1} \right) v_s$$

The overall gain is

$$\frac{v_o}{v_s} = \frac{R_1 + R_2}{R_1}$$

This circuit is a noninverting op amp circuit used to amplify the source input. Amplifiers are used in most clinical instrumentation, including ECK, EEG, and EOG, etc.   ■

The next example describes a summing op amp circuit.

Example Problem 8.20

Find the overall gain for the following circuit.

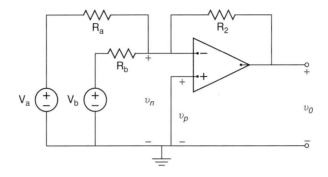

Solution

As before, we start the solution with $v_n = v_p$ and note that the noninverting input is connected to ground, yielding $v_n = v_p = 0$ V. Applying KCL at the inverting input node gives

$$-\frac{V_a}{R_a} - \frac{V_b}{R_b} - \frac{v_o}{R_2} = 0$$

or

$$v_o = -\left(\frac{R_2}{R_a}V_a + \frac{R_2}{R_b}V_b\right)$$

This circuit is a weighted summation of the input voltages. We can add additional source resistor inputs, so that in general

$$v_o = -\left(\frac{R_2}{R_a}V_a + \frac{R_2}{R_b}V_b + \ldots + \frac{R_2}{R_m}V_m\right)$$ ∎

Our next op amp circuit provides an output proportional to the difference of two input voltages. This op amp is often referred to as a differential amplifier.

Example Problem 8.21

Find the overall gain for the following circuit.

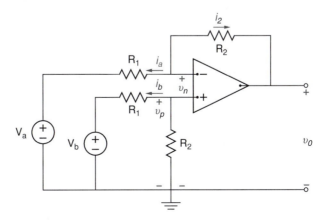

Solution

Assuming an ideal op amp, we note no current flows into the input terminals and that $v_n = v_p$. Applying KCL at the inverting input terminal gives

$$i_a = -i_2$$

or

$$\frac{v_n - V_a}{R_1} + \frac{v_n - v_o}{R_2} = 0$$

and

$$(R_1 + R_2)v_n - R_2V_a = R_1v_o$$

The previous equation involves two unknowns, thus we need another equation easily found by applying voltage divider at the noninverting input.

$$v_p = \frac{R_2}{R_1 + R_2}v_b = v_n$$

Substituting this result for v_n into the KCL equation at the inverting terminal gives

$$R_2V_b - R_2V_a = R_1v_o$$

or

$$v_o = \frac{R_2}{R_1}(V_b - V_a)$$

As shown, this op amp circuit, also known as the differential amplifier, subtracts the weighted input signals. This amplifier is used for bipolar measurements involving ECG and EEG because the typical recording is obtained between two bipolar input terminals. Ideally, the measurement contains only the signal of interest uncontaminated by noise from the environment. The noise is typically called common-mode signal. Common-mode signal comes from lighting, 60-Hz power line signals, inadequate grounding, and power supply leakage. A differential amplifier with appropriate filtering can reduce the impact of the common-mode signal.   ∎

The response of a differential amplifier can be decomposed into differential-mode and common-mode components,

$$v_{dm} = v_b - v_a$$

and

$$v_{cm} = \frac{(v_a + v_b)}{2}$$

As described, the common-mode signal is the average of the input voltages. Using the two previous equations, one can solve v_a and v_b in terms of v_{dm} and v_{cm} as

$$v_a = v_{cm} - \frac{v_{dm}}{2}$$

and

$$v_b = v_{cm} + \frac{v_{dm}}{2}$$

which, when substituted into the response in Example Problem 8.21 gives

$$v_0 = \left(\frac{R_1 R_2 - R_1 R_2}{R_1(R_1 + R_2)}\right)v_{cm} + \left(\frac{R_2(R_1 + R_2) + R_2(R_1 + R_2)}{2R_1(R_1 + R_2)}\right)v_{dm} = A_{cm}v_{cm} + A_{dm}v_{dm}$$

Notice the term multiplying, v_{cm}, A_{cm}, is zero, which is characteristic of the ideal op amp which amplifies only the differential mode of the signal. Since real amplifiers are not ideal and resistors are not truly exact, the common-mode gain is not zero. So when one designs a differential amplifier, the goal is to keep A_{cm} as small as possible and A_{dm} as large as possible.

The rejection of the common-mode signal is called common-mode rejection, and the measure of how ideal is the differential amplifier is called the common-mode rejection ratio, given as

$$CMRR = 20 \log_{10}\left|\frac{A_{dm}}{A_{cm}}\right|$$

where the larger the value of *CMRR* the better. Values of *CMRR* for a differential amplifier for EEG, ECG, and EMG are 100 to 120 db.

The general approach to solving op amp circuits is to first assume that the op amp is ideal and that $v_p = v_n$. Next we apply KCL or KVL at the two input terminals. In more complex circuits, we continue to apply our circuit analysis tools to solve the problem, as the next example illustrates.

Example Problem 8.22

Find v_0 for the following circuit.

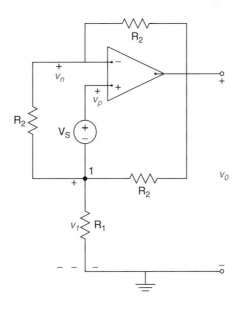

Solution

With $v_n = v_p$, we apply KCL at the inverting input

$$\frac{v_n - v_1}{R_2} + \frac{v_n - v_o}{R_2} = 0$$

and

$$2v_n - v_1 - v_o = 0$$

Next we apply KVL from ground to node 1 to the noninverting input and back to ground giving

$$-v_1 - V_s + v_p = 0$$

and with $v_n = v_p$ we have $v_n - v_1 = V_s$.

Now we apply KCL at node 1, noting no current flows into the noninverting input terminal

$$\frac{v_1}{R_1} + \frac{v_1 - v_o}{R_2} + \frac{v_1 - v_n}{R_2} = 0$$

Combining like terms in the previous equation gives

$$-R_1 v_n + (2R_1 + R_2)v_1 - R_1 v_o = 0$$

With three equations and three unknowns, we first eliminate v_1 by subtracting the inverting input KCL equation by the KVL equation, giving

$$v_1 = v_o - 2V_s$$

Next we eliminate v_n by substituting v_1 into the inverting input KCL equation as follows:

$$v_n = \frac{1}{2}(v_1 + v_o)$$
$$= \frac{1}{2}(v_o - 2V_s + v_o)$$
$$= v_o - V_s$$

Finally, we substitute the solutions for v_1 and v_n into the node 1 KCL equation, giving

$$-R_1 v_n + (2R_1 + R_2)v_1 - R_1 v_o = 0$$
$$-R_1(v_o - V_s) + (2R_1 + R_2)(v_o - 2V_s) - R_1 v_o = 0$$

After simplification, we have

$$v_o = \frac{(3R_1 + 2R_2)}{R_2} V_s \qquad \blacksquare$$

The next two examples illustrate an op amp circuit that differentiates and integrates by using a capacitor.

Example Problem 8.23

Find v_0 for the following circuit.

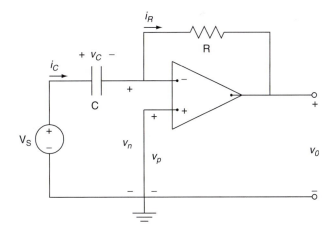

Solution

With the noninverting input connected to ground, we have $v_p = 0 = v_n$. From KVL

$$v_C = V_s$$

and it follows that

$$i_C = C\frac{dv_C}{dt} = C\frac{dV_s}{dt}$$

Since no current flows into the op amp, $i_C = i_R$. With

$$i_R = \frac{v_n - v_o}{R} = -\frac{v_o}{R}$$

and

$$i_C = C\frac{dV_s}{dt} = i_R = -\frac{v_o}{R}$$

we have

$$v_o = -RC\frac{dV_s}{dt}$$

If $R = \frac{1}{C}$, the circuit in this example differentiates the input, $v_o = -\frac{dV_s}{dt}$. ∎

Example Problem 8.24

Find v_0 for the following circuit.

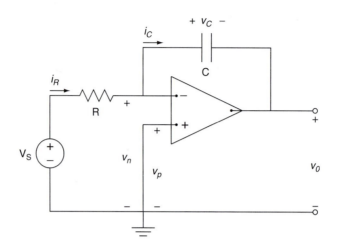

Solution

It follows that

$$v_n = v_p = 0$$

and

$$i_C = i_R = \frac{V_s}{R}$$

Therefore

$$v_C = \frac{1}{C} \int_{-\infty}^{t} i_C d\lambda = \frac{1}{C} \int_{-\infty}^{t} \frac{V_s}{R} d\lambda$$

From KVL, we have

$$v_C + v_o = 0$$

and

$$v_o = -\frac{1}{RC} \int_{-\infty}^{t} V_s d\lambda$$

With $R = \frac{1}{C}$, the circuit operates as an integrator

$$v_o = - \int_{-\infty}^{t} V_s d\lambda$$ ∎

8.11.1 Voltage Characteristics of the Op Amp

In the past examples involving the op amp, we have neglected to consider the supply voltage (shown in Fig 8.29) and that the output voltage of an ideal op amp is constrained to operate between the supply voltages V^+ and V^-. If analysis determines v_0 is greater than V^+, v_0 saturates at V^+. If analysis determines v_0 is less than V^-, v_0 saturates at V^-. The output voltage characteristics are shown in Figure 8.32.

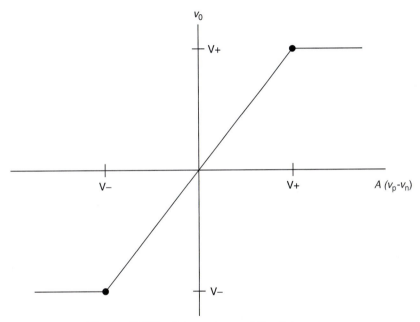

Figure 8.32 Voltage characteristics of an op amp.

Example Problem 8.25

For the circuit shown in Example Problem 8.22, let $V^+ = +10\,\text{V}$ and $V^- = -10\,\text{V}$. Graph the output voltage characteristics of the circuit.

Solution

The solution for Example Problem 8.22 is

$$v_o = \left(\frac{3R_1 + 2R_2}{R_2}\right)V_s$$

which saturates whenever v_0 is less than V^- and greater than V^+ as shown in the following graph.

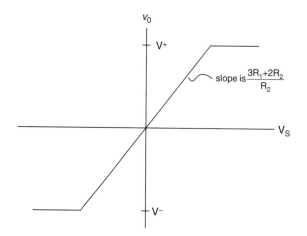

8.12 TIME-VARYING SIGNALS

An alternating current (*a-c*) or sinusoidal source of 50 or 60 Hz is common through-out the world as a power source supplying energy for most equipment and other devices. While most of this chapter has focused on the transient response when dealing with sinusoidal sources, attention is now focused on the steady-state or forced response. In bioinstrumentation, analysis in the steady state simplifies the design by focusing only on the steady-state response, which is where the device actually oper-ates. A sinusoidal voltage source is a time-varying signal given by

$$v_s = V_m \cos{(\omega t + \phi)} \tag{8.30}$$

where the voltage is defined by angular frequency (ω in radians/s), phase angle (ϕ in radians or degrees), and peak magnitude (V_m). The period of the sinusoid T is related to frequency f (Hz or cycles /s) and angular frequency by

$$\omega = 2\Pi f = \frac{2\Pi}{T} \tag{8.31}$$

An important metric of a sinusoid is its *rms value* (square *r*oot of the *m*ean value of the *s*quared function), given by

$$V_{rms} = \sqrt{\frac{1}{T} \int_0^T V_m^2 \cos^2{(\omega t + \phi)} dt} \tag{8.32}$$

which reduces to $V_{rms} = \frac{V_m}{\sqrt{2}}$.

To appreciate the response to a time-varying input, $v_s = V_m \cos{(\omega t + \phi)}$, consider the circuit shown in Figure 8.33, in which the switch is closed at $t = 0$ and there is no initial energy stored in the inductor.

Applying KVL to the circuit gives

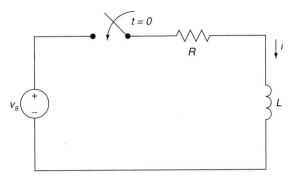

Figure 8.33 An RL circuit with sinusoidal input.

$$L\frac{di}{dt} + iR = V_m \cos(\omega t + \phi)$$

and after some work, the solution is

$$i = i_n + i_f$$

$$= \frac{-V_m}{\sqrt{R^2 + \omega^2 L^2}}\cos\left(\phi - \frac{\omega L}{R}\right)e^{-\frac{R}{L}t} + \frac{V_m}{\sqrt{R^2 + \omega^2 L^2}}\cos\left(\omega t + \phi - \frac{\omega L}{R}\right)$$

The first term is the natural response that goes to zero as t goes to infinity. The second term is the forced response that has the same form as the input (i.e., a sinusoid with the same frequency ω, but a different phase angle and maximum amplitude). If all you are interested in is the steady-state response as in most bioinstrumentation applications, then the only unknowns are the response amplitude and phase angle. The remainder of this section deals with techniques involving the *phasor* to efficiently find these unknowns.

8.12.1 Phasors

The phasor is a complex number that contains amplitude and phase angle information of a sinusoid, and for the signal in Eq. 8.30 is expressed as:

$$\mathbf{V} = V_m e^{j\phi} = V_m \underline{|\phi}\qquad(8.33)$$

In Eq. 8.30, by practice, the angle in the exponential is written in radians, and in the $\underline{|\phi}$ notation, in degrees. Work in the phasor domain involves the use of complex algebra in moving between the time and phasor domain; therefore, the rectangular form of the phasor is also used, given as

$$\mathbf{V} = V_m(\cos\phi + j\sin\phi)\qquad(8.34)$$

8.12.2   Passive Circuit Elements in the Phasor Domain

To use phasors with passive circuit elements for steady-state solutions, the relationship between voltage and current is needed for the resistor, inductor, and capacitor. Assume that

$$i = I_m \cos(\omega t + \theta)$$
$$I = I_m \lfloor \theta = I_m e^{j\theta}$$

For a resistor,

$$v = IR = RI_m \cos(\omega t + \theta)$$

and the phasor of v is

$$\mathbf{V} = RI_m \lfloor \theta = R\mathbf{I} \tag{8.35}$$

Note that there is no phase shift for the relationship between the phasor current and voltage for a resistor.

For an inductor,

$$v = L\frac{di}{dt} = -\omega LI_m \sin(\omega t + \theta) = -\omega LI_m \cos(\omega t + \theta - 90°)$$

and the phasor of v is

$$\mathbf{V} = -\omega LI_m \lfloor \theta - 90° = -\omega LI_m e^{j(\theta - 90°)}$$
$$= -\omega LI_m e^{j\theta} e^{-j90°} = -\omega LI_m e^{j\theta}(-j)$$
$$= j\omega LI_m e^{j\theta} \tag{8.36}$$
$$= j\omega L\mathbf{I}$$

Note that inductor current and voltage are out of phase by 90°—that is, current lags behind voltage by 90°.

For a capacitor, define $v = V_m \cos(\omega t + \theta)$ and $\mathbf{V} = V_m \lfloor \theta$. Now

$$i = C\frac{dv}{dt} = C\frac{d}{dt}(V_m \cos(\omega t + \theta))$$
$$= -CV_m \omega \sin(\omega t + \theta) = -CV_m \omega \cos(\omega t + \theta - 90°)$$

and the phasor for i is

$$\mathbf{I} = -\omega CV_m \lfloor \theta - 90° = -\omega CV_m e^{j\theta} e^{-j90°}$$
$$= -\omega CV_m e^{j\theta}(\cos(90°) - j\sin(90°))$$
$$= j\omega CV_m e^{j\theta}$$
$$= j\omega C\mathbf{V}$$

or

$$V = \frac{1}{j\omega C}I = \frac{-j}{\omega C}I \qquad (8.37)$$

Note that capacitor current and voltage are out of phase by 90°—that is, voltage lags behind current by 90°.

Eqs. 8.35 through 8.37 all have the form of $V = ZI$, where Z represents the impedance of the circuit element and is, in general, a complex number, with units of ohms. The impedance for the resistor is R, the inductor, $j\omega L$, and the capacitor, $\frac{-j}{\omega C}$. The impedance is a complex number and not a phasor, even though it may look like one. The imaginary part of the impendence is called reactance.

The final part to working in the phasor domain is to transform a circuit diagram from the time to phasor domain. For example, the circuit shown in Figure 8.34 is transformed into the phasor domain, shown in Figure 8.35, by replacing each circuit element with their impedance equivalent and sources by their phasor. For the voltage source, we have $V_s = 100 \sin 500t = 100 \cos (500t - 90°) \text{ mV} \leftrightarrow 500 \underline{|-90°}$
For the capacitor, we have

$$0.5 \ \mu F \leftrightarrow \frac{-j}{\omega C} = -j4000 \ \Omega$$

For the resistor, we have

$$1000 \ \Omega \leftrightarrow 1000 \ \Omega$$

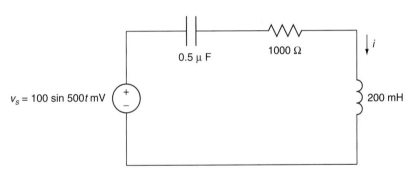

Figure 8.34   A circuit diagram.

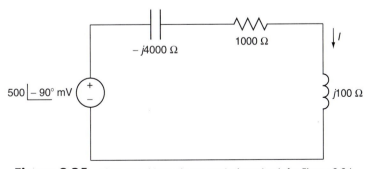

Figure 8.35   Phasor and impedance equivalent circuit for Figure 8.34.

For the inductor, we have

$$200\,\text{mH} \leftrightarrow j\omega L = j100\,\Omega$$

Each of the elements is replaced by its phasor and impedance equivalents as shown in Figure 8.35.

8.12.3 Kirchhoff's Laws and Other Techniques in the Phasor Domain

It is fortunate that all of the material presented so far in this chapter involving Kirchhoff's current and voltage laws, and all the other techniques, applies to phasors. That is, for KVL, the sum of phasor voltages around any closed path is zero:

$$\sum \mathbf{V}_i = 0 \tag{8.38}$$

and for KCL, the sum of phasor currents leaving any node is zero:

$$\sum \mathbf{I} = 0 \tag{8.39}$$

Impedances in series are given by

$$Z = Z_1 + \cdots + Z_n \tag{8.40}$$

Impedances in parallel are given by

$$Z = \frac{1}{\frac{1}{Z_1} + \cdots + \frac{1}{Z_n}} \tag{8.41}$$

The node-voltage method, as well as superposition and Thévenin equivalent circuits, are also applicable in the phasor domain. The following two examples illustrate the process, with the most difficult aspect involving complex algebra.

Example Problem 8.26

For the circuit shown in Figure 8.35, find the steady-state response i.

Solution

The impedance for the circuit is

$$Z = -j4000 + 1000 + j100 = 1000 - j3900\,\Omega$$

Using Ohm's law,

$$\mathbf{I} = \frac{\mathbf{V}}{Z} = \frac{0.5\,\lfloor -90^\circ}{1000 - j3900} = \frac{0.5\,\lfloor -90^\circ}{4026\,\lfloor -76^\circ} = 124\,\lfloor -14^\circ\,\mu\text{A}$$

Returning to the time domain, the steady-state current is

$$i = 124\cos(500t - 14^\circ)\mu\text{A}$$

∎

Example Problem 8.27

Find the steady-state response v using the node-voltage method for the following circuit.

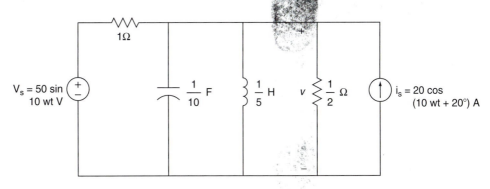

Solution

The first step is to transform the circuit elements into their impedances, which for the capacitor and inductor are:

$$\frac{1}{10}F \leftrightarrow \frac{-j}{\omega C} = -j\Omega$$

$$\frac{1}{5}H \leftrightarrow j\omega L = j2\,\Omega$$

The phasors for the two sources are

$$v_s = 50\sin\omega tV \leftrightarrow V_s = 50\,\underline{|-90°}V$$
$$i_s = 20\cos(\omega t + 20°)A \leftrightarrow I_s = 20\,\underline{|20°}$$

Since the two resistors retain their values, the phasor drawing of the circuit is shown in the following figure with the ground at the lower node.

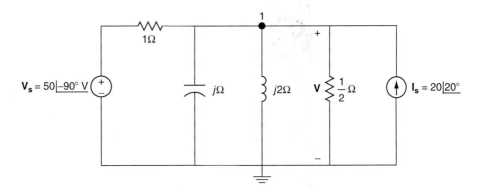

Writing the node-voltage equation for node 1 gives

$$\mathbf{V} - 50\lfloor -90^\circ + \frac{\mathbf{V}}{-j} + \frac{\mathbf{V}}{j2} + 2\mathbf{V} - 20\lfloor 20^\circ = 0$$

Collecting like terms, converting to rectangular form and converting to polar form gives

$$\mathbf{V}\left(3 + \frac{j}{2}\right) = 50\lfloor -90^\circ + 20\lfloor 20^\circ$$

$$\mathbf{V}\left(3 + \frac{j}{2}\right) = -50j + 18.8 + j6.8 = 18.8 - j43.2$$

$$\mathbf{V} \times 3.04\lfloor 9.5^\circ = 47.1\lfloor -66.5^\circ$$

$$\mathbf{V} = \frac{47.1\lfloor -66.5^\circ}{3.04\lfloor 9.5^\circ} = 15.5\lfloor -76^\circ$$

The steady-state solution is

$$v = 15.6\cos{(10t - 76^\circ)}\text{V}$$ ■

8.13   ACTIVE ANALOG FILTERS

This section presents several active analog filters involving the op amp. Passive analog filters use passive circuit elements: resistors, capacitors, and inductors. To improve performance in a passive analog filter, the resistive load at the output of the filter is usually increased. By using the op amp, fine control of the performance is achieved without increasing the load at the output of the filter. Filters are used to modify the measured signal by removing noise. A filter is designed in the frequency domain so that the measured signal to be retained is passed through and noise is rejected.

Shown in Figure 8.36 are the frequency characteristics of four filters: low-pass, high-pass, band-pass, and notch filters. The signal that is passed through the filter is indicated by the frequency interval called the pass-band. The signal that is removed by the filter is indicated by the frequency interval called the stop-band. The magnitude of the filter, $|H(j\omega)|$, is 1 in the pass-band and zero in the stop-band. The low-pass filter allows slowly changing signals with frequency less than ω_1 to pass through the filter, and eliminates any signal or noise above ω_1. The high-pass filter allows quickly changing signals with frequency greater than ω_2 to pass through the filter, and eliminates any signal or noise with frequency less than ω_2. The band-pass filter allows signals in the frequency band greater than ω_1 and less than ω_2 to pass through the filter, and eliminates any signal or noise outside this interval. The notch filter allows signals in the frequency band less than ω_1 and greater than ω_2 to pass through the filter, and eliminates any signal or noise outside this interval. The frequencies ω_1 and ω_2 are typically called cutoff frequencies for the low-pass and high-pass filters.

In reality, any real filter cannot possibly have these ideal characteristics, but instead has a smooth transition from the pass-band to the stop-band, as shown, for example, in Figure 8.37; the reason for this behavior is described in Chapter 10. Further, it is sometimes convenient to include both amplification and filtering in the same circuit,

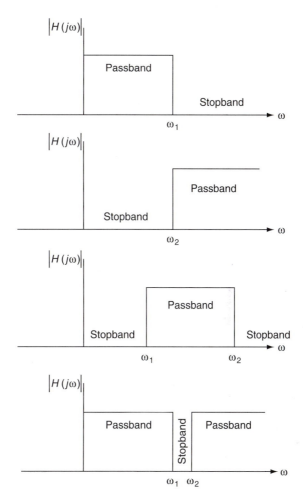

Figure 8.36 Ideal magnitude-frequency response for four filters: from top to bottom, low-pass, high-pass, band-pass, and notch.

so the maximum of the magnitude does not need to be 1, but can be a value of M specified by the needs of the application.

To determine the filter's performance, the filter is driven by a sinusoidal input. One varies the input over the entire spectrum of interest (at discrete frequencies) and records the output magnitude. The critical frequencies are when $|H(j\omega)| = \frac{M}{\sqrt{2}}$.

Example Problem 8.28

Using the low-pass filter in the following circuit, design the filter to have a gain of 5 and a cutoff frequency of $500\,\frac{\text{rad}}{s}$.

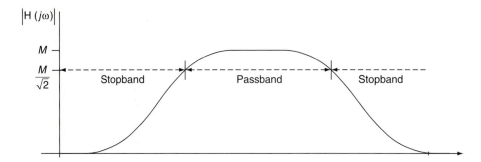

Figure 8.37   A realistic magnitude-frequency response for a band-pass filter. Note that the magnitude M does not necessarily need to be 1. The pass-band is defined as the frequency interval at which the magnitude is greater than $\frac{M}{\sqrt{2}}$.

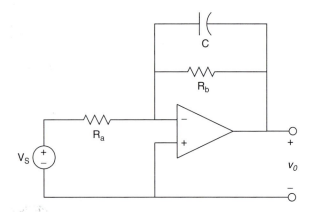

Solution

By treating the op amp as ideal, note that the noninverting input is connected to ground, and therefore, the inverting input is also connected to ground. The operation of this filter is readily apparent: at low frequencies the capacitor acts like an open circuit, reducing the circuit to an inverting amplifier that passes low-frequency signals, and at high frequencies the capacitor acts like a short circuit, which connects the output terminal to the inverting input and ground.

The phasor method will be used to solve this problem by first transforming the circuit into the phasor domain as shown in figure on the following page.

Summing the currents leaving the inverting input gives

$$-\frac{\mathbf{V}_s}{R_a} - \frac{\mathbf{V}_0}{\frac{1}{j\omega C}} - \frac{\mathbf{V}_0}{R_b} = 0$$

Collecting like terms and rearranging yields

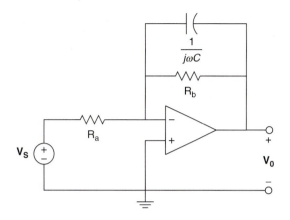

$$-V_0 \left(\frac{1}{\frac{1}{j\omega C}} + \frac{1}{R_b} \right) = \frac{V_s}{R_a}$$

After further manipulation,

$$\frac{V_0}{V_s} = -\frac{1}{R_a} \left(\frac{1}{\frac{1}{\frac{1}{j\omega C}} + \frac{1}{R_b}} \right) = -\frac{1}{R_a} \left(\frac{1}{j\omega C + \frac{1}{R_b}} \right)$$

$$\frac{V_0}{V_s} = -\frac{1}{R_a C} \left(\frac{1}{j\omega + \frac{1}{R_b C}} \right)$$

Similar to the reasoning for the characteristic equation for a differential equation, the cutoff frequency is defined as $\omega_c = \frac{1}{R_b C}$, (i.e., the denominator term, $j\omega + \frac{1}{R_b C}$ is set equal to zero). With the cutoff frequency set at $\omega_c = 500 \frac{\text{rad}}{s}$, then $\frac{1}{R_b C} = 500$. The cutoff frequency is also defined as $|H(j\omega)| = \frac{M}{\sqrt{2}}$ where $M = 5$. The magnitude of V_0 / V_s is given by

$$\left| \frac{V_0}{V_s} \right| = \frac{\frac{1}{R_a C}}{\sqrt{\omega^2 + \left(\frac{1}{R_b C} \right)^2}}$$

and at the cutoff frequency, $\omega_c = 500 \frac{\text{rad}}{s}$,

$$\frac{5}{\sqrt{2}} = \frac{\frac{1}{R_a C}}{\sqrt{\omega_c^2 + \left(\frac{1}{R_b C} \right)^2}}$$

With $\frac{1}{R_b C} = 500$, the magnitude is

$$\frac{5}{\sqrt{2}} = \frac{\frac{1}{R_a C}}{\sqrt{\omega_c^2 + \left(\frac{1}{R_b C} \right)^2}} = \frac{\frac{1}{R_a C}}{\sqrt{500^2 + 500^2}} = \frac{\frac{1}{R_a C}}{500\sqrt{2}}$$

which gives

$$R_a C = \frac{1}{2500}$$

Since we have three unknowns and two equations ($R_a C = \frac{1}{2500}$ and $\frac{1}{R_b C} = 500$), there are an infinite number of solutions. Therefore, one can select a convenient value for one of the elements, say $R_a = 20\,k\Omega$, and the other two elements are determined as

$$C = \frac{1}{2500 \times R_a} = \frac{1}{2500 \times 20,000} = 20\,nF$$

and

$$R_b = \frac{1}{500 \times C} = \frac{1}{500 \times 20 \times 10^{-9}} = 100\,k\Omega$$

A plot of the magnitude vs. frequency is shown in the following figure. As can be seen, the cutoff frequency gives a value of magnitude equal to 3.53 at 100 Hz, which is the design goal.

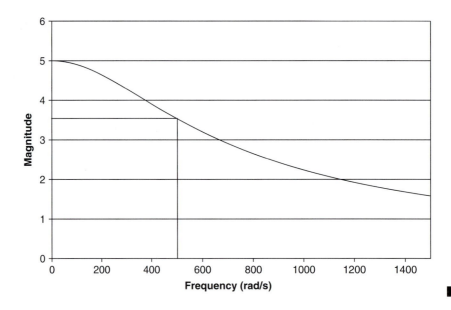

Example Problem 8.29

Using the high-pass filter in the following circuit, design the filter to have a gain of 5 and a cutoff frequency of $100\,\frac{rad}{s}$.

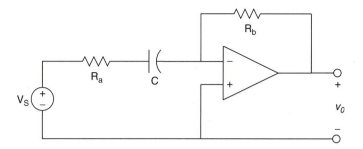

Solution

Since the op amp is assumed to be ideal and the noninverting input is connected to ground, the inverting input is also connected to ground. The operation of this filter is readily apparent. At low frequencies the capacitor acts like an open circuit, and so no input voltage is seen at the noninverting input. Since there is no input, then the output is zero. At high frequencies the capacitor acts like a short circuit, which reduces the circuit to an inverting amplifier that passes through high-frequency signals. As before, the phasor method will be used to solve this problem by first transforming the circuit into the phasor domain as shown in the following figure.

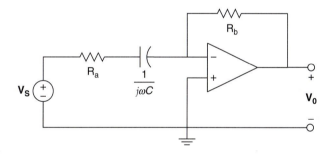

Summing the currents leaving the inverting input gives

$$-\frac{V_s}{R_a + \frac{1}{j\omega C}} - \frac{V_0}{R_b} = 0$$

Rearranging yields

$$\frac{V_0}{V_s} = -\frac{R_b}{R_a + \frac{1}{j\omega C}} = -\frac{R_b}{R_a} \frac{j\omega}{j\omega + \frac{1}{R_a C}}$$

At cutoff frequency $\omega_c = 100 \frac{\text{rad}}{s} = \frac{1}{R_a C}$. The magnitude of $\frac{V_0}{V_s}$ is given by

$$\left|\frac{V_0}{V_s}\right| = \frac{R_b}{R_a} \frac{\omega}{\sqrt{\omega^2 + \left(\frac{1}{R_a C}\right)^2}}$$

and at the cutoff frequency,

$$\frac{5}{\sqrt{2}} = \frac{R_b}{R_a} \frac{\omega_c}{\sqrt{\omega_c^2 + \left(\frac{1}{R_a C}\right)^2}}$$

With $\frac{1}{R_a C} = 100$ and $\omega_c = 100 \frac{rad}{s}$, gives

$$\frac{5}{\sqrt{2}} = \frac{R_b}{R_a} \frac{\omega_c}{\sqrt{\omega_c^2 + \left(\frac{1}{R_a C}\right)^2}} = \frac{R_b}{R_a} \frac{\frac{1}{R_a C}}{\sqrt{100^2 + 100^2}} = \frac{R_b}{R_a} \frac{100}{100} = \frac{R_b}{\sqrt{2} R_a}$$

Thus $\frac{R_b}{R_a} = 5$. Since we have three unknowns and two equations, one can select a convenient value for one of the elements, say $R_b = 20\,k\Omega$, and the other two elements are determined as

$$R_a = \frac{R_b}{5} = \frac{20,000}{5} = 4\,k\Omega$$

and

$$C = \frac{1}{100 R_a} = \frac{1}{100 \times 4000} = 2.5\,\mu F$$

A plot of the magnitude vs. frequency is shown in the following figure. As can be seen, the cutoff frequency gives a value of magnitude equal to 3.53 at 500 Hz, which is the design goal.

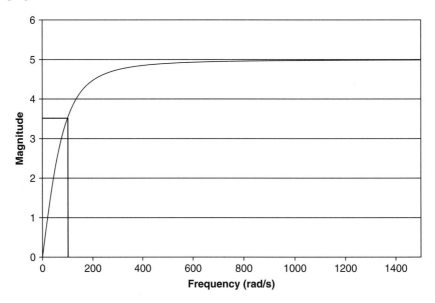

The next example demonstrates the technique to create a band-pass filter (which requires two cutoff frequencies).

Example Problem 8.30

Using the band-pass filter in the following circuit, design the filter to have a gain of 5 and pass through frequencies from 100 to 500 $\frac{rad}{s}$.

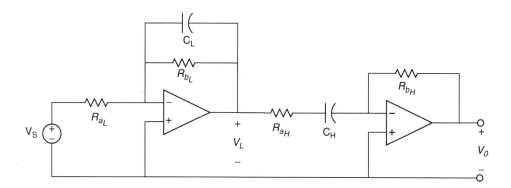

Solution

As usual, the design of the filter is done in the phasor domain, and makes use of work done in the previous two examples. Note the elements around the op amp on the left are the low-pass filter circuit elements, and those on the right, the high-pass filter. In fact, when working with op amps, filters can be cascaded together to form other filters; thus a low-pass and high-pass filter cascaded together form a band-pass. The phasor domain circuit is given in the next figure.

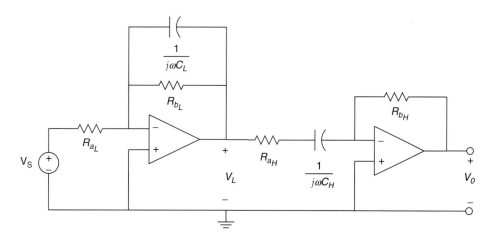

As before, the noninverting input to the op amps are connected to ground, which means that the inverting input is also connected to ground. Summing the currents leaving the inverting input for each op amp gives

$$-\frac{V_s}{R_{a_L}} - \frac{V_L}{\frac{1}{j\omega C_L}} - \frac{V_L}{R_{b_L}} = 0$$

$$-\frac{V_L}{R_{a_H} + \frac{1}{j\omega C_H}} - \frac{V_0}{R_{b_H}} = 0$$

Solving the first equation for V_L gives

$$V_L = -\frac{1}{R_{a_L} C_L}\left(\frac{1}{j\omega + \frac{1}{R_{b_L} C_L}}\right) V_s$$

Solving the second equation for V_0 gives

$$V_0 = -\frac{R_{b_H}}{R_{a_H}} \frac{j\omega}{j\omega + \frac{1}{R_{a_H} C_H}} V_L$$

Substituting V_L into the previous equation yields

$$V_0 = \frac{R_{b_H}}{R_{a_H}} \frac{j\omega}{j\omega + \frac{1}{R_{a_H} C_H}} \times \frac{1}{R_{a_L} C_L}\left(\frac{1}{j\omega + \frac{1}{R_{b_L} C_L}}\right) V_s$$

The form of the solution is simply the product of each filter. The magnitude of the filter is

$$\left|\frac{V_0}{V_s}\right| = \frac{R_{b_H}}{R_{a_H}} \frac{\omega}{\sqrt{\omega^2 + \left(\frac{1}{R_{a_H} C_H}\right)^2}} \frac{\frac{1}{R_{a_L} C_L}}{\sqrt{\omega^2 + \left(\frac{1}{R_{b_L} C_L}\right)^2}}$$

Since there are two cutoff frequencies, two equations evolve,

$$\omega_{c_H} = \frac{1}{R_{a_H} C_H} = 100 \frac{\text{rad}}{s}$$

and

$$\omega_{c_L} = \frac{1}{R_{b_L} C_L} = 500 \frac{\text{rad}}{s}$$

At either cutoff frequency, the magnitude is $\frac{5}{\sqrt{2}}$, such that at $\omega_{c_H} = 100 \frac{\text{rad}}{s}$

$$\frac{5}{\sqrt{2}} = \frac{R_{b_H}}{R_{a_H}} \frac{\omega_{c_H}}{\sqrt{\omega_{c_H}^2 + \left(\frac{1}{R_{a_H} C_H}\right)^2}} \frac{\frac{1}{R_{a_L} C_L}}{\sqrt{\omega_{c_H}^2 + \left(\frac{1}{R_{b_L} C_L}\right)^2}}$$

$$= \frac{R_{b_H}}{R_{a_H}} \frac{100}{\sqrt{100^2 + 100^2}} \frac{\frac{1}{R_{a_L} C_L}}{\sqrt{500^2 + 100^2}}$$

Therefore,

$$500\sqrt{26} = \frac{R_{b_H}}{R_{a_H}R_{a_L}C_L}$$

The other cutoff frequency gives the same result as the previous equation. There are now three equations ($\frac{1}{R_{a_H}C_H} = 100$, $\frac{1}{R_{b_L}C_L} = 500$ and $500\sqrt{26} = \frac{R_{b_H}}{R_{a_H}R_{a_L}C_L}$), and six unknowns. For convenience, set $R_{b_L} = 100\,\text{k}\Omega$ and $R_{a_H} = 100\,\text{k}\Omega$, which gives $C_L = \frac{1}{500R_{b_L}} = 20\,\text{nF}$ and $C_H = \frac{1}{100R_{a_H}} = 0.1\,\mu\text{F}$. Now from $500\sqrt{26} = \frac{R_{b_H}}{R_{a_H}}\frac{1}{R_{a_L}C_L}$,

$$\frac{R_{b_H}}{R_{a_L}} = 500\sqrt{26}C_LR_{a_H} = 5.099$$

Once again, one can specify one of the resistors, say $R_{a_L} = 10\,\text{k}\Omega$, giving $R_{b_H} = 50.099\,\text{k}\Omega$.

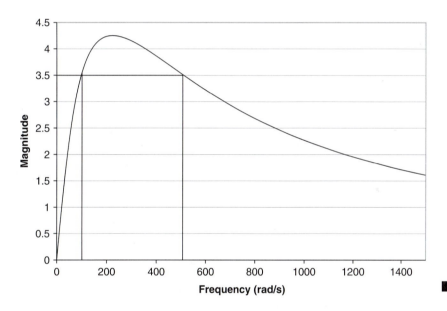

A plot of the magnitude versus frequency is shown in the following figure. As can be seen, the cutoff frequency gives a value of magnitude equal to 3.53 at 500 Hz, which is the design goal.

None of the filters in Example Problems 8.28–8.30 have the ideal characteristics of Figure 8.36. To improve the performance from the pass-band to stop-band in a low-pass filter with a sharper transition, one can cascade identical filters together (i.e., connect the output of the first filter to the input of the next filter, and so on). The more cascaded filters, the better the performance. The magnitude of the overall filter is the product of the individual filter magnitudes.

While this approach is appealing for improving the performance of the filter, the overall magnitude of the filter does not remain a constant in the pass-band. Better

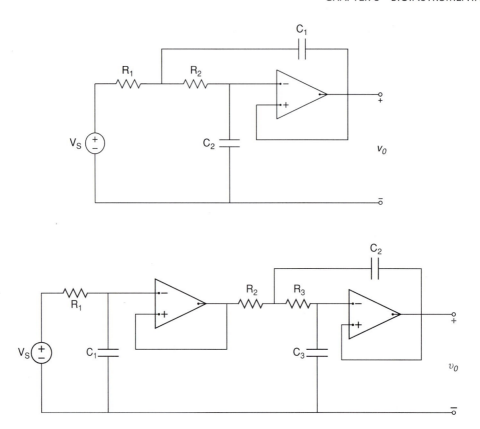

Figure 8.38   (Top) Second-order Butterworth low-pass filter. (Bottom) Third-order Butterworth low-pass filter.

filters are available with superior performance, such as a Butterworth filter. Two Butterworth filters are shown in Figure 8.38. Analysis of these filters is carried out in Exercises 54 and 55.

8.14   BIOINSTRUMENTATION DESIGN

Figure 8.2 described the various elements needed in a biomedical instrumentation system. The purpose of this type of instrument is to monitor the output of a sensor or sensors and to extract information from the signals that are produced by the sensors. Biomedical sensors are described in Chapter 9 and the processing of the measured signals is described in Chapter 10.

Acquiring a discrete-time signal and storing this signal in computer memory from a continuous-time signal is accomplished with an analog-to-digital (A/D) converter. The A/D converter uniformly samples the continuous-time waveform and transforms it into a sequence of numbers, one every t_k seconds. The A/D converter also transforms the continuous-time waveform into a digital signal (i.e., the amplitude takes one of 2^n

discrete values) which is converted into computer words and stored in computer memory. To adequately capture the continuous-time signal, the sampling instants t_k must be selected carefully so that information is not lost. As discussed in Chapter 10, the minimum sampling rate is twice the highest frequency content of the signal (based on the sampling theorem from communication theory). Realistically, we often sample at five to ten times the highest frequency content of the signal to achieve better accuracy by reducing aliasing error.

8.14.1 Noise

Measurement signals are always corrupted by noise in a biomedical instrumentation system. Interference noise occurs when unwanted signals are introduced into the system by outside sources such as power lines and transmitted radio and television electromagnetic waves. This kind of noise is effectively reduced by careful attention to the circuit's wiring configuration to minimize coupling effects.

Interference noise is introduced by power lines (50 or 60 Hz), fluorescent lights, AM/FM radio broadcasts, computer clock oscillators, laboratory equipment, cellular phones, and so forth. Electromagnetic energy radiating from noise sources is injected into the amplifier circuit or into the patient by capacitive and/or inductive coupling. Even the action potentials from nerve conduction in the patient generate noise at the sensor/amplifier interface. Filters are used to reduce the noise and to maximize the signal-to-noise (S/N) ratio at the input of the A/D converter.

Low-frequency noise (amplifier DC offsets, sensor drift, temperature fluctuations, etc.) is eliminated by a high-pass filter with the cutoff frequency set above the noise frequencies and below the biological signal frequencies. High-frequency noise (nerve conduction, radio broadcasts, computers, cellular phones, etc.) is reduced by a low-pass filter with the cutoff set below the noise frequencies and above the frequencies of the biological signal that is being monitored. Power-line noise is a very difficult problem in biological monitoring since the 50- or 60-Hz frequency is usually within the frequency range of the biological signal that is being measured. Band-stop filters are commonly used to reduce power-line noise. The notch frequency in these band-stop filters is set to the power-line frequency of 50 or 60 Hz, with the cutoff frequencies located a few Hertz to either side.

The second type of corrupting signal is called inherent noise. Inherent noise arises from random processes that are fundamental to the operation of the circuit's elements and, hence, is reduced by good circuit design practice. Although inherent noise can be reduced, it can never be eliminated. Low-pass filters can be used to reduce high-frequency components. However, noise signals within the frequency range of the biosignal being amplified cannot be eliminated by this filtering approach.

8.14.2 Computers

Computers consist of three basic units: the central processing unit (CPU), the arithmetic and logic unit (ALU), and memory. The CPU directs the functioning of all other units and controls the flow of information among the units during processing procedures. It is controlled by program instructions. The ALU performs all arithmetic calculations (add, subtract, multiply, and divide) as well as logical operations (AND, OR, NOT) that compare one set of information to another.

Computer memory consists of read only memory (ROM) and random access memory (RAM). ROM is permanently programmed into the integrated circuit that forms the basis of the CPU and cannot be changed by the user. RAM stores information temporarily and can be changed by the user. RAM is where user-generated programs, input data, and processed data are stored.

Computers are binary devices that use the presence of an electrical signal to represent 1 and the absence of an electrical pulse to represent 0. The signals are combined in groups of 8 bits, a byte, to code information. A word is made up of 2 bytes. Most desktop computers available today are 32-bit systems, which means that they can address 4.29×10^9 locations in memory. The first microcomputers were 8-bit devices that could interact with only 256 memory locations.

Programming languages relate instructions and data to a fixed array of binary bits so that the specific arrangement has only one meaning. Letters of the alphabet and other symbols (e.g., punctuation marks) are represented by special codes. ASCII stands for the American Standard Code for Information Exchange. ASCII provides a common standard that allows different types of computers to exchange information. When word processing files are saved as text files, they are saved in ASCII format. Ordinarily, word processing files are saved in special program-specific binary formats, but almost all data analysis programs can import and export data in ASCII files.

The lowest level of computer languages is machine language, which consists of the 0s and 1s that the computer interprets. Machine language represents the natural language of a particular computer. At the next level, assembly languages use English-like abbreviations for binary equivalents. Programs written in assembly language can manipulate memory locations directly. These programs run very quickly and are often used in data acquisition systems that must rapidly acquire a large number of samples, perhaps from an array of sensors, at a very high sampling rate.

Higher-level languages (e.g. FORTRAN, PERL, and C++) contain statements that accomplish tasks that require many machine or assembly language statements. Instructions in these languages often resemble English and contain commonly used mathematical notations. Higher-level languages are easier to learn than machine and assembly languages. Program instructions are designed to tell computers when and how to use various hardware components to solve specific problems. These instructions must be delivered to the CPU of a computer in the correct sequence to give the desired result.

When computers are used to acquire physiological data, programming instructions tell the computer when data acquisition should begin, how often samples should be taken from how many sensors, how long data acquisition should continue, and where the digitized data should be stored. The rate at which a system can acquire samples is dependent upon the speed of the computer's clock (e.g., 233 MHz) and the number of computer instructions that must be completed in order to take a sample. Some computers can also control the gain on the input amplifiers so that signals can be adjusted during data acquisition. In other systems, the gain of the input amplifiers must be manually adjusted.

EXERCISES

1. Find the power absorbed for the circuit element in Fig. 8.7 if

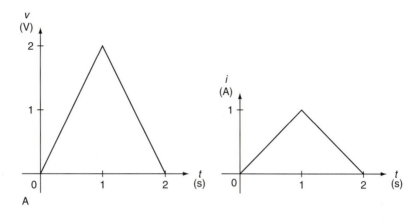

A

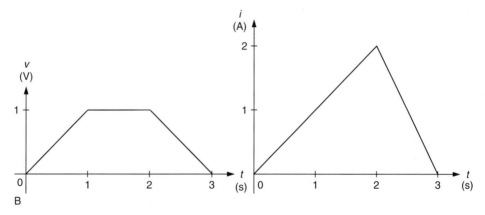

B

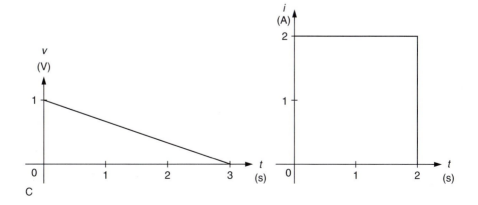

C

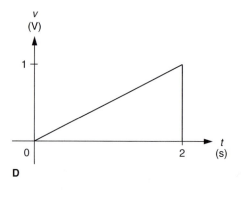

D

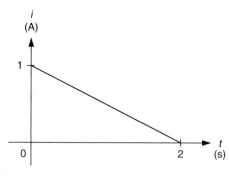

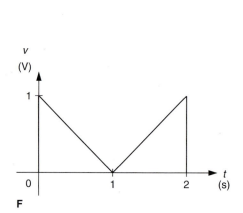

F

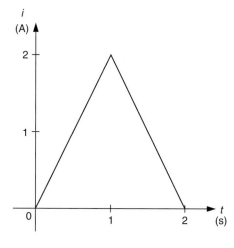

2. The voltage and current at the terminals in Fig. 8.7 are

$$v = te^{-10,000t}u(t) \text{ V}$$
$$i = (t + 10)e^{-10,000t}u(t) \text{ A}$$

    a. Find the time when the power is at its maximum.
    b. Find the maximum power.
    c. Find the energy delivered to the circuit at $t = 1 \times 10^{-4}$ s.
    d. Find the total energy delivered to the circuit element.

3. For the following circuit find (a) v_1, (b) the power absorbed and delivered.

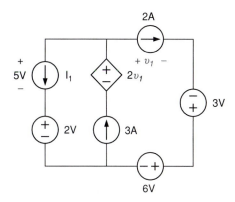

4. For the following circuit, find the power in each circuit element.

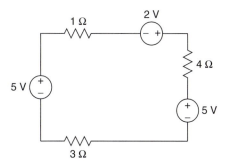

5. Find i_2 in the following circuit.

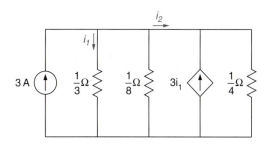

6. Find i_b for the following circuit.

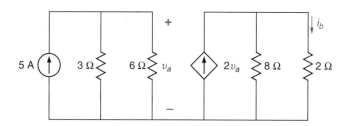

7. Find the equivalent resistance R_{ab} for the following circuit.

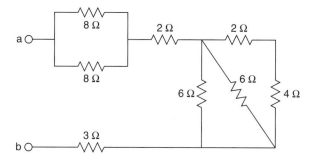

8. Find the equivalent resistance R_{ab} for the following circuit.

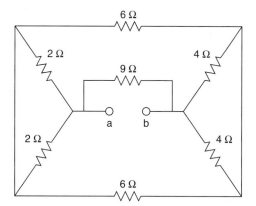

9. Find i_1 and v_1 for the following circuit.

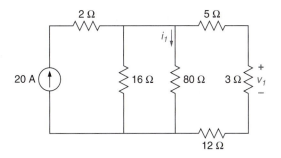

10. Use the node voltage method to determine v_1 and v_2.

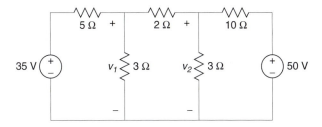

11. Use the node voltage method to determine v_1 and v_2.

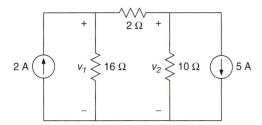

12. Use the node voltage method to determine v_1 and v_2.

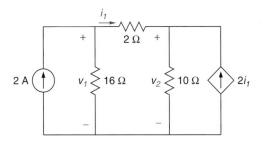

13. Use the node voltage method to determine v_1 and v_2.

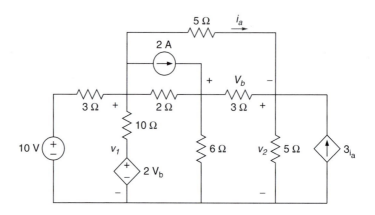

14. Use the node voltage method to determine v_1 and v_2.

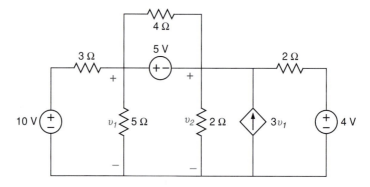

15. Use the node voltage method to determine v_1 and v_2.

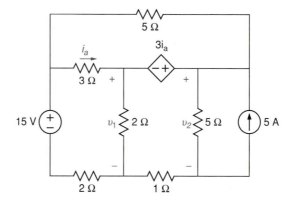

16. Use the superposition method to find v_o.

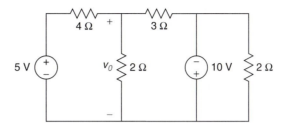

17. Use the superposition method to find v_o.

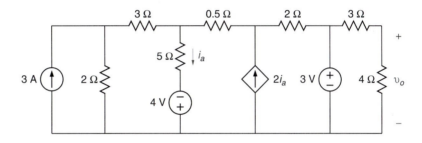

18. Find the Thévenin equivalent with respect to terminals a and b.

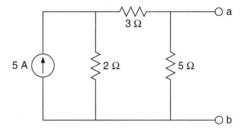

19. Find the Thévenin equivalent with respect to terminals a and b.

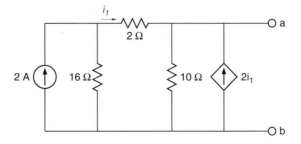

20. A current pulse given by $i(t) = (2 + 10e^{-2t})u(t)$ is applied through a 10-mH inductor. (a) Find the voltage across the inductor. (b) Sketch the current and voltage. (c) Find the power as a function of time.

21. The voltage across an inductor is given by the following figure. If $L = 30\,\text{mH}$ and $i(0) = 0\,\text{A}$, find $i(t)$ for t $\geq$ 0.

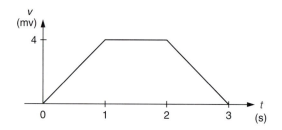

22. The voltage across a 4 μF capacitor is $v(t) = (200{,}000t - 50{,}000)e^{-2000t}u(t)$ V. Find (a) current through the capacitor, (b) power as a function of time, (c) energy.

23. The current through a 5 μF capacitor is

$$i(t) = \begin{cases} 0\,\text{mA} & t < 0 \ \text{ms} \\ 5t^2\,\text{mA} & 0 \leq t < 1 \ \text{ms} \\ 5(2 - t^2)\text{mA} & t \geq 1 \ \text{ms} \end{cases}$$

Find the voltage across the capacitor.

24. The switch has been in position a for a long time. At $t = 0$, the switch instantaneously moves to position b. Find i_L and v_1 for $t > 0$.

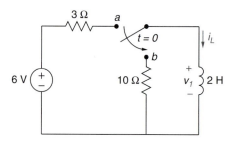

25. The switch has been in position a for a long time. At $t = 0$, the switch instantaneously moves to position b. Find i_L, i_1 and v_1 for $t > 0$.

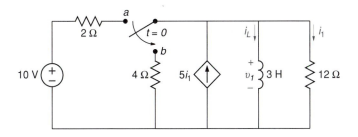

26. The switch has been in position a for a long time. At $t = 0$, the switch instantaneously moves to position b. Find v_c and i_1 for $t > 0$.

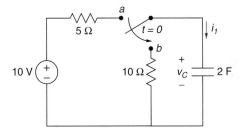

27. The switch has been in position a for a long time. At $t = 0$, the switch instantaneously moves to position b. Find v_c and i_1 for $t > 0$.

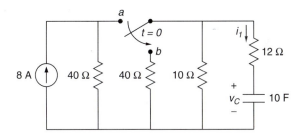

28. The switch has been in position a for a long time. At $t = 0$, the switch instantaneously moves to position b. Find v_c and i_1 for $t > 0$.

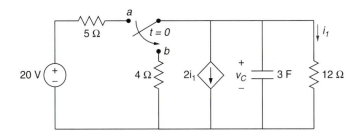

29. Find i_L and v_c for $t > 0$ for the following circuit if: (a) $i_s = 3u(t)$ A; (b) $i_s = 1 + 3u(t)$ A.

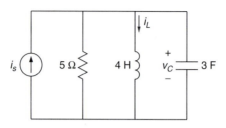

30. Find i_L and v_c for $t > 0$ for the following circuit if: (a) $v_s = 5u(t)$ V; (b) $v_s = 5u(t) + 3$ V.

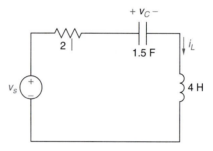

31. Find i_L and v_c for $t > 0$ for the following circuit if: (a) $i_s = 3u(t)$ A; (b) $i_s = 3u(t) - 1$ A.

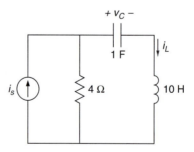

32. For the following circuit we are given that $v_b(0) = 2$ A, $i_{L_2}(0) = 5$ A, $v_{c_1}(0) = 2$ V, $v_{c_2}(0) = -3$ V, and $i_s = 2e^{-2t}u(t)$ A. Use the node-voltage method to find v_b for $t > 0$.

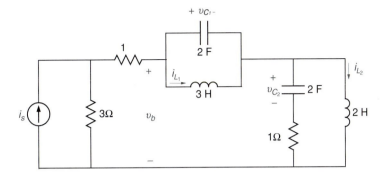

33. Use the node-voltage method to find v_{c_1} for $t > 0$ for the following circuit if: (a) $v_s = 2e^{-3t}u(t)$ V; (b) $v_s = 3\cos(2t)u(t)$ V; and (c) $v_s = 3u(t) - 1$ V.

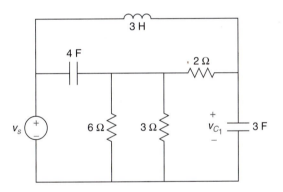

34. The operational amplifier shown in the following figure is ideal. Find v_o and i_o.

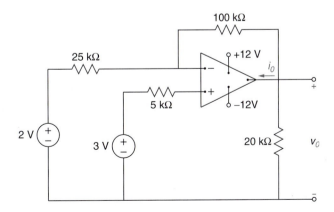

35. The operational amplifier shown in the following figure is ideal. Find v_o.

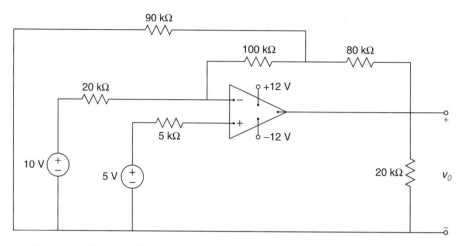

36. Find the overall gain for the following circuit if the operational amplifier is ideal. Draw a graph of v_o versus V_s if V_s varies between 0 to 10 V.

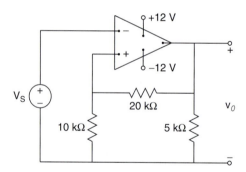

37. Find v_o in the following circuit if the operational amplifier is ideal.

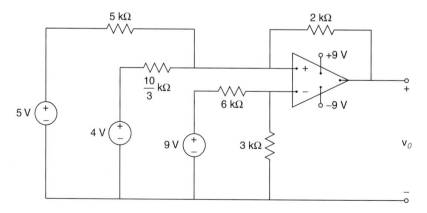

38. Find i_o in the following circuit if the operational amplifiers are ideal.

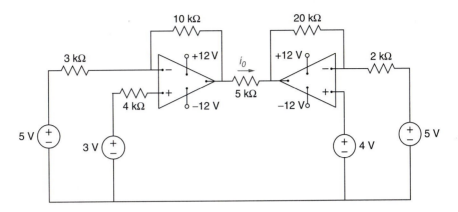

39. Suppose the input V_s is given as a triangular waveform as shown in the following figure. If there is no stored energy in the following circuit with an ideal operational amplifier, find v_o.

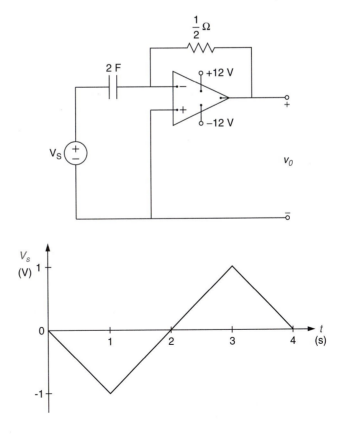

40. Suppose the input V_s is given in the following figure. If there is no stored energy in the following circuit with an ideal operational amplifier, find v_o.

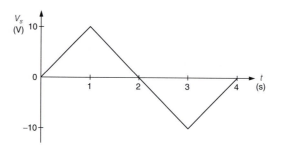

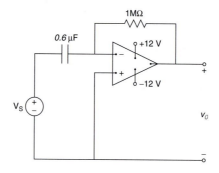

41. Suppose the input V_s is given in the following figure. If there is no stored energy in the following circuit with an ideal operational amplifier, find v_o.

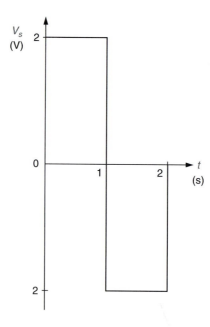

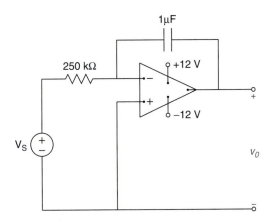

42. Suppose the input V_s is given in the following figure. If there is no stored energy in the following circuit with an ideal operational amplifier, find v_o.

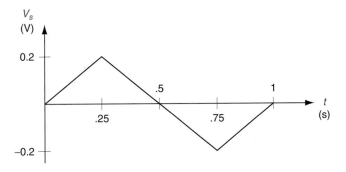

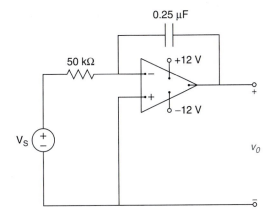

43. The following circuit is operating in the sinusoidal steady state. Find the steady-state expression for i_L if $i_s = 30 \cos 20t$ A.

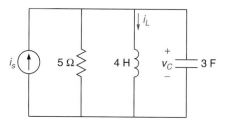

44. The following circuit is operating in the sinusoidal steady state. Find the steady-state expression for v_c if $v_s = 10 \sin 1000t$ V.

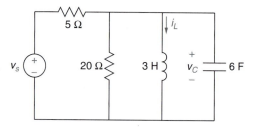

45. The following circuit is operating in the sinusoidal steady state. Find the steady-state expression for i_L if $i_s = 5 \cos 500t$ A.

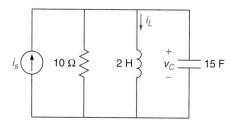

46. The following circuit is operating in the sinusoidal steady state. Find the steady-state expression for v_c if $i_s = 25 \cos 4000t$ V.

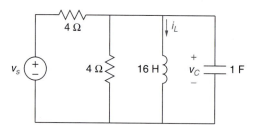

47. Design a low-pass filter with a magnitude of 10 and a cutoff frequency of 250 $\frac{rad}{s}$.

48. Design a high-pass filter with a magnitude of 20 and a cutoff frequency of 300 $\frac{rad}{s}$.

49. Design a band-pass filter with a gain of 15 and pass-through frequencies from 50 to 200 $\frac{rad}{s}$.

50. Design a low-pass filter with a magnitude of 5 and a cutoff frequency of 200 $\frac{rad}{s}$.

51. Design a high-pass filter with a magnitude of 10 and a cutoff frequency of 500 $\frac{rad}{s}$.

52. Design a band-pass filter with a gain of 10 and pass-through frequencies from 20 to 100 $\frac{rad}{s}$.

53. Suppose the operational amplifier in the following circuit is ideal. (The circuit is a low-pass first-order Butterworth filter.) Find the magnitude of the output v_o as a function of frequency.

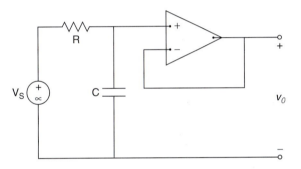

54. With an ideal operational amplifier, the following circuit is a second-order Butterworth low-pass filter. Find the magnitude of the output v_o as a function of frequency.

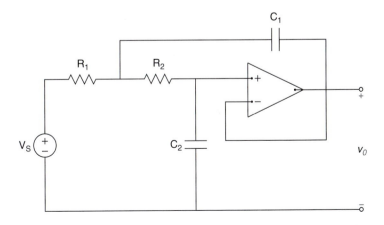

55. A third-order Butterworth low-pass filter is shown in the following circuit with an ideal operational amplifier. Find the magnitude of the output v_o as a function of frequency.

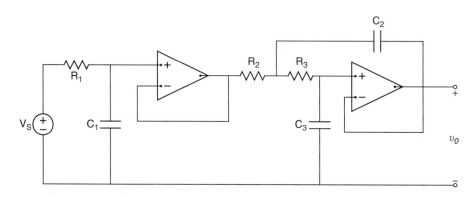

SUGGESTED READING

Aston, R. (1990). *Principles of Biomedical Instrumentation and Measurement*. Macmillan, New York.

Bronzino, J.D. (1986). *Biomedical Engineering and Instrumentation: Basic Concepts and Applications*. PWS Engineering, Boston.

Bronzino, J.D., Smith, V.H. and Wade, M.L. (1990). *Medical Technology and Society: An Interdisciplinary Perspective*. MIT Press, Cambridge, MA.

Carr J.J. and Brown, J.M. (2000). *Introduction to Biomedical Equipment Technology*, 4th Ed. Prentice Hall, Upper Saddle River, NJ.

Dempster, J. (1993). *Computer Analysis of Electrophysiological Signals*. Academic, London.

Johns, D.A. and Martin, K. (1997). *Analog Integrated Circuit Design*. John Wiley, New York.

Nilsson J.W. and Riedel, S. (2004). *Electric Circuits*, 7th Ed. Prentice Hall, Upper Saddle River, NJ.

Northrop, R.B. (2001). *Noninvasive Instrumentation and Measurement in Medical Diagnosis*. CRC, Boca Raton, FL.

Northrop, R.B. (1997). *Introduction to Instrumentation and Measurements*. CRC, Boca Raton, FL.

Perez, R. (2002). *Design of Medical Electronic Devices*. Academic, San Diego, CA.

Rosen A. and Rosen, H.D. (Eds.). (1995). *New Frontiers in Medical Device Technology*. John Wiley, New York.

Webster, J.G. (Ed.). (2003). *Bioinstrumentation*. John Wiley, New York.

Webster, J.G. (Ed.). (1997). *Medical Instrumentation—Application and Design*, 3rd Ed. John Wiley, New York.

Welkowitz, W., Deutsch, S. and Akay, M. (1992). *Biomedical Instruments—Theory and Design*, 2nd Ed. Academic, San Diego, CA.

Wise, D.E. (Ed.). (1991). *Bioinstrumentation and Biosensors*. Marcel Dekker, New York.

Wise, D.E. (Ed.). (1990). *Bioinstrumentation: Research, Developments, and Applications*. Butterworth, Stoneham, MA.

9 BIOMEDICAL SENSORS

Yitzhak Mendelson, PhD

Chapter Contents

At the conclusion of this chapter, the reader will be able to:

- Describe the different classifications of biomedical sensors.

- Describe the characteristics that are important for packaging materials associated with biomedical sensors.

- Calculate the half-cell potentials generated by different electrodes immersed in an electrolyte solution.

- Describe the electrodes that are used to record the ECG, EEG, and EMG and those that are used for intracellular recordings.

- Describe how displacement transducers, airflow transducers, and thermistors are used to make physical measurements.

- Describe how blood gases and blood pH are measured.

- Described how enzyme-based and microbial biosensors work and some of their uses.

- Explain how optical biosensors work and describe some of their uses.

9.1   INTRODUCTION

Biomedical sensors are used routinely in clinical medicine and biological research for measuring a wide range of physiological variables. They are often called biomedical transducers and are the main building blocks of diagnostic medical instrumentation found in physicians' offices, clinical laboratories, and hospitals. These sensors are routinely used in vivo to perform continuous invasive and noninvasive monitoring of critical physiological variables as well as in vitro to help clinicians in various diagnostic procedures. Some biomedical sensors are also used in nonmedical applications such as environmental monitoring, agriculture, bioprocessing, food processing, and the petrochemical and pharmacological industries.

Increasing pressures to lower health care costs, optimize efficiency, and provide better care in less expensive settings without compromising patient care are shaping the future of clinical medicine. As part of this ongoing trend, clinical testing is rapidly being transformed by the introduction of new tests that will revolutionize the way physicians diagnose and treat diseases in the future. Among these changes, patient self-testing and physician office screening are the two most rapidly expanding areas. This trend is driven by the desire of patients and physicians alike to have the ability to perform some types of instantaneous diagnosis and to move the testing apparatus from an outside central clinical laboratory to the point of care.

Biomedical sensors play an important role in a range of diagnostic medical applications. Depending on the specific needs, some sensors are used primarily in clinical laboratories to measure in vitro physiological quantities such as electrolytes, enzymes, and other biochemical metabolites in blood. Other biomedical sensors for measuring pressure, flow, and the concentrations of gases such as oxygen and carbon dioxide are used in vivo to follow continuously (monitor) the condition of a patient. For real-time continuous in vivo sensing to be worthwhile, the target analytes must vary rapidly and most often unpredictably.

The need for accurate medical diagnostic procedures places stringent requirements on the design and use of biomedical sensors. Usually, the first step in developing a biomedical sensor is to assess *in vitro* the accuracy,[1] operating range,[2] response time,[3] sensitivity,[4] resolution,[5] and reproducibility[6] of the sensor. Later, depending on the intended application, similar in vivo tests may be required to confirm the specifications of the sensor and to assure that the measurement remains sensitive, stable, safe, and cost-effective.

9.1.1 Sensor Classifications

Biomedical sensors are usually classified according to the quantity to be measured and are typically categorized as physical, electrical, or chemical depending on their specific applications. *Biosensors*, which can be considered a special subclassification of biomedical sensors, refers to a group of sensors that have two distinct components: (1) a biological recognition element such as a purified enzyme, antibody, or receptor, which functions as a mediator and provides the selectivity that is needed to sense the chemical component (usually referred to as the analyte) of interest, and (2) a supporting structure, which also acts as a transducer and is in intimate contact with the biological component. The purpose of the transducer is to convert the biochemical reaction into the form of an optical, electrical, or physical signal that is proportional to the concentration of a specific chemical. Thus, a blood pH sensor is not a biosensor according to this classification, although it measures a biologically important variable. It is simply a chemical sensor that can be used to measure a biological quantity.

9.1.2 Sensor Packaging

Packaging of certain biomedical sensors, primarily sensors for in vivo applications, is an important consideration during the design, fabrication, and use of the device. Obviously, the sensor must be safe and remain functionally reliable. In the development of implantable biosensors, an additional key issue is the long operational lifetime and biocompatibility of the sensor. Whenever a sensor comes into contact

---

[1]The ratio (expressed as a percentage) between the true value minus measured value and the true value.
[2]The maximum and minimum values that can be accurately measured.
[3]The time to reach 90% of the final value measured.
[4]The ratio of the incremental sensor output to the incremental input quantity.
[5]The smallest incremental quantity that the sensor can measure with certainty.
[6]The ability of the sensor to produce the same output when the same quantity is measured repeatedly.

with body fluids, the host itself may affect the function of the sensor or the sensor may affect the site in which it is implanted. For example, protein absorption and cellular deposits can alter the permeability of the sensor packaging which is designed to both protect the sensor and allow free chemical diffusion of certain analytes between the body fluids and the biosensor. Improper packaging of implantable biomedical sensors could lead to drift and a gradual loss of sensor sensitivity and stability over time. Furthermore, inflammation of tissue, infection, or clotting in a vascular site may produce harmful adverse effects (see Chapter 6). Hence, the materials used in the construction of the sensor's outer body must be nonthrombogenic and nontoxic since they play a critical role in determining the overall performance and longevity of an implantable sensor. One convenient strategy is to utilize various polymeric covering materials and barrier layers to minimize leaching of potentially toxic sensor components into the body. It is also important to keep in mind that once the sensor is manufactured, common sterilization practices by steam, ethylene oxide, or gamma radiation must not alter the chemical diffusion properties of the sensor packaging material.

This chapter will examine the operation principles of biomedical sensors including examples of invasive and noninvasive sensors for measuring biopotentials and other physical and biochemical variables encountered in different clinical and research applications.

9.2   BIOPOTENTIAL MEASUREMENTS

Biopotential measurements are made using different kinds of specialized electrodes. The function of these electrodes is to couple the ionic potentials generated inside the body to an electronic instrument. Biopotential electrodes are classified either as noninvasive (skin surface) or invasive (e.g., microelectrodes or wire electrodes).

9.2.1   The Electrolyte/Metal Electrode Interface

When a metal is placed in an electrolyte (i.e., an ionizable) solution, a charge distribution is created next to the metal/electrolyte interface as illustrated in Figure 9.1. This localized charge distribution causes an electric potential, called a half-cell potential, to be developed across the interface between the metal and the electrolyte solution.

The half-cell potentials of several important metals are listed in Table 9.1. Note that the hydrogen electrode is considered to be the standard electrode against which the half-cell potentials of other metal electrodes are measured.

Example Problem 9.1

Silver and zinc electrodes are immersed in an electrolyte solution. Calculate the potential drop between these two electrodes.

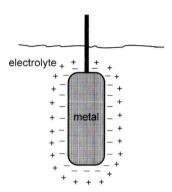

Figure 9.1 Distribution of charges at a metal/electrolyte interface.

TABLE 9.1 Half-cell Potentials of Important Metals

Primary Metal and Chemical Reaction			Half-cell Potential
Al	$\longrightarrow$	$Al^{3+} + 3e^-$	-1.706
Cr	$\longrightarrow$	$Cr^{3+} + 3e^-$	-0.744
Cd	$\longrightarrow$	$Cd^{2+} + 2e^-$	-0.401
Zn	$\longrightarrow$	$Zn^{2+} + 2e^-$	-0.763
Fe	$\longrightarrow$	$Fe^{2+} + 2e^-$	-0.409
Ni	$\longrightarrow$	$Ni^{2+} + 2e^-$	-0.230
Pb	$\longrightarrow$	$Pb^{2+} + 2e^-$	-0.126
H_2	$\longrightarrow$	$2H^+ + 2e^-$	0.000 (standard by definition)
Ag	$\longrightarrow$	$Ag^+ + e^-$	$+0.799$
Au	$\longrightarrow$	$Au^{3+} + 3e^-$	$+1.420$
Cu	$\longrightarrow$	$Cu^{2+} + 2e^-$	$+0.340$
$Ag + Cl^-$	$\longrightarrow$	$AgCl + 2e^-$	$+0.223$

Solution

From Table 9.1, the half-cell potentials for the silver and zinc electrodes are 0.799 and -0.763 V, respectively. Therefore, the potential drop between these two metal electrodes is equal to

$$0.799 - (-0.763) = 1.562\,\text{V}$$

Typically, biopotential measurements are made by utilizing two similar electrodes composed of the same metal. Therefore, the two half-cell potentials for these electrodes would be equal in magnitude. For example, two similar biopotential electrodes can be taped to the chest near the heart to measure the electrical potentials generated by the heart (electrocardiogram, or ECG). Ideally, assuming that the skin-to-electrode interfaces are electrically identical, the differential amplifier attached to these two electrodes would amplify the biopotential (ECG) signal but the half-cell potentials would be canceled out. In practice, however, disparity in electrode material or skin contact resistance could cause a significant DC offset voltage that would cause a current to flow through the two electrodes. This current will produce a voltage drop across the body. The offset voltage will appear superimposed at the output of the amplifier and may cause instability or base line drift in the recorded biopotential.   ■

Example Problem 9.2

Silver and aluminum electrodes are placed in an electrolyte solution. Calculate the current that will flow through the electrodes if the equivalent resistance of the solution is equal to 2 kΩ.

Solution

$$0.799 - (-1.706) = 2.505 \, \text{V}$$

$$2.505 \, \text{V}/2 \, \text{k}\Omega = 1.252 \, \text{mA}$$ ■

9.2.2   ECG Electrodes

Examples of different types of noninvasive biopotential electrodes used primarily for ECG recording are shown in Figure 9.2.

A typical flexible biopotential electrode for ECG recording is composed of certain types of polymers or elastomers which are made electrically conductive by the addition of a fine carbon or metal powder. These electrodes (Fig. 9.2a) are available with prepasted AgCl gel for quick and easy application to the skin using a double-sided peel-off adhesive tape.

The most common type of biopotential electrode is the "floating" silver/silver chloride electrode (Ag/AgCl), which is formed by electrochemically depositing a very

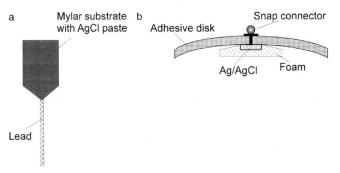

Figure 9.2   Biopotential skin surface ECG electrodes: (a) flexible Mylar electrode, and (b) disposable snap-type Ag/AgCl electrode.

thin layer of silver chloride onto a silver electrode (Fig. 9.2b). These electrodes are recessed and imbedded in foam that has been soaked with an electrolyte paste to provide good electrical contact with the skin. The electrolyte saturated foam is also known to reduce motion artifacts which could be produced, for example, during stress testing when the layer of the skin moves relative to the surface of the Ag/AgCl electrode. This motion artifact could cause large interference in the recorded biopotential and, in extreme cases, could severely degrade the measurement.

9.2.3 EMG Electrodes

A number of different types of biopotential electrodes are used in recording electromyographic (EMG) signals from different muscles in the body. The shape and size of the recorded EMG signals depends on the electrical property of these electrodes and the recording location. For noninvasive recordings, proper skin preparation, which normally involves cleansing the skin with alcohol or the application of a small amount of an electrolyte paste, helps to minimize the impedance of the skin–electrode interface and improve the quality of the recorded signal considerably. The most common electrodes used for surface EMG recording and nerve conduction studies are circular discs, about 1 cm in diameter, that are made of silver or platinum. For direct recording of electrical signals from nerves and muscle fibers, a variety of percutaneous needle electrodes are available, as illustrated in Figure 9.3. The most common type of needle electrode is the concentric bipolar electrode shown in Figure 9.3a. This electrode is made from thin metallic wires encased inside a larger canula or hypodermic needle. The two wires serve as the recording and reference electrodes. Another type of percutaneous EMG electrode is the unipolar needle electrode (Fig. 9.3b). This electrode is made of a thin wire that is mostly insulated by a thin layer of Teflon except about 0.3 mm near the distal tip. Unlike a bipolar electrode, this electrode requires a second unipolar reference electrode to form a closed electrical circuit. The second recording electrode is normally placed either adjacent to the recording electrode or attached to the surface of the skin.

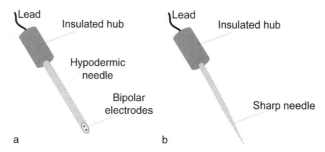

Figure 9.3 Intramascular biopotential electrodes: (a) bipolar and (b) unipolar configuration.

9.2.4   EEG Electrodes

The most commonly used electrodes for recording electroencephalographic (EEG) signals from the brain are cup electrodes and subdermal needle electrodes. Cup electrodes are made of platinum or tin and are approximately 5–10 mm in diameter. These cup electrodes are filled with a conducting electrolyte gel and can be attached to the scalp with an adhesive tape.

Recording of electrical potentials from the scalp is difficult because hair and oily skin impede good electrical contact. Therefore, clinicians sometimes prefer to use subdermal EEG electrodes instead of metal surface electrodes for EEG recording. These are basically fine platinum or stainless-steel needle electrodes about 10 mm long by 0.5 mm wide, which are inserted under the skin to provide a better electrical contact.

9.2.5   Microelectrodes

Microelectrodes are biopotential electrodes with an ultra-fine tapered tip that can be inserted into individual biological cells. These electrodes serve an important role in recording action potentials from single cells and are commonly used in neurophysiological studies. The tip of these electrodes must be small with respect to the dimensions of the biological cell to avoid cell damage and at the same time sufficiently strong to penetrate the cell wall. Figure 9.4 illustrates the construction of three typical types of microelectrodes: (a) glass micropipettes, (b) metal microelectrodes, and (c) solid-state microprobes.

In Figure 9.4a, a hollow glass capillary tube, typically 1 mm in diameter, is heated and softened in the middle inside a small furnace and then quickly pulled apart from both ends. This process creates two similar microelectrodes with an open tip that has a diameter on the order of 0.1 to 10 μm. The larger end of the glass tube (the stem) is then filled with a 3 M KCl electrolyte solution. A short piece of Ag/AgCl wire is inserted through the stem to provide an electrical contact with the electrolyte solution. When the tip of the microelectrode is inserted into an electrolyte solution, such as the intracellular cytoplasm of a biological cell, ionic current can flow through the fluid junction at the tip of the microelectrode. This establishes a closed electrical circuit between the Ag/AgCl wire inside the microelectrode and the biological cell.

A different kind of microelectrode made from a small-diameter strong metal wire (e.g., tungsten or stainless steel) is illustrated in Figure 9.4b. The tip of this microelectrode is usually sharpened down to a diameter of a few micrometers by an electrochemical etching process. The wire is then insulated up to its tip.

Solid-state microfabrication techniques commonly used in the production of integrated circuits can be used to produce microprobes for multichannel recordings of biopotentials or for electrical stimulation of neurons in the brain or spinal cord. An example of such a microsensor is shown in Figure 9.4c. The probe consists of a precisely micromachined silicon substrate with four exposed recording sites. One of the major advantages of this fabrication technique is the ability to mass produce very

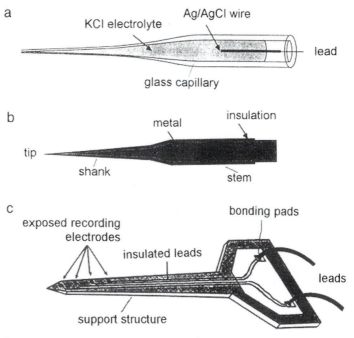

Figure 9.4 Biopotential microelectrodes: (a) capillary glass microelectrode, (b) insulated metal microelectrode, and (c) solid-state multisite recording microelectrode.

small and highly sophisticated microsensors with highly reproducible electrical and physical properties.

9.3 PHYSICAL MEASUREMENTS

9.3.1 Displacement Transducers

Inductive displacement transducers are based on the inductance L of a coil given by

$$L = n^2 \times G \times \mu \qquad (9.1)$$

where G is a geometric form constant, n is the number of coil turns, and, μ is the permeability of the magnetically susceptible medium inside the coil. These types of transducers measure displacement by changing either the self-inductance of a single coil or the mutual inductance coupling between two or more stationary coils, typically by the displacement of a ferrite or iron core in the bore of the coil assembly. A widely used inductive displacement transducer is the linear variable differential transformer (LVDT) illustrated in Figure 9.5.

This device is essentially a three-coil mutual inductance transducer that is composed of a primary coil (P) and two secondary coils (S_1 and S_2) connected in series but opposite in polarity in order to achieve a wider linear output range. The mutual

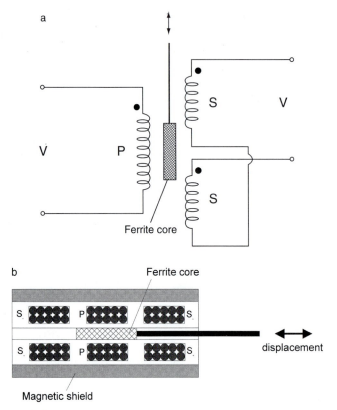

a

V

P

S

V

S

Ferrite core

b

Ferrite core

S   P   S

S   P   S

displacement

Magnetic shield

Figure 9.5   LVDT transducer: (a) electric diagram and (b) cross-section view.

inductance coupled between the coils is changed by the motion of a high-permeability slug. The primary coil is usually excited by passing an AC current. When the slug is centered symmetrically with respect to the two secondary coils, the primary coil induces an alternating magnetic field in the secondary coils. This produces equal voltages (but of opposite polarities) across the two secondary coils. Therefore, the positive voltage excursions from one secondary coil will cancel out the negative voltage excursions from the other secondary coil, resulting in a zero net output voltage. When the core moves toward one coil, the voltage induced in that coil is increased in proportion to the displacement of the core while the voltage induced in the other coil is decreased proportionally, leading to a typical voltage-displacement diagram as illustrated in Figure 9.6. Since the voltages induced in the two secondary coils are out of phase, special phase-sensitive electronic circuits must be used to detect both the position and the direction of the core's displacement.

Blood flow through an exposed vessel can be measured by means of an electro-magnetic flow transducer. It can be used in research studies to measure blood flow in major blood vessels near the heart including the aorta at the point where it exits from the heart.

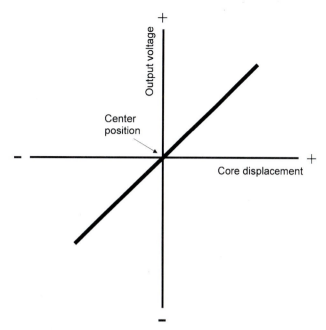

Figure 9.6   Output voltage versus core displacement of a typical LVDT transducer.

Consider a blood vessel of diameter l filled with blood flowing with a uniform velocity $\vec{u}$. If the blood vessel is placed in a uniform magnetic field $\vec{B}$ that is perpendicular to the direction of blood flow, the negatively charged anion and positively charged cation particles in the blood will experience a force $\vec{F}$ that is normal to both the magnetic field and blood flow directions and is given by

$$\vec{F} = q(\vec{u} \times \vec{B}) \qquad (9.2)$$

where q is the elementary charge (1.6×10^{-19} C). As a result, these charged particles will be deflected in opposite directions and will move along the diameter of the blood vessels according to the direction of the force vector $\vec{F}$. This movement will produce an opposing force $\vec{Fo}$ which is equal to

$$\vec{Fo} = q \times \vec{E} = q \times \frac{V}{l} \qquad (9.3)$$

where $\vec{E}$ is the net electrical field produced by the displacement of the charged particles and V is the potential produced across the blood vessel. At equilibrium, these two forces will be equal. Therefore, the potential difference V is given by

$$V = B \times l \times u \qquad (9.4)$$

and is proportional to the velocity of blood through the vessel.

Example Problem 9.3

Calculate the voltage induced in a magnetic flow probe if the probe is applied across a blood vessel with a diameter of 0.5 cm and the flow rate of blood is 5 cm/s. Assume that the magnitude of the magnetic field, B, is equal to 1.5×10^{-5} Wb/m^2.

Solution

From Eq. 9.4,

$$V = B \times l \times u = (1.5 \times 10^{-5} \text{ Wb/m}^2) \times (0.5 \times 10^{-2} \text{ m}) \times (5 \times 10^{-2} \text{ m/s})$$
$$= 3.75 \times 10^{-9} \text{ V}$$

(Note: [Wb] = [V × S])   ∎

    Practically, this device consists of a clip-on probe that fits snugly around the blood vessel as illustrated in Figure 9.7. The probe contains electrical coils to produce an electromagnetic field that is transverse to the direction of blood flow. The coil is usually excited by an AC current. A pair of very small biopotential electrodes are attached to the housing and rest against the wall of the blood vessel to pick up the induced potential. The flow-induced voltage is an AC voltage at the same frequency as the excitation voltage. Using an AC method instead of DC excitation helps to remove any offset potential error due to the contact between the vessel wall and the biopotential electrodes.

    A potentiometer is a resistive-type transducer that converts either linear or angular displacement into an output voltage by moving a sliding contact along the surface of a resistive element. Figure 9.8 illustrates linear and angular-type potentiometric transducers. A voltage, V_i, is applied across the resistor R. The output voltage, Vo, between the sliding contact and one terminal of the resistor is linearly proportional to the displacement. Typically, a constant current source is passed through the variable resistor, and the small change in output voltage is measured by a sensitive voltmeter using Ohm's law (i.e., $I = V/R$).

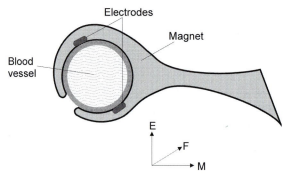

Figure 9.7   Electromagnetic blood-flow transducer.

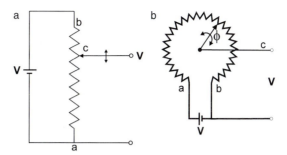

Figure 9.8   Linear translational (a) and angular (b) displacement transducers.

Example Problem 9.4

Calculate the change in output voltage of a linear potentiometer transducer that undergoes a 20% change in displacement.

Solution

Assuming that the current flowing through the transducer is constant, from Ohm's law,

$$\Delta V = I \times \Delta R$$

Hence, since the resistance between the sliding contact and one terminal of the resistor is linearly proportional to the displacement, a 20% change in displacement will produce a 20% change in the output voltage of the transducer.   ■

In certain clinical situations, it is desirable to measure changes in the peripheral volume of a leg when the venous outflow of blood from the leg is temporarily occluded by a blood pressure cuff. This volume-measuring method is called plethysmography and can indicate the presence of large venous clots in the legs. The measurement can be performed by wrapping an elastic resistive transducer around the leg and measuring the rate of change in resistance of the transducer as a function of time. This change corresponds to relative changes in the blood volume of the leg. If a clot is present, it will take more time for the blood stored in the leg to flow out through the veins after the temporary occlusion is removed. A similar transducer can be used to follow a patient's breathing pattern by wrapping the elastic band around the chest.

An elastic resistive transducer consists of a thin elastic tube filled with an electrically conductive material as illustrated in Figure 9.9. The resistance of the conductor inside the flexible tubing is given by

$$R = \rho \times \frac{l}{A} \tag{9.5}$$

where ρ is the resistivity of the electrically conductive material, l is the length, and A is the cross-sectional area of the conductor.

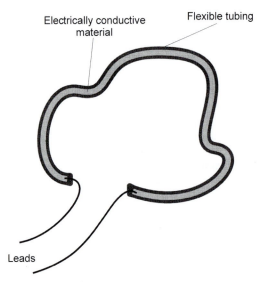

Electrically conductive material

Flexible tubing

Leads

Figure 9.9   Elastic resistive transducer.

Example Problem 9.5

A 10-cm long elastic resistive transducer with a resting resistance of 0.5 kΩ is wrapped around the chest. Assume that the chest diameter during exhalation is 33 cm. Calculate the resistance of the transducer after it has been applied to the chest.

Solution

After the transducer is stretched around the chest, its new length will increase from 10 to 103.7 cm. Assuming that the cross-sectional area of the transducer remains unchanged after it is stretched, the resistance will increase to

$$R_{stretched} = 0.5 \text{ k}\Omega \times \left(\frac{103.7 \text{ cm}}{10 \text{ cm}} \right) = 5.18 \text{ k}\Omega$$ ∎

Example Problem 9.6

Calculate the change in voltage that is induced across the elastic transducer in Example Problem 9.5. Assume that normal breathing produces a 10% change in chest circumference and a constant current of 5 mA is flowing through the transducer.

Solution

From Ohm's law (V = I × R),

$$V = 5 \text{ mA} \times 5.18 \text{ k}\Omega = 25.9 \text{ V}$$

If R changes by 10%, then

$$V = 5\,\text{mA} \times 1.1 \times 5.18\,\text{k}\Omega = 28.5\,\text{V}$$

$$\Delta V = 2.6\,\text{V}$$ ■

Strain gauges are displacement-type transducers that measure changes in the length of an object as a result of an applied force. These transducers produce a resistance change that is proportional to the fractional change in the length of the object, also called strain, S, which is defined as

$$S = \frac{\Delta l}{l} \tag{9.6}$$

where Δl is the fractional change in length and l is the initial length of the object. Examples include resistive wire elements and certain semiconductor materials.

To understand how a strain gauge works, consider a fine wire conductor of length l, cross-sectional area A, and resistivity ρ. The resistance of the unstretched wire is given by Eq. 9.5. Now suppose that the wire is stretched within its elastic limit by a small amount, Δl, such that its new length becomes $(l + \Delta l)$. Because the volume of the stretched wire must remain constant, the increase in the wire length results in a smaller cross-sectional area, $A_{\text{stretched}}$. Thus,

$$lA = (l + \Delta l) \times A_{\text{stretched}} \tag{9.7}$$

The resistance of the stretched wire is given by

$$R_{\text{stretched}} = \rho \times \frac{l + \Delta l}{A_{\text{stretched}}} \tag{9.8}$$

The increase in the resistance of the stretched wire ΔR is

$$\Delta R = R_{\text{stretched}} - \rho \times \frac{l}{A} \tag{9.9}$$

Substituting Eq. 9.8 and the value for $A_{\text{stretched}}$ from Eq. 9.7 into Eq. 9.9 gives

$$\Delta R = \rho \times \frac{(l + \Delta l)^2}{l \times A} - \rho \times \frac{l}{A} = \frac{\rho \times (l^2 + 2l\Delta l + \Delta l^2 - l^2)}{l \times A} \tag{9.10}$$

Assume that for small changes in length, $\Delta l \ll l$, this relationship simplifies to

$$\Delta R = \rho \times \frac{2 \times \Delta l}{A} = \frac{2 \times \Delta l}{l} \times R \tag{9.11}$$

The fractional change in resistance $(\Delta R/R)$ divided by the fractional change in length $(\Delta l/l)$ is called the gauge factor, G. For a common metal wire strain gauge made of constantan, G is approximately equal to 2. Semiconductor strain gauges made of silicon have a gauge factor about 70 to 100 times higher and are therefore much more sensitive than metallic wire strain gauges.

Example Problem 9.7

Calculate the strain in a metal wire gauge for a fractional change in resistance of 10%.

Solution

Combine Eqs. (9.6) and (9.11) to obtain

$$\frac{\Delta R}{R} = \frac{2 \times \Delta l}{l} = 2 \times S$$

$$\frac{0.1}{R} = 2 \times S$$

$$S = \frac{0.05}{R}$$ ■

Strain gauges typically fall into two categories, bonded or unbonded. A bonded strain gauge has a folded thin wire cemented to a semiflexible backing material, as illustrated in Fig. 9.10. An unbonded strain gauge consists of multiple resistive wires (typically four) stretched between a fixed and a movable rigid frame. In this configuration, when a deforming force is applied to the structure, two of the wires are stretched, and the other two are shortened proportionally. This configuration is used in blood pressure transducers, as illustrated in Figure 9.11. In this arrangement, a diaphragm is coupled directly by an armature to a movable frame that is inside the transducer. Blood in a peripheral vessel is coupled through a thin fluid-filled (saline) catheter to a disposable

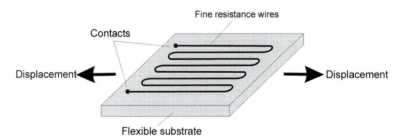

Figure 9.10   Bonded-type strain gauge transducer.

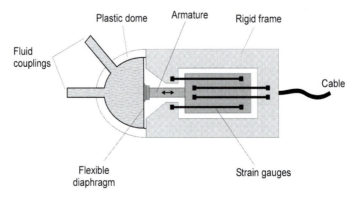

Figure 9.11   Resistive strain gauge (unbonded type) blood pressure transducer.

dome that is sealed by the flexible diaphragm. Changes in blood pressure during the pumping action of the heart apply a force on the diaphragm that causes the movable frame to move from its resting position. This movement causes the strain gauge wires to stretch or compress and results in a cyclical change in resistance that is proportional to the pulsatile blood pressure measured by the transducer.

In general, the change in resistance of a strain gauge is quite small. In addition, changes in temperature can also cause thermal expansion of the wire and thus lead to large changes in the resistance of a strain gauge. Therefore, very sensitive electronic amplifiers with special temperature compensation circuits are typically used in applications involving strain gauge transducers.

The capacitance C between two equal-size parallel plates of cross-sectional area A separated by a distance d is given by

$$C = \varepsilon_o \times \varepsilon_r \times \frac{A}{d} \tag{9.12}$$

where ε_o is the dielectric constant of free space (8.85×10^{-12} F/m) and ε_r is the relative dielectric constant of the insulating material placed between the two plates. The method that is most commonly employed to measure displacement is to change the separation distance d between a fixed and a movable plate as illustrated in Figure 9.12a. This arrangement can be used to measure force, pressure, or acceleration. Alternatively, it is possible to add a third plate and form a differential-type capacitance

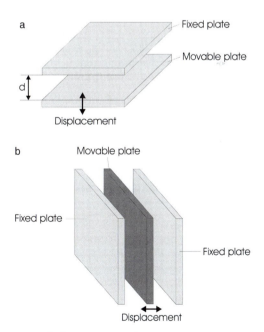

Figure 9.12 Capacitive displacement transducer: (a) single capacitance and (b) differential capacitance.

transducer (Fig. 9.12b). In this configuration, two of the plates are stationary whereas the middle plate can be moved freely relative to the position of the other plates, thus creating two variable-size capacitors. Accordingly, movement of the middle plate which will change the initial distance d by $\pm\Delta d$ will change the distance between two adjacent plates such that one capacitor will increase while the other will decrease in value. This double-capacitor arrangement provides improved sensitivity and can be incorporated into a Wheatstone bridge configuration (see Chapter 8). Capacitance sensors can be mass produced using solid-state microfabrication techniques that are commonly employed in making integrated circuits.

Example Problem 9.8

Two metal plates with an area of $4\,\text{cm}^2$ and separation distance of $0.1\,\text{mm}$ are used to form a capacitance transducer. If the material between the two plates has a dielectric constant $\varepsilon_r = 2.5$, calculate the capacitance of the transducer.

Solution

$$C = \varepsilon_o \times \varepsilon_r \frac{A}{d} = 8.85 \times 10^{-12}\text{F/m} \times 2.5 \times 4 \times 10^{-4}\text{m}^2/(0.1 \times 10^{-3}\text{m}) = 0.885\,\text{F} \quad \blacksquare$$

Capacitive displacement transducers can be used to measure respiration or movement by attaching multiple transducers to a mat that is placed on a bed. A capacitive displacement transducer can also be used as a pressure transducer by attaching the movable plate to a thin diaphragm that is in contact with a fluid or air. By applying a voltage across the capacitor and amplifying the small AC signal generated by the movement of the diaphragm, it is possible to obtain a signal that is proportional to the applied external pressure source.

Piezoelectric transducers are used in cardiology to listen to heart sounds (phonocardiography), in automated blood pressure measurements, and for measurement of physiological forces and accelerations. They are also commonly employed in generating ultrasonic waves (high-frequency sound waves typically above $20\,\text{kHz}$) which are used for measuring blood flow or imaging internal soft structures in the body (see Chapter 16).

A piezoelectric transducer consists of a small crystal (e.g., quartz) that contracts if an electric field (usually in the form of a short voltage impulse) is applied across its plates, as illustrated in Figure 9.13. Conversely, if the crystal is mechanically strained, it will generate a small electric potential. Besides quartz, several other ceramic materials, such as barium titanate and lead zirconate titanate, are also known to produce a piezoelectric effect.

The piezoelectric principle is based on the phenomenon that when an asymmetrical crystal lattice is distorted by an applied force F the internal negative and positive charges are reoriented. This causes an induced surface charge Q on the opposite

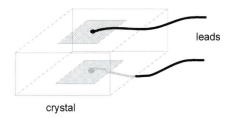

Figure 9.13 Ultrasonic transducer.

sides of the crystal. The induced charge is directly proportional to the applied force and is given by

$$Q = k \times F \qquad (9.13)$$

where k is a proportionality constant for the specific piezoelectric material. By assuming that the piezoelectric crystal acts like a parallel plate capacitor, the voltage across the crystal, V, is given by

$$\Delta V = \frac{\Delta Q}{C} \qquad (9.14)$$

where C is the equivalent capacitance of the crystal.

Example Problem 9.9

Derive a relationship for calculating the output voltage across a piezoelectric transducer that has a thickness d and area A in terms of an applied force F.

Solution

The capacitance of a piezoelectric transducer can be approximated by Eq. 9.12. Eq. 9.14 is combined with the relationship given by Eq. 9.13 and 9.14 to give

$$\Delta V = \frac{\Delta Q}{C} = \frac{k \times F}{C} = \frac{k \times F \times d}{\varepsilon_0 \times \varepsilon_r \times A} \qquad \blacksquare$$

Since the crystal has an internal leakage resistance, any steady charge produced across its surfaces will eventually be dissipated. Consequently, these piezoelectric transducers are not suitable for measuring a steady or low-frequency DC force. Instead, they are used either as variable force transducers or as mechanically resonating devices to generate high frequencies (typically from 1 to 10 MHz) either in crystal-controlled oscillators or as ultrasonic pulse transducers.

Piezoelectric transducers are commonly used in biomedical applications to measure the thickness of an object or in noninvasive blood pressure monitors. For instance, if two similar crystals are placed across an object (e.g., a blood vessel), one crystal can be excited to produce a short burst of ultrasound. The time it takes for this sound to reach the other transducer can be measured. Assuming that the velocity of sound propagation in soft tissue, c_t, is known (typically 1500 m/s), the time, t, it takes the ultrasonic pulse to propagate across the object can be measured and used to calculate the separation distance, d, of the two transducers from the following relationship

$$d = c_t \times t \qquad\qquad (9.15)$$

9.3.2   Airflow Transducers

One of the most common airflow transducers is the Fleish pneumotachometer, illustrated in Figure 9.14. The device consists of a straight short tube section with a fixed screen obstruction in the middle that produces a slight pressure drop as the air is passed through the tube. The pressure drop created across the screen is measured by a differential pressure transducer. The signal produced by the pressure transducer is proportional to the air velocity. The tube is normally shaped in a cone to generate a laminar airflow pattern. A small heater heats the screen so that water vapor does not condense on it over time and produce an artificially high pressure drop. Fleish type pneumotachometers are used to monitor volume, flow, and breathing rates of patients on mechanical ventilators.

9.3.3   Temperature Measurement

Body temperature is one of the most tightly controlled physiological variables and one of the four basic vital signs used in the daily assessment of a patient's health. The interior (core) temperature in the body is remarkably constant, about 37°C for a healthy person, and is normally maintained within ±0.5°C. Therefore, elevated body temperature is a sign of disease or infection, whereas a significant drop in skin temperature may be a clinical indication of shock.

There are two distinct areas in the body where temperature is measured routinely: the surface of the skin under the armpit or inside a body cavity such as the mouth or the rectum. The two most commonly used devices to measure body temperature are thermistors, which require direct contact with the skin or mucosal tissues, and

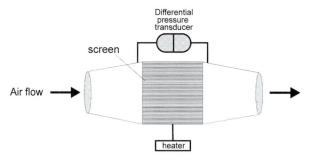

Figure 9.14   Fleish airflow transducer.

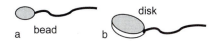

Figure 9.15 Common forms of thermistors.

non-contact thermometers, which measure body core temperature inside the auditory canal.

Thermistors are temperature-sensitive transducers made of compressed sintered metal oxides (such as nickel, manganese, or cobalt) that change their resistance with temperature. Commercially available thermistors range in shape from small beads to large disks, as illustrated in Figure 9.15.

Mathematically, the resistance-temperature characteristic of a thermistor can be approximated by

$$R_T = R_0 \times \exp\left[\beta \times \left(\frac{1}{T} - \frac{1}{T_0}\right)\right] \qquad (9.16)$$

where R_0 is the resistance at a reference temperature, T_0 (expressed in degrees K), R_T is the resistance at temperature, T (expressed in degrees K), and β is a material constant, typically between 2500 and 5500 K. A typical resistance-temperature characteristic of a thermistor is shown in Figure 9.16. Note that unlike metals and conventional resistors which have a positive temperature coefficient (as the temperature increases, the resistance increases), thermistors have a nonlinear relationship between temperature and resistance and a negative temperature coefficient. Increasing the temperature decreases the resistance of the thermistor.

Example Problem 9.10

A thermistor with a material constant β of 4500 K is used as a thermometer. Calculate the resistance of this thermistor at 25°C. Assume that the resistance of this thermistor at body temperature (37°C) is equal to 85 Ω.

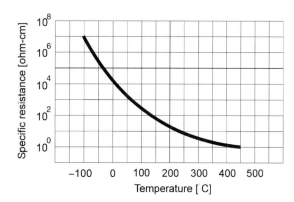

Figure 9.16 Resistivity versus temperature characteristics of a typical thermistor.

Solution

Using the resistance-temperature characteristic of a thermistor (Eq. 9.16) gives

$$R_T = 85 \times \exp\left[4500 \times \left(\frac{1}{298} - \frac{1}{310}\right)\right] = 152.5 \ \Omega \qquad \blacksquare$$

The size and mass of a thermistor probe in a medical thermometer must be small in order to produce a rapid response time to temperature variations. The probe is normally covered with a very thin sterile plastic sheet that is also disposable to prevent cross contamination between patients.

A thermistor sensor can be employed in a Swan-Ganz thermodilution technique for measuring cardiac output (the volume of blood ejected by the heart each minute) and assessing ventricular function. The procedure is normally performed in the operating room or the intensive care unit. It involves a rapid bolus injection of a cold indicator solution, usually 3–5 ml of a sterile saline or dextrose solution kept at 0°C, into the right atrium via a flexible pulmonary artery catheter (Fig. 9.17a).

The 5 or 7 French-size thermodilution catheter contains a small balloon (Fig. 9.17b) and is normally inserted into either the femoral or internal jugular veins. The catheter is constructed of a radiopaque material to enable easy visualization by an x-ray machine. It contains three lumina: a balloon inflation port to guide the flexible tip to the right location, a proximal central venous port, and a distal pulmonary artery port. After the balloon is inflated, the tip of the flexible catheter is passed across the tricuspid valve through the right ventricle, across the pulmonary valve, and into the pulmonary artery. The proximal and distal ports can be connected to pressure transducers that measure blood pressures inside the right side of the heart while the catheter is advanced into the right atrium.

After the catheter is inserted, a cold bolus is injected into the right atrium through the proximal lumen of the catheter. The bolus solution mixes with the venous blood in the right atrium and causes the blood to cool slightly. The cooled blood is ejected by the right ventricle into the pulmonary artery, where it contacts a thermistor that is located in the wall of the catheter near its distal tip. The thermistor measures the change in blood temperature as the blood passes on to the lungs. An instrument computes the cardiac output by integrating the change in blood temperature immediately following the bolus injection, which is inversely proportional to cardiac output.

Noncontact thermometers measure the temperature of the ear canal near the tympanic membrane, which is known to track the core temperature by about 0.5–1.0°C. Basically, as illustrated in Figure 9.18, infrared radiation from the tympanic membrane is channeled to a heat-sensitive detector through a metal waveguide that has a gold-plated inner surface for better reflectivity. The detector, which is either a thermopile or a pyroelectric sensor that converts heat flow into an electric current, is normally maintained at a constant temperature environment to minimize inaccuracies due to fluctuation in ambient temperature. A disposable speculum is used on the probe to protect patients from cross contamination.

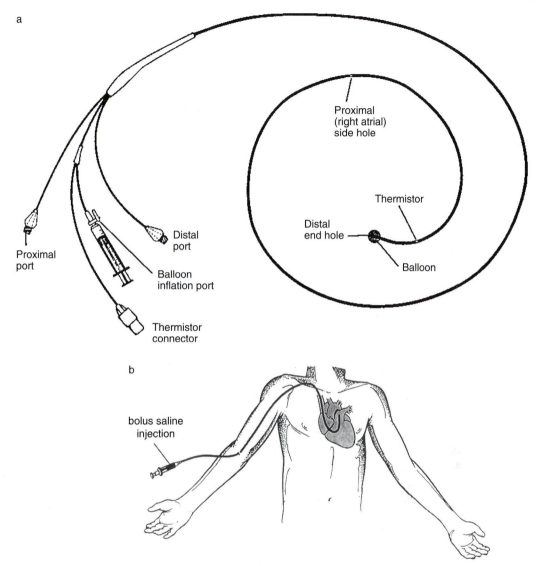

Figure 9.17 A Swan-Ganz thermodilution catheter.

9.4 **BLOOD GASES AND pH SENSORS**

Measurements of arterial blood gases (pO_2 and pCO_2) and pH are frequently performed on critically ill patients in both the operating room and the intensive care unit.
They are used by the physician to adjust mechanical ventilation or administer pharmacological agents. These measurements provide information about the respiratory and

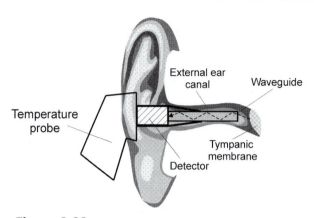

Figure 9.18   Noncontact-type infrared ear thermometer.

metabolic imbalances in the body and reflect the adequacy of blood oxygenation and CO_2 elimination.

Traditionally, arterial blood gas analysis has been performed by withdrawing blood from a peripheral artery. The blood sample is then transported to a clinical laboratory for analysis. The need for rapid test results in the management of unstable, critically ill patients has led to the development of newer methods for continuous noninvasive blood gas monitoring. This allows the physician to follow trends in the patient's condition and receive immediate feedback on the adequacy of certain therapeutic interventions.

Noninvasive sensors for measuring O_2 and CO_2 in arterial blood are based on the discovery that gases such as O_2 and CO_2 can easily diffuse through the skin. Diffusion occurs due to a partial pressure difference between the blood in the superficial layers of the skin and the outermost surface of the skin. This concept has been used to develop two types of noninvasive electrochemical sensors for transcutaneous monitoring of pO_2 and pCO_2. Furthermore, the discovery that blood changes its color depending on the amount of oxygen chemically bound to the hemoglobin in the erythrocytes has led to the development of several optical methods to measure the oxygen saturation in blood.

9.4.1 Oxygen Measurement

A quantitative method for measuring blood oxygenation is of great importance in assessing the circulatory and respiratory condition of a patient. Oxygen is transported by the blood from the lungs to the tissues in two distinct states. Under normal physiological conditions, approximately 2% of the total amount of oxygen carried by the blood is dissolved in the plasma. This amount is linearly proportional to the blood pO_2. The remaining 98% is carried inside the erythrocytes in a loose reversible chemical combination with hemoglobin (Hb) as oxyhemoglobin (HbO_2). Thus, there are two options for measuring blood oxygenation: either using a polarographic pO_2

sensor or measuring oxygen saturation (the relative amount of HbO_2 in the blood) by means of an optical oximeter.

A pO_2 sensor, also widely known as a Clark electrode, is used to measure the partial pressure of O_2 gas in a sample of air or blood. This sensor is categorized as an amperometric sensor and requires an external polarizing bias source. The measurement is based on the principle of polarography as illustrated in Figure 9.19. The electrode utilizes the ability of O_2 molecules to react chemically with H_2O in the presence of electrons to produce hydroxyl (OH^-) ions. This electrochemical reaction, called an oxidation/reduction or redox reaction, generates a small current and requires an externally applied constant polarizing voltage source of about 0.6V.

Oxygen is reduced (consumed) at the surface of a noble metal (e.g., platinum or gold) cathode (the electrode connected to the negative side of the voltage source) according to the following chemical reaction:

$$O_2 + 2H_2O + 4e^- \longleftrightarrow 4OH^-$$

In this reduction reaction, an O_2 molecule takes four electrons and reacts with two water molecules, generating four hydroxyl ions. The resulting OH^- ions migrate and react with a reference Ag/AgCl anode (the electrode connected to the positive side of the voltage source), causing a two-step oxidation reaction to occur as follows:

$$Ag \leftrightarrow Ag^+ + e^-$$
$$Ag^+ + Cl^- \leftrightarrow AgCl \downarrow$$

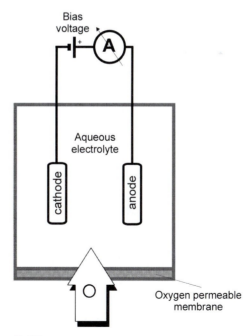

Figure 9.19   Principle of a polarographic Clark-type pO_2 sensor.

In this oxidation reaction, silver from the electrode is first oxidized to silver ions, and electrons are liberated to the anode. These silver ions are immediately combined with chloride ions to form silver chloride which precipitates on the surface of the anode. The current flowing between the anode and the cathode in the external circuit produced by this reaction is directly (i.e., linearly) proportional to the number of O_2 molecules constantly reduced at the surface of the cathode. The electrodes in the polarographic cell are immersed in an electrolyte solution of potassium chloride and surrounded by an O_2-permeable Teflon or polypropylene membrane that permits gases to diffuse slowly into the electrode. Thus, by measuring the change in current between the cathode and the anode, the amount of oxygen that is dissolved in the solution can be determined.

With a rather minor change in the configuration of a polarographic pO_2 sensor, it is also possible to measure the pO_2 transcutaneously. Figure 9.20 illustrates a cross section of a Clark-type transcutaneous pO_2 sensor. This sensor is essentially a standard polarographic pO_2 electrode that is attached to the surface of the skin by double-sided adhesive tape. It measures the partial pressure of oxygen that diffuses from the blood through the skin into the Clark electrode similar to the way it measures the pO_2 in a sample of blood. However, since the diffusion of O_2 through the skin is normally very low, a miniature heating coil is incorporated into the housing of this electrode to cause gentle vasodilatation (increased local blood flow) of the capillaries in the skin. By raising the local skin temperature to about $43°C$, the pO_2 measured by the transcutaneous sensor approximates that of the underlying arterial blood. This electrode has been used extensively in monitoring newborn babies in the intensive care unit. However, as the skin becomes thicker and matures in adult patients, the gas

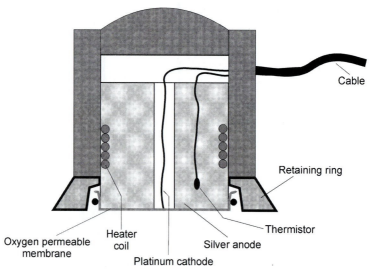

Figure 9.20   Transcutaneous pO_2 sensor.

diffusion properties of the skin change significantly and cause large errors that result in inconsistent readings.

Various methods for measuring the oxygen saturation, SO_2 (the relative amount of oxygen carried by the hemoglobin in the erythrocytes), of blood in vitro or in vivo in arterial blood (S_aO_2) or mixed venous blood (S_vO_2), have been developed. This method, referred to as oximetry, is based on the light absorption properties of blood and, in particular, the relative concentration of Hb and HbO_2 since the characteristic color of deoxygenated blood is blue, whereas fully oxygenated blood has a distinct bright red color.

The measurement is performed at two specific wavelengths: a red wavelength, λ_1, where there is a large difference in light absorbance between Hb and HbO_2 (e.g., 660 nm), and a second wavelength, λ_2, in the near-infrared region of the spectrum. The second wavelength can be either isobestic (a region of the spectrum around 805 nm where the absorbencies of Hb and HbO_2 are equal) or around 940–960 nm, where the absorbance of Hb is slightly smaller than that of HbO_2. Figure 9.21 shows the optical absorption spectra of blood in the visible and near-infrared region.

The measurement is based on Beer–Lambert's law that relates the transmitted light power, P_t, to the incident light power, P_0, according to the following relationship:

$$P_t = P_0 \times 10^{-abc} \qquad (9.17)$$

where a is a wavelength-dependent constant called the extinction coefficient (or molar absorptivity) of the sample, b is the light path length through the sample, and c is the concentration of the sample.

Assuming for simplicity that (1) $\lambda_1 = 660$ nm and $\lambda_2 = 805$ nm (i.e., isobestic), (2) the hemolyzed blood sample (blood in which the erythrocytes have been ruptured,

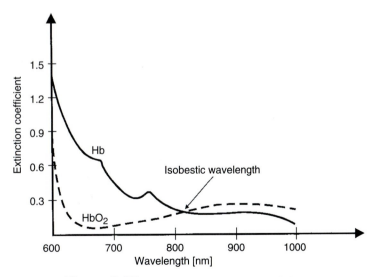

Figure 9.21 Optical properties of Hb and HbO_2.

i.e., the hemoglobin has been released and uniformly mixed with the plasma) consists of a two-component mixture of Hb and HbO_2, and (3) the total light absorbance by the mixture of these two components is additive, a simple mathematical relationship can be derived for computing the oxygen saturation of blood:

$$SO_2 = A - B \times \left[\frac{OD(\lambda_1)}{OD(\lambda_2)} \right] \tag{9.18}$$

where A and B are two coefficients that are functions of the specific absorptivity of Hb and HbO_2, OD (or absorbance) is defined as the optical density [i.e., $\log_{10}(1/T)$], where T represents the light transmission through the sample and is given by P_t/P_0, and SO_2 is defined as $c_{HB}/(c_{HB} + c_{HBO_2})$.

The measurement of SO_2 in blood can be performed either in vitro or in vivo. In vitro measurement using a bench top oximeter requires a sample of blood, usually withdrawn from a peripheral artery. The sample is transferred into an optical cuvette (a parallel-wall glass container which holds the sample) where it is first hemolyzed and then illuminated sequentially by light from an intense white source after proper wavelength selection using narrow-band optical filters. SO_2 can also be measured in vivo using a noninvasive pulse oximeter. Noninvasive optical sensors for measuring S_aO_2 by a pulse oximeter consist of a pair of small and inexpensive light emitting diodes (LEDs)—typically a red (R) LED around 660 nm and an infrared (IR) LED around 940–960 nm—and a single, highly sensitive silicon photodetector. These components are typically mounted inside a reusable spring-loaded clip or a disposable adhesive wrap (Fig. 9.22). Electronic circuits inside the pulse oximeter generate digital switching signals to turn on and off the two LEDs in a sequential manner and synchronously measure the photodetector output when the corresponding LEDs are turned on. The sensor is usually attached either to the finger tip or earlobe so that the tissue is sandwiched between the light source and the photodetector.

Pulse oximetry relies on the detection of a photoplethysmographic signal, as illustrated in Figure 9.23. This signal is caused by changes in the arterial blood volume associated with periodic contraction of the heart during systole. The magnitude of this signal depends on the amount of blood ejected from the heart into the peripheral vascular bed with each cardiac cycle, the optical absorption of the blood, the composition and color of the skin and underlying tissues, and the wavelengths used to illuminate the blood. S_aO_2 is derived by analyzing the magnitude of the red and infrared photoplethysmograms measured by the photodetector. Electronic circuits separate the red and infrared photopletysmograms into their pulsatile (AC) and nonpulsatile (DC) signal components. An algorithm inside the pulse oximeter performs a mathematical normalization by which the AC signal at each wavelength is divided by the corresponding DC component that is mainly due to the light absorbed by the bloodless tissue, residual arterial blood when the heart is in diastole, venous blood, and skin pigmentation. Since it is assumed that the AC portion in the photoplethysmogram results only from the pulsatile arterial blood component, this scaling process provides a normalized red/infrared ratio, R, which is highly dependent on the color of the arterial blood and is therefore related to S_aO_2 but is largely independent

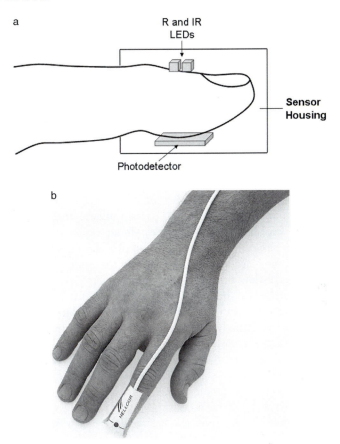

Figure 9.22   (a) Transmission-type pulse oximeter finger probe, and (b) disposable finger sensor (courtesy of Nellcor Puritan Bennett, Inc., Pleasanton, CA).

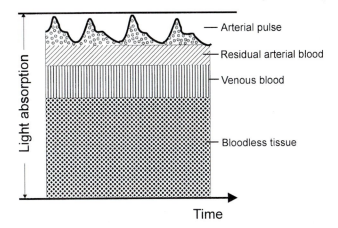

Figure 9.23   Time dependence of light absorption by a peripheral vascular tissue bed illustrating the effect of arterial pulsation.

of the volume of arterial blood entering the tissue during systole, skin pigmentation, and thickness. Hence, the instrument does not need to be recalibrated for measurements on different patients. The mathematical relationship between S_aO_2 and R is programmed by the manufacturer into the pulse oximeter.

9.4.2   pH Electrodes

pH describes the balance between acid and base in a solution. Acid solutions have an excess of hydrogen ions (H^+), whereas basic solutions have an excess of hydroxyl ions (OH^-). In a dilute solution, the product of these ion concentrations is a constant (1.0×10^{-14}). Therefore, the concentration of either ion can be used to express the acidity or alkalinity of a solution. All neutral solutions have a pH of 7.0.

The measurement of blood pH is fundamental to many diagnostic procedures. In normal blood, pH is maintained under tight control and is typically around 7.40 (slightly basic). By measuring blood pH, it is possible to determine whether the lungs are removing sufficient CO_2 gas from the body or how well the kidneys regulate the acid–base balance.

pH electrodes belong to a group of potentiometric sensors that generate a small potential difference without the need to polarize the electrochemical cell. A pH electrode essentially consists of two separate electrodes: a reference electrode and an active (indicator) electrode, as illustrated in Figure 9.24. The two electrodes are typically made of an Ag/AgCl wire dipped in a KCl solution and encased in a glass container. A salt bridge, which is essentially a glass tube containing an electrolyte enclosed in a membrane that is permeable to all ions, maintains the potential of the reference electrode at a constant value regardless of the solution under test. Unlike the reference electrode, the active electrode is sealed with hydrogen-impermeable

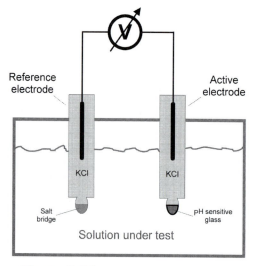

Figure 9.24   Principle of a pH electrode.

glass except at the tip. The reference electrode may also be combined with the indicator electrode in a single glass housing.

The boundary separating two solutions has a potential proportional to the hydrogen ion concentration of one solution and, at a constant temperature of 25°C, is given by

$$V = -59\,\text{mV} \times \log_{10}[\text{H}^+] + C \tag{9.19}$$

where C is a constant. Since pH is defined as

$$\text{pH} = -\log_{10}[\text{H}^+] \tag{9.20}$$

the potential of the active pH electrode V is proportional to the pH of the solution under test and is equal to

$$V = 59\,\text{mV} \times \text{pH} + C \tag{9.21}$$

The value of C is usually compensated for electronically when the pH electrode is calibrated by placing the electrode inside different buffer solutions with known pH values.

9.4.3 Carbon Dioxide Sensors

Electrodes for measurement of partial pressure of CO_2 in blood or other liquids are based on measuring the pH as illustrated in Figure 9.25. The measurement is based on the observation that, when CO_2 is dissolved in water, it forms a weakly dissociated

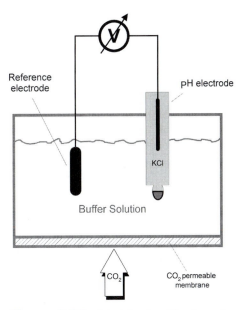

Figure 9.25 Principle of a pCO₂ electrode.

carbonic acid (H_2CO_3) that subsequently forms free hydrogen and bicarbonate ions according to the following chemical reaction:

$$CO_2 + H_2O \leftrightarrow H_2CO_3 \leftrightarrow H^+ + HCO_3^-$$

As a result of this chemical reaction, the pH of the solution is changed. This change generates a potential between the glass pH and a reference (e.g., Ag/AgCl) electrode that is proportional to the negative logarithm of the pCO_2.

9.5   BIOANALYTICAL SENSORS

The number of analytes that can be measured with electrochemical sensors can be increased significantly by adding biologically specific mediators (reagents that either undergo reactions or act as catalysts) to the semipermeable membrane structure. Several biosensors that have been constructed and used mainly for research applications have different enzymes and microorganisms as the primary sensing elements. Although these biosensors have been used successfully *in vitro* to demonstrate unique medical and industrial applications, further technical improvements are necessary to make these sensors robust and reliable enough to fulfill the demanding requirements of routine analytical and clinical applications. Examples of some interesting sensor designs are given in the following sections.

9.5.1   Enzyme-Based Biosensors

Enzymes constitute a group of more than 2000 proteins having so-called biocatalytic properties. These properties give the enzymes the unique and powerful ability to accelerate chemical reactions inside biological cells. Most enzymes react only with specific substrates even though they may be contained in a complicated mixture with other substances. It is important to keep in mind, however, that soluble enzymes are very sensitive both to temperature and pH variations and they can be inactivated by many chemical inhibitors. For practical biosensor applications, these enzymes are normally immobilized by insolubilizing the free enzymes via entrapment into an inert and stable matrix such as starch gel, silicon rubber, or polyacrylamide. This process is important to ensure that the enzyme retains its catalytic properties and can be reusable.

The action of specific enzymes can be utilized to construct a range of different biosensors. A typical example of an enzyme-based sensor is a glucose sensor that uses the enzyme glucose oxidase. Glucose plays an important role in metabolic processes. In patients suffering from diabetes mellitus, the pancreas does not produce sufficient amounts of insulin to control adequately the level of glucose in their blood. Therefore, to manage the disease, these patients must monitor and regulate their blood glucose level on a regular basis by medication and insulin injections. Currently available glucose sensors are based on an immobilized enzyme, such as glucose oxidase, which acts as a catalyst. Glucose is detected by measuring electrochemically either the amount of gluconic acid or hydrogen peroxide (H_2O_2) produced or by measuring the amount of oxygen consumed, according to the following chemical reaction:

$$\text{Glucose} + O_2 + H_2O \overset{\text{glucose oxidase}}{\longleftrightarrow} \text{gluconic acid} + H_2O_2$$

A glucose sensor is similar to a pO_2 sensor and is shown in Figure 9.26. Glucose and oxygen enter through the outside membrane to allow glucose to interact with the glucose oxidase enzyme. The remaining oxygen penetrates through the second oxygen-permeable membrane and is measured by the oxygen electrode.

Biocatalytic enzyme-based sensors generally consist of an electrochemical gas-sensitive transducer or an ion-selective electrode with an enzyme immobilized in or on a membrane that serves as the biological mediator. The analyte diffuses from the bulk sample solution into the biocatalytic layer where an enzymatic reaction takes place. The electroactive product that is formed (or consumed) is usually detected by an ion-selective electrode. A membrane separates the basic sensor from the enzyme if a gas is consumed (such as O_2) or is produced (such as CO_2 or NH_3). Although the concentration of the bulk substrate drops continuously, the rate of consumption is usually negligible. The decrease is detected only when the test volume is very small or

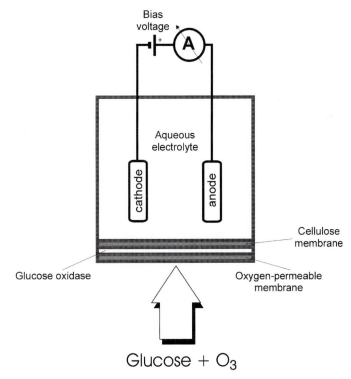

Figure 9.26   Principle of a glucose sensor.

when the area of the enzyme membrane is large enough. Thus, this electrochemical analysis is nondestructive, and the sample can be reused. Measurements are usually performed at a constant pH and temperature either in a stirred medium solution or in a flowthrough solution.

9.5.2 Microbial Biosensors

A number of microbial sensors have been developed, mainly for online control of biochemical processes in various environmental, agricultural, food, and pharmaceutical applications. Microbial biosensors typically involve the assimilation of organic compounds by the microorganisms, followed by a change in respiration activity (metabolism) or the production of specific electrochemically active metabolites, such as H_2, CO_2, or NH_3, that are secreted by the microorganism.

A microbial biosensor is composed of immobilized microorganisms that serve as specific recognition elements and an electrochemical or optical sensing device that is used to convert the biochemical signal into an electronic signal that can be processed. The operation of a microbial biosensor can be described by the following five-step process: (1) the substrate is transported to the surface of the sensor, (2) the substrate diffuses through the membrane to the immobilized microorganism, (3) a reaction occurs at the immobilized organism, (4) the products formed in the reaction are transported through the membrane to the surface of the detector, and (5) the products are measured by the detector.

Examples of microbial biosensors include ammonia (NH_3) and nitrogen dioxide (NO_2) sensors that utilize nitrifying bacteria as the biological sensing component. An ammonia biosensor can be constructed based on nitrifying bacteria, such as *Nitrosomonas* sp., that use ammonia as a source of energy and oxidize ammonia as follows:

$$NH_3 + 1.5O_2 \xrightarrow{\textit{Nitrosomonas sp.}} NO_2 + H_2O + H^+$$

This oxidation process proceeds at a high rate, and the amount of oxygen consumed by the immobilized bacteria can be measured directly by a polarographic oxygen electrode placed behind the bacteria.

Nitric oxide (NO) and NO_2 are the two principal pollution gases of nitrogen in the atmosphere. The principle of a NO_2 biosensor is shown in Figure 9.27. When a sample of NO_2 gas diffuses through the gas-permeable membrane, it is oxidized by the *Nitrobacter* sp. bacteria as follows:

$$2NO_2 + O_2 \xrightarrow{\textit{Nitrobacter sp.}} 2NO_3$$

Similar to an ammonia biosensor, the consumption of O_2 around the membrane is determined by an electrochemical oxygen electrode.

The use of microbial cells in electrochemical sensors offers several advantages over enzyme-based electrodes, the principal one being the increased electrode lifetime of several weeks. On the other hand, microbial sensors may be less favorable compared with enzyme electrodes with respect to specificity and response time.

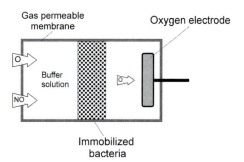

Figure 9.27   Principle of a NO_2 microbial-type biosensor.

9.6 OPTICAL BIOSENSORS

Optical biosensors play an important role in the development of highly sensitive and selective methods for biochemical analysis. The fundamental principle employed is based on the change in optical properties of a biological or physical medium. The change produced can be the result of intrinsic changes in absorbance, reflectance, scattering, fluorescence, polarization, or refractive index of the biological medium.

9.6.1   Optical Fibers

Optical fibers can be used to transmit light from one location to another. They are typically made from two concentric and transparent glass or plastic materials as illustrated in Figure 9.28. The center piece is known as the core and the outer layer, which serves as a coating material, is called the cladding.

The core and cladding of an optical fiber have a different index of refraction, n. The index of refraction is a number that expresses the ratio of the light velocity in free space to its velocity in a specific material. For instance, the refractive index for air is equal to 1.0, whereas the refractive index for water is equal to 1.33. Assuming that the refractive index of the core material is n_1 and the refractive index of the cladding is n_2, (where $n_1 > n_2$), according to Snell's law,

$$n_1 \times \sin \phi_1 = n_2 \times \sin \phi_2 \qquad (9.22)$$

where ϕ is the angle of incidence as illustrated in Figure 9.29.

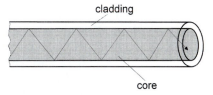

Figure 9.28   Principle of optical fibers.

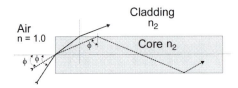

Figure 9.29   Optical fiber illustrating the incident and refracted light rays. Solid line shows the light ray escaping from the core into the cladding. The dashed line shows the ray undergoing total internal reflection inside the core.

Accordingly, any light passing from a lower refractive index to a higher refractive index is bent toward the line perpendicular to the interface of the two materials. For small incident angles, ϕ_1, the light ray enters the fiber core and bends inwards at the first core/cladding interface. For larger incident angles, ϕ_2, the ray exceeds a minimum angle required to bend it back into the core when it reaches the core/cladding boundary. Consequently, the light escapes into the cladding. By setting $\sin \phi_2 = 1.0$, the critical angle, ϕ_{cr}, is given by

$$\sin \phi_{cr} = \frac{n_2}{n_1} \tag{9.23}$$

Any light rays entering the optical fiber with incidence angles greater than ϕ_{cr} are internally reflected inside the core of the fiber by the surrounding cladding. Conversely, any entering light rays with incidence angles smaller than ϕ_{cr} escape through the cladding and are therefore not transmitted by the core.

Example Problem 9.11

Assume that a beam of light passes from a layer of glass with a refractive index $n_1 = 1.47$ into a second layer of glass with a refractive index of $n_2 = 1.44$. Using Snell's law, calculate the critical angle for the boundary between these two glass layers.

Solution

$$\phi_{cr} = \arcsin \times \left(\frac{n_2}{n_1}\right) = \arcsin(0.9796)$$

$$\phi_{cr} = 78.4°$$

Therefore, light that strikes the boundary between these two glasses at an angle greater than 78.4° will be reflected back into the first layer.   ■

The propagation of light along an optical fiber is not confined to the core region. Instead, the light penetrates a characteristic short distance (on the order of one wavelength) beyond the core surface into the less optically dense cladding medium. This effect causes the excitation of an electromagnetic field, called the evanescent wave, that depends on the angle of incidence and the incident wavelength. The intensity of the evanescent wave decays exponentially with distance according to Beer–Lambert's law. It starts at the interface and extends into the cladding medium.

9.6.2 Sensing Mechanisms

Two major optical techniques are commonly available to sense the optical change across a biosensor interface. These are usually based on evanescent wave spectroscopy, which plays a major role in fiber optic sensors, and a surface plasmon resonance principle.

Optical fibers can be used to develop a whole range of sensors for biomedical applications. These optical sensors are small, flexible, and free from electrical interference. They can produce an instantaneous response to microenvironments that surround their surface.

Commercial fiber optic sensors for blood gas monitoring became available at the end of the twentieth century. While many different approaches have been taken, they all have some features in common as illustrated in Figure 9.30. First, optical sensors are typically interfaced with an optical module. The module supplies the excitation light, which may be from a monochromatic source such as a diode laser or from a broad band source (e.g., quartz-halogen) that is filtered to provide a narrow bandwidth of excitation. Typically, two wavelengths of light are used: one wavelength is sensitive to changes in the species to be measured, whereas the other wavelength is unaffected by changes in the analyte concentration. This wavelength serves as a reference and is used to compensate for fluctuations in source output and detector

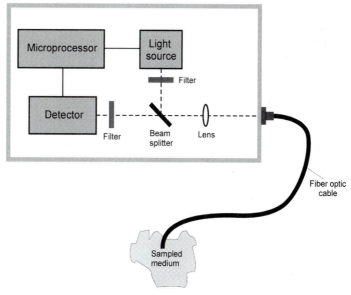

Figure 9.30 General principle of a fiber optic–based sensor.

stability. The light output from the optic module is coupled into a fiber optic cable through appropriate lenses and an optical connector.

Several sensing mechanisms can be utilized to construct optical fiber sensors. In fluorescence-based sensors, the incident light excites fluorescence emission, which changes in intensity as a function of the concentration of the analyte to be measured. The emitted light travels back down the fiber to the monitor where the light intensity is measured by a photodetector. In other types of fiber optic sensors, the light absorbing properties of the sensor chemistry change as a function of analyte chemistry. In the absorption-based design, a reflective surface near the tip or some scattering material within the sensing chemistry itself is usually used to return the light back through the same optical fiber. Other sensing mechanisms exploit the evanescent wave interaction with molecules that are present within the penetration depth distance and lead to attenuation in reflectance related to the concentration of the molecules. Because of the short penetration depth and the exponentially decaying intensity, the evanescent wave is absorbed by compounds that must be present very close to the surface. The principle has been used to characterize interactions between receptors that are attached to the surface and ligands that are present in solution above the surface.

The key component in the successful implementation of evanescent wave spectroscopy is the interface between the sensor surface and the biological medium. Receptors must retain their native conformation and binding activity and sufficient binding sites must be present for multiple interactions with the analyte. In the case of particularly weak absorbing analytes, sensitivity can be enhanced by combining the evanescent-wave principle with multiple internal reflections along the sides of an unclad portion of a fiber optic tip. Alternatively, instead of an absorbing species, a fluorophore (a compound that produces a fluorescent signal in response to light) can also be used. Light that is absorbed by the fluorophore emits detectable fluorescent light at a higher wavelength, thus providing improved sensitivity.

9.6.3   Indicator-Mediated Fiber Optic Sensors

Since only a limited number of biochemical substances have an intrinsic optical absorption or fluorescence property that can be measured directly with sufficient selectivity by standard spectroscopic methods, indicator-mediated sensors have been developed to use specific reagents that are immobilized either on the surface or near the tip of an optical fiber. In these sensors, light travels from a light source to the end of the optical fiber where it interacts with a specific chemical or biological recognition element. These transducers may include indicators and ion-binding compounds (ionophores) as well as a wide variety of selective polymeric materials. After the light interacts with the biological sample, it returns either through the same optical fiber (in a single-fiber configuration) or a separate optical fiber (in a dual-fiber configuration) to a detector, which correlates the degree of light attenuation with the concentration of the analyte.

Typical indicator-mediated sensor configurations are shown schematically in Figure 9.31. The transducing element is a thin layer of chemical material that is

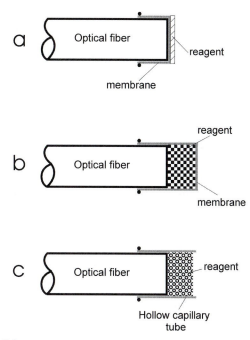

Figure 9.31 Different indicator-mediated fiber optic sensor configurations.

placed near the sensor tip and is separated from the blood medium by a selective membrane. The chemical-sensing material transforms the incident light into a return light signal with a magnitude that is proportional to the concentration of the species to be measured. The stability of the sensor is determined by the stability of the photosensitive material that is used and also by how effectively the sensing material is protected from leaching out of the probe. In Figure 9.31a, the indicator is immobilized directly on a membrane that is positioned at the end of the fiber. An indicator in the form of a powder can also be physically retained in position at the end of the fiber by a special permeable membrane as illustrated in Figure 9.31b, or a hollow capillary tube as illustrated in Figure 9.31c.

9.6.4 Immunoassay Sensors

The development of immunosensors is based on the observation of ligand-binding reaction products between a target analyte and a highly specific binding reagent. The key component of an immunosensor is the biological recognition element, which typically consists of antibodies or antibody fragments. Immunological techniques offer outstanding selectivity (the sensor's ability to detect a specific substance in a mixture containing other substances) and sensitivity through the process of antibody–antigen interaction. This is the primary recognition mechanism by which the immune system detects and fights foreign matter, which has allowed the measurement of many

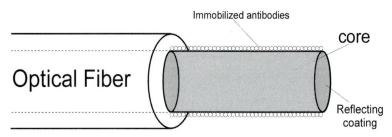

Figure 9.32   Principle of a fiber optic immunoassay biosensor.

important compounds at micromolar and even picomolar concentrations in complex biological samples.

Evanescent-type biosensors can be used in immunological diagnostics to detect antibody–antigen binding. Figure 9.32 shows a conceptual diagram of an immuno-assay biosensor. The immobilized antibody on the surface of the unclad portion of the fiber captures the antigen from the sample solution, which is normally introduced into a small flow-through chamber where the fiber tip is located. The sample solution is then removed and a labeled antibody is added into the flow chamber. A fluorescent signal is excited and measured when the labeled antibody binds to the antigen that is already immobilized by the antibody.

9.6.5   Surface Plasmon Resonance Sensors

When monochromatic polarized light (e.g., from a laser source) impinges on a transparent medium having a conducting metallized surface (e.g., Ag or Au), there is a charge density oscillation at the interface. When light at an appropriate wavelength interacts with the dielectric-metal interface at a defined angle, called the resonance angle, there is a match of resonance between the energy of the photons and the electrons at the metal interface. As a result, the photon energy is transferred to the surface of the metal as packets of electrons, called plasmons, and the light reflection from the metal layer will be attenuated. This results in a phenomenon known as surface plasmon resonance (SPR) which is illustrated schematically in Figure 9.33. The resonance is observed as a sharp dip in the reflected light intensity when the incident angle is varied. The resonance angle depends on the incident wavelength, the type of metal, the polarization state of the incident light, and the nature of the medium in contact with the surface. Any change in the refractive index of the medium will produce a shift in the resonance angle, and thus provide a highly sensitive means of monitoring surface interactions.

SPR is generally used for sensitive measurement of variations in the refractive index of the medium immediately surrounding the metal film. For example, if an antibody is bound to or absorbed into the metal surface, a noticeable change in the resonance angle can be readily observed because of the change of the refraction index at the surface if all other parameters are kept constant. The advantage of this concept

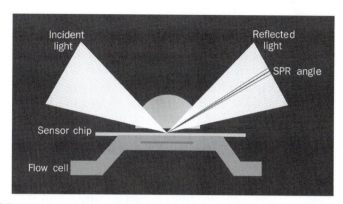

Figure 9.33 Principle of a surface plasmon resonance (SPR) detection system (courtesy of Biacore AB, Uppsala, Sweden).

is the improved ability to detect the direct interaction between antibody and antigen as an interfacial measurement.

EXERCISES

1. Two identical silver electrodes are placed in an electrolyte solution. Calculate the potential drop between the two electrodes.
2. Cadmium and zinc electrodes are placed in an electrolyte solution. Calculate the current that will flow through the electrodes if the equivalent resistance of the solution is equal to $8 \, k\Omega$.
3. By how much would the inductance of an inductive displacement transducer coil change if the number of coil turns increases by a factor of 4?
4. Determine the ratio between the cross-sectional areas of two blood vessels, assuming that the voltage ratio induced in identical magnetic flow probes is equal to 2:3 and the ratio of blood flows through these vessels is 1:5.
5. A $4.5 \, k\Omega$ linear rotary transducer is used to measure the angular displacement of the knee joint. Calculate the change in output voltage for a $165°$ change in the angle of the knee. Assume that a constant current of $7 \, mA$ is supplied to the transducer.
6. Provide a step-by-step derivation of Eq. 9.11.
7. An elastic resistive transducer with an initial resistance R_o and length l_o is stretched to a new length. Assuming that the cross-sectional area of the transducer changes during stretching, derive a mathematical relationship for the change in resistance ΔR as a function of the initial length, l_o; the change in length Δl; the volume of the transducer V; and the resistivity ρ.

8. The area of each plate in a differential capacitor sensor is equal to $4\,cm^2$. Calculate the equilibrium capacitance in air for each capacitor assuming that the equilibrium displacement for each capacitor is equal to 2 mm.

9. Plot the capacitance (y axis) versus displacement (x axis) characteristics of a capacitance transducer.

10. Calculate the sensitivity of a capacitive transducer (i.e., $\Delta C/\Delta d$) for small changes in displacements.

11. A capacitive transducer is used in a mattress to measure changes in breathing patterns of an infant. During inspiration and expiration, the rate of change (i.e., dV/dt) in voltage across the capacitor is equal to ± 1V/s, and this change can be modeled by a triangular waveform. Plot the corresponding changes in current flow through this transducer.

12. Derive the relationship for the current through the capacitor equivalent piezoelectric crystal as a function of V and C.

13. Two identical ultrasonic transducers are positioned across a blood vessel as shown in Figure 9.34. Calculate the diameter of the blood vessel if it takes 250 ns for the ultrasonic sound wave to propagate from one transducer to the other.

14. Calculate the resistance of a thermistor at 98°F, assuming that the resistance of this thermistor at 12°C is equal to $3.5\,k\Omega$ and $\beta = 4600$.

15. The resistance of a thermistor with a $\beta = 5500$ measured at 18°C is equal to $250\,\Omega$. Find the temperature of the thermistor when the resistance is doubled.

16. Calculate the β of a thermistor, assuming that it has a resistance of $3.2\,k\Omega$ at 21°C (room temperature) and a resistance of $1.85\,k\Omega$ when the room temperature increases by 10%.

17. Sketch the current (y axis) versus pO_2 (x axis) characteristics of a typical polarographic Clark electrode.

18. Explain why the value of the normalized ratio (R) in a pulse oximeter is independent of the volume of arterial blood entering the tissue during systole.

19. Explain why the value of the normalized ratio (R) in a pulse oximeter is independent of skin pigmentation.

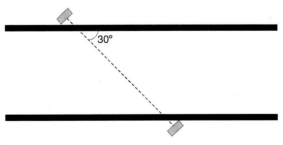

Figure 9.34 Two identical ultrasonic transducers positioned across a blood vessel.

20. Explain the difference between a potentiometric and amperometric sensor.

21. A pH electrode is attached to a sensitive voltmeter which reads $0.652\,V$ when the electrode is immersed in a buffer solution with a pH of 5.25. After the pH electrode is moved to an unknown buffer solution, the reading of the voltmeter is decreased by 25%. Calculate the pH of the unknown buffer solution.

22. Plot the optical density, OD, of an absorbing solution (y axis) versus the concentration of this solution (x axis). What is the slope of this curve?

23. An unknown sample solution whose concentration is $1.55 \times 10^{-3}\,g/L$ is placed in a 1cm clear holder and found to have a transmittance of 44%. The concentration of this sample is changed such that its transmittance has increased to 57%. Calculate the new concentration of the sample.

24. Calculate the angle of the refracted light ray if an incident light ray passing from air into water has a $62°$ angle with respect to the normal.

25. Explain why fiber optic sensors typically require simultaneous measurements using two wavelengths of light.

26. A chemical sensor is used to measure the pH of a dye with an absorbance spectrum shown in Figure 9.35. Assume that the absorbance of each form of the dye is linearly related to its pH. Devise a method to measure the pH of the dye.

27. Explain the difference between absorption-based and fluorescence-based measurements.

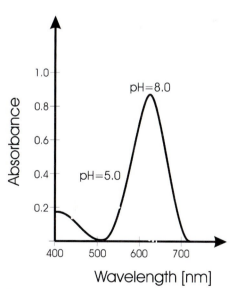

Figure 9.35   Optical absorbance spectra of a dye in its acid (pH = 5.0) and base (pH = 8.0) forms.

SUGGESTED READING

Allocca, J.A. and Stuart A. (1984). *Transducers: Theory and Applications*. Reston Publishing, Reston, VA.

Aston R. (1990). *Principles of Biomedical Instrumentation and Measurement*. MacMillan, New York.

Buerk, D. (1995). *Biosensors: Theory and Applications*. CRC, Boca Raton, FL.

Cobbold R.S.C. (1974). *Transducers for Biomedical Measurement: Principles and Applications*. Wiley, New York.

Cooper, J. and Cass A.E.G. (2004). *Biosensors*, Oxford University Press.

Cromwell, L., Weibell. F.J. and Pfeiffer, E.J. (1980). *Biomedical Instrumentation and Measurements*. Prentice Hall, Englewood Cliffs, NJ.

Eggins, B. (1997). *Biosensors: An Introduction*. Wiley, New York.

Eggins, B.R. (2002). *Chemical Sensors and Biosensors for Medical and Biological Applications*. John Wiley, New York.

Geddes L.A. and Baker L.E. (1989). *Principles of Applied Biomedical Instrumentation*, 3rd Ed. Wiley-Interscience, New York.

Hall, E.A.H. (1991). *Biosensors*. Prentice Hall, Englewood Cliffs.

Harsanyi, G. (2000). *Sensors in Biomedical Applications: Fundamental Technology and Applications*. CRC, Boca Raton, FL.

Neuman M.R. (1999). Biomedical Sensors. In *The Biomedical Engineering Handbook*, 2nd Ed. (J. D. Bronzino, Ed.). CRC/IEEE, Boca Raton, FL.

Togawa T., Tamura T. and Oberg P.A. (1997). *Biomedical Transducers and Instruments*. CRC, Boca Raton, FL.

Webster, J.G. (1988). *Encyclopedia of Medical Devices and Instrumentation. John* Wiley, New York.

Webster, J.G. (1998). *Medical Instrumentation: Application and Design*, 3rd Ed. John Wiley, New York.

Wise, D.L. (1991). *Bioinstrumentation and Biosensors*. Marcel Dekker, New York.

10   BIOSIGNAL PROCESSING

Monty A. Escabí, PhD*

Chapter Contents

*With contributions by Susan Blanchard, Carol Lucas, and Melanie T. Young.

At the end of this chapter, students will be able to:

- Describe the different origins and types of biosignals.

- Distinguish between deterministic, periodic, transient, and random signals.

- Explain the process of A/D conversion.

- Define the sampling theorem.

- Describe the main purposes and uses of the Fourier transforms.

- Define the Z transform.

- Describe the basic properties of a linear system.

- Describe the concept of filtering and signal averaging.

- Explain the basic concepts and advantages of fuzzy logic.

- Describe the basic concepts of artificial neural networks.

10.1   INTRODUCTION

Biological signals, or biosignals, are space, time, or space–time records of a biological event such as a beating heart or a contracting muscle. The electrical, chemical, and mechanical activity that occurs during these biological event often produces signals that can be measured and analyzed. Biosignals, therefore, contain useful information that can be used to understand the underlying physiological mechanisms of a specific biological event or system, and which may be useful for medical diagnosis.

Biological signals can be acquired in a variety of ways (e.g., by a physician who uses a stethoscope to listen to a patient's heart sounds or with the aid of technologically advanced biomedical instruments). Following data acquisition, biological signals are analyzed in order to retrieve useful information. Basic methods of signal analysis (e.g., amplification, filtering, digitization, processing, and storage) can be applied to many biological signals. These techniques are generally accomplished with simple electronic circuits or with digital computers. In addition to these common procedures, sophisticated digital processing methods are quite common and can significantly improve the quality of the retrieved data. These include signal averaging, wavelet analysis, and artificial intelligence techniques.

10.2 PHYSIOLOGICAL ORIGINS OF BIOSIGNALS

10.2.1 Bioelectric Signals

Nerve and muscle cells generate bioelectric signals that are the result of electrochemical changes within and between cells (see Chapter 5). If a nerve or muscle cell is stimulated by a stimulus that is strong enough to reach a necessary threshold, the cell will generate an action potential. The action potential, which represents a brief flow of ions across the cell membrane, can be measured with intracellular or extracellular electrodes. Action potentials generated by an excited cell can be transmitted from one cell to adjacent cells via its axon. When many cells become activated, an electric field is generated that propagates through the biological tissue. These changes in extracellular potential can be measured on the surface of the tissue or organism by using surface electrodes. The electrocardiogram (ECG), electrogastrogram (EGG), electroencephalogram (EEG), and electromyogram (EMG) are all examples of this phenomenon (Figure 10.1).

10.2.2 Biomagnetic Signals

Different organs, including the heart, brain, and lungs, also generate weak magnetic fields that can be measured with magnetic sensors. Typically, the strength of the magnetic field is much weaker than the corresponding physiological bioelectric signals. Biomagnetism is the measurement of the magnetic signals that are associated with specific physiological activity and that are typically linked to an accompanying electric field from a specific tissue or organ. With the aid of very precise magnetic sensors or SQUID magnetometers (superconducting quantum interference device) it is possible to directly monitor magnetic activity from the brain (magnetoencephalography, MEG), peripheral nerves (magnetoneurography, MNG), gastrointestinal tract (magnetogastrography, MGG), and the heart (magnetocardiography, MCG).

10.2.3 Biochemical Signals

Biochemical signals contain information about changes in concentration of various chemical agents in the body. The concentration of various ions, such as calcium and potassium, in cells can be measured and recorded. Changes in the partial pressures of oxygen (pO_2) and carbon dioxide (pCO_2) in the respiratory system or blood are often measured to evaluate normal levels of blood oxygen concentration. All of these constitute biochemical signals. These biochemical signals can be used for a variety of purposes such as determining levels of glucose, lactate, and metabolites and providing information about the function of various physiological systems.

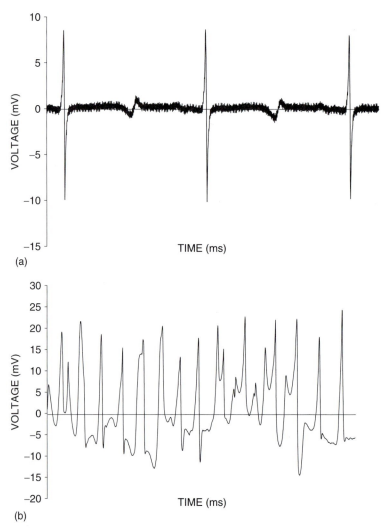

Figure 10.1   (a) Electrogram recorded from the surface of a pig's heart during normal sinus rhythm. (b) Electrogram recorded from the surface of the same pig's heart during ventricular fibrillation (VF) (sampled at 1000 samples/s).

10.2.4   Biomechanical Signals

Mechanical functions of biological systems, which include motion, displacement, tension, force, pressure, and flow, also produce measurable biological signals. Blood pressure, for example, is a measurement of the force that blood exerts against the walls of blood vessels. Changes in blood pressure can be recorded as a waveform (Fig. 10.2). The upstrokes in the waveform represent the contraction of

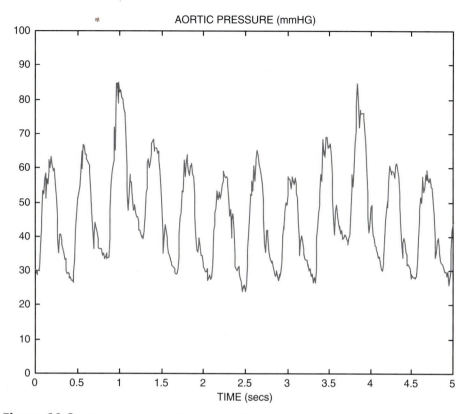

Figure 10.2   Blood pressure waveform recorded from the aortic arch of a 4-year-old child (sampled at 200 samples/s).

the ventricles of the heart as blood is ejected from the heart into the body and blood pressure increases to the systolic pressure, the maximum blood pressure (Chapter 3). The downward portion of the waveform depicts ventricular relaxation as the blood pressure drops to the minimum value, better known as the diastolic pressure.

10.2.5   Bioacoustic Signals

Bioacoustic signals are a special subset of biomechanical signals that involve vibrations (motion). Many biological events produce acoustic noise. For instance, the flow of blood through the valves in the heart has a distinctive sound. Measurements of the bioacoustic signal of a heart valve can be used to determine whether it is operating properly. The respiratory system, joints, and muscles also generate bioacoustic signals that propagate through the biological medium and can often be measured at the skin surface by using acoustic transducers such as microphones and accelerometers.

10.2.6   Biooptical Signals

Biooptical signals are generated by the optical, or light-induced, attributes of biological systems. Biooptical signals may occur either naturally or the signals may be introduced to measure a biological parameter with an external light medium. For example, information about the health of a fetus may be obtained by measuring the fluorescence characteristics of the amniotic fluid. Estimates of cardiac output can be made by using the dye dilution method which involves monitoring the concentration of a dye as it recirculates through the bloodstream. Finally, red and infrared light are used in various applications, such as obtaining precise measurements of blood oxygen levels by measuring the light absorption across the skin or a particular tissue.

Example Problem 10.1

What types of biosignals would the muscles in your lower legs produce if you were to sprint across a paved street?

Solution

Motion of the muscles and external forces imposed as your feet hit the pavement produce biomechanical signals. Muscle stimulation by nerves and the contraction of muscle cells produces bioelectric signals. Metabolic processes in the muscle tissue could be measured as biochemical signals.   ■

10.3   CHARACTERISTICS OF BIOSIGNALS

Biological signals can be classified according to various characteristics of the signal, including the waveform shape, statistical structure, and temporal properties. Two broad classes of signals that are commonly encountered include continuous and discrete signals. *Continuous* signals are defined over a continuum of time or space and are described by continuous variable functions. The notation $x(t)$ is used to represent a continuous time signal x that varies as a function of continuous time t. Signals that are produced by biological phenomena are almost always continuous signals. Some examples include voltage measurements from the heart (see Fig. 10.1), arterial blood pressure measurements (see Fig. 10.2), and measurements of electrical activity from the brain.

Discrete signals represent another signal class commonly encountered in today's clinical setting. Unlike continuous signals, which are defined along a continuum of points in space or time, discrete signals are defined only at a subset of regularly spaced points in time and/or space. Discrete signals are therefore represented by arrays or sequences of numbers. The notation $x(n)$ is used to represent a discrete sequence x that exists only at a subset of points in discrete time n. Here, $n = 0, 1, 2, 3 \ldots$ is always an integer that represents the nth element of the discrete sequence. Although most biological signals are not discrete per se, discrete signals play an important role due to today's advancements in digital technology. Sophisticated medical instruments are commonly used to convert continuous signals from the human body to discrete digital sequences (see Chapter 7) that can be analyzed and interpreted with a computer. CAT scans or computer axial tomography, for instance, take digital samples from continuous x-ray images of a patient that are obtained from different perspective angles (see Chapter 15). These digitized or discrete image slices are then digitally enhanced, manipulated, and processed to generate a full three-dimensional computer model of a patient's internal organs. Such technologies serve as indispensable tools for clinical diagnosis.

Biological signals can also be classified as being either *deterministic* or *random*. Deterministic signals can be described by mathematical functions or rules. *Periodic* and *transient* signals make up a subset of all deterministic signals. Periodic signals are usually composed of the sum of different sine waves or sinusoid components (see Sections 10.5.1 through 10.5.3) and can be expressed as

$$x(t) = x(t + kT) \tag{10.1}$$

where $x(t)$ is the signal, k is an integer, and T is the period. The period represents the distance along the time axis between successive copies of the periodic signal. Periodic signals have a stereotyped waveform with a duration of T units that repeats indefinitely. Transient signals are nonzero or vary only over a finite time interval and subsequently decay to a constant value as time progresses. The sine wave, shown in Figure 10.3a, is a simple example of a periodic signal since it repeats indefinitely with a repetition interval of 1 second. The product of a decaying exponential and a sine wave, as shown in Figure 10.3b, is a transient signal since the signal amplitude approaches zero as time progresses.

Real biological signals almost always have some unpredictable noise or change in parameters and, therefore, are not entirely deterministic. The ECG of a normal beating heart at rest is an example of a signal that appears to be almost periodic but has a subtle unpredictable component. The basic waveform shape consists of the P wave, QRS complex, and T wave and repeats (Figure 3.22). However, the precise shapes of the P waves, QRS complexes, and T wave are somewhat irregular from one heartbeat to the other. The length of time between QRS complexes, which is known as the R–R interval, also changes over time as a result of heart rate variability (HRV). HRV is used as a diagnostic tool to predict the health of a heart that has experienced a heart attack. The extended outlook for patients with low HRV is generally worse than it is for patients with high HRV.

Random signals, also called stochastic signals, contain uncertainty in the parameters that describe them. Because of this uncertainty, mathematical functions cannot be used to precisely describe random signals. Instead, random signals are most often analyzed using statistical techniques that require the treatment of the random parameters of the signal with probability distributions or simple statistical measures such as the mean and standard deviation. The electromyogram (EMG), an electrical recording of electrical activity in skeletal muscle that is used for the diagnosis of neuromuscular disorders, is a random signal. Stationary random signals are signals

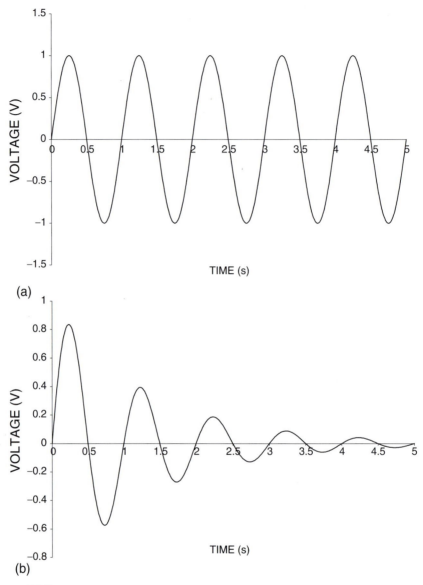

Figure 10.3 (a) Periodic sine wave signal $x(t) = \sin(\omega t)$ with a period of 1 Hz. (b) Transient signal $y(t) = e^{-0.75t}\sin(\omega t)$ for the same 1 Hz sine wave.

for which the statistics or frequency spectra remain constant over time. Conversely, nonstationary random signals have statistical properties or frequency spectra that vary with time. In many instances, the identification of stationary segments of random signals is important for proper signal processing, pattern analysis, and clinical diagnosis.

Example Problem 10.2

Ventricular fibrillation (VF) is a cardiac arrhythmia in which there are no regular QRS complexes, T waves, or rhythmic contractions of the heart muscle (Figure 10.1b). VF often leads to sudden cardiac death, which is one of the leading causes of death in the United States. What type of biosignal would most probably be recorded by an ECG when a heart goes into VF?

Solution

An ECG recording of a heart in ventricular fibrillation will be a random, continuous, bioelectric signal. ∎

10.4 SIGNAL ACQUISITION

10.4.1 Overview of Biosignal Data Acquisition

Biological signals are often very small and typically contain unwanted interference or noise. Such interference has the detrimental effect of obscuring relevant information that may be available in the measured signal. Noise can be extraneous in nature, arising from outside the body from sources such as thermal noise in sensors or 60-cycle noise in the electronic components of the acquisition system that can be caused by lighting systems. Noise can also be intrinsic to the biological media, meaning that noise can arise from adjacent tissues or organs within the desired measurement location. ECG measurements from the heart, for instance, can be affected by bio-electric activity from adjacent muscles.

To extract meaningful information from a signal that may be crucial in understanding a particular biological system or event, sophisticated data acquisition techniques and equipment are commonly used. High-precision low-noise equipment is often necessary to minimize the effects of unwanted noise. A diagram of the basic components in a bioinstrumentation system is shown in Figure 10.4.

Throughout the data acquisition procedure, it is critical that the information and structure of the original biological signal of interest be faithfully preserved. Since these signals are often used to aid the diagnosis of pathological disorders, the procedures of amplification, analog filtering, and A/D conversion should not generate misleading or untraceable distortions. Distortions in a signal measurement could lead to an improper diagnosis.

10.4.2 Sensors, Amplifiers, and Analog Filters

Signals are first detected in the biological medium such as a cell or on the skin's surface by using a sensor (see Chapter 6). A sensor converts a physical measurand into an electric output and provides an interface between biological systems and electrical recording instruments. The type of biosignal determines what type of sensor will be

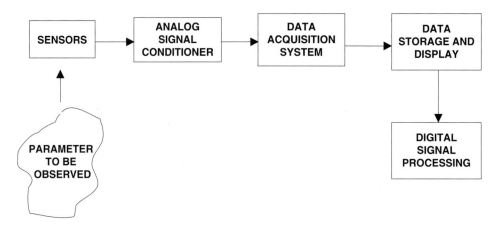

Figure 10.4   Sensors adapt the signal that is being observed into an electrical analog signal that can be measured with a data acquisition system. The data acquisition system converts the analog signal into a calibrated digital signal that can be stored. Digital signal processing techniques are applied to the stored signal to reduce noise and extract additional information that can improve understanding of the physiological meaning of the original parameter.

used. ECGs, for example, are measured with electrodes that have a silver/silver chloride (Ag/AgCl) interface attached to the body that detects the movement of ions. Arterial blood pressure is measured with a sensor that detects changes in pressure. It is very important that the sensor used to detect the biological signal of interest does not adversely affect the properties and characteristics of the signal it is measuring.

After the biosignal has been detected with an appropriate sensor, it is usually amplified and filtered. Operational amplifiers are electronic circuits that are used primarily to increase the amplitude or size of a biosignal. Bioelectric signals, for instance, are often faint and require up to a thousandfold boosting of their amplitude with such amplifiers. An analog filter may then be used to remove noise or to compensate for distortions caused by the sensor. Amplification and filtering of the biosignal may also be necessary to meet the hardware specifications of the data acquisition system. Continuous signals may need to be limited to a certain band of frequencies before the signal can be digitized with an analog-to-digital converter, prior to storing in a digital computer.

10.4.3   A/D Conversion

Analog-to-digital (A/D) converters are used to transform biological signals from continuous analog waveforms to digital sequences. An A/D converter is a computer-controlled voltmeter, which measures an input analog signal and gives a numeric representation of the signal as its output. Figure 10.5a shows an analog signal and Figure 10.5b shows a digital version of the same signal. The analog waveform, originally detected by the sensor and subsequently amplified and filtered, is a continu-

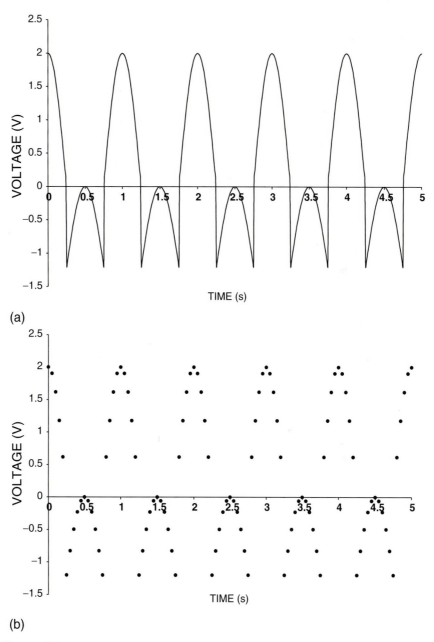

Figure 10.5   (a) Analog version of a periodic signal. (b) Digital version of the analog signal.

ous signal. The A/D converter transforms the continuous analog signal into a discrete digital signal. The discrete signal consists of a sequence of numbers that can easily be stored and processed on a digital computer. A/D conversion is particularly important in that, due to advances in computer technology, the storage and analysis of biosignals is becoming increasingly computer based.

The digital conversion of an analog biological signal does not produce an exact replica of the original signal. The discrete digital signal is a digital approximation of the original analog signal that is generated by repeatedly sampling the amplitude level of the original signal at fixed time intervals. As a result, the original analog signal is represented as a sequence of numbers—the digital signal.

The two main processes involved in A/D conversion are sampling and quantization. *Sampling* is the process by which a continuous signal is first converted into a discrete sequence in time. If $x(t)$ is an analog signal, sampling involves recording the amplitude value of $x(t)$ every T seconds. The amplitude value is denoted as $x(kT)$ where $k = 0, 1, 2, 3, \ldots$ is an integer that denotes the position or the sample number from the sample set or data sequence. T represents the sampling interval or the time between adjacent samples. In real applications, finite data sequences are generally used in digital signal processing. Therefore, the range of a data points is $k = 0, 1, \ldots N - 1$ where N is the total number of discrete samples. The sampling frequency, f_s, or the sampling rate, is equal to the inverse of the sampling period, $1/T$, and is measured in units of Hertz (s^{-1}).

Several digital sequences of particular importance are:

- The unit-sample of impulse sequence:

$$\delta(k) = \begin{cases} 1 \text{ if } k = 0 \\ 0 \text{ if } k \neq 0 \end{cases}$$

- The unit-step sequence:

$$u(k) = \begin{cases} 1 \text{ if } k \geq 0 \\ 0 \text{ if } k < 0 \end{cases}$$

- The exponential sequence:

$$a^k u(k) = \begin{cases} a^k \text{ if } k \geq 0 \\ 0 \ \text{ if } k < 0 \end{cases}$$

The sampling rate used to discretize a continuous signal is critical for the generation of an accurate digital approximation. If the sampling rate is too low, distortions will occur in the digital signal. Nyquist's theorem states that the minimum sampling rate used, f_s, should be at least twice the maximum frequency of the original signal to preserve all of the information of the analog signal. The Nyquist rate is calculated as

$$f_{\text{nyquist}} = 2 \cdot f_{\text{max}} \tag{10.2}$$

where f_{max} is the highest frequency present in the analog signal. The Nyquist theorem therefore states that f_s must be greater than or equal to $2 \cdot f_{\text{max}}$ to fully represent the analog signal by a digital sequence. Practically, sampling is usually done at five to ten times the highest frequency, f_{max}.

The second step in the A/D conversion process involves signal quantization. *Quantization* is the process by which the continuous amplitudes of the discrete signal are digitized by a computer. In theory, the amplitudes of a continuous signal can be any of an infinite number of possibilities. This makes it impossible to store all the values, given the limited memory in computer chips. Quantization overcomes this by reducing the number of available amplitudes to a finite number of possibilities that the computer can handle.

Since digitized samples are usually stored and analyzed as binary numbers on computers, every sample generated by the sampling process must be quantized. During quantization, the series of samples from the discretized sequence are transformed into binary numbers. The resolution of the A/D converter determines the number of bits that are available for storage. Typically, most A/D converters approximate the discrete samples with 8, 12, or 16 bits. If the number of bits is not sufficiently large, significant errors may be incurred in the digital approximation. Quantization errors occur when the sampled binary numbers are significantly different from the original sample values.

A/D converters are characterized by the number of bits used to generate a digital approximation. A quantizer with N bits is capable of representing a total of 2^N possible amplitude values. Therefore, the resolution of an A/D converter increases as the number of bits increases. A 16-bit A/D converter has better resolution than an 8-bit A/D converter since it is capable of representing a total of 65,536 amplitude levels, compared to 256 for the 8-bit converter. The resolution of an A/D converter is determined by the voltage range of the input analog signal divided by the numeric range (the possible number of amplitude values) of the A/D converter.

Example Problem 10.3

Find the resolution of an 8-bit A/D converter when an input signal with a 10 V range is digitized.

Solution

$$\frac{\text{input voltage range}}{2^N} = \frac{10\,\text{V}}{256} = 0.0391\,\text{V} = 39.1\,\text{mV} \qquad \blacksquare$$

Example Problem 10.4

The frequency content of an analog EEG signal is 0.5–100 Hz. What is the lowest rate at which the signal can be sampled to produce an accurate digital signal?

Solution

$$\text{Highest frequency in analog signal} = 100\,\text{Hz.}$$

$$f_{\text{nyquist}} = 2 \cdot f_{\text{max}} = 2 \cdot 100\,\text{Hz} = 200\,\text{samples/second} \qquad \blacksquare$$

Another problem often encountered is determining what happens if a signal is not sampled at a rate high enough to produce an accurate signal. One form of the sampling theorem is the statement: all frequencies of the form $[f - kf_s]$, where $-\infty \le k \le \infty$ and $f_s = 1/T$, look the same once they are sampled.

Example Problem 10.5

A 360 Hz signal is sampled at 200 samples/second. What frequency will the "aliased" digital signal look like?

Solution

According to the formula, $f_s = 200$, and the pertinent set of frequencies that look alike is in the form of $[360 - k200] = [\ldots 360\ 160\ -40\ -240 \ldots]$. The only signal in this group that will be accurately sampled is the signal at 40 Hz since the sampling rate is more than twice this value. $-40\,\text{Hz}$ and $+40\,\text{Hz}$ look alike for real signals [i.e., $\cos(-\omega t) = \cos(\omega t)$ and $\sin(-\omega t) = -\sin(\omega t)$]. Thus the sampled signal will look as if it had been a 40 Hz signal. The process is illustrated in Figure 10.6. $\qquad \blacksquare$

10.5 FREQUENCY DOMAIN REPRESENTATION OF BIOLOGICAL SIGNALS

In the early nineteenth century, Joseph Fourier laid out one of the most exquisite mathematical theories on the field of function approximation. At the time, his result was applied towards the problem of thermodynamic propagation of heat in solids, but it has since gained a much broader appeal. Today, Fourier's findings provide a general theory for approximating complex waveforms with simpler functions that has numerous applications in mathematics, physics, and engineering. This section summarizes the Fourier transform and variants of this technique that play an important conceptual role in the analysis and interpretation of biological signals.

10.5.1 Periodic Signal Representation: The Trigonometric Fourier Series

As an artist mixes oil paints on a canvas, a scenic landscape is meticulously recreated by combining various colors on a pallette. It is well known that all shades of the color spectrum can be recreated by simply mixing prime colors—red, green, and blue (RGB)—in the correct proportions. Television and computer displays

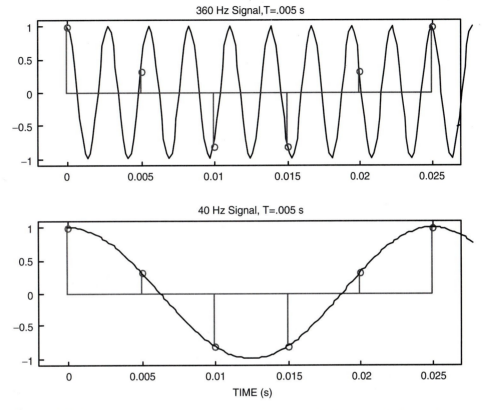

Figure 10.6 A 36 Hz sine wave is sampled every 5 ms (i.e., at 200 samples/s). This sampling rate will adequately sample a 40 Hz sine wave, but not a 36 Hz sine wave.

often transmit signals as RGB, and these signals are collated together to create colors much as a master painter would on a canvas. In fact, the human visual system takes exactly the opposite approach. The retina decomposes images and scenery from the outside world into purely red–green–blue signals that are independently analyzed and processed by our brains. Despite this, we perceive a multitude of colors and shades.

This simple color analogy is at the heart of Fourier's theory, which states that a complex waveform can be approximated to any degree of accuracy with simpler functions. In 1807, Fourier showed that an arbitrary periodic signal of period T can be represented mathematically as a sum of trigonometric functions. Conceptually, this is achieved by summing or mixing sinusoids while simultaneously adjusting their amplitudes and frequency as illustrated for a square wave function in Figure 10.7. If the amplitudes and frequencies are chosen appropriately, the trigonometric signals add constructively, thus recreating an arbitrary periodic signal. This is akin to combining prime colors in precise ratios to recreate an

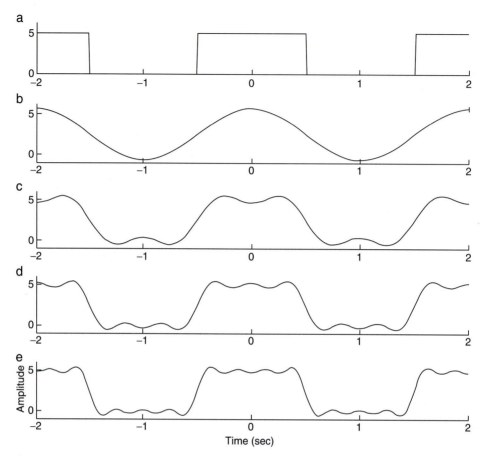

Figure 10.7 A square wave signal (a) is approximated by adding sinusoids (b–e): (b) 1 sinusoid, (c) 2 sinusoids, (d) 3 sinusoids, (e) 4 sinusoids. Increasing the number of sinusoids improves the quality of the approximation.

arbitrary color and shade. RGB are the building blocks for more elaborate colors much as sinusoids of different frequencies serve as the building blocks for more complex signals. All of these elements (the color and the required proportions; the frequencies and their amplitudes) have to be precisely adjusted to achieve a desired result. For the example, a first-order approximation of the square wave is achieved by fitting the square wave to a single sinusoid of appropriate frequency and amplitude. Successive improvements in the approximation are obtained by adding higher-frequency sinusoid components, or *harmonics*, to the first-order approximation. If this procedure is repeated indefinitely, it is possible to approximate the square wave signal with infinite accuracy.

The Fourier series summarizes this result

$$x(t) = a_0 + \sum_{m=1}^{\infty} (a_m \cos m\omega_0 t + b_m \sin m\omega_0 t) \qquad (10.3a)$$

where $\omega_0 = 2\pi/T$ is the fundamental frequency of $x(t)$ in units of radians/s, and the coefficients a_m and b_m determine the amplitude of each cosine and sine term at a specified frequency $\omega_m = m\omega_0$. Eq. 10.3a tells us that the periodic signal, $x(t)$, is precisely replicated by summing an infinite number of sinusoids. The frequencies of the sinusoid functions always occur at integer multiples of ω_0 and are referred to as "harmonics" of the fundamental frequency. If we know the coefficients a_m and b_m for each of the corresponding sine or cosine terms, we can completely recover the signal $x(t)$ by evaluating the Fourier series. How do we determine a_m and b_m for an arbitrary signal?

The coefficients of the Fourier series correspond to the amplitude and are related to the energy of each sine and cosine. These are determined as

$$a_0 = \frac{1}{T} \int_T x(t)dt \qquad (10.3b)$$

$$a_m = \frac{2}{T} \int_T x(t) \cos (m\omega_0 t)dt \qquad (10.3c)$$

$$b_m = \frac{2}{T} \int_T x(t) \sin (m\omega_0 t)dt \qquad (10.3d)$$

where the integrals are evaluated over a single period, T, of the waveform.

Example Problem 10.6

Find the trigonometric Fourier series of the square wave signal shown in Figure 10.7a and implement the result in MATLAB for the first 10 components. Plot the time waveform and the Fourier coefficients.

Solution

First note that

$$T = 2 \text{ and } \omega_0 = \frac{2\pi}{T} = \pi$$

To simplify the analysis, integration for a_m and b_m is carried out over the first period of the waveform (from -1 to 1)

$$a_0 = \frac{1}{T} \int\limits_{-1}^{1} x(t)dt = \frac{1}{2} \int\limits_{-1/2}^{1/2} 5dt = \frac{5}{2}$$

$$a_m = \frac{2}{T} \int\limits_{-1}^{1} x(t) \cos(m\omega_0 t)dt = \int\limits_{-1/2}^{1/2} 5 \cdot \cos(m\pi t)dt$$

$$= -5 \left. \frac{\sin(m\pi t)}{m\pi} \right|_{-1/2}^{1/2} = -5 \frac{\sin(m\pi/2)}{m\pi/2} = 5 \cdot \mathrm{sinc}(m\pi/2)$$

$$b_m = \frac{2}{T} \int_{-1}^{1} x(t) \sin(m\omega_0 t)dt = \int_{-1/2}^{1/2} 5 \cdot \sin(m\pi t)dt = -5 \cdot \left. \frac{\cos(m\pi t)}{m\pi} \right|_{-1/2}^{1/2} = 0$$

where by definition $\mathrm{sinc}(x) = \sin(x)/x$. Substituting the values for a_0, a_m, and b_m into Eq. 10.3a gives

$$x(t) = \frac{5}{2} + 5 \cdot \sum_{m=1}^{\infty} \frac{\sin(m\pi/2)}{m\pi/2} \cos(m\pi t)$$

MATLAB Implementation

```
%Plotting Fourier Series Approximation
subplot(211)
time=−2:0.01:2; %Time Axis
x = 5/2; %Initializing Signal
for m=1:10
   x = x + 5*sin (m*pi/2)/m/pi *2cos (m*pi*time);
end
plot(time,x,'k') %Plotting and Labels
xlabel('Time (sec)')
ylabel('Amplitude')
set(gca,'Xtick',[−2:2])
set(gca,'Ytick',[0 5])
set(gca,'Box','off')

%Plotting Fourier Magnitudes
subplot(212)
m=1:10;
Am=[5/2 5*sin(m* pi/2)./m/pi*2]; %Fourier Magnitudes
Faxis=(0:10)*.5; %Frequency Axis
plot(Faxis,Am,'k.') %Plotting
axis([0 5 −2 4])
set(gca,'Box','off')
xlabel('Frequency (Hz)')
ylabel('Fourier Amplitudes')
```

■

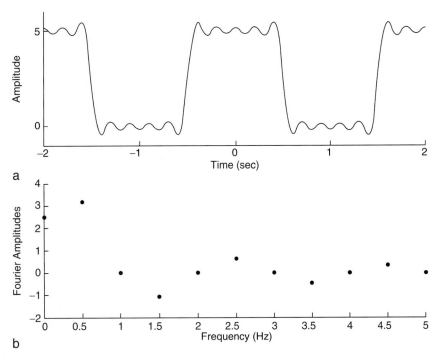

Figure 10.8 (a) MATLAB result showing the first 10 terms of Fourier series approximation for the periodic square wave of Fig. 10.7a. (b) The Fourier coefficients are shown as a function of the harmonic frequency.

Note that the approximation of summing the first 10 harmonics (Figure 10.8a) closely resembles the desired square wave. The Fourier coefficients, a_m, for the first 10 harmonics are shown as a function of the harmonic frequency in Figure 10.8b. To fully replicate the sharp transitions of the square wave, an infinite number of harmonics is required.

10.5.2 Compact Fourier Series

The trigonometric Fourier series provides a direct approach for fitting and analyzing various types of biological signals such as the repetitive beating of a heart or the cyclic oscillations produced by the vocal folds as one speaks. Despite its utility, alternative forms of the Fourier series are sometimes more appealing because they are easier to work with mathematically and because signal measurements often can be interpreted more readily. The most widely used counterparts for approximating and modeling biological signals are the *exponential* and *compact* Fourier series.

The compact Fourier series is a close cousin of the standard Fourier series. This version of the Fourier series is obtained by noting that the sum of sinusoids and cosines can be rewritten by a single cosine term with the addition of a phase constant $a_m \cos m\omega_0 t + b_m \sin m\omega_0 t = A_m \cos (m\omega_0 t + \phi_m)$, which leads to the compact form of the Fourier series:

$$x(t) = \frac{A_0}{2} + \sum_{m=1}^{\infty} A_m \cos (m\omega_0 t + \phi_m) \tag{10.4a}$$

The amplitude for each cosine, A_m, is related to the Fourier coefficients through

$$A_m = \sqrt{a_m^2 + b_m^2} \tag{10.4b}$$

and the cosine phase is obtained from a_m and b_m as

$$\phi_m = \tan^{-1}\left(\frac{-b_m}{a_m}\right) \tag{10.4c}$$

Example Problem 10.7

Convert the standard Fourier series for the square pulse function of Example Problem 10.5 to compact form and implement in MATLAB.

Solution

We first need to determine the magnitude, A_m, and phase, ϕ_m, for the compact Fourier series. The magnitude is obtained as:

$$A_m = \sqrt{a_m^2 + b_m^2} = \sqrt{(5 \cdot \text{sinc}(m\pi/2))^2 + (0)^2} = 5\frac{|\sin (m\pi/2)|}{\pi m/2}$$

Since

$$|\sin (m\pi/2)| = \begin{cases} 1 & m = \text{odd} \\ 0 & m = \text{even} \end{cases}$$

we have

$$A_m = \begin{cases} 10/m\pi & m = \text{odd} \\ 0 & m = \text{even} \end{cases}$$

Unlike a_m or b_m in the standard Fourier series, note that A_m is strictly a positive quantity for all m. The phase term is determined as:

$$\phi_m = \tan^{-1}\left(\frac{-b_m}{a_m}\right) = \tan^{-1}\left(\frac{0}{5 \cdot \text{sinc}(m\pi/2)}\right) = \begin{cases} 0 & m = 0, 1, 4, 5, 8, 9 \cdots \\ & \text{for} \\ \pi & m = 2, 3, 6, 7, 10, 11 \cdots \end{cases}$$

Combining results

$$x(t) = \frac{5}{2} + \sum_{m=1}^{\infty} \frac{10}{m\pi} \cos(m\omega_0 t + \phi_m)$$

where ϕ_m is as defined previously. An interesting point regards the similarity of standard and compact versions of the Fourier series for this square wave example. In the standard form, the coefficient a_m alternates between positive and negative values whereas for the compact form the Fourier coefficient, A_m, is identical in magnitude to a_m but it is always a positive quantity. The sign (+ or −) of the standard Fourier coefficient is now consumed in the phase term which alternates between 0 and π. This forces the cosine to alternate in its external sign because $-\cos(x) = \cos(x + \pi)$. The two equations are therefore mathematically identical, differing only in the way that the trigonometric functions are written out.

MATLAB Implementation

```
%Plotting Fourier Series Approximation
time=−2:0.01:2; %Time Axis
x=5/2; %Initializing Signal
m=1:10;
A=(10*sin *(m*pi/2)./m/pi); %Fourier Coefficients
P=angle(A); %Phase Angle
A=abs(A); %Fourier Magnitude
for m=1:10
   x=x+A(m)* cos(m*pi* time+P(m));
end
subplot(211)
plot(time,x,'k') %Plotting and Labels
xlabel('Time (sec)')
ylabel('Amplitude')
set(gca,'Xtick',[−2:2])
set(gca,'Ytick',[0 5])
set(gca,'Box','off')

%Plotting Fourier Magnitudes
subplot(212)
m=1:10;
A=[5/2 A]; %Fourier Magnitudes
Faxis=(0:10)*.5; %Frequency Axis
plot(Faxis,A,'k.') %Plotting
axis([0 5 −2 4])
set(gca,'Box','off')
xlabel('Frequency (Hz)')
ylabel('Fourier Amplitudes')
```

The results are identical to those shown in Figure 10.8a and b. ■

10.5.3   Exponential Fourier Series

The main result from Fourier series analysis is that an arbitrary periodic signal can approximate by summing individual cosine terms with specified amplitudes and phases. This result serves as much of the conceptual and theoretical framework for the field of signal analysis. In practice, the Fourier series is a useful tool for modeling various types of quasi-periodic signals.

An alternative and somewhat more convenient form of this result is obtained by noting that complex exponential functions are directly related to sinusoids and cosines through Euler's identities: $\cos(\theta) = (e^{j\theta} + e^{-j\theta})/2$ and $\sin(\theta) = (e^{j\theta} - e^{-j\theta})/2j$ where $j = \sqrt{-1}$. By applying Euler's identity to the compact trigonometric Fourier series, an arbitrary periodic signal can be expressed as a sum of complex exponential functions:

$$x(t) = \sum_{m=-\infty}^{+\infty} c_m e^{jm\omega_0 t} \tag{10.5a}$$

This equation represents the exponential Fourier series of a periodic signal. The coefficients c_m are complex numbers that are related to the trigonometric Fourier coefficients

$$c_m = \frac{a_m - jb_m}{2} = \frac{A_m}{2} e^{j\phi_m} \tag{10.5b}$$

The proof for this result is beyond the scope of this text; however, it is important to realize that the trigonometric and exponential Fourier series are intimately related, as can be seen by comparing their coefficients. The exponential coefficients can also be obtained directly by integrating $x(t)$,

$$c_m = \frac{1}{T} \int_T x(t) e^{-jm\omega_0 t} dt \tag{10.5c}$$

over one cycle of the periodic signal. As for the trigonometric Fourier series, the exponential form allows us to approximate a periodic signal to any degree of accuracy by adding a sufficient number of complex exponential functions. A distinct advantage of the exponential Fourier series, however, is that it requires that one only compute a single integral (Eq. 10.5c), compared to the trigonometric form, which requires three separate integrations.

Example Problem 10.8

Find the exponential Fourier series for the square wave of Figure. 10.7a and implement in MATLAB for the first 10 terms. Plot the time waveform and the Fourier series coefficients.

Solution

As for Problem 10.6, the Fourier coefficients are obtained by integrating from -1 to 1. Because a single cycle of the square wave signal has nonzero values between $-1/2$ to $+1/2$, the integral can be simplified by evaluating it between these limits:

$$c_m = \frac{1}{T}\int_T x(t)e^{-jm\omega_0 t}\,dt = \frac{1}{2}\int_{-1/2}^{1/2} 5e^{-jm\pi t}\,dt = \frac{5}{2}\cdot\frac{e^{-jm\pi t}}{-jm\pi}\bigg|_{-1/2}^{1/2}$$

$$\frac{5}{2}\cdot\frac{e^{+jm\pi/2}-e^{-jm\pi/2}}{jm\pi} = \frac{5}{2}\cdot\frac{\sin(m\pi/2)}{m\pi/2}$$

Therefore,

$$x(t) = \sum_{m=-\infty}^{+\infty} c_m e^{jk\omega_0 t} = \sum_{m=-\infty}^{\infty}\frac{5}{2}\cdot\frac{\sin(m\pi/2)}{m\pi/2}\cdot e^{jm\pi t}$$

MATLAB Implementation

```
%Plotting Fourier Series Approximation
subplot(211)
time=-2:0.01:2; %Time Axis
x=0; %Initialize Signal
for m=-10:10
   if m==0
   x=x+5/2; %Term for m=0
   else
   x=x+5/2 *sin(m* pi/2)/m/pi*2 **exp(j*m *pi*time);
   end
end
plot(time,x,'k') %Plotting and Labels
xlabel('Time (sec)')
ylabel('Amplitude')
set(gca,'Xtick',[-2:2])
set(gca,'Ytick',[0 5])
set(gca,'Box','off')

%Plotting Fourier Magnitudes
subplot(212)
m=(-10:10)+1E-10;
A=[5/2*sin (m*pi/2)./m/ pi*2]; %Fourier Magnitudes
Faxis=(-10:10)*.5; %Frequency Axis
plot(Faxis,A,'k.') %Plotting
axis([-5 5 -2 4])
set(gca,'Box','off')
xlabel('Frequency (Hz)')
ylabel('Fourier Amplitudes')
```

Note that we now require positive and negative frequencies in the approximation. Results showing the MATLAB output are illustrated in Figure 10.9. ∎

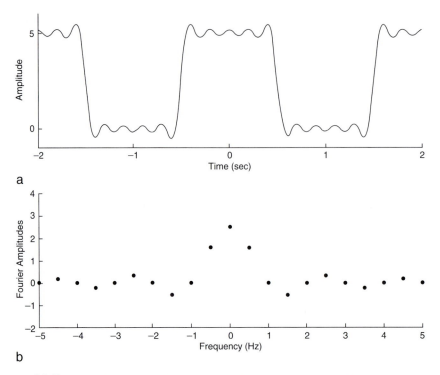

Figure 10.9   (a) MATLAB result showing the first 10 terms of exponential Fourier series approximation for the square wave of Fig. 10.7a. (b) The compact Fourier coefficients are shown as a function of the harmonic frequency. Note that both negative and positive frequencies are now necessary to approximate the square wave signal.

In practice, many periodic or quasi-periodic biological signals can be accurately approximated with only a few harmonic components. Figures 10.10 and 10.11 illustrate a harmonic reconstruction of an aortic pressure waveform obtained by applying a Fourier series approximation. Figure 10.10 plots the coefficients for the cosine series representation as a function of the harmonic number. Note that the low frequency coefficients are large in amplitude whereas the high frequency coefficients contain little energy and do not contribute substantially to the reconstruction. The amplitude coefficients, A_m, are plotted on a $\log_{10}$ scale so the smaller values are magnified and are therefore visible. Figure 10.11 shows several levels of harmonic reconstruction. The mean plus the first and second harmonics provide the basis for the general systolic and diastolic shape, since the amplitudes of these harmonics are large and contribute substantially to the reconstructed waveform. Additional harmonics add fine details but do not contribute significantly to the raw waveform.

10.5.4   Fourier Transform

In many instances, conceptualizing a signal in terms of its contributing cosine or sine functions has various advantages. The concept of frequency domain is an abstraction

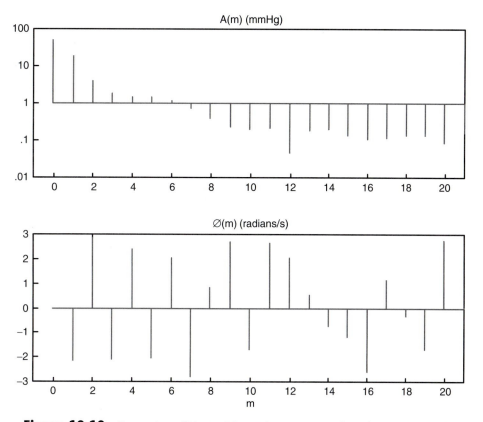

Figure 10.10 Harmonic coefficients of the aortic pressure waveform shown in Fig. 10.2.

that is born out of the Fourier series representation for a periodic signal. A signal can be expressed either in the "time domain" by the signal's time function, $x(t)$, or alternatively in the "frequency domain" by specifying the Fourier coefficient and phase, A_m and ϕ_m, as a function of the signal's harmonic frequencies, $\omega_m = m\omega_0$. Thus, if we know the Fourier coefficients and the frequency components that make up the signal, we can fully recover the periodic signal $x(t)$.

One of the disadvantages of the Fourier series is that it applies only to periodic signals, and many biological signals are not periodic. In fact, a broad class of biological signals includes signals that are continuous functions of time but that never repeat in time. Luckily, the concept of Fourier series for approximating a signal with cosines, sines, or complex exponentials can also be extended for signals that are not periodic. The Fourier integral, also referred to as the Fourier transform, is used to decompose a continuous aperiodic signal into its constituent frequency components

$$X(\omega) = \int_{-\infty}^{\infty} x(t)e^{-j\omega t}\,dt \tag{10.6}$$

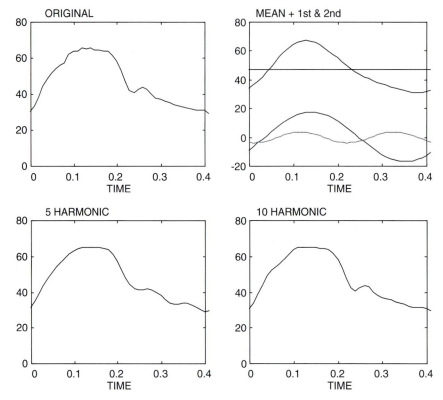

Figure 10.11   Harmonic reconstruction of the aortic pressure waveform shown in Fig. 10.2.

much as the Fourier series decomposes a periodic signal into its corresponding trigonometric components. $X(\omega)$ is a complex valued function of the continuous frequency, ω, and is analogous to the coefficients of the complex Fourier series, c_m. A rigorous proof for this relationship is beyond the scope of this text, but it is useful to note that the Fourier integral is derived directly from the exponential Fourier series by allowing the period, T, to approach infinity. The coefficients c_m of the trigonometric series approaches $X(\omega)$ as $T \to \infty$. Conceptually, a function that repeats at infinity can be considered as aperiodic since you will never observe it repeating. Tables of Fourier transforms for many common signals can be found in most digital signal processing textbooks.

As for the Fourier series, a procedure for converting the frequency-domain version of the signal, $X(\omega)$, to its time-domain expression is desired. The time-domain signal, $x(t)$, can be completely recovered from the Fourier transform with the inverse Fourier transform (IFT):

$$x(t) = \frac{1}{2\pi} \int_{-\infty}^{\infty} X(\omega)e^{j\omega t}d\omega \qquad (10.7)$$

These two representations of a signal are interchangeable, which means we can always go back and forth between the time-domain version of the signal, $x(t)$, and the frequency-domain version obtained with the Fourier transform, $X(\omega)$. The frequency domain expression therefore provides all of the necessary information for the signal and allows one to analyze and manipulate biological signals from a different perspective.

Example Problem 10.9

Find the Fourier Transform (FT) of the rectangular pulse signal

$$x(t) = \begin{cases} 1, & |t| \leq a \\ 0, & |t| > a \end{cases}$$

Solution

Equation 10.6 is used.

$$X(\omega) = \int_{-a}^{a} e^{-j\omega t} dt = \left. \frac{e^{-j\omega t}}{-j\omega} \right|_{-a}^{a} = \frac{2 \sin \omega a}{\omega} = 2a \cdot \text{sinc}(\omega a) \qquad \blacksquare$$

As for the Fourier series representation of a signal, the magnitude and the phase are important attributes of the Fourier transform. As stated previously, $X(\omega)$ is a complex valued function, meaning that it has a real, $\text{Re}\{X(\omega)\}$, and imaginary, $\text{Im}\{X(\omega)\}$, component and can be expressed as

$$X(\omega) = \text{Re}\{X(\omega)\} + j\text{Im}\{X(\omega)\} \qquad (10.8)$$

As for the Fourier series, the magnitude determines the amplitude of each complex exponential function (or equivalent cosine) required to reconstruct the desired signal, $x(t)$, from its Fourier transform

$$|X(\omega)| = \sqrt{\text{Re}\{X(\omega)\}^2 + \text{Im}\{X(\omega)\}^2} \qquad (10.9)$$

In contrast, the phase, determines the time shift of each cosine signal relative to a reference of time zero. It is determined as:

$$\theta(\omega) = \tan^{-1}\left(\frac{\text{Im}\{X(\omega)\}}{\text{Re}\{X(\omega)\}}\right) \qquad (10.10)$$

Note the close similarity for determining the magnitude and phase from the trigonometric and compact forms of the Fourier series (Eqs. 10.4a, b, c). The magnitude of the Fourier transform, $|X(\omega)|$, is analogous to A_m, whereas a_m and b_m are analogous to $\text{Re}\{X(\omega)\}$ and $\text{Im}\{X(\omega)\}$, respectively. The equations are identical in all other respects.

Example Problem 10.10

Find the magnitude and phase of the signal with Fourier transform

$$X(\omega) = \frac{1}{1 + j\omega}$$

Solution

The signal has to be put in a recognizable form similar to Equation 10.10. To achieve this,

$$X(\omega) = \frac{1}{1 + j\omega} \cdot \frac{1 - j\omega}{1 - j\omega} = \frac{1 - j\omega}{1 + \omega^2} = \frac{1}{1 + \omega^2} - j\frac{\omega}{1 + \omega^2}$$

Therefore

$$\mathrm{Re}\{X(\omega)\} = \frac{1}{1 + \omega^2} \text{ and } \mathrm{Im}\{X(\omega)\} = -\frac{\omega}{1 + \omega^2}$$

Using Eqs. 10.9 and 10.10, the magnitude is

$$|X(\omega)| = \frac{1}{1 + \omega^2}$$

and the phase

$$\theta(\omega) = \tan^{-1}(-\omega) = -\tan^{-1}(\omega) \qquad\qquad \blacksquare$$

10.5.5   Properties of the Fourier Transform

In practice, computing Fourier transforms (FT) for complex signals may be somewhat tedious and time consuming. When working with real-world problems, it is therefore useful to have tools available that help simplify calculations. The FT has several properties that help simplify frequency domain transformations. Some of these are summarized in the following paragraphs.

Let $x_1(t)$ and $x_2(t)$ be two signals in the time domain. The FTs of $x_1(t)$ and $x_2(t)$ are represented as $X_1(\omega) = F\{x_1(t)\}$ and $X_2(\omega) = F\{x_2(t)\}$.

Linearity

The Fourier transform is a linear operator. Therefore, for any constants a_1 and a_2,

$$F\{a_1 x_1(t) + a_2 x_2(t)\} = a_1 X_1(\omega) + a_2 X_2(\omega) \qquad\qquad (10.11)$$

This result demonstrates that the scaling and superposition properties defined for a liner system also hold for the Fourier transform.

Time Shifting/Delay

If $x_1(t - t_0)$ is a signal in the time domain that is shifted in time, the Fourier transform can be represented as

$$F\{x_1(t - t_0)\} = e^{-j\omega t_0} X_1(\omega) \qquad\qquad (10.12)$$

In other words, shifting a signal in time corresponds to multiplying its Fourier transform by a phase factor, $e^{-j\omega t_0}$.

Frequency Shifting

If $X_1(\omega - \omega_0)$ is the Fourier transform of a signal, shifted in frequency, the inverse Fourier transform is

$$F^{-1}\{X_1(\omega - \omega_0)\} = e^{j\omega_0 t}x(t) \tag{10.13}$$

Convolution Theorem

The convolution between two signals in the time domain is defined as

$$x(t) = \int_{-\infty}^{\infty} x_1(\tau)x_2(t - \tau)d\tau = x_1(t) * x_2(t) \tag{10.14}$$

and has an equivalent expression in the frequency domain

$$X(\omega) = F\{x(t)\} = F\{x_1(t) * x_2(t)\} = X_1(\omega)X_2(\omega). \tag{10.15}$$

Convolution in the time domain, which is relatively difficult to compute, is a straightforward multiplication in the frequency domain.

Next, consider the convolution of two signals, $X_1(\omega)$ and $X_2(\omega)$, in the frequency domain. The convolution integral in the frequency domain is expressed as

$$X(\omega) = \int_{-\infty}^{\infty} X_1(\nu)x_2(\omega - \nu)d\nu = X_1(\omega) * X_2(\omega) \tag{10.16}$$

The IFT of $X(\omega)$ is

$$x(t) = F^{-1}\{X(\omega)\} = F^{-1}\{X_1(\omega) * X_2(\omega)\} = 2\pi x_1(t)x_2(t) \tag{10.17}$$

Consequently, the convolution of two signals in the frequency domain is 2π times the product of the two signals in the time domain. As we will see subsequently, convolution is an important property for the filtering of biosignals.

Example Problem 10.11

What is the FT of $3\sin(25t) + 4\cos(50t)$? Express your answer only in a symbolic equation. Do not evaluate the result.

Solution

$$F\{3\sin(25t) + 4\cos(50t)\} = 3F\{\sin(25t)\} + 4F\{\cos(50t)\} \qquad \blacksquare$$

10.5.6   Discrete Fourier Transform

In digital signal applications, continuous biological signals are first sampled by an analog-to-digital converter and then transferred to a computer where they can be further analyzed and processed. Since the Fourier transform applies only to continuous signals of time, analyzing discrete signals in the frequency domain requires that we first modify the Fourier transform equations so that they are structurally compatible with the digital samples of a continuous signal.

The discrete Fourier transform (DFT)

$$X(m) = \sum_{k=0}^{N-1} x(k)e^{-j\frac{2\pi mk}{N}}; \; m = 0, \; 1, \; \ldots, \; N/2 \qquad (10.18)$$

provides the tool necessary to analyze and represent discrete signals in the frequency domain. The DFT is essentially the digital version of the Fourier transform. The index m represents the digital frequency index, $x(k)$ is the sampled approximation of $x(t)$, k is the discrete time variable, N is an even number that represents the number of samples for $x(k)$, and $X(m)$ is the DFT of $x(k)$.

The inverse discrete Fourier transform (IDFT) is the discrete-time version of the inverse Fourier transform. The inverse discrete Fourier transform (IDFT) is represented as

$$x(k) = \frac{1}{N}\sum_{m=0}^{N-1} X(m)e^{j\frac{2\pi mk}{N}}; \; k = 0, \; 1, \; \ldots, \; N-1 \qquad (10.19)$$

As for the FT and IFT, the DFT and IDFT represent a Fourier transform pair in the discrete domain. The DFT allows one to convert a set of digital time samples to its frequency domain representation. In contrast, the IDFT can be used to invert the DFT samples, allowing one to reconstruct the signal samples $x(k)$ directly from its frequency domain form, $X(m)$. These two equations are thus interchangeable, since either conveys all of the signal information.

Example Problem 10.12

Find the discrete Fourier transform of the signal $x(k) = 0.25^k$ for $k = 0\!:\!15$.

Solution

$$X(m) = \sum_{k=0}^{N-1} x(k)e^{-j\frac{2\pi mk}{N}} = \sum_{k=0}^{15} 0.25^k e^{-j\frac{2\pi mk}{16}} = \sum_{k=0}^{15} \left(0.25 \cdot e^{-j\frac{2\pi m}{N}}\right)^k = \sum_{k=0}^{15} a^k$$

Note that the preceding is a geometric sum in which $a = 0.25 \cdot e^{-j\frac{2\pi m}{N}}$. Since for a geometric sum

$$\sum_{k=M}^{N} a^k = \frac{a^{N+1} - a^M}{a-1}$$

we obtain

$$X(m) = \frac{a^{16} - a^0}{a-1} = \frac{0.25^{16}e^{-j\frac{32 m\pi}{N}} - 1}{0.25e^{-j\frac{2 m\pi}{N}} - 1}$$

An efficient computer algorithm for calculating the DFT is the fast Fourier transform (FFT). The output of the FFT and DFT algorithms are the same; however, the FFT has a much faster execution time than the DFT (proportional to $N \cdot \log_2(N)$ versus N^2 operations). The ratio of computing time for the DFT and FFT is therefore

$$\frac{DFT \text{ computing time}}{FFT \text{ computing time}} = \frac{N^2}{N \cdot \log_2 N} = \frac{N}{\log_2 N} \qquad (10.20)$$

For the FFT to be efficient, the number of data samples, N, must be a power of two. If $N = 1024$ signal samples, the FFT algorithm is approximately $1024/\log_2(1024) = 10$ times faster than the direct DFT implementation. If N is not a power of two, alternate DFT algorithms are usually used.

Figure 10.12 shows a signal and the corresponding DFT, which was calculated using the FFT algorithm. The signal shown in Figure 10.12a is a sine wave with a frequency of 100 Hz. Figure 10.12b shows the FFT of the 100 Hz sine wave. Notice that the peak of the FFT occurs at the frequency of the sine wave, 100 Hz. Figure 10.13a shows a 100 Hz sine wave that was corrupted with random noise that was added to the waveform. The frequency of the signal is not distinct in the time domain. After transforming this signal to the frequency domain, the signal (Figure 10.13b) reveals a definite frequency component at 100 Hz, which is marked by the large peak in the FFT.

Example Problem 10.13

Find and plot the magnitude of the discrete Fourier transform of the signal $x[n] = \sin(\pi/4 \cdot n) + 2 \cdot \cos(\pi/3 \cdot n)$ in MATLAB.

Solution

```
n = 1: 1024; %Discrete Time Axis
x = sin (pi/4*n) + 2* cos (pi/3*n); %Generating the signal
X = fft(x, 1024*16)/1024; %Computing 16k point Fast Fourier Transform
Freq = (1: 1024*16)/(1024*16)*2*pi; %Normalizing Frequencies between 0 – 2*pi
plot(Freq,abs(X),'k') %Plotting
axis([0.7 1.15 0 1.2])
xlabel('Frequency (rad/s)')
ylabel('Fourier Magnitude')
```

Results are shown in Figure 10.14. ∎

10.5.7 The *Z* Transform

The z transform provides an alternative tool for analyzing discrete signals in the frequency domain. This transform is essentially a variant of the DFT, which converts a discrete sequence into its z domain representation. In most applications, the z transform is somewhat easier to work with than the DFT because it does not require

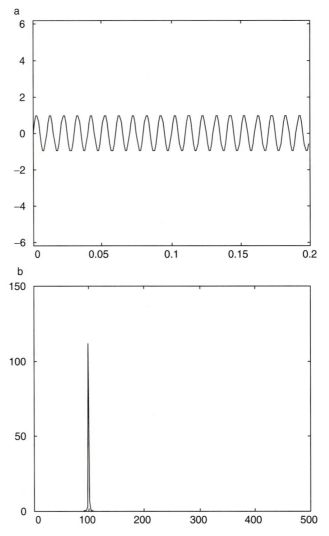

Figure 10.12   (a) 100 Hz sine wave. (b) Fast Fourier transform (FFT) of 100 Hz sine wave.

the use of complex numbers. The z transform plays a similar role for digital signals as the Laplace transform does for the analysis of continuous signals.

If a discrete sequence $x(k)$ is represented by x_k, the (one-sided) z transform of the discrete sequence is expressed by

$$X(z) = \sum_{k=0}^{\infty} x_k z^{-k} = x_0 + x_1 z^{-1} + x_2 z^{-2} + \ldots \qquad (10.21)$$

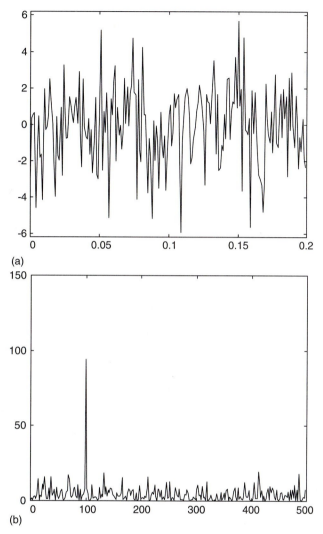

Figure 10.13 (a) 100 Hz sine wave corrupted with noise. (b) Fast Fourier transform (FFT) of the noisy 100 Hz sine wave.

Note that the z transform can be obtained directly from the DFT by allowing $N \to \infty$ and replacing $z = e^{-j\frac{2\pi m}{N}}$ in Equation 10.18. In most practical applications, sampled biological signals are represented by a data sequence with N samples so that the z transform is estimated for $k = 0 \ldots N - 1$ only. Tables of common z transforms and their inverse transforms can be found in most digital signal processing textbooks.

After a continuous signal has been sampled into a discrete sequence, its z transform is found quite easily. Since the data sequence of a sampled signal is represented as

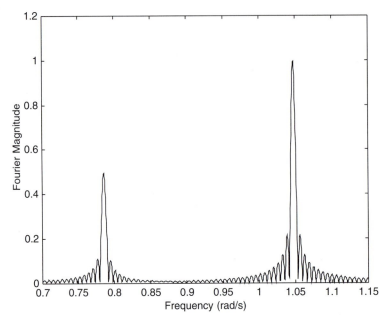

Figure 10.14   Fast Fourier transform magnitude for the sum of two sinusoids. Dominant energy peaks are located at the signal frequencies p/3 and p/4 rad/s.

$$x = [x(0),\ x(T),\ x(2T),\ \dots, x(kT)] \tag{10.22}$$

its z transform is obtained by applying Equation 10.21 to its samples

$$X(z) = x(0) + x(T)z^{-1} + x(2T)z^{-2} + \dots + x(kT)z^{-k} \tag{10.23}$$

where T is the sampling period or sampling interval.

A sampled signal is a data sequence with each sample separated from its neighboring samples by precisely one sampling period. In the z transform, the value of the multiplier, $x(kT)$, is the value of the data sample. The terms z^{-k} have an intuitive graphical explanation. The power k corresponds to the number of sampling periods following the start of the sampling process at time zero; z^{-k} can therefore be thought of as a "shift operator" which delays the sample by exactly k sampling periods or kT. The variable z^{-1}, for instance, represents a time separation of one period T following the start of the signal at time zero. In Equation 10.18, $z(0)$ is the value of the sampled data at $t = 0$ and $x(T)$ is the value of the sampled data that was obtained after the first sampling period. The z transform is an important method for describing the sampling process of an analog signal.

Example Problem 10.14

The discrete unit impulse function is represented as the sequence $\mathbf{x} = [1,\ 0,\ 0,\ 0,\ \dots, 0]$. Find the z transform of this sequence.

Solution

$$X(z) = 1 + 0z^{-1} + 0z^{-2} + \ldots + 0z^{-k} = 1 + 0 + 0 + \ldots + 0 = 1 \qquad \blacksquare$$

Example Problem 10.15

An A/D converter is used to convert a recorded signal of the electrical activity inside a nerve into a digital signal. The first five samples of the biological signal are $[-60.0, -49.0, -36.0, -23.0, -14.0]$mV. What is the z transform of this data sequence? How many sample periods after the start of the sampling process was the data sample -23.0 recorded?

Solution

$$Y(z) = -60.0 - 49.0z^{-1} - 36.0z^{-2} - 23.0z^{-3} - 14.0z^{-4}$$

The value of the negative exponent of the -23.0 mV z-term is 3. Therefore, the data sample with the value of -23.0 was recorded 3 sampling periods after the start of sampling. $\qquad \blacksquare$

10.5.8 Properties of the *Z* Transform

The z transform obeys many of the same rules and properties that we've already shown for the Fourier transform. These properties can significantly simplify the process of evaluating z transforms for complex signals. The following are some of the properties of the z transform. Note the close similarity to the properties for Equations 10.11, 10.12, and 10.14.

Let $x_1(k)$ and $x_2(k)$ be two digital signals with corresponding z transforms, $X_1(z)$ and $X_2(z)$.

Linearity

The z transform is a linear operator. For any constants a_1 and a_2,

$$Z\{a_1 x_1(k) + a_2 x_2(k)\} = \sum_{k=0}^{\infty} [a_1 x_1(k) + a_2 x_2(k)]z^{-k} = a_1 X_1(z) + a_2 X_2(z) \qquad (10.24)$$

Delay

Let $x_1(k - n)$ be the original signal that is delayed by n samples. The z transform of the delayed signal is

$$Z\{x_1(k - n)\} = \sum_{k=0}^{\infty} x_1(k - n)z^{-k} = \sum_{k=0}^{\infty} x_1(k)z^{-(k+n)} = z^{-n}X_1(z) \qquad (10.25)$$

As described previously, note that the operator z^{-n} represents a shift of n samples or precisely nT seconds.

Convolution

Let $x(k)$ be the discrete convolution between $x_1(k)$ and $x_2(k)$,

$$x(k) = x_1(k) * x_2(k)$$

$X(z)$, the z transform of $x(k)$ is calculated as

$$X(z) = Z\{x(k)\} = X_1(z)X_2(z) \tag{10.26}$$

As with the Fourier transform, this result demonstrates that convolution between two sequences is performed by simple multiplication in the z domain.

10.6   LINEAR SYSTEMS

A system is a process, machine, or a device that takes a signal as an input and manipulates it to produce an output that is related to, but is distinctly different from its input. A schematic showing the graphical representation of a system block diagram is shown in Figure 10.15.

   Biological organs and components are very often modeled as systems. The heart, for instance, is a large-scale system that takes oxygen deficient blood from the veins (the input) and pumps it through the lungs. This produces a blood output via the main arteries of the heart that is rich in oxygen content. Neurons in the brain can also be thought of as a simple microscopic system that takes electrical nerve impulses from various neurons as the input and sums these impulses to produce a single action

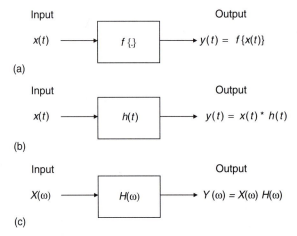

Figure 10.15   (a) Block diagram representation of a system. The input signal, $x(t)$, passes through the system transformation $f\{\cdot\}$ to produce an output, $y(t)$. (b) Time-domain representation of a linear system. The output of the linear system is represented by the convolution of the input and impulse response. (c) Frequency-domain representation of a linear system. The output corresponds to the product of the input and the system transfer function.

potential response—the output. Linear systems are a special class of systems with a unique set of properties that make them easy to analyze.

10.6.1 Linear System Properties

Although biological systems are not linear per se, very often they can be approximated by a linear system model. This is desired because it makes their analysis and the subsequent interpretation more tractable.

All linear systems are characterized by the principles of superposition (or additivity) and scaling. The superposition property states that the sum of two independent inputs produces an output that is the sum or superposition of the outputs for each individual input. The scaling property tells us that the change in the size of the input produces a comparable change at the output. Mathematically, if we know the outputs for two separate inputs, i.e.,

$$\begin{aligned} &\text{Input} \quad \text{Output} \\ &x_1(t) \rightarrow y_1(t) \\ &x_2(t) \rightarrow y_2(t) \end{aligned}$$

we can easily determine the output to any arbitrary combination of these inputs. More generally, a linear superposition and scaling of the input signals produces a linear superposition and scaling of the output signals

$$\begin{array}{cc} \text{Input} & \text{Output} \\ k_1 \cdot x_1(t) + k_2 \cdot x_2(t) \rightarrow k_1 \cdot y_1(t) + k_2 \cdot y_2(t) \end{array} \tag{10.27}$$

where k_1 and k_2 are arbitrary amplitude scaling constants. These constants scale the input amplitudes by making them larger ($k > 1$) or smaller ($k < 1$). This produces a comparable change in the net outputs, which are likewise scaled by the same constants.

Example Problem 10.16

The following information is given for a linear system:

$$\begin{array}{cc} \text{Input} & \text{Output} \\ x_1(t) = \cos(t) & \rightarrow y_1(t) = \cos(t + \pi/2) \\ x_2(t) = \cos(t) + \sin(2t) & \rightarrow y_2(t) = \cos(t + \pi/2) + 5\sin(2t) \\ x_3(t) = \cos(3t) & \rightarrow y_3(t) = 2\cos(3t) \end{array}$$

Find the output if the input is: $x(t) = 3\sin(2t) + 1/2\cos(3t)$.

Solution

The input is represented as superposition of x_1, x_2, and x_3:

$$x(t) = 3(x_2(t) - x_1(t)) + 1/2 x_3(t) = 3x_2(t) - 3x_1(t) + 1/2 x_3(t)$$

Applying the superposition and scaling properties produces an output:

$$y(t) = 3y_2(t) - 3y_1(t) + 1/2y_3(t) = 3(\cos(t + \pi/2) + 5\sin(2t)) - 3(\cos(t + \pi/2))$$
$$+ 1/2(2\cos(3t)) = 15 \cdot \sin(2t) + \cos(3t) \qquad \blacksquare$$

Example Problem 10.17

Consider the system given by the expression

$$y(t) = f\{x(t)\} = A \cdot x(t) + B$$

Determine if this is a linear system.

Solution

To solve this problem, consider a superposition of two separate inputs, $x_1(t)$ and $x_2(t)$, that independently produce outputs $y_1(t)$ and $y_2(t)$. Apply the input $x(t) = k_1 \cdot x_1(t) + k_2 \cdot x_2(t)$. If the system is linear, the output obeys

$$y_{Lin}(t) = k_1 \cdot y_1(t) + k_2 \cdot y_2(t) = k_1 \cdot (A \cdot x_1(t) + B) + k_2 \cdot (A \cdot x_2(t) + B)$$
$$= A(k_1 \cdot x_1(t) + k_2 \cdot x_2(t)) + (k_1 + k_2)B$$

The true system output, however, is determined as

$$y(t) = f\{x(t)\} = f\{k_1 \cdot x_1(t) + k_2 \cdot x_2(t)\} = A \cdot (k_1 \cdot x_1(t) + k_2 \cdot x_2(t)) + B$$

We need to compare our expected linear system output, $y_{Lin}(t)$, with the true system output, $y(t)$. Note that $y(t) \neq y_{Lin}(t)$ and therefore the system is not linear. $\blacksquare$

The superposition principle takes special meaning when applied to periodic signals. Because periodic signals are expressed as a sum of cosine or complex exponential functions with the Fourier series, their output must also be expressed as a sum of cosine or exponential functions. Thus if a linear system is stimulated with a periodic signal, its output is also a periodic signal with identical harmonic frequencies. The output, $y(t)$, of a linear system to a periodic input, $x(t)$, is related by

Input   Output

$$x(t) = \frac{A_0}{2} + \sum_{m=-\infty}^{+\infty} A_m \cos(m\omega_0 t + \phi_m) \Rightarrow y(t) = \frac{B_0}{2} + \sum_{m=-\infty}^{+\infty} B_m \cos(m\omega_0 t + \theta_m)$$

(10.28)

The input and output contain cosines with identical frequencies, $m\omega_0$, and are expressed by equations with similar form. A similar form of this expression is also obtained for the exponential Fourier series:

Input   Output

$$x(t) = \sum_{m=-\infty}^{+\infty} c_m e^{jm\omega_0 t} \Rightarrow y(t) = \sum_{m=-\infty}^{+\infty} b_m e^{jm\omega_0 t}$$

(10.29)

where the input and output coefficients, c_m and b_m, are explicitly related to A_m and B_m via Eq. 10.5b.

From Eqs. 10.28 and 10.29, the input and output of a linear system to a periodic input differ in two distinct ways. First, the amplitudes of each cosine are selectively scaled by different constants, A_m for the input and B_m for the output. These constants are uniquely determined by the linear system properties. Similarly, the phases angle of the input components, ϕ_m, are different from the output components, θ_m, meaning that the input and output components are shifted in time in relationship to each other. As for the amplitudes, the phase difference between the input and output is a function of the linear system. Thus, if we know the mathematical relationship of how the input components are amplitude scaled and phase shifted between the input and output, we can fully describe the linear system. This relationship is described by the system *transfer function*, H_m. This function fully describes how the linear system manipulates the amplitude and phases of the input to produce a specific output. This transformation is described by two separate components, the *magnitude* and the *phase*.

The magnitude of H_m is given by the ratio of the output to the input for the mth component

$$|H_m| = \frac{B_m}{A_m} \qquad (10.30)$$

Note that if we know the input magnitudes we can determine the output Fourier coefficients by multiplying the transfer function magnitude by the input Fourier coefficients: $B_m = |H_m| \cdot A_m$. The phase angle of the transfer function describes the phase relationship between the input and output for the mth frequency component

$$\angle H_m = \theta_m - \phi_m \qquad (10.31)$$

If we know the input phase, the output phase is determined as $\theta_m = \angle H_m + \phi_m$. Equations 10.30 and 10.31 are the two critical pieces of information that are necessary to fully describe a linear system. If these two properties of the transformation are known, it is possible to determine the output relationship for any arbitrary input.

10.6.2 Time-Domain Representation of Linear Systems

The relationship between the input and output of a linear system can be described by studying its behavior in the time domain (Figure 10.15b). The *impulse response* function, $h(t)$, is a mathematical description of the linear system that fully characterizes its behavior. As we will see subsequently, the impulse response of a linear system is directly related to the systems transfer function as outlined for the periodic signal. If one knows $h(t)$, one can readily compute the output, $y(t)$, to any arbitrary input, $x(t)$, using the *convolution integral*

$$y(t) = h(t) * x(t) = \int_{-\infty}^{\infty} h(\tau)x(t - \tau)d\tau \qquad (10.32)$$

The symbol * is shorthand for the convolution between the input and the system impulse response. Integration is performed with respect to the dummy integration variable τ. For the discrete case, the output of a discrete linear system is determined with the convolution sum

$$y(k) = h(k) * x(k) = \sum_{m=-\infty}^{\infty} h(m)x(k-m) \qquad (10.33)$$

where $h(m)$ is the impulse response of the discrete system. A detailed treatment of the convolution integral is found in many signal processing textbooks and is beyond the scope of this text. As shown in a subsequent section, a simpler treatment of the input-output relationship of a linear system is obtained by analyzing it in the frequency domain.

Example Problem 10.18

A cytoplasmic current injection $i(t) = u(t)$ to a cell membrane produces an intracellular change in the membrane voltage, $v(t)$. The membrane of a cell is modeled as a linear system with impulse response $h(t) = A \cdot e^{-t/\tau} \cdot u(t)$ where A is a constant in units $V/s/A$ and τ is the cell membrane time constant (units: seconds). Find the cell membrane voltage output.

Solution

The input, $i(t)$, and output, $v(t)$, are related by the convolution integral (Eq. 10.32):

$$v(t) = h(t) * i(t) = \int_{-\infty}^{\infty} h(\varsigma)i(t-\varsigma)d\varsigma = \int_{-\infty}^{\infty} A \cdot e^{-\varsigma/\tau} u(\varsigma)u(t-\varsigma)d\varsigma$$

where we use a dummy integration variable, ς, to distinguish it from the cell time constant, τ. The unit step functions inside the integral take values of one or zero, in which case they do not contribute to the integral. $u(\varsigma) = 1$ if $\varsigma > 0$ and $u(t - \varsigma) = 1$ if $t - \varsigma > 0$. Combining these two inequalities, we have that $0 < \varsigma < t$ and we can therefore change the limits of integration and replace the unit step function with 1,

$$v(t) = \int_{0}^{t} A \cdot e^{-\varsigma/\tau} d\varsigma = \frac{A}{\tau}(1 - e^{-t/\tau}) \qquad \blacksquare$$

10.6.3 Frequency-Domain Representation of Linear Systems

We have already considered the special case of linear systems in the frequency domain for periodic inputs. Recall that the system output of a linear system to a periodic stimulus is fully described by the system transfer function. The output of a linear system is also expressed in the time domain by the convolution integral (Figure 10.15b). The impulse response is the mathematical model that describes the linear system in the time domain. These two descriptions for the input-output relationship of a linear system are mutually related. Notably, for the aperiodic signal case, the transfer function is the Fourier transform of the impulse response

Time Domain		Frequency Domain	
$h(t)$	$\Leftrightarrow$	$H(\omega)$	(10.34)

where $H(\omega)$ is the system transfer function. Since the impulse response is a complete model of a linear system and since the Fourier transform is invertible [we can always go back and forth between $h(t)$ and $H(\omega)$] the transfer function contains all of the necessary information to fully describe the system. The advantage of the transfer function comes in its simplicity of use. Rather than performing a convolution integral, which can be quite intricate in many applications, the output of a linear system in the frequency domain is expressed as a product of the system input and its transfer function: $Y(\omega) = X(\omega)H(\omega)$. This result is reminiscent of the result for the Fourier series (Eqs. 10.30 and 10.31). Specifically, the *convolution property* of the Fourier transform states that a convolution in time corresponds to a multiplication in the frequency domain

Time Domain		Frequency Domain	
$y(t) = x(t) * h(t)$	$\Leftrightarrow$	$Y(\omega) = X(\omega)H(\omega)$	(10.35)

and thus the output of a linear system, $Y(\omega)$, is expressed in the frequency domain by the product of $X(\omega)$ and $H(\omega)$ (Figure 10.15c). Note that this result is essentially the convolution theorem (Eqs. 10.14 and 10.15; for proof see Example Problem 10.19) applied to the output of a linear system (Eq. 10.32). In many instances, this is significantly easier to compute than a direct convolution in the time domain.

Example Problem 10.19

Prove the convolution property of the Fourier transform.

Solution

$$Y(\omega) = FT\{y(t)\} = \int y(t)e^{-j\omega t}dt = \int\int x(\tau)h(t-\tau)d\tau e^{-j\omega t}dt$$

Make a change of variables, $u = t - \tau, \; du = dt$

$$= \int\int x(\tau)h(u)d\tau e^{-j\omega(u+\tau)}du = \int x(\tau)e^{-j\omega\tau}d\tau \cdot \int h(u)e^{-j\omega u}du = X(\omega)H(\omega) \qquad \blacksquare$$

Example Problem 10.20

Consider the cell membrane cytoplasmic current injection for Example Problem 10.18. Find the cell's output voltage in the Fourier domain.

Solution

The Fourier transform of the step current input is

$$I(\omega) = \int i(t)e^{-j\omega t}dt = \int_0^\infty 1 \cdot e^{-\omega t}dt = \left.\frac{e^{-\omega t}}{-j\omega}\right|_0^\infty = \frac{1}{j\omega}$$

The transfer function is determined as the Fourier transform of the impulse response:

$$H(\omega) = FT\{h(t)\} = \int A \cdot e^{-t/\tau} \cdot u(t)dt = \int_0^\infty A \cdot e^{-t/\tau}e^{-j\omega t}dt = \frac{A}{j\omega + 1/\tau}$$

The cell's voltage output in the frequency domain is determined as

$$V(\omega) = H(\omega)I(\omega) = \frac{A}{j\omega + 1/\tau} \cdot \frac{1}{j\omega} = \frac{A}{j\omega} - \frac{A}{j\omega + 1/\tau} \qquad\blacksquare$$

10.6.4   Analog Filters

Filters are a special class of linear systems that are widely used to manipulate the properties of a biological signal. Conceptually, a filter allows the user to selectively remove an undesired signal component while preserving or enhancing some component of interest. Although most of us are unaware of this, various types of filters are commonplace in everyday settings. Sunblock, for instance, is a type of filter that "removes" unwanted ultraviolet light from the sun in order to minimize the likelihood of sunburn and potentially reduce the risk of skin cancer. Filters are also found in many audio applications. Treble and bass control in an audio system are a special class of filter which the user selectively controls in order to boost or suppress the amount of high-frequency (treble) and low-frequency (bass) sound to a desired level and quality.

Filters play an important role in the analysis of biological signals, in part because signal measurements in clinical settings are often confounded by unwanted noise. Such noise distorts the signal waveform of interest, making it difficult to obtain a reliable diagnosis. If one could completely remove unwanted noise, one could significantly improve the quality of a signal and thus minimize the likelihood of an incorrect diagnosis.

Practically, most filters can be subdivided into three broad classes, according to how they modify the frequency spectrum of the desired signal. These broad classes include low-pass, high-pass, and band-pass filters. Low-pass filters work by removing high frequencies from a signal while selectively keeping the low frequencies (Figure 10.16a). This allows the low frequencies of the signal to pass through the filter

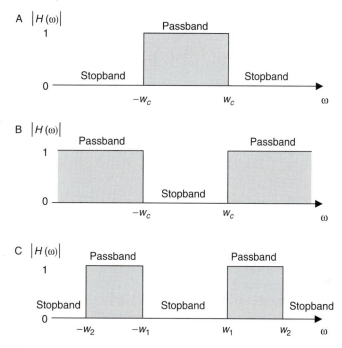

Figure 10.16 Frequency-domain magnitude response plot, $|H(\omega)|$, of the ideal (a) low-pass filter, (b) high-pass filter, and (c) band-pass filter. Signals in the shaded region—the pass-band—are preserved at the output whereas signals in the stop-band are selectively removed from the output.

uninterrupted, hence the name "low-pass." In some instances, low frequencies could be accentuated further by magnifying them while selectively removing the high frequencies. High-pass filters perform exactly the opposite function of a low-pass filter (see Figure 10.16b). They selectively pass the high frequencies but remove the low frequencies of the signal. The treble control in an audio system is a form of high-pass filter that accentuates the high frequencies, thus producing crisp and rich sound. In contrast, the bass control is a form of low-pass filter that selectively enhances low frequencies, or the "bass," creating a "warmer" sound quality. Band-pass filters fall somewhere in between the low-pass and high-pass filter. Rather than simply removing the low or high frequencies, band-pass filters remove both high and low frequencies but selectively keep a small "band" of frequencies (Figure 10.16c); hence its name. The function of a band-pass filter could be achieved by simply combining a low-pass and high-pass filter, as we will see subsequently.

Since filters are linear systems, the output of a filter is expressed by the convolution between the input and the filter's impulse response (Equation 10.32). Conversely, if the output is determined in the frequency domain, the output corresponds to the product of the filter transfer function and the input Fourier transform (Equation 10.35). The impulse response and transfer function of the ideal low-pass filter are

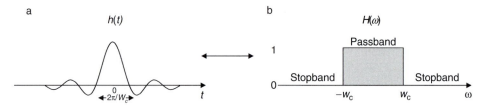

Figure 10.17   Time and frequency domain representation of the ideal low-pass filter. (a) The impulse response of the ideal low-pass filter, $h(t)$. (b) Transfer function of the ideal low-pass filter, $H(\omega)$.

shown in Figure 10.17. The transfer function of this filter takes a value of 1 within the filter pass-band and zero within the stop-band. Since the output of a linear system in the frequency domain is given as the product of the input Fourier transform and the signal transfer function, $Y(\omega) = H(\omega)X(\omega)$, any signal presented to this filter within its pass-band will pass through to the output uninterrupted because the frequency components are multiplied by one. In contrast, signals in the stop-band are removed at the output of the filter since the frequency components are multiplied by zero. The impulse response of the ideal low-pass analog filter is

$$h_{LP}(t) = \frac{W_c}{\pi}\,\text{sinc}(W_c t) \tag{10.36}$$

where $W_c = 2\pi f_c$ is the filter cutoff frequency. In the frequency domain, the ideal low-pass filter transfer function is

$$H_{LP}(\omega) = \begin{cases} 1 & |\omega| < \text{Wc} \\ 0 & |\omega| > \text{Wc} \end{cases} \tag{10.37}$$

This dual time-and-frequency-domain representation of the ideal low-pass filter is illustrated in Figure 10.17. Note that the transfer function takes a value of 1 only within the pass-band. At the cutoff frequency, the filter transfer function transitions from a value of 1 in the pass-band to a value of zero in the stop-band.

In the frequency domain, the ideal high-pass filter performs the exact opposite function of the low-pass filter:

$$H_{HP}(\omega) = \begin{cases} 0 & |\omega| < W_c \\ 1 & |\omega| > W_c' \end{cases} \tag{10.38}$$

that is, the pass-band exists for frequencies above the cutoff frequency, whereas the stop-band exists for frequencies below the filter's cutoff. This filter therefore preserves high-frequency signal components (above the cutoff frequency) and selectively removes low-frequency signals. The ideal high-pass filter transfer function can be easily derived from the ideal low-pass filter as

$$H_{HP}(\omega) = 1 - H_{LP}(\omega) \tag{10.39}$$

In the time domain, the ideal high-pass filter impulse response is obtained as

$$h_{HP}(t) = \delta(t) - h_{LP}(t) = \delta(t) - \frac{W_c}{\pi}\text{sinc}(W_c t) \tag{10.40}$$

A schematic depiction of the ideal high-pass filter transfer function is shown in Figure 10.16b.

The final class of filter we will consider is the band-pass filter. The prototypical band-pass filter is somewhat more complex than the low-pass and high-pass filters because it requires the definition of a lower and upper cutoff frequency, W_1 and W_2. Figure 10.16c illustrates the magnitude response of the ideal band-pass filter transfer function and impulse response. Only signals between the two cutoff frequencies are allowed to pass through to the output. All other signals are rejected. The transfer function of the ideal band-pass filter is given by

$$H_{BP}(\omega) = \begin{cases} 1 & W_1 < |\omega| < W_2, \\ 0 & \text{otherwise} \end{cases} \tag{10.41}$$

In the frequency domain, the ideal band-pass filter can be obtained by combining a high-pass filter with cutoff W_1 and a low-pass filter with cutoff W_2. The band-pass filter transfer function can therefore be expressed as the product of transfer functions for a low-pass and high-pass filter:

$$H_{BP}(\omega) = H_{HP}(\omega) \cdot H_{LP}(\omega) \tag{10.42}$$

In the time domain, the band-pass filter impulse response is obtained by the inverse Fourier transform of the filter transfer function:

$$h_{BP}(t) = h_{HP}(t) * h_{LP}(t) \tag{10.43}$$

This is done by applying the convolution theorem (Equations 10.14 and 10.15) to Equation 10.42.

Example Problem 10.21

An electromyographic (EMG) signal contains energy within the frequencies 25 and 100 Hz. Design a filter to remove unwanted noise.

Solution

We need to design a band-pass filter with pass-band frequencies 25 and 100 Hz. First determine the cutoff frequencies in rad/s. Since $W_c = 2\pi f_c$,

$$W_1 = 50\pi$$
$$W_2 = 200\pi$$

Next, we find the impulse response of the corresponding low-pass and high-pass filters.

$$h_{HP}(t) = \delta(t) - \frac{W_1}{\pi}\text{sinc}(W_1 t) = \delta(t) - 50\text{sinc}(50\pi t)$$

$$h_{LP}(t) = \frac{W_2}{\pi}\text{sinc}(W_2 t) = 200\text{sinc}(200\pi t)$$

The band-pass filter impulse response is

$$h_{BP}(t) = h_{BP}(t) * h_{LP}(t) = [\delta(t) - 50\text{sinc}(50\pi t)] * 200\text{sinc}(200\pi t)$$ ■

The described ideal analog filters provide a conceptual framework to aim for in various filter design applications. In practice, real analog filters cannot be implemented to achieve the strict specifications of the ideal filter because the impulse response of ideal filters is of infinite duration (extends from $-\infty$ to $+\infty$). Thus the ideal filters require an infinite amount of time to produce an output. Typically, most analog filters are designed with simple electronic circuits. Various approximations to the ideal low-pass, high-pass, and band-pass filter can be derived that are well suited for a variety of applications, including signal analysis of biomedical signals.

10.6.5 Digital Filters

Digital systems are described by difference equations, just like analog systems are described by differential equations. Difference equations are essentially discretized differential equations that have been sampled at a particular sampling rate. The general form of a real-time digital filter/difference equation is

$$y(k) = \sum_{m=0}^{M} b_m x(k-m) - \sum_{m=1}^{N} a_m y(k-m) \tag{10.44}$$

where the discrete sequence $x(k)$ corresponds to the input and $y(k)$ represents the output sequence of the discrete system. For instance, if $M = 2$ and $N = 2$, then

$$y(k) = b_0 x(k) + b_1 x(k-1) + b_2 x(k-2) - a_1 y(k-1) - a_2 y(k-2)$$

where $x(k)$ and $y(k)$ represent the input and output at time k, $x(k-1)$ and $y(k-1)$ represents the input and output one sample into the past, and similarly, $x(k-2)$ and $y(k-2)$ correspond to the input and output two samples into the past.

Digital systems, like analog systems, can also be defined by their impulse responses, $h(k)$, and the convolution sum (Equation 10.33). If the response has a finite number of nonzero points, the filter is called a finite impulse response (FIR) filter. If the response has an infinite number of nonzero points, the filter is called an infinite impulse response (IIR) filter. One positive quality of digital filters is the ease with which the output for any input can be calculated.

Example Problem 10.22

Find the impulse response for the digital filter

$$y(k) = \frac{1}{2}x(k) + \frac{1}{2}y(k-1)$$

Solution

Assume the system is at rest before input begins (i.e., $y(n) = 0$ *for* $n < 0$).

$$y(-2) = \frac{1}{2}\delta(-2) + \frac{1}{2}y(-3) = 0 + 0 = 0$$

$$y(-1) = \frac{1}{2}\delta(-1) + \frac{1}{2}y(-2) = 0 + 0 = 0$$

$$y(0) = \frac{1}{2}\delta(0) + \frac{1}{2}y(-1) = \frac{1}{2} + 0 = \frac{1}{2}$$

$$y(1) = \frac{1}{2}\delta(1) + \frac{1}{2}y(0) = 0 + \frac{1}{2}\left(\frac{1}{2}\right) = \left(\frac{1}{2}\right)^2$$

$$y(2) = \frac{1}{2}\delta(2) + \frac{1}{2}y(1) = 0 + \frac{1}{2}\left(\frac{1}{2}\right)^2 = \left(\frac{1}{2}\right)^3$$

$$y(3) = \frac{1}{2}\delta(3) + \frac{1}{2}y(2) = 0 + \frac{1}{2}\left(\frac{1}{2}\right)^3 = \left(\frac{1}{2}\right)^4$$

$$\cdots$$

$$y(k) = \left(\frac{1}{2}\right)^{k+1} u(k)$$

The impulse response for the filter is an exponential sequence. This is an IIR filter because the impulse response is of infinite duration. ■

Example Problem 10.23

Find the impulse response for the digital filter

$$y(k) = \frac{1}{3}x(k) + \frac{1}{3}x(k-1) + \frac{1}{3}x(k-2)$$

Solution

Assume the system is at rest before input begins (i.e., y(n) = 0 for n < 0).

$$y(-2) = \frac{1}{3}\delta(-2) + \frac{1}{3}\delta(-3) + \frac{1}{3}\delta(-4) = 0 + 0 + 0 = 0$$

$$y(-1) = \frac{1}{3}\delta(-1) + \frac{1}{3}\delta(-2) + \frac{1}{3}\delta(-3) = 0 + 0 + 0 = 0$$

$$y(0) = \frac{1}{3}\delta(0) + \frac{1}{3}\delta(-1) + \frac{1}{3}\delta(-2) = \frac{1}{3} + 0 + 0 = \frac{1}{3}$$

$$y(1) = \frac{1}{3}\delta(1) + \frac{1}{3}\delta(0) + \frac{1}{3}\delta(-1) = 0 + \frac{1}{3} + 0 = \frac{1}{3}$$

$$y(2) = \frac{1}{3}\delta(2) + \frac{1}{3}\delta(1) + \frac{1}{3}\delta(0) = 0 + 0 + \frac{1}{3} = \frac{1}{3}$$

$$y(3) = \frac{1}{3}\delta(3) + \frac{1}{3}\delta(2) + \frac{1}{3}\delta(1) = 0 + 0 + 0 = 0$$

$$y(4) = \frac{1}{3}\delta(4) + \frac{1}{3}\delta(3) + \frac{1}{3}\delta(2) = 0 + 0 + 0 = 0$$

$$\cdots$$

$$y(k) = 0; \ k \geq 3$$

This is an FIR filter with only three nonzero coefficients. ■

IIR filters are particularly useful for simulating analog systems. The main advantage of an IIR filter is that the desired job can usually be accomplished with fewer filter coefficients than would be required for an FIR filter (i.e., IIR filters tend to be more efficient). The main disadvantage of an IIR filter is that signals may be distorted in an undesirable way. FIR filters can be designed with symmetry to prevent undesired signal distortion. Methods for dealing with the distortion problem in FIR filters are outside of our discussions here.

Digital filters, as the name implies, are most often designed to perform specific filtering operations: low-pass filters, high-pass filters, band-pass filters, band-stop filters, notch filters, and so on. However, digital filters can be used to simulate most analog systems (e.g., to differentiate and to integrate). Many textbooks have been written on digital filter design. The key components of the process are described in the following paragraphs.

From Digital Filter to Transfer Function

The transfer function for the digital system, $H(z)$, can be obtained by rearranging the difference equation (10.23) and applying Equation 10.21. $H(z)$ is the quotient of the z transform of the output, $Y(z)$, divided by the z transform of the input, $X(z)$.

$$y(k) + a_1 y(k-1) + a_2 y(k-2) \ldots + a_N y(k-N) = b_0 x(k) + b_1 x(k-1) + \ldots + b_M x(k-M)$$

$$Y(z) + a_1 z^{-1} Y(z) + a_2 z^{-2} Y(z) \ldots + a_N z^{-N} Y(z) = b_0 X(z) + b_1 z^{-1} X(z) + b_2 z^{-2} X(z) \ldots + b_M z^{-M} X(z)$$

$$Y(z)(1 + a_1 z^{-1} + a_2 z^{-2} \ldots + a_N z^{-N}) = X(z)(b_0 + b_1 z^{-1} + b_2 z^{-2} \ldots + b_M z^{-M})$$

$$H(z) = \frac{Y(z)}{X(z)} = \frac{b_0 + b_1 z^{-1} + b_2 z^{-1} \ldots + b_M z^{-M}}{1 + a_1 z^{-1} + a_2 z^{-1} \ldots + a_N z^{-N}}$$

$$(10.45)$$

From Transfer Function to Frequency Response

The frequency response $(H'(\Omega))$ of a digital system can be calculated directly from $H(z)$ where Ω is in radians. If the data are samples of an analog signal as previously described, the relationship between ω and Ω is $\Omega = \omega T$:

$$H'(\Omega) = H(z)|_{z=e^{j\Omega}} \qquad (10.46)$$

For a linear system, an input sequence of the form

$$x(k) = A \sin(\Omega_0 k + \Phi)$$

will generate an output whose steady state sequence will fit into the following form

$$y(k) = \sin(\Omega_0 k + \varnothing)$$

Values for B and $\varnothing$ can be calculated directly

$$B = A|H'(\Omega_0)| \varnothing = \Phi + \ angle(H'(\Omega_0))$$

Example Problem 10.24

The input sequence for the digital filter used in Example Problem 10.22 is

$$x(k) = 100 \sin\left(\frac{\pi}{2}k\right)$$

What is the steady state form of the output?

Solution

$$y(k) - \frac{1}{2}y(k-1) = \frac{1}{2}x(k)$$

The difference equation is first converted into the z-domain:

$$Y(z) - \frac{1}{2}Y(z)z^{-1} = Y(z)\left[1 - \frac{1}{2}z^{-1}\right] = \frac{1}{2}X(z)$$

Solving for $H(z)$

$$H(z) = \frac{Y(z)}{X(z)} = \frac{\frac{1}{2}}{1 - \frac{1}{2}z^{-1}}$$

gives the filter transfer function. To determine the output, the transfer function is evaluated at the frequency of the input sinusoid ($z = e^{j\frac{\pi}{2}}$)

$$H'\left(\frac{\pi}{2}\right) = H(e^{j\frac{\pi}{2}}) = \frac{\frac{1}{2}}{1 - \frac{1}{2}e^{-j\frac{\pi}{2}}} = \frac{\frac{1}{2}}{1 + \frac{1}{2}j} = 0.4 - j0.2 = 0.45e^{-j0.15\pi}$$

This transfer function tells us that the output is obtained by scaling the input magnitude by 0.45 and shifting the signal by a phase factor of 0.15π rads. Therefore, the output is

$$y(k) = 45 \sin\left(\frac{\pi}{2}k - .15\pi\right)$$

Filter design problems begin with identifying the frequencies that are to be kept versus the frequencies that are to be removed from the signal. For ideal filters, $|H'(\Omega_{keep})| = 1$ *and* $|H'(\Omega_{remove})| = 0$. The filters in Example Problems 10.23 and 10.24 can both be considered as low-pass filters. However, their frequency responses, shown in Figure 10.18, show that neither is a particularly good low-pass filter. An ideal low-pass filter that has a cutoff frequency of $\pi/4$ with $|H'(\Omega)| = 1$ for $|\Omega| < \pi/4$ and $|H'(\Omega)| = 0$ for $\pi/4$ and $|\Omega| < \pi$ is superimposed for comparison. ∎

10.7 SIGNAL AVERAGING

Biological signal measurements are often confounded by measurement noise. Variability in the measurement of a signal often makes it difficult to determine the signal characteristics, making it nearly impossible to obtain a reliable clinical diagnosis.

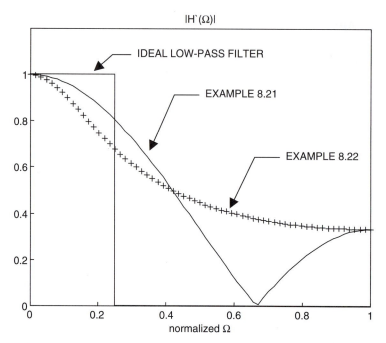

Figure 10.18   A frequency-domain comparison of low-pass filters described in Example Problems 10.23 and 10.24. An ideal low-pass filter with a cutoff frequency at p/4 rads, or 0.25 when normalized by p radians, is superimposed for comparison. The cutoff frequency of a low-pass filter is usually defined as the frequency at which the amplitude is equal to $1/\sqrt{2}$ or approximately 0.71, which matches Example Problem 10.12. Both digital filters have the same amplitude at f_{max} (i.e., where normalized $\Omega = 1$).

Many classes of biological signals are modeled as the sum of an ideal noiseless signal component, $x(t)$, and separate independent noise term, $n(t)$:

$$x_i(t) = x(t) + n(t). \tag{10.47}$$

The signal $x_i(t)$ corresponds to the "measured" ith trial or ith measurement of the signal. Note that the ith measurement contains both a deterministic component, $x(t)$, and a random or stochastic noise term, $n(t)$. Although the deterministic component of the signal is fixed from trial to trial, the noise term represents intrinsic variability, which may arise from a number of separate sources. The ith measurement can therefore exhibit significant trial-to-trial variability because the random component, $n(t)$, is different across consecutive trials. As an example, a measurement ECG (electrocardiogram) electrode can pick up extraneous signals from the muscles, lungs, and even from the internal electronics of the recording devices (e.g., 60-cycle noise from the power supply). The activity of these signals is unrelated to the activity of the beating heart and it therefore shows up in the signal measurement as noise. Other unpredictable changes in the activity of the heartbeat, such as from the caffeine jolt after taking a shot of espresso, could also show up in a measurement and be interpreted as noise.

We have already examined one possible way to separate out the signal term from the noise term by filtering the signal with an appropriately designed filter. Appropriate filtering allows one to clean up the signal, thus improving the quality of signal and the diagnostic reliability in clinical settings. If the spectrum of the noise and signal components do not overlap in the frequency domain, one can simply design a filter that keeps or enhances the desired signal term, $x(t)$, and discards the unwanted noise term, $n(t)$. While this is a simple and useful way of cleaning up a signal, this approach does not work in many instances because the biological signal and noise spectrums overlap.

Many biological signals are approximately periodic in nature. Signals associated with the beating heart—blood pressure, blood velocity, electrocardiogram—fall into this category. However, due to intrinsic natural variability, noise, and/or the influence of other functions such as respiration, beat-to-beat differences are to be expected. Figure 10.2 is an example of a blood pressure signal that has all of the described variability.

Blood pressure signals have many features that clinicians and researchers use to determine a patient's health. Some variables that are often measured include the peak pressure while the heart is ejecting blood (systolic phase), the minimum pressure achieved while the aortic valve is closed (diastolic phase), the peak derivative (dP/dt) during the early part of the systolic phase (considered an indication of the strength of the heart), and the time constant of the exponential decay during diastole (a function of the resistance and compliance of the blood vessels). One way to determine variables of interest is to calculate the variables or parameters for each beat in a series of beats and then report the means. This is often not possible because noise from individual measurements makes it very difficult to accurately determine the relevant biological parameters. An alternative approach is to first average the signal measurements from separate trials

$$\bar{x}(t) = \frac{1}{N}\sum_{i=1}^{N} x_i(t) \tag{10.48}$$

such that a representative beat is obtained. If the signal is discrete, this average is represented by:

$$\bar{x}(k) = \frac{1}{N}\sum_{i=1}^{N} x_i(k) \tag{10.49}$$

Here, $x_i(t)$, or $x_i(k)$ for the discrete case, represents the ith measured heart beat signal out of a total N measurements. The signal $\bar{x}(t)$, $\bar{x}(k)$ for the discrete case, represents the mean or average waveform obtained following the averaging procedure. Substituting Equation 10.48 into 10.47 leads to

$$\bar{x}(t) = x(t) + \frac{1}{N}\sum_{i=1}^{N} n(t) = x(t) + \epsilon(t) \tag{10.50}$$

If the noise term, $n(t)$, is purely random it can be shown that the measurement error term in Eq. 10.50, $\epsilon(t)$, which contains the influence of the noise, approaches 0 as $N \to \infty$. Thus, $\bar{x}(t) \approx x(t)$ for very large N whereas $\epsilon(t)$ tends to be small. This is a very powerful result! It tells us that we can effectively remove the noise by simply averaging measurements from many trials. Essentially, if we average a sufficiently large number of signal trials, the averaged signal closely approximates the true noiseless signal waveform.

Since the average heartbeat waveform closely approximates the true signal of interest, the variables can then be estimated based on the representative average heartbeat signal. Many biological acquisition systems are designed to calculate signal averages (Equations 10.48 and 10.49) as data are collected. The summation process is triggered by a signal or a signal-related feature. The ECG signal, which has many sharp features, is often used for heartbeat related data. Figure 10.19 shows a signal-averaged pressure waveform for the data shown in Figure 10.2. Figure 10.20 shows the signal averaging procedure for an auditory brainstem response (ABR) EEG measurement.

The preceding blood pressure example illustrates signal averaging in the time domain. For signals that are random in nature, signal averaging in the frequency

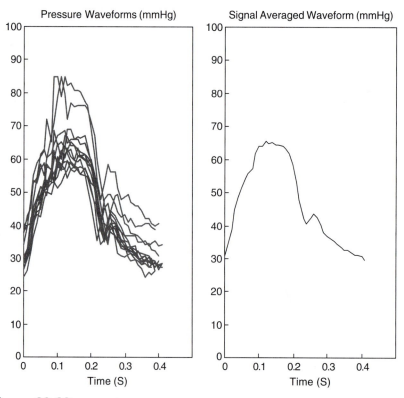

Figure 10.19   A signal-averaged pressure waveform for the data shown in Figure 10.2.

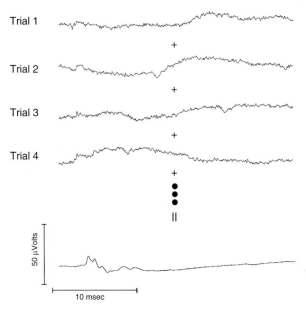

Figure 10.20 Single trials from auditory evoked response to a brief sound pulse (at time zero) measured on the temporal lobe. The auditory response from individual trials is obscured by random noise (shown first four out of one thousand). Averaged response of 1000 trials reveals the auditory response component (bottom trace).

domain is sometimes preferable. Figure 10.21 illustrates an EEG signal sampled over the occipital lobe of a patient. The sampling rate was 16 kHz. EEG analysis is usually done in the frequency domain since the presence of different frequencies is indicative of different brain states such as sleeping, resting, and alertness. The power at each frequency estimate, which can be approximated by the square of the Fourier transform, is the measurement of choice.

If a DFT is performed on the data to estimate the power of the frequencies in the signal, the expected noise in the measurement is of the same size as the measurement itself. To reduce the noise variance, a statistical approach must be undertaken. One popular approach is known as the Welch or periodogram averaging method. The signal is broken into L sections (disjoint if possible) of N points each. A DFT is performed on each of the L sections. The final result for the N frequencies is then the average at each frequency for the L sections.

The N data points in the ith segment are denoted as

$$x_i(k) = x(k + (i-1)N) \quad 0 \le k \le N-1, \ 1 \le i \le L$$

if the segments are consecutive and disjoint. The power estimate based on the DFT of an individual segment i is

$$\hat{P}_i(m) = \frac{1}{N}\left|\sum_{k=0}^{N-1} x_i(k)e^{-j\frac{2\pi mk}{N}}\right|^2 \quad \text{for } 0 \le m \le N-1 \tag{10.51}$$

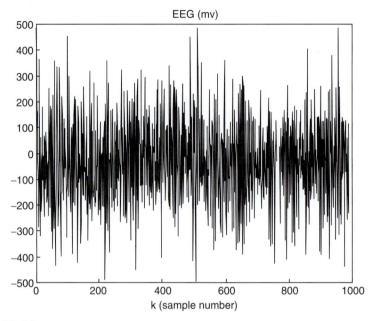

Figure 10.21   An EEG signal containing 1000 samples sampled at 16 kHz from the occipital area.

where m is associated with the power at a frequency of $\Omega = 2\pi m/N$ radians. The averaged signal spectrum is calculated by taking the mean at each frequency

$$\hat{P}(m) = \frac{1}{L}\sum_{i=1}^{L}\hat{P}_i(m) \qquad (10.52)$$

The selection of N is very important since N determines the resolution in the frequency domain. For example, if data are sampled at 500 samples/s and the resolution is desired at the 1 Hz level, at least 1 second or 500 samples ($N = 500$) should be included in each of the L sections. If resolution at the 10 Hz level is sufficient, only 0.1 seconds or 50 data points need to be included in each section. This process decreases the variance by a factor of $1/L$. This averaging process is demonstrated for the EEG data in Figure 10.22. Modifications to the procedure may include using overlapping segments if a larger value for L is needed and the number of available data points is not sufficient and/or multiplying each section by a window that forces continuity at the end points of the segments.

Example Problem 10.25

Consider the sinusoid signal

$$x(k) = \sin{(\pi/4k)} + n(k)$$

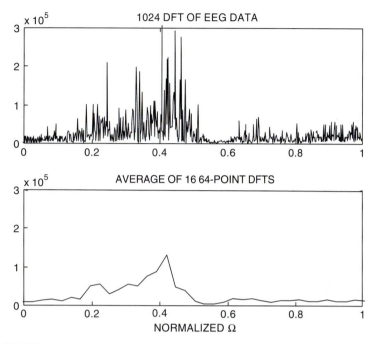

Figure 10.22   DFT averaging of an EEG. Top trace shows the raw DFT. Bottom trace shows the periodogram averaged DFT obtained with 16 64-point segments of the data.

that is corrupted by random noise, $n(k)$. Using MATLAB, show that averaging the signal removes the noise component and reveals the deterministic component. Show results for 1, 10, and 100 averages.

Solution

```
k=1:64; %Discrete Time Axis
for i=1:100 %Generating 100 signal Trials
x(i,:)=sin(pi/4*k)+randn(1,64); %i-th trial
end
X1=x(1,:); %1 Averages
X10=mean(x(1:10,:)); %10 Averages
X100=mean(x); %100 Averages
subplot(311) %Plotting Results, 1 Average
plot(k,X1,'k')
axis([1 64 −3 3])
title('1 Average')
ylabel('Amplitude')
subplot(312) %Plotting Results, 10 Averages
plot(k,X10,'k')
```

```
axis([1 64 −3 3])
title('10 Averages')
ylabel('Amplitude')
subplot(313) %Plotting Results, 100 Averages
plot(k,X100, 'k')
axis([1 64 −3 3])
title('100 Averages')
xlabel('Discrete Time')
ylabel('Amplitude')
```

Results are shown in Figure 10.23.   ■

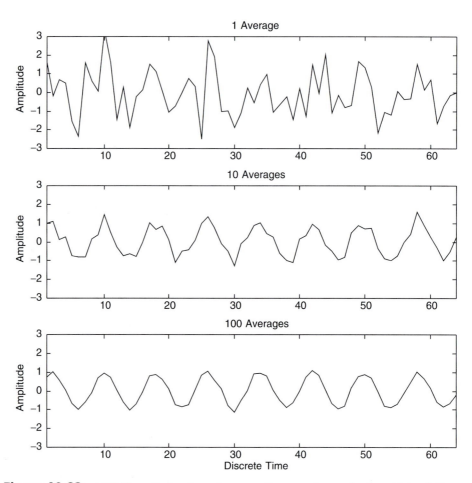

Figure 10.23   MATLAB results showing noise removal by averaging a noisy sinusoid signal. Shown for 1, 10, and 100 averages.

10.8   WAVELET TRANSFORM AND SHORT-TIME FOURIER TRANSFORM

The Fourier Transform (Equation 10.6) is a well-known signal processing tool for breaking a signal into constituent sinusoidal waveforms of different frequencies. For many applications, particularly those that change little over time, knowledge of the overall frequency content may be all that is desired. The Fourier Transform, however, does not delineate how a signal changes over time.

The short-time Fourier transform (STFT) and wavelet transform (WT) have been designed to help preserve the time-domain information. The STFT approach is to perform a Fourier transform on only a small section (window) of data at a time, thus mapping the signal into a 2D function of time and frequency. The transform is described mathematically as

$$X(\omega,\ a) = \int_{-\infty}^{\infty} x(t)g(t-a)e^{-j\omega t}dt \qquad (10.53)$$

where $g(t)$ may define a simple box or pulse function. The inverse of the STFT is given as

$$X(\omega,\ a) = K_g \int \int X(\omega,a)g(t-a)e^{j\omega t}dtda \qquad (10.54)$$

where K_g is a function of the window used.

To avoid the "boxcar" or "rippling" effects associated with a sharp window, the box may be modified to have more gradually tapered sides. Both designs are shown in Figure 10.24. The windows are superimposed on a totally periodic aortic pressure signal. For clarity, the windows have been multiplied by a factor of 100.

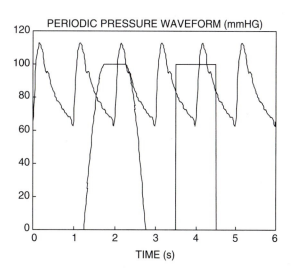

Figure 10.24   An example of two windows that might be used to perform an STFT on a perfectly periodic aortic pressure waveform. Each window approximates the width of one pulse. The tapered window on the left can help avoid the "boxcar" or "rippling" effects associated with the sharp window on the right. For clarity, the windows have been multiplied by a factor of 100.

The STFT amplitudes for three box window sizes, 1/2 period, 1 period, and 2 periods, are illustrated in Figure 10.25. The vertical lines in the top figure are indicative of longer periodicities than the window. The solid colored horizontal lines in the bottom two figures indicate that the frequency content is totally independent of time at that window size. This is expected since the window includes either one or two perfect periods. The dark (little or no frequency content) horizontal lines interspersed with the light lines in the bottom figure indicate that multiple periods exist within the window.

In contrast, Figure 10.26 shows an amplitude STFT spectrum for the aperiodic pressure waveform shown in Figure 10.2 with the window size matched as closely as possible to the heart rate. The mean has been removed from the signal so the variation

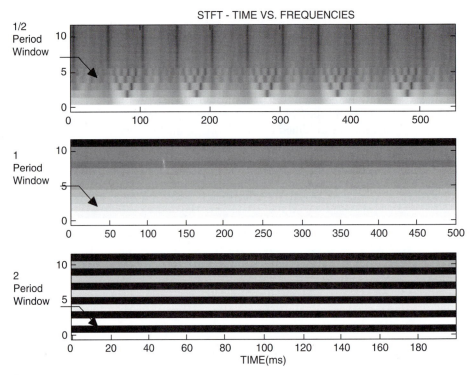

Figure 10.25 A 2-dimensional rendering of the STFT amplitude coefficients for three box window sizes—1/2 period, 1 period, and 2 periods—applied to the perfectly periodic data shown in Figure 10.16. The lighter the color, the higher the amplitude. For example, the 0th row corresponds to the mean term of the transform, which is the largest in all cases. Higher rows correspond to harmonics of the data, which, in general, decrease with frequency. The vertical lines in the top figure are indicative of longer periodicities than the window. The solid colored horizontal lines in the bottom two figures indicate that the frequency content is totally independent of time at that window size. The dark (little or no frequency content) horizontal lines interspersed with the light lines in the bottom figure indicate that multiple periods exist within the window.

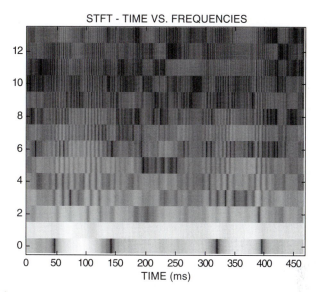

STFT - TIME VS. FREQUENCIES

Figure 10.26   A 2D rendering of the STFT of the aperiodic aortic pressure tracing shown in Figure 10.2. The window size was matched as closely as possible to the heart rate. The mean was removed from the signal so the variation in the lowest frequencies (i.e., frequency level 0) reflects changes with respiration.

in the lowest frequences (frequency level 0) reflects changes with respiration. The level of the heart rate (level 1) is most consistent across time, and the variability increases with frequency.

The main disadvantage of the STFT is that the width of the window remains fixed throughout the analysis. Wavelet analysis represents a change from both the FT and STFT in that the constituent signals are no longer required to be sinusoidal and the windows are no longer of fixed length. In wavelet analysis, the signals are broken up into shifted and scaled versions of the original or "mother" wavelet, $\psi(t)$. Figure 10.27 shows examples of two wavelets, the Haar on the left and one from the Daubechies (db2) series on the right. Conceptually, these mother wavelet functions are analogous to the impulse response of a band-pass filter. The sharp corners enable the transform to match up with local details that are not possible to observe using a Fourier transform. The notation for the 2D WT is

$$C(a, \ s) = \int\limits_{-\infty}^{\infty} x(t)\varphi(a, s, t)dt \qquad (10.55)$$

where $a =$ scale factor and $s =$ the position factor. C can be interpreted as the correlation coefficient between the scaled, shifted wavelet and the data. Figure 10.28 illustrates the db2 ($\varphi(t)$) wavelet at different scales and positions [e.g., φ (2,-100,t) = φ (2t-100)]. The inverse wavelet transform:

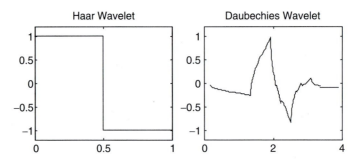

Figure 10.27   The general shape of two wavelets commonly used in wavelet analysis. The sharp corners enable the transform to match up with local details that cannot be observed when using a Fourier transform that matches only sinusoidal shapes.

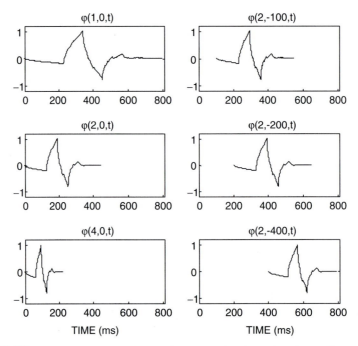

Figure 10.28   Illustrations of the db2 wavelet at several scales and positions. The upper lefthand corner illustrates the basic waveform φ(t). The notation for the illustrations is given in the form φ (scale, delay, t). Thus φ (t) = φ (1,0,t), φ (2t-100) = φ (2,-100,t), etc.

$$x(t) = K_\varphi \int \int C(a, s)\varphi(a, s, t)dtds \qquad (10.56)$$

can be used to recover the original signal, $x(t)$, from the wavelet coefficients, $C(a,s)$. K_φ is a function of the wavelet used.

In practice, wavelet analysis is performed on digitized signals using a subset of scales and positions (see MATLAB's Wavelet Toolbox). One computational process is to recursively break the signal into low-frequency ("high-scale" or "approximation") and high-frequency ("low scale" or "detail") components using digital low-pass and high-pass filters that are functions of the mother wavelet. The output of each filter will have the same number of points as the input. In order to keep the total number of data points the same at each level, every other data point of the output sequences is discarded. This is a process known as downsampling. Using upsampling and a second set of digital filters, called reconstruction filters, the process can be reversed, and the original data set is reconstructed. Remarkably, the inverse discrete wavelet transform does exist!

Athough this process will rapidly yield wavelet transform coefficients, the power of discrete wavelet analysis lies in its ability to examine waveform shapes at different resolutions and to selectively reconstruct waveforms using only the level of approximation and detail that is desired. Applications include detecting discontinuities and breakdown points, detecting long-term evolution, detecting self-similarity (e.g., fractal trees), identifying pure frequencies (similar to Fourier transform), and suppressing, de-noising, and/or compressing signals.

For comparison purposes, discrete Fourier transforms and discrete wavelet transforms are illustrated for the pressure waveforms shown in Figure 10.2. Figure 10.29 shows details of the DFT on the entire record of data. The beat-to-beat differences are reflected by the widened and irregular values around the harmonics of the heart rate. The respiration influence is apparent at the very low frequencies.

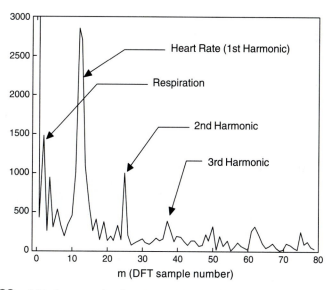

Figure 10.29   DFT of pressure data from Figure 10.2. The first, second, and third harmonics of the heart rate are clearly visible.

Finally, an example from the MATLAB Wavelet Toolbox is shown that uses the same pressure waveform. Figure 10.30 is a 2D rendering of the Wavelet Transform Coefficients. The *x* axis shows the positions and the *y* axis shows the scales with the low scales on the bottom and the high scales on the top. The top scale clearly shows

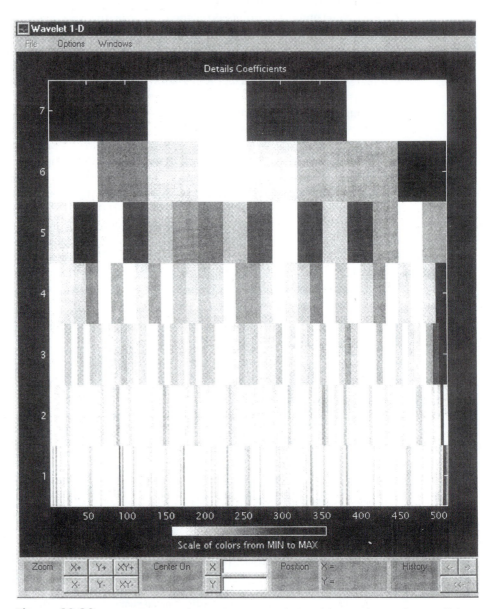

Figure 10.30   MATLAB was used to produce a 2D rendering of the wavelet transform coefficients with the Daubechies wavelet applied to the aortic pressure tracing in Figure 10.2. The *x* axis shows the positions and the *y* axis shows the scales, with the low scales on the bottom and the high scales on the top. The associated waveforms at selected levels of these scales are shown in Figure 10.31.

the two respiratory cycles in the signal. More informative than the transform coefficients, however, is a selective sample of the signal details and approximations. As the scale is changed from a1 to a7, the approximation goes from emphasizing the heart rate components to representing the respiration components. The details show that the noise at the heart rate levels is fairly random at the lower scales but moves to being quite regular as the heart rate data becomes the noise!

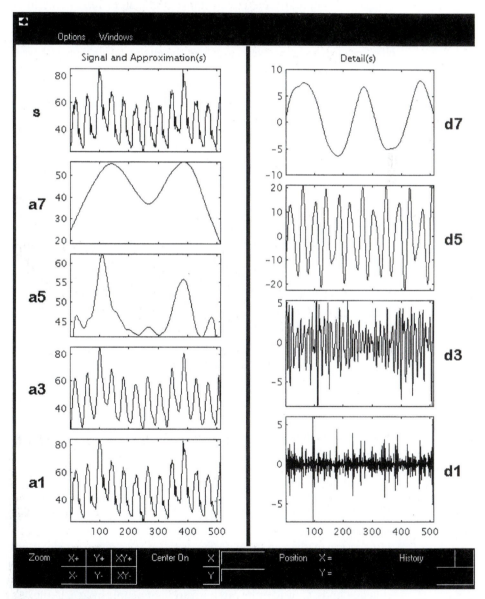

Figure 10.31   A selective sample of the signal details and approximations generated by MATLAB as part of the wavelet transform process.

10.9 ARTIFICIAL INTELLIGENCE TECHNIQUES

Artificial intelligence (AI) is a broad field that focuses on the application of computer systems that exhibit intelligent capabilities. AI systems can be built from a number of separate technologies, including fuzzy logic, neural networks, and expert systems. The principal aim of AI is to create intelligent machines that can function under adverse and unpredictable circumstances. The term *intelligent*, in regards to machines, indicates computer-based systems that can interact with their environment and adapt to changes in the environment. The adaptation is accomplished through self-awareness and perceived models of the environment that are based on qualitative and quantitative information. In other words, the basic goal of AI techniques is to produce machines that are more capable of humanlike reasoning, decision making, and adaptation.

The machine intelligence quotient (MIQ) is a measure of the intelligence level of machines. The higher the MIQ of a machine is, the higher the capacity of the machine for automatic reasoning and decision making. The MIQ of a wide variety of machines has risen significantly during the past few years. Many computer-based consumer products, industrial machinery, and biomedical instruments and systems are using more sophisticated artificial intelligence techniques. Advancements in the development of fuzzy logic, neural networks, and other soft computing techniques have contributed significantly to the improvement of the MIQ of many machines.

Soft computing is an alliance of complementary computing methodologies. These methodologies include fuzzy logic, neural networks, probabilistic reasoning, and genetic algorithms. Various types of soft computing often can be used synergistically to produce superior intelligent systems. The primary aim of soft computing is to allow for imprecision since many of the parameters that machines must evaluate do not have precise numeric values. Parameters of biological systems can be especially difficult to measure and evaluate precisely.

10.9.1 Fuzzy Logic

Fuzzy logic is based on the concept of using words, rather than numbers, for computing since words tend to be much less precise than numbers. Computing has traditionally involved calculations that use precise numerical values whereas human reasoning generally uses words. Fuzzy logic attempts to approximate human reasoning by using linguistic variables. *Linguistic variables* are words that are used to describe a parameter. For body temperature, linguistic variables that might be used are *high fever, above normal, normal, below normal,* and *frozen*. The linguistic variables are more ambiguous than the number of degrees Fahrenheit, such as 105.0, 98.9, 98.6, 97.0, and 27.5.

In classical mathematics, numeric sets called crisp sets are defined whereas the basic elements of fuzzy systems are called fuzzy sets. An example of a crisp set is A = [0, 20]. Crisp sets have precisely defined, numeric boundaries. Fuzzy sets do not have sharply defined bounds. Consider the categorization of people by age. Using crisp sets, the age groups could be divided as A= [0, 20], B = [30, 50], and C = [60, 80]. Figure 10.32a shows the characteristic function for the sets A, B, and C. The value of the function is either 0 or 1, depending on whether or not the age of a person is within the bounds of set A, B, or C. The scheme using crisp sets lacks flexibility. If a person is 25 years old or 37 years old, he or she is not categorized.

If the age groups were instead divided into fuzzy sets, the precise divisions between the age groups would no longer exist. Linguistic variables such as *young, middle-aged,* and *old* could be used to classify the individuals. Figure 10.32b shows the fuzzy sets for age categorization. Note the overlap between the categories. The words are basic descriptors, not precise measurements. A 30-year-old woman may seem old to a 6-year-old boy but quite young to an 80-year-old man. For the fuzzy sets, a value of 1 represents a 100% degree of membership to a set. A value of 0 indicates that there is no membership in the set. All numbers between 0 and 1 show the degree of membership to a group. A 35-year-old person, for instance, belongs 50% to the young set and 50% to the middle-aged set.

As with crisp sets from classical mathematics, operations are also defined for fuzzy sets. The fuzzy set operation of intersection is shown in Figure 10.33a. Figure 10.33b shows the fuzzy union operator, and Figure 10.33c shows the negation operator for fuzzy sets. The solid line indicates the result of the operator in each figure.

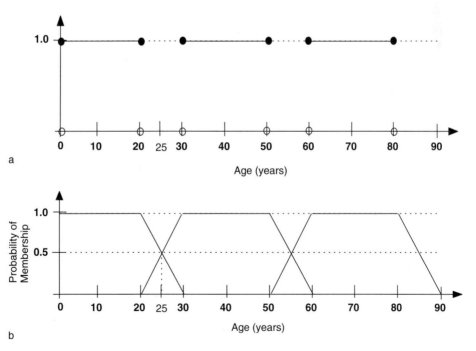

Figure 10.32 (a) Crisp sets for the classification of people by age; (b) fuzzy sets for the classification of people by age.

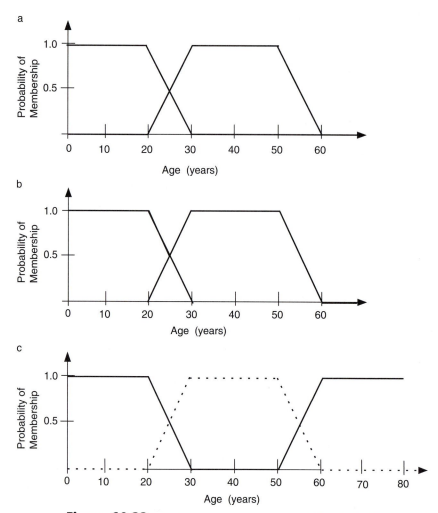

Figure 10.33 (a) Intersection of fuzzy sets: YOUNG AND MIDDLE-AGED. (b) Union of fuzzy sets: MIDDLE-AGED OR OLD. (c) Negation of fuzzy sets: NOT OLD.

Although it is easy to form fuzzy sets for a simple example such as age classification, fuzzy sets for more sophisticated applications are derived by using sophisticated calibration techniques. The linguistic variables are formulated mathematically and then can be processed by computers. Once the fuzzy sets have been established, rules are constructed. Fuzzy logic is a rule-based logic. Fuzzy systems are constructed by using a large number of rules. Most rules used in fuzzy logic computing are if/then statements that use linguistic variables. Two simple rules that use the fuzzy sets for age classification might be

- If patient is YOUNG, then use TREATMENT A.
- If patient is MIDDLE-AGED or OLD, then use TREATMENT B.

The degree of membership in a group helps determine which rule will be used and, consequently, the type of action that will be taken or, in the preceding example, the sort of treatment that will be used. Defuzzification methods are used to determine which rules will be used to produce the final output of the fuzzy system.

For many applications, fuzzy logic has significant advantages over traditional numeric computing methods. Fuzzy logic is particularly useful when information is too limited or too complex to allow for numeric precision since it tolerates imprecision. If an accurate mathematical model cannot be constructed, fuzzy logic may prove valuable. However, if a process can be described or modeled mathematically, then fuzzy logic will not generally perform better than traditional methods.

Biomedical engineering applications, which involve the analysis and evaluation of biosignals, often have attributes that confound traditional computing methods but are well suited to fuzzy logic. Biological phenomena often are not precisely understood and can be extremely complex. Biological systems also vary significantly from one individual to the next. In addition, many key quantities in biological systems cannot be measured precisely due to limitations in existing sensors and other biomedical measuring devices. Sensors may have the capability to measure biological quantities intermittently or in combination with other parameters but not independently. Blood glucose sensors, for example, are sensitive not only to blood glucose but also to urea and other elements in the blood. Fuzzy logic can be used to help compensate for the limitations of sensors.

Fuzzy logic is being used in a variety of biomedical engineering applications. Closed-loop drug delivery systems, which are used to automatically administer drugs to patients, have been developed by using fuzzy logic. In particular, fuzzy logic may prove valuable in the development of drug delivery systems for anesthetic administration since it is difficult to precisely measure the amount of anesthetic that should be delivered to an individual patient by using conventional computing methods. Fuzzy logic is also being used to develop improved neuroprosthetics for paraplegics. Neuroprosthetics for locomotion use sensors controlled by fuzzy logic systems to electrically stimulate necessary leg muscles and will, ideally, enable the paraplegic patient to walk.

Example Problem 10.26

A fuzzy system is used to categorize people by heart rates. The system is used to help determine which patients have normal resting heart rates, bradychardia, or tachycardia. Bradychardia is a cardiac arrhythmia in which the resting heart rate is less than 60 beats per minute, and tachycardia is defined as a cardiac arrhythmia in which the resting heart rate is greater than 100 beats per minute. A normal heart rate is considered to be in the range of 70–80 beats per minute. What are three linguistic variables that might be used to describe the resting heart rates of the individuals?

Solution

A variety of linguistic variables may be used. The names are important only in that they offer a good description of the categories and problem. *Slow, normal*, and *fast* might be used. Another possibility is simply *bradychardia, normal*, and *tachychardia*.

10.9.2   Artificial Neural Networks

Artificial neural networks (ANN) are the theoretical counterpart of real biological neural networks. The human brain is one of the most sophisticated biological neural networks, consisting of billions of brain cells (i.e., neurons) that are highly interconnected among each other. Such highly interconnected architecture of neurons allows for immense computational power, typically far beyond our most sophisticated computers. The brains of humans, mammals, and even simple invertebrate organisms (e.g., a fly) can easily learn through experience to recognize relevant sensory signals (e.g., sounds and images), and react to changes in the organisms' environment. Artifical neuronal networks are designed to mimic and attempt to replicate the function of real brains.

ANNs are simpler than biological neural networks. A sophisticated ANN contains only a few thousand neurons with several hundred connections. Although simpler than biological neural networks, the aim of ANNs is to build computer systems that have learning, generalized processing, and adaptive capabilities resembling those seen in real brains. Artificial neural networks can learn to recognize certain inputs and to produce a particular output for a given input. Therefore, artificial neural networks are commonly used for pattern detection and classification of biosignals.

ANNs consist of multiple, interconnected neurons. Different types of neurons can be represented in an ANN. Neurons are arranged in a layer, and the different layers of neurons are connected to other neurons and layers. The manner in which the neurons are interconnected determines the architecture of the ANN. There are many different ANN architectures, some of which are best suited for specific applications. Figure 10.34 shows a schematic of a simple ANN with three layers of neurons and a total of six neurons. The first layer is called the *input layer* and has two neurons, which accept the input to the network. The middle layer contains three neurons and is where much of the processing occurs. The *output layer* has one neuron which provides the result of the ANN.

Mathematical equations are used to describe the connections between the neurons. The diagram in Figure 10.35 represents a single neuron and a mathematical method for determining the output of the neuron. The equation for calculating the total input to the neuron is

$$x = (\text{Input}_1 \bullet \text{Weight}_1) + (\text{Input}_2 \bullet \text{Weight}_2) + \text{Bias Weight} \qquad (10.57)$$

The output for the neuron is determined by using a mathematical function, $g(x)$. Threshold functions and nonlinear sigmoid functions are commonly used. The output y of a neuron using the sigmoid function is calculated from the following simple equation:

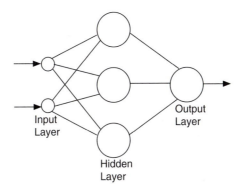

Figure 10.34 Schematic of a simple artificial neural network (ANN) with six neurons and three layers.

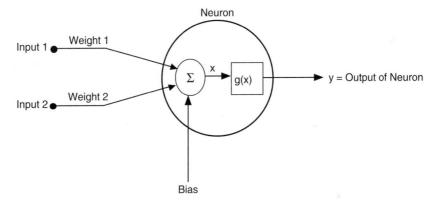

Figure 10.35 Diagram of a single neuron, showing mathematical input and output relationships.

$$y = 1/(1 + e^{-x}) \qquad (10.58)$$

In biosignal processing applications, the inputs to the first layer or input layer of the ANN can be raw data, a preprocessed signal, or extracted features from a biosignal. Raw data is generally a sample from a digitized signal. Preprocessed signals are biosignals that have been transformed, filtered, or processed using some other method before being input to the neural network. Features can also be extracted from biosignals and used as inputs for the neural network. Extracted features might include thresholds, a particular, recurring wave shape, or the period between waveforms.

The ANN must learn to recognize the features or patterns in an input signal, but this is not the case initially. For the ANN to learn, a training process must occur in which the user of the ANN presents the neural network with many different examples of important input. Each example is given to the ANN many times. Over time, after the ANN has been presented with all of the input examples several times, the ANN learns to produce particular outputs for specific inputs.

There are a variety of types of learning paradigms for ANNs. Learning can be broadly divided into two categories: unsupervised learning and supervised learning. In unsupervised learning, the outputs for the given input examples are not known. The ANN must perform a sort of self-organization. During unsupervised learning, the ANN learns to recognize common features in the input examples and produces a specific output for each different type of input. Types of ANNs with unsupervised learning that have been used in biosignal processing include the Hopfield network and self-organizing feature maps networks.

In supervised learning, the desired output is known for the input examples. The output which the ANN produces for a particular input or inputs is compared with the desired output or output function. The desired output is known as the target. The difference between the target and the output of the ANN is calculated mathematically for each given input example. A common training method for supervised learning is backpropagation. The multilayered perceptron trained with backpropagation is a type of a network with supervised learning that has been used for biosignal processing.

Backpropagation is an algorithm that attempts to minimize the error of the ANN. The error of the ANN can be regarded as simply the difference between the output of the ANN for an input example and the target for that same input example. Backpropagation uses a gradient-descent method to minimize the network error. In other words, the network error is gradually decreased down an error slope that is in some respects similar to how a ball rolls down a hill. The name *backpropagation* refers to the way by which the ANN is changed to minimize the error. Each neuron in the network is "credited" with a portion of the network error. The relative error for each neuron is then determined, and the connection strengths between the neurons are changed to minimize the errors. The weights, such as those that were shown in Figure 10.35, represent the connection strengths between neurons. The calculations of the neuron errors and weight changes propagate backwards through the ANN from the output neurons to the input neurons. Backpropagation is the method of finding the optimum weight values that produce the smallest network error.

ANNs are well suited for a variety of biosignal processing applications and may be used as a tool for nonlinear statistical analysis. They are often used for pattern recognition and classification. In addition, ANNs have been shown to perform faster and more accurately than conventional methods for signals that are highly complex or contain high levels of noise. ANNs also have the ability to solve problems that have no algorithmic solution—in other words, problems for which a conventional computer program cannot be written. Since ANNs learn, algorithms are not required to solve problems.

As advances are made in artificial intelligence techniques, ANNs are being used more extensively in biosignal processing and biomedical instrumentation. The viability of ANNs for applications ranging from the analysis of ECG and EEG signals to the interpretation of medical images and the diagnosis of a variety of diseases has been investigated. In neurology, research has been conducted by using ANNs to characterize brain defects that occur in disorders such as epilepsy, Parkinson's disease, and Alzheimer's disease. ANNs have also been used to characterize and classify ECG signals of cardiac arrhythmias. One study used an ANN in the emergency room to diagnose heart attacks. The results of the study showed that, overall, the ANN was able to diagnose heart attacks better than the emergency room physicians were. ANNs have the advantage of not being affected by fatigue, distractions, or emotional stress. As artificial intelligence technologies advance, ANNs may provide a superior tool for many biosignal processing tasks.

Example Problem 10.27

A neuron in a neural network has three inputs and uses a sigmoid function to calculate the output of the neuron. The three values of the inputs are 0.1, 0.9, and 0.1. The weights associated with these three inputs are 0.39, 0.72, and 0.26, and the bias weight is 0.48 after training. What is the output of the neuron?

Solution

Using Equation 10.57 to calculate the relative sum of the inputs gives

$$x = (\text{Input}_1 \bullet \text{Weight}_1) + (\text{Input}_2 \bullet \text{Weight}_2) + (\text{Input}_3 \bullet \text{Weight}_3) + \text{Bias Weight}$$
$$= (0.1)0.39 + (0.9)0.72 + (0.1)0.24 + 0.48$$
$$= 1.19$$

The output of the neuron is calculated using Equation 10.58

$$y = 1/(1 + e^{-x}) = 1/(1 + e^{-1.19}) = 0.77 \qquad \blacksquare$$

EXERCISES

1. What types of biosignals would the nerves in your legs produce during a sprint across the street?
2. What types of biosignals can be recorded with an EEG? Describe in terms of both origins and characteristics of the signal.
3. Describe the biosignal that the electrical activity of a normal heart would generate during a bicycle race.
4. A 16-bit A/D converter is used to convert an analog biosignal with a minimum voltage of $-30\,\text{mV}$ and a maximum voltage of $90\,\text{mV}$. What is the sensitivity?
5. An EMG recording of skeletal muscle activity has been sampled at 200–250 Hz and correctly digitized. What is the highest frequency of interest in the original EMG signal?
6. Two signals, $x_1(t)$ and $x_2(t)$, have the magnitude spectrum shown in Figure 10.36. Find the Nyquist rate for:
   a) $x_1(t)$

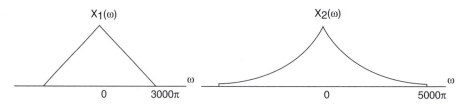

Figure 10.36

    b) $x_2(t)$
    c) $x(t) = x_1(t) * x_2(t)$ (Hint: apply the convolution theorem)
  7. Consider the signal

$$x(t) = 3 + \sin(2\pi 100t) + \cos(2\pi 250t + \pi/3)$$

Find the Nyquist frequency.

  8. A sinusoid with the frequency of 125 kHz is sampled at 70,000 samples per second. What is the apparent frequency of the sampled signal?

  9. An electroencephalographic (EEG) signal has a maximum frequency of 300 Hz. The signal is sampled and quantized into a binary sequence by an A/D converter.
    a) Determine the sampling rate if the signal is sampled at a rate 50% higher than the Nyquist rate.
    b) The samples are quantized into 2048 levels. How many binary bits are required for each sample

 10. Find the exponential Fourier series for the signal shown in Fig. 10.37a.

 11. Find the exponential Fourier series for the signal shown in Fig. 10.37b.

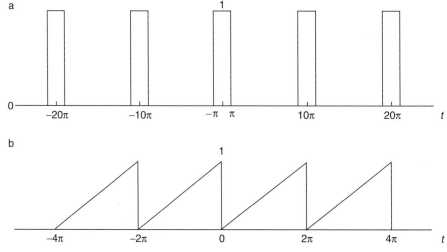

Figure 10.37

12. $f(t)$ is a periodic signal shown in Figure 10.38. Find its trigonometric Fourier series.

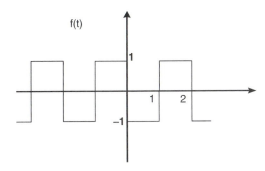

Figure 10.38

13. Consider the following trigonometric Fourier series:

$$f(t) = 3 + 3\cos(t) + 2\cos(2t) + 4\sin(2t) - 4\left(\frac{e^{j4t} + e^{-j4t}}{2}\right)$$

Write $f(t)$ in its compact trigonometric Fourier series form.

14. Explain why the exponential Fourier series requires negative frequencies.

15. Find the Fourier transform of
   a) $u(t)$
   b) $e^{-at}u(t)$
   c) $\cos(at)u(t)$

16. Find the Fourier transform of $f_1(t) = e^{-3t}u(t)$.

17. Find the Fourier transform of $f(t) = e^{-3|t|}$ and sketch its time and frequency domain representations. Hint: Find a few points on the curve by substituting values for the variable.

18. Prove the shift property of the Fourier transform.

19. Given $x(t) = e^{-at}u(t)$ and $h(t) = e^{-bt}u(t)$ where a and b are constants greater than zero, explain why it would be easier to evaluate the convolution $x(t) * h(t)$ in the frequency domain.

20. A brief current pulse of duration 50 ms and amplitude 1 mA is presented to a cell membrane with time constant 10 ms. Find the cell membrane voltage output.

21. The ion exchange process of a cell is estimated to have the following impulse response: $h(t) = e^{-4t}u(t)$.
   a) Explain what type of general information would be available to the researcher if this estimation of $h(t)$ were accurate.
   b) If sodium ions are injected into the system for two seconds in the form of a brief pulse approximated by the following equation, $x(t) = 3u(t) - 3u(t - 2)$, how would the cell respond to (e.g., pump out ions) such input? Find the answer using time-domain procedures. Hint: Convolve the input and the impulse response.

22. An ECG recording of the electrical activity of the heart during ventricular fibrillation is digitized and the signal begins with the following data sequence: -90.0, 10.0, -12.0, -63.0, 7.0, -22.0. The units of the data sequence are given in mV. What is the z transform of this data sequence of the biosignal?

23. For the systems described by the following equations, determine which of the systems is linear and which is not.
   a) $\frac{dy}{dt} + 2y(t) = f^2(t)$
   b) $\frac{dy}{dt} + 3ty(t) = t^2f(t)$
   c) $\frac{dy}{dt} + 2y(t) = f(t)\frac{df}{dt}$
   d) $y(t) = \int_{-\infty}^{t} f(\tau)d\tau$

24. Examine the characteristics of the digital filter

$$y(k) = \frac{1}{4}x(k) + \frac{1}{4}x(k-1) + \frac{1}{2}y(k-1)$$

Find the impulse response, $H(z)$ and $H'(\Omega)$. Use MATLAB to calculate and plot $|H'(\Omega)|$ for $0 < \Omega < \pi$. Observe the difference between this filter and the filter in Example Problem 10.12. Why is this a better low-pass filter? What is the output if the input sequence is $x(k) = 100\sin\left(\frac{\pi}{2}k + \frac{\pi}{8}\right)$? What is the output if the input sequence is $x(k) = 100u(k)$?

25. Find the z transform of
   a) $x(k) = u(k)$
   b) $x(k) = a^k u(k)$
   c) $x(k) = \cos(b \cdot k)u(k)$

26. Find the z transform of the following:
   a) $x[k] = \left(\frac{1}{2}\right)^k u(k)$
   b) $x[k] = (\cos \Omega k)u[k]$

27. Find the first four outputs of the following discrete system

$$y[k] - 3y[k-1] + 2y[k-2] = f[k-1]$$
$$\text{if } y[-1] = 2,\ y[-2] = 3,\ \text{and } f[k] = 3^k u[k].$$

28. Find the first four outputs of the following discrete system

$$y[k] - 2y[k-1] + 2y[k-2] = f[k-2]$$
$$\text{if } y[-1] = 1,\ y[-2] = 0,\ \text{and } f[k] = u[k].$$

29. In MATLAB, design a routine to show that averaging random noise across many trials approaches zero as the number of trials increases.

30. Accurate measurements of blood glucose levels are needed for the proper treatment of diabetes. Glucose is a primary carbohydrate, which circulates throughout the body and serves as an energy source for cells. In normal individuals the hormone insulin regulates the levels of glucose in the blood by promoting glucose transport out of the blood to skeletal muscle and fat

tissues. Diabetics suffer from improper management of glucose levels, and the levels of glucose in the blood can become too high. Describe how fuzzy logic might be used in the control of a system for measuring blood glucose levels. What advantages would the fuzzy logic system have over a more conventional system?

31. Describe three different biosignal processing applications for which artificial neural networks might be used. Give at least two advantages of artificial neural networks over traditional biosignal processing methods for the applications you listed.

32. The fuzzy sets in Example Problem 10.26 have been calibrated so that a person with a resting heart rate of 95 beats per minute has a 75% degree of membership in the normal category and a 25% degree of membership in the tachycardia category. A resting heart rate of 65 beats per minute indicates a 95% degree of membership in the normal category. Draw a graph of the fuzzy sets.

SUGGESTED READING

Akay, M. (1994). *Biomedical Signal Processing*. Academic, San Diego, CA.

Akay, M. (Ed.) (1998). *Time Frequency and Wavelets in Biomedical Signal Processing*. IEEE, New York.

Bauer, P., Nouak, S. and Winkler, T. (1996). A Brief Course in Fuzzy Logic and Fuzzy Control, (http://*www.flll*.uni-linz.ac.at/fuzzy). Fuzzy Logic Laboratorium Linz-Hagenberg, Linz, Austria.

Bishop, C.M. (1995) *Neural Networks for Pattern Recognition*. Oxford Univ. Press, New York.

Bruce, E.N. (2000). *Biomedical Signal Processing and Signal Modeling*. Wiley-Interscience, New York.

Ciaccio, E.J., Dunn, S.M. and Akay, M. (1993). Biosignal pattern recognition and interpretation systems: Part 1 of 4: Fundamental concepts. *IEEE Eng. Med. & Biol.* **12**, 810–897.
Ciaccio, E.J., Dunn, S.M. and Akay, M. (1993). Biosignal pattern recognition and interpretation systems: Part 2 of 4: Methods for feature extraction and selection. *IEEE Eng. Med. & Biol.* **12**, 106–113.

Ciaccio, E.J., Dunn, S.M. and Akay, M. (1994). Biosignal pattern recognition and interpretation systems: Part 3 of 4: Methods of classification. *IEEE Eng. Med. & Biol.* **12**, 269–279.

Ciaccio, E.J., Dunn, S.M. and Akay, M. (1994). Biosignal pattern recognition and interpretation systems: Part 4 of 4: Review of applications. *IEEE Eng. Med. & Biol.* **13**, 269–273.

Cohen, A. (1986). *Biomedical Signal Processing: Volume I Time and Frequency Domain Analysis*. CRC, Boca Raton, FL.

Cohen, A. (1986). *Biomedical Signal Processing: Volume II Compression and Automatic Recognition*. CRC, Boca Raton, FL.

Dempster, J. (1993). *Computer Analysis of Electrophysiological Signals*. Academic, San Diego, CA.

Devasahayam, S.R. (2000). *Signals and Systems in Biomedical Engineering: Signal Processing and Physiological Systems Modeling*. Kluwer Academic, New York.

Haykin, S. (1994). *Neural Networks—A Comprehensive Foundation*. Macmillan College, New York.

Northrop, R.B. (2003). *Signals and Systems Analysis in Biomedical Engineering*. CRC, Boca Raton, FL.

Onaral, B. (Ed.) (1995). Biomedical signal analysis. In *The Biomedical Engineering Handbook* (J.D. Bronzino, Ed.). CRC, Boca Raton, FL.

Oppenheim, A.V. and Schafer, R.W. (1975). *Digital Signal Processing*. Prentice-Hall, Englewood Cliffs, NJ.

Oppenheim, A.V., Willsky, A.S. and Young, I.T. (1983). *Signals and Systems*. Prentice-Hall, Englewood Cliffs, NJ.

Roberts R.A. and Mullis, C.T. (1987). *Digital Signal Processing*. Addison-Wesley, Reading, MA.

Smith, M. (1996). *Neural Networks for Statistical Modeling*. International Thomson Computer, Boston, MA.

Stearns, S.D. and David, R.A. (1993). *Signal Processing Algorithms in Fortran and C*. Prentice-Hall, Englewood Cliffs, NJ.

Thompkins, W.J. (1993). *Biomedical Digital Signal Processing*. Prentice-Hall, Englewood Cliffs, NJ.

Williams, C.S. (1993). *Designing Digital Filters*. Prentice-Hall, Englewood Cliffs, NJ.

Zadeh, L.A. (1987). *Fuzzy Sets and Applications*. John Wiley, New York.

Ziemer, R.E., Tranter, W.H. and Fannin, D.R. (1993). *Signals and Systems: Continuous and Discrete*, 3rd Ed. Macmillan, New York.

11 BIOELECTRIC PHENOMENA

John Enderle, PhD*

Chapter Contents

*With contributions by Joseph Bronzino.

11.7 Model of the Whole Neuron

Exercises

Suggested Reading

At the conclusion of this chapter, students will be able to:

- Describe the history of bioelectric phenomena.
- Qualitatively explain how signaling occurs among neurons.
- Calculate the membrane potential due to one or more ions.
- Compute the change in membrane potential due to a current pulse through a cell membrane.
- Describe the change in membrane potential with distance after stimulation.
- Explain the voltage clamp experiment and an action potential.
- Simulate an action potential using the Hodgkin–Huxley model.

11.1   INTRODUCTION

Chapter 3 briefly described the nervous system and the concept of a neuron. Here the description of a neuron is extended by examining its properties at rest and during excitation. The concepts introduced here are basic and allow further investigation of more sophisticated models of the neuron or groups of neurons by using GENESIS (a general neural simulation program—see suggested reading by J.M. Bower and D. Beeman) or extensions of the Hodgkin–Huxley model by using more accurate ion channel descriptions and neural networks. The models introduced here are an important first step in understanding the nervous system and how it functions.

Models of the neuron presented in this chapter have a rich history of development. This history continues today as new discoveries unfold that supplant existing theories and models. Much of the physiological interest in models of a neuron involves the neuron's use in transferring and storing information, whereas much engineering interest involves the neuron's use as a template in computer architecture and neural networks. To fully appreciate the operation of a neuron, it is important to understand the properties of a membrane at rest by using standard biophysics, biochemistry, and electric circuit tools. In this way, a more qualitative awareness of signaling via the generation of the action potential can be better understood.

The Hodgkin and Huxley theory that was published in 1952 described a series of experiments that allowed the development of a model of the action potential. This work was awarded a Nobel prize in 1963 (shared with John Eccles) and is covered in Section 11.6. It is reasonable to question the usefulness of covering the Hodgkin–Huxley model in a textbook today given all of the advances since 1952. One simple answer is that this model is one of the few timeless classics and should be covered. Another is that all current, and perhaps future, models have their roots in this model.

Section 11.2 describes a brief history of bioelectricity and can be easily omitted on first reading of the chapter. Following this, Section 11.3 describes the structure and provides a qualitative description of a neuron. Biophysics and biochemical tools useful in understanding the properties of a neuron at rest are presented in Section 11.4. An equivalent circuit model of a cell membrane at rest consisting of resistors, capacitors, and voltage sources is described in Section 11.5. Finally, Section 11.6 describes the Hodgkin–Huxley model of a neuron and includes a brief description of their experiments and the mathematical model describing an action potential.

11.2 HISTORY

11.2.1 The Evolution of a Discipline: The Galvani–Volta Controversy

In 1791, an article appeared in the *Proceedings of the Bologna Academy* reporting experimental results that, it was claimed, proved the existence of animal electricity. This now famous publication was the work of Luigi Galvani. At the time of its publication, this article caused a great deal of excitement in the scientific community and sparked a controversy that ultimately resulted in the creation of two separate and distinct disciplines: electrophysiology and electrical engineering. The controversy arose from the different interpretations of the data presented in this article. Galvani was convinced that the muscular contractions he observed in frog legs were due to some form of electrical energy emanating from the animal. On the other hand, Allesandro Volta, a professor of physics at the University of Padua, was convinced that the "electricity" described in Galvani's experiments originated not from the animal but from the presence of the dissimilar metals used in Galvani's experiments. Both of these interpretations were important. The purpose of this section, therefore, is to discuss them in some detail, highlighting the body of scientific knowledge available at the time these experiments were performed, the rationale behind the interpretations that were formed, and their ultimate effect.

11.2.2 Electricity in the Eighteenth Century

Before 1800, a considerable inventory of facts relating to electricity in general and bioelectricity in particular had accumulated. The Egyptians and Greeks had known that certain fish could deliver substantial shocks to an organism in their aqueous environment. Static electricity had been discovered by the Greeks, who produced it by rubbing resin (amber or, in Greek, *elektron*) with cat's fur or by rubbing glass with silk. For example, Thales of Miletus reported in 600 BC that a piece of amber, when vigorously rubbed with a cloth, responded with an "attractive power." Light particles such as chaff, bits of papyrus, and thread jumped to the amber from a distance and were held to it. The production of static electricity at that time became associated with an aura.

More than two thousand years elapsed before the English physician William Gilbert picked up where Thales left off. Gilbert showed that not only amber but also glass, agate, diamond, sapphire, and many other materials when rubbed exhibited the same attractive power described by the Greeks. However, Gilbert did not report that particles could also be repelled. It was not until a century later that electrostatic repulsion was noted by Charles DuFay (1698–1739) in France.

The next step in the progress of electrification was an improvement of the friction process. Rotating rubbing machines were developed to give continuous and large-scale production of electrostatic charges. The first of these frictional electric machines was developed by Otto von Guericke (1602–1685) in Germany. In the eighteenth century, electrification became a popular science and experimenters discovered many new attributes of electrical behavior. In England, Stephen Gray (1666–1736) proved that electrification could flow hundreds of feet through ordinary twine when suspended by silk threads. Thus, he theorized that electrification was a "fluid." Substituting metal wires for the support threads, he found that the charges would quickly dissipate. Thus, the understanding that different materials can either conduct or insulate began to take shape. The "electrics," such as silk, glass, and resin, held a charge. The "non-electrics," such as metals and water, conducted charges. Gray also found that electrification could be transferred by proximity of one charged body to another without direct contact. This was evidence of electrification by induction, a principle that was used later in machines that produced electrostatic charges.

In France, Charles F. DuFay, a member of the French Academy of Science, was intrigued by Gray's experiments. DuFay showed by extensive tests that practically all materials, with the exception of metals and those too soft or fluid to be rubbed, could be electrified. Later, however, he discovered that if metals were insulated they could hold the largest electric charge of all. DuFay found that rubbed glass would repel a piece of gold leaf whereas rubbed amber, gum, or wax attracted it. He concluded that there were two kinds of electric "fluids," which he labeled "vitreous" and "resinous." He found that while unlike charges attracted each other, like charges repelled. This indicated that there were two kinds of electricity.

In the American colonies, Benjamin Franklin (1706–1790) became interested in electricity and performed experiments that led to his hypothesis regarding the "one-fluid theory." Franklin stated that there was but one type of electricity and that the electrical effects produced by friction reflected the separation of electric fluid so that one body contained an excess and the other a deficit. He argued that "electrical fire" is a common element in all bodies and is normally in a balanced or neutral state. Excess or deficiency of charge, such as that produced by the friction between materials, created an imbalance. Electrification by friction was, thus, a process of separation rather than a creation of charge. By balancing a charge gain with an equal charge loss, Franklin had implied a law, namely that the quantity of the electric charge is conserved. Franklin guessed that when glass was rubbed the excess charge appeared on the glass, and he called that positive electricity. He thus established the direction of conventional current from positive to negative. It is now known that the electrons producing a current move in the opposite direction.

Out of this experimental activity came an underlying philosophy or law. Up to the end of the eighteenth century, the knowledge of electrostatics was mainly qualitative. There were means for detection, but not for measurement, and relationships between the charges had not been formulated. The next step was to quantify the phenomena of electrostatic charge forces.

For this determination, the scientific scene shifted back to France and the engineer-turned-physicist, Charles A. Coulomb (1726–1806). Coulomb demonstrated that a force is exerted when two charged particles are placed in the vicinity of one another. However, he went a step beyond experimental observation by deriving a general relationship that completely expressed the magnitude of this force. His inverse-square law for the force of attraction or repulsion between charged bodies became one of the major building blocks in understanding the effect of a fundamental property of matter—charge. However, despite this wide array of discoveries, it is important to note that before the time of Galvani and Volta, there was no source that could deliver a continuous flow of electric fluid, a term that we now know implies both charge and current.

In addition to a career as statesman, diplomat, publisher, and signer of the Declaration of Independence and the Constitution, Franklin was an avid experimenter and inventor. In 1743 at the age of 37, Franklin witnessed with excited interest a demonstration of static electricity in Boston and resolved to pursue the strange effects with investigations of his own. Purchasing and devising various apparati, Franklin became an avid electrical enthusiast. He launched into many years of experiments with electrostatic effects.

Franklin the scientist is most popularly known for his kite experiment during a thunderstorm in June 1752 in Philadelphia. Although various European investigators had surmised the identity of electricity and lightning, Franklin was the first to prove by an experimental procedure and demonstration that lightning was a giant electrical spark. Having previously noted the advantages of sharp metal points for drawing "electrical fire," Franklin put them to use as "lightning rods." Mounted vertically on rooftops they would dissipate the thundercloud charge gradually and harmlessly to the ground. This was the first practical application in electrostatics.

Franklin's work was well received by the Royal Society in London. The origin of such noteworthy output from remote and colonial America made Franklin especially marked. In his many trips to Europe as statesman and experimenter, Franklin was lionized in social circles and eminently regarded by scientists.

11.2.3 Galvani's Experiments

Against such a background of knowledge of the "electric fluid" and the many powerful demonstrations of its ability to activate muscles and nerves, it is readily understandable that biologists began to suspect that the "nervous fluid" or the "animal spirit" postulated by Galen to course in the hollow cavities of the nerves and mediate muscular contraction, and indeed all the nervous functions, was of an electrical nature. Galvani, an obstetrician and anatomist, was by no means the first to hold such a view, but his experimental search for evidence of the identity of the electric and nervous fluids provided the critical breakthrough.

Speculations that the muscular contractions in the body might be explained by some form of animal electricity were common. By the eighteenth century, experimenters were familiar with the muscular spasms of humans and animals that were subjected to the discharge of electrostatic machines. As a result, electric shock was viewed as a muscular stimulant. In searching for an explanation of the resulting muscular contractions, various anatomical experiments were conducted to study the possible relationship of "metallic contact" to the functioning of animal tissue. In 1750, Johann Sulzer (1720–1779), a professor of physiology at Zurich, described a chance discovery that an unpleasant acid taste occurred when the tongue was put between two strips of different metals, such as zinc and copper, whose ends were in contact. With the metallic ends separated, there was no such sensation. Sulzer ascribed the taste phenomenon to a vibratory motion set up in the metals that stimulated the tongue, and he used other metals with the same results. However, Sulzer's reports went unheeded for a half-century until new developments called attention to his findings.

The next fortuitous and remarkable discovery was made by Luigi Galvani (1737–1798), descendant of a very large Bologna family, who at age 25 was made professor of anatomy at the University of Bologna. Galvani had developed an ardent interest in electricity and its possible relation to the activity of the muscles and nerves. Dissected frog legs were convenient specimens for investigation, and in his laboratory Galvani used them for studies of muscular and nerve activity. In these experiments, he and his associates were studying the responses of the animal tissue to various stimulations. In this setting, Galvani observed that, while a freshly prepared frog leg was being probed by a scalpel, the leg jerked convulsively whenever a nearby frictional electrical machine gave off sparks.

Galvani, in writing of his experiments said:

> I had dissected and prepared a frog, and laid it on a table, on which there was an electrical machine. It so happened by chance that one of my assistants touched the point of his scalpel to the inner crural nerve of the frog; the muscles of the limb were suddenly and violently convulsed. Another of those who were helping to make the experiments in electricity thought that he noticed this happening only at the instant a spark came from the electrical machine. He was struck with the novelty of the action. I was occupied with other things at the time, but when he drew my attention to it I immediately repeated the experiment. I touched the other end of the crural nerve with the point of my scalpel, while my assistant drew sparks from the electrical machine. At each moment when sparks occurred, the muscle was seized with convulsions.

With an alert and trained mind, Galvani designed an extended series of experiments to resolve the cause of the mystifying muscle behavior. On repeating the experiments, he found that touching the muscle with a metallic object while the specimen lay on a metal plate provided the condition that resulted in the contractions.

Having heard of Franklin's experimental proof that a flash of lightning was of the same nature as the electricity generated by electric machines, Galvani set out to determine whether atmospheric electricity might produce the same results observed with his electrical machine. By attaching the nerves of frog legs to aerial wires and the feet to another electrical reference point known as electrical ground, he noted the same muscular response during a thunderstorm that he observed with the electrical machine. It was another chance observation during this experiment that led to further inquiry, discovery, and controversy.

Galvani also noticed that the prepared frogs, which were suspended by brass hooks through the marrow and rested against an iron trellis, showed occasional convulsions regardless of the weather. In adjusting the specimens, he pressed the brass hook against the trellis and saw the familiar muscle jerk occurring each time he completed the metallic contact. To check whether this jerking might still be from some atmospheric effect, he repeated the experiment inside the laboratory. He found that the specimen, laid on an iron plate, convulsed each time the brass hook in the spinal marrow touched the iron plate. Recognizing that some new principle was involved, he varied his experiments to find the true cause. In the process, he found that by substituting glass for the iron plate, the muscle response was not observed but using a silver plate restored the muscle reaction. He then joined equal lengths of two different metals and bent them into an arc. When the tips of this bimetallic arc touched the frog specimens, the familiar muscular convulsions were obtained. As a result, he concluded that not only was metal contact a contributing factor but also that the intensity of the convulsion varied according to the kinds of metals joined in the arc pair.

Galvani was now faced with trying to explain the phenomena he was observing. He had encountered two electrical effects for which his specimens served as indicator—one from the sparks of the electrical machine and the other from the contact of dissimilar metals. The electricity responsible for the action resided either in the anatomy of the specimens with the metals serving to release it or the effect was produced by the bimetallic contact with the specimen serving only as an indicator.

Galvani was primarily an anatomist and seized on the first explanation. He ascribed the results to "animal electricity" that resided in the muscles and nerves of the organism itself. Using a physiological model, he compared the body to a Leyden jar in which the various tissues developed opposite electrical charges. These charges flowed from the brain through nerves to the muscles. Release of electrical charge by metallic contact caused the convulsions of the muscles. "The idea grew," he wrote, "that in the animal itself there was an indwelling electricity. We were strengthened in such a supposition by the assumption of a very fine nervous fluid that during the phenomena flowed into the muscle from the nerve, similar to the electric current of a Leyden Jar." Galvani's hypothesis reflected the prevailing view of his day that ascribed the body activation to a flow of "spirits" residing in the various body parts.

In 1791, Galvani published his paper, *De Viribus Electricitatis In Motu Musculari*, in the proceedings of the Academy of Science in Bologna. This paper set forth his experiments and conclusions. Galvani's report created a sensation and implied to many a possible revelation of the mystery of the life force. Men of science and laymen alike, both in Italy and elsewhere in Europe, were fascinated and challenged by these findings. However, no one pursued Galvani's findings more assiduously and used them as a stepping stone to greater discovery than Allesandro Volta.

11.2.4   Volta's Interpretation

Galvani's investigations aroused a virtual furor of interest. Wherever frogs were found, scientists repeated his experiments with routine success. Initially, Galvani's explanation for the muscular contractions was accepted without question—even by the prominent physician Allesandro Volta who had received a copy of Galvani's paper and verified the phenomenon.

Volta was a respected scientist in his own right. At age 24, Volta published his first scientific paper, *On the Attractive Force of the Electric Fire*, in which he speculated about the similarities between electric force and gravity. Engaged in studies of physics and mathematics and busy with experimentation, Volta's talents were so evident that before the age of 30 he was named the professor of physics at the Royal School of Como. Here he made his first important contribution to science with the invention of the electrophorus or "bearer of electricity." This was the first device to provide a replenishable supply of electric charge by induction rather than by friction.

In 1782, Volta was called to the professorship of physics at the University of Padua. There he made his next invention, the condensing electrophorus, a sensitive instrument for detecting electric charge. Earlier methods of charge detection employed the "electroscope," which consisted of an insulated metal rod that had pairs of silk threads, pith balls, or gold foil suspended at one end. These pairs diverged by repulsion when the rod was touched by a charge. The amount of divergence indicated the strength of the charge and thus provided quantitative evidence for Coulomb's law.

By combining the electroscope with his electrophorus, Volta provided the scientific community with a detector for minute quantities of electricity. Volta continued to innovate, and made his condensing electroscope a part of a mechanical balance that made it possible to measure the force of an electric charge against the force of gravity. This instrument was called an electrometer and was of great value in Volta's later investigations of the electricity created by contact of dissimilar metals.

Volta expressed immediate interest on learning of Galvani's 1791 report to the Bologna Academy on the "Forces of Electricity in Their Relation to Muscular Motion." Volta set out quickly to repeat Galvani's experiments and initially confirmed Galvani's conclusions on "animal electricity" as the cause of the muscular reactions. Along with Galvani, he ascribed the activity to an imbalance between electricity of the muscle and that of the nerve, which was restored to equilibrium when a metallic connection was made. On continuing his investigations, however, Volta began to have doubts about the correctness of that view. He found inconsistencies in the balance theory. In his experiments, muscles would convulse only when the nerve was in the electrical circuit made by metallic contact.

In an effort to find the true cause of the observed muscle activity, Volta went back to an experiment previously performed by Sulzer. When Volta placed a piece of tinfoil on the tip and a silver coin at the rear of his tongue and connected the two with a copper wire, he got a sour taste. When he substituted a silver spoon for the coin and omitted the copper wire, he got the same result as when he let the handle of the spoon touch the foil. When using dissimilar metals to make contact between the tongue and the forehead, he got a sensation of light. From these results, Volta came to the conclusion that the sensations he experienced could not originate from the metals as conductors but must come from the ability of the dissimilar metals themselves to generate electricity.

After two years of experimenting, Volta published his conclusions in 1792. While crediting Galvani with a surprising original discovery, he disagreed with him on what produced the effects. By 1794, Volta had made a complete break with Galvani. He became an outspoken opponent of the theory of animal electricity and proposed the theory of "metallic electricity." Galvani, by nature a modest individual, avoided any direct confrontation with Volta on the issue and simply retired to his experiments on animals.

Volta's conclusive demonstration that Galvani had not discovered animal electricity was a blow from which the latter never recovered. Nevertheless, he persisted in his belief in animal electricity and conducted his third experiment, which definitely proved the existence of bioelectricity. In this experiment, he held one foot of the frog nerve–muscle preparation and swung it so that the vertebral column and the sciatic nerve touched the muscles of the other leg. When this occurred or when the vertebral column was made to fall on the thigh, the muscles contracted vigorously. According to most historians, it was his nephew Giovanni Aldini (1762–1834) who championed Galvani's cause by describing this important experiment in which he probably collaborated. The experiment conclusively showed that muscular contractions could be evoked without metallic conductors. According to Fulton and Cushing, Aldini wrote:

> Some philosophers, indeed, had conceived the idea of producing contractions in a frog without metals; and ingenious methods, proposed by my uncle Galvani, induced me to pay attention to the subject, in order that I might attain to greater simplicity. He made me sensible of the importance of the experiment and therefore I was long ago inspired with a desire of discovering that interesting process. It will be seen in the Opuscoli of Milan (No. 21), that I showed publicly, to the Institute of Bologna, contractions in a frog without the aid of metals so far back as the year 1794. The experiment, as described in a memoir addressed to M. Amorotti [sic] is as follows: I immersed a prepared frog in a strong solution of muriate of soda. I then took it from the solution, and, holding one extremity of it in my hand, I suffered the other to hang freely down. While in this position, I raised up the nerves with a small glass rod, in such a manner that they did not touch the muscles. I then suddenly removed the glass rod, and every time that the spinal marrow and nerves touched the muscular parts, contractions were excited. Any idea of a stimulus arising earlier from the action of the salt, or from the impulse produced by the fall of the nerves, may be easily removed. Nothing will be necessary but to apply the same nerves to the muscles of another prepared frog, not in a Galvanic circle; for, in this case, neither the salt, nor the impulse even if more violent, will produce muscular motion.

The claims and counterclaims of Volta and Galvani developed rival camps of supporters and detractors. Scientists swayed from one side to the other in their opinions and loyalties. Although the subject was complex and not well understood, it was on the verge of an era of revelation. The next great contribution to the field was made by Carlo Matteucci, who both confirmed Galvani's third experiment and made a new discovery. Matteucci showed that the action potential precedes the contraction of skeletal muscle. In confirming Galvani's third experiment, which demonstrated the injury potential, Matteucci noted,

> I injure the muscles of any living animal whatever, and into the interior of the wound I insert the nerve of the leg, which I hold, insulated with glass tube. As I move this nervous filament in the interior of the wound, I see immediately strong contractions in the leg. To always obtain them, it is necessary that one point of the nervous filament touches the depths of the wound, and that another point of the same nerve touches the edge of the wound.

By using a galvanometer, Matteucci found that the difference in potential between an injured and uninjured area was diminished during a tetanic contraction. The study of this phenomenon occupied the attention of all succeeding electrophysiologists. More than this, however, Matteucci made another remarkable discovery—that a transient bioelectric event, now designated the action potential, accompanies the contraction of intact skeletal muscle. He demonstrated this by showing that a contracting muscle is able to stimulate a nerve that, in turn, causes contraction of the muscle it innervates. The existence of a bioelectric potential was established through the experiments of Galvani and Matteucci. Soon thereafter, the presence of an action potential was discovered in cardiac muscle and nerves.

Volta, on the other hand, advocated that the source of the electricity was due to the contact of the dissimilar metals only, with the animal tissue acting merely as the indicator. His results differed substantially depending on the pairs of metals used. For example, Volta found that the muscular reaction from dissimilar metals increased in vigor depending on the metals that were used.

In an effort to obtain better quantitative measurements, Volta dispensed with the use of muscles and nerves as indicators. He substituted instead his "condensing electroscope." He was fortunate in the availability of this superior instrument because the contact charge potential of the dissimilar metals was minute, far too small to be detected by the ordinary gold-leaf electroscope. Volta's condensing electroscope used a stationary disk and a removable disk separated by a thin insulating layer of shellac varnish. The thinness of this layer provided a large capacity for accumulation of charge. When the upper disk was raised after being charged, the condenser capacity was released to give a large deflection of the gold leaves.

Volta proceeded systematically to test the dissimilar metal contacts. He made disks of various metals and measured the quantity of the charge on each disk combination by the divergence of his gold-foil condensing electroscope. He then determined whether the charge was positive or negative by bringing a rubbed rod of glass or resin near the electroscope. The effect of the rod on the divergence of the gold foil indicated the polarity of the charge.

Volta's experiments led him toward the idea of an electric force or electrical "potential." This, he assumed, resided in contact between the dissimilar metals. As Volta experimented with additional combinations, he found that an electrical potential also existed when there was contact between the metals and some fluids. As a result, Volta added liquids, such as brine and dilute acids, to his conducting system and classified the metal contacts as "electrifiers of the first class" and the liquids as electrifiers of the "second class."

Volta found that there was only momentary movement of electricity in a circuit composed entirely of dissimilar metals. However, when he put two dissimilar metals in contact with a separator soaked with a saline or acidified solution, there was a steady indication of potential. In essence, Volta was assembling the basic elements of an electric battery—two dissimilar metals and a liquid separator. Furthermore, he found that the overall electric effect could be enlarged by multiplying the elements. Thus, by stacking metal disks and the moistened separators vertically he constructed an "electric pile," the first electric battery. This was the most practical discovery of his career.

11.2.5 The Final Result

Considerable time passed before true explanations became available for what Galvani and Volta had done. Clearly, both demonstrated the existence of a difference in electric potential—but what produced it eluded them. The potential difference present in the experiments carried out by both investigators is now clearly understood. Although Galvani thought that he had initiated muscular contractions by discharging animal electricity resident in a physiological capacitor consisting of the nerve (inner conductor) and muscle surface (outer conductor), it is now known that the stimulus consists of an action potential which in turn causes muscular contractions.

It is interesting to note that the fundamental unit of the nervous system—the neuron—has an electric potential between the inside and outside of the cell even at rest. This membrane resting potential is continually affected by various inputs to the cell. When a certain potential is reached, an action potential is generated along its axon to all of its distant connections. This process underlies the communication mechanisms of the nervous system. Volta's discovery of the electrical battery provided the scientific community with the first steady source of electrical potential, which when connected in an electric circuit consisting of conducting materials or liquids results in the flow of electrical charge (i.e., electrical current). This device launched the field of electrical engineering.

11.3 NEURONS

A reasonable estimate of the human brain is that it contains about 10^{12} neurons partitioned into fewer than 1000 different types in an organized structure of rather uniform appearance. Though not important in this chapter, it is important to note that there are two classes of neuron: the nerve cell and the neuroglial cell. Even though there are 10 to 50 times as many neuroglial cells as nerve cells in the brain, attention is focused here on the nerve cell since the neuroglial cells are not involved in signaling and primarily provide a support function for the nerve cell. Therefore, the terms *neuron* and *nerve cell* are used interchangeably since the primary focus here is to better understand the signaling properties of a neuron. Overall, the complex abilities of the brain are best described by virtue of a neuron's interconnections with other

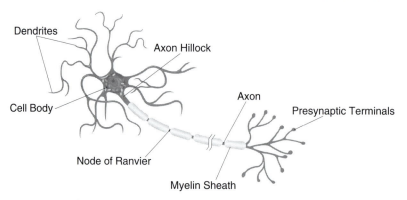

Figure 11.1   Diagram of a typical neuron.

neurons or the periphery and not as a function of the individual differences among neurons.

A typical neuron, as shown in Figure 11.1, is defined with four major regions: cell body, dendrites, axon, and presynaptic terminals. The cell body of a neuron contains the nucleus and other apparatus needed to nourish the cell and is similar to other cells. Unlike other cells, however, the neuron's cell body is connected to a number of branches called dendrites and a long tube called the axon that connects the cell body to the presynaptic terminals. Dendrites are the receptive surfaces of the neuron that receive signals from thousands of other neurons passively and without amplification. Located on the dendrite and cell body are receptor sites that receive input from presynaptic terminals from adjacent neurons. Neurons typically have 10^4 to 10^5 synapses. Communication between neurons, as previously described in Chapter 2, is through a neurotransmitter that changes membrane properties. Also connected to the cell body is a single axon that ranges in length from 1 meter in the human spinal cord to a few millimeters in the brain. The diameter of the axon also varies from less than 1 to 500 μm. In general, the larger the diameter of the axon, the faster the signal travels. Signals traveling in the axon range from 0.5 m/s to 120 m/s. The purpose of an axon is to serve as a transmission line to move information from one neuron to another at great speeds. Large axons are surrounded by a fatty insulating material called the myelin sheath and have regular gaps, called the nodes of Ranvier, that allow the action potential to jump from one node to the next. The action potential is most easily envisioned as a pulse that travels the length of the axon without decreasing in amplitude. Most of the remainder of this chapter is devoted to understanding this process. At the end of the axon is a network of up to 10,000 branches with endings called the presynaptic terminals. A diagram of the presynaptic terminal is shown in Figure 3.28. All action potentials that move through the axon propagate through each branch to the presynaptic terminal. The presynaptic terminals are the transmitting unit of the neuron which, when stimulated, release a neurotransmitter that flows across a gap of approximately 20 nanometers to an adjacent cell where it interacts with the postsynaptic membrane and changes its potential.

11.3.1 Membrane Potentials

The neuron, like other cells in the body, has a separation of charge across its external membrane. The cell membrane is positively charged on the outside and negatively charged on the inside as illustrated in Figure 11.2. This separation of charge, due to the selective permeability of the membrane to ions, is responsible for the membrane potential. In the neuron, the potential difference across the cell membrane is approximately 60 mV to 90 mV, depending on the specific cell. By convention, the outside is defined as 0 mV (ground), and the resting potential is $V_m = v_i - v_o = -60$ mV. This charge differential is of particular interest since most signaling involves changes in this potential across the membrane. Signals such as action potentials are a result of electrical perturbations of the membrane. By definition, if the membrane is more negative than resting potential (i.e., -60 to -70 mV), it is called hyperpolarization, and an increase in membrane potential from resting potential (i.e., -60 to -50 mV) is called depolarization.

To create a membrane potential of -60 mV does not require the separation of many positive and negative charges across the membrane. The actual number, however, can be found from the relationship $Cdv = dq$, or $C\Delta v = \Delta q$ (Δq = the number of charges times the electron charge of 1.6022×10^{-19} C). Therefore, with $C = 1\mu F/cm^2$ and $\Delta v = 60 \times 10^{-3}$, the number of charges equals approximately 1×10^8 per cm^2. These charges are located within 1 μm distance of the membrane.

Graded Response and Action Potentials

A neuron can change the membrane potential of another neuron to which it is connected by releasing its neurotransmitter. The neurotransmitter crosses the

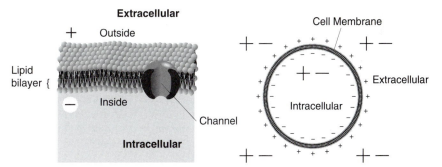

Figure 11.2   Diagrams illustrating separation of charges across a cell membrane. The figure on the left shows a cell membrane with positive ions along the outer surface of the cell membrane and negative ions along the inner surface of the cell membrane. The figure on the right further illustrates separation of charge by showing that only the ions along the inside and outside of the cell membrane are responsible for membrane potential (negative ions along the inside and positive ions along the outside of the cell membrane). Elsewhere the negative and positive ions are approximately evenly distributed as indicated with the large +− symbols for the illustration on the right. Overall, there is a net excess of negative ions inside the cell and a net excess of positive ions in the immediate vicinity outside the cell. For simplicity, the membrane shown on the right is drawn as a solid circle and ignores the axon and dendrites.

synaptic cleft or gap, interacts with receptor molecules in the postsynaptic membrane of the dendrite or cell body of the adjacent neuron, and changes the membrane potential of the receptor neuron (see Fig. 11.3).

The change in membrane potential at the postsynaptic membrane is due to a transformation from neurotransmitter chemical energy to electrical energy. The change in membrane potential depends on how much neurotransmitter is received and can be depolarizing or hyperpolarizing. This type of change in potential is typically called a graded response since it varies with the amount of neurotransmitter received. Another way of envisaging the activity at the synapse is that the neurotransmitter received is integrated or summed, which results in a graded response in the membrane potential. Note that a signal from a neuron is either inhibitory or excitatory, but specific synapses may be excitatory and others inhibitory, providing the nervous system with the ability to perform complex tasks.

The net result of activation of the nerve cell is the action potential. The action potential is a large depolarizing signal of up to 100 mV that travels along the axon and lasts approximately 1 to 5 ms. Figure 11.4 illustrates a typical action potential. The action potential is an all or none signal that propagates actively along the axon without decreasing in amplitude. When the signal reaches the end of the axon at the presynaptic terminal, the change in potential causes the release of a packet of neurotransmitter. This is a very effective method of signaling over large distances. Additional details about the action potential are described throughout the remainder of this chapter after some tools for better understanding this phenomenon are introduced.

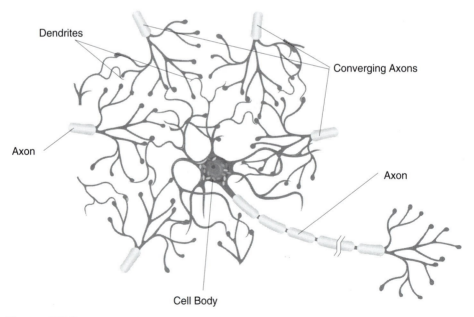

Figure 11.3   Diagram illustrating a typical neuron with presynaptic terminals of adjacent neurons in the vicinity of its dendrites.

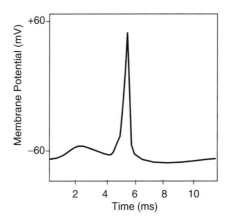

Figure 11.4 An action potential.

11.3.2 Resting Potential, Ionic Concentrations, and Channels

A resting membrane potential exists across the cell membrane because of the differential distribution of ions in and around the membrane of the nerve cell. The cell maintains these ion concentrations by using a selectively permeable membrane and, as described later, an active ion pump. A selectively permeable cell membrane with ion channels is illustrated in Figure 11.2. The neuron cell membrane is approximately 10 nm thick and, because it consists of a lipid bilayer (i.e., two plates separated by an insulator), has capacitive properties. The extracellular fluid is composed of primarily Na^+ and Cl^-, and the intracellular fluid (cytoplasm) is composed of primarily K^+ and A^-. The large organic anions (A^-) are primarily amino acids and proteins and do not cross the membrane. Almost without exception, ions cannot pass through the cell membrane except through a channel.

Channels allow ions to pass through the membrane, are selective, and are either passive or active. Passive channels are always open and are ion specific. Figure 11.5 illustrates a cross section of a cell membrane with passive channels only. As shown, a particular channel allows only one ion type to pass through the membrane and prevents all other ions from crossing the membrane through that channel. Passive channels exist for Cl^-, K^+, and Na^+. In addition, a passive channel exists for Ca^{++}, which is important in the excitation of the membrane at the synapse. Active channels, or gates, are either opened or closed in response to an external electrical or chemical stimulation. The active channels are also selective and allow only specific ions to pass through the membrane. Typically, active gates open in response to neurotransmitters and an appropriate change in membrane potential.

Figure 11.6 illustrates the concept of an active channel. Here, K^+ passes through an active channel and Cl^- passes through a passive channel. As will be shown, passive channels are responsible for the resting membrane potential, and active channels are responsible for the graded response and action potentials.

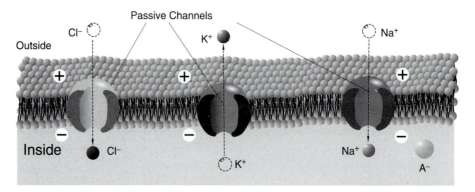

Figure 11.5   Idealized cross section of a selectively permeable membrane with channels for ions to cross the membrane. The thickness of the membrane and size of the channels are not drawn to scale. When the diagram is drawn to scale, the cell membrane thickness is 20 times the size of the ions and 10 times the size of the channels, and the spacing between the channels is approximately 10 times the cell membrane thickness. Note that a potential difference exists between the inside and outside of the membrane as illustrated with the + and − signs. The membrane is selectively permeable to ions through ion-specific channels; that is, each channel shown here allows only one particular ion to pass through it.

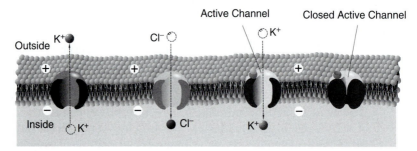

Figure 11.6   Passive and active channels provide a means for ions to pass through the membrane. Each channel is ion specific. As shown, the active channel on the left allows K^+ to pass through the membrane, but the active channel on the right is not open, preventing any ion from passing through the membrane. Also shown is a passive Cl^- channel.

11.4   BASIC BIOPHYSICS TOOLS AND RELATIONSHIPS

11.4.1   Basic Laws

Two basic biophysics tools and a relationship are used to characterize the resting potential across a cell membrane by quantitatively describing the impact of the ionic gradients and electric fields.

Fick's Law

The flow of particles due to diffusion is along the concentration gradient with particles moving from high-concentration areas to low ones. Specifically, for a cell membrane, the flow of ions across a membrane is given by

$$J(\text{diffusion}) = -D\frac{d[I]}{dx} \tag{11.1}$$

where J is the flow of ions due to diffusion, $[I]$ is the ion concentration, dx is the membrane thickness, and D is the diffusivity constant in m^2/s. The negative sign indicates that the flow of ions is from higher to lower concentration, and $\frac{d[I]}{dx}$ represents the concentration gradient.

Ohm's Law

Charged particles in a solution experience a force resulting from other charged particles and electric fields present. The flow of ions across a membrane is given by

$$J(\text{drift}) = -\mu Z[I]\frac{dv}{dx} \tag{11.2}$$

where J is the flow of ions due to drift in an electric field $\vec{E}$, $\mu =$ mobility in m^2/sV, $Z =$ ionic valence, $[I]$ is the ion concentration, v is the voltage across the membrane, and $\frac{dv}{dx}$ is $(-\vec{E})$. Note that Z is positive for positively charged ions (e.g., $Z = 1$ for Na^+ and $Z = 2$ for Ca^{++}) and negative for negatively charged ions (e.g., $Z = -1$ for Cl^-). Positive ions drift down the electric field and negative ions drift up the electric field.

Figure 11.7 illustrates a cell membrane that is permeable to only K^+ and shows the forces acting on K^+. Assume that the concentration of K^+ is that of a neuron with a higher concentration inside than outside and that the membrane resting potential is negative from inside to outside. Clearly, only K^+ can pass through the membrane, and Na^+, Cl^-, and A^- cannot move since there are no channels for them to pass through. Depending on the actual concentration and membrane potential, K^+ will pass through the membrane until the forces due to drift and diffusion are balanced. The chemical force due to diffusion from inside to outside decreases as K^+ moves through the membrane, and the electric force increases as K^+ accumulates outside the cell until the two forces are balanced.

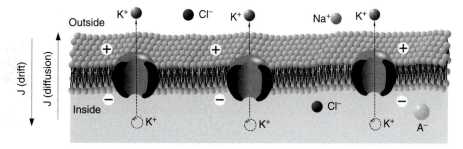

Figure 11.7 Diagram illustrating the direction of the flow of K^+ due to drift and diffusion across a cell membrane that is permeable only to K^+.

Einstein Relationship

The relationship between the drift of particles in an electric field under osmotic pressure, that is the relationship between diffusivity and mobility, is given by

$$D = \frac{KT\mu}{q} \tag{11.3}$$

where D is the diffusivity constant, μ is mobility, K is Boltzmann's constant, T is the absolute temperature in degrees Kelvin, and q is the magnitude of the electric charge (i.e., 1.60186×10^{-19} coulombs).

11.4.2 Resting Potential of a Membrane Permeable to One Ion

The flow of ions in response to concentration gradients is limited by the selectively permeable nerve cell membrane and the resultant electric field. As described, ions pass through channels that are selective for that ion only. For clarity, the case of a membrane permeable to one ion only is considered first and then the case of a membrane permeable to more than one ion follows. It is interesting to note that neuroglial cells are permeable to only K^+ and that nerve cells are permeable to K^+, Na^+, and Cl^-. As will be shown, the normal ionic gradient is maintained if the membrane is permeable only to K^+ as in the neuroglial cell.

Consider the cell membrane shown in Figure 11.7 that is permeable only to K^+ and assume that the concentration of K^+ is higher in the intracellular fluid than in the extracellular fluid. For this situation, the flow due to diffusion (concentration gradient) tends to push K^+ outside of the cell and is given by

$$J_K(\text{diffusion}) = -D\frac{d[K^+]}{dx} \tag{11.4}$$

The flow due to drift (electric field) tends to push K^+ inside the cell and is given by

$$J_K(\text{drift}) = -\mu Z[K^+]\frac{dv}{dx} \tag{11.5}$$

which results in a total flow

$$J_K = J_K(\text{diffusion}) + J_K(\text{drift}) = -D\frac{d[K^+]}{dx} - \mu Z[K^+]\frac{dv}{dx} \tag{11.6}$$

Using the Einstein relationship $D = \frac{KT\mu}{q}$, the total flow is now given by

$$J_K = -\frac{KT}{q}\mu\frac{d[K^+]}{dx} - \mu Z[K^+]\frac{dv}{dx} \tag{11.7}$$

From Equation 11.7, the flow of K^+ is found at any time for any given set of initial conditions. In the special case of steady state, that is, at equilibrium when the flow of K^+ into the cell is exactly balanced by the flow out of the cell or $J_K = 0$, Equation 11.7 reduces to

$$0 = -\frac{KT}{q}\mu\frac{d[K^+]}{dx} - \mu Z[K^+]\frac{dv}{dx} \tag{11.8}$$

With $Z = +1$, Equation 11.8 simplifies to

$$dv = -\frac{KT}{q[K^+]}d[K^+] \tag{11.9}$$

Integrating Equation 11.9 from outside the cell to inside yields

$$\int_{v_o}^{v_i} dv = -\frac{KT}{q}\int_{[K^+]_o}^{[K^+]_i}\frac{d[K^+]}{[K^+]} \tag{11.10}$$

where v_o and v_i are the voltages outside and inside the membrane and $[K^+]_o$ and $[K^+]_i$ are the concentrations of potassium outside and inside the membrane. Thus,

$$v_i - v_o = -\frac{KT}{q}\ln\frac{[K^+]_i}{[K^+]_o} = \frac{KT}{q}\ln\frac{[K^+]_o}{[K^+]_i} \tag{11.11}$$

Equation 11.11 is known as the Nernst equation, named after a German physical chemist Walter Nernst, and $E_K = v_i - v_o$ is known as the Nernst potential for K^+. At room temperature, $\frac{KT}{q} = 26\,mV$, and thus the Nernst equation for K^+ becomes

$$E_K = v_i - v_o = 26\ln\frac{[K^+]_o}{[K^+]_i}\,mV \tag{11.12}$$

Though Equation 11.12 is specifically written for K^+, it can be easily derived for any permeable ion. At room temperature, the Nernst potential for Na^+ is

$$E_{Na} = v_i - v_o = 26\ln\frac{[Na^+]_o}{[Na^+]_i}\,mV \tag{11.13}$$

and the Nernst potential for Cl^- is

$$E_{Cl} = v_i - v_o = -26\ln\frac{[Cl^-]_o}{[Cl^-]_i} = 26\ln\frac{[Cl^-]_i}{[Cl^-]_o}\,mV \tag{11.14}$$

The negative sign in Equation 11.14 is due to $Z = -1$ for Cl^-.

11.4.3 Donnan Equilibrium

In a neuron at steady state (equilibrium) that is permeable to more than one ion (for example K^+, Na^+, and Cl^-) the Nernst potential for each ion is calculated using Equations 11.12 to 11.14, respectively. The membrane potential, $V_m = v_i - v_o$, however, is due to the presence of all ions and is influenced by the concentration and permeability of each ion. In this section, the case in which two ions are permeable is presented. In the next section, the case in which any number of permeable ions are present is considered.

Suppose a membrane is permeable to both K^+ and Cl^-, but not to a large cation, R^+, as shown in Figure 11.8. For equilibrium, the Nernst potentials for both K^+ and Cl^- must be equal, that is $E_K = E_{Cl}$, or

$$E_K = \frac{KT}{q}\ln\frac{[K^+]_o}{[K^+]_i} = E_{Cl} = \frac{KT}{q}\ln\frac{[Cl^-]_i}{[Cl^-]_o} \tag{11.15}$$

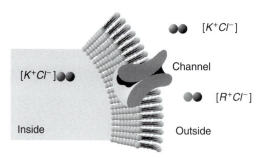

Figure 11.8   Membrane is permeable to both K^+ and Cl^-, but not to a large cation R^+.

After simplifying,

$$\frac{[K^+]_o}{[K^+]_i} = \frac{[Cl^-]_i}{[Cl^-]_o} \qquad (11.16)$$

Equation 11.16 is known as the Donnan equilibrium. An accompanying principle is space charge neutrality, which states that the number of cations in a given volume is equal to the number of anions. Thus, in the equilibrium state ions still diffuse across the membrane, but each K^+ that crosses the membrane must be accompanied by a Cl^- for space charge neutrality to be satisfied. If in Figure 11.8 R^+ were not present, then at equilibrium, the concentration of K^+ and Cl^- on both sides of the membrane would be equal. With R^+ in the extracellular fluid, the concentrations of [KCl] on both sides of the membrane are different as shown in the following example.

Example Problem 11.1

A membrane is permeable to K^+ and Cl^-, but not to a large cation R^+. Find the steady-state equilibrium concentration for the following initial conditions.

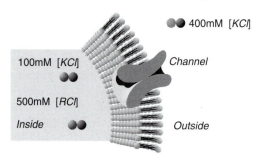

Solution

By conservation of mass,

$$[K^+]_i + [K^+]_o = 500$$

$$[Cl^-]_i + [Cl^-]_o = 1000$$

and by space charge neutrality,

$$[K^+]_i + 500 = [Cl^-]_i$$

$$[K^+]_o = [Cl^-]_o$$

From the Donnan equilibrium,

$$\frac{[K^+]_o}{[K^+]_i} = \frac{[Cl^-]_i}{[Cl^-]_o}$$

Substituting for $[K^+]_o$ and $[Cl^-]_o$ from the conservation of mass equations into the Donnan equilibrium equation gives

$$\frac{500 - [K^+]_i}{[K^+]_i} = \frac{[Cl^-]_i}{1000 - [Cl^-]_i}$$

and eliminating $[Cl^-]_i$ by using the space charge neutrality equations gives

$$\frac{500 - [K^+]_i}{[K^+]_i} = \frac{[K^+]_i + 500}{1000 - [K^+]_i - 500} = \frac{[K^+]_i + 500}{500 - [K^+]_i}$$

Solving the previous equation yields $[K^+]_i = 167\,\text{mM}$ at steady state. Using the conservation of mass equations and space charge neutrality equation gives $[K^+]_o = 333\,\text{mM}$, $[Cl^-]_i = 667\,\text{mM}$, and $[Cl^-]_o = 333\,\text{mM}$ at steady state. At steady state and at room temperature, the Nernst potential for either ion is 18 mV, as shown for $[K^+]$

$$E_K = v_i - v_o = 26\ln\frac{333}{167} = 18\,\text{mV}$$

Summarizing, at steady state

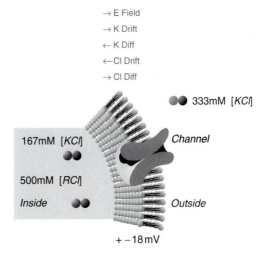

11.4.4   Goldman Equation

The squid giant axon resting potential is $-60\,\text{mV}$, which does not correspond to the Nernst potential for Na^+ or K^+. As a general rule, when V_m is affected by two or more ions, each ion influences V_m as determined by its concentration and membrane permeability. The Goldman equation quantitatively describes the relationship between V_m and permeable ions but applies only when the membrane potential or electric field is constant. This situation is a reasonable approximation for a resting membrane potential. Here the Goldman equation is first derived for K^+ and Cl^- and then extended to include K^+, Cl^-, and Na^+. The Goldman equation is used by physiologists to calculate the membrane potential for a variety of cells and, in fact, was used by Hodgkin, Huxley, and Katz in studying the squid giant axon.

Consider the cell membrane shown in Figure 11.9. To determine V_m for both K^+ and Cl^-, flow equations for each ion are derived separately under the condition of a constant electric field and then combined using space charge neutrality to complete the derivation of the Goldman equation.

Potassium Ions

The flow equation for K^+ with mobility μ_K is

$$J_K = -\frac{KT}{q}\mu_K\frac{d[K^+]}{dx} - \mu_K Z_K[K^+]\frac{dv}{dx} \qquad (11.17)$$

Under a constant electric field,

$$\frac{dv}{dx} = \frac{\Delta v}{\Delta x} = \frac{V}{\delta} \qquad (11.18)$$

Substituting Equation 11.18 into 11.17 with $Z_K = 1$ gives

$$J_K = -\frac{KT}{q}\mu_K\frac{d[K^+]}{dx} - \mu_K Z_K[K^+]\frac{V}{\delta} \qquad (11.19)$$

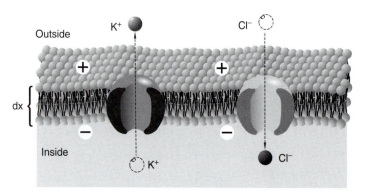

Figure 11.9   Diagram illustrating a cell membrane permeable to both K^+ and Cl^-. The width of the membrane is $dx = \delta$.

Let the permeability for K^+, P_K, equal

$$P_K = \frac{\mu_K KT}{\delta q} = \frac{D_K}{\delta} \tag{11.20}$$

Therefore, using Equation 11.20 in 11.19 gives

$$J_K = \frac{-P_K q}{KT} V[K^+] - P_K \delta \frac{d[K^+]}{dx} \tag{11.21}$$

Rearranging the terms in Equation 11.21 yields

$$dx = \frac{d[K^+]}{\frac{-J_K}{P_K \delta} - \frac{qV[K^+]}{KT\delta}} \tag{11.22}$$

Taking the integral of both sides, while assuming that J_K is independent of x, gives

$$\int_0^\delta dx = \int_{[K^+]_i}^{[K^+]_o} \frac{d[K^+]}{\frac{-J_K}{P_K \delta} - \frac{qV[K^+]}{KT\delta}} \tag{11.23}$$

resulting in

$$x\big|_0^\delta = -\frac{KT\delta}{qV} \ln\left(\frac{J_K}{P_K \delta} + \frac{qV[K^+]}{KT\delta}\right)\Bigg|_{[K^+]_i}^{[K^+]_o} \tag{11.24}$$

and

$$\delta = -\frac{KT\delta}{qV} \ln\left(\frac{\frac{J_K}{P_K \delta} + \frac{qV[K^+]_o}{KT\delta}}{\frac{J_K}{P_K \delta} + \frac{qV[K^+]_i}{KT\delta}}\right) \tag{11.25}$$

Removing δ from both sides of Equation 11.25, bringing the term $-\frac{KT}{qV}$ to the other side of the equation, and then taking the exponential of both sides yields

$$e^{-\frac{qv}{KT}} = \frac{\frac{J_K}{P_K \delta} + \frac{qV[K^+]_o}{KT\delta}}{\frac{J_K}{P_K \delta} + \frac{qV[K^+]_i}{KT\delta}} \tag{11.26}$$

Solving for J_K in Equation 11.26 gives

$$J_K = \frac{qVP_K}{KT}\left(\frac{[K^+]_o - [K^+]_i e^{\frac{-qV}{KT}}}{e^{\frac{-qV}{KT}} - 1}\right) \tag{11.27}$$

Chlorine Ions

The same derivation carried out for K^+ can be repeated for Cl^-, which yields

$$J_{Cl} = \frac{qVP_{Cl}}{KT}\left(\frac{[Cl^-]_o e^{\frac{-qV}{KT}} - [Cl^-]_i}{e^{\frac{-qV}{KT}} - 1}\right) \tag{11.28}$$

where P_{Cl} is the permeability for Cl^-.

Summarizing for Potassium and Chlorine Ions

From space charge neutrality, $J_K = J_{Cl}$, with Equations 11.27 and 11.28 gives

$$P_K\left([K^+]_o - [K^+]_i e^{\frac{-qV}{KT}}\right) = P_{Cl}\left([Cl^-]_o e^{\frac{-qV}{KT}} - [Cl^-]_i\right) \qquad (11.29)$$

Solving for the exponential terms yields

$$e^{\frac{-qV}{KT}} = \frac{P_K[K^+]_o + P_{Cl}[Cl^-]_i}{P_K[K^+]_i + P_{Cl}[Cl^-]_o} \qquad (11.30)$$

Solving for V gives

$$V = v_o - v_i = -\frac{KT}{q}\ln\left(\frac{P_K[K^+]_o + P_{Cl}[Cl^-]_i}{P_K[K^+]_i + P_{Cl}[Cl^-]_o}\right) \qquad (11.31)$$

or in terms of V_m

$$V_m = \frac{KT}{q}\ln\left(\frac{P_K[K^+]_o + P_{Cl}[Cl^-]_i}{P_K[K^+]_i + P_{Cl}[Cl^-]_o}\right) \qquad (11.32)$$

This equation is called the Goldman equation. Since sodium is also important in membrane potential, the Goldman equation for K^+, Cl^-, and Na^+ can be derived as

$$V_m = \frac{KT}{q}\ln\left(\frac{P_K[K^+]_o + P_{Na}[Na^+]_o + P_{Cl}[Cl^-]_i}{P_K[K^+]_i + P_{Na}[Na^+]_i + P_{Cl}[Cl^-]_o}\right) \qquad (11.33)$$

where P_{Na} is the permeability for Na^+. To derive Equation 11.33, first find J_{Na} and then use space charge neutrality $J_K + J_{Na} = J_{Cl}$. Equation 11.33 then follows. In general, when the permeability to one ion is exceptionally high, as compared with the other ions, then V_m predicted by the Goldman equation is very close to the Nernst equation for that ion.

Tables 11.1 and 11.2 contain the important ions across the cell membrane, the ratio of permeabilities, and Nernst potentials for the squid giant axon and frog skeletal muscle. The squid giant axon is extensively reported on and used in experiments due to its large size, lack of myelination, and ease of use. In general, the intracellular and extracellular concentration of ions in vertebrate neurons is approximately three to four times less than in the squid giant axon.

TABLE 11.1   Approximate Intracellular and Extracellular Concentrations of the Important Ions across a Squid Giant Axon, Ratio of Permeabilities, and Nernst Potentials

Ion	Cytoplasm (mM)	Extracellular Fluid (mM)	Ratio of Permeabilities	Nernst Potential (mV)
K^+	400	20	1	−74
Na^+	50	440	0.04	55
Cl^-	52	560	0.45	−60

Note that the permeabilities are relative, that is $P_K : P_{Na} : P_{Cl}$, and not absolute. Data were recorded at 6.3°C, resulting in KT/q approximately equal to 25.3 mV.

TABLE 11.2 Approximate Intracellular and Extracellular Concentrations of the Important Ions across a Frog Skeletal Muscle, Ratio of Permeabilities, and Nernst Potentials

Ion	Cytoplasm (mM)	Extracellular Fluid (mM)	Ratio of Permeabilities	Nernst Potential (mV)
K^+	140	2.5	1.0	−105
Na^+	13	110	0.019	56
Cl^-	3	90	0.381	−89

Data were recorded at room temperature, resulting in KT/q approximately equal to $26\,mV$.

Example Problem 11.2

Calculate V_m for the squid giant axon at 6.3°C.

Solution

Using Equation 11.33 and the data in Table 11.1, gives

$$V_m = 25.3 \times \ln\left(\frac{1 \times 20 + 0.04 \times 440 + 0.45 \times 52}{1 \times 400 + 0.04 \times 50 + 0.45 \times 560}\right) mV = -60\,mV \qquad \blacksquare$$

11.4.5 Ion Pumps

At rest, separation of charge and ionic concentrations across the cell membrane must be maintained, otherwise V_m changes. That is, the flow of charge into the cell must be balanced by the flow of charge out of the cell. For Na^+, the concentration and electric gradient creates a force that drives Na^+ into the cell at rest. At V_m, the K^+ force due to diffusion is greater than that due to drift and results in an efflux of K^+ out of the cell. Space charge neutrality requires that the influx of Na^+ be equal to the flow of K^+ out of the cell. Although these flows cancel each other and space charge neutrality is maintained, this process cannot continue unopposed. Otherwise, $[K^+]_i$ goes to zero as $[Na^+]_i$ increases, with subsequent change in V_m as predicted by the Goldman equation.

Any change in the concentration gradient of K^+ and Na^+ is prevented by the Na-K pump. The pump transports a steady stream of Na^+ out of the cell and K^+ into the cell. Removal of Na^+ from the cell is against its concentration and electric gradient and is accomplished with an active pump that consumes metabolic energy. Figure 11.10 illustrates a Na-K pump along with an active and passive channel.

The Na-K pump has been found to be electrogenic; that is, there is a net transfer of charge across the membrane. Nonelectrogenic pumps operate without any net transfer of charge. For many neurons, the Na-K ion pump removes three Na^+ ions for every two K^+ ions moved into the cell, which makes V_m slightly more negative than predicted with only passive channels.

In general, when the cell membrane is at rest, the active and passive ion flows are balanced and a permanent potential exists across a membrane only if

1. The membrane is impermeable to some ion(s).
2. An active pump is present.

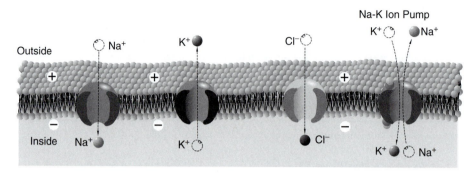

Figure 11.10   An active pump is illustrated along with a passive and active channel.

The presence of the Na-K pump forces V_m to a given potential based on the K^+ and Na^+ concentrations that are determined by the active pump. Other ion concentrations are determined by V_m. For instance, since Cl^- moves across the membrane only through passive channels, the Cl^- concentration ratio at rest is determined from the Nernst equation with $E_{Cl} = V_m$, or

$$\frac{[Cl^-]_i}{[Cl^-]_o} = e^{\frac{qV_m}{KT}} \tag{11.34}$$

Example Problem 11.3

Consider a membrane in which there is an active K^+ pump, passive channels for K^+ and Cl^-, and a nonequilibrium initial concentration of $[KCl]$ on both sides of the membrane. Find an expression for the active K^+ pump.

Solution

From space charge neutrality, $J_{Cl} = J_K$, or

$$J_K = J_p - \frac{KT}{q}\mu_K \frac{d[K^+]}{dx} - \mu_K Z_K [K^+] \frac{dv}{dx}$$

$$J_{Cl} = -\frac{KT}{q}\mu_{Cl} \frac{d[Cl^-]}{dx} - \mu_{Cl} Z_{Cl} [Cl^-] \frac{dv}{dx}$$

where J_p is the flow due to the active K^+ pump.

Solving for $\frac{dv}{dx}$ using the J_{Cl} equation with $Z_{Cl} = -1$ gives

$$\frac{dv}{dx} = \frac{KT}{q[Cl^-]} \frac{d[Cl^-]}{dx}$$

By space charge neutrality, $[Cl^-] = [K^+]$, which allows rewriting the previous equation as

$$\frac{dv}{dx} = \frac{KT}{q[K^+]} \frac{d[K^+]}{dx}$$

At equilibrium, both flows are zero and with $Z_K = 1$, the J_K equation with $\frac{dv}{dx}$ substitution is given as

$$J_K = 0 = J_p - \mu_K[K^+]\frac{dv}{dx} - \frac{KT\mu_K}{q}\frac{d[K^+]}{dx} = J_p - \mu_K[K^+]\frac{KT}{q[K^+]}\frac{d[K^+]}{dx} - \frac{KT\mu_K}{q}\frac{d[K^+]}{dx}$$

$$= J_p - \frac{2KT\mu_K}{q}\frac{d[K^+]}{dx}$$

Moving J_p to the left side of the equation, multiplying both sides by dx, and then integrating yields,

$$-\int_0^\delta J_p dx = -\frac{2KT\mu_K}{q}\int_{[K^+]_i}^{[K^+]_o} d[K^+]$$

or

$$J_p = \frac{2KT\mu_K}{q\delta}\left([K^+]_o - [K^+]_i\right) \qquad \blacksquare$$

Note: In this example, if no pump were present, then at equilibrium, the concentration on both sides of the membrane would be the same.

11.5 EQUIVALENT CIRCUIT MODEL FOR THE CELL MEMBRANE

In this section, an equivalent circuit model is developed using the tools previously developed. Creating a circuit model is helpful when discussing the Hodgkin–Huxley model of an action potential in the next section, a model that introduces voltage- and time-dependent ion channels. As described in Sections 11.3 and 11.4, the nerve cell has three types of passive electrical characteristics: electromotive force, resistance, and capacitance. The nerve membrane is a lipid bilayer that is pierced by a variety of different types of ion channels, where each channel is characterized as being passive (always open) or active (gates that can be opened). Each ion channel is also characterized by its selectivity. In addition, there is the active Na-K pump that maintains V_m across the cell membrane.

11.5.1 Electromotive, Resistive, and Capacitive Properties

Electromotive Force Properties

The three major ions K^+, Na^+, and Cl^- are differentially distributed across the cell membrane at rest and across the membrane through passive ion channels as illustrated in Figure 11.5. This separation of charge exists across the membrane and results in a voltage potential V_m as described by the Goldman Equation 11.33.

Across each ion-specific channel, a concentration gradient exists for each ion that creates an electromotive force, a force that drives that ion through the channel at a constant rate. The Nernst potential for that ion is the electrical potential difference across the channel and is easily modeled as a battery, as is illustrated in Figure 11.11 for K^+. The same model is applied for Na^+ and Cl^- with values equal to the Nernst potentials for each.

Resistive Properties

In addition to the electromotive force, each channel also has resistance; that is, it resists the movement of electrical charge through the channel. This is mainly due to collisions with the channel wall where energy is given up as heat. The term *conductance*, G, measured in Siemens (S), which is the ease with which the ions move through the membrane, is typically used to represent resistance. Since the conductances (channels) are in parallel, the total conductance is the total number of channels, N, times the conductance for each channel, G'

$$G = N \times G'$$

It is usually more convenient to write the conductance as resistance $R = \dfrac{1}{G}$, measured in ohms (Ω).

An equivalent circuit for the channels for a single ion is now given as a resistor in series with a battery as shown in Figure 11.12.

Conductance is related to membrane permeability, but they are not interchangeable in a physiological sense. Conductance depends on the state of the membrane, varies with ion concentration, and is proportional to the flow of ions through a membrane. Permeability describes the state of the membrane for a particular ion. Consider the case in which there are no ions on either side of the membrane. No matter how many channels are open, $G = 0$ because there are no ions available to flow across the cell membrane (due to a potential difference). At the same time, ion permeability is constant and is determined by the state of the membrane.

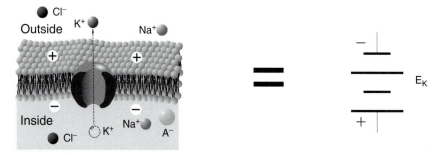

Figure 11.11 A battery is used to model the electromotive force for a K^+ channel with a value equal to the K^+ Nernst potential. The polarity of the battery is given with the ground on the outside of the membrane, in agreement with convention. From Table 11.1, note that the Nernst potential for K^+ is negative, which reverses the polarity of the battery, driving K^+ out of the cell.

Equivalent Circuit for Three Ions

Each of the three ions, K^+, Na^+, and Cl^-, is represented by the same equivalent circuit, as shown in Figure 11.12, with Nernst potentials and appropriate resistances. Combining the three equivalent circuits into one circuit with the extracellular fluid and cytoplasm connected by short circuits completely describes a membrane at rest (Fig. 11.13).

Example Problem 11.4

Find V_m for the frog skeletal muscle (Table 11.2) if the Cl^- channels are ignored. Use $R_K = 1.7 \, k\Omega$ and $R_{Na} = 15.67 \, k\Omega$.

Solution

The following diagram depicts the membrane circuit with mesh current I, current I_{Na} through the sodium channel, and current I_K through the potassium channel. Current I is found using mesh analysis:

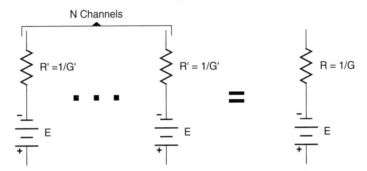

Figure 11.12 The equivalent circuit for N ion channels is a single resistor and battery.

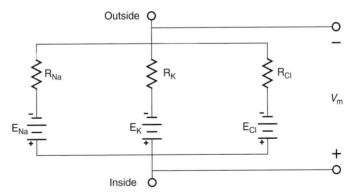

Figure 11.13 Model of the passive channels for a small area of nerve at rest with each ion channel represented by a resistor in series with a battery.

$$E_{Na} + IR_{Na} + IR_K - E_K = 0$$

and solving for I,

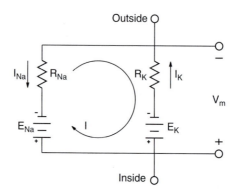

$$I = \frac{E_K - E_{Na}}{R_{Na} + R_K} = \frac{(-105 - 56) \times 10^{-3}}{(15.67 + 1.7) \times 10^3} = -9.27\,\mu A$$

yields

$$V_m = E_{Na} + IR_{Na} = -89\,mV$$

Notice that $I = -I_{Na}$ and $I = -I_K$, or $I_{Na} = I_K$ as expected. Physiologically, this implies that the inward Na^+ current is exactly balanced by the outward bound K^+ current.   ■

Example Problem 11.5

Find V_m for the frog skeletal muscle if $R_{Cl} = 3.125$ kΩ.

Solution

To solve, first find a Thévenin equivalent circuit for the circuit in Example Problem 11.4.

$$V_{Th} = V_m = -89\,mV$$

and

$$R_{Th} = \frac{R_{Na} \times R_K}{R_{Na} + R_K} = 1.534\,k\Omega$$

The Thévenin equivalent circuit is shown in the following figure.

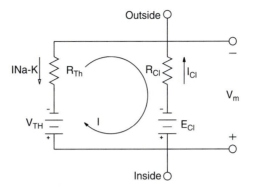

Since $E_{Cl} = V_{Th}$ according to Table 11.2, no current flows. This is the actual situation in most nerve cells. The membrane potential is determined by the relative conductances and Nernst potentials for K^+ and Na^+. The Nernst potentials are determined by the active pump that maintains the concentration gradient. Cl^- is usually passively distributed across the membrane. ∎

Na-K Pump

As shown in Example Problem 11.4 and Section 11.4, there is a steady flow of K^+ ions out of the cell and Na^+ ions into the cell even when the membrane is at the resting potential. Left unchecked, this would drive E_K and E_{Na} toward 0. To prevent this, current generators—the Na-K pump—are used that are equal and opposite to the passive currents and incorporated into the model as shown in Figure 11.14.

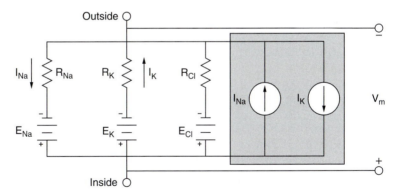

Figure 11.14   Circuit model of the three passive channels for a small area of the nerve at rest with each ion channel represented by a resistor in series with a battery. The Na-K active pump is modeled as two current sources within the shaded box.

11.5.2 Capacitive Properties

Capacitance occurs whenever electrical conductors are separated by an insulating material. In the neuron, the cytoplasm and extracellular fluid are the electrical conductors and the lipid bilayer of the membrane is the insulating material (Fig. 11.3). Capacitance for a neuron membrane is approximately $1\,\mu F/cm^2$. Membrane capacitance implies that ions do not move through the membrane except through ion channels.

The membrane can be modeled using the circuit in Figure 11.15 by incorporating membrane capacitance with the electromotive and resistive properties. A consequence of membrane capacitance is that changes in membrane voltage are not immediate but follow an exponential time course due to first-order time constant effects. To appreciate the effect of capacitance, the circuit in Figure 11.15 is reduced to Figure 11.16 by using a Thévenin equivalent for the batteries and the resistors with R_{Th} and V_{Th} given in Equations 11.35 and 11.36.

$$R_{Th} = \frac{1}{\frac{1}{R_K} + \frac{1}{R_{Na}} + \frac{1}{R_{Cl}}} \tag{11.35}$$

$$V_{Th} = -\frac{R_{Na}R_{Cl}E_k + R_K R_{Cl}E_{Na} + R_K R_{Na}E_{Cl}}{R_{Na}R_{Cl} + R_K R_{Cl} + R_K R_{Na}} \tag{11.36}$$

The time constant for the membrane circuit model is $\tau = R_{Th} \times C_m$, and at 5τ the response is within 1% of steady state. The range for τ is from 1 to 20 ms in a typical neuron. In addition, at steady state, the capacitor acts as an open circuit and $V_{Th} = V_m$, as it should.

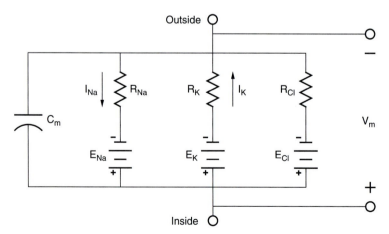

Figure 11.15 Circuit model of a small area of the nerve at rest with all of its passive electrical properties. The Na-K active pump shown in Fig. 11.14 is removed because it does not contribute electrically to the circuit.

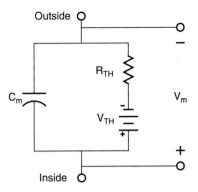

Figure 11.16 Thévenin equivalent circuit of the model in Fig. 11.15.

Example Problem 11.6

Compute the change in V_m due to a current pulse through the cell membrane.

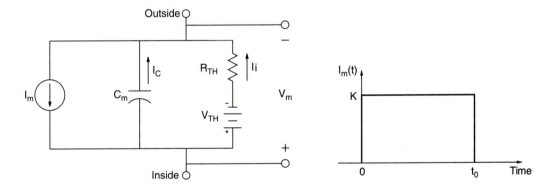

Solution

Experimentally, the stimulus current is a pulse passed through the membrane from an intracellular electrode to an extracellular electrode as depicted in the preceding circuit diagram. The membrane potential, V_m, due to a current pulse, I_m, with amplitude K and duration t_o applied at $t = 0$, is found by applying Kirchhoff's current law at the cytoplasm, yielding

$$-I_m + \frac{V_m - V_{Th}}{R_{Th}} + C_m \frac{dV_m}{dt} = 0$$

The Laplace transform of the node equation is

$$-I_m(s) + \frac{V_m(s)}{R_{Th}} - \frac{V_{Th}}{sR_{Th}} + sC_mV_m(s) - C_mV_m(0^+) = 0$$

Combining common terms gives

$$\left(s + \frac{1}{C_m R_{Th}}\right) V_m(s) = V_m(0^+) + \frac{I_m(s)}{C_m} + \frac{V_{Th}}{s C_m R_{Th}}$$

The Laplace transform of the current pulse is $I_m(s) = \frac{K}{s}(1 - e^{-t_o s})$. Substituting $I_m(s)$ into the node equation and rearranging terms yields

$$V_m(s) = \frac{V_m(0^+)}{\left(s + \frac{1}{C_m R_{Th}}\right)} + \frac{K(1 - e^{-t_o s})}{s C_m \left(s + \frac{1}{C_m R_{Th}}\right)} + \frac{V_{Th}}{s C_m R_{Th} \left(s + \frac{1}{C_m R_{Th}}\right)}$$

Performing a partial fraction expansion, and noting that $V_m(0^+) = V_{Th}$, gives

$$V_m(s) = K R_{Th} \left(\frac{1}{s} - \frac{1}{\left(s + \frac{1}{C_m R_{Th}}\right)}\right)(1 - e^{-t_o s}) + \frac{V_{Th}}{s}$$

Transforming back into the time domain yields the solution

$$V_m(t) = V_{Th} + R_{Th} K \left(1 - e^{-\frac{t}{R_{Th} C_m}}\right) u(t) - R_{Th} K \left(1 - e^{-\frac{t-t_0}{R_{Th} C_m}}\right) u(t - t_0)$$

The ionic current (I_I) and capacitive current (I_c) are shown in the following figure, where

$$I_i = \frac{V_m - V_{Th}}{R_{Th}}$$

$$I_c = C_m \frac{dV_m}{dt}$$

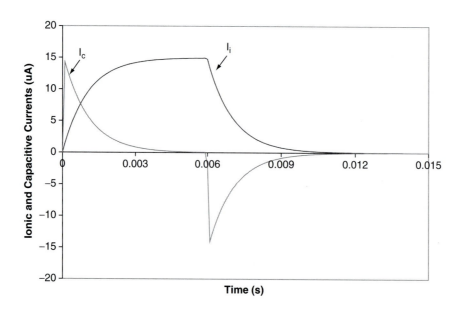

Shown in the following figures are graphs of V_m in response to a 15 μA current pulse of 6 ms (top) and 2 ms (bottom) using parameters for the frog skeletal muscle. The time constant is approximately 1 ms. Note that in the figure on the top, V_m reaches steady state before the current pulse returns to zero, and in the figure on the bottom, V_m falls short of the steady-state value reached on the top.

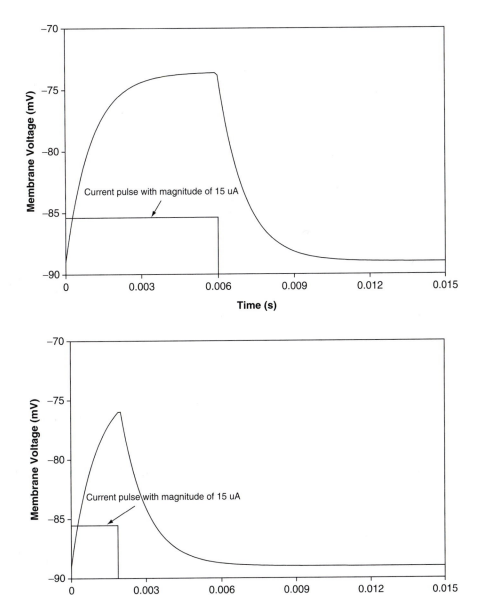

The value of the time constant is important in the integration of currents (packets of neurotransmitter) at the synapse. Each packet of neurotransmitter acts as a current pulse. Note, that the longer the time constant, the more time the membrane is excited. Most excitations are not synchronous, but because of τ, a significant portion of the stimulus is added together to cause signal transmission.

The following figure is due to a series of 15 μA current pulses of 6 ms duration with the onsets occurring at 0, 2, 4, 6, and 8 ms. Since the pulses occur within 5τ of the previous pulses, the effect of each on V_m is additive, allowing the membrane to depolarize to approximately −45 mV. If the pulses were spaced at intervals greater than 5τ, then V_m would be a series of pulse responses as previously illustrated.   ■

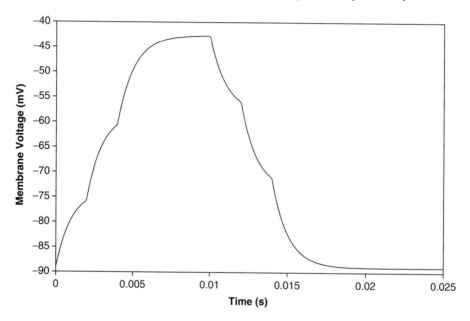

11.5.3   Change in Membrane Potential with Distance

The circuit model in Figure 11.15 or 11.16 describes a small area or section of the membrane. In Example Problem 11.6, a current pulse was injected into the membrane and resulted in a change in V_m. The change in V_m in this section of the membrane causes current to flow into the adjacent membrane sections, which causes a change in V_m in each section and so on, continuing throughout the surface of the membrane. Since the volume inside the dendrite is much smaller than the extracellular space, there is significant resistance to the flow of current in the cytoplasm from one membrane section to the next as compared with the flow of current in the extracellular space. The larger the diameter of the dendrite, the smaller the resistance to the spread of current from one section to the next. To model this effect, a resistor, R_a, is placed in the cytoplasm connecting each section together as shown

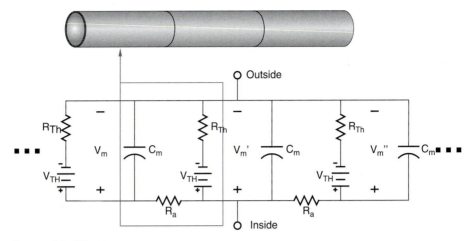

Figure 11.17 Equivalent circuit of series of membrane sections connected with axial resistance, R_a.

in the Figure 11.17. This model is actually a three-dimensional surface and continues in the x, y, and z directions. The outside resistance is negligible since it has a greater volume and is modeled as a short circuit.

Suppose a current is injected into a section of the dendrite as shown in Figure 11.18, similar to the situation in Example Problem 11.6 where t_o is large. At steady state, the transient response due to C_m has expired and only current through the resistance is important. Most of the current flows out through the section into which the current was injected since it has the smallest resistance (R_{Th}) in relation to the other sections. The next largest current flowing out of the membrane occurs in the next section since it has the next smallest resistance, $R_{Th} + R_a$. The change in V_m, ΔV_m, from the injection site is independent of C_m and depends solely on the relative values of R_{Th} and R_a. The resistance seen in n sections from the injection site is $R_{Th} + n \times R_a$. Since current decreases with distance from the injection site, then ΔV_m also decreases with distance from the injection site because it equals the current through that section times R_{Th}. The change in membrane potential, ΔV_m, decreases exponentially with distance and is given by

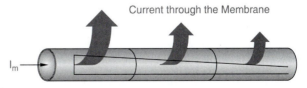

Figure 11.18 A diagram illustrating the flow of current through a dendrite at steady state. Since current seeks the path of least resistance, most of the current leaves the dendrite at the injection site, and becomes smaller with distance from the injection site.

$$\Delta V_m = V_o e^{\frac{-x}{\lambda}} \tag{11.37}$$

where $\lambda = \sqrt{\frac{R_{Th}}{R_a}}$ is the membrane length constant, x is the distance away from the injection site, and V_o is the change in membrane potential at the injection site. The range of values for λ is 0.1 to 1 mm. The larger the value of λ, the greater the effect of the stimulation along the length of the membrane.

11.6 HODGKIN–HUXLEY MODEL OF THE ACTION POTENTIAL

Hodgkin and Huxley published five papers in 1952 that described a series of experiments and an empirical model of an action potential in a squid giant axon. Their first four papers described the experiments that characterized the changes in the cell membrane that occurred during the action potential. The last paper presented the empirical model. The empirical model they developed is not a physiological model based on the laws and theory developed in this chapter but a model based on curve fitting by using an exponential function. In this section, highlights of the Hodgkin–Huxley experiments are presented along with the empirical model. All of the figures presented in this section were simulated using SIMULINK and the Hodgkin–Huxley empirical model parameterized with their squid giant axon data.

11.6.1 Action Potentials and the Voltage Clamp Experiment

The ability of nerve cells to conduct action potentials makes it possible for signals to be transmitted over long distances within the nervous system. An important feature of the action potential is that it does not decrease in amplitude as it is conducted away from its site of initiation. An action potential occurs when V_m reaches a value called the threshold potential at the axon hillock (see Fig. 11.1). Once V_m reaches threshold, time- and voltage-dependent conductance changes occur in the active Na^+ and K^+ gates that drive V_m toward E_{Na}, then back to E_K, and finally to the resting potential. These changes in conductance were first described by Hodgkin and Huxley (and Katz as a co-author on one paper and a collaborator on several others). Figure 11.19 illustrates a stylized action potential with the threshold potential at approximately -40 mV.

Stimulation of the postsynaptic membrane along the dendrite and cell body must occur for V_m to rise to the threshold potential at the axon hillock. As previously described, the greater the distance from the axon hillock, the smaller the contribution of postsynaptic membrane stimulation to the change in V_m at the axon hillock. Also, because of the membrane time constant, there is a time delay in stimulation at the postsynaptic membrane and the resultant change in V_m at the axon hillock. Thus, time and distance are important functions in describing the graded response of V_m at the axon hillock.

Once V_m reaches threshold, active Na^+ conductance gates are opened and an inward flow of Na^+ ions results, causing further depolarization. This depolarization increases Na^+ conductance, consequently inducing more Na^+ current. This iterative

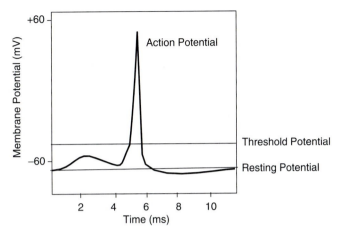

Figure 11.19 Stylized diagram of an action potential once threshold potential is reached at approximately 5 ms. The action potential is due to voltage and time-dependent changes in conductance. The action potential rise is due to Na^+ and the fall is due to K^+ conductance changes.

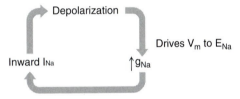

Figure 11.20   Illustration of the conductance gate for sodium.

cycle, shown in Figure 11.20, continues driving V_m to E_{Na} and concludes with the closure of the Na^+ gates. A similar, but slower change in K^+ conductance occurs that drives V_m back to the resting potential. Once an action potential is started, it continues until completion. This is called the "all or none" phenomenon. The active gates for Na^+ and K^+ are both functions of V_m and time.

The action potential moves through the axon at high speeds and appears to jump from one node of Ranvier to the next in myelinated neurons. This occurs because the membrane capacitance of the myelin sheath is very small, making the membrane appear only resistive with almost instantaneous changes in V_m possible.

To investigate the action potential, Hodgkin and Huxley used an unmyelinated squid giant axon in their studies because of its large diameter (up to 1 mm) and long survival time of several hours in seawater at 6.3°C. Their investigations examined the then existing theory that described an action potential as due to enormous changes in membrane permeability that allowed all ions to freely flow across the membrane, driving V_m to zero. As they discovered, this was not the case. The success of the Hodgkin–Huxley studies was based on two new experimental techniques, the space clamp and voltage clamp, and collaboration with Cole and Curtis from Columbia University.

The space clamp allowed Hodgkin and Huxley to produce a constant V_m over a large region of the membrane by inserting a silver wire inside the axon and thus eliminating R_a. The voltage clamp allowed the control of V_m by eliminating the effect of further depolarization due to the influx of I_{Na} and efflux of I_K as membrane permeability changed. Selection of the squid giant axon was fortunate for two reasons: (1) it was large and survived a very long time in seawater and (2) it had only two types of voltage–time-dependent permeable channels. Other types of neurons have more than two voltage–time-dependent permeable channels, which would have made the analysis extremely difficult or even impossible.

Voltage Clamp

To study the variable voltage–time-resistance channels for K^+ and Na^+, Hodgkin and Huxley used a voltage clamp to separate these two dynamic mechanisms so that only the time-dependent features of the channel were examined. Figure 11.21 illustrates the voltage clamp experiment by using the equivalent circuit model previously described. The channels for K^+ and Na^+ are represented using variable voltage–time resistances, and the passive gates for Na^+, K^+, and Cl^- are given by a leakage channel with resistance R_l (that is, the Thévenin equivalent circuit of the passive channels). The function of the voltage clamp is to suspend the interaction between Na^+ and K^+ channel resistance and the membrane potential as shown in Figure 11.22. If the membrane voltage is not clamped, then changes in Na^+ and K^+ channel resistance modify membrane voltage, which then changes Na^+ and K^+ channel resistance, and so on and so forth as previously described.

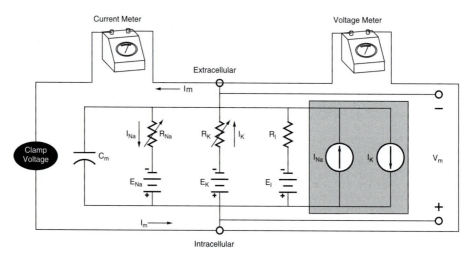

Figure 11.21   Equivalent circuit model of an unmyelinated section of squid giant axon under voltage clamp conditions. The channels for K^+ and Na^+ are now represented using variable voltage–time resistances, and the passive gates for Na^+, K^+, and Cl^- are given by a leakage channel with resistance R_l. The Na-K pump is illustrated within the shaded area of the circuit. In the experiment, the membrane is immersed in the seawater bath.

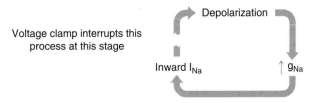

Figure 11.22 Voltage clamp experiment interrupts the cycle shown in Figure 11.20.

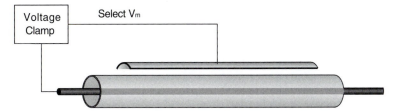

Figure 11.23 Physical setup for the voltage clamp experiment.

A voltage clamp is created by using two sets of electrodes as shown in Figure 11.23. In an experiment, one pair injects current, I_m, to keep V_m constant and another pair is used to observe V_m. To estimate the conductance in the Na$^+$ and K$^+$ channels, I_m is also measured during the experiment. Meters for recording V_m and I_m are illustrated in Figure 11.21. They are placed outside the seawater bath. Today, these would be connected to an analog-to-digital converter (ADC) with data stored in the hard disk of a computer. Back in 1952, these meters were strip chart recorders. The application of a clamp voltage, V_c, causes a change in Na$^+$ conductance that results in an inward flow of Na$^+$ ions. This causes the membrane potential to be more positive than V_c. The clamp removes positive ions from inside the cell, which results in no net change in V_m. The current, I_m, is the dependent variable in the voltage clamp experiment and V_c is the independent variable.

To carry out the voltage clamp experiment, the investigator first selects a clamp voltage and then records the resultant membrane current, I_m, that is necessary to keep V_m at the clamp voltage. Figure 11.24 shows the resulting I_m due to a clamp voltage of -20 mV. Initially, the step change in V_m causes a large current to pass through the membrane that is primarily due to the capacitive current. The clamp voltage also creates a constant leakage current through the membrane that is equal to

$$I_l = \frac{V_c - E_l}{R_l} \tag{11.38}$$

Subtracting both the capacitive and leakage current from I_m leaves only the Na$^+$ and K$^+$ currents. To separate the Na$^+$ and K$^+$ currents, Hodgkin and Huxley substituted a large impermeable cation for Na$^+$ in the external solution. This eliminated the Na$^+$ current and left only the K$^+$ current. Returning the Na$^+$ to the external solution allowed the Na$^+$ current to be estimated by subtracting the capacitive, leakage, and

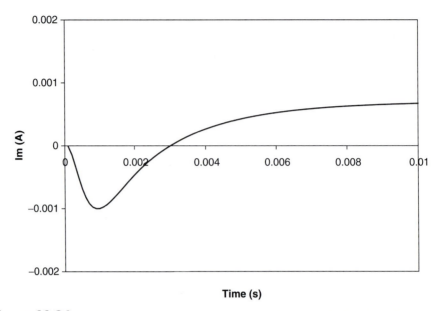

Figure 11.24   Diagram illustrating membrane current I_m due to a $-20\,\text{mV}$ voltage clamp.

K^+ currents from I_m. The Na^+ and K^+ currents due to a clamp voltage of $-20\,\text{mV}$ are illustrated in Figure 11.25. Since the clamp voltage in Figure 11.25 is above threshold, the Na^+ and K^+ channel resistances are engaged and follow a typical profile. The Na^+ current rises to a peak first and then returns to zero as the clamp voltage is maintained.

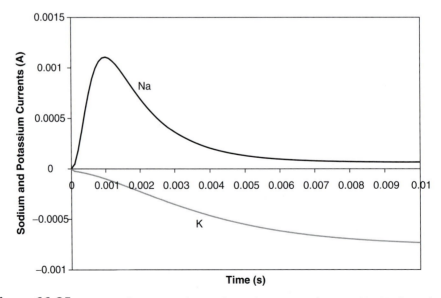

Figure 11.25   Diagram illustrating sodium and potassium currents due to a $-20\,\text{mV}$ voltage clamp.

The K^+ current falls to a steady-state current well after the Na^+ current peaks, and is maintained at this level until the clamp voltage is removed. This general pattern holds for both currents for all clamp voltages above threshold.

The Na^+ and K^+ channel resistance or conductance is easily determined by applying Ohm's law to the circuit in Figure 11.20 and the current waveforms in Figure 11.21

$$I_K = \frac{V_m - E_K}{R_K} = G_K(V_m - E_K) \tag{11.39}$$

$$I_{Na} = \frac{E_{Na} - V_m}{R_{Na}} = G_{Na}(E_{Na} - V_m) \tag{11.40}$$

These conductances are plotted as a function of clamp voltages ranging from $-50\,mV$ to $+20\,mV$ in Figure 11.26.

For all clamp voltages above threshold, the rate of onset for opening Na^+ channels is more rapid than for K^+ channels, and the Na^+ channels close after a period of time whereas K^+ channels remain open while the voltage clamp is maintained. Once the Na^+ channels close, they cannot be opened until the membrane has been hyperpolarized to its resting potential. The time spent in the closed state is called the refractory period. If the voltage clamp is turned off before the time course for Na^+ is complete (returns to zero), G_{Na} almost immediately returns to zero, and G_K returns to zero slowly regardless of whether or not the time course for Na^+ is complete.

Example Problem 11.7

Compute I_c and I_l through a cell membrane for a subthreshold clamp voltage.

Solution

Assume that the Na^+ and K^+ voltage–time-dependent channels are not activated because the stimulus is below threshold. This eliminates these gates from the analysis although this is not actually true, as shown in Example Problem 11.9. The cell membrane circuit is given by

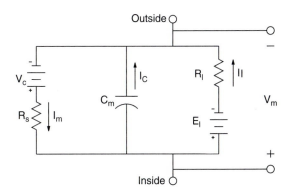

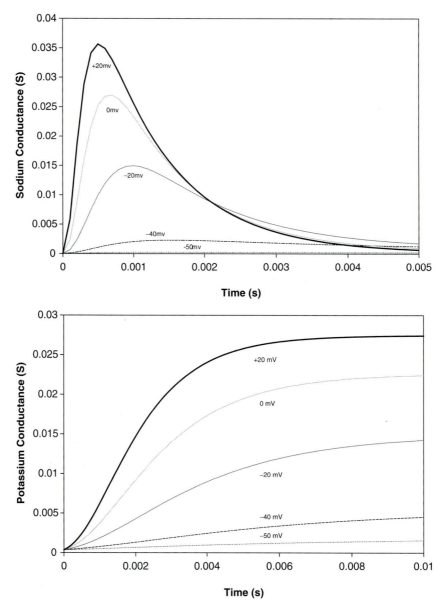

Figure 11.26   Diagram illustrating the change in Na$^+$ and K$^+$ conductance with clamp voltage ranging from −50 mV [below threshold] to +20 mV. Note that the time scales are different in the two conductance plots.

where R_s is the resistance of the wire. Applying Kirchhoff's current law at the cytoplasm gives

$$C_m \frac{dV_m}{dt} + \frac{V_m - E_l}{R_l} + \frac{V_m - V_c}{R_s} = 0$$

Rearranging the terms in the previous equation yields

$$C_m \frac{dV_m}{dt} + \frac{R_l + R_s}{R_l R_s} V_m = \frac{R_l V_c + R_s E_l}{R_l R_s}$$

With the initial condition $V_m(0) = E_l$, the solution is given by

$$V_m = \frac{R_l V_c + R_s E_l}{R_l + R_s} + \frac{R_l(E_l - V_c)}{R_l + R_s} e^{\frac{-(R_l + R_s)t}{R_l R_s C_m}}$$

Now

$$I_c = C_m \frac{dV_m}{dt} = \frac{E_l - V_c}{R_s} e^{-\frac{(R_l + R_s)t}{R_l R_s C_m}}$$

and $I_l = \frac{V_m - E_l}{R_l}$. At steady state, $I_l = \frac{V_c - E_l}{R_l + R_s}$.   ■

Reconstruction of the Action Potential

By analyzing the estimated G_{Na} and G_K from voltage clamp pulses of various amplitudes and durations, Hodgkin and Huxley were able to obtain a complete set of nonlinear empirical equations that described the action potential. Simulations using these equations accurately describe an action potential in response to a wide variety of stimulations. Before presenting these equations, it is important to qualitatively understand the sequence of events that occur during an action potential by using previously described data and analyses. An action potential begins with a depolarization above threshold that causes an increase in G_{Na} and results in an inward Na^+ current. The Na^+ current causes a further depolarization of the membrane, which then increases the Na^+ current. This continues to drive V_m to the Nernst potential for Na^+. As shown in Figure 11.26, G_{Na} is a function of both time and voltage and peaks and then falls to zero. During the time it takes for G_{Na} to return to zero, G_K continues to increase, which hyperpolarizes the cell membrane and drives V_m from E_{Na} to E_K. The increase in G_K results in an outward K^+ current. The K^+ current causes further hyperpolarization of the membrane, which then increases K^+ current. This continues to drive V_m to the Nernst potential for K^+, which is below resting potential. Figure 11.27 illustrates the changes in V_m, G_{Na}, and G_K during an action potential.

The circuit shown in Figure 11.16 is a useful tool for modeling the cell membrane during small subthreshold depolarizations. This model assumes that the K^+ and Na^+ currents are small enough to neglect. As illustrated in Example Problem 11.6, a current pulse sent through the cell membrane briefly creates a capacitive current, which decays exponentially and creates an exponentially increasing I_l. Once the current pulse is turned off, capacitive current flows again and exponentially decreases to zero. The leakage current also exponentially decays to zero.

As the current pulse magnitude is increased, depolarization of the membrane increases, causing activation of the Na$^+$ and K$^+$ voltage–time-dependent channels. For sufficiently large depolarizations, the inward Na$^+$ current exceeds the sum of the outward K$^+$ and leakage currents ($I_{Na} > I_K + I_l$). The value of V_m at this current is called threshold. Once the membrane reaches threshold, the Na$^+$ and K$^+$ voltage–time channels are engaged and run to completion as shown in Figure 11.27.

If a slow-rising stimulus current is used to depolarize the cell membrane, then the threshold will be higher. During the slow approach to threshold, inactivation of G_{Na} channels occurs and activation of G_K channels develops before threshold is reached. The value of V_m, where $I_{Na} > I_K + I_l$ is satisfied, is much larger than if the approach to threshold occurs quickly.

11.6.2   Equations Describing G$_{Na}$ and G$_K$

The empirical equation used by Hodgkin and Huxley to model G_{Na} and G_K is of the form

$$G(t) = \left(A + Be^{-Ct}\right)^D \tag{11.41}$$

Values for the parameters A, B, C, and D were estimated from the voltage clamp data that were collected on the squid giant axon. Not evident in Equation 11.41 is the voltage dependence of the conductance channels. The voltage dependence is captured

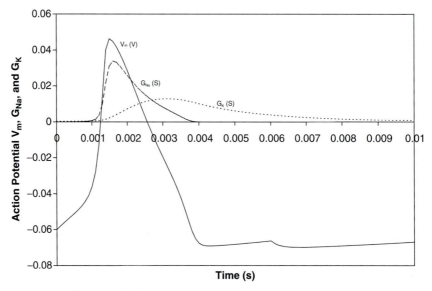

Figure 11.27 V_m, G_{Na}, and G_K during an action potential.

in the parameters as described in this section. In each of the conductance models, D is selected as 4 to give a best fit to the data. Figure 11.27 was actually calculated using SIMULINK, a simulation package that is part of MATLAB, and the parameter estimates found by Hodgkin and Huxley. Details concerning the simulation are covered later in this section.

Potassium

The potassium conductance waveform shown in Figure 11.26 is described by a rise to a peak while the stimulus is applied. This aspect is easily included in a model of G_K by using the general Hodgkin–Huxley expression as follows.

$$G_K = \bar{G}_K n^4 \qquad (11.42)$$

where $\bar{G}_K$ is maximum K$^+$ conductance and n is thought of as a rate constant and given as the solution to the following differential equation:

$$\frac{dn}{dt} = \alpha_n(1 - n) - \beta_n n \qquad (11.43)$$

where

$$\alpha_n = 0.01 \frac{V + 10}{e^{\frac{V+10}{10}} - 1}$$
$$\beta_n = 0.125 e^{\frac{V}{80}}$$
$$V = V_{rp} - V_m$$

V_{rp} is the membrane potential at rest without any membrane stimulation. Note that V is the displacement from resting potential and should be negative. Clearly, G_K is a time-dependent variable since it depends on Equation 11.43 and a voltage-dependent variable since n depends on voltage because of α_n and β_n.

Sodium

The sodium conductance waveform in Figure 11.26 is described by a rise to a peak with a subsequent decline. These aspects are included in a model of G_{Na} as the product of two functions, one describing the rising phase and the other describing the falling phase, and modeled as

$$G_{Na} = \bar{G}_{Na} m^3 h \qquad (11.44)$$

where $\bar{G}_{Na}$ is maximum Na$^+$ conductance and m and h are thought of as rate constants and given as the solutions to the following differential equations:

$$\frac{dm}{dt} = \alpha_m(1 - m) - \beta_m m \qquad (11.45)$$

where

$$\alpha_m = 0.1 \frac{V + 25}{e^{\frac{V+25}{10}} - 1}$$

$$\beta_m = 4e^{\frac{V}{18}}$$

and

$$\frac{dh}{dt} = \alpha_h(1 - h) - \beta_h h \qquad (11.46)$$

where

$$\alpha_h = 0.07e^{\frac{V}{20}}$$

$$\beta_h = \frac{1}{e^{\frac{V+30}{10}} + 1}$$

Note that m describes the rising phase and h describes the falling phase of G_{Na}. The units for the α_i's and β_i's in Equations 11.43, 11.45, and 11.46 are ms^{-1} while n, m, and h are dimensionless and range in value from 0 to 1.

Example Problem 11.8

Calculate G_K and G_{Na} at resting potential for the squid giant axon using the Hodgkin–Huxley model. Parameter values are $\bar{G}_K = 36 \times 10^{-3} S$ and $\bar{G}_{Na} = 120 \times 10^{-3} S$.

Solution

At resting potential, G_K and G_{Na} are constant with values dependent on n, m, and h. Since the membrane is at steady state $\frac{dn}{dt} = 0$, $\frac{dm}{dt} = 0$ and $\frac{dh}{dt} = 0$. Using Equations 11.43, 11.45, and 11.46, at resting potential and steady state

$$n = \frac{\alpha_n^0}{\alpha_n^0 + \beta_n^0}$$

$$m = \frac{\alpha_m^0}{\alpha_m^0 + \beta_m^0}$$

$$h = \frac{\alpha_h^0}{\alpha_h^0 + \beta_h^0}$$

where α_i^0 is α at $V = 0$ for $i = n$, m and h, and β_i^0 is β at $V = 0$ for $i = n$, m, and h. Calculations yield $\alpha_n^0 = 0.0582$, $\beta_n^0 = 0.125$, $n = 0.31769$, $\alpha_m^0 = 0.2236$, $\beta_m^0 = 4$, $m = 0.05294$, $\alpha_h^0 = 0.07$, $\beta_h^0 = 0.04742$, and $h = 0.59615$. Therefore, at resting potential and steady state

$$G_K = \bar{G}_K n^4 = 36.0 \times 10^{-3}(0.31769)^4 = 0.3667 \times 10^{-3} S$$

and

$$G_{Na} = \bar{G}_{Na} m^3 h = 120.0 \times 10^{-3} (0.05294)^3 \times 0.59615 = 0.010614 \times 10^{-3} S \quad \blacksquare$$

11.6.3 Equation for the Time Dependence of the Membrane Potential

Figure 11.28 shows a model of the cell membrane that is stimulated via an external stimulus, I_m, which is appropriate for simulating action potentials. Applying Kirchhoff's current law at the cytoplasm yields

$$I_m = G_K(V_m - E_K) + G_{Na}(V_m - E_{Na}) + \frac{(V_m - E_l)}{R_l} + C_m \frac{dV_m}{dt} \tag{11.47}$$

where G_K and G_{Na} are the voltage–time-dependent conductances given by Equations 11.42 and 11.44.

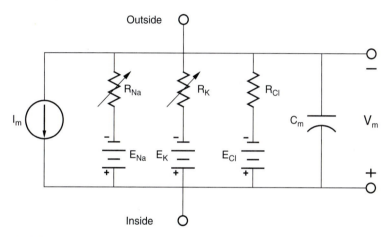

Figure 11.28 Circuit model of an unmyelinated section of squid giant axon. The channels for K^+ and Na^+ are represented using the variable voltage–time conductances given in Equations 11.42 and 11.44. The passive gates for Na^+, K^+, and Cl^- are given by a leakage channel with resistance, R_1, and Nernst potential, E_1. The Na-K pump is not drawn for ease in analysis since it does not contribute any current to the rest of the circuit.

Example Problem 11.9

For the squid giant axon, compute the size of the current pulse (magnitude and pulse width) necessary to raise the membrane potential from its resting value of -60 mV to -40 mV and then back to its resting potential. Neglect any changes in K^+ and Na^+ conductances from resting potential but include G_K and G_{Na} at resting potential in the analysis. Hodgkin–Huxley parameter values for the squid giant axon are $G_l = \frac{1}{R_l} = 0.3 \times 10^{-3}$ S, $\bar{G}_K = 36 \times 10^{-3}$ S, $\bar{G}_{Na} = 120 \times 10^{-3}$ S, $E_K = -72 \times 10^{-3}$ V, $E_l = -49.4 \times 10^{-3}$ V, $E_{Na} = 55 \times 10^{-3}$ V, and $C_m = 1 \times 10^{-6}$ F.

Solution

Let current I_m be given by

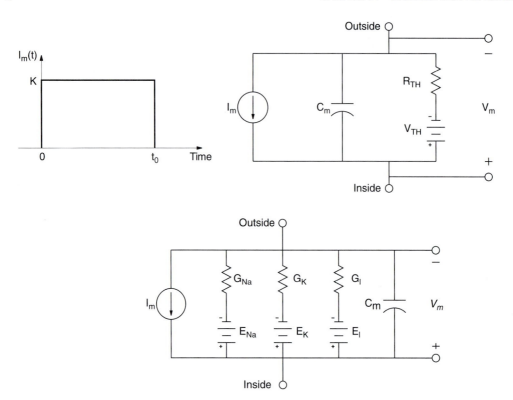

In Example Problem 11.8, the conductances at resting potential were calculated as $G_K = 0.3667 \times 10^{-3}\,S$ and $G_{Na} = 0.010614 \times 10^{-3}\,S$. Since G_K and G_{Na} remain constant for a subthreshold current stimulus in this problem, the circuit in Figure 11.28 reduces to the circuit shown above right. For ease in analysis, this circuit is replaced by the Thévenin equivalent circuit shown on the right with

$$R_{Th} = \frac{1}{G_{Na} + G_K + G_l} = 1.4764\,k\Omega$$

and $V_{Th} = -60\,mV$.

Since the solution in Example Problem 11.6 is the same as the solution in this problem,

$$V_m(t) = V_{Th} + R_{Th}K\left(1 - e^{-\frac{t}{R_{Th}C_m}}\right)u(t) - R_{Th}K\left(1 - e^{-\frac{t-t_o}{R_{Th}C_m}}\right)u(t - t_o)$$

For convenience, assume the current pulse $t_0 > 5\tau$. Therefore, for $t \le t_0$, $V_m = -40\,mV$ according to the problem statement at steady state, and from the previous equation, V_m reduces to

$$V_m = -0.040 = V_{Th} + K \times R_{Th} = -0.060 + K \times 1,476.4$$

which yields $K = 13.6\,\mu A$. Since $\tau = R_{Th}C_m = 1.47\,ms$, any value for t_0 greater than $5\tau = 7.35\,ms$ brings V_m to $-40\,mV$ with $K = 13.6\,\mu A$. Naturally, a larger current pulse magnitude is needed for an action potential because as V_m exponentially approaches threshold (reaching it with a duration of infinity), the Na^+ conductance channels become active and shut down.

To find V_m during an action potential, four differential equations (Equations 11.43 and 11.45 through 11.47) and six algebraic equations (α_i's and β_i's in Equations 11.43, 11.45, and 11.46) need to be solved. Since the system of equations is nonlinear due to the n^4 and m^3 conductance terms, an analytic solution is not possible. To solve for V_m, it is therefore necessary to simulate the solution. There are many computer tools that allow a simulation solution of nonlinear systems. SIMULINK, a general-purpose toolbox in MATLAB that simulates solutions for linear and nonlinear, continuous and discrete dynamic systems, is used in this textbook. SIMULINK is a popular and widely used simulation program with a user-friendly interface that is fully integrated within MATLAB. SIMULINK is interactive and works on most computer platforms. Analogous to an analog computer, programs for SIMULINK are developed based on a block diagram of the system.

The SIMULINK program for an action potential is shown in Figures 11.29 through 11.32. The block diagram is created by solving for the highest derivative term in Equation 11.47, which yields Equation 11.48. The SIMULINK program is then created by using integrators, summers, and so forth:

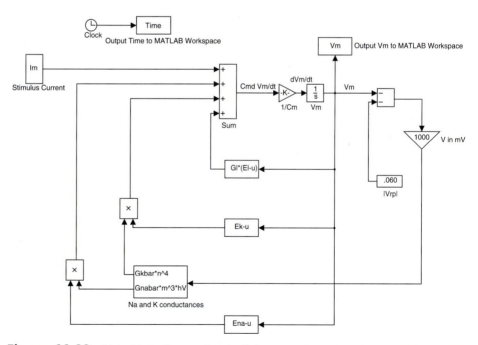

Figure 11.29 Main block diagram for simulating an action potential using SIMULINK. The stimulus current is a pulse created by subtracting two step functions as described in Figure 11.30. The Na^+ and K^+ conductance function blocks are described in Figures 11.31 and 11.32.

$$\frac{dV_m}{dt} = \frac{1}{C_m}(I_m + G_K(E_K - V_m) + G_{Na}(E_{Na} - V_m) + G_l(E_l - V_m)) \qquad (11.48)$$

Figure 11.25 shows the main block diagram. Figures 11.29–11.31 are subsystems that were created for ease in analysis. The Workspace output blocks were used to pass simulation results to MATLAB for plotting. Parameter values used in the simulation were based on the empirical results from Hodgkin and Huxley, with $G_l = \frac{1}{R_l} = 0.3 \times 10^{-3}S$, $\bar{G}_K = 36 \times 10^{-3}S$, $\bar{G}_{Na} = 120 \times 10^{-3}S$, $E_K = -72 \times 10^{-3}V$, $E_l = -49.4 \times 10^{-3}V$, $E_{Na} = 55 \times 10^{-3}V$, and $C_m = 1 \times 10^{-6}F$. Figure 11.23 is a SIMULINK simulation of an action potential. The blocks Gl*(El-u), Ek-u, and Ena-u are function blocks that were used to represent the terms $G_l(E_l - V_m)$, $(E_K - V_m)$, and $(E_{Na} - V_m)$ in Equation 11.48, respectively.

The stimulus pulse current was created by using the SIMULINK step function as shown in Figure 11.30. The first step function starts at $t = 0$ with magnitude K, and the other one starts at $t = t_0$ with magnitude $-K$. The current pulse should be sufficient to quickly bring V_m above threshold.

Figure 11.31 illustrates the SIMULINK program for the conductance channels for Na^+ and K^+. Function blocks Gkbar*u ^4, Gnabar*u ^3, and Gnabar*m ^3*h

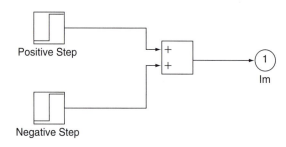

Figure 11.30   The stimulus current.

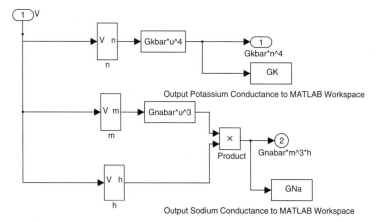

Figure 11.31   SIMULINK program for the K^+ and Na^+ conductance channels.

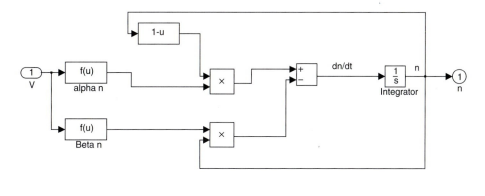

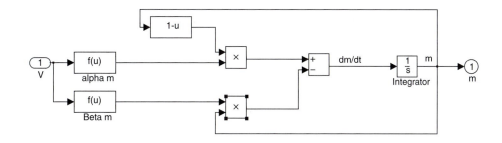

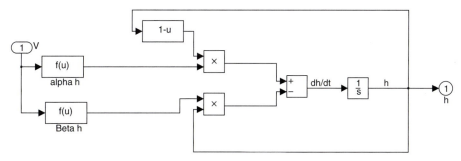

Figure 11.32   SIMULINK program for the alpha and beta terms in Equations 11.43, 11.45, and 11.46.

represent $\bar{G}_K n^4$, $\bar{G}_{Na} m^3$, and $\bar{G}_{Na} m^3 h$, respectively. The subsystems n, m, and h are described in Figure 11.32 and are based on six algebraic equations for α_i's and β_i's in Equations 11.43, 11.45, and 11.46.   ■

11.7   MODEL OF THE WHOLE NEURON

This section brings together the entire neuron, combing the dendrite, soma, axon, and presynaptic terminal. Dendrites and axons can be modeled as a series of cylindrical compartments, each connected together with an axial resistance as described in Section 11.5.3. Both the axon and dendrites are connected to the soma. Of course, real neurons have many different arrangements, such as the dendrite connected to the axon, which then connects to the soma. The basic neuron consists of many dendrites, one axon, and one soma. Note that the dendrite and axon do not have to have constant-diameter cylinders, but may narrow toward the periphery.

As described previously, Figure 11.17 illustrates a generic electrical dendrite compartment model with passive channels, and Figure 11.28 illustrates the axon compartment with active channels at the axon hillock and the node of Ranvier. To model the myelinated portion of the axon, a set of passive compartments, like the dendrite compartment, can be used with capacitance, passive ion channels, and axial resistance. Shown in Figure 11.33 is a portion of the axon with myelin sheath, with three passive channels, and an active component for the node of Ranvier. The structure in Figure 11.33 can be modified for any number of compartments as appropriate. The soma can be modeled as an active or passive compartment depending on the type of neuron.

To model the neuron in Figure 11.33, Kirchhoff's current law is applied for each compartment (i.e., each line in Eq. 11.49 is for a compartment), giving

$$
\begin{aligned}
\ldots + C_m \frac{dV_m}{dt} &+ \frac{(V_m - V_{TH})}{R_{TH}} + \frac{(V_m - V_m')}{R_a} \\
+ C_m \frac{dV_m'}{dt} &+ \frac{(V_m' - V_{TH})}{R_{TH}} + \frac{(V_m' - V_m'')}{R_a} \\
+ C_m \frac{dV_m''}{dt} &+ \frac{(V_m'' - V_{TH})}{R_{TH}} + \frac{(V_m'' - V_m''')}{R_a} \\
+ G_K(V_m''' - E_K) &+ G_{Na}(V_m''' - E_{Na}) + \frac{(V_m''' - E_l)}{R_l} + C_m \frac{dV_m'''}{dt} + \ldots
\end{aligned}
\tag{11.49}
$$

Because neurons usually have other channels in addition to the three of the squid giant axon, a model of the neuron should have the capability of including other channels,

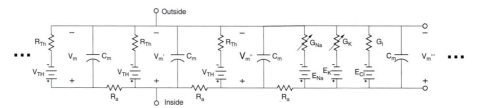

Figure 11.33   A segment of the axon with active and passive compartments.

such as a fast sodium channel, delayed potassium conductance, or high-threshold calcium conductance. Additional ion channels can be added for each compartment in Equation 11.49, by adding

$$\sum_{i=1}^{n} G_i(V_m - E_i)$$

for each compartment for channels $i = 1, n$. The values of C_m, R_{TH}, R_a, and G_i are dependent on the size of the compartment and the type of neuron modeled.

A complete model of the neuron can be constructed by including as many dendritic branches as needed, each described using Figure 11.17, each modeled by

$$\ldots + C_m \frac{dV_m}{dt} + \frac{(V_m - V_{TH})}{R_{TH}} + \frac{(V_m - V'_m)}{R_a} + C_m \frac{dV'_m}{dt} + \frac{(V'_m - V_{TH})}{R_{TH}} + \frac{(V'_m - V''_m)}{R_a} + \ldots$$

$$(11.50)$$

a soma with passive or active properties using either

$$C_m \frac{dV_m}{dt} + \frac{(V_m - V_{TH})}{R_{TH}} + \frac{(V_m - V'_m)}{R_a} \qquad (11.51)$$

or

$$G_K(V'''_m - E_K) + G_{Na}(V'''_m - E_{Na}) + \frac{(V'''_m - E_l)}{R_l} + C_m \frac{dV'''_m}{dt} \qquad (11.52)$$

and an axon using Equation 11.49 as described in Rodriguez and Enderle (2004). Except for the terminal compartment, two inputs are needed for the dendrite compartment; the input defined by the previous compartment's membrane potential and the next compartment's membrane potential. Additional neurons can be added using the same basic neuron, interacting with each other using the current from the adjacent neuron (presynaptic terminal) to stimulate the next neuron.

For illustration purposes, the interaction between two adjacent neurons is modeled using SIMULINK, shown in Figure 11.34, and the results shown in Figure 11.35. Three voltage-dependent channels for Na+, K+, and Ca2+, and also a leakage channel are used for the axon. We use a myelinated axon with four passive compartments between each node of Ranvier. The total axon consists of three active compartments and two myelinated passive segments. The dendrite consists of five passive compartments, and the soma is a passive spherical compartment. The stimulus is applied at the terminal end of the dendrite of the first neuron. It is modeled as an active electrode compartment. The size of each axon compartment is the same but different from the dendrite compartment. The input to the first neuron is shown in Figure 11.36.

Although this chapter has focused on the neuron, it is important to note that numerous other cells have action potentials that involve signaling or triggering.

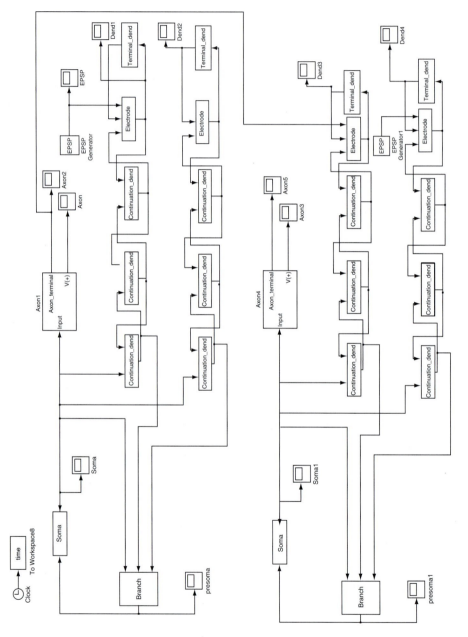

Figure 11.34   SIMULINK model for two adjacent neurons.

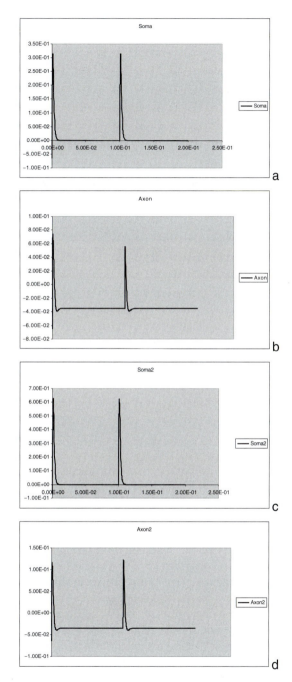

Figure 11.35   (a) Soma of first neuron, (b) axon of first neuron, (c) soma of second neuron, and (d) axon of second neuron.

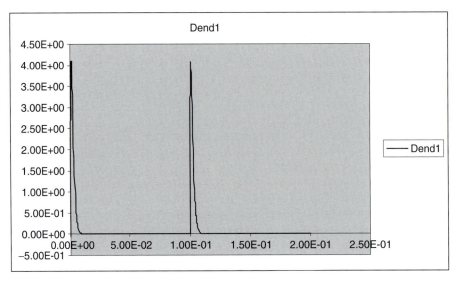

Figure 11.36   The stimulus to the first neuron.

Many of the principles discussed in this chapter also apply to these other cells but the action potential defining equations are different. For example, the cardiac action potential can be defined with a DiFranceso–Noble, Luo–Rudy, or other models rather than a Hodgkin–Huxley model of the neuron.

EXERCISES

1. Assume a membrane is permeable to only Ca^{++}. (a) Derive the expression for the flow of Ca^{++}; (b) Find the Nernst potential for Ca^{++}.
2. Assume that a membrane is permeable to Ca^{++} and Cl^{-} but not to a large cation R^{+}. The inside concentrations are $[RCl] = 100\,mM$ and $[CaCl_2] = 200\,mM$, and the outside concentration is $[CaCl_2] = 300\,mM$. (a) Derive the Donnan equilibrium. (b) Find the steady-state equilibrium concentration for Ca^{++}.
3. Assume that a membrane is permeable to Ca^{++} and Cl^{-}. The initial concentrations on the inside are different from the outside, and these are the only ions in the solution. (a) Write an equation for J_{Ca} and J_{Cl}. (b) Write an expression for the relationship between J_{Ca} and J_{Cl}. (c) Find the equilibrium voltage. (d) Find the relationship between the voltage across the membrane and $[CaCl_2]$ before equilibrium.
4. Assume that a membrane is permeable to only ion R^{+++}. The inside concentration is $[RCl_3] = 2\,mM$ and the outside concentration is $[RCl_3] = 1.4\,mM$.

(a) Write an expression for the flow of R^{+++}. (b) Derive the Nernst potential for R^{+++} at equilibrium.

5. Derive the Goldman equation for a membrane in which Na^+, K^+, and Cl^- are the only permeable ions.

6. Calculate V_m for the frog skeletal muscle at room temperature.

7. The following steady-state concentrations and permeabilities are given for a membrane.

Ion	Cytoplasm (mM)	Extracellular Fluid (mM)	Ratio of Permeabilities
K^+	140	2.5	1.0
Na^+	13	110	0.019
Cl^-	3	90	0.381

(a) Find the Nernst potential for K^+. (b) What is the resting potential predicted by the Goldman equation? (c) Explain whether space charge neutrality is satisfied. (d) Explain why the equilibrium membrane potential does not equal zero.

8. A membrane has the following concentrations and permeabilities.

Ion	Cytoplasm (mM)	Extracellular Fluid (mM)	Ratio of Permeabilities
K^+	?	4	?
Na^+	41	276	0.017
Cl^-	52	340	0.412

The resting potential of the membrane is -52 mV at room temperature. Find the K^+ cytoplasm concentration.

9. The following steady-state concentrations and permeabilities are given for a membrane. Note that A^+ is not permeable.

Ion	Cytoplasm (mM)	Extracellular Fluid (mM)	Ratio of Permeabilities
K^+	136	15	1.0
Na^+	19	155	0.019
Cl^-	78	112	0.381
A^+	64	12	-

(a) Find the Nernst potential for Cl^-. (b) What is the resting potential predicted by the Goldman equation? (c) Explain whether space charge neutrality is satisfied. (d) Explain why the equilibrium membrane potential does not equal zero. (e) Explain why the resting potential does not equal the Nernst potential of any of the ions.

10. A membrane is permeable to B^{+++} and Cl^-, but not to a large cation R^+. The following initial concentrations are given.

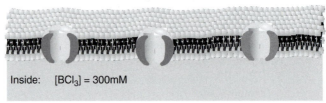

       (a) Derive the Donnan equilibrium. (b) Find the steady-state equilibrium concentration for B^{+++}.

11. The following membrane is permeable to Ca^{++} and Cl^-. (a) Write expressions for the flow of Ca^{++} and Cl^- ions. (b) Write an expression for the relationship between J_{Ca} and J_{Cl}. (c) Find the equilibrium voltage. (d) Find the relationship between voltage across the membrane and the concentration of $CaCl_2$ before equilibrium is reached.

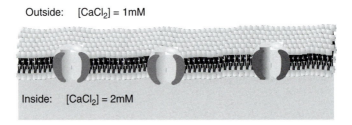

12. The following membrane has an active Ca^{++} pump. Assume that the membrane is permeable to both Ca^{++} and Cl^-, and the Ca^{++} pump flow is J_p. The width of the membrane is δ. Find the pump flow as a function of $[Ca^{++}]$.

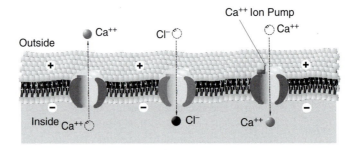

13. The membrane shown is permeable to K^+ and Cl^-. The active pump transports K^+ from the outside to the inside of the cell. The width of the membrane is δ.

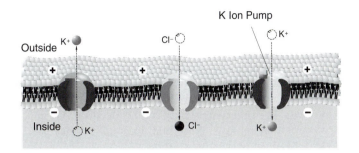

(a) Write an equation for the flow of each ion. (b) Find the flows at equilibrium. (c) Find the pump flow as a function of $([K^+]_i - [K^+]_o)$. (d) Qualitatively describe the ion concentration on each side of the membrane.

14. The following membrane is given with two active pumps. Assume that the membrane is permeable to Na^+, K^+, and Cl^-, and $J_p(K) = J_p(Na) = J_p$. The width of the membrane is δ. Solve for the quantity $([Cl^-]_i - [Cl^-]_o)$ as a function of J_p.

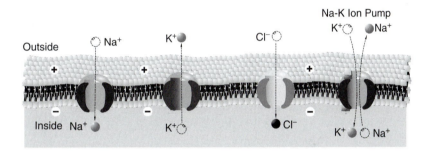

15. The following steady-state concentrations and permeabilities are given for a membrane. Note that A^+ is not permeable. The ion channel resistances are $R_K = 1.7\,k\Omega$, $R_{Na} = 9.09\,k\Omega$, and $R_{Cl} = 3.125\,k\Omega$.

Ion	Cytoplasm (mM)	Extracellular Fluid (mM)	Ratio of Permeabilities
K^+	168	6	1.0
Na^+	50	337	0.019
Cl^-	41	340	0.381
A^+	64	12	-

(a) Find the Nernst potential for each ion. (b) Draw a circuit model for this membrane (Hint: See Fig. 11.13). (c) Find the membrane resting potential using the circuit in part (b). (d) Find the Thévenin equivalent circuit for the circuit in part (b).

16. Suppose the membrane in Figure 11.13 is given with $R_K = 0.1\,k\Omega$, $R_{Na} = 2\,k\Omega$, $R_{Cl} = 0.25\,k\Omega$, $E_K = -74\,mV$, $E_{Na} = 55\,mV$, and $E_{Cl} = -68\,mV$. (a) Find V_m. (b) Find the Thévenin equivalent circuit.

17. Suppose a membrane has an active Na-K pump with $R_K = 0.1\,k\Omega$, $R_{Na} = 2\,k\Omega$, $R_{Cl} = 0.25\,k\Omega$, $E_K = -74\,mV$, $E_{Na} = 55\,mV$, and $E_{Cl} = -68\,mV$, as shown in Figure 11.14. Find I_{Na} and I_K for the active pump.

18. The following steady-state concentrations and permeabilities are given for a membrane. Note that A^+ is not permeable.

Ion	Cytoplasm (mM)	Extracellular Fluid (mM)	Ratio of Permeabilities
K^+	140	2.5	1.0
Na^+	13	110	0.019
Cl^-	3	90	0.381
A^+	64	12	-

(a) If $R_K = 1.7\,k\Omega$ and $R_{Cl} = 3.125\,k\Omega$, then find R_{Na}. (b) Find the Thévenin equivalent circuit model.

19. Suppose that a membrane that has an active Na-K pump with $R_K = 0.1\,k\Omega$, $R_{Na} = 2\,k\Omega$, $R_{Cl} = 0.25\,k\Omega$, $E_K = -74\,mV$, $E_{Na} = 55\,mV$, $E_{Cl} = -68\,mV$, and $C_m = 1\,\mu F$ as shown in Figure 11.15, is stimulated by a current pulse of $10\,\mu A$ for 6 ms. (a) Find V_m. (b) Find the capacitive current. (c) Calculate the size of the current pulse applied at 6 ms for 1 ms, necessary to raise V_m to $-40\,mV$. (d) If the threshold voltage is $-40\,mV$ and the stimulus is applied as in part (c), then explain whether an action potential occurs.

20. Suppose that a membrane that has an active Na-K pump with $R_K = 2.727\,k\Omega$, $R_{Na} = 94.34\,k\Omega$, $R_{Cl} = 3.33\,k\Omega$, $E_K = -72\,mV$, $E_{Na} = 55\,mV$, $E_{Cl} = -49.5\,mV$, and $C_m = 1\,\mu F$ and is shown in Figure 11.15 is stimulated by a current pulse of $13\,\mu A$ for 6 ms. Find (a) V_m, (b) I_K, and (c) the capacitive current.

21. Suppose a membrane has an active Na-K pump with $R_K = 1.75\,k\Omega$, $R_{Na} = 9.09\,k\Omega$, $R_{Cl} = 3.125\,k\Omega$, $E_K = -85.9\,mV$, $E_{Na} = 54.6\,mV$, $E_{Cl} = -9.4\,mV$, and $C_m = 1\,\mu F$ as shown in Figure 11.15. (a) Find the predicted resting membrane potential. (b) Find V_m if a small subthreshold current pulse is used to stimulate the membrane.

22. Suppose a membrane has an active Na-K pump with $R_K = 2.727\,k\Omega$, $R_{Na} = 94.34\,k\Omega$, $R_{Cl} = 3.33\,k\Omega$, $E_K = -72\,mV$, $E_{Na} = 55\,mV$, $E_{Cl} = -49.5\,mV$, and $C_m = 1\,\mu F$ as shown in Figure 11.15. Design a stimulus that will drive V_m to threshold at 3 ms. Assume that the threshold potential is $-40\,mV$. (a) Find the current pulse magnitude and duration. (b) Find and sketch V_m.

23. Suppose a current pulse of $20\,\mu A$ is passed through the membrane of a squid giant axon. The Hodgkin–Huxley parameter values for the squid giant axon are $G_l = \frac{1}{R_l} = 0.3 \times 10^{-3}\,S$, $\bar{G}_K = 36 \times 10^{-3}\,S$, $\bar{G}_{Na} = 120 \times 10^{-3}\,S$, $E_K = -72 \times 10^{-3}\,V$, $E_l = -49.4 \times 10^{-3}\,V$, $E_{Na} = 55 \times 10^{-3}\,V$ and $C_m = 1 \times 10^{-6}\,F$. Simulate the action potential. Plot (a) V_m, G_{Na}, and G_K versus time, (b) V_m, n, m, and h versus time, and (c) V_m, I_{Na}, I_K, I_c, and I_l versus time.

24. Suppose a current pulse of $20\,\mu A$ is passed through an axon membrane. The parameter values for the axon are $G_l = \frac{1}{R_l} = 0.3 \times 10^{-3}\,S$, $\bar{G}_K = 36 \times 10^{-3}\,S$, $\bar{G}_{Na} = 120 \times 10^{-3}\,S$, $E_K = -12 \times 10^{-3}\,V$, $E_l = 10.6 \times 10^{-3}\,V$, $E_{Na} = 115 \times 10^{-3}\,V$, and $C_m = 1 \times 10^{-6}\,F$. Assume that Equations 11.41 through 11.48 describe the axon. Simulate the action potential. Plot (a) V_m, G_{Na}, and G_K versus time, (b) V_m, n, m, and h versus time, and (c) V_m, I_{Na}, I_K, I_c, and I_l versus time.

25. This exercise examines the effect of the threshold potential on an action potential for the squid giant axon. The Hodgkin–Huxley parameter values for the squid giant axon are $G_l = \frac{1}{R_l} = 0.3 \times 10^{-3}\,S$, $\bar{G}_K = 36 \times 10^{-3}\,S$, $\bar{G}_{Na} = 120 \times 10^{-3}\,S$, $E_K = -72 \times 10^{-3}\,V$, $E_l = -49.4 \times 10^{-3}\,V$, $E_{Na} = 55 \times 10^{-3}\,V$, and $C_m = 1 \times 10^{-6}\,F$. (a) Suppose a current pulse of $-10\,\mu A$, which hyperpolarizes the membrane, is passed through the membrane of a squid giant axon for a very long time. At time $t = 0$, the current pulse is removed. Simulate the resultant action potential. (b) The value of the threshold potential is defined as when $I_{Na} > I_K + I_l$. Changes in threshold potential can be implemented easily by changing the value of 25 in the equation for α_m to a lower value. Suppose the value of 25 is changed to 10 in the equation, defining α_m and a current pulse of $-10\,\mu A$ (hyperpolarizes the membrane) is passed through the membrane of a squid giant axon for a very long time. At time $t = 0$, the current pulse is removed. Simulate the resultant action potential.

26. Simulate the plots shown in Figures 11.20–11.22 in the voltage clamp mode.

27. Select an input current waveform necessary to investigate the refractory period that follows an action potential. (Hint: Use a two-pulse current input.) What is the minimum refractory period? If the second pulse is applied before the minimum refractory period, how much larger is the stimulus magnitude that is needed to generate an action potential?

28. Explain whether the following circuit allows the investigator to conduct a voltage clamp experiment. The unmyelinated axon is sealed at both ends. A silver wire is inserted inside the axon that eliminates R_a. Clearly state any assumptions.

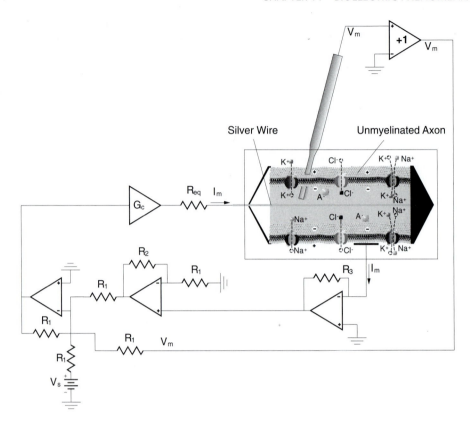

SUGGESTED READING

Bahill, A.T. (1981). *Bioengineering: Biomedical, Medical and Clinical Engineering.* Prentice-Hall, Englewood Cliffs, NJ.

Bronzino, J.D. (2000). *The Biomedical Engineering Handbook.* CRC, Boca Raton, FL.

Bower, J.M. and Beeman, D. (1998). *The Book of Genesis: Exploring Realistic Neural Models with the General Neural Simulation System.* Springer-Verlag, New York.

Deutsch, S. and Deutsch, A. (1993). *Understanding the Nervous System—An Engineering Perspective.* IEEE Press, New York.

DiFrancesco, D. and Noble, D. (1985). A model of cardiac electrical activity incorporating ionic pumps and concentration changes. *Phil Trans R. Soc London [B]* **307**, 307–353.

Enderle, J.D. (2002). Neural control of saccades. In J. Hyönä, D. Munoz, W. Heide, and R. Radach (Eds.), *The Brain's Eyes: Neurobiological and Clinical Aspects to Oculomotor Research, Progress in Brain Research, V. 140.* Elsevier, Amsterdam.

Fulton, J.F. and Cushing, H. (1936). A bibliographical study of the Galvani and Aldini writings on animal electricity. *Ann. Sci.* **1**, 239–268.

Guyton, A.C. and Hall, J. E. (1995). *Textbook on Medical Physiology,* 9th Ed. Saunders, Philadelphia.

Hille, B. (1992). *Ionic Channels of Excitable Membranes*, 2nd Ed. Sunderland, Massachusetts.

Hodgkin, A., Huxley, A. and Katz, B. (1952). Measurement of current-voltage relations in the membrane of the giant axon of *Loligo. J. Physiol. (London)* **116**, 424–448.

Hodgkin, A. and Huxley, A. (1952a). Currents carried by sodium and potassium ions through the membrane of the giant axon of *Loligo. J. Physiol. (London)* **116**, 449–472.

Hodgkin, A. and Huxley, A. (1952b). Currents carried by sodium and potassium ions through the membrane of the giant axon of *Loligo. J. Physiol. (london)* **116**, 49–472

Hodgkin, A. and Huxley, A. (1952c). The components of membrane conductance in the giant axon of *Loligo. J. Physiol. (London)* **116**, 473–496.

Hodgkin, A. and Huxley, A. (1952d). The dual effect of membrane potential on sodium conductance in the giant axon of *Loligo. J. Physiol. (London)* **116**, 497–506.

Hodgkin, A. and Huxley, A. (1952e). A quantitative description of membrane current and its application to conduction and excitation in nerve. *J. Physiol. (London)* **117**, 500–544.

Kandel, E.R., Schwartz, J.H. and Jessell, T.M. (2000). *Principles of Neural Science,* 5th Ed. McGraw-Hill, New York.

Keener, J. and Sneyd, J. (1998). *Mathematical Physiology.* Springer, New York.

Luo, C. and Rudy, Y. (1994). A dynamic model of the cardiac ventricular action potential: I. Simulations of ionic currents and concentration changes. *Circ. Res.* **74**, 1071.

Matthews, G.G. (1991). *Cellular Physiology of Nerve and Muscle.* Blackwell Scientific, Boston.

Nernst, W. (1889). Die elektromotorishe Wirksamkeit der Jonen. *Z. Physik. Chem.* **4**, 129–188.

Northrop, R. (2001). *Introduction to Dynamic Modeling of Neurosensory Systems.* CRC, Boca Raton.

Plonsey, R. and Barr, R.C. (1982). *Bioelectricity: A Quantitative Approach.* Plenum, New York.

Rinzel, J. (1990). Electrical excitability of cells, theory and experiment: Review of the Hodgkin-Huxley foundation and an update. *Bull. Math. Biology: Classics of Theoretical Biology* **52**, 5–23.

Rodriguez Campos, F. and Enderle, J.D. (2004). Porting Genesis to SIMULINK, *Proceedings of the 30th IEEE EMBS Annual International Conference*, September 2–5.

12 PHYSIOLOGICAL MODELING

John Enderle, PhD*

Chapter Contents

*With contributions by R.J. Fisher.

At the conclusion of this chapter, the student will be able to:

- Describe the process used to build a mathematical physiological model.

- Explain the concept of a compartment.

- Analyze a physiological system using compartmental analysis.

- Solve a nonlinear compartmental model.

- Qualitatively describe a saccadic eye movement.

- Describe the saccadic eye movement system with a second-order model.

- Explain the importance of the pulse-step saccadic control signal.

- Explain how a muscle operates using a nonlinear and linear muscle model.

- Simulate a saccade with a fourth-order saccadic eye movement model.

- Estimate the parameters of a model using system identification.

12.1 INTRODUCTION

A *quantitative* physiological model is a mathematical representation that approximates the behavior of an actual physiological system. *Qualitative* physiological models, most often used by biologists, describe the actual physiological system without the use of mathematics. Quantitative physiological models, however, are much more useful and are the subject of this chapter. Physiological systems are almost always dynamic and are characterized mathematically with differential equations. The modeling techniques developed in this chapter are intimately tied to many other interdisciplinary areas, such as physiology, biophysics, and biochemistry, and involve electrical and mechanical analogs. A model is usually constructed using basic and natural laws. This chapter extends this experience by presenting models that are more complex and involve larger systems.

Creating a model is always accompanied by carrying out an experiment and obtaining data. The best experiment is one that provides data that are related to variables used in the model. Consequently, the design and execution of an experiment is one of the most important and time-consuming tasks in modeling. A model constructed from basic and natural laws then becomes a tool for explaining the underlying processes that cause the experimental data and predicting the behavior of the system to other types of stimuli. Models serve as vehicles for thinking, organizing complex data, and testing hypotheses. Ultimately, modeling's most important goals are the generation of new knowledge, prediction of observations before they occur, and assistance in designing new experiments.

Figure 12.1 illustrates the typical steps in developing a model. The first step involves observations from an experiment or a phenomenon that lead to a conjecture or a verbal description of the physiological system. An initial hypothesis is formed via a mathematical model. The strength of the model is tested by obtaining data and testing the model against the data. If the model performs adequately, the model is satisfactory, and a solution is stated. If the model does not meet performance specifications, then the model is updated and additional experiments are carried out. Usually some of the variables in the model are observable and some are not. New experiments provide additional data that increase the understanding of the physiological system by providing information about previously unobservable variables, which improves the model. The process of testing the model against the data continues until a satisfactory solution is attained. Usually a statistical test is performed to test the goodness of fit

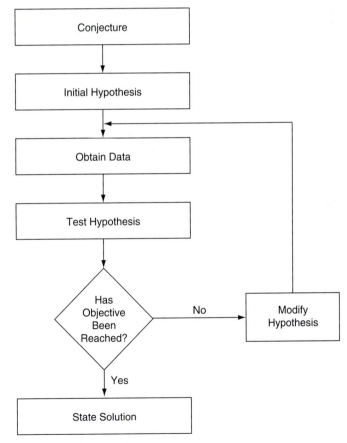

Figure 12.1   Flow chart for modeling.

between the model and the data. One of the characteristics of a good model is how well it predicts the future performance of the physiological system.

The introduction of the digital computer, programming languages, and simulation software have caused a rapid change in the use of physiological models. Before digital computers, mathematical models of biomedical systems were either oversimplified or involved a great deal of hand calculation as described in the Hodgkin–Huxley investigations published in 1952. Today, digital computers have become so common that the terms *modeling* and *simulation* have almost become synonymous. This has allowed the development of much more realistic or homeomorphic models that include as much knowledge as possible about the structure and interrelationships of the physiological system without any overriding concern about the number of calculations. Though models can continue to be made more complex, it is important to evaluate the value added with each stage of complexity—the model should be made as simple as possible to explain the data, but not so simple that it becomes meaningless. On the other hand, a model that is made too complex is also of little use.

12.1.1   Deterministic and Stochastic Models

A deterministic model is one that has an exact solution that relates the independent variables of the model to each other and to the dependent variable. For a given set of initial conditions, a deterministic model yields the same solution each and every time. A stochastic model involves random variables that are functions of time and include probabilistic considerations. For a given set of initial conditions, a stochastic model yields a different solution each and every time. Suffice it to say that solutions involving stochastic models are much more involved than the solution for a deterministic model.

It is interesting to note that all deterministic models include some measurement error. The measurement error introduces a probabilistic element into the deterministic model so that it might be considered stochastic. However, in this chapter, models are deterministic if their principle features lead to definitive predictions. On the other hand, models are stochastic if their principle features depend on probabilistic elements. This chapter is primarily concerned with deterministic models.

12.1.2   Solutions

There are two types of solutions available to the modeler. A closed-form solution exists for models that can be solved by analytic techniques such as solving a differential equation using the classical technique or by using Laplace transforms. For example, given the following differential equation

$$\ddot{x} + 4\dot{x} + 3x = 9$$

with initial conditions $x(0) = 0$ and $\dot{x}(0) = 1$, the solution is found as

$$x(t) = -4e^{-t} + e^{-3t} + 3$$

A numerical or simulation solution exists for models that have no closed-form solution. Consider the following function

$$x = \int_{-20}^{20} \frac{1}{33\sqrt{2\pi}} e^{-\frac{1}{2}\left(\frac{t-7}{33}\right)^2} dt$$

This function (the area under a Gaussian curve) has no closed-form solution and must be solved using an approximation technique such as the trapezoidal rule for integration. Most nonlinear differential equations do not have an exact solution and must be solved via an iterative method or simulation package such as SIMULINK. This was the situation in Chapter 11 when the Hodgkin–Huxley model was solved.

Inverse Solution

Engineers often design and build systems to a predetermined specification. They often use a model to predict how the system will behave because a model is efficient and economical. The model that is built is called a plant and consists of parameters that completely describe the system, the characteristic equation. The engineer selects the parameters of the plant to achieve a certain set of specifications such as rise time, settling time, or peak overshoot time.

In contrast, biomedical engineers involved with physiological modeling do not build the physiological system, but only observe the behavior of the system—the input and output of the system—and then characterize it with a model. Characterizing the model as illustrated in Figure 12.1 involves identifying the form or structure of the model, collecting data, and then using the data to estimate the parameters of the model. The goal of physiological modeling is not to design a system, but to identify the components (or parameters) of the system. Most often, data needed for building the model are not the data that can be collected using existing bioinstrumentation and biosensors as discussed in Chapters 8–10. Typically, the recorded data are transformed from measurement data into estimates of the variables used in the model. Collecting appropriate data is usually the most difficult aspect of the discovery process.

Model building typically involves estimating the parameters of the model that optimize, in a mean square error sense, the output of the model or model prediction, $\hat{x}_i$, and the data, x_i. For example, one metric for estimating the parameters of a model, S, is given by minimizing the sum of squared errors between the model prediction and the data

$$S = \sum_{i=1}^{n} \varepsilon_i^2 = \sum_{i=1}^{n} (x_i - \hat{x}_i)^2$$

where ε_i is the error between the data x_i and the model prediction $\hat{x}_i$. This technique provides an unbiased estimate with close correspondence between the model prediction and the data.

The first part of this chapter deals with compartmental models, a unified technique for modeling many systems of the body. To provide a feeling for the modeling process described in Figure 12.1, the latter part of this chapter focuses on one particular system—the fast eye movement system, the modeling of which began with early muscle modeling experiences in the 1920s and continues today with neural network models for the control of the fast eye movement system. This physiological system is probably the best understood of all systems in the body. Some of the reasons for this success are the relative ease in obtaining data, the simplicity of the system in relation to others, and the lack of feedback during dynamic changes in the system. Finally, the topic of system identification or parameter estimation closes the chapter.

12.2   COMPARTMENTAL MODELING

When analyzing systems of the body characterized by a transfer of solute from one compartment to another, such as the respiratory and circulatory systems, it is convenient to describe the system as a series of compartments. Compartmental modeling is based on metabolism of tracer-labeled compounds studies in the 1920s. Compartmental analysis differs from physiological modeling in that it is concerned with maintaining correct chemical levels in the body and their correct fluid volumes.

Some readily identifiable compartments are:

■ Cell volume that is separated from the extracellular space by the cell membrane
■ Interstitial volume that is separated from the plasma volume by the capillary walls that contain the fluid that bathes the cells
■ Plasma volume contained in the circulatory system that consists of the fluid that bathes blood cells

Variables tracked in compartmental analysis are quantity and concentration of a substance (solute), temperature, and pressure. Substances of interest are exogenous, such as a drug or tracer, or endogenous, such as glucose, or an enzyme or hormone such as insulin. The process of transfer of substance from one compartment to another is based on mass conservation, and compartmental models can be linear or nonlinear. As shown, compartmental analysis provides a uniform theory that can be systematically applied to many of the body's systems. An example of a compartment model was introduced in Chapter 11, with the whole neuron model that described the flow of current.

Although interest in compartmental analysis here is focused on the body, other scientists use this technique in studying pharmacokinetics, chemical reaction engineering, fluid transport, and even semiconductor design and fabrication. Before investigating large systems, Fick's law of diffusion is presented first from the viewpoint of compartmental analysis.

12.2.1   Transfer of Substances Between Two Compartments Separated by a Thin Membrane

In this section, the time course of the transfer of a solute between two compartments separated by a thin membrane is examined using Fick's law of diffusion, as given by

$$\frac{dq}{dt} = -DA\frac{dc}{dx} \tag{12.1}$$

where

q = quantity of solute
A = membrane surface area
c = concentration
D = diffusion coefficient
dx = membrane thickness

One typically works with concentrations rather than quantity since measurements are in concentration rather than quantity. Thus, the following relationship is used in moving between quantity and concentration.

$$\text{Concentration} = \frac{\text{Quantity}}{\text{Volume}}$$

Consider the system of two compartments shown in Fig. 12.2, where

V_1 and V_2 are the volumes of compartments I and II
q_1 and q_2 are the quantities of solute in compartments I and II
c_1 and c_2 are the concentrations of solute in compartments I and II

and an initial amount of solute, Q_{10}, is dumped into compartment I. The rate of change of solute in compartment I is given by

$$\frac{dq_1}{dt} = -DA\frac{dc_1}{dx} = -DA\frac{(c_1 - c_2)}{\Delta x} \qquad (12.2)$$

Next, quantity of solute is then converted into a concentration by

$$q_1 = V_1 c_1 \qquad (12.3)$$

and then after differentiating Eq. 12.3, gives

$$\frac{dq_1}{dt} = V_1 \frac{dc_1}{dt} \qquad (12.4)$$

Substituting Eq. 12.4 into Eq. 12.2 yields

$$V_1 \frac{dc_1}{dt} = \frac{-DA}{\Delta x}(c_1 - c_2) \qquad (12.5)$$

With the transfer rate K defined as

$$K = \frac{DA}{\Delta x}$$

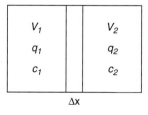

Figure 12.2 Two compartment model of diffusion, where dx = membrane thickness (*cm*).

when substituted into Eq. 12.5 yields

$$\frac{dc_1}{dt} = -\frac{K}{V_1}(c_1 - c_2) \tag{12.6}$$

Due to conservation of mass,

$$q_1 + q_2 = Q_{10}$$

and after converting to concentrations gives

$$V_1 c_1 + V_2 c_2 = V_1 C_{10} \tag{12.7}$$

where C_{10} is the initial concentration in compartment I due to the initial amount of solute dumped into the compartment.

The concentration in compartment II is found from Eq. 12.7 as

$$c_2 = \frac{C_{10}V_1 - V_1 c_1}{V_2} \tag{12.8}$$

which when substituted into Eq. 12.6 gives

$$\frac{dc_1}{dt} = \frac{-K}{V_1 V_2}[V_2 c_1 - V_1 C_{10} + V_1 c_1] = \frac{KC_{10}}{V_2} - \frac{Kc_1}{V_1 V_2}(V_1 + V_2)$$

or

$$\frac{dc_1}{dt} + K\left(\frac{V_1 + V_2}{V_1 V_2}\right)c_1 = \frac{KC_{10}}{V_2} \tag{12.9}$$

This is a first-order linear differential equation with forcing function

$$f(t) = \frac{KC_{10}}{V_2} \tag{12.10}$$

and initial condition $c_1(0) = C_{10}$ can be solved using standard techniques. Assume for simplicity that $V_1 = V_2$. Then Eq. 12.9 becomes

$$\frac{dc_1}{dt} + \frac{2K}{V_1}c_1 = \frac{KC_{10}}{V_1} \tag{12.11}$$

To solve Eq. 12.11, note that the root is $-\frac{2K}{V_1}$ and the natural solution is

$$c_{1_n} = B_1 e^{-\frac{2Kt}{V_1}} \tag{12.12}$$

where B_1 is a constant to be determined from the initial condition. The forced response has the same form as the forcing function in Eq. 12.9, $c_{1_f} = B_2$, which when substituted into Eq. 12.11 yields

$$\frac{2K}{V_1}B_2 = \frac{KC_{10}}{V_1}$$

or $B_2 = \frac{C_{10}}{2}$. Thus the complete response is

$$c_1 = c_{1_n} + c_{1_f} = B_1 e^{-\frac{2Kt}{V_1}} + \frac{C_{10}}{2} \tag{12.13}$$

To find B_1, the initial condition is used

$$c_1(0) = C_{10} = B_1 e^{-\frac{2Kt}{V_1}}\bigg|_{t=0} + \frac{C_{10}}{2} = B_1 + \frac{C_{10}}{2}$$

or $B_1 = \frac{C_{10}}{2}$. The complete solution is

$$c_1 = \frac{C_{10}}{2}\left(e^{-\frac{2Kt}{V_1}} + 1\right)$$

for $t \geq 0$. Note that the concentration in compartment II is found using Eq. 12.8 as

$$c_2 = \frac{V_1 C_{10} - V_1 c_1}{V_2} = \frac{C_{10}}{2}\left(1 - e^{-\frac{2Kt}{V_1}}\right)$$

The previous diffusion model is an example of a two compartment model and also applicable to describing the diffusion across a cell membrane, where dx represents the width of the cell membrane.

12.2.2 Compartmental Modeling Basics

Compartmental modeling involves describing a system with a finite number of compartments, each connected with a flow of solute from one compartment to another. Given a system described by a group of compartments, some exchange of solute (i.e., a radioactive tracer, a molecule such as glucose or insulin, or a gas such as oxygen or carbon dioxide) is expected between compartments by diffusion. Compartmental analysis predicts the concentrations of solutes under consideration in each compartment as a function of time using conservation of mass: accumulation equals input minus output. The model may be linear, nonlinear, continuous, discrete, and even have time-varying or stochastic parameters. If the model is continuous and linear, then the change in solute concentration is described as a sum of exponential terms. Boxes are used to define a compartment and the flow of solute between compartments defined by arrows.

The following assumptions are made when describing the transfer of a solute by diffusion between any two compartments:

1. The volume of each compartment remains constant.
2. Any solute q entering a compartment is instantaneously mixed throughout the entire compartment.
3. The rate of loss of a solute from a compartment is proportional to the amount of solute in the compartment times the transfer rate, K, given by Kq.

From a modeling perspective, identifying compartments and the number of compartments to describe a system is a difficult step. Acquiring measurement data for model facilitation is another difficult step because some compartments are inaccessible. Both of these steps are beyond the scope of this textbook; interested readers can examine books listed at the end of this chapter for more details.

The simplest compartment model consists of only one compartment. This can be considered a special case of a two-compartment model because the solute leaves the compartment and enters another compartment or space.

Example Problem 12.1

Consider the behavior of radioactive iodine (I^{131}) for the single compartment (plasma) shown in the following diagram.

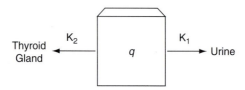

where q = quantity of iodine in the compartment, and K_1 and K_2 = transfer rates.
I^{131} is ingested through the digestive system and immediately placed in the plasma. I^{131} is removed from the plasma into the thyroid gland and excreted into the urine. Assume the initial quantity of I^{131} in the plasma is $q(0)$. Find the response to a single tracer dose of I^{131} (initial condition) in the compartment.

Solution

Using conservation of mass, the differential equation describing the rate of change of the quantity of I^{131} in the compartment is given by accumulation = input − output, where

$$\text{Accumulation} = \frac{dq}{dt}$$

$$\text{Input} = 0$$

$$\text{Output} = K_{1q} + K_{2q}$$

Thus

$$\frac{dq}{dt} = -(K_1 + K_2)q$$

The solution of the differential equation is similar to the previous section, giving

$$q = q(0)e^{-(K_1 + K_2)t}$$

for $t \geq 0$.   ■

The technique used in the previous example has a number of applications in diagnostic medicine and with a minor modification, some practical applications. One of the assumptions used in Ex. 12.1 is that the solute is immediately placed in the plasma compartment, a physical impossibility with ingestion. It is, however, a mathematical reality if the input is a delta function, $\delta(t)$, which can instantaneously change the initial conditions of the system. The closest one comes to the delta function input is via a rapid intravenous injection, which is referred to as a bolus.

Consider the more realistic compartmental model shown in Figure 12.3 with ingestion of a solute in the digestive system, and removal of solute via metabolism and excretion in urine. By including a digestive system component, the solute is not instantaneously delivered into the plasma, but is slowly released from the digestive system into the plasma. The conservation of mass is written for each compartment as

$$\frac{dq_1}{dt} = K_3 q_2 - (K_1 + K_2)q_1$$

$$\frac{dq_2}{dt} = -K_3 q_2$$

The second equation involves only q_2, with solution $q_2 = q_2(0)e^{-K_3 t}$, where $q_2(0)$ is the total quantity ingested. The solution for q_2 can be substituted into the first equation, giving

$$\frac{dq_1}{dt} = q_2(0)K_3 e^{-K_3 t} - (K_1 + K_2)q_1$$

and after rearranging

$$\frac{dq_1}{dt} + (K_1 + K_2)q_1 = q_2(0)K_3 e^{-K_3 t} \qquad (12.14)$$

This is a first-order differential equation with a forcing function $q_2(0)K_3 e^{-K_3 t}$. The natural solution is $q_{1_n} = B_1 e^{-(K_1 + K_2)t}$ and the forced response is $q_{1_f} = B_2 e^{-K_3 t}$, giving a total solution of

$$q_1 = q_{1_n} + q_{1_f} = B_1 e^{-(K_1 + K_2)t} + B_2 e^{-K_3 t} \qquad (12.15)$$

Substituting the forced response into Eq. 12.14 to determine B_2 gives

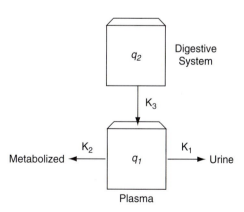

Figure 12.3 A compartmental model with realistic ingestion of solute with removal by metabolism and excretion in urine.

$$-K_3 B_2 e^{-K_3 t} + (K_1 + K_2) B_2 e^{-K_3 t} = q_2(0) K_3 e^{-K_3 t}$$

Solving for B_2 gives

$$B_2 = \frac{q_2(0) K_3}{K_1 + K_2 - K_3}$$

The complete response is now

$$q_1 = B_1 e^{-(K_1 + K_2)t} + \frac{q_2(0) K_3}{K_1 + K_2 - K_3} e^{-K_3 t} \tag{12.16}$$

and B_1 is found using the initial condition $q_1(0) = 0$

$$q_1(0) = 0 = \left[B_1 e^{-(K_1 + K_2)t} + \frac{q_2(0) K_3}{K_1 + K_2 - K_3} e^{-K_3 t} \right]_{t=0} = B_1 + \frac{q_2(0) K_3}{K_1 + K_2 - K_3}$$

giving

$$B_1 = -\frac{q_2(0) K_3}{K_1 + K_2 - K_3}$$

and

$$q_1 = \frac{q_2(0) K_3}{K_1 + K_2 - K_3} \left(e^{-K_3 t} - e^{-(K_1 + K_2)t} \right) \tag{12.17}$$

or in terms of concentration,

$$c_1 = \frac{1}{V_1} \frac{q_2(0) K_3}{(K_1 + K_2 - K_3)} \left(e^{-K_3 t} - e^{-(K_1 + K_2)t} \right) \tag{12.18}$$

for $t \geq 0$. To determine the time when maximum solute is in compartment I, Eq. 12.17 is differentiated with respect to t, set equal to zero, and solved as follows:

$$\begin{aligned}
\frac{dq_1}{dt} &= \frac{d}{dt} \left(\frac{q_2(0) K_3}{K_1 + K_2 - K_3} \left(e^{-K_3 t} - e^{-(K_1 + K_2)t} \right) \right) \\
&= \frac{q_2(0) K_3}{K_1 + K_2 - K_3} \left(-K_3 e^{-K_3 t} + (K_1 + K_2) e^{-(K_1 + K_2)t} \right)
\end{aligned} \tag{12.19}$$

Setting Eq. 12.19 equal to zero gives

$$\frac{q_2(0) K_3}{K_1 + K_2 - K_3} \left(-K_3 e^{-K_3 t} + (K_1 + K_2) e^{-(K_1 + K_2)t} \right) = 0$$

or

$$K_3 e^{-K_3 t} = (K_1 + K_2) e^{-(K_1 + K_2)t}$$

Multiplying both sides of the previous equation by $e^{(K_1 + K_2)t}$ and dividing by K_3 gives

$$e^{(K_1 + K_2)t} e^{-K_3 t} = e^{(K_1 + K_2 - K_3)t} = \frac{K_1 + K_2}{K_3}$$

and taking the logarithm of both sides gives

$$(K_1 + K_2 - K_3)t = \ln\left(\frac{K_1 + K_2}{K_3}\right)$$

Solving for t_{max} yields

$$t_{max} = \frac{\ln\left(\frac{K_1+K_2}{K_3}\right)}{(K_1 + K_2 - K_3)} \qquad (12.20)$$

It should be clear from Eq. 12.20, the smaller the term $K_1 + K_2$ is compared to K_3, the more time it takes to reach maximum quantity in the plasma.

Example Problem 12.2

Suppose 50 g of solute is ingested. Find the maximum amount of solute in the plasma if the compartmental model in Figure 12.2 is used with $K_1 + K_2 = 0.005$ min^{-1} and $K_3 = 0.02$ min^{-1}.

Solution

Using Eq. 12.20 gives

$$t_{max} = \frac{\ln\left(\frac{K_1+K_2}{K_3}\right)}{(K_1 + K_2 - K_3)} = \frac{\ln\left(\frac{0.005}{0.02}\right)}{0.005 - 0.02} = \frac{\ln(0.25)}{-0.015} = 92.42 \text{ min}$$

The maximum amount of solute in compartment I at t_{max} is therefore

$$q_1(t_{max}) = \frac{q_2(0)K_3}{K_1 + K_2 - K_3}\left(e^{-K_3 t} - e^{-(K_1+K_2)t}\right)\Big|_{t=92.42}$$

$$= \frac{50 \times 0.02}{0.005 - 0.02}(e^{-0.02 \times 92.42} - e^{-0.005 \times 92.42}) = 31.5 \text{ g} \qquad \blacksquare$$

12.2.3 Multicompartmental Models

As described previously, real models of the body involve many more compartments than described in the previous section, such as cell volume, interstitial volume, and plasma volume. Each of these volumes can be further compartmentalized. For instance, the interstitial volume can be defined with compartments including the GI tract, mouth, liver, kidneys, and other unidentified compartments. Each of these compartments has its own transfer rate for moving the solute from one compartment to another. In general, concern about how the solute moves from and into a compartment is not a focus, only the amount of solute that is transferred.

The concepts described in the previous section can be applied to a model with any number of compartments. Each compartment is characterized by a conservation of mass differential equation describing the rate of change of solute. Thus, for the case of N compartments, there are N equations of the general form

$$\frac{dq_i}{dt} = \text{input} - \text{output}$$

where q_i is the quantity of solute in compartment i. For a linear system, the transfer rates are constants.

Example Problem 12.3

Consider a two-compartment system for the distribution of creatinine in the body. Creatinine is a waste product of metabolism in the muscle which is cleared from the body through the urine. A sketch of the system is shown in the following figure. Assume creatinine production in the muscle is $\dot{q}_{in}$ and is given by a step input. Find the concentration of creatinine in the plasma compartment.

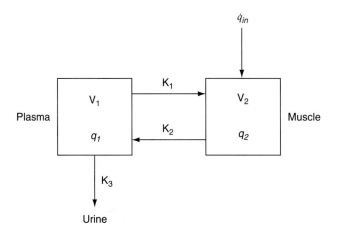

Solution

The differential equations describing the rate of change of creatinine in compartments A and B are written by using the conservation of mass equation as

$$\frac{dq_1}{dt} = K_2 q_2 - (K_1 + K_3)q_1$$

$$\frac{dq_2}{dt} = K_1 q_1 - K_2 q_2 + \dot{q}_{in}$$

These equations are changed to concentrations using the relationship

$$\text{Concentration} = \frac{\text{Quantity}}{\text{Volume}}$$

yielding

$$V_1 \frac{dc_1}{dt} = K_2 V_2 c_2 - (K_1 + K_3) V_1 c_1$$

$$V_2 \frac{dc_2}{dt} = K_1 V_1 c_1 - K_2 V_2 c_2 + \dot{q}_{in}$$

The D-operator is used to eliminate c_2, where in general, Dy is used to denote the differential operator $\frac{dy}{dt}$, $D^2 y$ denotes $\frac{d^2 y}{dt^2}$, and so on. The D-operator technique allows one to algebraically eliminate variables to write the set of differential equations as a single differential equation. From the single differential equation, the differential equation is then solved. Transforming the differential equations into the D-operator format gives

$$V_1 D c_1 = V_2 K_2 c_2 - V_1 (K_1 + K_3) c_1$$

$$V_2 D c_2 = V_1 K_1 c_1 - V_2 K_2 c_2 + \dot{q}_{in}$$

Collecting like terms gives

$$(V_1 D + V_1(K_1 + K_3)) c_1 - V_2 K_2 c_2 = 0$$

$$-V_1 K_1 c_1 + (V_2 D + V_2 K_2) c_2 = \dot{q}_{in}$$

To eliminate c_2, multiply the first equation by $(V_2 D + V_2 K_2)$ and the second equation by $V_2 K_2$, yielding

$$(V_2 D + V_2 K_2)(V_1 D + V_1(K_1 + K_3)) c_1 - (V_2 D + V_2 K_2) V_2 K_2 c_2 = 0$$

$$- V_2 K_2 V_1 K_1 c_1 + V_2 K_2 (V_2 D + V_2 K_2) c_2 = V_2 K_2 \dot{q}_{in}$$

Adding the equations together eliminates c_2, giving

$$(V_2 V_1 D^2 + (V_2 V_1 K_2 D + V_2 V_1(K_1 + K_3)D) + V_2 V_1 K_2 K_3) c_1 = K_2 V_2 \dot{q}_{in}$$

or

$$(D^2 + (K_1 + K_2 + K_3)D + K_2 K_3) c_1 = \frac{K_2 \dot{q}_{in}}{V_1}$$

Returning to the time domain gives

$$\ddot{c}_1 + (K_1 + K_2 + K_3) \dot{c}_1 + K_2 K_3 c_1 = \frac{K_2 \dot{q}_{in}}{V_1}$$

The roots of the characteristic equation are

$$s_{1,2} = \frac{-(K_1 + K_2 + K_3) \pm \sqrt{(K_1 + K_2 + K_3)^2 - 4 K_2 K_3}}{2}$$

The natural response is an overdamped response since K_1, K_2, and K_3 are greater than zero, giving

$$c_{1_n} = B_1 e^{s_1 t} + B_2 e^{s_2 t}$$

The forced response is a constant (B_3) because the input is a constant, and when substituted into the differential equation, yields

$$c_{1_f} = B_3 = \frac{\dot{q}_{in}}{K_3 V_1}$$

The complete response is

$$c_1 = B_1 e^{s_1 t} + B_2 e^{s_2 t} + \frac{\dot{q}_{in}}{K_3 V_1}$$

for $t \geq 0$. The constants B_1 and B_2 can be determined using the initial conditions.   ∎

12.2.4 Modified Compartmental Modeling

In the previous section, compartmental models of any order could be readily put together to describe rather complex systems, and with some effort, solved for the variable of interest. Many systems are not appropriately described by the compartmental analysis presented in the previous section because the transfer rates are not constant, but depend, for example, on the solute in a compartment. Compartmental analysis, now termed *modified compartmental analysis*, can still be applied to these systems by incorporating the nonlinearities in the model. Because of the nonlinearity, solution of the differential equation is usually not possible analytically, but can be easily simulated. Another method of handling the nonlinearity is to linearize the nonlinearity or invoke pseudostationary conditions.

Infectious Disease Models

In this section, the spread of an infection through a population is presented, illustrating modified compartmental modeling. The study of the occurrence of a disease and all of the factors that influence it is known as *epidemiology*. The ultimate purpose for the development of infectious disease models is to facilitate practical applications such as

1. Determining the cause or etiology of a disease
2. Controlling the spread of the disease

The history of infectious or communicable disease modeling dates to 1760 when D. Bernoulli studied the population dynamics of smallpox with a mathematical model. Little work was done until the early twentieth century, when Hamer and Soper presented mathematical models that described the spread of measles in Glasgow, Scotland. In 1928, Kermack and McKendrick (continuous time), and Reed and Frost (discrete time) presented extensions of the work of Hammer and Soper. Since the 1950s, when Abbey and Bailey presented their work, there has been an epidemic of work in this area. This is primarily due to the nonlinear nature of the models and the advent of computing facilities and algorithms necessary to solve differential equations.

Both the Kermack–McKendrick and Reed–Frost models are described in this section. For either, the course of an infection is described in two ways, as shown in the following timeline.

a

Infected

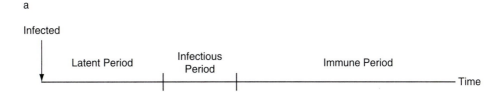

b

Infected

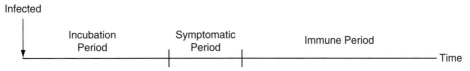

where

- *Latent period:* The time elapsed between contact and the actual discharge of the infectious agent.
- *Infectious period:* The time during which the contagious agent is spread to others.
- *Immune period:* The time during which a person no longer transmits the agent, and is either temporarily or permanently immune to the disease.
- *Incubation period:* The time elapsed between contact and the observation of symptoms.
- *Symptomatic period:* The time interval in which the person overtly displays signs of the illness.

Kermack–McKendrick Continuous Time Model

Consider a community of total size n, with x susceptibles, y infectives, and z immunes, whereby

$$n = x + y + z$$

The following assumptions are given in developing the model:

1. Uniform mixing among the population
2. Zero latent period
3. Population is closed and isolated
4. Negative exponential distribution for infectious period
5. β is the infectious rate
6. γ is the removal rate

The course of an acute epidemic in a closed population is a function of the number of susceptibles and the infective rate between susceptibles and infectives. A block diagram for the compartmental model is given in Figure 12.4. Arrows indicate a nonnegative transfer of individuals from one state to another, dependent on the infective rate β (infectives) and the removal rate γ.

Figure 12.4   Kermack–McKendrick compartmental model.

The Kermack–McKendrick model describes the transfer of x susceptibles, y infectives, and z immunes at time t from state to state. With βy as the infective rate, the differential equations describing the model are:

$$\frac{dx}{dt} = -\beta xy$$

$$\frac{dy}{dt} = \beta xy - \gamma y$$

$$\frac{dz}{dt} = \gamma y$$

Fortunately, one can solve this set of nonlinear equations analytically using a Taylor series approximate technique. Assume that $y_0 = y(0)$ and $z_0 = z(0)$ take on small values at time zero, and therefore $x_0 = x(0) \approx n$. Define the relative removal rate as

$$\rho = \frac{\gamma}{\beta}$$

To have $\frac{dy}{dt} > 0$ requires that $\rho < x_0$, or no epidemic will start. The relative removal rate of $\rho = x_0$ gives the threshold density of susceptibles.

To solve for x, eliminate y by dividing the first by the third equation, and then solve the differential equation:

$$\frac{\frac{dx}{dt}}{\frac{dz}{dt}} = \frac{-\beta xy}{\gamma y} = \frac{-x}{\rho}$$

and

$$x = x_0 e^{\frac{-z}{\rho}}$$

Next, use the differential equation for immunes and substitute $y = n - x - z$ and the solution for x.

$$\frac{dz}{dt} = \gamma y = \gamma(n - x - z) = \gamma\left(n - z - x_0 e^{\frac{-z}{\rho}}\right)$$

Using a Taylor series for x

$$\frac{dz}{dt} = \gamma\left\{ n - z - x_0 + \underbrace{\frac{x_0 z}{\rho} - \frac{x_0}{2}\frac{z^2}{\rho^2} + \cdots}_{\text{Taylor Series}} \right\}$$

and neglecting higher-order terms, gives

$$\frac{dz}{dt} = \gamma \left\{ (n - x_0) + \left(\frac{x_0}{\rho} - 1 \right) z - \frac{x_0}{2\rho^2} z^2 \right\}$$

Using $\tanh(x) = \frac{e^x - e^{-x}}{e^x + e^{-x}}$, and after much work, the solution for z is found as

$$z = \frac{\rho^2}{x_0} \left\{ \frac{x_0}{\rho} - 1 + \alpha \tanh\left(\frac{1}{2} \alpha \gamma t - \phi \right) \right\} \tag{12.21}$$

where

$$\alpha = \sqrt{\frac{(x_0 - 1)^2}{\rho} + \frac{2x_0 y_0}{\rho^2}}$$

and

$$\phi = \tanh^{-1} \frac{1}{\alpha} \left(\frac{x_0}{\rho} - 1 \right)$$

To determine the total size of an epidemic, evaluate $z(+\infty)$ using Eq. 12.21

$$z(+\infty) = \frac{\rho^2}{x_0} \left\{ \frac{x_0}{\rho} - 1 + \alpha \right\}$$

since $\tanh\left(\frac{1}{2} \alpha \gamma t - \phi \right)_{t \to \infty} = 1$. Note that if

$$\left(\frac{x_0}{\rho} - 1 \right)^2 \gg \frac{2x_0 y_0}{\beta}, \quad \text{then } \alpha = \left(\frac{x_0}{\rho} - 1 \right)$$

and

$$z(+\infty) \approx 2\rho \left(1 - \frac{\rho}{x_0} \right)$$

No epidemic will occur if $x_0 < \rho$.

Example Problem 12.4

Use SIMULINK to create an epidemic curve using the Kermack–McKendrick model for a population of 10,000 susceptibles, 5 infectives, and 0 immunes, and $\beta = .000100$ and $\gamma = 0.9$.

Solution

Before simulating, check to see whether an epidemic will occur via the relative removal rate ρ as

$$\rho = \frac{\gamma}{\beta} = \frac{0.90}{0.0001} = 9000 \le 10,000$$

Since ρ is less than the initial population of susceptibles, an epidemic occurs. The SIMULINK model is given in the following figure, with the simulation results after that.

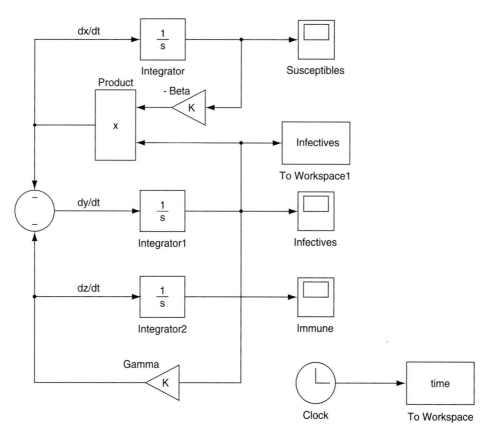

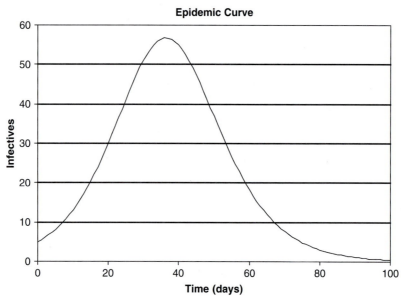

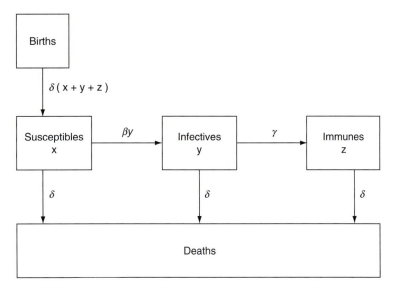

Figure 12.5   Kermack–McKendrick compartmental model with birth and death rates.

To include the possibility of following a population over long periods of time, it is necessary to include a birth and death rate as shown in Figure 12.5. The equations describing this model are

$$\frac{dx}{dt} = -\beta xy + \delta(z + y)$$

$$\frac{dy}{dt} = \beta yx - \gamma y - \delta y$$

$$\frac{dz}{dt} = \gamma y - \delta z$$ ∎

Unfortunately, including deaths as part of the model yields results that are endemic (a constant infection throughout the population at all times and no epidemics), as shown in the following example.

Example Problem 12.5

Use SIMULINK to create an epidemic curve using the Kermack–McKendrick model with a death rate for a population of 10,000 susceptibles, 5 infectives, and 0 immunes, and $\beta = .000100$, $\gamma = 0.9$, and $\delta = 0.01$.

Solution

The SIMULINK model is shown in the following figure.

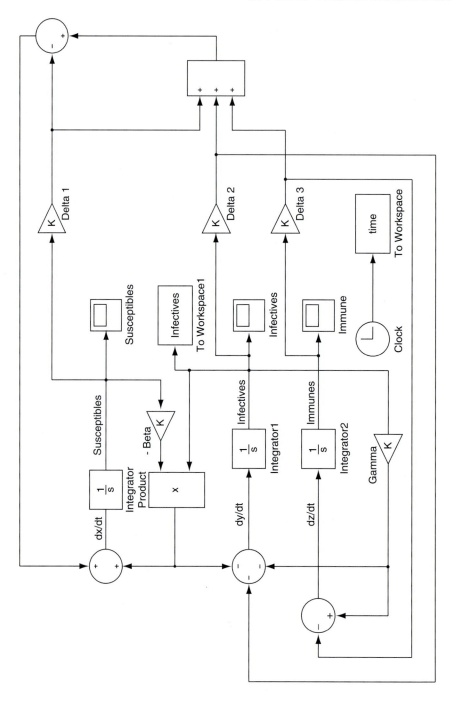

The simulation results are shown in the following figure illustrating the endemic nature of the epidemic. It is important to note that this type of epidemic does not follow the data. ■

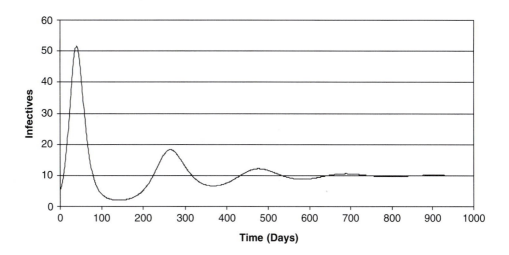

Reed–Frost Model

The Reed–Frost model is a deterministic discrete time model. One reason for utilizing a discrete time model rather than a continuous time model is that recorded data is measured at regular intervals. Another reason is that extensions to the simplest models are easily accomplished, such as adding a nonzero latent period with a precisely defined distribution. Assume

1. There is uniform mixing among the population.
2. There is a nonzero latent period.
3. The population is closed and at steady state.
4. Any susceptible individuals, after contact with an infectious person, develop the infection and are infectious to others only in the following period, after which they are immune.
5. Since the person can be infected at any instant during the time period, the average latent period is one-half of the time period, where the length of the time period represents the period of infectivity.
6. Each individual has a fixed probability of coming into adequate contact p with any other specified individual within one time period.

The structure of the Reed–Frost model is shown in Figure 12.6. Note that the probability of adequate contact p can be thought of as

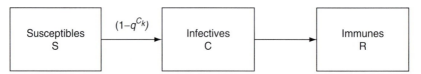

Figure 12.6   The Reed–Frost model.

$$p = \frac{\text{average number of adequate contacts}}{N}$$

With

$$q = 1 - p$$

the probability that a susceptible individual does not come into adequate contact is

$$q^{C_k}$$

The Reed–Frost model describes the transfer of S susceptibles, C infectives, and R immunes from state to state at discrete time $k + 1$. After adequate contact with an infective in a given time period, a susceptible will develop the infection and be infectious to others only during the subsequent time period, after which he or she becomes immune. With $(1 - q^{C_k})$ as the infective rate, the nonlinear difference equations describing the model are

$$C_{k+1} = S_k(1 - q^{C_k})$$
$$S_{k+1} = S_k - C_{K+1}$$
$$R_{k+1} = R_k + C_k$$

The time period T is understood to be the length of time an individual is infectious, so that the removal rate is equal to one.

Example Problem 12.6

Use SIMULINK to create an epidemic curve using the Reed–Frost model with 10,000 susceptibles, 10 infectives, and 0 immunes, and $q = 0.9998$.

Solution

The SIMULINK model is shown in the following figure.

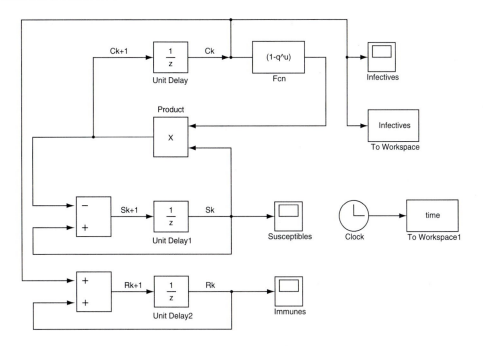

The simulation results are given in the next figure.

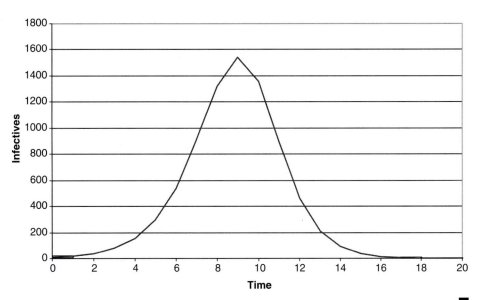

12.2.5 Transfer of Solutes Between Physiological Compartments by Fluid Flow

In this section, we use a modified compartmental model to consider the transfer of solutes between compartments by fluid flow. Figure 12.7 illustrates a compartment washed through by a stream of fluid where V_A = volume of compartment, $\frac{dq_{in}}{dt} = \dot{q}_{in}$ = rate at which the quantity of solute enters the compartment, $\frac{dq_{out}}{dt} = \dot{q}_{out}$ = rate at which the quantity of solute leaves the compartment.

To analyze a system washed by fluid flow, the conservation of mass approach is used. The rate at which the solute enters the compartment is

$$\dot{q}_{in} = \dot{V}c_{in}$$

where $\dot{V}$ = volume rate of fluid entering the compartment and c_{in} = concentration of solute in the fluid.

The rate at which the solute leaves the compartment is given by

$$\dot{q}_{out} = \dot{V}c_{out}$$

where C_{out} = concentration of solute in fluid leaving the compartment. The differential equation describing the rate of change of the quantity of solute in the compartment with time is given by:

$$\frac{dq_A}{dt} = \dot{q}_{in} - \dot{q}_{out} = \dot{V}(c_{in} - c_{out})$$

or writing this equation in terms of concentration

$$V_A \frac{dc_A}{dt} = \dot{V}(c_{in} - c_{out})$$

This basic equation is used in writing fluid flows through compartments. When describing the cardiovascular system, $\dot{Q}$ is typically used instead of $\dot{V}$.

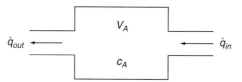

Figure 12.7   Compartmental model for the transfer of solutes between compartments by fluid flow.

Example Problem 12.7

Given in the following diagram is a three-compartment model used to describe the time course of steroid concentration in the arterial and venous blood. The steroid is injected as a bolus into the pulmonary artery. Assume all metabolism of steroids takes place in the liver, and that the metabolic rate of steroids is directly proportional to the concentration of steroids in the tissue compartment. $\dot{Q}$ is the cardiac output. If 5% of the total cardiac output flows into the hepatic artery and 20% of the cardiac output flows into the portal vein, then write the differential equations that describe this system, where

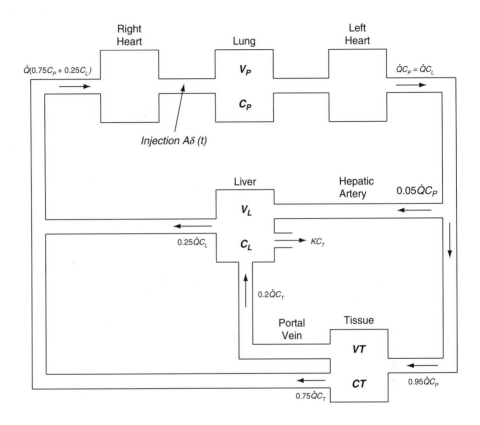

$$A\delta(t) = \text{bolus of steroid}$$
$$\dot{Q} = \text{cardiac output}$$
$$c_p = \text{concentration of steroid in lung}$$
$$c_L = \text{concentration of steroid in liver}$$
$$c_T = \text{concentration of steroid in tissue}$$
$$V_p = \text{volume of lung compartment}$$
$$V_L = \text{volume of liver compartment}$$
$$V_T = \text{volume of tissue compartment}$$
$$K = \text{metabolic rate factor}$$

Solution

Conservation of mass equations are written for each compartment as follows.
Lung compartment:

$$V_p \frac{dc_p}{dt} = \dot{Q}(.75c_T + .25c_L) + A\delta(t) - \dot{Q}c_p$$

Liver compartment:

$$V_L \frac{dc_L}{dt} = 0.05\dot{Q}c_p + .2\dot{Q}c_T - .25\dot{Q}c_L - Kc_T$$

Tissue compartment:

$$V_T \frac{dc_T}{dt} = .95\dot{Q}c_p - .95\dot{Q}c_T \qquad\blacksquare$$

12.2.6   Dye Dilution Model

Dye dilution studies are used to determine cardiac output, cardiac function, perfusion of organs, and the functional state of the vascular system. Usually the dye (e.g., Evans blue) is injected at one site in the cardiovascular system and observed at one or more sites as a function of time.

The need for a dye that is retained in the blood for long periods of time is evident. If some of the dye leaves the plasma, the measurements will be in error. The most widely used drug is Evans blue, which combines with the plasma albumin. Another is radioactive chromium Cr^{51}.

Example Problem 12.8

Compute the blood volume of a human being. A known quantity of dye is introduced into the bloodstream. After a few minutes (to allow for complete mixing), a blood sample is withdrawn.

Solution

$$\text{volume of blood} = \frac{\text{quantity of dye injected}}{\text{concentration}} \qquad\blacksquare$$

Example Problem 12.9

Estimate the cardiac output.

Solution

To model the system, focus on the heart–lung compartment as a single washed-through compartment with the dye injected as a step input, with magnitude K, as shown in the following figure.

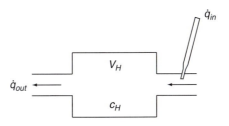

where V_H = the volume of the heart–lung compartment, $\dot{q}_{in}$ = the rate at which dye is injected, $\dot{q}_{out}$ = the rate at which dye leaves the compartment, c_H = the concentration of dye in the heart–lung compartment.

By conservation of mass

$$\frac{dq_H}{dt} = \dot{q}_{in} - \dot{q}_{out}$$

Substituting $\dot{q}_{out} = \dot{Q}c_H$ and $\dot{q}_{in} = Ku(t)$ for the step input gives

$$V_H \frac{dc_H}{dt} = \dot{q}_{in} - \dot{Q}c_H = Ku(t) - \dot{Q}c_H$$

The solution of the differential equation is given by

$$C_H(t) = \frac{K}{\dot{Q}} \left(1 - e^{-\frac{\dot{Q}t}{V_H}} \right)$$

for $t \geq 0$, which is shown in the following figure.

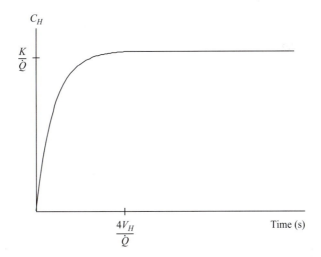

From the data, $\dot{Q}$ is experimentally calculated from steady state and V_H is experimentally calculated from the time constant. Unfortunately, the system does not achieve steady state before recirculation, thus this method is not satisfactory for estimating cardiac output.

If a pulse is used for the injection rather than a step, the solution for the dye concentration is

$$c_H(t) = \frac{K}{\dot{Q}}\left(1 - e^{-\frac{\dot{Q}t}{V_H}}\right)u(t) - \frac{K}{\dot{Q}}\left(1 - e^{-\frac{\dot{Q}(t-t_d)}{V_H}}\right)u(t - t_d)$$

Two solutions are possible, depending on the width of the pulse, as shown in the following figures.

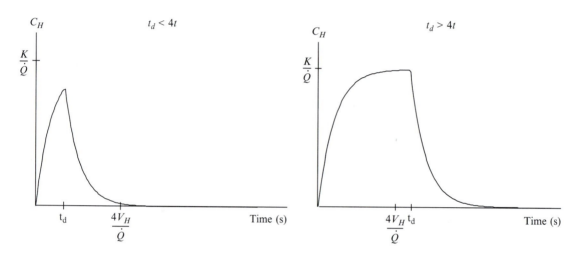

It is very difficult to determine $\dot{Q}$ from real data because of the recirculation effect. To rectify this, consider a pulse in which $t_d < 4\tau$ (i.e., no recirculation effect). One can estimate the integral of the concentration by calculating the area under the curve manually. Analytically,

$$S = \int_0^\infty c_H(t)dt$$

$$= \int_0^\infty \frac{K}{\dot{Q}}\left(1 - e^{-\frac{\dot{Q}t}{V_H}}\right)dt - \int_{t_d}^\infty \frac{K}{\dot{Q}}\left(1 - e^{-\frac{\dot{Q}(t-t_d)}{V_H}}\right)dt$$

$$= \frac{K}{\dot{Q}}\left(\int_0^{t_d} dt - \int_0^\infty e^{-\frac{\dot{Q}t}{V_H}}dt + \int_{t_d}^\infty e^{-\frac{\dot{Q}(t-t_d)}{V_H}}dt\right)$$

$$= \frac{K}{\dot{Q}}\left[t_d + \frac{V_H}{\dot{Q}} - \frac{V_H}{\dot{Q}}\right] = \frac{K}{\dot{Q}}t_d$$

So an estimate of $\dot{Q}$ is given by

$$\dot{Q} = \frac{Kt_d}{S}$$ ■

12.3   AN OVERVIEW OF THE FAST EYE MOVEMENT SYSTEM

A fast eye movement is usually referred to as a *saccade* and involves quickly moving the eye from one image to another image. *Saccade* is a French term that means to pull, which originated from the jerk of the reins on a horse. This type of eye movement is very common and is observed most easily while reading. When the end of a line is reached, the eyes are moved quickly to the beginning of the next line. The saccade system is part of the oculomotor system that controls all movements of the eyes due to any stimuli. The eyes are moved within the orbit by the oculomotor system to allow the individual to locate, see, and track objects in visual space. Each eye can be moved within the orbit in three directions: vertically, horizontally, and torsionally. These movements are due to three pairs of agonist–antagonist muscles. These muscles are called antagonistic pairs because their activity opposes each other and follows the principle of reciprocal innervation. Shown in Figure 12.8 is a diagram illustrating the muscles of the eye. The overall strategy of the system is to keep the central portion of the retina, called the fovea, on the target of interest.

The oculomotor system responds to visual, auditory, and vestibular stimuli, which results in one of five types of eye movements. Fast eye movements are used to locate or acquire targets. Smooth pursuit eye movements are used to track or follow a target. Vestibular ocular movements are used to maintain the eyes on the target during head movements. Vergence eye movements are used to track near and far targets. Optokinetic eye movements are reflex movements that occur when moving through a target-filled environment or to maintain the eyes on target during continuous head rotation. Except for the vergence eye movement, each of the other four movements is conjugate—that is, the eyes move together in the same direction and distance. The vergence eye movement system uses nonconjugate eye movements to keep the eyes on the

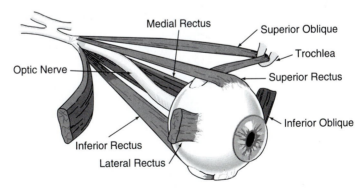

Figure 12.8 Diagram illustrating the muscles and optic nerve of the right eye. The left eye is similar except that the lateral and medial rectus muscles are reversed. The lateral and medial rectus muscles are used to move the eyes in a horizontal motion. The superior rectus, inferior rectus, superior oblique, and inferior oblique are used to move the eyes vertically and torsionally. The contribution from each muscle depends on the position of the eye. When the eyes are looking straight ahead, called primary position, the muscles are stimulated and under tension.

target. If the target moves closer, the eyes converge—farther away, they diverge. Each of these movements is controlled by a different neuronal system and uses the same final common pathway to the muscles of the eye. In addition to the five types of eye movements, these stimuli also cause head and body movements. Thus, the visual system is part of a multiple input–multiple output system.

Regardless of the input, the oculomotor system is responsible for movement of the eyes so that targets are focused on the central $\frac{1}{2}°$ region of the retina, known as the fovea (Fig. 12.9). Lining the retina are photoreceptive cells that translate images into neural impulses. These impulses are then transmitted along the optic nerve to the central nervous system via parallel pathways to the superior colliculus and the cerebral cortex. The fovea is more densely packed with photoreceptive cells than the retinal periphery; thus, a higher-resolution image (or higher visual acuity) is generated in the fovea than in the retinal periphery. The purpose of the fovea is to

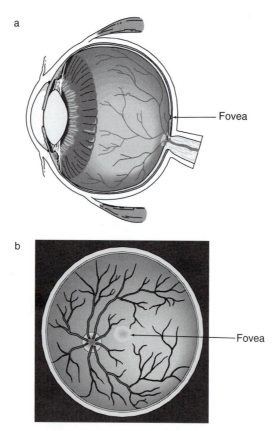

Figure 12.9   (a) Diagram illustrating a side view of the eye. The rear surface of the eye is called the retina. The retina is part of the central nervous system and consists of two photoreceptors, rods, and cones. (b) Front view looking at the rear inside surface (retina) of the eye. The fovea is located centrally and is approximately 1 mm in diameter. The oculomotor system maintains targets centered on the fovea.

allow us to *clearly* see an object and the purpose of the retinal periphery is to allow us to *detect* a new object of interest. Once a new object of interest is detected in the periphery, the saccade system redirects the eyes, as quickly as possible, to the new object. This type of saccade is typically called a goal-directed saccade.

During a saccade, the oculomotor system operates in an open-loop mode. After the saccade, the system operates in a closed-loop mode to ensure that the eyes reach the correct destination. The saccade system operates without feedback during a fast eye movement because information from the retina and muscle proprioceptors is not transmitted quickly enough during the eye movement for use in altering the control signal. The oculomotor plant and saccade generator are the basic elements of the saccadic system. The oculomotor plant consists of three muscle pairs and the eyeball. These three muscle pairs contract and lengthen to move the eye in horizontal, vertical, and torsional directions. Each pair of muscles acts in an antagonistic fashion due to reciprocal innervation by the saccade generator. For simplicity, the models described here involve only horizontal eye movements and one pair of muscles, the lateral and medial rectus muscle.

12.3.1   Saccade Characteristics

Saccadic eye movements, among the fastest voluntary muscle movements the human is capable of producing, are characterized by a rapid shift of gaze from one point of fixation to another. Shown in Figure 12.10 is a 10° saccade. The usual experiment for recording saccades is for a subject to sit before a horizontal target display of small light emitting diodes (LEDs). Subjects are instructed to maintain their eyes on the lit LED by moving their eyes as quickly as possible to avoid errors. A saccade is made by the subject when the active LED is switched off and another LED is switched on. Saccadic eye movements are conjugate and ballistic, with a typical duration of 30–100 ms and a latency of 100–300 ms. The subject was looking straight ahead when the target switched from the 0° position to the 10° position as illustrated in Figure 12.10. The subject then executed a saccade 150 ms later and completed the saccade at 200 ms. The latent period in Figure 12.10 is approximately 150 ms and is thought to be the time interval during which the central nervous system (CNS) determines whether to make a saccade and, if so, calculates the distance the eyeball is to be moved, transforming retinal error into transient muscle activity.

Generally, saccades are extremely variable, with wide variations in the latent period, time to peak velocity, peak velocity, and saccade duration. Furthermore, variability is well coordinated for saccades of the same size. Saccades with lower peak velocity are matched with longer saccade durations, and saccades with higher peak velocity are matched with shorter saccade durations. Thus, saccades driven to the same destination usually have different trajectories.

To appreciate differences in saccade dynamics, it is often helpful to describe them with saccade main sequence diagrams. The main sequence diagrams plot saccade peak velocity–saccade magnitude, saccade duration–saccade magnitude, and saccade latent

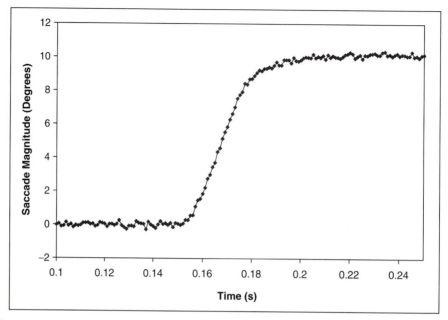

Figure 12.10   Sample saccadic eye movement of approximately 10 degrees. Data was collected with a sampling rate of 1000 samples/s.

period–saccade magnitude. Shown in Figure 12.11 are the main sequence characteristics for a subject executing 26 saccades. The subject actually executed 52 saccades in
both the positive and negative directions with only the results of the saccades in the
positive direction displayed in Figure 12.11 for simplicity. Notice that the peak
velocity–saccade magnitude starts off as a linear function and then levels off to a
constant for larger saccades. Many researchers have fit this relationship to an exponential function. The solid lines in Figure 12.11a include an exponential fit to the data
for positive eye movements. The lines in the first graph are fitted to the equation

$$V = \alpha\left(1 - e^{-\frac{x}{\beta}}\right) \tag{12.22}$$

where V is the maximum velocity, x the saccade size, and the constants α and β were
evaluated to minimize the summed error squared between the model and the data.
Note that α represents the steady state of the peak velocity–saccade magnitude curve
and β represents the "time constant" for the peak velocity–saccade magnitude
curve. For this data set, α equals 825, and β equals 9.3.

A similar pattern is observed with eye movements moving in the negative
direction, but the parameters α and β are typically different from the values computed
for the positive direction. The exponential shape of the peak velocity–saccade
amplitude relationship might suggest that the system is nonlinear if a step input
to the system is assumed. A step input provides a linear peak velocity–saccade

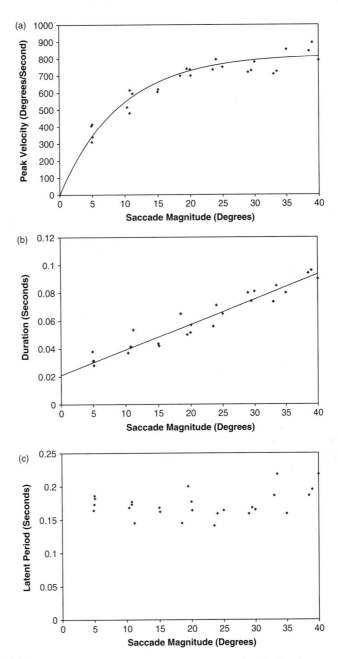

Figure 12.11   Main sequence diagrams for positive saccades. Similar shapes are observed for negative saccades. (a) Peak velocity–saccade magnitude, (b) saccade duration–saccade magnitude, and (c) latent period–saccade magnitude for 26 saccadic movements by a single subject (adapted from Enderle, Observations on pilot neurosensory control performance during saccadic eye movements. *Aviation, Space, and Environmental Medicine,* 59: 309, 1988).

amplitude relationship. In fact, the saccade system is not driven by a step input but rather by a more complex pulse-step waveform. Thus, the saccade system cannot be assumed to be nonlinear solely based on the peak velocity–saccade amplitude relationship. The input to the saccade system is discussed more fully in Section 12.4.

Shown in Figure 12.11b are data depicting a linear relationship between saccade duration–saccade magnitude. If a step input is assumed, then the dependence between saccade duration and saccade magnitude also might suggest that the system is nonlinear. A linear system with a step input always has a constant duration. Since the input is not characterized by a step waveform, the saccade system cannot be assumed to be nonlinear solely based on the saccade duration–saccade magnitude relationship. Shown in Figure 12.11c is the latent period–saccade magnitude data. It is quite clear that the latent period does not show any linear relationship with saccade size (i.e., the latent period's value appears independent of saccade size). However, some other investigators have proposed a linear relationship between the latent period and saccade magnitude. This feature is unimportant for the presentation in this chapter since in the development of the oculomotor plant models, the latent period is implicitly assumed within the model.

Because of the complexity of the eye movement system, attention is restricted to horizontal fast eye movements. In reality, the eyeball is capable of moving horizontally, vertically, and torsionally. An appropriate model for this system would include a model for each muscle and a separate controller for each muscle pair. The development of the horizontal eye movement models in this chapter is historical and is presented in increasing complexity with models of muscle introduced out of sequence so that their importance is fully realized. Not every oculomotor model is discussed. A few are presented for illustrative purposes.

12.4 WESTHEIMER SACCADIC EYE MOVEMENT MODEL

The first quantitative saccadic horizontal eye movement model, illustrated in Figure 12.12, was published by Westheimer in 1954. Based on visual inspection of a recorded 20° saccade and the assumption of a step controller, Westheimer proposed a second-order model (Eq. 12.23) that follows directly from Figure 12.12.

$$J\ddot{\theta} + B\dot{\theta} + K\theta = \tau(t) \tag{12.23}$$

Laplace analysis is used to analyze the characteristics of this model and compare it to data. Taking the Laplace transform of Equation 12.23 with zero initial conditions yields

$$s^2 J\theta + sB\theta + K\theta = \tau(s)$$
$$(s^2 J + sB + K)\theta = \tau(s) \tag{12.24}$$

The transfer function of Equation 12.24, written in standard form, is given by

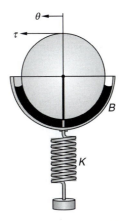

Figure 12.12 A diagram illustrating Westheimer's second-order model of the saccade system. The parameters J, B, and K are rotational elements for moment of inertia, friction, and stiffness, respectively, and represent the eyeball and its associated viscoelasticity. The torque applied to the eyeball by the lateral and medial rectus muscles is given by $\tau(t)$, and θ is the angular eye position. The radius of the eyeball is r.

$$H(s) = \frac{\theta(s)}{\tau(s)} = \frac{\frac{\omega_n^2}{K}}{s^2 + 2\zeta\omega_n s + \omega_n^2} \quad (12.25)$$

where $\omega_n = \sqrt{\frac{K}{J}}$, and $\zeta = \frac{B}{2\sqrt{KJ}}$. Based on the saccade trajectory for a $20°$ saccade, Westheimer estimated $\omega_n = 120$ radians per second, and $\zeta = 0.7$. With the input $\tau(s) = \frac{\gamma}{s}$, $\theta(t)$ is determined as:

$$\theta(t) = \frac{\gamma}{K}\left[1 + \frac{e^{-\zeta\omega_n t}}{\sqrt{1 - \zeta^2}}\cos(\omega_d t + \phi)\right] \quad (12.26)$$

where $\omega_d = \omega_n\sqrt{1 - \zeta^2}$ and $\phi = \pi + \tan^{-1}\frac{-\zeta}{\sqrt{1-\zeta^2}}$

Example Problem 12.10

Show the intermediate steps in going from Equation 12.25 to Equation 12.26.

Solution

Substituting the input $\tau(s) = \frac{\gamma}{s}$ into Equation 12.25 yields

$$\theta(s) = \frac{\gamma\omega_n^2}{Ks(s^2 + 2\zeta\omega_n s + \omega_n^2)}$$

Assuming a set of complex roots based on the estimates from Westheimer, a partial fraction expansion gives

$$\theta(s) = \frac{\gamma}{Ks} + \frac{\overbrace{\frac{\gamma}{2K\left((\zeta^2-1)-j\zeta\sqrt{1-\zeta^2}\right)}}}{s + \zeta\omega_n - j\omega_n\sqrt{1-\zeta^2}} + \frac{\overbrace{\frac{\gamma}{2K\left((\zeta^2-1)+j\zeta\sqrt{1-\zeta^2}\right)}}}{s + \zeta\omega_n + j\omega_n\sqrt{1-\zeta^2}}$$

$$= \frac{\gamma}{Ks} + \frac{|M|e^{j\phi}}{s + \zeta\omega_n - j\omega_n\sqrt{1-\zeta^2}} + \frac{|M|e^{-j\phi}}{s + \zeta\omega_n + j\omega_n\sqrt{1-\zeta^2}}$$

$|M|$ is the magnitude of the partial fraction coefficient (numerator of either of the complex terms)—that is,

$$|M| = \frac{\gamma}{2K\sqrt{(\zeta^2 - 1)^2 + \zeta^2(1 - \zeta^2)}} = \frac{\gamma}{2K\sqrt{1 - \zeta^2}}$$

ϕ, the phase angle, is found by first removing the imaginary term from the denominator of the first partial fraction coefficient by multiplying by the complex conjugate and then rearranging terms—that is,

$$\frac{\gamma}{2K\left((\zeta^2 - 1) - j\zeta\sqrt{1 - \zeta^2}\right)} = \frac{\gamma\left((\zeta^2 - 1) + j\zeta\sqrt{1 - \zeta^2}\right)}{2K\left((\zeta^2 - 1) - j\zeta\sqrt{1 - \zeta^2}\right)\left((\zeta^2 - 1) + j\zeta\sqrt{1 - \zeta^2}\right)}$$

$$= \frac{\gamma\left(-\left(\sqrt{1 - \zeta^2}\right)^2 + j\zeta\sqrt{1 - \zeta^2}\right)}{2K\left((\zeta^2 - 1) - j\zeta\sqrt{1 - \zeta^2}\right)\left((\zeta^2 - 1) + j\zeta\sqrt{1 - \zeta^2}\right)}$$

$$= \gamma\sqrt{1 - \zeta^2}\,\frac{(-\sqrt{1 - \zeta^2} + j\zeta)}{2K\left((\zeta^2 - 1) - j\zeta\sqrt{1 - \zeta^2}\right)\left((\zeta^2 - 1) + j\zeta\sqrt{1 - \zeta^2}\right)}$$

As shown in the figure in this solution, the phase angle of the partial fraction coefficient from the above equation is then given by

$$\phi = \pi + \psi = \pi + \tan^{-1}\frac{\zeta}{-\sqrt{1 - \zeta^2}}$$

and

$$\psi = \tan^{-1}\left(\frac{\zeta}{-\sqrt{1 - \zeta^2}}\right)$$

Note that the hypotenuse of the triangle in the following figure is 1. Returning to the time domain yields Equation 12.25 by noting that the form of the solution for the complex terms is

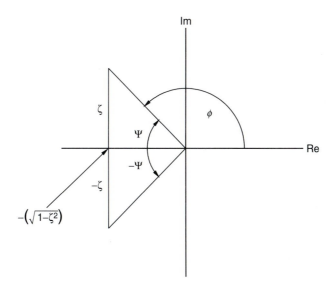

$$2|M|e^{-\zeta\omega_n t}\cos\left(\omega_n\sqrt{1-\zeta^2}t+\phi\right)$$

It is always helpful to check analysis results and one easy point to check is usually at time zero.

$$\theta(0)=\frac{\gamma}{K}\left[1+\frac{\cos(\phi)}{\sqrt{1-\zeta^2}}\right]=\frac{\gamma}{K}\left[1+\frac{-\sqrt{1-\zeta^2}}{\sqrt{1-\zeta^2}}\right]=0$$

as it should since the saccade starts at primary position or $\theta(0)=0$. ■

To fully explore the quality of a model, it is necessary to compare its performance against the data. For a saccade, convenient metrics are time to peak overshoot, which gives an indication of saccade duration, and peak velocity. These metrics were discussed previously in Section 12.3.1 when the main sequence diagram was described.

The time to peak overshoot of saccade model, T_p, is found by first calculating

$$\frac{\partial\theta}{\partial t}=\frac{\gamma e^{-\zeta\omega_n t}}{K\sqrt{1-\zeta^2}}[-\zeta\omega_n\cos(\omega_d t+\phi)-\omega_d\sin(\omega_d t+\phi)] \qquad (12.27)$$

using the chain rule on Equation 12.26 and then determining T_p from $\frac{\partial\theta}{\partial t}|_{t=T_p}=0$, yielding

$$T_p=\frac{\pi}{\omega_n\sqrt{1-\zeta^2}} \qquad (12.28)$$

With Westheimer's parameter values, $T_p=37$ ms for saccades of all sizes, which is independent of saccade magnitude and not in agreement with the experimental data which have a duration that increases as a function of saccade magnitude as presented in Figure 12.11.

Example Problem 12.11

Show that Equation 12.28 follows from Equation 12.27 set equal to zero. Find the value of $\theta(T_p)$.

Solution

With

$$\frac{\gamma e^{-\zeta\omega_n t}}{K\sqrt{1-\zeta^2}}[-\zeta\omega_n \cos(\omega_d t + \phi) - \omega_d \sin(\omega_d t + \phi)] = 0$$

the terms multiplying the sinusoids are removed since they do not equal zero. Therefore,

$$-\zeta\omega_n \cos(\omega_d t + \phi) = \omega_d \sin(\omega_d t + \phi) = \omega_n \sqrt{1-\zeta^2}\sin(\omega_d t + \phi)$$

which reduces to

$$\tan(\omega_d t + \phi) = \frac{-\zeta}{\sqrt{1-\zeta^2}} = \tan(\phi)$$

The last term in the preceding equation follows from Example Problem 12.9. Now

$$\tan(\omega_d t + \phi) = \tan(\phi)$$

only when $\omega_d t = n\pi$ or $t = \frac{n\pi}{\omega_d}$. The time to peak overshoot is the smallest value of n that satisfies $t = \frac{n\pi}{\omega_d}$ which is $n = 1$. Thus with $t = T_p$

$$T_p = \frac{\pi}{\omega_d} = \frac{\pi}{\omega_n\sqrt{1-\zeta^2}}$$

With T_p substituted into Equation 12.26, the size of the saccade at time of peak overshoot is

$$\theta(T_p) = \frac{\gamma}{K}\left[1 + \frac{e^{-\zeta\omega_n\frac{\pi}{\omega_n\sqrt{1-\zeta^2}}}}{\sqrt{1-\zeta^2}}\cos\left(\omega_d\frac{\pi}{\omega_n\sqrt{1-\zeta^2}} + \phi\right)\right]$$

$$= \frac{\gamma}{K}\left[1 + \frac{e^{-\zeta\omega_n\frac{\pi}{\omega_n\sqrt{1-\zeta^2}}}}{\sqrt{1-\zeta^2}}\cos(\pi + \phi)\right]$$

$$= \frac{\gamma}{K}\left[1 + \frac{e^{-\zeta\omega_n\frac{\pi}{\omega_n\sqrt{1-\zeta^2}}}}{\sqrt{1-\zeta^2}}\sqrt{1-\zeta^2}\right]$$

$$= \frac{\gamma}{K}\left(1 + e^{-\zeta\frac{\pi}{\sqrt{1-\zeta^2}}}\right)$$
■

The predicted saccade peak velocity, $\dot{\theta}(t_{pv})$, is found by first calculating

$$
\frac{\partial^2 \theta}{\partial t^2} = \frac{-\gamma e^{-\zeta\omega_n t}}{K\sqrt{1-\zeta^2}}(-\zeta\omega_n\zeta\omega_n \cos(\omega_d t + \phi) + \omega_d \sin(\omega_d t + \phi))
$$

$$
+ (-\zeta\omega_n\omega_d \sin(\omega_d t + \phi) + \omega_d^2 \cos(\omega_d t + \phi))
$$

$$(12.29)$$

and then determining time at peak velocity, t_{pv}, from $\frac{\partial^2 \theta}{\partial t^2}|_{t=t_{pv}} = 0$, yielding

$$
t_{pv} = \frac{1}{\omega_d} \tan^{-1}\left(\frac{\sqrt{1-\zeta^2}}{\zeta}\right)
$$

$$(12.30)$$

Substituting t_{pv} into Equation 12.27 gives the peak velocity $\dot{\theta}(t_{pv})$. Using Westheimer's parameter values with any arbitrary saccade magnitude given by $\Delta\theta = \frac{\gamma}{K}$ and Equation 12.27 gives

$$
\dot{\theta}(t_{pv}) = 55.02\Delta\theta
$$

$$(12.31)$$

Equation 12.31 indicates that peak velocity is directly proportional to saccade magnitude. As illustrated in the main sequence diagram shown in Figure 12.11, experimental peak velocity data have an exponential form and do not represent a linear function as predicted by the Westheimer model. Both Equation 12.28 and 12.31 are consistent with linear systems theory. That is, for a step input to a linear system, the duration (and time to peak overshoot) stays constant regardless of the size of the input, and the peak velocity increases with the size of the input.

Example Problem 12.12

Show that Equation 12.30 follows from 12.29.

Solution

With

$$
\frac{-\gamma e^{-\zeta\omega_n t}}{K\sqrt{1-\zeta^2}}(-\zeta\omega_n\zeta\omega_n \cos(\omega_d t + \phi) + \omega_d \sin(\omega_d t + \phi))
$$

$$
+ (-\zeta\omega_n\omega_d \sin(\omega_d t + \phi) + \omega_d^2 \cos(\omega_d t + \phi)) = 0
$$

the terms multiplying the sinusoids are removed since they do not equal zero. Therefore,

$$
((\omega_d^2 - \zeta^2\omega_n^2) \cos(\omega_d t + \phi)) - (2\zeta\omega_n\omega_d \sin(\omega_d t + \phi)) = 0
$$

which reduces to

$$
\tan(\omega_d t + \phi) = \frac{\omega_d^2 - \zeta^2\omega_n^2}{2\zeta\omega_n\omega_d} = \frac{\omega_n^2(1 - \zeta^2) - \zeta^2\omega_n^2}{2\zeta\omega_n\omega_n\sqrt{1-\zeta^2}} = \frac{1 - 2\zeta^2}{2\zeta\sqrt{1-\zeta^2}} = \left(1 - \frac{1}{2\zeta^2}\right)\tan(\psi)
$$

The last term in the previous expression makes use of $\tan(\psi) = \frac{-\zeta}{\sqrt{1-\zeta^2}}$ as given in Example Problem 12.10. Also note that $\tan(\psi) = \tan(\phi)$ as evident from the figure in Example Problem 12.10. Now the trigonometric identity

$$\tan(\omega_d t + \phi) = \frac{\tan(\omega_d t) + \tan(\phi)}{1 - \tan(\omega_d t)\tan(\phi)}$$

is used to simplify the above expression involving $\tan(\omega_d t + \phi)$, which after substituting yields

$$\tan(\omega_d t) + \tan(\phi) = (1 - \tan(\omega_d t)\tan(\phi)) \times \left(1 - \frac{1}{2\zeta^2}\right)\tan(\psi)$$

$$= \tan(\psi) - \frac{1}{2\zeta^2}\tan(\psi) - \left(1 - \frac{1}{2\zeta^2}\right)\tan(\psi)\tan(\omega_d t)\tan(\phi)$$

Using $\tan(\psi) = \tan(\phi)$ and rearranging the previous equation gives

$$\tan(\omega_d t)\left[1 + (\tan(\phi))^2\left(1 - \frac{1}{2\zeta^2}\right)\right] = \frac{-\tan(\phi)}{2\zeta^2}$$

or

$$\tan(\omega_d t) = \frac{-\tan(\phi)}{2\zeta^2\left(1 + (\tan(\phi))^2\left(\frac{2\zeta^2-1}{2\zeta^2}\right)\right)} = \frac{-\tan(\phi)}{2\zeta^2 + (\tan(\phi))^2(2\zeta^2 - 1)}$$

With $\tan(\phi) = \frac{-\zeta}{\sqrt{1-\zeta^2}}$, the previous expression reduces to

$$\tan(\omega_d t) = \frac{\frac{\zeta}{\sqrt{1-\zeta^2}}}{2\zeta^2 + \frac{\zeta^2(2\zeta^2-1)}{1-\zeta^2}} = \frac{\sqrt{1-\zeta^2}}{\zeta}$$

or

$$t_{pv} = \frac{1}{\omega_d}\tan^{-1}\left(\frac{\sqrt{1-\zeta^2}}{\zeta}\right)$$ ∎

Westheimer noted the differences between saccade duration–saccade magnitude and peak velocity–saccade magnitude in the model and the experimental data and inferred that the saccade system was not linear because the peak velocity–saccade magnitude plot was nonlinear. He also noted that the input was not an abrupt step function. Overall, this model provided a satisfactory fit to the eye position data for a saccade of 20°, but not for saccades of other magnitudes. Interestingly, Westheimer's second-order model proves to be an adequate model for saccades of all sizes if a different input function, as described in the next section, is assumed. Due to its simplicity, the Westheimer model of the oculomotor plant is still popular today.

12.5 THE SACCADE CONTROLLER

One of the challenges in modeling physiological systems is the lack of data or information about the input to the system. For instance, in the fast eye movement system the input is the neurological signal from the CNS to the muscles connected to the eyeball. Information about the input is not available in this system since it involves thousands of neurons firing at a very high rate. Recording the signal would involve invasive surgery and instrumentation that is not yet available. The difficulty in modeling this system—and most physiological systems—is the lack of information about the input. Often, however, it is possible to obtain information about the input via indirect means as described in this section for the fast eye movement system.

In 1964, Robinson performed an experiment in an attempt to measure the input to the eyeballs during a saccade. To record the measurement, one eye was held fixed using a suction contact lens, while the other eye performed a saccade from target to target. Since the same innervation signal is sent to both eyes during a saccade, Robinson inferred that the input, recorded through the transducer attached to the fixed eyeball, was the same input driving the other eyeball. He proposed that muscle tension driving the eyeballs during a saccade is a pulse plus a step, or simply, a pulse-step input (Figure 12.13).

Microelectrode studies have been carried out to record the electrical activity in oculomotor neurons. Figure 12.14 illustrates a micropipet being used to record the activity in the oculomotor nucleus, an important neuron population responsible for driving a saccade. Additional experiments on oculomotor muscle have been carried out to learn more about the saccade controller since Robinson's initial study. In 1975, for instance, Collins and his coworkers reported using a miniature "C"-gauge force transducer to measure muscle tension in vivo at the muscle tendon during unrestrained human eye movements. This type of study has allowed a better understanding of the tensions exerted by each muscle, rather than the combined effect of both muscles as shown in Figure 12.14.

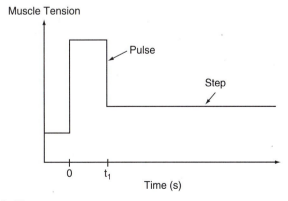

Figure 12.13   Diagram illustrating the muscle tension recorded during a saccade.

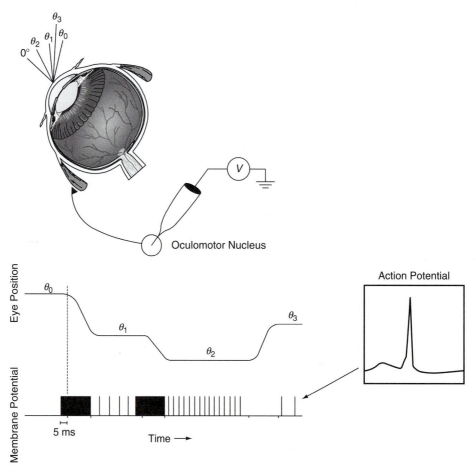

Figure 12.14   Diagram of recording of a series of saccades using a micropipet and the resultant electrical activity in a single neuron. Spikes in the membrane potential indicate that an action potential occurred with dynamics as described in Chapter 11. Saccade neural activity initiates with a burst of neural firing approximately 5 ms before the eye begins to move and continues until the eye has almost reached its destination. Relative position of the eye is shown at the top with angles θ_0 through θ_3. Initially the eye starts in position θ_0, a position in the extremity in which the muscle is completely stretched with zero input. To move the eye from θ_0 to θ_1, neural burst firing occurs. To maintain the eye at θ_1, a steady firing occurs in the neuron. The firing rate for fixation is in proportion to the shortness of the muscle. Next, the eye moves from θ_1 to θ_2. This saccade moves much more slowly than the first saccade with approximately the same duration as the first. The firing level is also approximately at the same level as the first. The difference in input corresponds to fewer neurons firing to drive the eye to its destination, which means a smaller input than the first saccade. Because the muscle is shorter after completing this saccade, the fixation firing rate is higher than the previous position at θ_1. Next, the eye moves in the opposite direction to θ_3. Since the muscle is lengthening, the input to the muscle is zero, that is, no action potentials are used to stimulate the muscle. The fixation firing level θ_3 is less than that for θ_1 because the muscle is longer.

It is important to distinguish between the tension or force generated by a muscle, called muscle tension, and the force generator within the muscle, called the active-state tension generator. The active-state tension generator creates a force within the muscle that is transformed through the internal elements of the muscle into the muscle tension. Muscle tension is external and measurable, and the active-state tension is internal and unmeasurable. Active-state tension follows most closely the neural input to the muscle. From Figure 12.14, a pattern of neural activity is observed as follows:

1. The muscle that is being contracted (agonist) is stimulated by a pulse, followed by a step to maintain the eyeball at its destination.
2. The muscle that is being stretched (antagonist), is unstimulated during the saccade (stimulated by a pause or a negative pulse to zero), followed by a step to maintain the eyeball at its destination.

Figure 12.15 quantifies these relationships for the agonist neural input, N_{ag}, and the antagonist neural input, N_{ant}. The pulse input is required to get the eye to the target as soon as possible, and the step is required to keep the eye at that location.

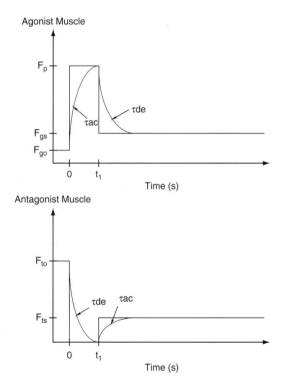

Figure 12.15 Agonist, N_{ag}, and antagonist, N_{ant}, control signals (solid lines) and the agonist, F_{ag}, and antagonist, F_{ant}, active-state tensions (dashed lines). Note that the time constant for activation, τ_{ac}, is different from the time constant for deactivation, τ_{de}. The time interval, t_1, is the duration of the pulse.

It has been reported that the active-state tensions are not identical to the neural controllers but can be described by low-pass filtered pulse-step waveforms. The active-state tensions are shown in Figure 12.15 as dashed lines with time-varying time constants τ_{ac} and τ_{de}. It is thought that the low-pass filtering involves the movement of Ca^{++} across the cell membrane. Some investigators have reported a different set of time constants for the agonist and antagonist activity, and others have noted a firing frequency–dependent agonist activation time constant. Others suggest that the agonist activation time constant is a function of saccade magnitude. For simplicity in this textbook, activation and deactivation time constants are assumed to be identical for both agonist and antagonist activity.

In 1964, Robinson described a model for fast eye movements (constructed from empirical considerations) that simulated saccades over a range of 5° to 40° by changing the amplitude of the pulse-step input. These simulation results were adequate for the position–time relationship of a saccade, but the velocity–time relationship was inconsistent with physiological evidence. To correct this deficiency of the model, physiological and modeling studies of the oculomotor plant were carried out during the 1960s through the 1990s that allowed the development of a more homeomorphic oculomotor plant. Essential to this work was the construction of oculomotor muscle models.

12.6   DEVELOPMENT OF AN OCULOMOTOR MUSCLE MODEL

It is clear that an accurate model of muscle is essential in the development of a model of the horizontal fast eye movement system that is driven by a pair of muscles (lateral and medial rectus muscles). In fact, the Westheimer model does not include any muscle model and relies solely on the inertia of the eyeball, friction between the eyeball and socket, and elasticity due to the optic nerve and other attachments as the elements of the model. In this section, the historical development of a muscle model is discussed as it relates to the oculomotor system. Muscle model research involves a broad spectrum of topics, ranging from the micro models that deal with the sarcomeres to macro models in which collections of cells are grouped into a lumped parameter system and described with ordinary mechanical elements. Here the focus is on a macro model of the oculomotor muscle based on physiological evidence from experimental testing. The model elements, as presented, consist of an active-state tension generator (input), elastic elements, and viscous elements. Each element is introduced separately and the muscle model is incremented in each subsection. It should be noted that the linear muscle model presented at the end of this section completely revises the subsections before it. The earlier subsections were presented because of their historical significance and to appreciate the current muscle model.

12.6.1   Passive Elasticity

Consider the experiment of stretching an unexcited muscle and recording tension to investigate the passive elastic properties of muscle. The data curve shown in

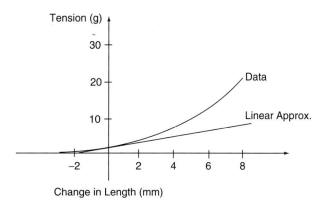

Figure 12.16 Diagram illustrating the tension-displacement curve for unexcited muscle. The slope of the linear approximation to the data is muscle passive elasticity, K_{pe}.

Figure 12.16 is a typical recording of the tension observed in an eye rectus muscle. The tension required to stretch a muscle is a nonlinear function of distance. Thus, to precisely model this element, a nonlinear spring element should be used. Note that the change in length at 0 refers to the length of the muscle at primary position (looking straight ahead). Thus the eye muscles are stretched, approximately 3 mm, when the eye movement system is at rest in primary position. At rest, the muscle length is approximately 37 mm.

To be useful in a linear model of muscle, Figure 12.16 must be linearized in the vicinity of an operating point. The operating point should be somewhat centered in the region in which the spring operates. In Figure 12.16, a line tangent to the curve at primary position provides a linear approximation to the spring's behavior in this region. For ease in analysis, the following relationships hold for a sphere representing the eyeball radius of 11 mm:

$$1\,gm = 9.806 \times 10^{-3}N$$

$$1° = 0.192\,mm = 1.92 \times 10^{-4}m$$

The slope of the line, K_{pe}, is approximately

$$K_{pe} = 0.2\frac{g}{°} = 0.2\frac{g}{°} \times \frac{9.806 \times 10^{-3}N}{1g} \times \frac{1°}{1.92 \times 10^{-4}m} = 10.2\frac{N}{m}$$

and represents the elasticity of the passive elastic element. The choice of the operating region is of vital importance concerning the slope of the curve. At this time, a point in the historical operating region of rectus muscle is used. In most of the oculomotor literature, the term K_{pe} is typically subtracted out of the analysis and is not used. Later in this section, the operating point will be revisited and this element will be completely removed from the model.

12.6.2   Active-State Tension Generator

In general, a muscle produces a force in proportion to the amount of stimulation. The element responsible for the creation of force is the active-state tension generator. Note that this terminology is used so that there is no confusion with the force created within the muscle when the tension created by the muscle is discussed. The active tension generator is included along with the passive elastic element in the muscle model as shown in Figure 12.17. The relationship between tension, T, active-state tension, F, and elasticity is given by

$$T = F - K_{pe}x \qquad (12.32)$$

Isometric (constant length) experiments have been performed on humans over the years to estimate the active-tension generator at different levels of stimulation. These experiments were usually performed in conjunction with strabismus surgery when muscles were detached and reattached to correct crossed eyes. Consider the tension created by a muscle when stimulated as a function of length as shown in Figure 12.18. The data were collected from the lateral rectus muscle that was detached from one eyeball while the other unoperated eyeball fixated at different locations in the nasal (N) and temporal (T) directions from $-45°$ to $45°$. This experiment was carried out under the assumption that the same neural input is sent to each eyeball (Hering's law of equal innervation), thus the active-state tension in the freely moving eyeball should be the same as that in the detached lateral rectus muscle. At each fixation point, the detached lateral rectus muscle was stretched and tension data were recorded at each of the points indicated on the graph. The thick line represents the muscle tension at that particular eye position under normal conditions. The curve for $45°$ T is the zero stimulation case and represents the passive mechanical properties of muscle. The results are similar to those reported in Figure 12.10. Note that the tension generated is a nonlinear function of the length of the muscle.

To compare the model in Figure 12.17 against the data in Figure 12.18, it is convenient to subtract the passive elasticity in the data (represented by the $45°$ T

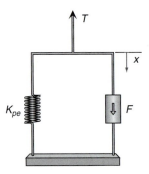

Figure 12.17   Diagram illustrating a muscle model consisting of an active-state tension generator, F, and a passive elastic element, K_{pe}. Upon stimulation of the active-state tension generator a tension, T, is exerted by the muscle.

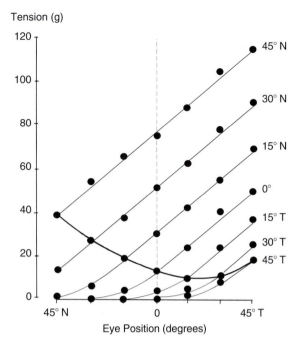

Figure 12.18 Length-tension curves for lateral rectus muscle at different levels of activation. Dots represent tension data recorded from the detached lateral rectus muscle during strabismus surgery while the unoperated eyeball fixated at targets from −45° to 45° (adapted from Collins, O'Meara, and Scott, 1975).

curve) from each of the data curves 30° T through 45° N, leaving only the hypothetical active-state tension. Shown in the graph on the left in Figure 12.19 is one such calculation for 15° N with the active-state tension given by the dashed line. The other curves in Figure 12.18 give similar results and have been omitted because of the clutter. The dashed line should represent the active-state tension, which appears to be a function of length. If this was a pure active-state tension element, the subtracted curve should be a horizontal line indicative of the size of the input. One such input is shown for the active-state tension in the graph on the right in Figure 12.19. The result in Figure 12.19 implies that either the active-state tension's effect is a nonlinear element (i.e., there may be other nonlinear or linear elements missing in the model), or perhaps some of the assumptions made in the development of the model are wrong. For the moment, consider that the analysis is correct and assume that some elements are missing. This topic will be revisited at the end of this section.

12.6.3 Elasticity

With the effects of the passive elasticity removed from the tension data as shown in Figure 12.19, a relationship still exists between length and tension as previously

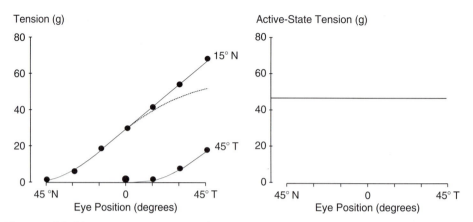

Figure 12.19 (Left) Length-tension curves for extraocular muscle at two levels of activation corresponding to the 15° N and 45° T positions. Dots represent tension data recorded from the detached lateral rectus muscle during strabismus surgery while the unoperated eyeball fixated at targets. Dashed line is the 15° curve with the 45° curve subtracted from it. The resultant dashed curve represents the active-state tension as a function of eye position. (Right) The theoretical graph for active-state tension vs. eye position as given by Equation 7.18 (adapted from Collins, O'Meara, and Scott, 1975).

described. To account for the relationship between length and tension, a new elastic element is added into the model as shown in Figure 12.20 and described by Equation 12.33.

$$T = F - K_{pe}x - Kx \qquad (12.33)$$

The new elastic element, K, accounts for the slope of the subtracted curve shown with dashes in the graph on the left in Figure 12.19. The slope of the line, K, at primary position is approximately $0.8\,\text{g}/° = 40.86\,\text{N/m}$ (a value typically reported in the literature). The slope for each of the curves in Figure 12.18 can be calculated in the same manner at primary position with the resultant slopes all approximately equal to the same value as the one for 15° N. At this time, the introduction of additional experiments will provide further insight to the development of the muscle model.

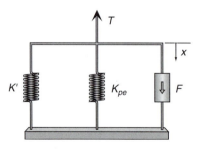

Figure 12.20 Diagram illustrating a muscle model consisting of an active-state tension generator, F, passive elastic element, K_{pe}, and elastic element, K. Upon stimulation of the active-state tension generator, a tension T is exerted by the muscle.

Series Elastic Element

Experiments carried out by Levin and Wyman in 1927, and Collins in 1975 indicated the need for a series elasticity element in addition to the other elements previously presented in the muscle model. The experimental setup and typical data from the experiment are shown in Figure 12.21. The protocol for this experiment, called the quick release experiment, is as follows. (1) A weight is hung onto a muscle. (2) The muscle is fully stimulated at time t_1. (3) The weight is released at time t_2. At time t_2, the muscle changes length almost instantaneously when the weight is released. The only element that can instantaneously change its length is a spring. Thus to account for this behavior, a spring, called the series elastic element, K_{se}, is connected in series to the active-state tension element. The updated muscle model is shown in Figure 12.22.

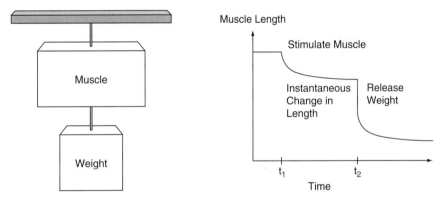

Figure 12.21   Diagram illustrating the quick release experiment. Figure on left depicts the physical setup of the experiment. Figure on the right shows typical data from the experiment. At time t_1 the muscle is fully stimulated and at time t_2 the weight is released.

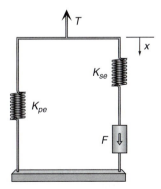

Figure 12.22   Diagram illustrating a muscle model consisting of an active-state tension generator F, passive elastic element K_{pe} and series elastic element K_{se}. Upon stimulation of the active-state tension generator F, a tension T is exerted by the muscle.

Based on the experiment carried out by Collins in 1975 on rectus eye muscle, an estimate for K_{se} was given as 125 N/m (2.5 gm/°). Since the value of K_{se} does not equal the value of K, another elastic element is needed to account for this behavior.

Length–Tension Elastic Element

Given the inequality between K_{se} and K, another elastic element, called the length–tension elastic element, K_{lt}, is placed in parallel with the active-state tension element as shown in the illustration on the left in Figure 12.23. For ease of analysis, K_{pe} is subtracted out (removed) using the graphical technique shown in Figure 12.19. To estimate a value for K_{lt}, the muscle model shown on the right in Figure 12.23 is analyzed and reduced to an expression involving K_{lt}. Analysis begins by summing the forces acting on nodes 1 and 2.

$$T = K_{se}(x_2 - x_1) \tag{12.34}$$

$$F = K_{lt}x_2 + K_{se}(x_2 - x_1) \quad \rightarrow \quad x_2 = \frac{F + K_{se}x_1}{K_{se} + K_{lt}} \tag{12.35}$$

Substituting x_2 from Equation 12.35 into 12.34 gives

$$T = \frac{K_{se}}{K_{se} + K_{lt}}(F + K_{se}x_1) - K_{se}x_1 = \frac{K_{se}}{K_{se} + K_{lt}}F - \frac{K_{se}K_{lt}}{K_{se} + K_{lt}}x_1 \tag{12.36}$$

Equation 12.36 is an equation for a straight line with y-intercept $\frac{K_{se}}{K_{se}+K_{lt}}$ and slope $\frac{K_{se}K_{lt}}{K_{se}+K_{lt}}$. The slope of the length–tension curve in Figure 12.17 is given by $K = 0.8\,\text{g}/° = 40.86\,\text{N/m}$. Therefore,

$$K = \frac{K_{se}K_{lt}}{K_{se} + K_{lt}} = 40.86\frac{N}{m} \tag{12.37}$$

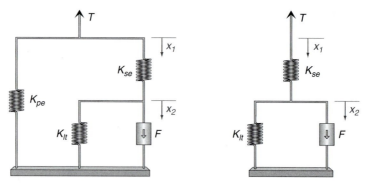

Figure 12.23   Diagram on left illustrates a muscle model consisting of an active-state tension generator F in parallel to a length-tension elastic element K_{lt}, connected to a series elastic element K_{se}, all in parallel with the passive elastic element K_{pe}. Upon stimulation of the active-state tension generator F, a tension T is exerted by the muscle. The diagram on the right is the same muscle model except that K_{pe} has been removed.

Solving Equation 12.37 for K_{lt} yields

$$K_{lt} = \frac{K_{se}K}{K_{se} - K} = 60.7\frac{N}{m}$$ (12.38)

12.6.4 Force–Velocity Relationship

Early experiments indicated that muscle had elastic as well as viscous properties. Muscle was tested under isotonic (constant force) experimental conditions as shown in Figure 12.24 to investigate muscle viscosity. The muscle and load were attached to a lever with a high lever ratio. The lever reduced the gravity force (mass × gravity) of the load at the muscle by one over the lever ratio, and the inertial force (mass × acceleration) of the load by one over the lever ratio squared. With this arrangement, it was assumed that the inertial force exerted by the load during isotonic shortening could be ignored. The second assumption was that if mass was not reduced enough by the lever ratio (enough to be ignored), then taking measurements at maximum velocity provided a measurement at a time when acceleration is zero, and therefore inertial force equals zero. If these two assumptions are valid, then the experiment would provide data free of the effect of inertial force as the gravity force is varied.

According to the experimental conditions, the muscle is stretched to its optimal length at the start of the isotonic experiment. The isotonic experiment begins by attaching a load M, stimulating the muscle, and recording position. The two curves in Figure 12.25 depict the time course for the isotonic experiment for a small and large load. Notice that the duration of both responses are approximately

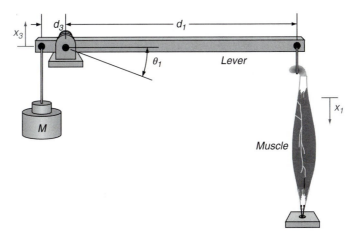

Figure 12.24 Drawing of the classical isotonic experiment with inertial load and muscle attached to the lever. The muscle is stretched to its optimal length according to experimental conditions and attached to ground.

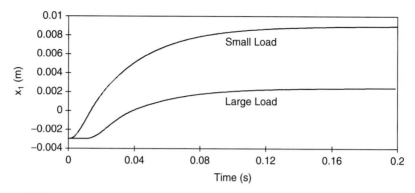

Figure 12.25   Diagram illustrating typical response of a muscle stimulated with a large and small load.

equal regardless of the load, in spite of the apparent much longer time delay associated with the large load. Next notice that the heavier the load, the less the total shortening. Maximum velocity is calculated numerically from the position data. To estimate muscle viscosity, this experiment is repeated with many loads at the same stimulation level and maximum velocity is calculated. Figure 12.26 illustrates the typical relationship between load ratio (P/P_o) and maximum velocity, where $P = Mg$ and P_o is the isometric tension (the largest weight that the muscle can move) for maximally stimulated muscle. This curve is usually referred to as the force–velocity curve.

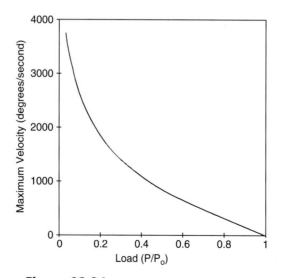

Figure 12.26   Illustrative force-velocity curve.

Clearly, the force–velocity curve is nonlinear and follows a hyperbolic shape. If a smaller stimulus than maximum is used to stimulate the muscle, then a family of force–velocity curves results as shown in Figure 12.27. Each curve is generated with a different active-state tension as indicated. The force–velocity characteristics in Figure 12.27 are similar to those shown in Figure 12.26. In particular, the slope of the force–velocity curve for a small value of active-state tension is quite different than that for a large value of the active-state tension in the operating region of the eye muscle (i.e., approximately 800°/s).

To include the effects of viscosity from the isotonic experiment in the muscle model, a viscous element is placed in parallel with the active-state tension generator and the length–tension elastic element as shown in Figure 12.28. The impact of this element is examined by analyzing the behavior of the model in Example Problem 12.12 by simulating the conditions of the isotonic experiment. At this stage, it is assumed that the viscous element is linear in this example. For simplicity, the lever is removed along with the virtual acceleration term $\ddot{M}x_1$. A more thorough analysis including the lever is considered later in this chapter. For simplicity, the passive elastic element K_{pe} is removed from the diagrams and analysis.

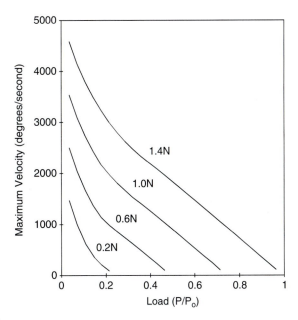

Figure 12.27 Illustrative family of force-velocity curves for active state tensions ranging from 1.4 to 0.2 N.

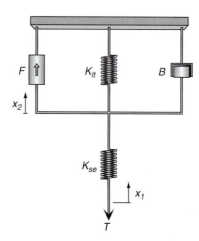

Figure 12.28   Diagram illustrates a muscle model consisting of an active-state tension generator F in parallel with a length-tension elastic element K_{lt} and viscous element B, connected to a series elastic element K_{se}. The passive elastic element K_{pe} has been removed from the model for simplicity. Upon stimulation of the active-state tension generator F, a tension T is exerted by the muscle.

Example Problem 12.13

Consider the system shown in the following figure that represents a model of the isotonic experiment. Assume that the virtual acceleration term $M\ddot{x}_1$ can be ignored. Calculate and plot maximum velocity as a function of load.

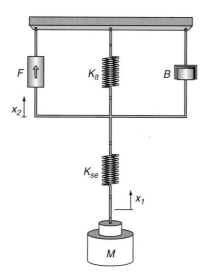

Solution

Assume that $\dot{x}_2 > \dot{x}_1$ and that the mass is supported so that $x_1 > 0$. Let the term $K_{st} = K_{se} + K_{lt}$. Summing the forces acting on nodes 1 and 2 gives

$$Mg = K_{se}(x_2 - x_1) \quad \rightarrow \quad x_1 = x_2 - \frac{Mg}{K_{se}}$$

$$F = B\dot{x}_2 + K_{lt}x_2 + K_{se}(x_2 - x_1)$$

Substituting x_1 into the second equation yields

$$F = B\dot{x}_2 + K_{lt}x_2 + Mg$$

Solving the previous equation for x_2 and $\dot{x}_2$ gives

$$x_2(t) = \frac{F - Mg}{K_{lt}}\left(1 - e^{-\frac{K_{lt}t}{B}}\right)$$

$$\dot{x}_2(t) = \frac{F - Mg}{B}e^{-\frac{K_{lt}t}{B}}$$

Maximum velocity, $V_{\max}$, for all loads is given by $V_{\max} = \frac{F-Mg}{B}$ and $\dot{x}_1 = \dot{x}_2$ since $\dot{x}_1 = \frac{d}{dt}\left(x_2 - \frac{Mg}{K_{se}}\right)$. The following graph depicts a linear relationship between maximum velocity and load, and the data.

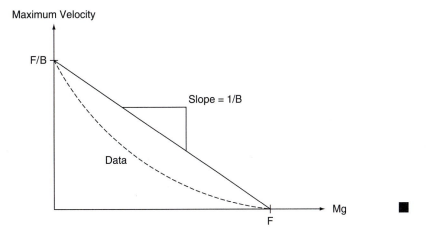

The assumption of a linear viscosity element appears to be in error since the analysis in Example Problem 12.12 predicts a linear relationship between load and maximum velocity (according to the assumptions of the solution) and the data from the isotonic experiment shown in Figure 12.26 is clearly nonlinear. Thus, a reasonable assumption is that the viscosity element is nonlinear.

Traditionally, muscle viscosity is characterized by the nonlinear Hill hyperbola, given by

$$V_{\max}(P + a) = b(P_0 - P) \tag{12.39}$$

where V_{max} is the maximum velocity, P is the external force, P_o the isometric tension, and a and b are the empirical constants representing the asymptotes of the hyperbola. As described previously, P_o represents the isometric tension, which is the largest weight that the muscle can move, and P is the weight Mg. Hill's data suggest that

$$a = \frac{P_0}{4} \quad \text{and} \quad b = \frac{V_{max}}{4}$$

Therefore, with these values for a and b, the Hill equation is rewritten from Equation 12.39 as

$$P = P_0 - \frac{V_{max}(P_0 + a)}{b + V_{max}} = P_0 - BV_{max} \tag{12.40}$$

where

$$B = \frac{P_0 + a}{b + V_{max}} \tag{12.41}$$

The term B represents the viscosity of the element. Clearly, the force due to viscosity is nonlinear due to the velocity term, V_{max}, in the denominator of Equation 12.41.

In oculomotor models, V_{max} is usually replaced by $\dot{x}_2$, P is replaced by muscle tension, T, and P_o is replaced by the active-state tension, F, as defined from Figure 12.28. Therefore, Equations 12.40 and 12.41 are rewritten as

$$T = F - BV \tag{12.42}$$

where

$$B = \frac{F + a}{b + \dot{x}_2} \tag{12.43}$$

Example Problem 12.14

Write an equation for the tension created by the muscle model shown in Figure 12.28.

Solution

Assuming that $\dot{x}_2 > \dot{x}_1$, the forces acting at nodes 1 and 2 give

$$T = K_{se}(x_2 - x_1) \quad \rightarrow \quad x_2 = \frac{T}{K_{se}} + x_1$$

$$F = B\dot{x}_2 + K_{lt}x_2 + K_{se}(x_2 - x_1)$$

Eliminating x_2 in the second equation by substituting from the first equation yields

$$F = B\dot{x}_2 + (K_{se} + K_{lt})\left(\frac{T}{K_{se}} + x_1\right) - K_{se}x_1$$

Solving for T gives

$$T = \frac{K_{se}}{K_{se}+K_{lt}}F - \frac{K_{se}K_{lt}}{K_{se}+K_{lt}}x_1 - \frac{K_{se}B}{K_{se}+K_{lt}}\dot{x}_2$$

Equation 12.43 can be substituted for parameter B in the preceding equation to give a nonlinear model of oculomotor muscle. ∎

Some oculomotor investigators have reported values for a and b in the Hill equation that depend on whether the muscle is being stretched or contracted. There is some evidence to suggest that stretch dynamics are different from contraction dynamics. However, the form of the viscosity expression for muscle shortening or lengthening is given by Equation 12.43, with values for a and b parameterized appropriately. For instance, Hsu and coworkers described the viscosity for shortening and lengthening for oculomotor muscles as

$$B_{ag} = \frac{F_{ag}+AG_a}{\dot{x}_2+AG_b} \tag{12.44}$$

$$B_{ant} = \frac{F_{ant}-ANT_a}{\dot{x}_2+ANT_b} \tag{12.45}$$

where AG_a, AG_b, ANT_a, and ANT_b are parameters based on the asymptotes for contracting (agonist) or stretching (antagonist), respectively.

Equations 12.44 and 12.45 define B_{ag} and B_{ant} as nonlinear functions of velocity. To develop a linear model of muscle, both B_{ag} and B_{ant} can be linearized by making a straight-line approximation to the hyperbolic functions shown in Figure 12.27. Antagonist activity is typically at the 5% level ($F = 0.2\,\text{N}$) and agonist activity is at the 100% level ($F = 1.4\,\text{N}$).

12.7 A LINEAR MUSCLE MODEL

This section reexamines the static and dynamic properties of muscle in the development of a linear model of oculomotor muscle. The updated linear model for oculomotor muscle is shown in Figure 12.29. Each of the elements in the model is linear and is supported with physiological evidence. The muscle is modeled as a parallel combination of viscosity B_2 and series elasticity K_{se}, connected to the parallel combination of active state tension generator F, viscosity element B_1, and length–tension elastic element K_{lt}. Variables x_1 and x_2 describe the displacement from the equilibrium for the stiffness elements in the muscle model. The only structural difference between this model and the previous oculomotor muscle model is the addition of viscous element B_2 and the removal of passive elasticity K_{pe}. As will be described, the viscous element B_2 is vitally important to describe the nonlinear force–velocity characteristics of the muscle, and the elastic element K_{pe} is unnecessary.

The need for two elastic elements in the linear oculomotor muscle model is supported through physiological evidence. As described previously, the use and value of the series elasticity K_{se} was determined from the isotonic–isometric quick release experiment by Collins. Length–tension elasticity K_{lt} was estimated in a slightly

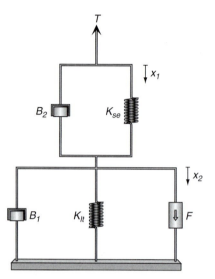

Figure 12.29   Diagram illustrates an updated linear muscle model consisting of an active-state tension generator F in parallel with a length–tension elastic element K_{lt} and viscous element B_1, connected to a series elastic element K_{se} in parallel with a viscous element B_2. Upon stimulation of the active-state tension generator F, a tension T is exerted by the muscle.

different fashion than before from the slope of the length–tension curve. Support for the two linear viscous elements is based on the isotonic experiment and estimated from simulation results presented in this section.

12.7.1   Length–Tension Curve

The basis for assuming nonlinear elasticity is the nonlinear length–tension relation for excited and unexcited muscle for tensions below 10 g as shown in Figure 12.18. Using a miniature "C"-gauge force transducer, Collins in 1975 measured muscle tension in vivo at the muscle tendon during unrestrained human eye movements. Data shown in Figure 12.18 were recorded from the rectus muscle of the left eye by measuring the isometric tensions at different muscle lengths, ranging from eye positions of $-45°$ to $45°$, and different levels of innervation, established by directing the subject to look at the corresponding targets with the unhampered right eye from $-45°$ temporal (T) to $45°$ nasal (N). The change in eye position during this experiment corresponds to a change in muscle length of approximately 18 mm. Collins described the length–tension curves as "straight, parallel lines above about 10 g. Below the 10 g level, the oculorotary muscles begin to go slack." He also reported that the normal range of tensions for the rectus muscle during all eye movements never falls below 10 g into the slack region when the in vivo force transducer is used.

In developing a muscle model for use in the oculomotor system, it is imperative that the model accurately exhibits the static characteristics of rectus eye muscle within the normal range of operation. Thus, any oculomotor muscle model must have length–tension characteristics consisting of straight, parallel lines above 10 g tension. Since oculomotor muscles do not operate below 10 g, it is unimportant that the linear behavior of the model does not match this nonlinear portion in the length–tension curves observed in the data as was done in the development of the muscle model earlier. As demonstrated in this section, by concentrating on the operational region of the oculomotor muscles, accurate length–tension curves are obtained from the muscle model using just series elastic and length–tension elastic elements, even when active-state tension is zero. Thus, there is no need to include a passive elastic element in the muscle model as previously required.

Since the rectus eye muscle is not in equilibrium at primary position (looking straight ahead, $0°$) within the oculomotor system, it is necessary to define and account for the equilibrium position of the muscle. Equilibrium denotes the unstretched length of the muscle when the tension is zero, with zero input. It is assumed that the active-state tension is zero on the $45°$ T length–tension curve. Typically, the equilibrium position for rectus eye muscle is found from within the slack region, where the $45°$ T length–tension curve intersects the horizontal axis. Note that this intersection point was not shown in the data collected by Collins (Figure 12.18) but is reported to be approximately $15°$ (3 mm short of primary position), a value that is typical of those reported in the literature.

Since the muscle does not operate in the slack region during normal eye movements, using an equilibrium point calculated from the operational region of the muscle provides a much more realistic estimate for the muscle. Here, the equilibrium point is defined according to the straight-line approximation to the $45°$ T length–tension curve above the slack region. The value at the intersection of the straight-line approximation with the horizontal axis gives an equilibrium point of $-19.3°$. By use of the equilibrium point at $-19.3°$, there is no need to include an additional elastic element K_{pe} to account for the passive elasticity associated with unstimulated muscle as others have done.

The tension exerted by the linear muscle model shown in Figure 12.29 is given by

$$T = \frac{K_{se}}{K_{se} + K_{lt}}F - \frac{K_{se}K_{lt}}{K_{se} + K_{lt}}x_1 \tag{12.46}$$

With the slope of the length–tension curve equal to $0.8\,\text{g/}° = 40.86\,\text{N/m}$ in the operating region of the muscle (nonslack region), $K_{se} = 2.5\,\text{g/}° = 125\,\text{N/m}$, and Equation 12.46 has a slope of

$$\frac{K_{se}K_{lt}}{K_{se} + K_{lt}} \tag{12.47}$$

K_{lt} is evaluated as $1.2\,\text{g/}° = 60.7\,\text{N/m}$. To estimate the active-state tension for fixation at all locations, Equation 12.46 is used to solve for F for each innervation level straight-line approximation, yielding

$$F = 0.4 + 0.0175\theta \ \text{N} \quad \text{for } \theta > 0° \ (N \text{ direction}) \tag{12.48}$$

and

$$F = 0.4 + 0.012\theta \ \text{N} \quad \text{for } \theta \leq 0° \ (T \text{ direction}) \tag{12.49}$$

where θ is the angle that the eyeball is deviated from the primary position measured in degrees, and $\theta = 5208.7 \times (x_1 - 3.705)$. Note that $5208.7 = \frac{180}{\pi r}$, where r equals the radius of the eyeball.

Figure 12.30 displays a family of static length–tension curves obtained using Equations 12.46 through 12.49, which depicts the length–tension experiment. No attempt is made to describe the activity within the slack region since the rectus eye muscle does not normally operate in that region. The length–tension curves shown in this graph are in excellent agreement with the data shown in Figure 12.18 within the operating region of the muscle.

12.7.2 Force–Velocity Relationship

The original basis for assuming nonlinear muscle viscosity is that the expected linear relation between external load and maximum velocity (see Example Problem 12.12) was not found in early experiments by Fenn and Marsh. As Fenn and Marsh reported, "If the muscle is represented accurately by a viscous elastic system this force–velocity

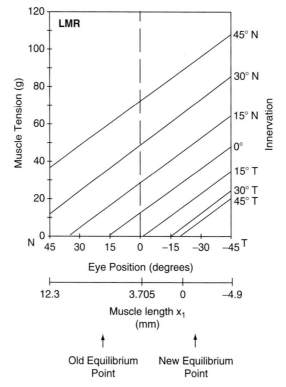

Figure 12.30 Length-tension curve generated using Equations 7.46–7.49 derived from the linear muscle model and inputs: $F = 130\,g$ for 45°N, $F = 94.3\,g$ for 30°N, $F = 64.9\,g$ for 15° N, $F = 40.8\,g$ for 0°, $F = 21.7\,g$ for 15°T, $F = 5.1\,g$ for 30°T, and $F = 0\,g$ for 45°T.

curve should have been linear, the loss of force being always proportional to the velocity. The slope of the curve would then represent the coefficient of viscosity." Essentially the same experiment was repeated for rectus eye muscle by Close and Luff in 1974 with similar results.

The classical force–velocity experiment was performed to test the viscoelastic model for muscle as described in Section 12.6.4. Under these conditions, it was first assumed that the inertial force exerted by the load during isotonic shortening could be ignored. The second assumption was that if mass was not reduced enough by the lever ratio (enough to be ignored), then taking measurements at maximum velocity provided a measurement at a time when acceleration is zero, and therefore inertial force equals zero. If these two assumptions are valid, then the experiment would provide data free of the effect of inertial force as the gravity force is varied. Both assumptions are incorrect. The first assumption is wrong since the inertial force is never minimal (minimal would be zero) and therefore has to be taken into account. The second assumption is wrong since, given an inertial mass not equal to zero, then maximum velocity depends on the forces that act prior to the time of maximum velocity. The force–velocity relationship is carefully reexamined with the inertial force included in the analysis in this section.

The dynamic characteristics for the linear muscle model are described with a force–velocity curve calculated via the lever system presented in Figure 12.24 and according to the isotonic experiment. For the rigid lever, the displacements x_1 and x_3 are directly proportional to the angle θ_1 and to each other, such that

$$\theta_1 = \frac{x_1}{d_1} = \frac{x_3}{d_3} \tag{12.50}$$

The equation describing the torques acting on the lever is given by

$$Mgd_3 + Md_3^2\ddot{\theta}_1 = d_1 K_{se}(x_2 - x_1) + d_1 B_2(\dot{x}_2 - \dot{x}_1) \tag{12.51}$$

The equation describing the forces at node 2, inside the muscle, is given by

$$F = K_{lt}x_2 + B_1\dot{x}_2 + B_2(\dot{x}_2 - \dot{x}_1) + K_{se}(x_2 - x_1) \tag{12.52}$$

Equation 12.51 is rewritten by removing θ_1 using Equation 12.50, hence

$$Mg\frac{d_3}{d_1} + M\left(\frac{d_3}{d_1}\right)^2\ddot{x}_1 = K_{se}(x_2 - x_1) + B_2(\dot{x}_2 - \dot{x}_1) \tag{12.53}$$

Ideally, to calculate the force–velocity curve for the lever system, $x_1(t)$ is found first. Then $\dot{x}_1(t)$ and $\ddot{x}_1(t)$ are found from $x_1(t)$. Finally, the velocity is found from $V_{max} = \dot{x}_1(T)$, where time T is the time it takes for the muscle to shorten to the stop, according to the experimental conditions of Close and Luff. While this velocity may not be maximum velocity for all data points, the symbol V_{max} is used to denote the velocities in the force–velocity curve for ease in presentation. Note that this definition of velocity differs from the Fenn and Marsh definition of velocity. Fenn and Marsh denoted maximum velocity as $V_{max} = \dot{x}_1(T)$, where time T is found when $\ddot{x}_1(T) = 0$.

It should be noted that this is a third-order system and the solution for $x_1(t)$ is not trivial and involves an exponential approximation (for an example of an exponential approximation solution for V_{max} from a fourth-order model, see the paper by Enderle et al., 1988). It is more expedient, however, to simply simulate a solution for $x_1(t)$ and then find V_{max} as a function of load.

Using a simulation to reproduce the isotonic experiment, elasticities estimated from the length–tension curves as previously described, and data from rectus eye muscle, parameter values for the viscous elements in the muscle model are found as $B_1 = 2.0\,Ns/m$ and $B_2 = 0.5\,Ns/m$ as demonstrated by Enderle and coworkers in 1991. The viscous element B_1 is estimated from the time constant from the isotonic time course. The viscous element B_2 is calculated by trial and error so that the simulated force–velocity curve matches the experimental force–velocity curve.

Shown in Figure 12.31 are the force–velocity curves using the model described in Equation 12.53 (with triangles), plotted along with an empirical fit to the data (solid line). It is clear that the force–velocity curve for the linear muscle model is hardly a straight line, and that this curve fits the data well.

The muscle lever model described by Equations 12.51 and 12.52 is a third-order linear system and is characterized by three poles. Dependent on the values of the parameters, the eigenvalues (or poles) consist of all real poles or a real and a pair of complex conjugate poles. A real pole is the dominant eigenvalue of the system. Through a sensitivity analysis, viscous element B_1 is the parameter that has the

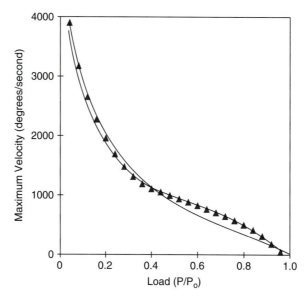

Figure 12.31   Force–velocity curve derived from simulation studies with the linear muscle model with an end stop. Shown with triangles indicating simulation calculation points and an empirical fit to the force-velocity data (solid line) as described by Close and Luff (1974) (adapted from Enderle, Engelken, and Stiles, 1991).

greatest effect on the dominant eigenvalue or time constant for this system whereas viscous element B_2 has very little effect on the dominant eigenvalue. Thus, viscous element B_1 is estimated so that the dominant time constant of the lever system model (approximately B_1/K_{lt} when $B_1 > B_2$) matches the time constant from the isotonic experimental data. For rectus eye muscle data, the duration of the isotonic experiment is approximately 100 ms. A value for $B_1 = 2.0\,Ns/m$ yields a simulated isotonic response with approximately the same duration. For skeletal muscle experimental data, the duration of the isotonic experiment is approximately 400 ms, and a value for $B_1 = 6.0\,Ns/m$ yields a simulated isotonic response with approximately the same duration. It is known that fast and slow muscle have differently shaped force–velocity curves and that the fast muscle force–velocity curve data has less curvature. Interestingly, the changes in the parameter values for B_1 as suggested here gives differently shaped force–velocity curves consistent with fast (rectus eye muscle) and slow (skeletal muscle) muscle.

The parameter value for viscous element B_2 is selected by trial and error so that the shape of the simulation force–velocity curve matches the data. As the value for B_2 is decreased from 0.5 Ns/m, the shape of the force–velocity curves changes to a more linear shaped function. Moreover, if the value of B_2 falls below approximately 0.3 Ns/m, strong oscillations appear in the simulations of the isotonic experiment, which are not present in the data. Thus, the viscous element B_2 is an essential component in the muscle model. Without it, the shape of the force–velocity curve is more linear and the time course of the isotonic experiment does not match the characteristics of the data.

Varying the parameter values of the lever muscle model changes the eigenvalues of the system. For instance, with $M = 0.5$ kg, the system's nominal eigenvalues (as defined with the parameter values previously specified) are a real pole at -30.71 and a pair of complex conjugate poles at $-283.9 \pm j221.2$. If the value of B_2 is increased, three real eigenvalues describe the system. If the value of B_2 is decreased, a real pole and a pair of complex conjugate poles continue to describe the system. Changing the value of B_1 does not change the eigenvalue composition, but does significantly change the value of the dominant eigenvalue from -292 with $B_1 = .1$ to -10 with $B_1 = 6$.

12.8   A LINEAR HOMEOMORPHIC SACCADIC EYE MOVEMENT MODEL

Based on physiological evidence, Bahill and coworkers in 1980 presented a linear fourth-order model of the horizontal oculomotor plant that provides an excellent match between model predictions and horizontal eye movement data. This model eliminates the differences seen between velocity and acceleration predictions of the Westheimer and Robinson models and the data. For ease in this presentation, the 1984 modification of this model by Enderle and coworkers is used.

Figure 12.32 illustrates the mechanical components of the oculomotor plant for horizontal eye movements, the lateral and medial rectus muscle and the eyeball. The agonist muscle is modeled as a parallel combination of an active-state tension generator F_{ag}, viscosity element B_{ag}, and elastic element K_{lt} connected to a series elastic

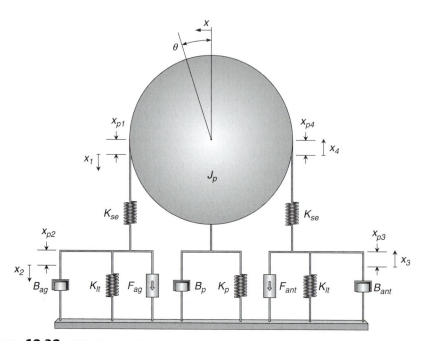

Figure 12.32　This diagram illustrates the mechanical components of the oculomotor plant. The muscles are shown extended from equilibrium, a position of rest, at the primary position (looking straight ahead), consistent with physiological evidence. The average length of the rectus muscle at the primary position is approximately 40 mm, and at the equilibrium position is approximately 37 mm. θ is the angle the eyeball is deviated from the primary position, and variable x is the length of arc traversed. When the eye is at the primary position, both θ and x are equal to zero. Variables x_1 through x_4 are the displacements from equilibrium for the stiffness elements in each muscle. Values x_{p1} through x_{p4} are the displacements from equilibrium for each of the variables x_1 through x_4 at the primary position. The total extension of the muscle from equilibrium at the primary position is x_{p1} plus x_{p2} or x_{p3} plus x_{p4}, which equals approximately 3 mm. It is assumed that the lateral and medial rectus muscles are identical, such that x_{p1} equals x_{p4} and x_{p3} equals x_{p2}. The radius of the eyeball is r.

element K_{se}. The antagonist muscle is similarly modeled as a parallel combination of an active-state tension generator F_{ant}, viscosity element B_{ant}, and elastic element K_{lt} connected to a series elastic element K_{se}. The eyeball is modeled as a sphere with moment of inertia J_p, connected to viscosity element B_p, and elastic element K_p. The passive elasticity of each muscle is included in spring K_p for ease in analysis. Each of the elements defined in the oculomotor plant is ideal and linear.

Physiological support for the muscle model is based on the model presented in Section 12.6 using linear approximations to the force–velocity curve and elasticity curves. Eyeball moment of inertia, elasticity, and viscosity in Figure 12.32 are supported from the studies by Robinson and coworkers during strabismus surgery. Passive elasticity of the eyeball is usually from a combination of the effects due to the four other oculomotor muscles and the optic nerve. Viscous effects are attributed to the friction of the eyeball within the eye socket. Moment of inertia is due to the eyeball.

By summing the forces at junctions 2 and 3 (the equilibrium positions for x_2 and x_3) and the torques acting on the eyeball, using Laplace variable analysis about the operating point, the linear homeomorphic model, as shown in Figure 12.32, is derived as

$$\delta \left(K_{se} \left(K_{st} \left(F_{ag} - F_{ant} \right) + B_{ant} \dot{F}_{ag} - B_{ag} \dot{F}_{ant} \right) \right) = \ddddot{\theta} + C_3 \dddot{\theta} + C_2 \ddot{\theta} + C_1 \dot{\theta} + C_0 \theta \quad (12.54)$$

where

$$K_{st} = K_{se} + K_{lt}, \quad J = \frac{57.296 J_p}{r^2}, \quad B = \frac{57.296 B_p}{r^2}, \quad K = \frac{57.296 K_p}{r^2}, \quad \delta = \frac{57.296}{r J B_{ANT} B_{AG}},$$

$$C_3 = \frac{J K_{st} (B_{ag} + B_{ant}) + B B_{ant} B_{ag}}{J B_{ant} B_{ag}}$$

$$C_2 = \frac{J K_{st}^2 + B K_{st} (B_{ag} + B_{ant}) + B_{ant} B_{ag} (K + 2K_{se})}{J B_{ant} B_{ag}}$$

$$C_1 = \frac{B K_{st}^2 + (B_{ag} + B_{ant})(K K_{st} + 2K_{se} K_{st} - K_{se}^2)}{J B_{ant} B_{ag}}$$

$$C_0 = \frac{K K_{st}^2 + 2K_{se} K_{st} K_{lt}}{J B_{ant} B_{ag}}$$

The agonist and antagonist active-state tensions follow from Figure 12.15, which assume no latent period, and are given by the following low-pass filtered waveforms:

$$\dot{F}_{ag} = \frac{N_{ag} - F_{ag}}{\tau_{ag}} \quad \text{and} \quad \dot{F}_{ant} = \frac{N_{ant} - F_{ant}}{\tau_{ant}} \quad (12.55)$$

where N_{ag} and N_{ant} are the neural control inputs (pulse-step waveforms), and

$$\tau_{ag} = \tau_{ac}(u(t) - u(t - t_1)) + \tau_{de} u(t - t_1) \quad (12.56)$$

$$\tau_{ant} = \tau_{de}(u(t) - u(t - t_1)) + \tau_{ac} u(t - t_1) \quad (12.57)$$

are the time-varying time constants.

Based on an analysis of experimental evidence, a set of parameter estimates for the oculomotor plant are: $K_{se} = 125 \, \text{Nm}^{-1}$, $K_{lt} = 32 \, \text{Nm}^{-1}$, $K = 66.4 \, \text{Nm}^{-1}$, $B = 3.1 \, \text{Nsm}^{-1}$, $J = 2.2 \times 10^{-3} \, \text{Ns}^2 \text{m}^{-1}$, $B_{ag} = 3.4 \, \text{Nsm}^{-1}$, $B_{ant} = 1.2 \, \text{Nsm}^{-1}$, $\tau_{ac} = 0.009 \, \text{s}$, $\tau_{de} = 0.0054 \, \text{s}$, and $\delta = 5.80288 \times 10^5$, and the steady-state active-state tensions as:

$$F_{ag} = \begin{cases} 0.14 + 0.0185\theta & \text{N for } \theta < 14.23° \\ 0.0283\theta & \text{N for } \theta \geq 14.23° \end{cases} \quad (12.58)$$

$$F_{ant} = \begin{cases} 0.14 - 0.00980\theta & \text{N for } \theta < 14.23° \\ 0 & \text{N for } \theta \geq 14.23° \end{cases} \quad (12.59)$$

Since saccades are highly variable, estimates of the dynamic active-state tensions are usually carried out on a saccade by saccade basis. One method to estimate the active-state tensions is to use the system identification technique, a conjugate gradient search program carried out in the frequency domain discussed later in this chapter. Estimates for agonist pulse magnitude are highly variable from saccade to saccade, even for saccades of the same size. Agonist pulse duration is closely coupled with pulse amplitude. As the pulse amplitude increases, the pulse duration decreases for saccades of the same magnitude. Reasonable values for the pulse amplitude for this model range from about 0.6 N to 1.4 N. The larger the magnitude of the pulse amplitude, the larger the peak velocity of the saccade.

Example Problem 12.15

Show that Equation 12.54 follows from Figure 12.32.

Solution

From Figure 12.32, the following relationships exist among variables x_1, x_4, θ, and x:

$$x_1 = x - x_{p1}$$
$$x_4 = x + x_{p4}$$
$$\theta = 57.296 \, x/r$$

Assuming that $\dot{x}_2 > \dot{x}_1 \geq \dot{x}_4 > \dot{x}_3$, the forces acting at nodes 2 and 3 and torques acting on the eyeball give

$$F_{ag} = B_{ag}\dot{x}_2 + K_{se}(x_2 - x_1) + K_{lt}x_2$$
$$rK_{se}(x_2 - x_1) - rK_{se}(x_4 - x_3) = J_p\ddot{\theta} + B_p\dot{\theta} + K_p\theta$$
$$K_{se}(x_4 - x_3) = F_{ant} + K_{lt}x_3 + B_{ant}\dot{x}_3$$

The previous three equations could be simplified and written as one equation by taking the Laplace transform of each equation and eliminating all variables except x. However, initial conditions must be considered in the transforms. An alternative method is to reduce the equations to include only changes from the initial eye position. Introduce

$$\hat{x} = x - x(0)$$
$$\hat{\theta} = \theta - \theta(0)$$
$$\hat{x}_1 = x_1 - x_1(0)$$
$$\hat{x}_2 = x_2 - x_2(0)$$
$$\hat{x}_3 = x_3 - x_3(0)$$
$$\hat{x}_4 = x_4 - x_4(0)$$
$$\hat{F}_{ag} = F_{ag} - F_{ag}(0)$$
$$\hat{F}_{ant} = F_{ant} - F_{ant}(0)$$

Node equations 2 and 3 and the equation for the torques acting on the eyeball are rewritten in terms of the new variables as

$$\hat{F}_{ag} = K_{st}\hat{x}_2 + B_{ag}\dot{\hat{x}}_2 - K_{se}\hat{x}$$
$$\hat{F}_{ant} = K_{se}\hat{x} - K_{st}\hat{x}_3 - B_{ant}\dot{\hat{x}}_3$$
$$K_{se}(\hat{x}_2 + \hat{x}_3 - 2\hat{x}) = J\ddot{\hat{x}} + B\dot{\hat{x}} + K\hat{x}$$

where

$$K_{st} = K_{se} + K_{lt}$$
$$J = \frac{57.296 J_p}{r^2}$$
$$B = \frac{57.296 B_p}{r^2}$$
$$K = \frac{57.296 K_p}{r^2}$$

Note that variables $\hat{x}_1$ and $\hat{x}_4$ are eliminated from the previous equations since they are both equal to $\hat{x}$ and $\hat{\theta} = \frac{57.296\hat{x}}{r}$. The Laplace transform of node equations 2 and 3 and the equation for the torques acting on the eyeball are given by

$$L\{\hat{F}_{ag}\} = (K_{st} + sB_{ag})\hat{x}_2 - K_{se}\hat{x}$$
$$L\{\hat{F}_{ant}\} = K_{se}\hat{x} - (K_{st} + sB_{ag})\hat{x}_3$$
$$K_{se}(\hat{x}_2 + \hat{x}_3 - 2\hat{x}) = (Js^2 + Bs + K)\hat{x}$$

These three equations are written as one equation by first eliminating $\hat{x}_2$ and $\hat{x}_3$ from the node equations—that is,

$$\hat{x}_2 = \frac{L\{\hat{F}_{ag}\} + K_{se}\hat{x}}{K_{st} + sB_{ag}} \quad \text{and} \quad \hat{x}_3 = \frac{K_{se}\hat{x} - L\{\hat{F}_{ant}\}}{K_{st} + sB_{ant}}$$

and upon substituting into the torque equation yields

$$K_{se}\left(\frac{L\{\hat{F}_{ag}\} + K_{se}\hat{x}}{K_{st} + sB_{ag}} + \frac{K_{se}\hat{x} - L\{\hat{F}_{ant}\}}{K_{st} + sB_{ant}} - 2\hat{x}\right) = (s^2J + sB + K)\hat{x}$$

Multiplying the preceding equation by $(sB_{ant} + K_{st}) \times (sB_{ag} + K_{st})$ reduces it to

$$K_{se}(sB_{ant} + K_{st})L\{\hat{F}_{ag}\} - K_{se}(sB_{ag} + K_{st})L\{\hat{F}_{ant}\} = (P_4s^4 + P_3s^3 + P_2s^2 + P_1s + P_0)\hat{x}$$

where

$$P_4 = JB_{ant}B_{ag}$$
$$P_3 = JK_{st}(B_{ag} + B_{ant}) + BB_{ant}B_{ag}$$
$$P_2 = JK_{st}^2 + BK_{st}(B_{ag} + B_{ant}) + B_{ant}B_{ag}(K + 2K_{se})$$
$$P_1 = BK_{st}^2 + (B_{ag} + B_{ant})(KK_{st} + 2K_{se}K_{st} - K_{se}^2)$$
$$P_0 = KK_{st}^2 + 2K_{se}K_{st}K_{lt}$$

Transforming back into the time domain yields

$$K_{se}\left(K_{st}(\hat{F}_{ag} - \hat{F}_{ant}) + B_{ant}\dot{\hat{F}}_{ag} - B_{ag}\dot{\hat{F}}_{ant}\right) = P_4\ddddot{\hat{x}} + P_3\dddot{\hat{x}} + P_2\ddot{\hat{x}} + P_1\dot{\hat{x}} + P_0\hat{x}$$

Note that with $\dot{F}_{ag} = \dot{\hat{F}}_{ag}$, $\dot{F}_{ant} = \dot{\hat{F}}_{ant}$, and derivatives of $\hat{x}$ equal to x, the preceding equation is rewritten as

$$K_{se}(K_{st}(\hat{F}_{ag} - \hat{F}_{ant}) + B_{ant}\dot{F}_{ag} - B_{ag}\dot{F}_{ant}) = P_4\ddddot{x} + P_3\dddot{x} + P_2\ddot{x} + P_1\dot{x} + P_0\hat{x}$$

In general, it is not necessary for $F_{ag}(0) = F_{ant}(0)$. With $\hat{x} = x - x(0)$ and $\hat{F}_{ag} - \hat{F}_{ant} = F_{ag} - F_{ag}(0) - F_{ant} + F_{ant}(0)$, the steady-state difference of the two node equations is

$$F_{ag}(0) - F_{ant}(0) = K_{st}(x_2(0) + x_3(0)) - K_{se}(x_1(0) + x_4(0))$$

With $x_1(0) = x(0) - x_{p1}$ and $x_4(0) = x(0) + x_{p1}$,

$$F_{ag}(0) - F_{ant}(0) = K_{st}(x_2(0) + x_3(0)) - 2K_{se}x(0)$$

From the node 1 equation at time 0^-,

$$K_{se}(x_2(0) + x_3(0) - 2x(0)) = Kx(0)$$

or

$$x_2(0) + x_3(0) = \left(\frac{K}{K_{se}} + 2\right)x(0)$$

This gives

$$F_{ag}(0) - F_{ant}(0) = \left(K_{st}\left(\frac{K}{K_{se}} + 2\right) - 2K_{se}\right)x(0)$$

Therefore,

$$K_{se}(K_{st}(F_{ag} - F_{ant}) - K_{st}(F_{ag}(0) - F_{ant}(0)) + B_{ant}\dot{F}_{ag} - B_{ag}\dot{F}_{ant})$$
$$= P_4\ddddot{x} + P_3\dddot{x} + P_2\ddot{x} + P_1\dot{x} + P_0(x - x(0))$$

From the previous expression for $F_{ag}(0) - F_{ant}(0)$, it follows from multiplying both sides by $K_{se}K_{st}$ that

$$K_{se}K_{st}(F_{ag}(0) - F_{ant}(0)) = K_{se}K_{st}\left(K_{st}\left(\frac{K}{K_{se}} + 2\right) - 2K_{se}\right)x(0)$$

The right-hand side of the previous equation reduces to

$$K_{se}K_{st}(F_{ag}(0) - F_{ant}(0)) = (K_{st}^2K + 2K_{se}K_{st}^2 - 2K_{se}^2K_{st})x(0)$$
$$= (K_{st}^2K + 2K_{se}K_{st}K_{lt})x(0) = P_0x(0)$$

Substituting the above into the differential equation describing the system gives

$$K_{se}(K_{st}(F_{ag} - F_{ant}) + B_{ant}\dot{F}_{ag} - B_{ag}\dot{F}_{ant}) = P_4\ddddot{x} + P_3\dddot{x} + P_2\ddot{x} + P_1\dot{x} + P_0x$$

With $x = \theta r/57.296$ substituted into the previous result, Equation 12.52 follows. ∎

12.9 A TRUER LINEAR HOMEOMORPHIC SACCADIC EYE MOVEMENT MODEL

The linear model of the oculomotor plant presented in Section 12.8 is based on a nonlinear oculomotor plant model by Hsu and coworkers using a linearization of the force–velocity relationship and elasticity curves. Using the linear model of muscle described in Section 12.6, it is possible to avoid the linearization and to derive a truer linear homeomorphic saccadic eye movement model.

The linear muscle model in Section 12.7 has the static and dynamic properties of rectus eye muscle, a model without any nonlinear elements. As presented, the model has a nonlinear force–velocity relationship that matches eye muscle data using linear viscous elements, and the length–tension characteristics are also in good agreement with eye muscle data within the operating range of the muscle. Some additional advantages of the linear muscle model are that a passive elasticity is not necessary if the equilibrium point $x_e = -19.3°$, rather than $15°$, and muscle viscosity is a constant that does not depend on the innervation stimulus level.

Figure 12.33 illustrates the mechanical components of the updated oculomotor plant for horizontal eye movements: the lateral and medial rectus muscle and the eyeball. The agonist muscle is modeled as a parallel combination of viscosity B_2 and series elasticity K_{se}, connected to the parallel combination of active-state tension generator F_{ag}, viscosity element B_1, and length–tension elastic element K_{lt}. For simplicity, agonist viscosity is set equal to antagonist viscosity. The antagonist muscle is similarly modeled with a suitable change in active-state tension to F_{ant}. Each of the elements defined in the oculomotor plant is ideal and linear.

The eyeball is modeled as a sphere with moment of inertia J_p, connected to a pair of viscoelastic elements connected in series. The update of the eyeball model is based on observations by Robinson, presented in 1981, and the following discussion. In the model of the oculomotor plant described in Section 12.8, passive elasticity K_{pe} was combined with the passive elastic orbital tissues. In the new linear model muscle presented in Section 12.6, the elastic element K_{pe} was no longer included in the muscle model. Thus, the passive orbital tissue elasticity needs to be updated due to the elimination of K_{pe} and the new observations by Robinson. As reported by Robinson in 1981, "When the human eye, with horizontal recti detached, is displaced and suddenly released, it returns rapidly about 61% of the way with a time constant of about 0.02 sec, and then creeps the rest of the way with a time constant of about 1 sec." As suggested according to this observation, there are at least two viscoelastic elements. Here it is proposed that these two viscoelastic elements replace the single viscoelastic element of the previous oculomotor plant. Connected to the sphere are $B_{p1}||K_{p1}$ connected in series to $B_{p2}||K_{p2}$. As reported by Robinson, total orbital elasticity is equal to $12.8 \times 10^{-7} g/°$ (scaled for this model). Thus with the time constants previously described, the orbital viscoelastic elements are evaluated as

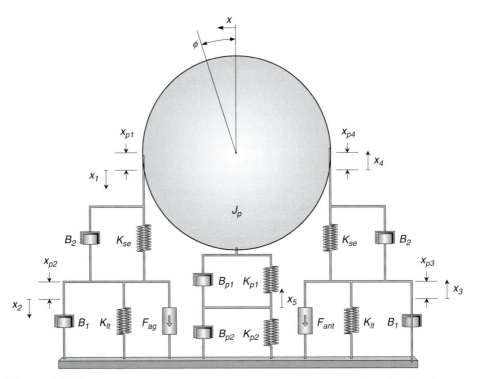

Figure 12.33   This diagram illustrates the mechanical components of the updated oculomotor plant. The muscles are shown to be extended from equilibrium, a position of rest, at the primary position (looking straight ahead), consistent with physiological evidence. The average length of the rectus muscle at the primary position is approximately 40 mm, and at the equilibrium position is approximately 37 mm. θ is the angle the eyeball is deviated from the primary position, and variable x is the length of arc traversed. When the eye is at the primary position, both θ and x are equal to zero. Variables x_1 through x_4 are the displacements from equilibrium for the stiffness elements in each muscle, and θ_5 is the rotational displacement for passive orbital tissues. Values x_{p1} through x_{p4} are the displacements from equilibrium for each of the variables x_1 through x_4 at the primary position. The total extension of the muscle from equilibrium at the primary position is x_{p1} plus x_{p2} or x_{p3} plus x_{p4}, which equals approximately 3 mm. It is assumed that the lateral and medial rectus muscles are identical, such that x_{p1} equals x_{p4} and x_{p3} equals x_{p2}. The radius of the eyeball is r.

$K_{p1} = 1.28 \times 10^{-6} g/^{\circ}$, $K_{p2} = 1.98 \times 10^{-6} g/^{\circ}$, $B_{p1} = 2.56 \times 10^{-8} gs/^{\circ}$, and $B_{p2} = 1.98 \times 10^{-6} gs/^{\circ}$. For modeling purposes, θ_5 is the variable associated with the change from equilibrium for these two pairs of viscoelastic elements. Both θ and θ_5 are removed from the analysis for simplicity using the substitution $\theta = 57.296x/r$ and $\theta_5 = 57.296x_5/r$.

By summing the forces acting at junctions 2 and 3 and the torques acting on the eyeball and junction 5, a set of four equations is written to describe the oculomotor plant.

$$F_{ag} = K_{lt}x_2 + B_1\dot{x}_2 + K_{se}(x_2 - x_1) + B_2(\dot{x}_2 - \dot{x}_1)$$

$$B_2(\dot{x}_4 - \dot{x}_3) + K_{se}(x_4 - x_3) = F_{ant} + K_{lt}x_3 + B_1\dot{x}_3$$

$$B_2(\dot{x}_2 + \dot{x}_3 - \dot{x}_1 - \dot{x}_4) + K_{se}(x_2 + x_3 - x_1 - x_4) = J\ddot{x} + B_3(\dot{x} - \dot{x}_5) + K_1(x - x_5)$$

$$K_1(x - x_5) + B_3(\dot{x} - \dot{x}_5) = B_4\dot{x}_5 + K_2x_5$$

$$(12.60)$$

where

$$J = \frac{57.296}{r^2}J_p, \ B_3 = \frac{57.296}{r^2}B_{p1}, \ B_4 = \frac{57.296}{r^2}B_{p2}, \ K_1 = \frac{57.296}{r^2}K_{p1}, \ K_2 = \frac{57.296}{r^2}K_{p2}$$

Using Laplace variable analysis about an operating point similar to the analysis used in Example Problem 12.15 yields

$$K_{se}K_{12}(F_{ag} - F_{ant}) + (K_{se}B_{34} + B_2K_{12})(\dot{F}_{ag} - \dot{F}_{ant}) + B_2B_{34}(\ddot{F}_{ag} - \ddot{F}_{ant})$$
$$= C_4\ddddot{x} + C_3\dddot{x} + C_2\ddot{x} + C_1\dot{x} + C_0x$$

$$(12.61)$$

where

$$B_{12} = B_1 + B_2, \ B_{34} = B_3 + B_4, \ K_{12} = K_1 + K_2$$

$$C_4 = JB_{12}B_{34}$$

$$C_3 = B_3B_4B_{12} + 2B_1B_2B_{34} + JB_{34}K_{st} + JB_{12}K_{12}$$

$$C_2 = 2B_1B_{34}K_{se} + JK_{st}K_{12} + B_3B_{34}K_{st} + B_3B_{12}K_{12} + K_1B_{12}B_{34}$$
$$\qquad - B_3^2K_{st} - 2K_1B_3B_{12} + 2B_2K_{lt}B_{34} + 2B_1K_{12}B_2$$

$$C_1 = 2K_{lt}B_{34}K_{se} + 2B_1K_{12}K_{se} + B_3K_{st}K_2 + K_1B_{34}K_{st} + K_1B_{12}K_{12} - K_{st}K_1B_3$$
$$\qquad - K_1^2B_{12} + 2B_2K_{lt}K_{12}$$

$$C_0 = 2K_{lt}K_{se}K_{12} + K_1K_{st}K_2$$

Converting from x to θ gives

$$\delta(K_{se}K_{12}(F_{ag} - F_{ant}) + (K_{se}B_{34} + B_2K_{12})(\dot{F}_{ag} - \dot{F}_{ant}) + B_2B_{34}(\ddot{F}_{ag} - \ddot{F}_{ant}))$$
$$= \ddddot{\theta} + P_3\dddot{\theta} + P_2\ddot{\theta} + P_1\dot{\theta} + P_0\theta$$

$$(12.62)$$

where

$$\delta = \frac{57.296}{rJB_{12}B_{34}}, \ P_3 = \frac{C_3}{C_4}, \ P_2 = \frac{C_2}{C_4}, \ P_1 = \frac{C_1}{C_4}, \ P_0 = \frac{C_0}{C_4}$$

Based on the updated model of muscle and length–tension data presented in Section 12.6, steady-state active-state tensions are determined as:

$$F = \begin{cases} 0.4 + 0.0175\theta & \text{N for } \theta \geq 0° \\ 0.4 + 0.0125\theta & \text{N for } \theta < 0° \end{cases}$$

$$(12.63)$$

The agonist and antagonist active-state tensions follow from Figure 12.15, which assume no latent period, and are given by the following low-pass filtered waveforms:

$$\dot{F}_{ag} = \frac{N_{ag} - F_{ag}}{\tau_{ag}} \quad \text{and} \quad \dot{F}_{ant} = \frac{N_{ant} - F_{ant}}{\tau_{ant}} \tag{12.64}$$

where N_{ag} and N_{ant} are the neural control inputs (pulse-step waveforms), and

$$\tau_{ag} = \tau_{ac}(u(t) - u(t - t_1)) + \tau_{de}u(t - t_1) \tag{12.65}$$

$$\tau_{ant} = \tau_{de}(u(t) - u(t - t_1)) + \tau_{ac}u(t - t_1) \tag{12.66}$$

are the time-varying time constants. Based on an analysis of experimental data, parameter estimates for the oculomotor plant are: $K_{se} = 125\,\text{Nm}^{-1}$, $K_{lt} = 60.7\,\text{Nm}^{-1}$, $B_1 = 2.0\,\text{Nsm}^{-1}$, $B_2 = 0.5\,\text{Nsm}^{-1}$, $J = 2.2 \times 10^{-3}\,\text{Ns}^2\text{m}^{-1}$, $B_3 = 0.538\,\text{Nsm}^{-1}$, $B_4 = 41.54\,\text{Nsm}^{-1}$, $K_1 = 26.9\,\text{Nm}^{-1}$, $K_2 = 41.54\,\text{Nm}^{-1}$, and $r = 0.0118\,\text{m}$.

Saccadic eye movements simulated with this model have characteristics that are in good agreement with the data, including position, velocity, and acceleration, and the main sequence diagrams. As before, the relationship between agonist pulse magnitude and pulse duration is tightly coupled.

Example Problem 12.16

Using the oculomotor plant model described in Equation 12.62, parameters given after this model, and the steady-state input from Equation 12.63, simulate a 10° saccade. Plot agonist and antagonist active-state tension, position, velocity, and acceleration versus time. Compare the simulation with the main sequence diagram in Figure 12.11.

Solution

The solution to this example involves selecting a set of parameters (F_p, t_1, τ_{ac}, and τ_{de}) that match the characteristics observed in the main sequence diagram shown in Figure 12.11. There is a great deal of flexibility in simulating a 10° saccade. The only constraints for the 10° saccade simulation results are that the duration is approximately 40 to 50 ms and peak velocity is in the 500 to 600°/s range. For realism, a latent period of 150 ms has been added to the simulation results. A SIMULINK block diagram of Equation 12.62 is shown in the following figures.

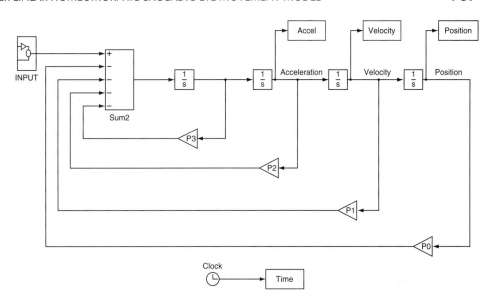

where the input block is given by

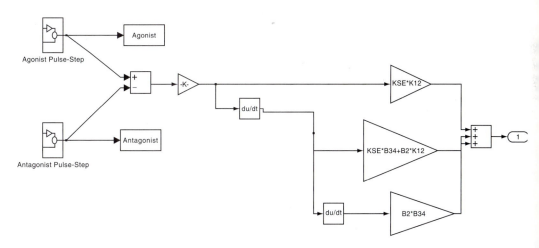

the agonist pulse-step block by

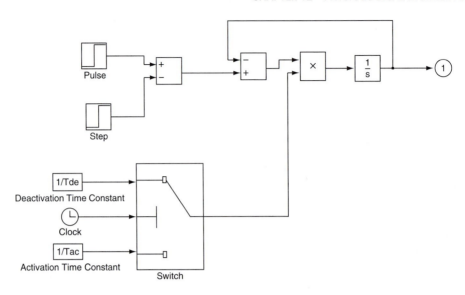

and the antagonist pulse-step block by

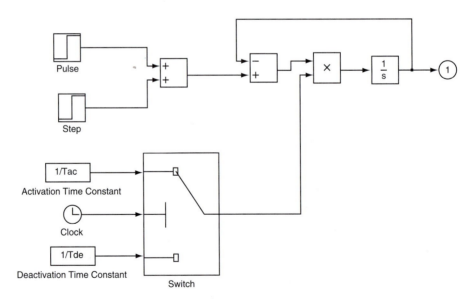

With $F_p = 1.3$ N, $t_1 = 0.10$ s, $\tau_{ac} = 0.018$ s, and $\tau_{de} = 0.018$ s, the following simulation results.

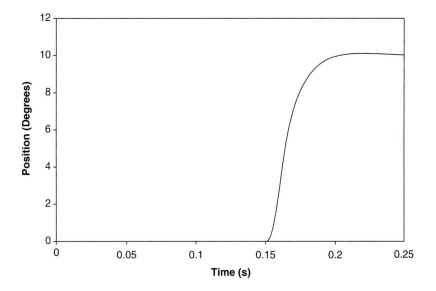

These 10° simulation results have the main sequence characteristics with a peak velocity of 568°/s and a duration of 45 ms.

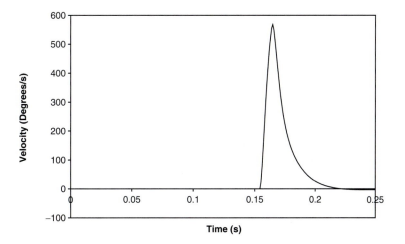

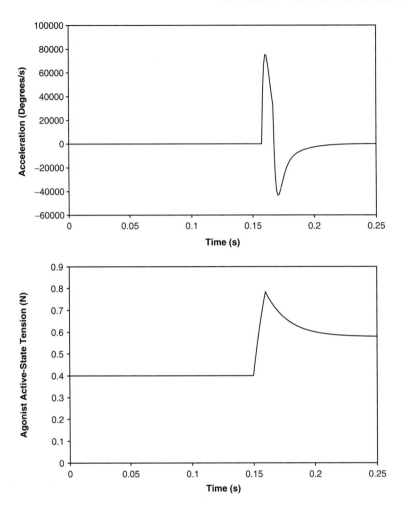

Many other parameter sets can also simulate a $10°$ saccade. For instance, consider reducing τ_{de} to $.009\,s$. Because the antagonist active-state tension activity goes toward zero more quickly than in the last case, a greater total active-state tension $(F_{ag} - F_{ant})$ results. Therefore, to arrive at $10°$ with the appropriate main sequence characteristics, F_p needs to be reduced to $1.0\,N$ if τ_{ac} remains at $0.018\,s$ and t_1 equals 0.0115 s. This $10°$ simulation is shown in the following figures.

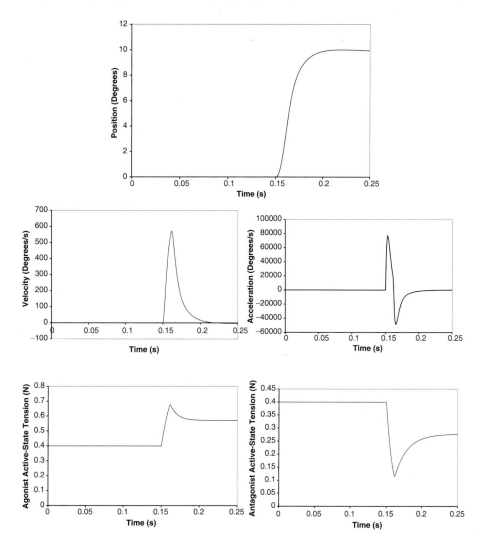

To simulate larger saccades with main sequence characteristics, the time constants for the agonist and antagonist active-state tensions can be kept at the same values as the $10°$ saccades or made functions of saccade amplitude (see Bahill, 1981 for several examples of amplitude-dependent time constants). Main sequence simulations for $15°$ and $20°$ saccades are obtained with $F_p = 1.3\,N$ and the time constants both fixed at $0.018\,s$ (the first case) by changing t_1 to 0.0155 and $0.0223\,s$, respectively. For example, the $20°$ simulation results are shown in the following figures with a peak velocity of $682°/s$ and a duration of $60\,ms$.

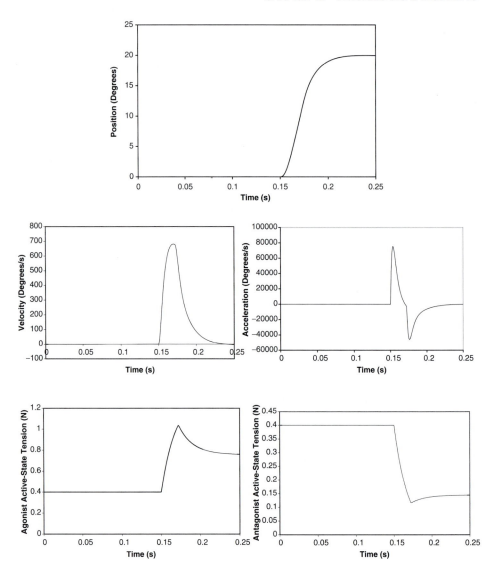

In general, as F_p increases, t_1 decreases to maintain the same saccade amplitude. Additionally, peak velocity increases as F_p increases. For the saccade amplitude to remain a constant as either or both time constants increase, F_p should also increase. ∎

12.10 SYSTEM IDENTIFICATION

In traditional applications of electrical, mechanical, and chemical engineering, the main application of modeling is as a design tool to allow the efficient study of the effects of parametric variation on system performance as a means of cost containment. In modeling physiological systems, the goal is not to design a system, but to identify the parameters and structure of the system. Ideally, the input and the output of the physiological system are known and some information about the internal dynamics of the system are available (Figure 12.34). In many cases, either the input or output is not measurable or observable but is estimated from a remote signal, and no information about the system is known. System identification is the process of creating a model of a system and estimating the parameters of the model. This section introduces the concept of system identification in both the frequency and time domain.

A variety of signals is available to the biomedical engineer as described earlier. Those produced by the body include action potentials, EEGs, EKGs, EMGs, EOGs and pressure transducer output. Additional signals are available through ultrasound, x-ray tomography, MRI, and radiation. From these signals a model is built and parameters estimated according to the modeling plan in Figure 12.1. Before work on system identification begins, understanding the characteristics of the input and output signals is important, that is, knowing the voltage range, frequency range, whether the signal is deterministic or stochastic, and if coding (i.e., neural mapping) is involved. Most biologically generated signals are low frequency and involve some coding. For example, EEGs have an upper frequency of 30 Hz and eye movements have an upper frequency under 100 Hz. The saccadic system uses neural coding that transforms burst duration into saccade amplitude. After obtaining the input and output signals, these signals must be processed. A fundamental block is the amplifier, which is characterized by its gain and frequency as described in Chapter 8. Note that the typical amplifier is designed as a low-pass filter (LPF) since noise amplification is not desired. Interestingly, most amplifiers have storage elements (i.e., capacitors and inductors), so the experimenter must wait until the transient response of the amplifier has been completed before any useful information can be extracted. An important point to remember is that the faster the cutoff of the filter, the longer the transient response of the amplifier.

In undergraduate classes, a system (the transfer function or system description) and input are usually provided and a response or output is requested. Though this seems difficult, it is actually much easier than trying to determine the parameters of a physiological system when all that is known are the input (and perhaps not the direct input as described in the saccadic eye movement system) and noisy output

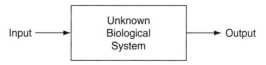

Figure 12.34 Block diagram of typical physiological system without feedback.

characteristics of the model. In the ideal case, the desired result here is the transfer function, which can be determined from

$$H(s) = \frac{V_o(s)}{V_i(s)} \qquad (12.67)$$

12.10.1 Classical System Identification

The simplest and most direct method of system identification is sinusoidal analysis. A source of sinusoidal excitation is needed which usually consists of a sine wave generator, a measurement transducer, and a recorder to gather frequency response data. Many measurement transducers are readily available for changing physical variables into voltages as described in Chapter 9. Devices that produce the sinusoidal excitation are much more difficult to obtain and are usually designed by the experimenter. A recording of the frequency response data can easily be obtained from an oscilloscope. Figure 12.35 illustrates the essential elements of sinusoidal analysis.

The experiment to identify model parameters using sinusoidal analysis is simple to carry out. The input is varied at discrete frequencies over the entire spectrum of interest and the output magnitude and phase are recorded for each input. To illustrate this technique, consider the following analysis. From Figure 12.35, it is clear that the transfer function is given by

$$H(j\omega) = \frac{V_o(j\omega)}{V_i(j\omega)} \qquad (12.68)$$

The Fourier transform of the input is

$$V_i(j\omega) = F\{A\cos(\omega_x t + \theta)\} = A \int e^{-j\omega\lambda} \cos(\omega_x\lambda + \theta)d\lambda = A \int e^{-j\omega\left(-\frac{\theta}{\omega_x}+\tau\right)} \cos(\omega_x\tau)d\tau$$

$$(12.69)$$

by substituting $\lambda = \tau - \frac{\theta}{\omega_x}$. Factoring out the terms not involving τ, gives

$$V_i(j\omega) = Ae^{\frac{j\omega\theta}{\omega_x}} \int e^{-j\omega\tau} \cos\omega_x\tau d\tau = Ae^{\frac{j\omega\theta}{\omega_x}}[\pi\delta(\omega - \omega_x) + \pi\delta(\omega + \omega_x)] \qquad (12.70)$$

Similarly,

$$V_o(j\omega) = Be^{\frac{j\omega\phi}{\omega_x}}[\pi\delta(\omega - \omega_x) + \pi\delta(\omega + \omega_x)] \qquad (12.71)$$

According to Equation 12.68

$$H(j\omega) = \frac{V_o(j\omega)}{V_i(j\omega)} = \frac{Be^{\frac{j\omega\phi}{\omega_x}}}{Ae^{\frac{j\omega\theta}{\omega_x}}} = \frac{B}{A}e^{\frac{j\omega(\phi-\theta)}{\omega_x}} \qquad (12.72)$$

Figure 12.35 Impulse response block diagram.

At steady state with $\omega = \omega_x$, Equation 12.72 reduces to

$$H(j\omega) = \frac{B}{A} e^{j(\phi - \theta)} \tag{12.73}$$

Each of these quantities in Equation 12.73 is known (i.e., B, A, ϕ, and θ), so the magnitude and phase angle of the transfer function is also known. Thus, ω_x can be varied over the frequency range of interest to determine the transfer function.

In general, a transfer function, $G(s)|_{s=j\omega} = G(j\omega)$, is composed of the following terms:

1. Constant term K
2. M poles or zeros at the origin of the form $(j\omega)^M$
3. P poles of the form $\prod_{p=1}^{P}(1 + j\omega\tau_p)$ or Z zeros of the form $\prod_{z=1}^{Z}(1 + j\omega\tau_z)$. Naturally, the poles or zeros are located at $-\frac{1}{\tau}$.
4. R complex poles of the form $\prod_{r=1}^{R}\left(1 + \left(\frac{2\zeta_r}{\omega_{n_r}}\right)j\omega + \left(\frac{j\omega}{\omega_{n_r}}\right)^2\right)$ or S zeros of the form

$$\prod_{s=1}^{S}\left(1 + \left(\frac{2\zeta_s}{\omega_{n_s}}\right)j\omega + \left(\frac{j\omega}{\omega_{n_s}}\right)^2\right).$$

5. Pure time delay $e^{-j\omega T_d}$

where M, P, Z, R, S, and T_d are all positive integers. Incorporating these terms, the transfer function is written as

$$G(j\omega) = \frac{K \times e^{-j\omega T_d} \times \left(\prod_{z=1}^{Z}(1 + j\omega\tau_z)\right) \times \left(\prod_{s=1}^{S}\left(1 + \left(\frac{2\zeta_s}{\omega_{n_s}}\right)j\omega + \left(\frac{j\omega}{\omega_{n_s}}\right)^2\right)\right)}{(j\omega)^M \times \left(\prod_{p=1}^{P}(1 + j\omega\tau_p)\right) \times \left(\prod_{r=1}^{R}\left(1 + \left(\frac{2\zeta_r}{\omega_{n_r}}\right)j\omega + \left(\frac{j\omega}{\omega_{n_r}}\right)^2\right)\right)} \tag{12.74}$$

This equation is used as a template when describing the data with a model. To determine the value of the unknown parameters in the model, the logarithm and asymptotic approximations to the transfer function are used. In general, the logarithmic gain, in dB, of the transfer function template is

$$20\log|G(j\omega)| = 20\log K + 20\sum_{z=1}^{Z}\log|1 + j\omega\tau_z| + 20\sum_{s=1}^{S}\log\left|1 + \left(\frac{2\zeta_s}{\omega_{n_s}}\right)j\omega + \left(\frac{j\omega}{\omega_{n_s}}\right)^2\right|$$

$$-20\log|(j\omega)^M| - 20\sum_{p=1}^{P}\log|1 + j\omega\tau_p| - 20\sum_{r=1}^{R}\log\left|1 + \left(\frac{2\zeta_r}{\omega_{n_r}}\right)j\omega + \left(\frac{j\omega}{\omega_{n_r}}\right)^2\right| \tag{12.75}$$

and the phase, in degrees, is

$$\phi(\omega) = -\omega T_d + \sum_{z=1}^{Z} \tan^{-1}(\omega\tau_z) + \sum_{s=1}^{S} \tan^{-1}\left(\frac{2\zeta_s\omega_{n_s}\omega}{\omega_{n_s}^2 - \omega^2}\right) - M \times (90°)$$

$$- \sum_{p=1}^{P} \tan^{-1}(\omega\tau_p) - \sum_{r=1}^{R} \tan^{-1}\left(\frac{2\zeta_r\omega_{n_r}\omega}{\omega_{n_r}^2 - \omega^2}\right)$$

(12.76)

where the phase angle of the constant is $0°$ and the magnitude of the time delay is 1. Evident from these expressions is that each term can be considered separately and added together to obtain the complete Bode diagram. The asymptotic approximations to the logarithmic gain for the poles and zeros are given by

- **Poles at the origin**
  Gain: $-20 \log |(j\omega)| = -20 \log \omega$. The logarithmic gain at $\omega = 1$ is 0 (i.e., the line passes through 0 dB at $\omega = 1$ radian/s).
  Phase: $\phi = -90°$. If there is more than one pole, the slope of the gain changes by $M \times (-20)$ and the phase by $M \times (-90°)$.

- **Pole on the real axis**

$$Gain: -20 \log |1 + j\omega\tau_p| = \begin{cases} 0 & \text{for } \omega < \frac{1}{\tau_p} \\ -20 \log(\omega\tau_p) & \text{for } \omega \geq \frac{1}{\tau_p} \end{cases}$$

  Phase: An asymptotic approximation to $-\tan^{-1}(\omega\tau_p)$ is drawn with a straight line from $0°$ at one decade below $\omega = \frac{1}{\tau_p}$ to $-90°$ at 1 decade above $\omega = \frac{1}{\tau_p}$.
  The pole is located at $-\frac{1}{\tau_p}$.

- **Zero on the real axis**

$$Gain: 20 \log |1 + j\omega\tau_z| = \begin{cases} 0 & \text{for } \omega < \frac{1}{\tau_z} \\ 20 \log(\omega\tau_z) & \text{for } \omega \geq \frac{1}{\tau_z} \end{cases}$$

- *Phase:* An asymptotic approximation to $\tan^{-1}(\omega\tau_z)$ is drawn with a straight line from $0°$ at one decade below $\omega = \frac{1}{\tau_z}$ to $+90°$ at 1 decade above $\omega = \frac{1}{\tau_z}$.

The zero is located at $-\frac{1}{\tau_z}$.

- **Complex poles**

$$Gain: -20 \log \left|1 + \left(\frac{2\zeta_r}{\omega_{n_r}}\right)j\omega + \left(\frac{j\omega}{\omega_{n_r}}\right)^2\right| = \begin{cases} 0 & \text{for } \omega < \omega_{n_r} \\ -40 \log\left(\frac{\omega}{\omega_{n_r}}\right) & \text{for } \omega \geq \omega_{n_r} \end{cases}$$

A graph of the actual magnitude frequency is shown in Figure 12.36 with $\omega_n = 1.0$ and ζ ranging from 0.05 to 1.0. Notice that as ζ decreases from 1.0, the magnitude peaks at correspondingly larger values. As ζ approaches zero, the magnitude approaches infinity at $\omega = \omega_{n_r}$. For values of $\zeta > 0.707$ there is no resonance.

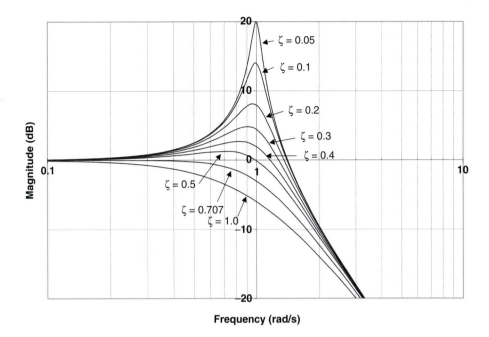

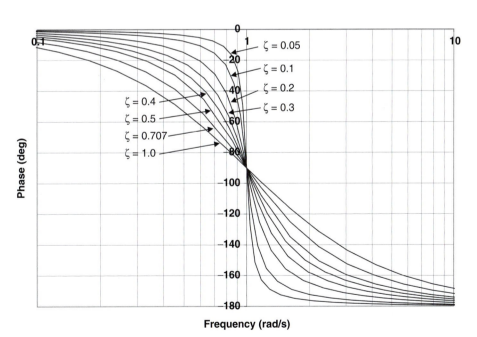

Figure 12.36   Bode plot of complex poles with $\omega_n = 1.0$ rad/s.

Phase: Depending on the value of ζ_r, the shape of the curve is quite variable but in general is 0° at one decade below $\omega = \omega_{n_r}$ and $-180°$ at 1 decade above $\omega = \omega_{n_r}$.

A graph of the actual phase frequency is shown in Figure 12.36 with $\omega_n = 1.0$ and ζ ranging from 0.05 to 1.0. Notice that as ζ decreases from 1.0, the phase changes more quickly from 0° to 180° over a smaller frequency interval.

The poles are located at $-\zeta_r \omega_{n_r} \pm j\omega_{n_r}\sqrt{1 - \zeta_r^2}$.

- **Complex zeros**

$$\text{Gain: } 20\log\left|1 + \left(\frac{2\zeta_s}{\omega_{n_s}}\right)j\omega + \left(\frac{j\omega}{\omega_{n_s}}\right)^2\right| = \begin{cases} 0 & \text{for } \omega < \omega_{n_s} \\ 40\log\left(\frac{\omega}{\omega_{n_s}}\right) & \text{for } \omega \geq \omega_{n_s} \end{cases}$$

Phase: Depending on the value of ζ_s, the shape of the curve is quite variable but in general is 0° at one decade below $\omega = \omega_{n_s}$ and $+180°$ at 1 decade above $\omega = \omega_{n_s}$.

Both the magnitude and phase follow the two previous graphs and discussion with regard to the complex poles with the exception that the slope is $+40$ dB/decade rather than -40 dB/decade.

The zeros are located at $-\zeta_s \omega_{n_s} \pm j\omega_{n_s}\sqrt{1 - \zeta_s^2}$.

- **Time delay**
  Gain: 1 for all ω
  Phase: $-\omega T_d$
- **Constant K**
  Gain: $20\log K$
  Phase: 0

The frequency at which the slope changes in a Bode magnitude-frequency plot is called a break or corner frequency. The first step in estimating the parameters of a model involves identifying the break frequencies in the magnitude-frequency and/or phase-frequency responses. This simply involves identifying points at which the magnitude changes slope in the Bode plot. Poles or zeros at the origin have a constant slope of -20 or $+20$ dB/decade, respectively, from $-\infty$ to ∞. Real poles or zeros have a change in slope at the break frequency of -20 or $+20$ dB/decade, respectively. The value of the pole or zero is the break frequency. Estimating complex poles or zeros is much more difficult. The first step is to locate the break frequency ω_n, the point at which the slope changes by 40 dB/decade. To estimate ζ, use the actual magnitude frequency (size of the peak) and phase frequency (quickness of changing 180°) curves in Figure 12.35 to closely match the data.

The error between the actual logarithmic gain and straight-line asymptotes at the break frequency is 3 dB for a pole on the real axis (the exact curve equals the asymptote -3 dB). The error drops to 0.3 dB one decade below and above the break frequency. The error between the real zero and the asymptote is similar except the exact curve equals the asymptote $+3$ dB. At the break frequency for the complex poles or zeros, the error between the actual logarithmic gain and straight-line asymptotes depends on ζ and can be quite large as observed from Figure 12.36.

Example Problem 12.17

Sinusoids of varying frequencies were applied to an open loop system and the following results were measured. Construct a Bode diagram and estimate the transfer function.

Frequency (radians/s)	0.01	0.02	0.05	0.11	0.24	0.53	1.17	2.6	5.7	12.7	28.1	62	137	304	453	672	1000
$20 \log \lvert G \rvert$ (dB)	58	51	44	37	30	23	15	6	−7	−20	−34	−48	−61	−75	−82	−89	−96
Phase (degrees)	−90	−91	−91	−93	−97	−105	−120	−142	−161	−171	−176	−178	−179	−180	−180	−180	−180

Solution

Bode plots of gain and phase versus frequency for the data given are shown in the following graphs. From the phase-frequency graph, it is clear that this system has two more poles than zeros because the phase angle approaches $-180°$ as $\omega \to \infty$. Also, note that there is no peaking observed in the gain-frequency graph or sharp changes in the phase-frequency graph so there does not appear to be any lightly damped ($\zeta < 0.5$) complex poles. However, this does not imply that there are no heavily damped complex poles at this time.

For frequencies in the range 0.01 to 1 radians/s, the slope of the gain-frequency graph is -20 dB per decade. Since it is logical to assume that the slope remains at -20 dB for frequencies less than 0.01 radians/s, a transfer function with a pole at the origin provides such a response. If possible, it is important to verify that the magnitude-frequency response stays at -20 dB per decade for frequencies less than 0.01 radians/s to the limits of the recording instrumentation.

For frequencies in the range 2 to 1000 radians/s, the slope of the gain-frequency graph is -40 dB per decade. Since a pole has already been identified in the previous frequency interval, it is reasonable to conclude that there is another pole in this interval. Other possibilities exist such as an additional pole and zero that are closely spaced, or a complex pole and a zero. However, in the interests of simplicity and because these possibilities are not evident in the graphs, the existence of a pole is all that is required to describe the data.

Summarizing at this time, the model contains a pole at the origin and another pole somewhere in the region above 1 radians/s. It is also safe to conclude that the model does not contain heavily damped complex poles since the slope of the gain-frequency graph is accounted for completely.

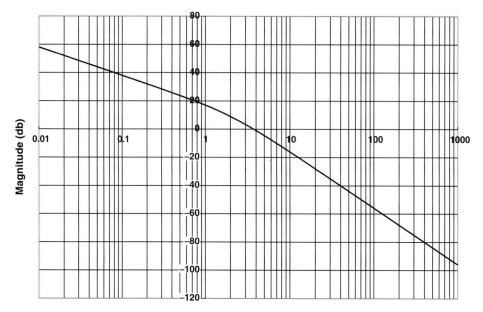

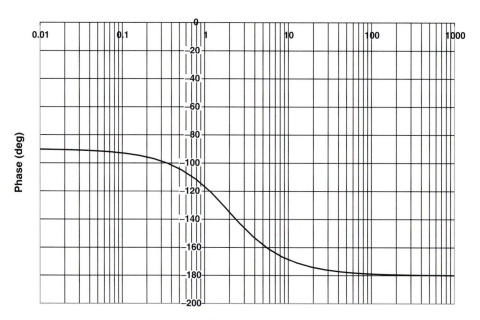

To estimate the unknown pole, straight lines are drawn tangent to the magnitude-frequency curve as shown in the following figure. The intersection of the two lines gives the break frequency for the pole at approximately 2 radians/s. Notice that the actual curve is approximately 3 dB below the asymptotes as discussed previously and observed in the following figure. The model developed thus far is

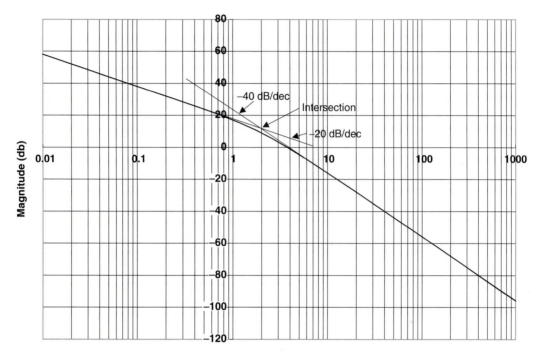

Frequency (rad/s)

$$G(j\omega) = \frac{1}{j\omega\left(\frac{j\omega}{2} + 1\right)}$$

To determine the constant K, the magnitude at $\omega = 1$ radian/s is investigated. The pole at zero contributes a value of 0 toward the logarithmic gain. The pole at 2 contributes a value of -1 dB$\left(-20\log\left|\frac{j\omega}{2} + 1\right| = -20\log\left(\sqrt{\left(\frac{1}{2}\right)^2 + 1^2}\right) = -1$ dB$\right)$ toward the logarithmic gain. The reason that the contribution of the pole at 2 was computed was that the point 1 radian/s was within the range of $\pm$ a decade of the break frequency. At $\omega = 1$ radian/s, the nonzero terms are

$$20\log|G(\omega)| = 20\log K - 20\log\left|\frac{j\omega}{2} + 1\right|$$

From the gain-frequency graph, $20\log|G(\omega)| = 17$ dB. Therefore

$$17 = 20 \log K - 1$$

or $K = 8$. The model now consists of

$$G(j\omega) = \frac{8}{j\omega\left(\frac{j\omega}{2} + 1\right)}$$

The last term to investigate is whether there is a time delay in the system. At the break frequency $\omega = 2.0$ radians/s, the phase angle from the current model should be

$$\phi(\omega) = -90 - \tan^{-1}\left(\frac{\omega}{2}\right) = -90 - 45 = -135°$$

This value is approximately equal to the data, and thus there does not appear to be a time delay in the system.   ∎

Example Problem 12.17 illustrated a process of thinking in determining the structure and parameters of a model. Carrying out an analysis in this fashion on complex systems is extremely difficult if not impossible. Software packages that automatically carry out estimation of poles, zeros, a time delay, and gain of a transfer function from data are available, such as the System Identification toolbox in MATLAB. There are also other programs that provide more flexibility than MATLAB in analyzing complex systems, such as the FORTRAN program written by Seidel (1975).

There are a variety of other inputs that one can use to stimulate the system to elicit a response. These include such transient signals as a pulse, step, and ramp, and noise signals such as white noise and pseudo random binary sequences. The reason these other techniques might be of interest is that not all systems are excited via sinusoidal input. One such system is the fast eye movement system. Here we typically use a step input to analyze the system.

In analyzing the output data obtained from step input excitation to determine the transfer function, we use a frequency response method using the Fourier transform and the fast Fourier transform. The frequency response of the input is known (s^{-1}). The frequency response of the output is calculated via a numerical algorithm called the fast Fourier transform (FFT). To calculate the FFT of the output, the data must first be digitized using an A/D converter and stored in disk memory. Care must be taken to anti-alias (low-pass) filter any frequency content above the highest frequency of the signal, and sample at a rate of at least 2.5 times the highest frequency. The transfer function is then calculated according to Equation 12.47.

12.10.2   Identification of a Linear First-Order System

Another type of identification technique specifically for a first-order system is presented here using a time-domain approach. Assume that the system of interest is a first-order system that is excited with a step input. The response to the input is

$$v_0(t) = v_{ss} + Ke^{-\frac{t}{\tau}}u(t) \tag{12.77}$$

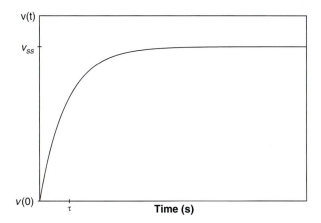

Figure 12.37 First-order system response to a step input.

where $K = -(v_{ss} - v(0))$, and is shown in Figure 12.37. Note that at $t = \tau$, the response is 63% of the way from the initial to the steady-state value. Similarly, at $t = 4\tau$, the response is 98% of the way from the initial to the steady-state value.

Suppose step input data are collected from an unknown first-order system. To describe the system, the parameters of Equation 12.76 need to be estimated. One way to estimate the system time constant is from the initial slope of the response, and a smoothed steady-state value (via averaging). That is, the time constant τ is found from

$$\dot{v}(t) = \frac{1}{\tau}(v_{ss} - v(0))e^{-\frac{t}{\tau}} \quad \rightarrow \tau = \frac{v_{ss} - v(0)}{\dot{v}(t)}e^{-\frac{t}{\tau}} \tag{12.78}$$

At $t = 0$, $\tau = \frac{v_{ss} - v(0)}{\dot{v}(0)}$, where $\dot{v}(0)$ is the initial slope of the response. The equation for estimating τ is nothing more than the equation of a straight line. This technique is illustrated in Figure 12.38.

Example Problem 12.18

The following data were collected for the step response for an unknown first-order system. Find the parameters that describe the model.

t	0	0.05	0.1	0.15	0.2	0.25	0.3	0.4	0.5	0.7	1.0	1.5	2.0
$v(t)$	0.0	0.56	0.98	1.30	1.54	1.73	1.86	2.04	2.15	2.24	2.27	2.28	2.28

Solution

The model under consideration is described by Equation 12.76, $v_o(t) = v_{ss} + Ke^{-\frac{t}{\tau}}u(t)$ with unknown parameters, v_{ss}, K, and τ. Clearly $v_{ss} = 2.28$ and $K = -2.28$. The data

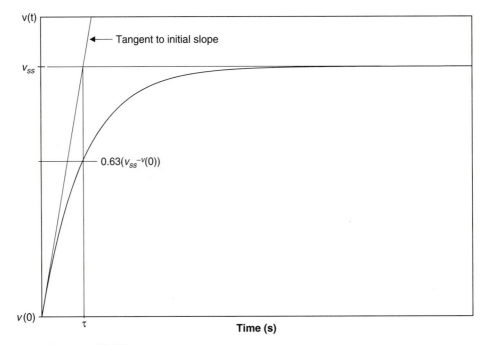

Figure 12.38   Estimating the time constant from the initial slope of the response.

are plotted in the following figure along with the tangent to $\dot{v}(0)$. From the graph, $\tau = 0.17\,s$.

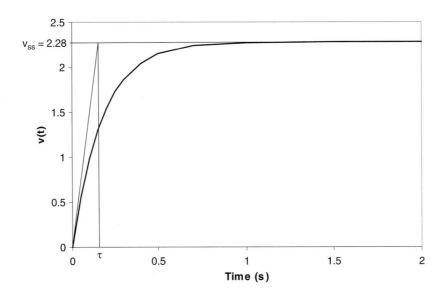

The model is given by

$$v(t) = 2.28 \left(1 - e^{-\frac{t}{0.17}} \right) u(t)$$

or by

$$V(s) = \frac{v_{ss}}{s\tau + 1} = \frac{2.28}{0.17s + 1} = \frac{13.4}{s + 5.9}$$ ∎

12.10.3 Identification of a Linear Second-Order System

Consider estimating the parameters of a second-order system using a time-domain approach. For ease in analysis, suppose the following system is given.

The differential equation describing the system is given by $f(t) = M\ddot{x} + B\dot{x} + Kx$. It is often convenient to rewrite the original differential equation in the standard form for ease in analysis.

$$\frac{f(t)}{M} = \ddot{x} + 2\zeta\omega_n\dot{x} + \omega_n^2 x \tag{12.79}$$

where ω_n is the undamped natural frequency and ζ is the damping ratio, which for the system in Figure 12.39 is given by

$$\zeta = \frac{B}{2\sqrt{KM}} \quad \text{and} \quad \omega_n = \sqrt{\frac{K}{M}}$$

The roots of the characteristic equation are

$$s_{1,2} = -\zeta\omega_n \pm \omega_n\sqrt{\zeta^2 - 1} = -\zeta\omega_n \pm j\omega_n\sqrt{1 - \zeta^2} = -\zeta\omega_n \pm j\omega_d \tag{12.80}$$

By holding ω_n constant and varying ζ, the roots move about the complex plane as illustrated in Figure 12.40. A system with $0 < \zeta < 1$ is called underdamped, with $\zeta = 1$ critically damped, and with $\zeta > 1$ overdamped. The natural or homogenous solution of this system can be solved using the classical approach via the roots of characteristic equation using the following equations.

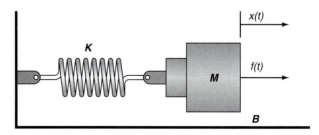

Figure 12.39 A simple mechanical system.

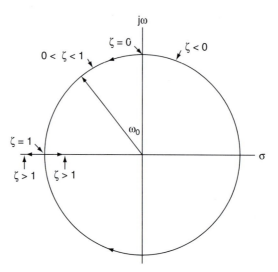

Figure 12.40   Root trace as ζ is varied while ω_n is held constant. The circle radius is ω_n.

Damping	Natural Response Equations
Overdamped	$x(t) = x_{ss} + A_1 e^{s_1 t} + A_2 e^{s_2 t}$
Underdamped	$x(t) = x_{ss} + B_1 e^{-\zeta \omega_n t} \cos(\omega_d t + \phi)$
Critically damped	$x(t) = x_{ss} + (C_1 + C_2 t) e^{-\zeta \omega_n t}$

where x_{ss} is the steady-state value of $x(t)$ and A_1, A_2, B_1, ϕ, C_1, and C_2 are the constants which describe the system evaluated from the initial conditions of the system.

To estimate ζ and ω_n, a step input of magnitude γ is applied to the system, and the data are collected. After visually inspecting a plot of the data, one of the three types of responses is selected that describes the system. The parameters are then estimated from the plot. For instance, consider the following step response for an unknown system.

It appears that a suitable model for the system might be a second-order underdamped model (i.e., $0 < \zeta < 1$), with a solution similar to the one carried out in Example Problem 12.9

$$x(t) = C\left[1 + \frac{e^{-\zeta \omega_n t}}{\sqrt{1 - \zeta^2}} \cos(\omega_d t + \phi)\right] \tag{12.81}$$

where C is the steady-state response x_{ss}.

The following terms illustrated in Figure 12.41 are typically used to describe, quantitatively, the response to a step input:

- Rise time, T_r: The time for the response to rise from 10% to 90% of steady state.
- Settling time, T_s: The time for the response to settle within $\pm 5\%$ of the steady-state value.

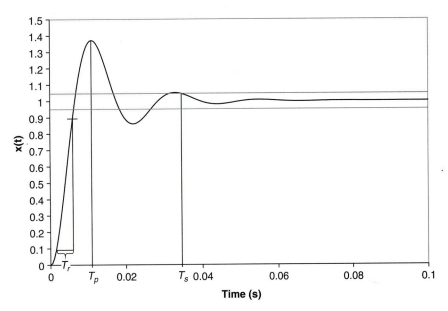

Figure 12.41 Sample second-order response with graphical estimates of rise time, time to 21 peak overshoot, and settling time.

- Peak overshoot time, T_p: The time for the response to reach the first peak overshoot.

Using values graphically determined from the data for the above quantities as shown in Figure 12.41 provides estimates for the parameters of the model. In Example Problem 12.10, the peak overshoot time was calculated for a second-order under-damped system as

$$T_p = \frac{\pi}{\omega_n \sqrt{1 - \zeta^2}} \tag{12.82}$$

and the response at T_p as

$$x(T_p) = C\left(1 + e^{-\zeta \frac{\pi}{\sqrt{1-\zeta^2}}}\right) \tag{12.83}$$

With the performance estimates calculated from the data and Equations 12.82 and 12.83, it is possible to estimate ζ and ω_n. First find ζ by using the Equation 12.83. Then using the solution for ζ, substitute this value into Equation 12.82 to find ω_n. The phase angle ϕ in Equation 12.81 is determined using Equation 12.26 and the estimate for ζ, that is

$$\phi = \pi + \tan^{-1} \frac{-\zeta}{\sqrt{1 - \zeta^2}}$$

Example Problem 12.19

Find ζ and ω_n for the data in Figure 12.41.

Solution

From the data in Figure 12.41, $C = 1.0$, $T_p = 0.011$, and $x(T_p) = 1.37$. Therefore,

$$x(T_p) = C\left(1 + e^{-\zeta\frac{\pi}{\sqrt{1-\zeta^2}}}\right) \quad \rightarrow \quad \zeta = \sqrt{\frac{\frac{(\ln(x(T_p)-1))^2}{\pi^2}}{1 + \frac{(\ln(x(T_p)-1))^2}{\pi^2}}} = 0.3$$

$$T_p = \frac{\pi}{\omega_n\sqrt{1-\zeta^2}} \quad \rightarrow \quad \omega_n = \frac{\pi}{T_p\sqrt{1-\zeta^2}} = 300 \text{ radians/s} \quad \blacksquare$$

EXERCISES

1. Suppose 500 mg of dye was introduced into the plasma compartment. After reaching steady state, the concentration in the blood is $0.0893 \frac{mg}{cm^3}$. Find the volume of the plasma compartment.
2. For the compartmental system that follows, a radioactively labeled bolus of solute was injected into compartment II. Write a single differential equation involving only variable (a) q_1; (b) q_2; (c) q_3.

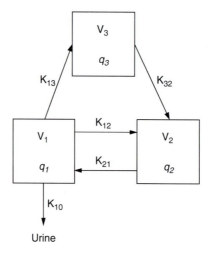

3. Two compartments, with equal volumes of 0.0572 cm^3, are separated by a thin membrane. If all of the solute is initially dumped into one compartment, then find the transfer rate if the time constant equals $27 \times 10^3 \text{ s}^{-1}$.

4. Suppose the volume of compartment I equals $0.0572 \, \text{cm}^3$ and is twice as large as compartment II. If 100 moles of solute are dumped into compartment II, then solve for the concentration in both compartments.

5. A radioactively labeled bolus of substance is injected intravenously into the plasma. The time dependence of the concentration of the substance in the plasma was found to fit

$$c_1 = A_1 e^{\lambda_1 t} + A_2 e^{\lambda_2 t}$$

where

$c_1 = $ concentration of substance in plasma (mg/100 ml)

$A_1 = 143 \, \text{mg}/100 \, \text{ml}$

$A_2 = 57 \, \text{mg}/100 \, \text{ml}$

$\lambda_1 = -1.6 \, \text{day}^{-1}$

$\lambda_2 = -2.8 \, \text{day}^{-1}$

The amount of substance injected equals 10K mg. The following model depicts the compartmental model.

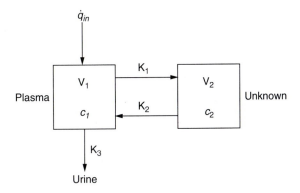

(a) Find the volume of the plasma compartment. (b) Determine the transfer rates K_1, K_2, and K_3.

6. Suppose the step response in the dye-dilution test for a $15 \, \frac{\text{mg}}{\text{s}}$ magnitude step input has a steady-state value of $\frac{15}{70} \, \frac{\text{mg}}{\text{cm}^3}$ and the time constant is $\frac{10}{7} \, \text{s}$. (a) Find the cardiac flow rate and the volume of the heart–lung compartment. (b) Suppose the input is a pulse of $15 \, \frac{\text{mg}}{\text{s}}$ applied for 1 s. Find the concentration in the heart–lung compartment.

7. Consider the following three-compartment model.

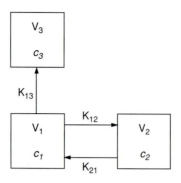

A 10 mg radioactively labeled bolus is injected into compartment I. The time dependence of the concentration of solute in compartment I is estimated as

$$c_1 = 57e^{-2.8t} + 143e^{-1.6t} \; \frac{\text{mg}}{100 \, \text{ml}}$$

where t is in days. (a) What is the volume of compartment I? (b) Determine the transfer rates K_1, K_2, and K_3.

8. Shown in the following figure is data for the pulse response in a dye-dilution test (the input is superimposed on the concentration graph for clarity). It is known that the pulse width is less than 4 time constants and that $\int_0^{+\infty} c_H dt = 0.2 \, \frac{\text{mg-s}}{\text{cm}^3}$. Find the cardiac output and the volume of the heart–lung compartment.

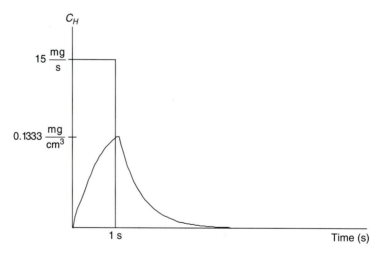

9. A radioactive bolus of I^{131} is injected into the plasma compartment. The time dependence of the concentration of I^{131} in the plasma is estimated as $c_1 = 143e^{-1.6t} \; \frac{\text{mg}}{100 \, \text{ml}}$. The amount of I^{131} is 10 K mg. Assuming the compartmental model in Exercise 12.1, find: (a) the volume of the plasma compartment, (b) $K = K_1 + K_2$.

10. An unknown quantity of radioactive iodine (I^{131}) is instantaneously passed into the plasma. The time dependence of the quantity of I^{131} in the plasma is an exponential decay from 100 mg with a time constant of 1 day, and in the urine as an exponential rise from zero to 75 mg with a time constant of 1 day. Assuming the compartment model in Exercise 12.1, determine K_1 and K_2.

11. What is the main sequence diagram? How do results from the Westheimer model in Section 12.4 compare with the main sequence diagram?

12. Simulate a 20° saccade with the Westheimer model in Section 12.4 with $\zeta = 0.7$ and $\omega_n = 120$ radians/s using SIMULINK. Assume that $K = 1$ N-m. Repeat the simulation for a 5, 10, and 15° saccade. Compare these results with the main sequence diagram in Figure 12.11.

13. Suppose the input to the Westheimer model in Section 12.4 is a pulse-step waveform as described in Section 12.5, and $\zeta = 0.7$, $\omega_n = 120$ radians/s, and $K = 1$ N-m. (a) Estimate the size of the step necessary to keep the eyeball at 20°. (b) Using SIMULINK, find the pulse magnitude that matches the main sequence diagram in Figure 12.11 necessary to drive the eyeball to 20°. (c) Repeat part (b) for saccades of 5, 10, and 15°. (d) Compare these results with those of the Westheimer model.

14. A model of the saccadic eye movement system is characterized by the following equation.

$$\tau = 1.74 \times 10^{-3}\ddot{\theta} + 0.295\dot{\theta} + 25\theta$$

where τ is the applied torque. Suppose $\tau = 250u(t)$ and the initial conditions are zero. Use Laplace transforms to solve for $\theta(t)$. Sketch $\theta(t)$.

15. Suppose a patient ingested a small quantity of radioactive iodine (I^{131}). A simple model describing the removal of I^{131} from the bloodstream into the urine and thyroid is given by a first-order differential equation. The rate of transfer of I^{131} from the bloodstream into the thyroid is given by K_1 and the urine by K_2. The time constant for the system is $\tau = \frac{1}{K_1 + K_2}$. (a) Sketch the response of the system. (b) Suppose the thyroid is not functioning and does not take up any I^{131}; sketch the response of the system and compare to the result from (a).

16. Given a population of 50,000 susceptibles, 10 infectives, and 0 immunes, and $\beta = .001$ and $\gamma = 0.8$, simulate an epidemic using the Kermack–McKendrick model.

17. Use SIMULINK to create an epidemic curve using the Kermack–McKendrick model with a birth and death rate for a population of 50,000 susceptibles, 10 infectives, and 0 immunes, and $\beta = .001$, $\gamma = 0.8$ and $\delta = 0.01$.

18. Use SIMULINK to create an epidemic curve using the Reed–Frost model with 50,000 susceptibles, 20 infectives, and 0 immunes, and q = 0.98896.

19. Consider the following system defined with $M_1 = 2$ kg, $M_2 = 1$ kg, $B_1 = 1$ N − s/m, $B_2 = 2$ N − s/m, $K_1 = 1$ N/m, and $K_2 = 1$ N/m. Let $f(t)$

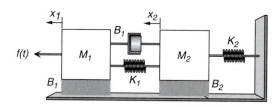

be the applied force, and x_1 and x_2 be the displacements from rest. (a) Find the transfer function. (b) Use MATLAB to draw the Bode diagram.

20. Consider the system illustrated in the following diagram defined with $M_1 = 1\,\text{kg}$, $M_2 = 2\,\text{kg}$, $B_2 = 2\,\text{N} - \text{s/m}$, $K_1 = 1\,\text{N/m}$, and $K_2 = 1\,\text{N/m}$. Let $f(t)$ be the applied force, and x_1 and x_2 be the displacements from rest. The pulley is assumed to have no inertia or friction. (a) Write the differential equations that describe this system. (b) If the input is a step with magnitude $10\,\text{N}$, simulate the solution with SIMULINK.

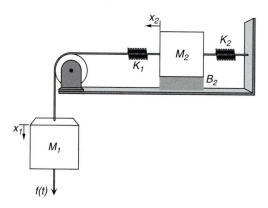

21. Consider the system in the following diagram, where there is a viscous element B_2, K_1 is a translational element, and K_2 is a rotational element. Let $\tau(t)$ be the applied torque, x_1 the displacement of M from rest, and θ the angular displacement of the element J from rest (i.e., when the springs are neither stretched or compressed). The pulley has no inertia or friction, and the cable does not stretch. (a) Write the differential equations that describe the system. (b) Write the state variable equations that describe the system.

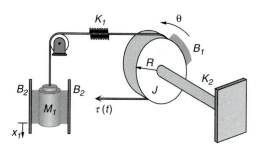

22. Consider the system in the following diagram, where there are two viscous elements B_1 and B_2, K_1 is a translational element and K_2 is a rotational element. Let $f(t)$ be the applied force, x_1 the displacement of M_1, x_2 the displacement of M_2 from rest, and θ the angular displacement of the element J from rest (i.e., when the springs are neither stretched nor compressed). The pulley has no inertia or friction, and the cable does not stretch. (a) Write the differential equations that describe the system. (b) Write the state variable equations that describe the system.

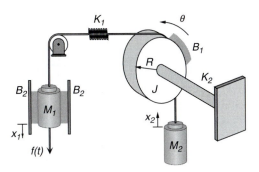

23. Consider the system in the following diagram defined with $M_3 = 0.5$ kg, $K_3 = 4$ N/m (translational), $J_1 = 0.5$ kg$-$m^2, $B_1 = 1$ N$-$s$-$m, $K_1 = 2$ N $-$ m, $J_2 = 2$ kg $-$ m^2, $K_2 = 1$ N $-$ m, $R_1 = 0.2$ m, and $R_2 = 1.0$ m. Let $\tau(t)$ be the applied torque, x_3 be the displacement of M_3 from rest, and θ_1 and θ_2 be the angular displacement of the elements J_1 and J_2 from rest (i.e., when the springs are neither stretched nor compressed). (a) Write the differential equations that describe this system. (b) If the input is a step with magnitude 10 N, simulate the solution with SIMULINK.

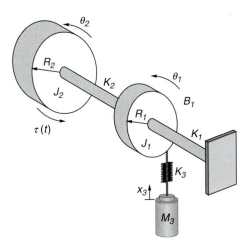

24. Given the Westheimer model described in Section 12.4 with $\zeta = \frac{1}{\sqrt{2}}$, $\omega_n = 100\,\text{radians/s}$, and $K = 1$, solve for the general response with a pulse-step input as described in Figure 12.11. Examine the change in the response as the pulse magnitude is increased and the duration of the pulse, t_1, is decreased while the steady-state size of the saccade remains constant.

25. With the Westheimer model described in Section 12.4, separately estimate ζ and ω_n for a 5, 10, 15 and 20-degree saccades using information in the main sequence diagram in Figure 12.11. Assume that peak overshoot, $x(T_p)$, is 1 degree greater than the saccade size. Simulate the four saccades. Develop a relationship between ζ and ω_n as a function of saccade size that matches the main sequence diagram. With these relationships, plot T_p and peak velocity as a function of saccade size. Compare these results to the original Westheimer main sequence results and those in Figure 12.11.

26. Consider an unexcited muscle model as shown in Figure 12.28 with $K_{lt} = 32\,\text{Nm}^{-1}$, $K_{se} = 125\,\text{Nm}^{-1}$, and $B = 3.4\,\text{Nsm}^{-1}$ ($F = 0$ for the case of an unexcited muscle). (a) Find the transfer function $H(j\omega) = \frac{X_1}{T}$. (b) Use MATLAB to draw the Bode diagram.

27. Consider an unexcited muscle model in Figure 12.29 with $K_{lt} = 60.7\,\text{Nm}^{-1}$, $K_{se} = 125\,\text{Nm}^{-1}$, $B_1 = 2\,\text{Nsm}^{-1}$, and $B_2 = 0.5\,\text{Nsm}^{-1}$ ($F = 0$ for the case of an unexcited muscle). (a) Find the transfer function $H(j\omega) = \frac{X_1}{T}$. (b) Use MATLAB to draw the Bode diagram.

28. Simulate a 20° saccade using the linear homeomorphic saccadic eye movement model from Section 12.8 using SIMULINK.

29. Simulate a 20° saccade using the linear homeomorphic saccadic eye movement model from Section 12.9 using SIMULINK.

30. Consider the linear homeomorphic saccadic eye movement model given in Equation 12.41. (a) Find the transfer function. (b) Use MATLAB to draw the Bode diagram.

31. Verify the length–tension curves in Figure 12.30.

32. Consider the following model of the passive orbital tissues driven by torque τ with $K_p = 0.5\,\text{g/}°$, $B_p = 0.06\,\text{gs/}°$, and $J_p = 4.3 \times 10^{-5}\,\text{gs}^2/°$. All elements

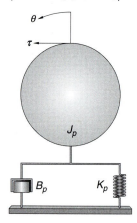

are rotational. (a) Find the transfer function $\frac{\theta(s)}{\tau(s)}$. (b) Use MATLAB to draw the Bode diagram.

33. Verify the force–velocity curve for the muscle model in Figure 12.31. Hint: Use SIMULINK to calculate peak velocity for each value of M.

34. Consider the following model of the passive orbital tissues driven by torque τ with $J_p = 4.308 \times 10^{-5} \,\text{gs}^2/°$, $K_{p1} = 0.5267 \,\text{g}/°$, $K_{p2} = 0.8133 \,\text{g}/°$, $B_{p1} = 0.010534 \,\text{gs}/°$, and $B_{p2} = 0.8133 \,\text{gs}/°$. All elements are rotational. (a) Find the transfer function $\frac{\theta_1(s)}{\tau(s)}$. (b) Use MATLAB to draw the Bode diagram.

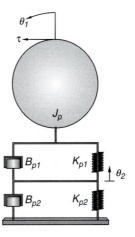

35. Consider the following model of the eye movement system. The elements are all rotational and $f_K(\theta) = K_1\theta^2$ (a nonlinear rotational spring). (a) Write the nonlinear differential equation that describes this system. (b) Write a linearized differential equation using a Taylor series first-order approximation about an operating point.

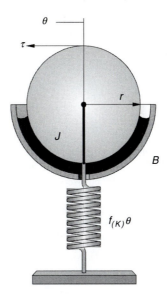

36. Suppose the passive elasticity of unexcited muscle is given by the following nonlinear translational force-displacement relationship

$$f(x) = |x|x$$

where x is the displacement from equilibrium position. Determine a linear approximation for this nonlinear element in the vicinity of the equilibrium point.

37. Sinusoids of varying frequencies were applied to an open loop system and the following results were measured. Construct a Bode diagram and estimate the transfer function.

Frequency (radians/s)	0.6	1.6	2.6	3.6	5.5	6.1	7.3	9.8	12.7	32.9	62.1	100
Magnitude Ratio, $|G|$	2.01	2.03	2.09	2.17	2.37	2.43	2.49	2.16	1.39	0.18	0.05	0.018
Phase (degrees)	−3.3	−8.9	−15.0	−21.8	−38.1	−44.8	−60.0	−93.6	−123.2	−164.2	−170.0	−175.0

38. Sinusoids of varying frequencies were applied to an open loop system and the following results were measured. Construct a Bode diagram and estimate the transfer function.

Frequency (radians/s)	0.001	0.356	1.17	2.59	8.53	12.7	18.9	41.8	62.1	137	304	1000
$20 \log |G|$ (dB)	6.02	6.02	5.96	5.74	3.65	1.85	−0.571	−6.64	−9.95	−16.8	−23.6	−34.0
Phase (degrees)	−0.086	−3.06	−10.0	−22.0	−64.9	−88.1	−116.0	−196.0	−259.0	−479.0	−959.0	−2950.

39. Sinusoids of varying frequencies were applied to an open loop system and the following results were measured. Construct a Bode diagram and estimate the transfer function.

Frequency (radians/s)	0.11	0.24	0.53	1.17	2.6	5.7	12.7	28.1	62	137	304	453	672	1000
Magnitude Ratio, $|G|$	2.0	2.0	2.0	2.0	1.93	1.74	1.24	0.67	0.32	0.15	0.07	0.044	0.03	0.02
Phase (degrees)	−0.62	−1.37	−3.03	−6.7	−14.5	−29.8	−51.8	−70.4	−80.9	−85.8	−88.1	−88.7	−89.1	−89.4

40. Sinusoids of varying frequencies were applied to an open loop system and the following results were measured. (Data is from Seidel, 1975.) Construct a Bode diagram for the data.

Frequency (radians/s)	1	3	7	10	15	20	25	30	35	40	50	60	70	80	90	100	110	120	130	140
Magnitude Ratio, $|G|$	1.	.95	.77	.7	.67	.63	.6	.53	.48	.44	.35	.31	.33	.35	.32	.32	.3	.29	.27	.26
Phase (radians)	−.035	−.227	−.419	−.541	−.611	−.768	−.995	−1.08	−1.24	−1.31	−1.52	−1.92	−1.61	−1.83	−2.08	−2.23	−2.53	−2.72	−2.9	−3.0

Estimate the transfer function if it consists of (a) two poles; (b) a pole and a complex pole pair; (c) two poles, a zero, and a complex pole pair; (d) three poles, a zero and a complex pole pair. Hint: It may be useful to solve this program using the MATLAB System Identification toolbox or Seidel's program.

41. The following data were collected for the step response for an unknown first-order system. Find the parameters that describe the model.

T	0.0	0.005	0.01	0.015	0.02	0.025	0.03	0.035	0.04	0.045	0.05	0.055	0.06	0.1
$v(t)$	0.00	3.41	5.65	7.13	8.11	8.75	9.18	9.46	9.64	9.76	9.84	9.90	9.93	10.

42. Suppose a second-order underdamped system response to a step is given by Equation 12.81 and has $C = 10$, $T_p = 0.050$, and $x(T_p) = 10.1$. Find ζ and ω_n.

43. A stylized $10°$ saccade is shown in the following figure. Estimate ζ and ω_n for the Westheimer model described in Section 12.4. Calculate the time to peak velocity and peak velocity.

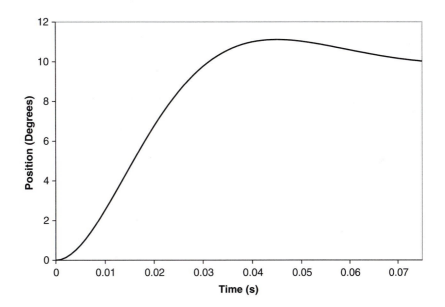

SUGGESTED READING

Ackerman, E. and Gatewood, L.C. (1979). *Mathematical Models in the Health Sciences: A Computer-aided Approach*. Univ. Minnesota Press, Minneapolis.

Bahill, A.T., Latimer, J.R. and Troost, B.T. (1980). Linear homeomorphic model for human movement. *IEEE Trans. Biomed. Eng.* **27**, 631–639.

Bahill, A.T. (1981). *Bioengineering: Biomedical, Medical and Clinical Engineering*. Prentice-Hall, Englewood Cliffs, NJ.

Carpenter, R.H.H. (1988). *Movements of the Eyes*. Pion, London.

Carson, E. and Cobelli, C. (2001). *Modeling Methodology for Physiology and Medicine*. Academic, San Diego.

Close, R.I. and Luff, A.R. (1974). Dynamic properties of inferior rectus muscle of the rat. *J. Physiol.* **236**, 259–270.

Collins, C.C. (1975). The human oculomotor control system. In *Basic Mechanisms of Ocular Motility and Their Clinical Implications* (G. Lennerstrand and P. Bach-y-Rita, Eds.). Pergamon, Oxford, 145–180.

Collins, C.C., O'Meara, D., and Scott, A.B. (1975). Muscle tension during unrestrained human eye movements. *J. Physiol.* **245**: 351–369.

Enderle, J.D., Wolfe, J.W. and Yates, J.T. (1984). The linear homeomorphic saccadic eye movement model—A modification. *IEEE Trans. Biomed. Eng.* **31**(11) 717–720.

Enderle, J.D. and Wolfe, J.W. (1987). Time-optimal control of saccadic eye movements. *IEEE Trans. Biomed. Eng.* **BME-34(1):** 43–55.

Enderle, J.D. and Wolfe, J.W. (1988). Frequency response analysis of human saccadic eye movements: estimation of stochastic muscle forces. *Computers in Biol. & Med.* **18(3),** 195–219.

Enderle, J.D., Engelken, E.J. and Stiles, R.N. (1991). A comparison of static and dynamic characteristics between rectus eye muscle and linear muscle model predictions. *IEEE Trans. Biomed. Eng.* **BME-38(12),** 1235–1245.

Enderle, J.D. (2000). The fast eye movement control system. In *The Biomedical Engineering Handbook Volume 2*, 2nd Ed. (J. Bronzino, Ed.). CRC, Boca Raton, FL, 166-1–166-2.

Enderle, J.D. (2002). Neural control of saccades. In *The Brain's Eyes: Neurobiological and Clinical Aspects to Oculomotor Research, Progress in Brain Research, V. 140* (J. Hyönä, D. Munoz, W. Heide and R. Radach, Eds.). Elsevier, Amsterdam, 21–50.

Fenn, W.O. and Marsh, B.S. (1935). Muscular force at different speeds of shortening. *J. Physiol. (London)* **35,** 277–297.

Hill, A.V. (1938). The heat of shortening and the dynamic constants of muscle. *Proc. R. Soc. London Ser. B* **126,** 136–195.

Hsu, F.K., Bahill, A.T. and Stark, L. (1976). Parametric sensitivity of a homeomorphic model for saccadic and vergence eye movements. *Computer Programs in Biomed.* **6,** 108–116.

Jacquez, J. (1996). *Compartmental Analysis in Biology and Medicine.* BioMedware, Ann Arbor, MI.

Keener, J. and Sneyd, J. (1998). *Mathematical Physiology.* Springer-Verlag, New York.

Khoo, M.C. (2000). *Physiological Control Systems: Analysis, Simulation, and Estimation.* IEEE Press, Piscataway, NJ.

Kuo, B.C. (1991). *Automatic Control Systems.* Prentice-Hall, Englewood Cliffs, NJ.

Leigh, R.J. and Zee, D.S. (1999). *The Neurology of Eye Movements*, 3rd Ed. Oxford Univ. Press, New York.

Levin, A. and Wyman, A. (1927). The viscous elastic properties of muscle. *Proc. Roy. Soc. (London)* **B101,** 218–243.

Robinson, D.A. (1964). The mechanics of human saccadic eye movement. *J. Physiol.* (London) **174,** 245–264.

Robinson, D.A. (1975). Oculomotor control signals. In *Basic Mechanisms of Ocular Motility and their Clinical Implication* (G. Lennerstrand and P. Bach-y-Rita, Eds.). Pergamon, Oxford.

Robinson, D.A. (1981). Models of mechanics of eye movements. In *Models of Oculomotor Behavior and Control* (B. L. Zuber, Ed.). CRC, Boca Raton, FL.

Scudder, C.A. (1988). A new local feedback model of the saccadic burst generator. *J. Neurophys.* **59(4),** 1454–1475.

Seidel, R.C. (1975, September). Transfer-function-parameter estimation from frequency response data-A FORTRAN program. *NASA Technical Memorandum NASA TM X-3286.*

Sparks, D.L. (1986). Translation of sensory signals into commands for control of saccadic eye movements: role of the primate superior colliculus. *Physiol. Rev.* **66,** 118–171.

Van Gisbergen, J.A.M., Robinson, D.A. and Gielen, S. (1981). A quantitative analysis of generation of saccadic eye movements by burst neurons. *J. Neurophys.* **45,** 417–442.

Westheimer, G. (1954). Mechanism of saccadic eye movements. *AMA Arch. Ophthalmol.* **52,** 710–724.

Wilkie, D.R. (1968). *Muscle: Studies in Biology*, Vol. 11. Edward Arnold, London.

13 GENOMICS AND BIOINFORMATICS

Spencer Muse, PhD

Chapter Contents

At the conclusion of this chapter, the reader will be able to:

- Discuss the basic principles of molecular biology regarding genome science.
- Describe the major types of data involved in genome projects, including technologies for collecting them.

- Describe practical applications and uses of genomic data.
- Understand the major topics in the field of bioinformatics and DNA sequence analysis.
- Use key bioinformatics databases and web resources.

13.1   INTRODUCTION

In April 2003, sequencing of all three billion nucleotides in the human genome was declared complete. This landmark of modern science brought with it high hopes for the understanding and treatment of human genetic disorders. There is plenty of evidence to suggest that the hopes will become reality—1631 human genetic diseases are now associated with known DNA sequences, compared to the less than 100 that were known at the initiation of the Human Genome Project (HGP) in 1990. The success of this project (it came in almost 3 years ahead of time and 10% under budget, while at the same time providing more data than originally planned) depended on innovations in a variety of areas: breakthroughs in basic molecular biology to allow manipulation of DNA and other compounds; improved engineering and manufacturing technology to produce equipment for reading the sequences of DNA; advances in robotics and laboratory automation; development of statistical methods to interpret data from sequencing projects; and the creation of specialized computing hardware and software systems to circumvent massive computational barriers that faced genome scientists. Clearly, the HGP served as an incubator for interdisciplinary research at both the basic and applied levels.

The human genome was not the only organism targeted during the genomic era. As of June 2004, the complete genomes were available for 1557 viruses, 165 microbes, and 26 eukaryotes ranging from the malaria parasite *Plasmodium falciparum* to yeast, rice, and humans. Continued advances in technology are necessary to accelerate the pace and to reduce the expense of data acquisition projects. Improved computational and statistical methods are needed to interpret the mountains of data. The increase in the rate of data accumulation is outpacing the rate of increases in computer processor speed, stressing the importance of both applied and basic theoretical work in the mathematical and computational sciences.

In this chapter, the key technologies that are being used to collect data in the laboratory, as well as some of the important mathematical techniques that are being used to analyze the data, are surveyed. Applications to medicine are used as examples when appropriate.

13.1.1   The Central Dogma: DNA to RNA to Protein

Understanding the applications of genomic technologies requires an understanding of three key sets of concepts: how genetic information is stored, how that information is processed, and how that information is transmitted from parent to offspring.

In most organisms, the genetic information is stored in molecules of DNA, deoxyribonucleic acid (Fig. 13.1). Some viruses maintain their genetic data in RNA, but no emphasis will be placed on such exceptions. The size of genomes, measured in counts of nucleotides or base pairs, varies tremendously, and a curious observation is that genome size is only loosely associated with organismal complexity (Table 13.1). Most of the known functional units of genomes are called genes. For purposes of this chapter, a gene can be defined as a contiguous block of nucleotides operating for a single purpose. This definition is necessarily vague, for there are a number of types of genes, and even within a given type of gene, experts have difficulty agreeing on precisely where the beginning and ending boundaries of those genes lie. A structural gene is a gene that codes instructions for creating a protein (Fig. 13.2). A second category of genes with many members is the collection of RNA genes. An RNA gene does not contain protein information; instead, its function is determined by its ability to fold into a specific three-dimensional configuration, at which point it is able to interact with other molecules and play a part in a biochemical process. A common RNA gene found in most forms of life is the tRNA gene illustrated in Figure 13.3.

Structural genes are the entities most scientists envision when the word "gene" is mentioned, and from this point on, the term *gene* will be used to mean "structural gene" unless specified otherwise. The number and variety of genes in organisms is a current topic of importance for genome scientists. Gene number in organisms ranges from tiny (470 in mycoplasma) to enormous (60,000 or more in plants).

Figure 13.1 The DNA molecule consists of two strands of nucleotides arranged in a double helix. Complementary pairs of nucleotides (A,T and C,G) form chemical bonds that hold the helix intact.

TABLE 13.1 Genome Size and Genome Content (most from Taft and Mattick, 2003)

Organism	Genome Size (x10^6)	Gene Number
HIV1	0.009	9
SARS	0.03	14
M. genitalium	0.58	470
N. equitans	0.49	552
H. helaticus	1.80	1875
E. coli	4.64	4288
P. falciparum	22.85	5268
S. cerevisiae	12.10	6000
D. melanogaster	122	13,600
C. elegans	97	19,049
A. thaliana	115	25,000
H. sapiens	3200	30,000
M. musculus	2500	37,000

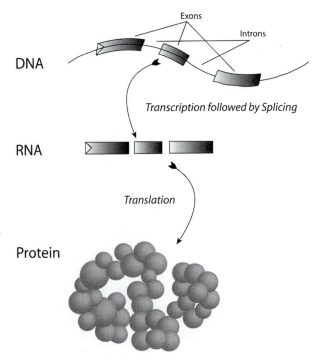

Figure 13.2 The central dogma of molecular biology. Genomic DNA is first transcribed into an RNA copy that contains introns and exons. The introns are spliced from the RNA, and the remaining exons are concatenated. The mature RNA sequence is then translated to produce a protein.

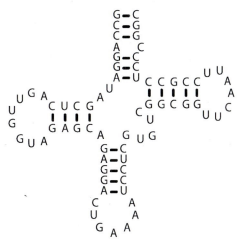

Figure 13.3   The transfer RNA (tRNA) is an example of a non-protein-coding gene. Its function is the result of the specific two- and three-dimensional structures formed by the RNA sequence itself.

Non-free-living organisms have even smaller gene numbers (the HIV virus contains only nine). The number of genes in a typical human genome has been estimated to be about 30,000, perhaps the single most surprising finding from the Human Genome Project. This number was thought to be as large as 120,000 as recently as 1998. The confusion over this number arose in part because there is a not a "one gene, one protein" rule in humans, or indeed, in many eukaryotic organisms. Instead, a single gene region can contain the information needed to produce multiple proteins. To understand this fact, the series of steps involved in creating a functional protein from the underlying DNA sequence instructions must be understood. The central dogma of molecular biology states that genetic information is stored in DNA, copied to RNA, and then interpreted from the RNA copy to form a functional protein (Fig. 13.2).

The process of copying the genetic information in DNA into an RNA copy is known as transcription (see Chapter 3). The process is thought by many to be a remnant of an early RNA world, in which the earliest life forms were based on RNA genomes. It is at the level of transcription that gene expression is regulated, determining where and when a particular gene is turned on or off.

The transcription of a gene occurs when an enzyme known as RNA polymerase binds to the beginning of a gene and proceeds to create a molecule of RNA that matches the DNA in the genome. It is this molecule of messenger RNA (mRNA) that will serve as a template for producing a protein. However, it is necessary for organisms to regulate the expression of genes to avoid having all genes being produced in all cells at all times. Transcription factors interact with either the genomic DNA or the polymerase molecule to allow delicate control of the gene expression process. A feedback loop is created whereby an environmental stimulus such as a drug leads to the production of a transcription factor, which triggers the expression of a gene. In addition to this example of a positive control mechanism, negative control is also possible. An emerging theme is that sets of genes are often coregulated by a single or small group of transcription factors. These sets of genes often share a short upstream DNA sequence that serves as a binding site for the transcription factor.

One of the earliest surprises of the genomic era was the discovery that many eukaryotic gene sequences are not contiguous, but are instead interrupted by DNA sequences known as introns. As shown in Figure 13.2, introns are physically cut, or spliced, from the mRNA sequence before the RNA is converted into a protein. The presence of introns helps to explain the phenomenon that there are more proteins produced in an organism than there are genes present. The process of alternative splicing allows for exons to be assembled in a combinatoric fashion, resulting in a multitude of potential proteins. For example, consider a gene sequence with exons E1, E2, and E3 interrupted by introns I1 and I2. If both introns are spliced, the resulting protein would be encoded as E1-E2-E3. However, it is also possible to splice the gene in a way that produces protein E1-E3, skipping exon E2. Much like transcription factors regulate gene expressions, there are factors that help to regulate alternative splicing. A common theme is to find a single gene that is spliced in different ways to produce isoforms that are expressed in specific tissues.

The process of reading the template in an mRNA molecule and using it to produce a protein is known as translation. Conceptually, this process is much more simple than the transcription and splicing processes. A structure known as a ribosome binds to the mRNA molecule. The ribosome then moves along the RNA in units of 3 nucleotides. Each of these triplets, or codons, encodes one of 20 amino acids. At each codon the ribosome interacts with tRNAs to interpret a codon and add the proper amino acid to the growing chain before moving along to the next codon in the sequence (see Chapter 3).

13.2   CORE LABORATORY TECHNOLOGIES

Genome science is heavily driven by new technological advances that allow for the rapid and inexpensive collection of various types of data. It has been said that the field is data-driven rather than hypothesis-driven, a reflection of the tendency for researchers to collect large amounts of genomic data with the (realistic) expectation that subsequent data analyses, along with the experiments they suggest, will lead to better understanding of genetic processes. Although the list of important biotechnologies changes on an almost daily basis, there are three prominent data types in today's environment: (1) genome sequences provide the starting point that allows scientists to begin understanding the genetic underpinnings of an organism; (2) measurements of gene expression levels facilitate studies of gene regulation, which, among other things, help us to understand how an organism's genome interacts with its environment; and (3) genetic polymorphisms are variations from individual to individual within species, and understanding how these variations correlate with phenotypes such as disease susceptibility is a crucial element of modern biomedical research.

13.2.1 Gene Sequencing

The basic principles for obtaining DNA sequences have remained rather stable over the past few decades, although the specific technologies have evolved dramatically. The most widely used sequencing techniques rely on attaching some sort of "reporter" to each nucleotide in a DNA sequence, then measuring how quickly or how far the nucleotide migrates through a medium. The principles of Sanger sequencing, originally developed in 1974, are illustrated in Figure 13.4.

DNA sequences have an orientation. The 5′ end of a sequence can be considered to be the left end, and the 3′ end is on the right. Sanger sequencing begins by creating all possible subsequences of the target sequence that begin at the same 5′ nucleotide. A reporter, originally radioactive but now fluorescent, is attached to the final 3′ nucleotide in each subsequence. By using a unique reporter for each of the four

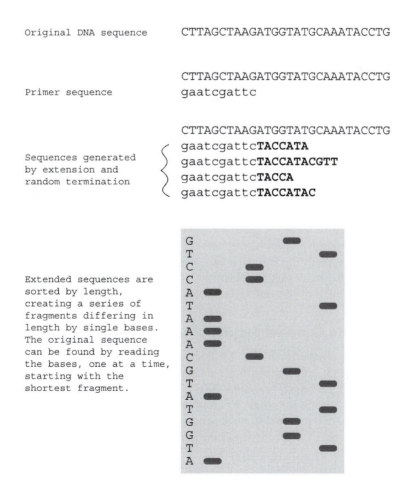

Figure 13.4   The principles of Sanger sequencing of DNA.

nucleotides, it is possible to identify the final 3′ nucleotide in each of the subsequences. Consider the task of sequencing the DNA molecule AGGT. There are four possible subsequences that begin with the 5′ A: A, AG, AGG, and AGGT. The technology of Sanger sequencing produces each of those four sequences and attaches the reporter to the final nucleotide. The subsequences are sorted from shortest to longest based on the rate at which they migrate through a medium. The shortest sequence would correspond to the subsequence A; its reporter tells us that the final nucleotide is an A. The second shortest subsequence is AG, with a final nucleotide of G. By arranging the subsequences in a "ladder" from shortest to longest, the sequence of the complete target sequence can be found simply by reading off the final nucleotide of each subsequence.

A series of new advances have allowed Sanger sequencing to be applied in a high-throughput way, paving the way for sequencing of entire genomes, including that of the human. Radioactive reporters have been replaced with safer and cheaper fluorescent dyes, and automatic laser-based systems now read the sequence of fluorescent-labeled nucleotides directly as they migrate. Early versions of Sanger sequencing only allowed for reading a few hundred nucleotides at a time; modern sequencing devices can read sequences of 800 nucleotides or more. Perhaps most important has been the replacement of "slab gel" systems with capillary sequencers. The older system required much labor and a steady hand; capillary systems, in conjunction with the development of necessary robotic devices for manipulating samples, have allowed almost completely automated sequencing pipelines to be developed. Not to be ignored in the series of technological advances is the development of automated base-calling algorithms. A laser reads the intensities of each of the four fluorescent reporter dyes as each nucleotide passes it. The resulting graph of those intensities is a chromatogram. Statistical algorithms, including the landmark program phred, are able to accept chromatograms as input and output DNA sequences with very high levels of accuracy, reducing the need for laborious human intervention. By assessing the relative levels of the four curves, the base-calling algorithms not only report the most likely nucleotide at each position, but they also provide an error probability for each site. A single state-of-the-art DNA sequencing machine can currently produce upwards of one million nucleotides per day.

13.2.1 Whole Genome Sequencing

Large regions of DNA are not sequenced in single pieces. Instead, larger contigs of DNA are fragmented into multiple, short, overlapping sequences. The emergence of shotgun sequencing (Fig. 13.5), pioneered by Dr. Craig Venter, has revolutionized approaches for obtaining complete genome sequences. The fundamental approach to shotgun sequencing of a genome is simple: (1) create many identical copies of a genome; (2) randomly cut the genomes into millions of fragments, each short enough to be sequenced individually; (3) align the overlapping fragments by identifying matching nucleotides at the ends of fragments, and finally; (4) read the complete genome sequence by following a gap-free path through the fragments. Until Venter's work, the idea of shotgun sequencing was considered unfeasible for a variety of

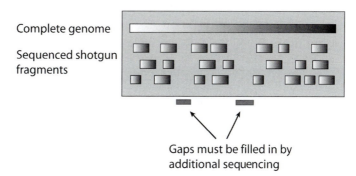

Figure 13.5 Shotgun sequencing of genomes or other large fragments of DNA proceeds by cutting the original DNA into many smaller segments, sequencing the smaller fragments, and assembling the sequenced fragments by identifying overlapping ends.

reasons. Perhaps most daunting was the computational task of aligning the millions of fragments generated in the shotgun process. Specialized hardware systems and associated algorithms were developed to handle these problems.

13.2.3 Gene Expression

Following in the footsteps of high-throughput genome sequencing came technology that allowed scientists to survey the relative abundance of thousands of individual gene products. These technologies are, in essence, a modern high-throughput replacement of the Northern blot procedure. For each member in a collection of several thousand genes, the assays provide a quantitative estimate of the number of mRNA copies of each gene being produced in a particular tissue. Two technologies, cDNA and oligonucleotide microarrays, currently dominate the field, and they have opened the door to many exploratory analyses that were previously impossible.

As a first example, consider taking two samples of cells from an individual cancer patient: one sample from a tumor and one from normal tissue. A microarray experiment makes it possible to identify the set of genes that are produced at different levels in the two tissue types. It is likely that this set of differentially expressed genes contains many genes involved in biological processes related to tumor formation and proliferation (Fig. 13.6).

A second common type of study is a time course experiment. Microarray data is collected from the same tissues at periodic intervals over some length of time. For instance, gene expression levels may be measured in 6-hour increments following the administration of a drug. For each gene, the change in gene expression level can be plotted against time (Fig. 13.7). Groups of coregulated genes will be identified as having points in time where they all experience either an increase or decrease in expression levels. A likely cause for this behavior is that all genes in the coregulated set are governed by a single transcription factor.

A final important medical application of microarray technologies involves diagnosis. Suppose that a physician obtains microarray data from tumor cells of a patient.

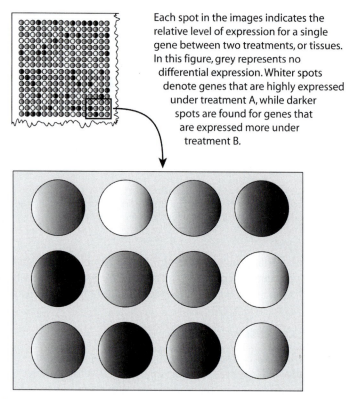

Each spot in the images indicates the relative level of expression for a single gene between two treatments, or tissues. In this figure, grey represents no differential expression. Whiter spots denote genes that are highly expressed under treatment A, while darker spots are found for genes that are expressed more under treatment B.

Figure 13.6   Microarray technology allows genome scientists to measure the relative expression levels of thousands of genes simultaneously. A typical microarray slide can hold 5000 or more spots.

The data, consisting of the relative levels of gene expression for a suite of many genes, can be compared to similar data collected from tumors of known types. If the patient's gene expression profile matches the profile of one of the reference samples, the patient can be diagnosed with that tumor type. The advent of microarray techniques has rapidly improved the accuracy of this type of diagnosis in a variety of cancers.

cDNA microarrays were the first, and are still the most widely used, form of high-throughput gene expression methods. The procedure begins by attaching the DNA sequences of thousands of genes onto a microscope slide in a pattern of spots, with each spot containing only DNA sequences of a single gene. A variety of technologies have emerged for creating such slides, ranging from simple pin spotting devices to technologies using laser jet printing techniques. The RNA of expressed genes is next collected from the target cell population. Through the process of reverse transcription, a cDNA version of each RNA is created. A cDNA molecule is complementary to the genomic DNA sequence in the sense that complementary base pairs will physically bind to one another. For example, a cDNA reading GTTAC could physically bind to the genomic DNA sequence CAATG. During the process of creating the cDNA collection, each cDNA is labeled with a fluorescent dye. The collection of labeled

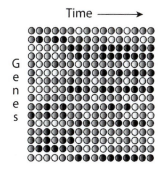

In the original image at left, each column contains the expression levels for all genes on an entire microarray at a specific time point. All of the spots in each row are expression values for a single gene. By moving from left to right along a single row, the change in that gene's expression over time can be traced. However, it is difficult to identify any similarities among genes.

By applying a clustering algorithm to the data from different genes, it is possible to identify genes with similar patterns of expression across the time course. The image on the right contains exactly the same data as the previous one; however, in this image the ordering of genes (i.e., rows) has been altered to match the results of a clustering algorithm. With genes from the same clusters sitting above one another, it is easy to see that the genes fall into three distinct groups with regard to their expression patterns over time.

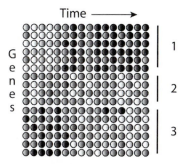

Figure 13.7 Time course expression studies examine patterns of gene expression over time. By finding genes that have similar expression profiles, sets of coregulated genes can be identified.

cDNAs is poured over the microscope slide and its set of attached DNA molecules. The cDNAs that match a DNA on the slide physically bind to their mates, and unbound cDNAs are washed from the slide. Finally, the number of bound molecules at each spot (genes) can be read by measuring the fluorescence level at each spot. Highly expressed genes will create more RNA, which results in more labeled cDNAs binding to those spots.

A more common variant on the basic cDNA approach is illustrated in Figure 13.8. In this experiment, RNA from two different tissues or individuals is collected, labeled with two different dyes, and competitively hybridized on a single slide. The relative abundance of the two dyes allows the scientist to state, for instance, that a particular gene is expressed fivefold times more in one tissue than in the other.

Oligonucleotide arrays take a slightly different approach to assaying the relative abundance of RNA sequences. Instead of attaching full-length DNAs to a slide, oligonucleotide systems make use of short oligonucleotides chosen to be specific to individual genes. For each gene included in the array, approximately 10 to 20 different oligonucleotides of length 20–25 nucleotides are designed and printed onto a chip. The use of multiple oligonucleotides for each gene helps to reduce the effects of a variety of potential errors. Fluorescently labeled RNA (rather than cDNA) is collected from the target tissue and hybridized against the oligonucleotide array. One limitation

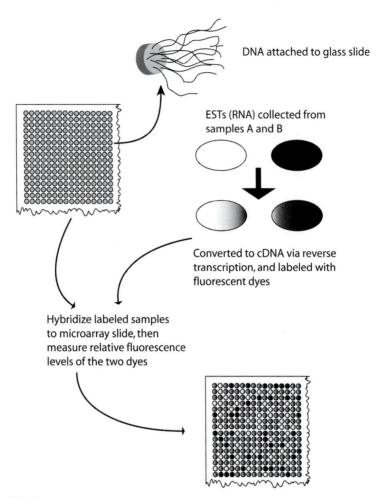

DNA attached to glass slide

ESTs (RNA) collected from samples A and B

Converted to cDNA via reverse transcription, and labeled with fluorescent dyes

Hybridize labeled samples to microarray slide, then measure relative fluorescence levels of the two dyes

Figure 13.8   A cDNA microarray slide is created by (1) attaching DNA to spots on a glass slide, (2) collecting expressed RNA sequences (expressed sequence tags, ESTs) from tissue samples, (3) converting the RNA to DNA and labeling the molecules with fluorescent dyes, (4) hybridizing the labeled DNA molecules to the DNA bound to the slide, and (5) extracting the quantity of each expressed sequence by measuring the fluorescence levels of the dyes.

of the oligonucleotide approach is that only a single sample can be assayed on a single chip—competitive hybridization is not possible.

Although oligonucleotide and cDNA approaches to assaying gene expression rely on the same basic principles, each has its own advantages and disadvantages. As already noted, competitive hybridization is currently only possible in cDNA systems. The design of oligonucleotide arrays requires that the sequences of genes for the chip are already available. The design phase is very expensive, and oligonucleotide systems are only available for commercially important and model organisms. In contrast, cDNA arrays can be developed fairly quickly even in organisms without sequenced genomes. In their favor, oligonucleotide arrays allow for more genes to be spotted in a given area (thus allowing more measurements to be made on a single chip) and tend to offer higher repeatability of measurements. Both of these facts reduce the overall level of experimental error rate in oligonucleotide arrays relative to cDNA micro-arrays, although at a higher per observation cost. Because of the trade-off between obtaining many cheap noisy measurements versus a smaller number of more precise but expensive measurements, it is not clear that either technology has an obvious cost advantage.

Both techniques share the same major disadvantage: only measurements of RNA levels are found. These measurements are used as surrogates for the much more desirable and useful quantities of the amount of protein produced for each gene. It appears that RNA levels are correlated with protein levels, but the extent and strength of this relationship is not understood well. The near future promises a growing role for protein microarray systems, which are currently seeing limited use because of their very high costs.

13.2.4 Polymorphisms

The "final draft" of the human genome was announced in April 2003. It included roughly 2.9 billion nucleotides, with some 30,000 to 40,000 genes spread across 23 pairs of chromosomes. The next phase of major data acquisition on the human genome is to discover how differences, both large and small, from individual to individual, result in variation at the phenotypic level. Toward this end, a major effort has been made to find and document genetic polymorphisms. Polymorphisms have long been important to studies of genetics. Variations of the banding patterns in polytene chromosomes, for instance, have been studied for many decades. Allozyme assays, based on differences in the overall charge of amino acid sequences, were popular in the 1960s. Most modern studies of genetic polymorphisms, though, focus on identifying variation at the individual nucleotide level.

The international SNP Consortium (http://snp.cshl.org) is a collaboration of public and private organizations that discovered and characterized approximately 1.8 million single nucleotide polymorphisms (SNPs) in human populations. In medicine, the expectation is that knowledge of these individual nucleotide variants will accelerate the description of genetic diseases and the drug development process. Pharmaceutical companies are optimistic that surveys of variation will be of use for selecting the proper drug for individual patients and for predicting likely side effects on an individual-to-individual basis.

Most SNPs (pronounced "snips") are the result of a mutation from one nucleotide to another, whereas a minority are insertions and deletions of individual nucleotides. Surveys of SNPs have demonstrated that their frequencies vary from organism to organism and from region to region within organisms. In the human genome, a SNP is found about every 1000 to 1500 nucleotides. However, the frequency of SNPs is much higher in noncoding regions of the genome than in coding regions, the result of natural selection eliminating deleterious alleles from the population. Furthermore, synonymous or silent polymorphisms, which do not result in a change of the encoded amino acids, are more frequent than nonsynonymous or replacement polymorphisms. The fields of population genetics and molecular evolution provide many empirical surveys of SNP variation, along with mathematical theory, for analyzing and predicting the frequencies of SNPs under a variety of biologically important settings.

Simple sequence repeats (SSRs) consist of a moderate (10–50) number of tandemly repeated copies of the same short sequence of 2 to 13 nucleotides. SSRs are an important class of polymorphisms because of their high mutation rates, which lead to SSR loci being highly variable in most populations. This high level of variability makes SSR markers ideal for work in individual identification. SSRs are the markers typically employed for DNA fingerprinting in the forensics setting. In human populations, an SSR locus usually has 10 or more alleles and a per generation mutation rate of 0.001. The FBI uses a set of 13 tetranucleotide repeats for identification purposes, and experts claim that no two unrelated individuals have the exact same collection of alleles at all 13 of those loci.

13.3   CORE BIOINFORMATICS TECHNOLOGIES

As the technology for collecting genomic data has improved, so has the need for new methods for management and analysis of the massive amounts of accumulated data. The term *bioinformatics* has evolved to include the mathematical, statistical, and computational analysis of genomic data. Work in bioinformatics ranges from database design to systems engineering to artificial intelligence to applied mathematics and statistics, all with an underlying focus on genomic science.

A variety of bioinformatics topics may be illustrated using the core technologies described in the preceding section. It is necessary to carry out sequence alignments in order to assemble sequence fragments. All of these sequences, along with the vital information about their sources, functions, and so on, must be stored in databases, which must be readily available to users in a variety of locations. Once a sequence has been obtained, it is necessary to annotate its function. One of the most fundamental annotation tasks is that of computational gene finding, in which a genome or chromosome sequence is input to an algorithm that subsequently outputs the predicted location of genes. A gene sequence, whether predicted or experimentally determined, must have its function predicted, and many bioinformatics tools are available for this task. Once microarray data are available, it is necessary to identify subsets of coregulated genes and to identify genes that are differentially expressed between two or more treatments or tissue types. Polymorphism data from SNPs are used to search for correlations with, for example, the presence or absence of a disease in family pedigrees. These questions are all of fundamental importance and draw on many different fields. By necessity, bioinformatics is a highly multidisciplinary field.

13.3.1   Genomics Databases

Genome projects involve far-reaching collaborations among many researchers in many fields around the globe, and it is critical that the resulting data be easily available both to project members and to the general scientific community. In light of this requirement, a number of key central data repositories have emerged. In addition to providing storage and retrieval of gene sequences, several of these databases also offer advanced sequence analysis methods and powerful visualization tools.

In the United States, the primary public genomics resource is that of the National Center for Biotechnology Information (NCBI). The NCBI website (http://www.ncbi.nlm.nih.gov) provides a seemingly endless collection of data and data analysis tools. Perhaps the most important element of the NCBI collection is the GenBank database of DNA and RNA sequences. NCBI provides a variety of tools for searching GenBank, and elements in GenBank are linked to other databases, both within and outside of NCBI. Figure 13.9 shows some results from a simple query of the GenBank nucleotide database.

GenBank data files contain a wealth of information. Figure 13.10 shows a simple GenBank file for a prion sequence from duck. The accession number, AF283319, is

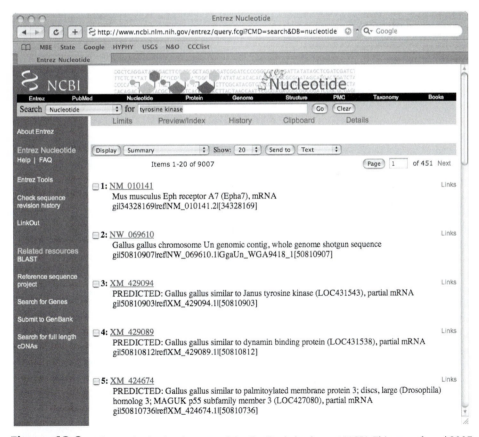

Figure 13.9 The result of a simple query of the GenBank database at NCBI. This query found 9007 entries in the GenBank nucleotide database containing the term "tyrosine kinase." Each entry can be clicked to find additional information.

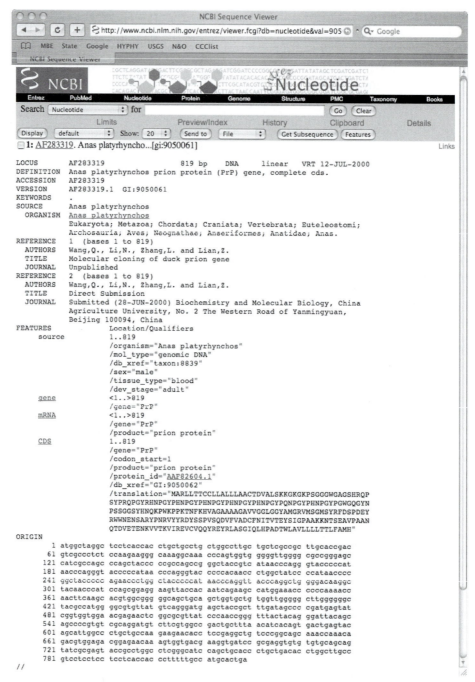

Figure 13.10   A simple GenBank file containing the DNA sequence for a prion protein gene.

the unique identifier for this entry. The GenBank file contains a DNA sequence of 819 nucleotides, its predicted amino acid sequence, and a citation to the Chinese laboratory that obtained the data. The "Links" icon in the upper right provides access to related information found in other databases. It is essential for those working in genomics or bioinformatics to become familiar with GenBank and the content of GenBank files.

NCBI is also the home of the BLAST database searching tool. BLAST uses algorithms for sequence alignment (described later in this chapter) to find sequences in GenBank that are similar to a query sequence provided by the user. To illustrate the use of BLAST, consider a study by Professor Eske Willerslev at the University of Copenhagen. Willerslev and his colleagues collected samples from Siberian permafrost that included a variety of preserved plant and animal material estimated to be 300,000–400,000 years old. They were able to extract short DNA sequences from the *rbcL* gene. These short sequences were used as input to the BLAST algorithm, which reported a list of similar sequences. It is likely that the most similar sequences come from close relatives of the organisms that provided the ancient DNA.

The European Bioinformatics Institute (EBI, http://www.ebi.ac.uk) is the European "equivalent" of NCBI. Users who explore the EBI website will find much of the same type of functionality as provided by NCBI. Of particular note is the Ensembl project (http://www.ensembl.org), a joint venture between EBI and the Sanger Institute. Ensembl has particularly nice tools for exploring genome project data through its Genome Browser. Figure 13.11 shows a portion of the display for a region of human chromosome 7. Ensembl provides comparisons to other completed genome sequences (rat, mouse, and chimpanzee), along with annotations of the locations of genes and other interesting features. Most of the items in the display are clickable and provide links to more detailed information on each display component.

Many other databases and Web resources play important roles in the day-to-day working of genome scientists. Table 13.2 includes a selection of these resources, along with short descriptions of their unique features.

13.3.2 Sequence Alignment

The most fundamental computational algorithm in bioinformatics is that of pairwise sequence alignment. Not only is it of immediate practical value, but the underlying dynamic programming algorithm also serves as a conceptual framework for many other important bioinformatics techniques. The goal of sequence alignment is to accept as input two or more DNA, RNA, or amino acid sequences; identify the regions of the sequences that are similar to one another according to some measure; and output the sequences with the similar positions aligned in columns. An alignment of six sequences from HIV strains is shown in Figure 13.12.

Sequence alignments have numerous uses. Alignments of pairs of sequences help us to determine whether or not they have the same or similar functions. Regions of alignments with little sequence variation likely correspond to important structural or functional regions of protein coding genes. By studying patterns of similarity in an alignment of genes from several species, it is possible to infer the evolutionary history

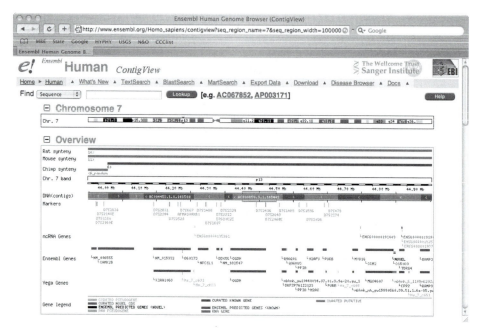

Figure 13.11 A view of a portion of human chromosome 7 in the Ensembl genome browser.

TABLE 13.2 Examples of Online Genome Resources

Comprehensive Genome Resources

www.ncbi.nlm.nih.gov	NCBI: GenBank and many others
www.tigr.org	TIGR: Microbial genomes: Rich software collection
www.ebi.ac.uk	European Bioinformatics Institute: Databases and analysis tools

Microbial Genomes

www.tigr.org/tigr-scripts/CMR2/CMRHomePage.spl	Comprehensive Microbial Resource (CMR) Home Page
www.ncbi.nlm.nih.gov/genomes/MICROBES/ Complete.html	NCBI Microbial Genomes Databases

Model Organism Genomic Databases and Resources

www.arabidopsis.org	TAIR: The Arabidopsis Information Resource
www.wormbase.org	*C. elegans* database
flybase.bio.indiana.edu	FlyBase: A database of the *Drosophila* genome
www.yeastgenome.org	*Saccharomyces* Genome Database
genome.ucsc.edu	Genome browsers for a variety of completed genomes

Annotation and Analysis Resources

www.sanger.ac.uk/Software/Pfam	PFAM: Protein family database and HMM software
genes.mit.edu/GENSCAN.html	GENSCAN HMM-based gene finder
www.ebi.ac.uk/arrayexpress	Microarray databases and software
www.ncbi.nlm.nih.gov/BLAST	NCBI BLAST database search server
www-hto.usc.edu/software/seqaln	Sequence alignment software and server
www.cellbiol.com	Resources and software indices

```
Strain1   ATAGTAATTA GATCTGAAAA CTTCTCGAAC AATGCTAAAA CCATAATAGT ACAGCTAAAT
Strain2   GTAGTAATTA GATCTGAAAA CTTCTCGAAC AATGCTAAAA CCATAATAGT ACAGCTAAAT
Strain3   ATAGTA---- --TCTGAAAA CTTCACGAAC AATGCTAAAA CCATAATAGT ACATCTAAAT
Strain4   GTAGTAATTA GATCT---AA CTTCACGAAC AATGCTAAGA CCATAATAGT ACAGCTAAAT
Strain5   ATAGTAATTA GATCT---AA CTTCACGAAC AATGCTAAAA CCATAATAGT ACAGCTAAAG
Strain6   ATAGTAATCA GATCT---AA CTTCTCGGAC AATGCTAAAA CCATAATAGT ACAGCTAAAC
```

Figure 13.12   A multiple sequence alignment of six HIV sequences. Indels, which are insertions or deletions of nucleotides into or from existing sequences, are indicated with the symbol "-".

of the species, and even to reconstruct DNA or amino acid sequences that were present in the ancestral organisms. Many methods for annotation, including assigning protein function and identifying transcription factor binding sites, rely on multiple sequence alignments as input.

To illustrate the principles underlying sequence alignment, consider the special case of aligning two DNA sequences. If the two sequences are similar, it is most likely because they have evolved from a common ancestral sequence at some time in the past. As illustrated in Figure 13.13, the sequences differ from the ancestral sequence and from each other because of past mutations. Most mutations fall into one of two classes: nucleotide substitutions, which result in these two sequences being different at the location of the mutation (Fig. 13.13a), and insertions or deletions of short sequences (Fig. 13.13b). The term *indel* is often used to denote an insertion or deletion mutation. Figure 13.13b shows that indels lead to one sequence having nucleotides present at certain positions, whereas the second sequence has no nucleotides at those positions. To align two sequences without error, it would be necessary to have knowledge of the entire collection of mutations in the history of the two sequences. Since this information is not available, it is necessary to rely on computational algorithms for reconstructing the likely locations of the various mutation events. A score function is chosen to evaluate alignment quality, and the algorithms attempt to find the pairwise alignment that has the highest numerical score among all possible alignments.

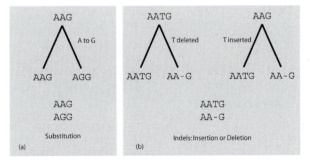

Figure 13.13   Types of mutations. A substitution (a) involves the replacement of one nucleotide with another. Note that a single alignment might be consistent with either an insertion or deletion mutation (b).

Consider aligning the two short sequences CAGG and CGA. It can be shown that there are 129 possible ways to align these two sequences, several of which are shown in Figure 13.14. How does one determine which of the 129 possibilities is best? Alignments (a) and (b) each have two positions with matching nucleotides; however, alignment (b) includes three columns with indels, whereas (a) has only one. On the other hand, alignment (a) has one mismatch to (b)'s zero. There is no definitive answer to the question of which alignment is best; however, it makes sense that "good" alignments will tend to have more matches and fewer mismatches and indels. It is possible to quantify that intuition by invoking a scoring scheme in which each column receives a score, s_i, according to the formula

$$s_i = \begin{cases} m, & \text{the bases at column } i \text{ match} \\ d, & \text{the bases at column } i \text{ do not match} \\ i, & \text{there is an indel at column } i \end{cases}$$

Using this scheme with match score $m = 5$, mismatch score $d = -1$, and indel score $i = -3$, the alignment in Figure 13.14a would receive a score of $5 - 3 + 5 - 1 = 6$. Similarly, the alignment in Figure 13.14b has a score of $5 - 3 - 3 + 5 - 3 = 1$. The remaining alignments in Figure 13.14 have scores of 0, 0, 1, 0, -5, and 1, respectively. Alignment (a) is considered best under the standards of this scoring scheme, and, in fact, it has the best score of all 129 possible alignments. This example suggests an algorithm for finding the best scoring alignment of any two sequences: enumerate all possible alignments, calculate the score for each, and select the alignment with the highest score. Unfortunately, it turns out that this approach is not practical for real data. It can be shown that the number of possible alignments of 2 sequences of length n is approximately $2^{2n}/\sqrt{2\pi n}$ when n is large. Even for a pair of short sequences of length 100, the number of alignments is 6×10^{58}, 35 orders of magnitude larger than Avogadro's number! Techniques such as the Needleman–Wunsch and Smith–Waterman algorithms, which allow for computationally efficient identifications of the optimal alignments, are important practical and theoretical components of bioinformatics.

```
CAGG        CAGG-       CAGG        CAGG
+  +        +   +       +            +
C-GA        C--GA       CGA-        -CGA

(a)         (b)         (c)         (d)


CAGG-       CAGG        CA-GG       CAG-G
+  +        +            +         +  +
C-G-A       CG-A        --CGA       C-GA-

(e)         (f)         (g)         (h)
```

Figure 13.14   Eight of the 129 possible alignments of the sequences CAGG and CGA.

Conceptually, the task of aligning three or more sequences is essentially the same as that of aligning pairs of sequences. The computational task, however, becomes enormously more complex, growing exponentially with the number of sequences to be aligned. No practically useful solutions have been found, and the problem has been shown to belong to a class of fundamentally hard computational problems said to be NP-complete. In addition to the increased computation, there is one important new concept that arises when shifting from pairwise alignment to multiple alignment. Scoring columns in the pairwise case was simple; that is not the case for multiple sequences. Complications arise because the evolutionary tree relating the sequences to be aligned is typically unknown, which makes assigning biologically plausible scores difficult. This problem is often ignored, and columns are scored using a sum of pairs scoring scheme in which the score for a column is the sum of all possible pairwise scores. For example, the score for a column containing the three nucleotides CGG, again using the scores $m = 5$, $d = -1$, and $i = -3$, is $-1 - 1 + 5 = 3$. Other algorithms, such as the popular CLUSTALW program, use an approach known as progressive alignment to circumvent this issue.

Almost all widely used methods for finding sequence alignments rely on a scoring scheme similar to the one used in the preceding paragraphs. Clearly, this formula has very little biological basis. Furthermore, how does one select the scores for matches, mismatches, and indels? Considerable work has addressed these issues with varying degrees of success. The most important improvement is the replacement of the simple match and mismatch scores with scoring matrices obtained from empirical collections of amino acid sequences. Rather than assigning, for example, all mismatches a value of -1, the BLOSUM and PAM matrices provide a different penalty for each possible pair of amino acids. Since these penalties are derived from actual data, mismatches between chemically similar amino acids such as leucine and isoleucine receive smaller penalties than mismatches between chemically different ones.

A second area of improvement is in the assignment of indel penalties. The alignments in Figure 13.14b and 13.14e each have a total of three sites with indels. However, the indels at sites 2 and 3 of Figure 13.14b could have been the result of a single insertion or deletion event. Recognizing this fact, it is common to use separate open and extension penalties for indels. If the open penalty is $o = -5$ and the extension penalty is $e = -1$, then a series of three consecutive indels would receive a score of $-5 - 1 - 1 = -7$.

13.3.3 Database Searching

The most common bioinformatics task is searching a molecular database such as GenBank for sequences that are similar to a query sequence of interest. For example, the query sequence may be a gene sequence from a newly isolated viral outbreak, and the search task may be to find out if any known viral sequences are similar to this new one. It turns out that this type of database searching is a special case of pairwise sequence alignment. Essentially, all sequences in the database are concatenated end to end, and this new "supersequence" is aligned to our query sequence. Since the supersequence is many, many times longer than the query sequence, the resulting pairwise alignment would consist mostly of gaps and provide relatively little useful information. A more useful procedure is to ask if the supersequence contains a short subsequence that aligns well with the query sequence. This problem is known as local sequence alignment, and it can be solved with algorithms very similar to those for the basic alignment problem. The Smith–Waterman algorithm is guaranteed to find the best such local alignment.

Even though the Smith–Waterman algorithm provides a solution to the database search problem for many applications, it is still too slow for high-volume installations such as NCBI, where multiple query requests are handled every second. For these settings, a variety of heuristic searches have been developed. These tools, including BLAST and FASTA, are not guaranteed to find the best local alignment, but they usually do and are, therefore, valuable research tools.

It is no exaggeration to claim that BLAST (http://www.ncbi.nlm.nih.gov/BLAST) is one of the most influential research tools of any field in the history of science. The algorithm has been cited in upwards of 30,000 studies to date. In addition to providing a fast and effective method for database searching, the use of BLAST spread rapidly because of the statistical theory developed to accompany it. When searching a very large database with a short sequence, it is very likely that one or more instances of the query sequence will be found in the database simply by chance alone. When BLAST reports a list of database matches, it sorts them according to an E-value, which is the number of matches of that quality expected to be found by chance. An E-value of 0.001 indicates a match that would only be found once every 1000 searches, and it suggests that the match is biologically interesting. On the other hand, an E-value of 2.0 implies that two matches of the observed quality would be found every search simply by chance, and therefore, the match is probably not of interest.

Consider an effort to identify the virus responsible for SARS. The sequence of the protease gene, a ubiquitous viral protein, was isolated and stored under GenBank accession number AY609081. If that sequence is submitted to the tblastx variant of BLAST at NCBI, the best matching non-SARS entries in GenBank (remember that the SARS entries would not have been in the database at the time) all belong to corona-viruses, providing strong evidence that SARS is caused by a coronavirus. This type of comparative genomic approach has become invaluable in the field of epidemiology.

13.3.4   Hidden Markov Models

Much work in genomic science and bioinformatics focuses on problems of identifying biologically important regions of very long DNA sequences, such as chromosomes or genomes. Many important regions such as genes or binding sites come in the form of relatively short contiguous blocks of DNA. Hidden Markov models (HMMs) are a class of mathematical tools that excel at identifying this type of feature. Historically, HMMs have been used in problems as diverse as finding sources of pollution in rivers, formal mathematical descriptions of written languages, and speech recognition, so there is a rich body of existing theory. Predictably, many successful applications of HMMs to new problems in genomic science have been seen in recent years. HMMs have proven to be excellent tools for identifying genes in newly sequenced genomes, predicting the functional class of proteins, finding boundaries between introns and exons, and predicting the higher-order structure of protein and RNA sequences.

To introduce the concept of an HMM in the context of a DNA sequence, consider the phenomenon of isochores, regions of DNA with base frequencies unique from neighboring regions. Data from the human genome demonstrate that regions of a million or more bases have G+C content varying from 20% to 70%, a much higher range than one would expect to see if base composition were homogeneous across the entire genome. A simple model of the genome assigns each nucleotide to one of three possible classes (Fig. 13.15a): a high G+C class (H), a low G+C class (L), or a normal G+C class (N). In the normal class, each of the four bases A, C, G, and T is used with equal frequency (25%). In the high G+C class, the frequencies of the four bases are 15% A, 35% C, 35% G, and 15% T, and in the low G+C class, the frequencies are 30% A, 20% C, 20% G, and 30% T. In the parlance of HMMs, these three classes are called hidden states, since they are not observed directly. Instead, the emitted characters A, C, G, and T are the observations. Thus, this simple model of a genome consists of successive blocks of nucleotides from each of the three classes (Fig. 13.15b).

The formal mathematical details of HMMs will not be discussed, but it is useful to understand the basic components of the models (Fig. 13.16). Each hidden state in an HMM is able to emit characters, but the emission probabilities vary among hidden states. The model must also describe the pattern of hidden states, and the transition probabilities determine both the expected lengths of blocks of a single hidden state and the likelihood of one hidden state following another (e.g., is it likely for a block of high G+C to follow a block of low G+C?). The transition probabilities play important roles in applications such as gene finding.

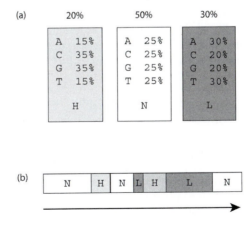

Figure 13.15   (a) Three hidden states: H, N, and L, along with their emission probabilities. In normal (N) regions of the genome, each base appears 25% of the time. Fifty percent of the genome is in a normal region. (b) The genome consists of patches of the three types of hidden states: H, N, and L.

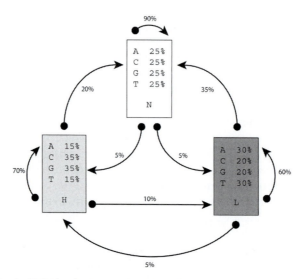

Figure 13.16   An HMM for the states in Fig. 13.15. Transition probabilities govern the chance that one hidden state follows another. For example, an N state is followed by another N state 90% of the time, by an L state 5% of the time, and by an H state 5% of the time. Emission probabilities control the frequency of the four nucleotides found at each type of hidden state. In the hidden state L, there is 30% A, 20% C, 20% G, and 30% T.

With the components of the model in place, it is possible to take a string of nucleotides and apply existing statistical methods to answer questions such as

- Where are the boundaries between the three hidden states?
- Are nucleotides 1200–1500 in a high G+C region?
- What is the most likely hidden state for nucleotide 3872?
- What is the chance of seeing a block of high G+C nucleotides shorter than 5000?

These types of questions will be addressed in the examples discussed in the next section.

13.3.5   Gene Prediction

The task of gene prediction is conceptually simple to describe: given a very long sequence of DNA, identify the locations of genes. Unfortunately, the solution of the problem is not quite as simple. As a first pass, one might simply find all pairs of start (ATG) and stop (TAG, TGA, TAA) codons. Blocks of sequence longer than, say, 300 nucleotides that are flanked by start and stop codons and that have lengths in multiples of three are likely to be protein coding genes. Although this simple method will be likely to find many genes, it will probably have a high false positive rate (incorrectly predict that a sequence is a gene), and it will certainly have a high false negative rate (fail to predict real genes). For instance, the method fails to consider the possibility of introns, and it is unable to predict short genes. Gene finding algorithms rely on a variety of additional information to make predictions, including the known structure of genes, the distribution of lengths of introns and exons in known genes, and the consensus sequences of known regulatory sequences.

HMMs turn out to be exceptionally well-suited for gene finding, and the basic structure of a simple gene finding HMM is shown in Figure 13.17. Note that the HMM includes hidden states for promoter regions, the start and stop codons, exons, introns, and the noncoding DNA falling between different genes. Also note that not all hidden states are connected to one another. This fact reflects an understanding of gene and genome structure. The sequence of states Start – Intron – Stop – Exon – Promoter is not biologically possible, whereas the series Noncoding – Promoter – Start – Exon – Intron – Exon – Intron – Exon – Stop – Noncoding is. Good HMMs incorporate this type of knowledge extensively.

To put the HMM of Figure 13.17 to use, the model must first be trained. The training step involves taking existing sequences of known genes and estimating all of the transition and emission probabilities for each of the model's hidden states. For example, if a training data set included 150 introns, the observed frequency of C in those intron sequences could be used to come up with the emission probability for C in the hidden state intron. The average lengths of introns and exons would be used to estimate the transition probabilities to and from the Exon and Intron hidden states. Once the training step is complete, the HMM machinery can be used to predict the locations of genes in a long sequence of DNA, along with their intron/exon boundaries, promoter sites, and so forth.

Gene finding algorithms in actual use are much more complex than the one shown in Figure 13.17, but they retain the same basic structure. The performance of gene

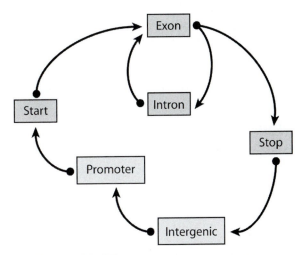

Figure 13.17 A simple HMM for gene finding.

finders continues to get better and better as more genomes are studied and the quality of the underlying HMMs is improved. In bacteria, modern gene finding algorithms are rarely incorrect. Upwards of 95% of the predicted genes are subsequently found to be actual genes, and only 1–2% of true genes are missed by the algorithms. The situation is not as rosy for eukaryotic gene prediction, however. Eukaryotic genomes are much larger, and the gene structure is more complex (most notably, eukaryotic genes have introns). The effectiveness of gene finding algorithms is usually measured in terms of sensitivity and specificity. If these quantities are measured on a per nucleotide basis, an algorithm's sensitivity, S_n, is defined to be the percentage of nucleotides in real genes that are actually predicted to be in genes. The specificity, S_p, is the percentage of nucleotides predicted to be in genes that truly are in genes. Good gene predictors have high sensitivity and high specificity. The best gene eukaryotic gene finders today have sensitivities and specificities around 90% at the individual nucleotide. If the quantities are measured at the level of entire exons (e.g., did the algorithm correctly predict the location of the entire exon or not?), the values drop to around 70%.

An emerging and powerful approach for predicting the location of genes uses a comparative genomics approach. The entire human genome sequence is now available, and the locations of tens of thousands of genes are known. Suppose that a laboratory now sequences the genome of the cheetah. Since humans and cheetahs are both mammals, they should have reasonably similar genomes. In particular, most of the gene sequences should be quite similar. Gene prediction can proceed by doing a pairwise sequence alignment of the two genomes and then predicting that positions in the cheetah genome corresponding to locations of known human genes are also genes in the cheetah. This approach is remarkably effective, although it will obviously miss genes that are unique to one species or the other. The degree of relatedness of the two organisms also has a major impact on the utility of this approach. The human genome could be used to predict genes in the gorilla genome much better than it could be used to predict genes in the sunflower or paramecium genomes.

In addition to HMMs and comparative genomics approaches, a variety of other techniques are being used for gene prediction. Neural networks and other artificial intelligence methods have been used effectively. Perhaps most intriguing, as more and more genomes become available, are hybrid methods that integrate, for example, HMMs with comparative genomic data from two or more genomes.

13.3.6   Functional Annotation

Once a genome is sequenced and its genes are found or predicted, the next step in the bioinformatics pipeline is to determine the biological function of the genes. Ideally, molecular biological work would be carried out in the laboratory to study each gene's function, but clearly that approach is not feasible. Two basic computational approaches will be described, one using comparative genomics and the other using HMMs.

Comparative genomics approaches to assigning function to genes rely on a simple logical assumption: if a gene in species A is very similar to a gene in species B, then the two genes most likely have the same or related functions. This logic has long been applied at higher biological levels (e.g., the kidneys of different species have the same basic biological function even though the exact details may differ in the two species). At the level of genes, the inference is less accurate, especially if the species involved in the comparison are not closely related, but the approach is nonetheless useful and usually effective.

Simple database searches are the most straightforward comparative genomic approach to functional annotation. A newly discovered gene sequence that returns matches to cytochrome oxidase genes when input to BLAST is likely to be a cytochrome oxidase gene itself. Complications arise when matches are to distantly related species, when the matching regions are very short, or when the sequence matches members of a multigene family. In the first case, the functions of the genes may have changed during the tens or hundreds of millions of years since the two organisms shared a common ancestor. However, if two or more such distantly related organisms have gene sequences that are nearly identical, a strong argument can be made that the gene is critical in both organisms and that the same function has been maintained throughout evolutionary history. Short matches may arise simply as a result of elementary protein structure. For example, two sequences may have regions that match simply because they both encode alpha helical regions. Such matches provide useful structural information, but the stronger inference of shared function is not justified. Multigene families are the result of gene duplications followed by functional divergence. Examples include the globin and amylase families of genes. At some point in the past, a single gene in one organism was completely duplicated in the genome. At that point, the duplicated copy was free to evolve a new, but often related, function. Subsequent duplications allow for the growth and diversification of such families. Because of their shared ancestry, all members of a gene family tend to have similar DNA sequences. This fact makes it difficult to assign function with high accuracy when matches appear in database searches, but it often provides a general class of functions for the query sequence.

Efforts have been made to classify all known proteins into functional groups using comparative genomics. Suppose that the GenBank protein database is queried with protein sequence A and the result is that its closest match is protein sequence B. If the database is next queried using sequence B and the closest match for B is found to be sequence A, then these two proteins are said to be reciprocal best matches, and they are likely to have the same function. Likewise, if the best match to sequence A is B, the best match to B is C, and the best match to C is A, then A, B, and C are likely to have the same function. This general principle has been used to create clusters of genes that are predicted to have similar or identical functions. The COGs (Clusters of Orthologous Groups of proteins) database at NCBI (http://www.ncbi.nlm.nih.gov/COG) represents a comprehensive clustering of the entire GenBank protein database using this type of scheme.

There are many known examples of proteins or individual protein domains that have the same function or structure. The PFAM (Protein Family) database (http://www.sanger.ac.uk/Software/Pfam) includes multiple sequence alignments of almost 7500 such protein families. Using the sequence data for each alignment, the PFAM project members created a special type of HMM called a profile HMM. This database makes it possible to take a query sequence and, for each of the 7500 families and their associated profile HMMs, ask the question, "Is the query sequence a member of this gene family?" A query to the PFAM results in a probability assigned to each of the included protein families, providing not only the best matches but also indications of the strength of the matches. Currently, about 74% of the proteins in GenBank have a match in PFAM, indicating a fairly high likelihood of any newly discovered protein having a PFAM match. PFAM is of interest not only because of its effectiveness, but also because of its theoretical approach of combining comparative genomic and HMM components.

13.3.7 Identifying Differentially Expressed Genes

A common experiment is to use microarray or oligonucleotide array technology to measure the expression level for several thousand genes under two different "treatments." It is often the case that one treatment is a control while the other is an environmental stimulus such as a drug, chemical, or change in a physical variable such as temperature or pH. Other possibilities include comparisons between two tissue types (e.g., brain vs. heart), between diseased and undiseased tissues (e.g., tumor vs. normal), or between samples at two developmental phases (e.g., embryo vs. adult). One of the primary reasons to carry out such an experiment is to identify the genes that are differentially expressed between the two treatments.

The basic format of the data from a simple two-treatment microarray experiment is the following:

Spot	Trt 1	Trt 2	Ratio
1	350	250	1.4
2	1200	1250	0.96
3	900	300	3.00
4	14,000	52,200	0.27

Spot: Location on microarray
Trt 1, Trt 2: The raw expression levels measured for each of the two treatments
Ratio: Ratio of the treatment 1 measurement to the treatment 2 measurement

Each spot on a microarray corresponds to a single gene, and in competitive hybridization experiments, a single spot usually provides measurements of gene expression under two different treatments. Note that the first column has been intentionally labeled "Spot" instead of "Gene." It is important that the same gene be used and measured multiple times; therefore, a number of different spots will typically correspond to the same gene. The final column of data is the most important for interpreting this experiment. The most extreme difference in *relative* expression levels is found at spot 4, where the gene is expressed almost fourfold higher under treatment 2. The question now becomes, "How large (or small) must the ratio be to say that the expression levels are really different?" This question is one of variability and of statistical significance. Phrased differently, would a ratio near 0.27 for spot 4 be likely if the experiment were repeated? The data in the table do not provide the necessary information to answer this question, and this fact points out the importance of replication in experimental design. Whenever quantitative measurements are to be compared, replication is needed in order to estimate the variance of the measurements. This fundamental tenet of experimental design was largely ignored during the early history of microarray studies. Fortunately, recent work has included careful attention to experimental design and proper analysis using the analysis of variance (ANOVA). Typical experiments now include five or more replicate measurements of each gene. In order to detect very small treatment effects on levels of expression, even larger amounts of replication are needed.

13.3.8 Clustering Genes with Shared Expression Patterns

A second type of microarry experiment is designed not to find differentially expressed genes, but to identify sets of genes that respond to two or more treatments in the same manner. This type of study is best illustrated with a time course study in which expression levels are measured at a series of time intervals. Examples of such studies might involve measuring expression levels in laboratory mice each hour following exposure to a toxic chemical, expression levels in a mother or fetus at each trimester of a pregnancy, or expression levels in patients each year following infection with HIV. If plots of expression levels (y axis) against time (x axis) for each gene are overlaid as shown in Figure 13.18, it is possible to visually compare the expression profiles of genes. The desired pattern is a group of genes that tend to increase or decrease their expression levels in unison. In Figure 13.18 it appears that genes 2 and 5 have very similar expression profiles, as do genes 1 and 4.

The similarity between the expression profiles of two genes can be described using the correlation coefficient,

$$r_{X,Y} = \frac{1}{n} \sum_{i=1}^{n} \left(\frac{X_i - \bar{X}}{s_X} \right) \left(\frac{Y_i - \bar{Y}}{s_Y} \right),$$

where X_i and Y_i are the expression levels of genes X and Y at time point i. Values near 1 or -1 indicate that the two genes have very similar profiles. When faced with thousands of profiles, the task becomes a bit more problematic. A common theme is to cluster genes on the basis of the similarity in their profiles, and many algorithms for carrying out the clustering have been published. All of these algorithms share the objective of assigning genes to clusters so that there is little variation among profiles within clusters, but considerable variation between clusters. Top down clustering begins with all genes in a single cluster, then recursively partitions the genes into smaller and smaller clusters. Bottom up methods start with each gene in its own cluster and progressively merge smaller clusters into larger ones. Clustering algorithms may also be supervised, meaning that the user specifies ahead of time the final number of clusters, or unsupervised, in which case the algorithm determines the final number of clusters.

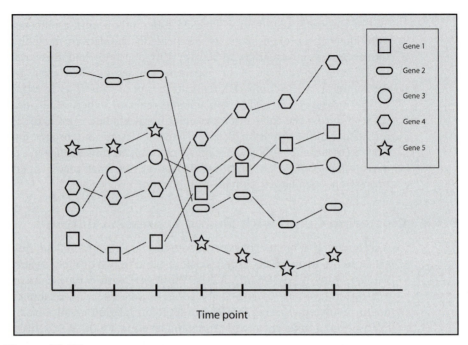

Figure 13.18   Overlaid expression profiles for 5 genes. Note that genes 2 and 5, as well as genes 10 and 4, have similar profiles, suggesting that each pair may be regulated by the same transcription factors.

13.4   CONCLUSION

The emergence of genomic science has not simply provided a rich set of tools and data for studying molecular biology. It has been the catalyst for an astounding burst of interdisciplinary research, and it has challenged long-established hierarchies found in most institutions of higher learning. The next generation of biologists will need to be as comfortable at a computer workstation as they are at the lab bench. Recognizing this fact, many universities have already reorganized their departments and their curricula to accommodate the demands of genomic science.

From a more practical point of view, the results of genomic research will begin to trickle into medicine. Already, diagnostic procedures are changing rapidly as a result of genomics. The next phase of genomics will focus on relating genotypes to complex phenotypes, and as those connections are uncovered, new therapies and drugs will follow. Consider, for example, a drug that is of significant benefit to 99% of users, but causes serious side effects in the remaining 1%. Such drugs currently have difficulty remaining in the marketplace. However, the use of genetic screens to identify the patients likely to suffer side effects should make it possible for these drugs to be used safely and effectively. Less imminent, but certainly in the foreseeable future, are gene therapies that will allow for repair of genetic defects. The continued interplay of biology, engineering, and the mathematical sciences will be responsible for exploration of these frontiers.

EXERCISES

1. How many possible proteins could be formed by a gene region containing four exons?

2. In general, eukaryotes have introns, whereas prokaryotes do not. What are possible advantages and disadvantages of introns?

3. Most amino acids are encoded by more than a single codon. If one of these synonymous codons is energetically more efficient for the organism to use, what effect would that have on the organism's genome content? How might this fact be used in gene finding algorithms?

4. What is the chance that a 20-nucleotide oligonucleotide matches a sequence other than the one it was designed to match? Assume for simplicity that all nucleotides have frequency 25%. How many matches to that oligonucleotide would one expect to find in the human genome?

5. If each of the 13 SSR markers used by the FBI for identification purposes has 20 equally frequent alleles, what is the chance that two randomly chosen individuals have the same collection of alleles at those 13 markers?

6. How many mammalian genomes have been completely sequenced? What are they?

7. What is the size of the *Anopheles gambiae* genome? How many chromosomes does it have? How many genes does it have?

8. What is the length of the *Drosphila melanogaster* alcohol dehydrogenase gene?

9. Consider the following two alignments for the sequences CGGTCA and CAGCA:

> C-GGTCA   C-GGTCA
> CA-G-CA   CA-GCA.

> **a.** Find the score of each alignment using a match score of 5, mismatch penalty of −2, and gap penalty of −4.
> **b.** Find the score of each if the gap penalty is −5 for opening and −1 for extending.

10. Suppose a computer can calculate the scores for one million alignments per second. How long would it take to find the best alignment of two 1000 bp sequences by exhaustive search?

11. Find an example of a zinc finger gene sequence using GenBank. Use BLAST to discover how many GenBank sequences are similar to the sequence you found. What does the result tell you about zinc finger genes?

12. What are some additional features that might be added to the simple gene finding HMM of Fig. 13.17?

13. Draw a diagram of a simple gene finding HMM that might be useful for prokaryotes. The HMM should contain hidden states for exons and intergenic regions, and it should guarantee that exons have lengths that are multiples of three.

14. Use the PFAM website to give a brief description of the structure and function of members of the hamartin gene family.

15. In Fig. 13.18, gene 2 seems to be expressed at higher levels than gene 5. Justify the claim that the two genes have similar profiles and might be coregulated.

16. The expression levels for three genes measured at four times are:

Gene 1	5	7	20	22
Gene 2	15	12	18	17
Gene 3	10	15	50	45

Compute the correlation coefficient for each pair of genes. Do any of them have similar profiles?

17. The expression levels for two genes measured at four times are:

Gene 1	12	16	24	28
Gene 2	64	16	16	64

Compute the correlation coefficient for this pair of genes. Make a plot of the gene 1 levels against the gene 2 levels. Is the information in the plot consistent with the implication of the correlation coefficient? Comment on potential hazards of using the correlation coefficient based on your findings.

18. Often, the gene sequences placed on microarray slides are of unknown function. Suppose that an experiment identifies such a gene as being important for formation of a particular type of tumor. Explain how you might use resources at NCBI to identify the gene's function.

19. In Fig. 13.7, genes in a time course study are clustered according to their expression profiles. Develop an algorithm for assigning genes to clusters.

20. When carrying out a database search using BLAST with a protein coding gene as the query sequence, there are two possible approaches. First, it is possible to query using the original DNA sequence. Second, one could translate the coding DNA and query using the amino acid sequence of the encoded protein. Which approach is likely to be better? What are the advantages and disadvantages of each?

SUGGESTED READING

Altschul, S., Gish, W., Miller, W., Myers, E. and Lipman, D.J. (1990). Basic local alignment search tool. *J. Mol. Biol.* **215**, 403–410.

Bernardi, G. (2000). Isochores and the evolutionary genomics of vertebrates. *Gene* **241**, 3–17.

Brown, P.O. and Botstein, D. (1999). Exploring the new world of the genome with DNA microarrays. *Nat. Genetics* **21 (suppl)**, 33–37.

Burge, C. and Karlin, S. (1997). Prediction of complete gene structures in human genomic DNA. *J. Mol. Biol.* **268**, 78–94.

Burris, J., Cook-Deegan, R. and Alberts, B. (1998). The Human Genome Project after a decade: Policy issues. *Nat. Genetics* **20**, 333–335.

Cantor, C.R. and Smith, C.L. (1999). *Genomics: The Science and Technology Behind the Human Genome Project.* John Wiley and Sons, New York, NY.

Collins, F.S. et al. (1998). New goals for the US Human Genome Project: 1998–2003. *Science* **262**, 682–689.

Fraser, C.M. et al. (1995). The minimal gene complement of *Mycoplasma genitalium. Science* **270**, 397–403.

Gibson, G. and Muse, S.V. (2002). *A Primer of Genome Science.* Sinauer Associates, Sunderland, MA.

Hartl, D. and Clark, A. (1998). *Principles of Population Genetics,* 3rd Ed. Sinauer Associates, Sunderland, MA.

Henikoff, S. and Henikoff, J. (1992). Amino acid substitution matrices from protein blocks. *Proc. Natl. Acad. Sci. USA* **89**, 10915–10919.

International Human Genome Sequencing Consortium. (2001). Initial sequencing and analysis of the human genome. *Nature* **409**, 860–921.

International SNP Map Working Group. (2001). A map of human genome sequence variation containing 1.42 million single-nucleotide polymorphisms. *Nature* **409**, 928–933.

Kerr, M.K., Martin, M. and Churchill, G. (2000). Analysis of variance for gene expression in microarray data. *J. Comput. Biol.* **7**, 819–837.

Lee, C., Weindruch, R. and Prolla, T. (2000). Gene-expression profile of the aging brain in mice. *Nat. Genetics* **25**, 294–297.

Mount, D.W. (2001). *Bioinformatics: Sequence and genome analysis.* Cold Spring Harbor Laboratory Press, Cold Spring Harbor, NY.

Needleman, S. and Wunsch, C. (1970). A general method applicable to the search for similarities in the amino acid sequences of two proteins. *J. Mol. Biol.* **48**, 443–453.

Scherf, U. et al. (2000). A gene expression database for the molecular pharmacology of cancer. *Nat. Genetics* **24**, 236–244.

Smith, T. and Waterman, M. (1981). Identification of common molecular subsequences. *J. Mol. Biol.* **147**, 195–197.

Sonnhammer, E., Eddy, S., Birney, E., Bateman, A. and Durbin, R. (1998). Pfam: Multiple sequence alignments and HMM profiles of protein domains. *Nucl. Acids Res.* **26**, 320–322.

Taft, R.J. and Mattick, J.S. (2003). Increasing biological complexity is positively correlated with the relative genome-wide expansion of non-protein-coding DNA sequences. *Genome Biology* **5**, P1.

Venter, J.C., Adams, M., Sutton, G., Kerlavage, A., Smith, H. and Hunkapiller, M. (1998). Shotgun sequencing of the human genome. *Science* **280**, 1540–1542.

Venter, J.C. et al. (2001). The sequence of the human genome. *Science* **291**, 1304–1351.

Waterman, M.S. (1995). *Introduction to Computational Biology.* Chapman and Hall, London.

Wheeler, D.L. et al. (2000). Database resources of the National Center for Biotechnology Information. *Nucleic Acids Res.* **28**, 10–14.

Willerslev, E. et al. (2003). Diverse plant and animal genetic records from Holocene and Pleistocene sediments. *Science* **300**, 791–795.

Zhang, L. et al. (1997). Gene expression profiles in normal and cancer cells. *Science* **276**, 1268–1272.

14 COMPUTATIONAL CELL BIOLOGY AND COMPLEXITY

Charles Coward, MD[*]
Banu Onaral, PhD[*]

Chapter Contents

At the end of this chapter, the reader will be able to:

- Describe how computational modeling is used to model cellular processes.

- Develop simple models of cellular control.

- Describe the basic concepts defining complexity theory.

- Describe how complexity theory applies to cellular biology.

[*]With contributions by Susan Blanchard and Anne-Marie Stomp.

14.1   COMPUTATIONAL BIOLOGY

Computational biology involves modeling, measuring, or classifying the processes within a cell, such as metabolic or control pathways, or the proteins that are active within those pathways. Specifically, computational biology includes the use of computer systems to search the genome/proteome through the use of the huge genomic and proteomic databases or computer systems and can be used to analyze cellular messenger RNA (mRNA) levels within a tissue through microarray testing. Computational biology tools can be used to model pathways to understand how pathways interact or to find drug targets to treat specific diseases. These systems can bridge the purely theoretical through modeling or utilize or generate experimental data. Ultimately, these different tools and approaches are used to generate information about cellular behavior and to integrate the information to develop an understanding about how cells function within an organism.

14.1.1   Computational Modeling of Cellular Processes

Modeling of cellular processes can be divided into two areas, modeling the internal control of the cell or modeling the metabolic pathways within the cell. Both types of processes will, in general, interact with the environment around the cell. Control of the cell can be affected by stimuli from other cells nearby, from tissues distant from the cell of interest, and from the extracellular matrix to which the cell is attached. Metabolite concentrations may be affected by glucose concentrations in the environment around the cell. Both of the areas (control and metabolic modeling) may therefore consider the exterior of the cell. Cellular modeling may also require simulating the effects of different compartments (see Chapter 12) that may restrict the interactions between different control paths or interactions or metabolic concentrations. Examples of different compartments within a eukaryotic cell are cytoplasm, mitochondria, and nucleus. Other compartments may also be considered such as Golgi apparatus or sarcoplasmic reticulum.

14.1.2   Modeling Control Mechanisms within the Cell

Control within a cell is distributed throughout the entire cell. Receptors interact with the exterior of the cell and initiate effects within the cell. Some receptors interact with DNA to stimulate or inhibit production of mRNA that leads to production of proteins. These receptors may function on the cell membrane or within the cytoplasm and exert their influence on the mRNA/protein production through cascades of protein messengers or interact with second messengers, ultimately influencing protein concentrations within the cell. Some of these proteins may stimulate the initial conversion to cancer. These types of receptors are called proto-oncogenes and may be tyrosine kinases. Alternatively, receptors exist within the nucleus for steroid hormones which can travel through cell membranes that enclose the cytoplasm and separate the cytoplasm from the interior of the nucleus. Receptors may also facilitate absorption of substrate for metabolic pathways, such as GLUT receptors which allow a cell to absorb glucose from the bloodstream.

14.1.3 Modeling Metabolic Pathways within the Cell

Metabolic pathways within the cell provide the ability to convert energy within various molecules (such as glucose, ketoacids, and certain amino acids) into energy needed by the cell to perform various maintenance functions, cell division, or the functions required by the tissue to maintain the organism as a whole. The energy within these molecules is stored in chemical bonds between specific atoms within the molecules of interest. These metabolic pathways may involve a process called *respiration* that utilizes the characteristics of oxygen molecules to assist in the chemical conversion and liberation of energy from the fuel molecules. Energy can also be extracted from glucose without the need of oxygen, although this is significantly less efficient. Energy generated in this manner produces lactic acid, a molecule that can change the pH and adversely affect the activity of enzymes within the cell.

The process of energy extraction or energy conversion is a pathway requiring many enzymes, with the corresponding complexity due to many control points within the pathway, and will cross into several compartments within the cell. The citric acid cycle (Kreb's cycle) functions in part in the cytoplasm and in part within the mitochondria. If the glycolysis pathway and the Kreb's cycle are to be modeled, the cytoplasm and mitochondria must be modeled. Provisions must be made to allow metabolites within the Kreb's cycle to cross into the mitochondria, and adenosine triphosphate (ATP) must be permitted to exit the mitochondria.

14.2 THE MODELING PROCESS

In developing a model of cellular processes, whether control or metabolic, the purpose of the model must be established first. Is the purpose of the model to simply understand the processes, to identify the key control points within a process (for a drug discovery process), or to attempt to develop information or data for future experimentation? This question is important because it can influence how the model is developed and which assumptions are made in constructing the model. The types of assumptions may influence decisions on which computational platforms or software packages are used to execute the model. If a preexisting software package is used for the model (or simulation), then the assumptions made during its design and implementation should be evaluated to determine if they are consistent with the goals of the model.

The goals of the model may also determine the computational time required for the model to perform its simulation with simpler models requiring several minutes on a personal computer or workstation and larger models requiring hours to days of processing on supercomputers or Linux clusters.

14.2.1   Methods of Modeling

Computational models can be built using existing higher-level software packages such as MATLAB/SIMULINK®, V-Cell (http://www.nrcam.uchc.edu/, Fig. 14.1), E-Cell (http://www.e-cell.org/), Jig Cell (http://jigcell.biol.vt.edu/), LabVIEW®, and Microsoft® Excel, or through programming languages such as C/C++, Java, or Basic. The higher-level software packages provide an environment that allows the user to quickly create a model or simulation with reduced debugging time and with predefined graphics capabilities. Some of the packages may allow the user to simply click and drag various icons onto the workspace to simulate processes. The disadvantage of these packages is that it may not be possible to simulate desired features, and the simulations may have considerable overhead that will lead to extended processing times for complicated simulations. The advantages of using programming languages include higher processing speeds and additional flexibility to create a detailed simulation that meets the requirements. The disadvantages of using programming languages are the additional time needed to debug the programs and the difficulty in developing

Figure 14.1   V-Cell (http://www.nrcam.uchc.edu/), which allows a user to simulate cellular processes. The colored circle in the center of the screen indicates diffusion into a cell at a specific time. The time can be changed with controls.

output that demonstrates the model's results. Usually, a greater understanding of a programming language is required to create a model than to create a simulation using a software package such as V-Cell or E-Cell. The higher-level software packages can be combined with the models developed in one of the programming languages to allow a simulation to have the speed and flexibility afforded by the language and the graphics capabilities of the software packages.

Computational models are built using equations to represent the various aspects that define the simulation. Usually, models or simulations (the terms are used here interchangeably) are based on modeling changes within a cell or tissue. Modeling the changes within the cell allows the modeler to represent a signal pathway (cellular control) or an ion/metabolite concentration or flux increase within the cytoplasm. Change can be spatial (concentration gradient between different compartments within the cell) or temporal (change in calcium concentration within the cytoplasm in a muscle cell during contraction). Changes are usually modeled with ordinary or partial differential equations, but they can be modeled with other types of methods or equations (perhaps, for example, a Boolean network with delays). The model could consist of the closed form solution to a solvable differential equation or a numerical solution of a more difficult, ordinary, or partial differential equation.

An example of a simple model using a closed form solution to an ordinary differential equation would be a simple exponential solution. The coefficients could be modified to compare results using different characteristics.

14.2.2 Equations of Modeling

As mentioned in Section 14.2.1, the equations of modeling principally involve change. This change will be dependent on many factors, such as enzyme concentrations, substrate concentrations, initiating factors, concentration gradients, electrical gradients, decay of complex molecules (mRNA, proteins, hormones), and spatio-temporal dimensions. First-order or second-order differential equations are used to model the change within the cell. When a simulation is based on a differential equation, a method of numerical solution must be used to integrate the equation within the software. If the solution of differential equations is numerical, a grid or coordinate system is usually defined. This coordinate system is used to define points spatially for arrays that contain the values of the concentrations throughout the cell. By using an array (one-, two-, or three- dimensional) to store the concentrations, spatial change can be modeled and concentration gradients can be established.

In solving first-order differential equations, the prototypical method of solution is the Euler approximation. In this approximation, the first-order derivative is based on the general definition of a derivative, as shown for the following differential equation:

$$\frac{df}{dx} = f(x, y) \tag{14.1}$$

Using the definitions of derivatives:

$$\frac{df(x)}{dx} = \frac{\Delta f(x, y)}{\Delta x} = (f(x_2, y) - f(x_1, y))/\Delta x \tag{14.2}$$

the equation can be manipulated to generate the Euler method:

$$x_{n+1} = x_n + \Delta x f(x_n, y_n) \tag{14.3}$$

This approximation is relatively inaccurate and, in some equations, may lead to instability of the solution. The Euler method evaluates the function at the current (x_{n+1}) position by using the value of the function at the previous position (x_n), so for a rapidly changing function, the results can lead to the inaccuracy mentioned earlier. Alternatively, a second-order or fourth-order Runge–Kutta method or one of several other methods could be used to model the derivative, and this would lead to a more stable and accurate solution.

The second-order Runge–Kutta approximation is:

$$k_1 = hf(x_n, y_n) \tag{14.4}$$

$$k_2 = hf(x_n + \frac{1}{2}h, y_n + \frac{1}{2}k_1) \tag{14.5}$$

$$x_{n+1} = x_n + k_2 + O(h^3) \tag{14.6}$$

The Runge–Kutta method evaluates the function at the midpoint, so the results more closely reflect the behavior of the function itself. All of these approaches may have problems if the rate of change of the function changes over an interval (i.e., the function is slowly changing and then begins to change quickly) and these differential equations are said to be "stiff." Stiff ordinary differential equations simulating this type of changing behavior may occur in biological models when signal transduction paths are modeled. A common approach to solving stiff ordinary differential equations is to change the integration step (decrease Δx) when the function begins to change rapidly with respect to x. Many of the higher-level simulation packages allow the user to select which method should be used to solve the first-order differential equations required in the simulation.

The Michaelis–Menten equation provides one of the basic first-order differential equations necessary for analysis of metabolic networks. The Michaelis–Menten equation relates the rate of production of a specific molecule to the product formation velocity (V_{max}), substrate concentration [S], and substrate concentration when the product formation velocity is one-half of V_{max}. The equation reflects a binding of the enzyme (E) and the substrate (S). During the binding the enzyme speeds the transition of the substrate to the product. The transition speed is increased by lowering the activation energy required for the conversion to occur. Without the enzyme, the transition from substrate to product would occur at a very low rate, so the enzyme, by lowering the activation energy, increases the probability and speed of conversion. With every enzymatic reaction, there is the possibility that the substrate will be "released" from the enzyme without making a conversion to the product, and this is reflected in the constant k_m. The k_m constant is equal to the ratio of the rates of breakdown of the enzyme-substrate complex to the rate of the creation of the enzyme-substrate complex and therefore reflects the stability of the enzyme-substrate complex. This equation assumes that enzyme concentration is insignificant when compared to the substrate concentration. The equation mathematically represents the following chemical equation:

$$E + S \underset{k-1}{\overset{k1}{\rightleftharpoons}} ES \overset{k_2}{\longrightarrow} E + P$$

The Michaelis–Menten equation is

$$\frac{d[P]}{dt} = V_{\max}[S]/(k_m + [S]) \tag{14.7}$$

where

$$k_m = (k - 1 + k_2)/k_1 \tag{14.8}$$

and

$$V_{\max} = k_2[E_o] \tag{14.9}$$

This velocity ($V_{\max}$) (maximal rate of conversion) occurs when all of the enzymes are saturated with substrate and is not a function of the rates of enzyme-substrate complex creation or dissociation. [E_o] is the enzyme concentration, [S] is the substrate concentration, k_m is the substrate concentration at which product formation velocity equals $\frac{1}{2} V_{\max}$, $\frac{d[P]}{dt}$ = rate of increase of the product, k_1 = rate for synthesis of E and S to ES, $k-1$ = rate for catabolism of ES to E and S, k_2 = rate for conversion of ES to E and P (production of P), k_2 is also called k_{cat}, the turnover number, and $\frac{d[P]}{dt} = -\frac{d[S]}{dt}$ is the rate of decrease of the substrate.

The Michaelis–Menten equation demonstrates both zero-order and first-order kinetics with respect to the substrate concentration. Initially, as the substrate concentration is low, the equation demonstrates first-order kinetics when the rate of conversion increases as the substrate concentration increases. Eventually, as the substrate concentration increases, the enzymes in the system become saturated and the rate of increase of the product decreases (second time derivative of [P] is negative), leading to the asymptotic approach to $V_{\max}$. In this region the equation demonstrates zero-order kinetics, in which the velocity is no longer a linear function of the substrate concentration. This relationship exists because the second step in the chemical equation is assumed to be essentially irreversible, meaning that the enzymes in the system do not catalyze the conversion of the product into the substrate. The equation can be modified to reflect multiple reactions prior to the irreversible reaction and to incorporate the rates of those reactions.

To model spatial change, the diffusion equation is used. The equation relates spatial changes in concentrations to temporal changes and can be modified to include one-, two-, or three-dimensional change. The following equation is the one-dimensional diffusion equation in Cartesian coordinates:

$$\frac{\partial u}{\partial t} = D \frac{\partial^2 u}{\partial x^2} \tag{14.10}$$

The following equation is the three-dimensional diffusion equation:

$$\frac{\partial u}{\partial t} = D\left(\frac{\partial^2 u}{\partial x^2} + \frac{\partial^2 u}{\partial y^2} + \frac{\partial^2 u}{\partial z^2}\right) \tag{14.11}$$

This equation can be used to model concentration gradients that exist across cell membranes or to simulate the diffusion that exists due to interstitial pressure within tissues.

Signal Transduction Example

A common control mechanism in both control and metabolic processes in a cell is phosphorylation (Fig. 14.2). In the mechanism, an enzyme receives a phosphate (PO_4^{3-}) group from a member of a class of proteins called kinases. When the enzyme is phosphorylated, the behavior of the enzyme is changed. In many cases, the enzyme is activated. A differential equation can be created that models the process of phosphorylation. The equation will reflect the rates of change from an unphosphorylated form to a phosphorylated form and back. Each direction will have a different rate constant and will be dependent upon the stimulus.

The quantity of R_p, which may be either concentration or amount depending on the model, will provide the output used to interface to another process.

The general form of the equation is:

$$\frac{dR_p}{dt} = k_1 S(R - R_p) - k_2 R_p \tag{14.12}$$

k_1 = rate of phosphorylation of enzyme R by a kinase

k_2 = rate of loss of the phosphorylation of enzyme R by a phphatase

S = stimulus or signal

R = amount or concentration of enzyme R in the unphosphorylated form

R_p = amount or concentration of enzyme R in the phosphorylated form

The solution to this equation is a hyperbolic (Fig. 14.3). Equation 14.12 reflects the rates of the phosphorylation and loss of phosphorylation of the enzymes and the quantities of each form of the enzyme.

A more accurate representation of the reaction uses the following equation:

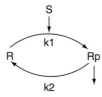

Figure 14.2   A diagram of the phosphorylation of enzyme R_p caused by stimulus S. The phosphorylation is completed by a kinase and the phosphorus is then transferred from R_p to another enzyme.

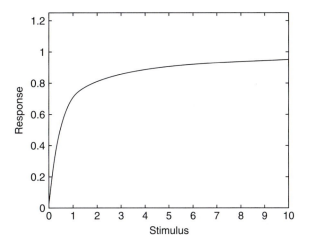

Figure 14.3 Hyperbolic curve defined by Equation 14.12.

$$\frac{dR_p}{dt} = [V_1 S(R - R_p)]/[k_1 + R - R_p] - [V_2 R_p]/[k_2 + R_p] \qquad (14.13)$$

k_1 = Michaelis constant of phosphorylated enzyme R_1

k_2 = Michaelis constant of phosphorylated enzyme R_2

V_1 = max rate of phosphorylation

V_2 = max rate of loss of phosphorylation

S = stimulus or signal

R = amount or concentration of enzyme R in the unphosphorylated form

R_p = amount or concentration of enzyme R in the phosphorylated form

This approach is more realistic because these enzymes follow Michaelis–Menten kinetics (Fig. 14.4).

If phosphorylation of one enzyme leads to phosphorylation of another enzyme, then the signal is transmitted through a cascade. In this example (Figs. 14.5 and 14.6), enzyme R is a kinase.

$$\frac{dR_p}{dt} = [V_1 S(R - R_p)]/[k_1 + R - R_p] - [V_2 R_p]/[k_2 + R_p] \qquad (14.14)$$

$$\frac{dR_1 p}{dt} = [V_3 R_p (R_1 - R_1 p)]/[k_3 + R_1 - R_1 p] - [V_4 R_1 p]/[k_4 + R_1 p] \qquad (14.15)$$

In the cascade shown in Figure 14.7, three phosphorylations lead to control of a system. The characteristics of the curves are determined by the rate constants (Fig. 14.8). A third phosphorylation equation is added to model the last step in the cascade:

$$\frac{dR_2 p}{dt} = [V_5 R_p (R_2 - R_2 p)]/[k_5 + R_2 - R_2 p] - [V_6 R_2 p]/[k_6 + R_6 p] \qquad (14.16)$$

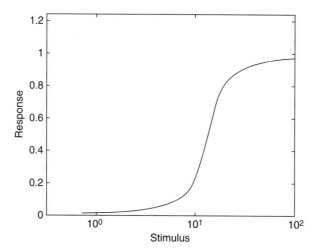

```
function sigmoid
options = odeset('RelTol',le-4, 'AbsTol',[1e-4 le-4 le-5]);
Y = zeros(3,1); % a column vector
[t,Y] = ode23(@s_model,[0 100],[0 0 0],options);
figure, semilogx(t,Y(:,1),'k-'); % Plots fig. 14.4
axis([0 100 0 1.25]);
xlabel('Stimulus')
ylabel('Response')
figure, semilogx(t,Y(:,1),'k-',t,Y(:,2), 'k--'); % Plots fig. 14.6
axis([0 100 0 1.25])
xlabel('Stimulus')
ylabel('Response')
figure, semilogx(t,Y(:,1),'k-',t,Y(:,2), 'k--',t,Y(:,3),'k.') % Plots fig. 14.8
axis([1 100 0 1.25])
xlabel('Stimulus')
ylabel('Response')

function dy = s_model(t,y)
dy = zeros(3,1);
dy(1) = (0.05*t*(1 − y(1)))/(0.2 + (1 − y(1))) − (0.50*y(1))/(0.10 + y(1));
dy(2) = (0.87*y(1)*(1 − y(2)))/(0.10 + (1 − y(2))) − (0.10*y(2))/(0.1 + y(2));
dy(3) = (0.85*y(2)*(1 − y(3)))/(0.08 + (1 − y(3))) − (0.08*y(3))/(0.1 + y(3));
```

Figure 14.4   Sigmoidal curve defined by Equation 14.13.

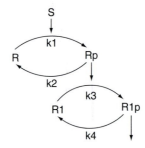

Figure 14.5   A diagram of a cascade of phosphorylation of a series of enzymes.

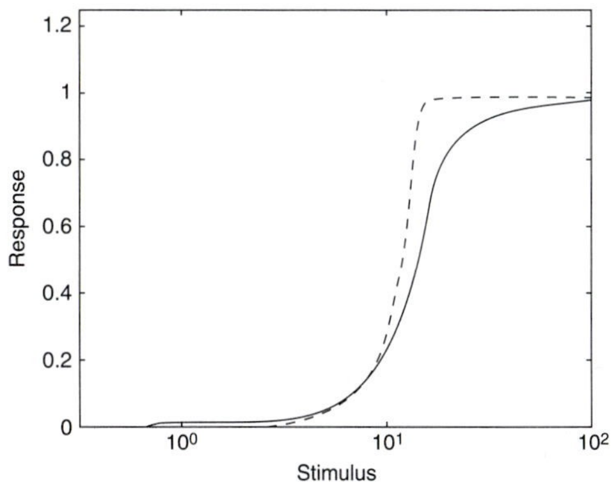

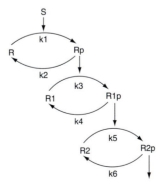

```
function sigmoid
options = odeset('RelTol',le-4, 'AbsTol',[1e-4 le-4 le-5]);
Y = zeros(3,1); % a column vector
[t,Y] = ode23(@s_model,[0 100],[0 0 0],options);
figure, semilogx(t,Y(:,1),'k-'); % Plots fig. 14.4
axis([0 100 0 1.25]);
xlabel('Stimulus')
ylabel('Response')
figure, semilogx(t,Y(:,1),'k-',t,Y(:,2), 'k–'); % Plots fig. 14.6
axis([0 100 0 1.25])
xlabel('Stimulus')
ylabel('Response')
figure, semilogx(t,Y(:,1),'k-',t,Y(:,2), 'k–',t,Y(:,3),'k.') % Plots fig. 14.8
axis([1 100 0 1.25])
xlabel('Stimulus')
ylabel('Response')

function dy = s_model(t,y)
dy = zeros(3,1);
dy(1) = (0.05*t*(1 − y(1)))/(0.2 + (1 − y(1))) − (0.50*y(1))/(0.10 + y(1));
dy(2) = (0.87*y(1)*(1 − y(2)))/(0.10 + (1 − y(2))) − (0.10*y(2))/(0.1 + y(2));
dy(3) = (0.85*y(2)*(1 − y(3)))/(0.08 + (1 − y(3))) − (0.08*y(3))/(0.1 + y(3));
```

Figure 14.6 Plot of Equations 14.14 and 14.15. In these sigmoidal curves, the stimulus leads to the first phosphorylation (solid line), which in turn causes the second phosphorylation (dashed line).

Figure 14.7 A diagram of a cascade of phosphorylation of a series of enzymes.

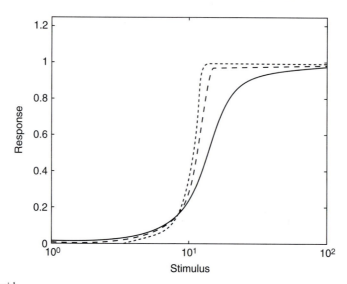

```
function sigmoid
options = odeset('RelTol',le-4, 'AbsTol',[1e-4 le-4 le-5]);
Y = zeros(3,1); % a column vector
[t,Y] = ode23(@s_model,[0 100],[0 0 0],options);
figure, semilogx(t,Y(:,1),'k-'); % Plots fig. 14.4
axis([0 100 0 1.25]);
xlabel('Stimulus')
ylabel('Response')
figure, semilogx(t,Y(:,1),'k-',t,Y(:,2), 'k–'); % Plots fig. 14.6
axis([0 100 0 1.25])
xlabel('Stimulus')
ylabel('Response')
figure, semilogx(t,Y(:,1),'k-',t,Y(:,2), 'k–',t,Y(:,3),'k.') % Plots fig. 14.8
axis([1 100 0 1.25])
xlabel('Stimulus')
ylabel('Response')

function dy = s_model(t,y)
dy = zeros(3,1);
dy(1) = (0.05*t*(1 − y(1)))/(0.2 + (1 − y(1))) − (0.50*y(1))/(0.10 + y(1));
dy(2) = (0.87*y(1)*(1 − y(2)))/(0.10 + (1 − y(2))) − (0.10*y(2))/(0.1 + y(2));
dy(3) = (0.85*y(2)*(1 − y(3)))/(0.08 + (1 − y(3))) − (0.08*y(3))/(0.1 + y(3));
```

Figure 14.8   Plots of Eqs. 14.14, 14.15, and 14.16. In these sigmoidal curves, the stimulus leads to the first phosphorylation (solid line), which in turn causes the second phosphorylation (dashed line). The second phosphorylation then effects the third phosphorylation (dotted line). The constants in the phosphorylation equations do not reflect the behavior of the actual MAPK cascade but are for illustration only.

This model is a simplified approximation of the mitogen-activated protein kinase (MAPK) pathway (Fig. 14.9), which is involved in many aspects of cellular regulation, including growth, differentiation, inflammation, and apoptosis. The MAPK pathway is a conserved cascade made up of kinases which receive stimuli from a variety of sources, such as integrins, G protein-coupled receptors (GPCRs), and tyrosine kinase receptors (RTKs). From integrins, the MAPK cascade receives signals from the extra-cellular matrix about the stress or strain imparted on the cell through the basement membrane. G protein-coupled receptors will, generally, provide signals from extra-cellular nonsteroid hormones about the status of the tissues farther from the cell to the MAPK cascade and, finally, tyrosine kinase receptors will provide signals from extra-cellular growth factors (e.g., Epidermal Growth Factor (EGF)) for the cascade. The MAPK cascade integrates these signals (along with others) to determine if gene expression will occur. Through this process, cellular regulation is initiated and con-trolled. The MAPK cascade family includes several pathways within a cell, such as ERK1/2, ERK5, p38, and JNK. In this pathway, a signal stimulates the initial phos-phorylation which leads to a cascade. Through the cascade, the response to the stimulus develops bistable (switch-like) behavior (on/off). This simplified model is based on papers by Bhalla and Iyengar (1999), Huang and Ferrell (1996), Kholodenko (2000), and Shvartsman et al. (2001).

Protein Production Example

In this example, the phosphorylation equation is used to simulate the production of mRNA (transcription), which in turn leads to translation. In this model, the first

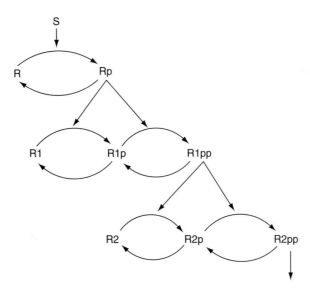

Figure 14.9 The MAPK pathway. This signal pathway is conserved in many species. It maintains tight control over specific cellular processes. The signal transduction cascade involves multiple phos-phorylations to transmit the signal through the enzymes. The first phosphorylation is made to an MAPKKK (*R*) in response to the stimulus. MAPKKK then phosphorylates MAPKK (1) twice. In turn, MAPKK phosphorylates MAPK (*R2*).

equation models a promoter that enables transcription to proceed. The promoter is phosphorylated and transcription begins. After transcription begins, mRNA is produced and begins to decay. Existing mRNA enables translation, and protein levels increase and then begin to decay. The rate of decay of mRNA is faster than the rate of decay of protein (Fig. 14.10).

$$\frac{dR_p}{dt} = \frac{V_1 S(R - R_p)}{k_1 + (R - R_p)} - \frac{V_2 R_p}{k_1 + R_p} \tag{14.17}$$

$$\frac{d[mRNA]}{dx} = k_3^*[R_p] - k_4^*[mRNA] \tag{14.18}$$

$$\frac{d[protein]}{dx} = k_5^*[mRNA] - k_6^*[protein] \tag{14.19}$$

k_3 = rate of production of mRNA

k_4 = rate of decay of mRNA

k_5 = rate of production of protein

k_6 = rate of decay of protein

The model reflects the initial production of protein and represents a model for the production of a protein that is only needed periodically for cellular processes. To model proteins that are needed at some nominal concentration to maintain the viability of the cell, the stimulus must be received at a periodic rate which is related to the decay rate. Figure 14.11 reflects the model with periodic stimulus to produce necessary proteins.

14.3 BIONETWORKS

The previous examples illustrated a small series of reactions that represented several steps within a metabolic pathway. A bionetwork is a pathway within the cell that performs a specific purpose. This pathway involves many steps that require different proteins and may be connected to other pathways at many points. At any of these contact points, the pathway of interest may receive molecules (source) or export molecules (sink). Examples of the various bionetworks are glycolysis, the citric acid cycle, and ß-oxidation. The challenge of this approach is that all of the various bionetwork examples interact either through substrates, products, and/or the various molecules that establish the state of the cell (ATP, NADH, NAD, etc.). This interaction adds high levels of complexity to this type of analysis.

In the network shown in Figure 14.12, each step can be defined by a single chemical equation that relates the substrates necessary for the creation of the product. Each product then becomes a substrate for the next chemical reaction in the pathway. Thus a series of linear equations can be written to define the stoichiometry of the system. This series of linear equations can be combined to create a stoichiometric matrix, and

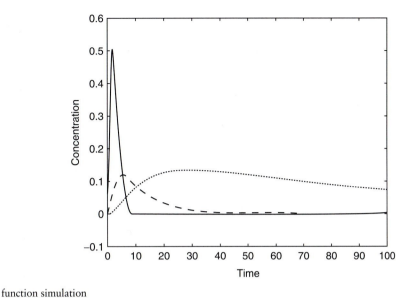

```
function simulation
options = odeset('RelTol',le-4, 'AbsTol',[1e-4 le-4 le-5]);
Y = zeros(3,1); % a column vector
[t,Y] = ode23(@simmodel,[0 100],[0 0 0],options);
figure, plot(t, Y(:,1),'k-',t,Y(:,2), 'k- -',t,Y(:,3),'k.');
xlabel('time')
ylabel('Concentration')

function dy = simmodel(t,y)
dy = zeros(3,1);% a column vector

k11 = 0.15; % Scaling purposes only
k1 = 0.2;
k2 = 0.1;
k3 = 0.1;
k4 = 0.1;
k5 = 0.1;
k6 = 0.01;
v1 = 0.6;
v2 = 0.15;

if t <2 % Used to model a limited stimulus
s = 5; % Length of stimuli
else( t>2)
s = 0;
end
dy(1) = (v1*k11*s*(1 − y(1)))/(k1 + (1 − y(1))) − (v2*y(1))/(k2 + y(1));
dy(2) = k3*y(1) − k4*y(2);
dy(3) = k5*y(2) − k6*y(3);
```

Figure 14.10 A simple model of promoter leading to transcription and then to translation. The solid line is the level of the phosphorylated promoter, the dashed line is the level of mRNA, and the dotted line is the level of the protein produced.

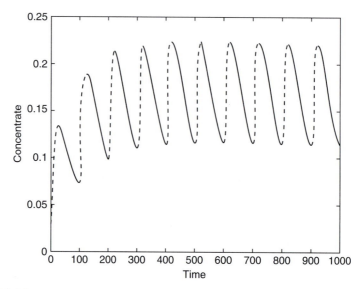

Figure 14.11   Protein levels. With simple modifications to the MATLAB code, leading to periodic stimuli (every 100 time units), protein production can compensate for the protein decay. In this manner, protein levels can remain above a required threshold.

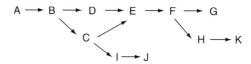

Fig. 14.12   Schematic of a simple network.

the matrix can be evaluated to derive various qualities of the system, including gene networks that are part of the mechanism that exerts control on the cellular machinery. By modeling multiple pathways that interact, the influence of inactivating particular proteins or reducing kinetic activity can be modeled. When the bionetwork reflects the changes that occur due to disease, the outcome of the inactivation of pathways or portions of pathways can reflect drugs that are competitive inhibitors of specific proteins. A library of many metabolic and regulatory pathways can be found in the Kyoto Encyclopedia of Genes and Genomes (http://www.genome.jp/kegg/). Bionetworks, both metabolic and control, demonstrate complex behavior due to the redundancy of pathways, the interaction of various pathways, and distribution of the control of cellular behavior over the various pathways.

14.4 **INTRODUCTION TO COMPLEXITY THEORY**

Complexity theory is the study of how systems interact and evolve. Complexity theory is applied to the study of biological systems to understand or characterize how various biological mechanisms function to maintain life. The study of complexity is based on observing the behavior of a system due to both the system response to specific stimuli and as a result of the interactions and functions of the various component subsystems. Through the study of complex systems, common characteristics have been identified that both characterize the systems and are required to maintain system viability.

Typically, complex systems have the following qualities:

- There can be rich interactions between a large numbers of local subsystems.
- Interactions between subsystems can be inhibitory, stimulatory, competitive, or cooperative and may be dependent upon conditions or other subsystems.
- Due to the resulting subsystem interactions, a large variety of possible behaviors (many degrees of freedom) exist.
- The possible behaviors will change over space and time (temporal and spatial degrees of freedom) (Fig. 14.13).
- Global behavior of the system can be characterized more by the interactions between subsystems and less by the behavior of the subsystems themselves.
- The global observed behavior is nonlinear, both due to the interactions of subsystems and due to the qualities of the subsystems themselves.
- Subsystems are not controlled by a central process but have control distributed over the interactions that exist among the subsystems (self-organization).
- The emergent global behavior is determined by self-organization.

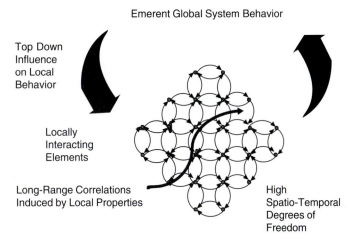

Figure 14.13   Schematic of the interactions between varying scales and emergence of a systemic behavior.

- A complex system is adaptive and stable with relatively small changes in the environment.
- A complex system demonstrates critical phenomena; it can undergo an abrupt change leading to a different state of the system.

Complex systems are systems that are composed of richly interacting subsystems (Fig. 14.14). In essence, they are systems that are composed of subsystems that interact in many ways and affect each other and may effect change in the outcome of other subsystems. These systems have many degrees of freedom and are spatially distributed, which leads to system effects that have both (or either) temporal and spatial degrees of freedom. This suggests that the events "caused" by stimuli will change over time and space. So the observed global behavior is determined more by the interaction of the subsystems than by the behavior of the individual subsystems; it "emerges" from the interactions. The observed global behavior is, in effect, established by the spatially and temporally distributed interactions, leading to distributed "self-organization." Systemic stability and adaptive behavior are related to the distributed self-organization. The interactions and the behavior of the individual subsystems are nonlinear.

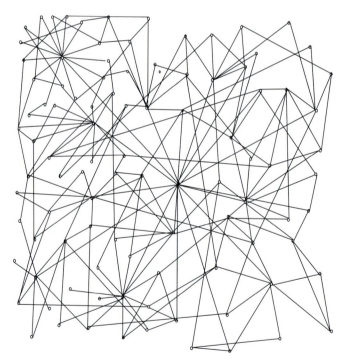

Figure 14.14   This is an example of an interaction map. Each node could indicate a protein, transcription site, etc. The chart indicates that each node interacts with other nodes and that any one node does not interact with all other nodes. The interactions that are distributed over the nodes determine which processes are active and define the system behavior.

14.4.1 Complexity in Metabolic and Control Networks

Although knowledge about biological functionality can be gained from modeling individual metabolic or control pathways by using linear differential equations, this reductionist approach will not demonstrate the full complexity of a biological organism. Both individual metabolic and control pathways interact with other pathways within a cell. These interactions between multiple metabolic pathways may lead to substrate being directed down other paths and used for other purposes. Interactions between control pathways (and metabolic paths in certain cases) communicate the status of the cell to other control processes/pathways. This communication can provide information about the environment surrounding the cell (e.g., ion concentrations and paracrine or autocrine interactions), information about distant conditions of the organism or tissue (e.g., endocrine), or status information about the cell itself. Because the status of the environment (outside and inside of the cell) is communicated to other pathways, this mechanism provides a methodology for distributed "decision making" in terms of which other pathways should be active or passive. The cellular behavior dictated by the active and passive pathways will direct the cell to perform a specific role within the organism, it's phenotype, and this system behavior can be defined to be adaptive, another important quality of complexity. Examples of the cellular activities include: secretion of hormones, ions, or proteins; contraction; growth; mitosis; apoptosis; or senescence. This distributed interaction provides a mechanism for the emergence of behavior, one of the hallmarks of complexity theory. *Emergence of behavior* is a quality that can be defined colloquially as "the whole is greater than the sum of the parts" and specifically as "the interaction of multiple pathways or systems that produces system behavior that cannot be predicted from the study of the pathways/systems individually."

An example of the communication between pathways is the generation of ATP within the cell. Because it is the currency of energy within the cell, its cellular concentration will also indicate whether the cell is in an energetic state or not. ATP, at higher cellular concentrations, which is indicative of an energetic cellular status, will have a greater probability of being used to provide the phosphate (PO_4^{3-}) group that activates existing cellular proteins, either activating or inactivating specific metabolic or control pathways. Similar examples can be found with other intracellular messengers such as GTP and cyclic AMP.

With this complex interaction between pathways leading to distributed "decision making," the cell also has to demonstrate stability. Stability occurs if a system exposed to small stimuli or perturbations in the environment results in small changes in the system's state (e.g., with small changes in extracellular sodium or potassium concentration, the cell remains viable). Large changes in the environment may or may not lead to catastrophic changes in the system (e.g., the cell may not be viable). In contrast, an unstable system will exhibit catastrophic and unpredictable results when exposed to a small stimulus or perturbation.

Specifically, an organism has to demonstrate stability in metabolic, control, and reproductive functions. Though stability is important to maintain a particular organism, evolution requires change. So, whereas a cell must be stable in terms of its metabolism and distributed control, it may be advantageous for nonzero error rates to occur in copying of DNA during the S-phase (synthesis) of the cell cycle. In this case, mutations in the DNA or RNA allow changes in functional components (e.g., proteins) which may increase the fitness of the cell. The specific organism then competes within the environment and is subsequently selected by the environment and reproduces according to its acquired fitness. This type of change probably led to sickle cell anemia.

During modeling, the processes are generally assumed to be linear systems. This assumption is made to simplify the solution and will yield a solution and an approximation of the actual process. Most systems are actually nonlinear, and this is another quality of complexity.

Although this discussion has been directed to the ways that cellular behavior demonstrates complexity, in fact, complexity exists on all levels of life from subcellular to population and ecological systems. Cellular systems demonstrate complex behavior through the interactions of various metabolic and control pathways that lead to an ability to respond to internal and external stimuli and maintain viability (i.e., they are adaptive and stable). Cellular systems demonstrate a behavior that is different from what would be predicted if systems were studied individually; they demonstrate emergence of behavior.

Through the use of computer systems to identify the purpose of particular proteins within the cell, analyze and cluster mRNA expression levels, and model metabolic or control networks, research is continuing to develop new understanding of cellular functionality and to identify drug targets to treat disease. These three methodologies can be combined to provide access to existing experimental information, develop a theoretical understanding through analysis of various pathways, and then analyze experiments to validate those hypotheses. Computer systems are able to emulate some of the cellular complexity through various metabolic and control simulations, and these models will continue to demonstrate greater complexity as computer hardware becomes more powerful, computer simulation designs and architectures develop, and molecular biology provides greater understanding of the interaction between cellular functions and additional quantitative cellular data.

EXERCISES

1. The process described by the Michaelis–Menten equation can be represented by a series of first-order differential equations. These differential equations define the rate of change of each substance to be equal to the rate constant multiplied by the concentration of each molecule in the chemical equation. Develop the four equations and describe the meaning of each term.

2. Develop the four first-order ordinary differential equations that model the Michaelis–Menten equation when the transition step from ES to $E + P$ is reversible (not assumed to be irreversible).

3. The characteristics of a sigmoidal curve are dependent upon the constants and the stimulus. Change the constants and observe the effects. What range of values leads to a curve that appears to be hyperbolic?

4. The MAPK cascade is an important cellular signal transduction pathway. Which diseases are implicated in changes to the MAPK cascade?

5. In the MAPK cascade example, feedback was not modeled. What cellular mechanisms would allow negative or positive feedback to control the cascade?

6. The NF-κB signal pathway is active in inflammation. Several models of the pathway have been developed (Hoffmann et al., 2002; Lipniacki et al., 2004). What are the important proteins that are active in the pathway? How is the pathway controlled? Is there feedback in the pathway? What is the result of the active NF-κB pathway? Which diseases are attributed to the NF-κB pathway?

7. Cell division models have demonstrated cellular protein oscillations. Which proteins are included in the models? What role does phosphorylation play in the models? How do the proteins regulate the cell cycle?

8. The combination of stimuli frequency, production, and decay rates determine protein levels within a cell. Modify the protein production code to increase the time of the model by making the following changes to functions simulation and simmodel.

Simulation:
[t,Y] = ode23(@simmodel,[0 **1000**],[0 0 0],options);
Simmodel:
if **mod(t,100)** < 2 % Used to model a limited stimulus
s = 5;
else(**mod(t,100)** > 2)
s = 0;
end

Increase and decrease the stimuli frequency, production, and decay rates and compare the result with Fig. 14.11.

9. Relating to the phosphorylation cascade example, what are the Goldbeter–Koshland function and ultrasensitivity? How would they affect the example?

10. In the phosphorylation cascade example, the cell was assumed to be in a high energy state. How is the energy state exhibited in a cell? If the cell was in a low energy state how would the curves change?

11. The phosphorylation cascade example can also be described by a series of first-order ordinary differential equations, but some of these equations reflect the input stimulus. Develop the ordinary differential equations for each of the molecules in the cascade. (Each phosphorylated molecule is different from a nonphosphorylated molecule.)

12. Emergent behavior is present in normal cellular and pathologic processes. How does emergent behavior lead to disease?

13. One of the metabolic enzymes in a cell is phosphofructokinase (PFK-1). In which pathways is it active? (See the Kyoto Encyclopedia of Genes and Genomes.) Phosphorylation controls PFK-1. How does phosphorylated PFK-1 activate and deactivate metabolic pathways? How does this relate to the energy state of the cell? What other molecules regulate PFK-1 activity and how?

14. Find example pathways for the following pathway components (Tyson et al., 2003):
a. Feed-forward loops
b. Perfect adaptation
c. Positive feedback
d. Negative feedback
e. Substrate-depletion oscillators
f. Activator-inhibitor oscillators

15. Sickle cell anemia is thought to increase the fitness (resistance to specific environmental conditions) of individuals with the mutation. What is the environmental pressure that performs the selection? What other disease also increases the fitness of individuals with the mutations against the same environmental conditions?

SUGGESTED READINGS

Allman, E.S. and Rhodes, J.A. (2004). *Mathematical Models in Biology.* Cambridge, Cambridge, U.K.

Bhalla, U.S. and Iyengar, R. (1999). Emergent properties of networks of biological signaling pathways. *Science* **283**, 381–387.

Fall, C.P., Marland, E.S., Wagner, J.M. and Tyson, J.J. (Eds.). (2002). *Computational Cell Biology.* Springer-Verlag, New York.

Hoffmann, A., Levchenko, A., Scott, M.L. and Baltimore, D. (2002). The IκB-NF-κB signaling module: Temporal control and selective gene activation. *Science* **298**, 1241–1245.

Huang, C.Y. and Ferrell, J.E. (1996). Ultrasensitivity in the mitogen-activated protein kinase cascade. *Proc. Nat. Acad. Sci. USA* **93**, 10078–10083.

Keener, J. and Sneyd, J. (1998). *Mathematical Physiology.* Springer-Verlag, New York.

Kholodenko, B.N. (2000). Negative feedback and ultrasensitivity can bring about oscillations in the mitogen-activated protein kinase cascades. *Eur. J. Biochem.* **267**, 1583–1588.

Nelson, D.L. and Cox, M.M. (2005). *Lehninger Principles of Biochemistry.* W.H. Freeman, New York.

Lipniacki, T., Paszek, P., Brasier, A.R., Luxon, B. and Kimmel, M. (2004). Mathematical model of NF-κB module. *J. Theoret. Biol.* **228**, 195–215.

Lodish, H. et al. (1999). *Molecular Cell Biology, W.H. Freeman & Co.*, New York.

Markevich, N.I., Hoek, J.B. and Kholodenko, B.N. (2004). Signaling switches and bistability arising from multisite phosphorylation in protein kinase cascades. *J Cell Biol.* **164**, 353–359 (plus supplemental materials).

Press, W.H., Teukolsky, S.A., Vetterling, W.T. and Flannery, B.P. (2002). *Numerical Recipies in C++: The Art of Scientific Computing*, 2nd Ed. Cambridge Univ. Press, Cambridge, U.K.

Sauro, H.M. and Kholodenko, B.N. (2004). Quantitative analysis of signaling networks. *Prog. Biophys. & Molecular Biol.* 86, 5–43.

Shvartsman, S.Y. et al. (2001). Autocrine loops with positive feedback enable context-dependent cell signaling. *Am. J. Physiol. Cell Physiol.* **282**, C545–C559.

Stephanopoulos, G.N., Aristidou, A.A. and Nielsen, J. (1998). *Metabolic Engineering Principles and Methodologies,* Academic, San Diego, CA.

Tyson, J.J. (1991). Modeling the cell division cycle: cdc2 and cyclin interactions. *Proc. Nat. Acad. Sci. USA* **88**, 7328–7332.

Tyson, J.J., Chen, K.C. and Novak, B. (2003). Sniffers, buzzers, toggles, and blinkers: Dynamics of regulatory and signaling pathways in the cell. *Curr. Op. Cell Biol.* **15**, 221–231.

Weng, G., Bhalla, U.S. and Iyengar, R. (1999). Complexity in biological signaling systems. *Science* **284**, 92–96.

Wolf, D.M. and Arkin, A.P. (2003). Motifs, modules and games in bacteria. *Curr. Op. Microbiol.* **6**, 125–134.

15 RADIATION IMAGING

Joseph Bronzino, PhD, PE

Chapter Contents

At the conclusion of this chapter, students will be able to:

- Understand the fundamental principles of radioactivity.

■ Understand that ionizing radiation, as generally employed for medical imaging, is either externally produced and detected after it passes through the patient, or introduced into the body, making the patient the source of radiation emissions.

15.1   INTRODUCTION

Throughout history, a persistent goal of medicine has been the development of procedures for determining the basic cause of a patient's distress. As a result, the search for tools capable of "looking into" the human organism with minimal harm to the patient has always been considered important. However, not until the later part of the twentieth century have devices capable of providing images of normal and diseased tissue within a patient's body been available. Today, modern imaging devices, based on fundamental concepts in physical science (e.g., x-ray and nuclear physics, optics, acoustics, etc.) and incorporating the latest innovations in computer technology and data processing techniques, have not only proved extremely useful in patient care, they have revolutionized health care.

Over a century ago (1895), the first Nobel laureate, physicist Wilhelm Conrad Roentgen, described a new type of radiation, x-rays, that ultimately led to the birth of a new medical specialty, radiology, and the medical imaging industry. Initially, these radiographic imaging systems were very rudimentary, primarily providing images of broken bones or contrast-enhanced structures such as the urinary or gastrointestinal systems. However, since the 1970s, advances in imaging techniques and, in particular, the use of the computer have revolutionized the application of radiographic imaging techniques in medical diagnosis. The same excitement that surrounded Roentgen followed Allan Macleod Cormack and Godfrey Newbold Hounsfield. Cormack in the physics department at Tufts University and Hounsfield in the research laboratories of the British company EMI Limited, worked independently of each other, but obviously under the spell of a common dream, and accomplished quite different things. Cormack elegantly demonstrated the mathematical rudiments of image reconstruction in a remarkable paper published in 1963. Less than a decade later, Hounsfield unveiled an incredible engineering achievement: the first commercial instrument capable of obtaining digital axial images with high-contrast resolution for medical purposes. In recognition of the significant advances made possible by the development of computerized tomography, Cormack and Hounsfield shared the award of the Nobel prize in physiology and medicine in 1979. The excitement of their discovery has not yet subsided.

Ionizing radiation (i.e., radiation capable of producing ion pairs) as generally employed for medical imaging is either (1) radiation that is introduced into the body, thereby making the patient the "source" of radiation emissions, or (2) externally produced radiation, which passes through the patient and is detected by radiation-sensitive devices "behind" the patient.

During cellular or organ system function studies, gamma or x-rays are emitted and detected outside the patient, providing physiological (rate of decay or "washout") and anatomical (imaging) information. More than 90% of the diagnostic procedures in nuclear medicine use emission-imaging techniques. On the other hand, externally produced radiation, which passes through the patient, is the basis of operation for x-ray machines and computerized tomography. These devices and the procedures involving them are housed within departments of radiology. The purpose of this chapter is to present the fundamental principles of operation for each of these radiation imaging modalities.

15.2 EMISSION IMAGING SYSTEMS

Nuclear medicine is that branch of medicine that employs emission scanning for the purpose of helping physicians arrive at a proper diagnosis. An outgrowth of the atomic age and ushered in by the advances made in nuclear physics and technology during World War II, nuclear medicine emerged as a powerful and effective approach in detecting and treating specific physiological abnormalities.

The field of nuclear medicine is a classic example of a medical discipline that has embraced and utilized the concepts developed in the physical sciences. Conceived as a "joint venture" between the clinician and the physical scientist, it has evolved into an interdisciplinary field of activity with its own body of knowledge and techniques. In the process, the domain of nuclear medicine has grown to include studies pertaining to:

- The creation and proper utilization of radioactive tracers (or radiopharmaceuticals) that can be safely administered into the body
- The design and application of nuclear instrumentation devices and systems to detect and display the activity of these radioactive elements
- The determination of the relationship between the activity of the radioactive tracer and specific physiological processes

To better understand this radiation imaging modality, it is necessary to discuss radioactivity, its detection, and the instruments available to monitor the activity of radioactive materials.

15.2.1 Basic Concepts

In 1895, when Roentgen announced the discovery of a new type of penetrating radiation which he called x-rays, he opened a new realm of scientific inquiry. X-rays are a form of electromagnetic (EM) energy just like radio waves and light. The main difference between x-rays and light or radio waves, however, is in their frequency or wavelength. Table 15.1 shows the EM radiation spectrum. It will be noted that x-rays typically have a wavelength from 100 nm to 0.01 nm, which is much shorter than radio or light waves.

Spurred on by Roentgen's discovery, the French physicist Henri Becquerel investigated the possibility that known fluorescent or phosphorescent substances produced a type of radiation similar to the x-rays discovered by Roentgen. In 1896, Becquerel announced that certain uranium salts also radiated, that is, emitted penetrating radiations, whether or not they were fluorescent. These results were startling and presented the world with a new and entirely unexpected property of matter.

TABLE 15.1   Electromagnetic Wave Spectrum

Energy (eV)	Frequency (Hz)		Wavelength (m)
4×10^{-11}	10^4		10^4
4×10^{-10}	10^5	AM radio waves	10^3
4×10^{-9}	10^6		10^2
4×10^{-8}	10^7	Short radio waves	10^1
		FM radio waves and TV	
4×10^{-7}	10^8		10^0
4×10^{-6}	10^9		10^{-1}
4×10^{-5}	10^{10}	Microwaves and radar	10^{-2}
4×10^{-4}	10^{11}		10^{-3}
4×10^{-3}	10^{12}	Infrared light	10^{-4}
4×10^{-2}	10^{13}		10^{-5}
4×10^{-1}	10^{14}	Visible light	10^{-6}
4×10^{0}	10^{15}	Ultraviolet light	10^{-7}
4×10^{1}	10^{16}		10^{-8}
4×10^{2}	10^{17}		10^{-9}
4×10^{3}	10^{18}	X–ray	10^{-10}
4×10^{4}	10^{19}		10^{-11}
4×10^{5}	10^{20}		10^{-12}
4×10^{4}	10^{21}	Gamma ray	10^{-13}
4×10^{7}	10^{22}	Cosmic ray	10^{-14}

Urged on by scientific curiosity, Becquerel convinced Marie Curie, one of the most promising young scientists at the École Polytechnique, to investigate exactly what it was in the uranium that caused the radiation he had observed. Madame Curie, who subsequently was to coin the term *radioactivity* to describe the emitting property of radioactive materials, and her husband Pierre, an established physicist by virtue of his studies of piezoelectricity, became engrossed in their search for the mysterious substance. In the course of their studies, the Curies discovered a substance far more radioactive than uranium. They called this substance radium and announced its discovery in 1898.

While the Curies were investigating radioactive substances, there was a tremendous flurry of activity among English scientists who were beginning to identify the constituent components of atoms. Since the turn of the nineteenth century, Dalton's chemical theory of atoms, in which he postulated that all matter is composed of atoms that are indivisible, reigned supreme as the accepted view of the internal composition of matter. However, in 1897, when J.J. Thompson identified the electron as a negatively charged particle having a much smaller mass than the lightest atom, a new concept of the basic elements comprising matter had to be formulated. Our present image of atoms was developed in 1911 by Ernest Rutherford, who demonstrated that the principal mass of an atom was concentrated in a dense, positively charged nucleus surrounded by a cloud of negatively charged electrons. This idea became incorporated into the planetary concept of the atom, a nucleus surrounded by very light orbiting electrons, which was conceived by Niels Bohr in 1913.

With this insight into the elementary structure of atoms, Rutherford was the first to recognize that radioactive emissions involve the spontaneous disintegration of atoms. After the general acceptance of this basic concept, Rutherford was awarded a Nobel prize in chemistry in 1908, and the "mystery" began to unfold. By observing the behavior of these radioactive emissions in a magnetic field, the Curies discovered that there are three distinct types of active radiation emitted from radioactive material. The three, arbitrarily called *alpha, beta,* and *gamma* by Rutherford, are now known to be (1) alpha (α) particles, which are positively charged and identical to the nucleus of the helium atom; (2) beta (β) particles, which are negatively charged electrons; and (3) gamma (γ) rays, which are pure electromagnetic radiation with zero mass and charge.

15.2.2 Elementary Particles

By the 1930s, the research in this field clearly identified three elementary particles: the electron, the proton, and the neutron, usually considered the building blocks of atoms. Consider the arrangement of these particles as shown in Figure 15.1, a common view of the atom. The atom includes a number of particles: (1) one or more electrons, each having a mass of about 9.1×10^{-31} kg and a negative electrical charge of 1.6×10^{-19} coulombs; (2) at least one proton with a mass of 1.6×10^{-27} kg, which is approximately 1800 times that of the electron; and (3) perhaps neutrons, which have the same mass as protons but possess no charge.

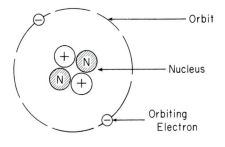

Figure 15.1 Planetary view of atomic structures. The primary mass is the nucleus, which contains protons and neutrons. The nucleus, which has a net positive charge, is surrounded by smaller orbiting electrons. In stable atoms, the net charge of electrons in orbit is equal and opposite to that of the nucleus. The atom illustrated is helium.

Since the mass of an individual nuclear particle is very small and, when expressed in grams, involves the use of unwieldy negative exponents, the system of atomic mass units (amu) has been developed, which uses carbon 12 ($^{12}_{6} \leftarrow$ C) as a reference atom. The arbitrary value of 12 mass units has been assigned to carbon 12. The masses of all other atoms are based on a unit, which is 1/12 the mass of the carbon 12 atom. The mass of the lightest isotope of hydrogen is thus approximately 1 amu.

The mass in grams of an isotope numerically equal to its atomic mass is called a gram-atomic mass. A gram-atomic mass of a substance is also referred to as a mole or equivalent mass of the substance. Since gram-atomic masses have magnitudes proportional to the actual masses of the individual atoms, it follows that one mole of any substance contains a definite number of atoms. The number of atoms in one gram-atomic mass is given by Avogadro's number (N_A) and is equal to 6.023×10^{23}.

The mass of any atom in grams can thus be found by dividing the gram-atomic mass of the isotope by Avogadro's number. For example, the mass of an atom of carbon-12 is its gram-atomic mass (12 grams) divided by Avogadro's number (N_A):

$$\text{mass of} {}^{12}\text{C} = \frac{12\,\text{gm}}{6.023 \times 10^{23}}$$
$$= 1.99 \times 10^{-23}\,\text{gm}$$

Since ^{12}C is equal to 12 atomic mass units (amu), the mass of 1 amu would be 1/12 of the mass of a carbon-12 atom.

$$1\,\text{amu} = \frac{1.99 \times 10^{-23}\,\text{gm}/{}^{12}\text{C atoms}}{12\,\text{amu}/{}^{12}\text{C atom}}$$
$$1\,\text{amu} = 1.66 \times 10^{-24}\,\text{gm}$$

The electrical charge carried by the electron is a fundamental property of matter, as is the mass of a particle. Since this is the smallest amount of electricity that can exist, it is usually expressed as a negative unit charge (-1). When expressing a unit charge in the metric system of units (meters, kilograms, and seconds), one obtains the value 1.6×10^{-19} coulombs. The charges carried by all other atomic particles are therefore some integral multiple of this value. Consequently, it is impossible for a particle to have, for example, a charge equal to two and one-half times that of the electron.

Neutrons and protons exist together in the nucleus and have been given the collective name of nucleons. The total number of nucleons in the nucleus of an element is called the atomic mass, or mass number, and is represented by the symbol (A), whereas the number of protons alone is referred to as the atomic number (Z). Different atoms having the same A are called isobars whereas those with the same number of neutrons are called isotones.

The various combinations of neutrons and protons that may exist in nature are illustrated by examining the composition of the nuclei of the three types of hydrogen atom (hydrogen, deuterium, and tritium). Although hydrogen has a nucleus consisting of a single proton, the combination of one proton and one neutron exists as a single particle called a demtron and is the nucleus of the atom called heavy hydrogen, or deuterium. Extending this concept further, the combination of two protons and two neutrons forms a stable particle—the alpha particle—which in nature exists as the nucleus of the helium atom. Alpha radiation emitted from radioactive substances consists essentially of a stream of such particles. Tritium, the third atom of hydrogen, on the other hand, has a nucleus consisting of only one proton and two neutrons.

These three types of hydrogen are examples of atoms whose nuclei have the same number of protons [i.e., have the same atomic number (Z)] and at the same time may have a different number of neutrons and, thereby, a different atomic mass (A). Atoms exhibiting this characteristic were given the name *isotope* (from the Greek, meaning "same place"). The term *isotope* has been widely used to refer to any atom, particularly a radioactive one. However, current usage favors the word *nuclide* to refer to a particular combination of neutrons and protons. Thus isotopes are nuclides that have the same atomic number. All elements with an atomic number (Z) greater than 83, and/or atomic mass (A) greater than 209 are radioactive—that is, they decay spontaneously into other elements, and this decay causes the emission of active particles.

To specifically designate each individual atom, certain symbols have been developed. In the process of reading the literature in this field, it is therefore necessary to become somewhat familiar with them. In the United States, it was the practice to place the atomic number as a subscript before and the atomic mass as a superscript after the chemical symbol of the atom ($_{53}I^{131}$). Since the chemical symbol itself also specifies the atomic number, one often omits it and simply writes I^{131}. In Europe, on the other hand, it was customarily written as a superscript prior to the chemical symbol (^{131}I). In an effort to achieve international standards, it was agreed in 1964 that the atomic mass should be placed as a superscript preceding the chemical symbol (^{131}I). When superscripts are not used, a more literal form of designation, such as cobalt 60, is commonly used. Referring to the helium atom, one simply writes ^4He, where 4 is equal to the atomic weight (because of the aggregation of protons and neutrons in the nucleus).

15.2.3 Atomic Structure and Emissions

As already mentioned, Bohr's model suggested the existence of an atomic structure analogous to the planetary system. In this system electrons rotated in discrete orbits or shells around the nucleus, and the orbital diameters were determined by a quantum number *(n)* having integer values. These orbits were then represented by K, L, M, and N, corresponding to an increasing number of *n*, a nomenclature still in use today. This model was further refined in 1925 by Wolfgang Pauli in terms of quantum mechanical principles. Pauli's work on atomic structure explained various observed phenomena, including the estimates for binding energy of the electrons at various orbits of an atom. Pauli further observed that an atom can be defined by four quantum numbers: (1) *n* is the principal quantum number, which is an integer and scalar quantity; (2) *l* is the angular momentum quantum number, a vector quantity which has integral values ranging from 0 to $n - 1$; (3) m_1 is the magnetic quantum number with integral values ranging from -1 to $+1$; and (4) m_s is the spin magnetic quantum number, which has the values of $+1/2$ and $-1/2$. According to the Pauli exclusion principle, no two electrons in an atom can have the same set of quantum numbers.

In an electrically neutral atom, the number of orbital electrons exactly balances the number of positive charges in the nucleus. The chemical properties of an atom are determined by the orbital electrons since they are predominantly responsible for molecular bonding, light spectra, fluorescence, and phosphorescence. Electrons in the inner shells, on the other hand, are more tightly bound and may be removed from their orbits only by considerable energy such as by radiation interaction.

The amount of energy required to eject an electron from an orbit is equivalent to the binding energy for that shell, which is highest for the electrons at the innermost shell. The energy required to move an electron from an inner shell to an outer shell is equal to the difference in binding energies between the two shells. This energy requirement represents one of the natural characteristics of an element. When this characteristic energy is released as a photon, in the case of transition of an electron from an outer shell to an inner shell, it is known as a characteristic x-ray. However, if instead of the emission of a photon, the energy is transferred to another orbital electron, called an Auger electron, it will be ejected from orbit. The probability for the yield of characteristic x-ray in such a transition is known as fluorescent yield.

Protons and neutrons (also known as nucleons) experience a short-range nuclear force that is far greater than the electromagnetic force of repulsion between the protons. The movement of nucleons is often described by a shell model, analogous to orbital electrons. However, only a limited number of motions are allowed, and they are defined by a set of nuclear quantum numbers. The most stable arrangement is known as the ground state. The other two broad arrangements are: (1) the metastable state, when the nucleus is unstable, but has a relatively long lifetime before transforming into another state; and (2) the excited state, when the nucleus is so unstable that it has only a transient existence before transforming into another state. Thus, an atomic nucleus may have separate existence at two energy levels, known as isomers (both have the same Z and the same A). An unstable nucleus ultimately transforms itself to a more stable condition, either by absorbing or releasing energy (photons or particles) to a nucleus at ground state. This process is known as radioactive transformation or decay. As stated previously, naturally occuring heavier elements, having Z greater than 83, are all unstable.

Assessment of nuclear binding energy is important in determining the relative stability of a nuclide. This binding energy represents the minimum amount of energy necessary to overcome the nuclear force required to separate the individual nucleons. This can be assessed on the basis of mass–energy equivalence as represented by $E = mc^2$, where E, m, and c represent energy, mass, and speed of light, respectively. This has led to the common practice of referring to masses in terms of electron-volts (eV). The mass of an atom is always found to be less than the sum of the masses of the individual components (neutrons, protons, and electrons). This apparent loss of mass (Δm), often called mass defect or deficiency, is responsible for the binding energy of the nucleus and is equivalent to some change in energy ($\Delta\, mc^2$). As mentioned previously, the mass of a neutral carbon-12 atom has been accepted as 12.0 atomic mass units (amu). The sum of the masses of the components of ^{12}C, however, is 12.10223 amu. The difference in masses (0.10223 amu) is equivalent to 95.23 mega-electron-volts (MeV) of binding energy for this nucleus, or 7.936 MeV/nucleon (obtained by dividing MeV by $A = 12$) for carbon 12. For nuclei with atomic mass numbers greater than 11, the binding energy per nucleon ranges between 7.4 and 8.8 MeV. One atomic mass unit (1 amu), therefore, is equal to $1.6605655 \times 10^{-24}$ g, or, using the mass–energy relation, is equivalent to 931.502 MeV. The resting mass of an electron, on the other hand, is very small (i.e., only 0.511 MeV).

Example Problem 15.1

Verify that the energy released by 1 amu is 931 MeV.

Solution

The conversion can be expressed by Einstein's equation

$$E = mc^2$$

where m is the mass in grams, and c is the velocity of light in cm/sec. Then, the energy equivalent (E) of one atomic mass unit is given by

$$E = (1.66 \times 10^{-24} \, \text{gm})(3 \times 10^{10} \, \text{cm/sec})^2$$

$$E = 1.49 \times 10^{-3} \, \text{gm-cm}^2/\text{sec}^2$$

The unit, $\text{gm-cm}^2/\text{sec}^2$, is frequently encountered in physics and is termed the *erg*, thus:

$$E = 1.49 \times 10^{-3} \, \text{ergs}$$

Another convenient unit of energy is the electron volt. An electron volt is defined as the amount of kinetic energy acquired by an electron when it is accelerated in an electric field produced by a potential difference of one volt. Since the work done by a difference of potential V acting on a charge e is Ve, and the charge on one electron is 1.6×10^{-19} coulombs, it is possible to calculate the amount of energy in one electron volt as follows:

$$1 \text{eV} = 1.6 \times 10^{-19} \text{coulomb} \times 1 \text{ volt}$$

$$1 \text{eV} = 1.6 \times 10^{-19} \text{joules}$$

or

$$1 \text{eV} = 1.6 \times 10^{-12} \text{ergs}$$

Since the electron volt is a very small amount of energy, it is more commonly expressed in thousands of electron volts (keV) or millions of electron volts (MeV).

$$1 \text{ MeV} = 1,000,000 \text{ eV}$$

$$1 \text{ keV} = 1000 \text{ eV}$$

It is now possible to express the atomic mass unit in MeV.

$$1 \, \text{amu} = 1.49 \times 10^{-3} \, \text{ergs, and}$$

$$1 \, \text{MeV} = 1.60 \times 10^{-6} \, \text{ergs; therefore,}$$

$$1 \, \text{amu} = 931 \, \text{MeV}$$

Therefore, if one amu could be completely transformed into energy, 931 MeV would result. To gain a concept of the magnitude of the electron volt, one million electron volts (MeV) is enough to lift only a milligram weight one millionth of a centimeter. ■

An unstable nuclide, commonly known as a radionuclide, eventually comes to a stable condition with the emission of ionizing radiation after a specific probability of life expectancy. In general, there are two classifications of radionuclides: natural and artificial. Naturally occurring radionuclides are those nuclides that emit radiation spontaneously and therefore require no additional energy from external sources. Artificial radionuclides, on the other hand, are essentially man-made and are produced by bombarding so-called stable nuclides with high-energy particles. Both types of radionuclides play an important role in emission scanning and nuclear medicine. The average life or half-life, the mode of transformation or decay, and the nature of emission (type and energy of the ionizing radiation) constitute the basic characteristics of a radionuclide.

All radioactive materials, whether they occur in nature or are artificially produced, decay by the same types of processes (i.e., they emit alpha, beta, and/or gamma radiations). As just discussed, the emission of alpha and beta particles involves the disintegration of one element, often called the parent, into another, the daughter. The modes or phases of transformation in the process of a radionuclide passage from an unstable to stable condition may be divided into six different categories: (1) α (alpha) decay/emission, (2) β- (negatron) decay, (3) β+ (positron) decay, (4) electron capture (EC), (5) isomeric transition (IT), and (6) fission.

Alpha (α) Decay

Alpha particles are ionized helium atoms ($_2^4$He) moving at a high velocity. If a nucleus emits an alpha particle, it loses two protons and two neutrons, thereby reducing Z by 2 and A by 4. A number of naturally occurring heavy elements undergo such decay. For example, a parent nucleus $_{92}^{238}$U emits an alpha particle, thereby changing to a daughter nucleus $_{90}^{234}$Th (Thorium). In symbolic form this process may be written as follows:

$$_{92}^{238}\text{U} \rightarrow _{90}^{234}\text{Th} + _2^4\text{He}$$

Note the following facts about this reaction: (1) the atomic number (number of protons) on the left is the same as on the right ($92 = 90 + 2$) because charge must be conserved, and (2) the mass number (protons plus neutrons) on the left is the same as on the right ($238 = 234 + 4$).

When one element changes into another, as in alpha decay, the process is called transmutation. For alpha emission to occur, the mass of the parent must be greater than the combined mass of the daughter and the alpha particle. In the decay process, this excess mass is converted into energy and appears in the form of kinetic energy in the daughter nucleus and the alpha particle. Most of the kinetic energy is carried away by the alpha particle because it is much less massive than the daughter nucleus. That is, because momentum must be conserved in the decay process, the lighter alpha particle recoils with a much higher velocity than the daughter nucleus. Generally, light particles carry off most of the energy in nuclear decays.

Example Problem 15.2

Radium, $^{226}_{88}$ Ra, decays by alpha emission. What is the daughter element formed?

Solution

The decay can be written symbolically as

$$^{226}_{88} \text{Ra} \rightarrow \text{X} + ^4_2 \text{He}$$

where X is the unknown daughter element. Requiring that the mass numbers and atomic numbers balance on the two sides of the arrow, we find that the daughter nucleus must have a mass number of 222 and an atomic number of 86:

$$^{226}_{88} \text{Ra} \rightarrow ^{222}_{86} \text{X} + ^4_2 \text{He}$$

The periodic table shows that the nucleus with an atomic number of 86 is radon, Rn. ∎

Example Problem 15.3

In Example Problem 15.2, we showed that the $^{226}_{88}$ Ra nucleus undergoes alpha decay to $^{222}_{86}$ Rn. Calculate the amount of energy liberated in this decay. Take the mass of $^{226}_{88}$ Ra to be 226.025406 amu, that of $^{222}_{86}$ Rn to be 222.017574 amu, and that of $^4_2$ He to be 4.002603 amu.

Solution

After decay, the mass of the daughter, m_d, plus the mass of the alpha particle, m_α, is

$$m_d + m_\alpha = 222.017574 \text{ amu} + 4.002603 \text{ amu} = 222.020177 \text{ amu}$$

Thus, calling the mass of the parent nucleus M_p, we find that the mass lost during decay is

$$\Delta m = M_p - (m_d + m_\alpha) = 226.025406 \text{ amu} - 226.020177 \text{ amu} = 0.005229 \text{ amu}$$

Using the relationship 1 amu = 931.5 MeV, we find that the energy liberated is

$$E = (0.005229 \text{ amu})(931.50 \text{ MeV/amu}) = 4.87 \text{ MeV}$$ ∎

Negatron ($\beta-$) or ($\beta-$, γ) Decay

When a radioactive nucleus undergoes beta decay, the daughter nucleus contains the same number of nucleons as the parent nucleus but the atomic number (Z) is increased by 1. A typical beta decay event is

$$\,^{14}_{6}\text{C} \rightarrow \,^{14}_{7}\text{N} + \,^{0}_{-1}\text{e}$$

The superscripts and subscripts on the carbon and nitrogen nuclei follow our usual conventions, but those on the electron may need some explanation. The -1 indicates that the electron has a charge whose magnitude is equal to that of the proton but negative. The 0 used for the electron's mass number indicates that the mass of the electron is almost zero relative to that of carbon and nitrogen nuclei.

The emission of electrons from a nucleus is surprising because the nucleus is usually thought to be composed of protons and neutrons only. This apparent discrepancy can be explained by noting that the electron that is emitted is created in the nucleus by a process in which a neutron is transformed into a proton. This can be represented by the following equation:

$$\,^{1}_{0}\text{n} \rightarrow \,^{1}_{1}\text{p} + \,^{0}_{-1}\text{e}$$

Example Problem 15.4

Find the energy liberated in the beta decay of $\,^{14}_{6}\text{C}$ to $\,^{14}_{7}\text{N}$.

Solution

$\,^{14}_{6}\text{C}$ has a mass of 14.003242 amu and $\,^{14}_{7}\text{N}$ has a mass of 14.003074 amu. Here, the mass difference between the initial and final states is

$$\Delta m = 14.003242 \text{ amu} - 14.003074 \text{ amu} = 0.000168 \text{ amu}$$

This corresponds to an energy release of

$$E = (0.000168 \text{ amu})(931.50 \text{ MeV/amu}) = 0.156 \text{ MeV} \quad \blacksquare$$

Only a small number of electrons have this kinetic energy. Most of the emitted electrons have kinetic energies less than this predicted value. If the daughter nucleus and the electron are not carrying away this liberated energy, then the requirement that energy is conserved leads one to ask the question, "What accounts for the missing energy?"

In 1930, Pauli proposed that a third particle must be present to carry away the "missing" energy and to conserve momentum. Enrico Remi later named this particle the neutrino (little neutral one) because it had to be electrically neutral and have little or no resting mass. Although it eluded detection for many years, the neutrino (symbol v) was finally detected experimentally in 1950. The neutrino has the following properties:

1. It has zero electric charge.
2. It has a resting mass smaller than that of the electron.
3. It interacts very weakly with matter and is therefore very difficult to detect.

Phosphorus 32 is a typical example of a pure β− emitter (^{32}P is transformed to ^{32}S) that has been used for therapy.

Very often a nucleus that undergoes radioactive decay is left in an excited energy state. The nucleus can then undergo a second decay to an even lower energy state by emitting one or more photons. The process is very similar to the emission of light by an atom. An atom emits radiation to release some extra energy when an electron "jumps" from a state of high energy to a state of lower energy. Likewise, the nucleus uses essentially the same method to release any extra energy it may have following a decay or some other nuclear event. In nuclear de-excitation, the "jumps" that release energy are made by protons or neutrons in the nucleus as they move from a higher energy level to a lower level. The photons emitted in such a de-excitation process are called gamma rays and have very high energy relative to the energy of visible light. Most of the radionuclides undergoing β− decay also emit γ rays, almost simultaneously. For example, iodine 131 emits several β− and γ rays in this process.

The following sequence of events represents a typical situation in which gamma decay occurs:

$$^{12}_{5}B \rightarrow ^{12}_{6}C^* + ^{0}_{-1}e$$

$$^{12}_{6}C^* \rightarrow ^{12}_{6}C + \gamma$$

The first process represents a beta decay in which ^{12}B decays to ^{12}C*, where the asterisk is used to indicate that the carbon nucleus is left in an excited state. The excited carbon nucleus then decays to a ground state by emitting an x-ray. Note that an x-ray emission does not result in any change in Z or A.

Positron (β+) or (β+, γ) Decay

With the introduction of the neutrino, the beta decay process in its correct form can be written

$$^{14}_{6}C \rightarrow ^{14}_{7}N + ^{0}_{-1}e + \text{-}v$$

where the bar in the symbol -v indicates that this is an antineutrino. To explain what an antineutrino is, consider the following decay process:

$$^{12}_{7}N \rightarrow ^{12}_{6}C + ^{0}_{+1}e + v$$

When ^{12}N decays into ^{12}C, a particle is produced that is identical to the electron except that is has a positive charge of +e. This particle is called a positron. Because it is like the electron in all respects except charge, the positron is said to be antiparticle to the electron.

It should be noted that positrons (β+) are created in the nucleus, as if a proton was converted to a neutron and a positron. Positron decay produces a different element by decreasing Z by 1, with A being the same. For example, carbon 11 decays to the predominant stable isotope of boron (i.e., ^{11}C transforms to ^{11}B). Positrons (like negatrons) share their energy with neutrinos. Positron decay is also associated with γ-ray emission. The positron, once emitted, however, is often annihilated as a result of a collision with an electron within one nanosecond. This produces a pair of photons of 0.511 MeV that move in opposite directions. A minimum transition energy of 1.022 MeV is required for any positron decay.

Example Problem 15.5

^{12}N beta decays to an excited state of ^{12}C which subsequently decays to the ground state with an emission of 4.43 MeV gamma ray. What is the maximum energy of the emitted beta particle?

Solution

The decay process for positive beta emission is

$$^{12}_{7}\text{N} \rightarrow ^{12}_{6}\text{C}^* + ^{0}_{+1}\text{e} + \text{v} \rightarrow ^{12}_{6}\text{C} + \gamma$$

Conservation of energy tells us that a nucleus X will decay into a lighter nucleus X^1 with an emission of one or more particles (collectively designated as x) only if the mass of X is greater than the total mass $X^1 + x$. The excess mass energy is known as the Q value of the decay. In this process

$$Q = [m(^{12}_{7}\text{N}) - m(^{12}_{6}\text{C}^*) - 2m_e]c^2$$

To determine Q for this decay, we first need to find the mass of the product nucleus ^{12}C in its excited state. In the ground state ^{12}C has a mass of 12.000000 amu so its mass in the excited state is

$$12.000000 \text{ amu} + \frac{4.43 \text{ MeV}}{931.5 \text{ MeV/amu}} = 12.004756 \text{ amu}$$

$$\therefore \ Q = [12.018613 - 12.004756 \text{ amu} - 2 \times 0.000549 \text{ amu}]$$

$$931.5 \text{ MeV/amu} = 11.89 \text{ MeV} \qquad \blacksquare$$

Electron Capture or K Capture

An orbital electron, usually from the inner shell (K shell), may be captured by the nucleus (as if a proton captured an electron and converted itself to a neutron). An electron capture process produces a different element by decreasing Z by 1, with A being the same (similar to β+ decay). Indeed, some radionuclides have definite probabilities of undergoing either positron decay or electron capture (such as iron 52, which decays with about 42% EC and 58% β+ emission). Further, there could be an associated gamma emission with electron capture. For example, ^{51}Cr (chromium 51) transforms to ^{51}V (vanadium 51) with about 90% going directly to the ground state. The remaining 10% goes to an excited state of ^{51}V followed by transition to ground state with the emission of photons. An electron capture also causes a vacancy in the inner shell, which leads to the emission of a characteristic x-ray or Auger electron. Absence of high-energy electrons (beta particles) in EC causes low radiation absorbed close to the tissue.

Isomeric Transition and Internal Conversion

Radionuclides at a metastable state emit only γ rays. The element remains the same with no change in A (isomeric transition). The atomic mass number of the isomer is, therefore, denoted by Am. For example, ^{99}mTc (technetium 99m) decays to ^{99}Tc. However, there is a definite probability that instead of a photon coming out, the energy may be transferred to an inner orbital electron. This is known as internal conversion, and the internally converted electrons are close to monoenergetic beta particles. For example, barium 135m (decaying by IT) emits about 84% IC electrons. They also create a vacancy in the shell, consequently leading to the emission of characteristic x-rays and Auger electrons.

Nuclear Fission

Usually, a heavy nuclide may break up into two nuclides (more or less equal fragments). This may happen spontaneously, but is more likely with the capture of a neutron. The uranium fission products mostly range between atomic numbers 42 and 56. A number of medically useful radionuclides are produced as fission products, such as Xenon 133, which may be extracted by appropriate radiochemical procedures.

15.2.4 Radioactive Decay

The reduction of the number of atoms through disintegration of their nuclei is known as radioactive decay and is characteristic of all radioactive materials. Unaffected by changes in temperature, pressure, or chemical combination, the rate of the decay process remains constant with the same number of disintegrations occurring during each interval of time. Furthermore, this decay process is a random event. Consequently, every atom in a radioactive element has the same probability of disintegrating.

As the decay process continues, it is clear that fewer atoms will be available to disintegrate. This fraction of the remaining number of atoms that decay per unit of time is called the decay constant (λ). The half-life and decay constant are obviously related since the larger the value of the decay constant (λ), the faster the process of decay and consequently the shorter the half-life. In any event, the decay constant is an unchanging value throughout the decay process. Each radionuclide exhibits a distinctive disintegration process because of its inherent properties—that is, its decay constant (λ) and half-life ($T_{1/2}$). All nuclides decay in the same manner, but not at the same rate, since this is a parameter determined by the unique nature of the particular radioactive element in question.

Considering a number of atoms (N) of a specific type of radionuclide present at a time t, the transformation rate can be defined by $- dN/dt$ (the minus sign denotes the decay/decrease), which would be proportional to the number of atoms, or

$$dN/dt = \lambda N \tag{15.1}$$

where λ is the transformation constant.

Taking the initial number of atoms as $N_0 (N = N_0$ when $t = t_0$), we get by integration

$$N = N_0 e^{-\lambda_t} \tag{15.2}$$

where N is the number of radioactive nuclei present at time t, N_0 is the number present at time $t = 0$, and $e = 2.718\ldots$ is the base of the natural logarithm (see Figure 15.2).

The unit of activity is the curie (Ci), defined as

$$1 \text{ Ci} \equiv 3.7 \times 10^{10} \text{ decays/s}$$

This number of decay events per second was selected as the original activity unit because it is the approximate activity of 1 g of radium. The SI unit of activity is the becquerel (Bq):

$$1 \text{ Bq} = 1 \text{ decay/s}$$

Therefore, 1 Ci = 3.7×10^{10} Bq. The most commonly used units of activity are the millicurie (10^{-3} Ci) and the microcurie (10^{-6} Ci).

One of the most common terms encountered in any discussion of radioactive materials is half-life, $T_{1/2}$. This is because all radioactive substances follow the same general decay pattern. After a certain interval of time, half of the original number of

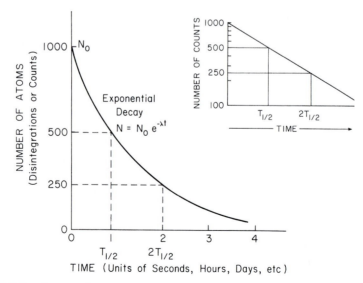

Figure 15.2   The rate of decay of the radioactive material is exponential. The natural logarithm (see insert) of this decay process is therefore a straight line.

nuclei in a sample will have decayed, then in a second time interval equal to the first, half of those nuclei remaining will have decayed, and so on.

The half-life is the time required for half of a given number of radioactive nuclei to decay. The half-life, or $T_{1/2}$ (the time corresponding to transformation of 50% of the nuclides, when $N = N_0/2$), therefore may be obtained by solving for $\lambda T_{1/2}$ in the equation (15.2). Therefore

$$N = N_0 e^{-\lambda_t}.$$

and

$$T_{1/2} = \ln 2/\lambda = 0.693/\lambda$$

Example Problem 15.6

Derive the equation for $T_{1/2}$.

Solution

Starting with $N = N_0 e^{-\lambda t}$. After a time interval equal to one half-life, $t = T_{1/2}$, the number of radioactive nuclei remaining is $N = N_0/2$. Therefore,

$$N_0/2 = N_0 e^{-\lambda T_{1/2}}$$

Dividing both sides by N_0 gives

$$1/2 = e^{-\lambda T_{1/2}}$$

Taking the natural logarithm of both sides of this equation eliminates the exponential factor on the right, since $\ln e = 1$:

$$\ln 1/2 = -\lambda T_{1/2}$$

Since $\ln 1/2 = -0.693$ (use your calculator to check this), we have

$$-0.693 = -\lambda T_{1/2}$$

$$T_{1/2} = \frac{0.693}{\lambda}$$

∎

Let us assume that the number of radioactive nuclei present at $t = 0$ is N_0. The number left after one half-life, $T_{1/2}$, passes is $N_0/2$. After the second half-life, the number remaining is again reduced by one half. Hence, after a time $2T_{1/2}$, the number remaining is $N_0/4$, and so forth. Half-lives range from about 10^{-22} s to 10^{21} years.

Example Problem 15.7

The half-life of the radioactive nucleus ^{286}Ra is 1.6×10^3 years. If a sample contains 3×10^{16} of such nuclei, determine the activity.

Solution

First, calculate the decay constant, λ, using the fact that

$$T_{1/2} = 1.6 \times 10^3 \text{ years} = (1.6 \times 10^3 \text{ years})(3.15 \times 10^7 \text{ s/year}) = 5.0 \times 10^{10} \text{ s}$$

Therefore,

$$\lambda = \frac{0.693}{T_{1/2}} = \frac{0.693}{5.0 \times 10^{10}\text{s}} = 1.4 \times 10^{-11}\text{s}^{-1}$$

The activity, or decay rate, of the sample at $t = 0$ is calculated using the form -$\underline{d}N_0 = dt$

R_0 is the decay rate at $t = 0$ and N_0 is the number of radioactive nuclei present at $t = 0$. Since $N_0 = 3 \times 10^{16}$, we have

$$R_0 = \lambda N_0 = (1.4 \times 10^{-11} \text{ s}^{-1})(3 \times 10^{16}) = 4.2 \times 10^5 \text{ decays/s}$$

Since $1\text{Ci} = 3.7 \times 10^{10}$ decays/s, the activity, or decay rate is

$$R_0 = 11.3 \ \mu\text{Ci} \qquad\qquad\qquad ■$$

Progress in emission scanning and nuclear medicine has been linked to the availability of radionuclides that could be used in human subjects in appropriate chemical forms. The choice of a radionuclide depends on its physical characteristics in relation to diagnostic and therapeutic applications and the possibility of incorporating it into an appropriate chemical compound suitable for biomedical investigation.

A radiopharmaceutical contains a specific chemical compound labeled with a radionuclide. This may be a simple inorganic salt or a complex organic molecule. It is well recognized that the chemical properties of an element reflect some of the possible biological behaviors, and the behavior of several elements within the same group of the periodic table appears similar. For example, strontium 85 has been used to represent calcium metabolism in bone. Table 15.2 summarizes some of the important radionuclides that are currently used in nuclear medicine.

15.2.5 Measurement of Radiation: Units

It should now be apparent that all radioactive substances decay and in the process emit various types of radiation (alpha, beta, and/or gamma) during this process. However, since they usually occur in various combinations, it is difficult to measure each type of radiation separately. As a result, measurement of radioactivity is usually accomplished by using one of the following techniques: (1) counting the number of disintegrations that occur per second in a radioactive material, (2) noting how effective this radiation is in producing atoms (ions) possessing a net positive or negative charge, or (3) measuring the energy absorbed by matter from the radiation penetrating it. Using these fundamental concepts, three kinds of radiation units have been established: the curie (Ci), the roentgen (R), and the radiation-absorbed dose (rad).

Attention should be paid to the fact that the curie defines the number of disintegrations per unit time and not the nature of the radiation, which may be alpha,

TABLE 15.2 Commonly Used Radionuclides in Nuclear Medicine

Radionuclide	Half-life	Transition	Production	Chemical Form
Carbon 11	20.38 min	β^+	Cyclotron	3-N-methylspiperone
Flourine 18	109.77 min	β^+	Cyclotron	Flourodeoxyglucose
Phosphorus 31	14.29 days	β^-	Reactor	Phosphates
Chromium 51	27.704 days	EC	Reactor	Sodium chromate
Cobalt 57	270.9 days	EC	Cyclotron	Cyanocobalamin
Gallium 67	78.26 h	EC	Cyclotron	Citrate complex
Molybdenum 99	66.0 h	β^-	Reactor	Molybdate in column
Technetium 99m	6.02 h	IT	Generator	TcO_4 and complexes
Indium 111	2.83 days	EC	Cyclotron	DTPA and oxine
Iodine 123	13.2 h	EC	Cyclotron	Mainly iodide
Iodine 123	60.14 days	EC	Reactor	Diverse proteins
Iodine 131	8.04 days	β^-	Reactor	Diverse compounds
Xenon 133	5.245 days	β^-	Reactor	Gas
Thallium 201	3.044 days	EC	Cyclotron	Thallous chloride

beta, or gamma rays. The curie is simply a measure of the activity of a radioactive source.

The roentgen and the rad, on the other hand, are units based on the effect of the radiation on an irradiated object. Thus, while the curie defines a source, the roentgen and rad define the effect of the source on an object. One of the major effects of x-ray or gamma radiation is the ionization of atoms, that is, the creation of atoms possessing a net positive or net negative charge (ion pair). The roentgen is determined by observing the total number of ion pairs produced by x-ray or gamma radiation in 1 cc of air at standard conditions (at 760 mm Hg and 0°C). Since each ion pair has an electrical charge, this can be related to electrical effects that can be detected by various instruments. Thus, one roentgen is defined as that amount of x-ray or gamma radiation that produces enough ion pairs to establish an electrical charge separation of 2.58×10^{-4} coulombs per kilogram of air. Thus the roentgen is a measure of radiation quantity, not intensity. The rad on the other hand, is based on the total energy absorbed by the irradiated material. One rad means that 0.01 joule (the unit of energy in the metric system) of energy is absorbed per kilogram of material.

Since human tissue is exposed to various types of radioactive materials in nuclear medicine, another unit of measure, the rem (roentgen equivalent man), is often used to specify the biological effect of radiation. Consequently, the rem is a unit of human biological dose resulting from exposure of the biological preparation to one or many types of ionizing radiation.

These radiation units are used in various situations in nuclear medicine. The roentgen, or more commonly its submultiple the milliroentgen (mr), is used as a value for most survey meter readings. The rad is used as a unit to describe the amount of exposure received, for example, by an organ of interest on injection of a radiopharmaceutical. The rem is the unit used to express exposure values of some personnel-monitoring devices such as film badges.

The classical method of dosimetry, however, has been replaced by a modern method of radiation dose calculation through the effort of the Medical Internal Radiation Dose (MIRD) Committee. The internationally accepted unit for radiation dose to tissue is the gray (Gy); 1 Gy is equivalent to the absorption of energy of one joule per kilogram of the tissue under consideration. Previously the radiation-absorbed dose was expressed as radiation-absorbed dose (rad); 1 rad is equivalent to the absorption of energy of 100 ergs per gram or 0.01 J per kilogram, which is equal to 0.01 Gy. Thus, an expression of 1 rad per millicurie (mCi) is equivalent to 0.27027 milligray per megabecquerel (mGy/MBq), or 1 mGy per megabecquerel is equivalent to 3.7 rad per millicurie. Most of the computations on radiation dosimetry that are available for nuclear medicine procedures, however, have been expressed in terms of rads.

Exposure of living organisms—mammals in particular—to a high level of radiation induces pathological conditions and even death. This feature is utilized in therapy for malignant diseases where the intention is to deliver a localized high radiation dose to destroy the undesirable tissue. Exposure to lower levels of radiation may not show any apparent effect, but it increases the risk of cancer as a long-term effect. It also increases the probability of genetic defects if the gonads are exposed. This imposes the need for minimization of radiation exposure to the worker, and specific guidelines for medical applications.

The United States Nuclear Regulatory Commission (NRC) has adopted standards that limit maximum exposure for the general public to 0.5 sievert (Sv) (Rem) per year. Limits for occupational exposure are 1.25 Sv/3 mo for the whole body and 18.75 Sv/3 mo for the extremities. Thus, one of the most important radiation safety procedures is to monitor both the personnel and the work area. Routine personnel monitoring is usually done with film badges and ring-type finger badges. Radiation survey and wipe tests are carried out at certain intervals and at times of incidental/accidental contamination. The common shielding material is lead, the thickness depending on the energy of gamma rays. Further, radioactive wastes are stored at assigned areas with appropriate shielding. The procedures used for radioactive waste disposal depend on the nature and half-life of the waste. All require a temporary storage facility, and they are finally disposed of locally according to a prescribed set of procedures.

15.3   INSTRUMENTATION AND IMAGING DEVICES

From the discussion of radiation measurement it should be clear that radiation energy can be measured only indirectly, that is, by measuring some effect caused by the radiation. Included among the various indirect techniques used to measure radioactivity are the following:

- *Photography*: The blackening of film when it is exposed to a specific type of radiation such as x-rays (which are the equivalent of gamma rays and will be discussed in the section on externally produced radiation).
- *Ionization*: The passage of radiation through a volume of gas established in the probe of a gas detector produces ion pairs. The function of this type of detector

depends on the collection of these ion pairs in such a way that they may be counted. This technique has been most effective in measuring alpha radiation and least effective in measuring gamma radiation.

■ *Luminescence*: The emission of light not due to incandescence. Since the flash of light produced by the bombardment of a certain type of material with penetrating-type radiation can be detected and processed, this technique is extremely useful. As a matter of fact, the fluorescent effect produced by ionizing radiation is the basis of the scintillation detector discussed in the next section. This type of indirect detection scheme is excellent for observing the presence of all three types of radiation.

15.3.1 Scintillation Detectors

Since the majority of modern detector systems in nuclear medicine utilize probes based on the scintillation principle, it will be described in detail. Scintillation detectors may be used for all types of radiation, depending on the particular type of scintillator used and its configuration. Regardless of the application, however, the general technique is the same for all scintillation probes. Certain materials, for example, zinc sulfide and sodium iodide, have the property of emitting a flash of light or scintillation when struck by ionizing radiation. The amount of light emitted is, over a wide range, proportional to the energy expended by the particle in this material. When the scintillator material is placed next to the sensitive surface of an electronic device called a photomultiplier, the light from the scintillator is then converted into a series of small electrical pulses whose height is directly proportional to the energy of the incident gamma ray. These electrical pulses can then be amplified and processed in such a way as to provide the operator with information regarding the amount and nature of the radioactivity striking the scintillation detector. Thus scintillators may be used for diagnostic purposes to determine the amount and/or distribution of radionuclides in one or more organs of a patient.

Figure 15.3 illustrates the basic scintillation detection system. It consists of (1) a detector, which usually includes the scintillation crystal, photomultiplier tubes, and preamplifier; (2) signal processing equipment such as the linear amplifier and the single-channel pulse analyzer; and (3) data display units such as the scaler, scanner, and oscilloscope. Once the radioactive event is detected by the crystal and an appropriate pulse is generated by the photomultiplier circuitry, the resulting voltage pulses are still very small. To avoid any serious loss of information caused by distortion from unwanted signals (such as noise) and to provide a strong enough signal to be processed and displayed, the amplifier is used to increase the amplitude of the pulses by a constant factor. This process is called linear amplification.

In such a system, it should be apparent that because of the wide variation in energies of gamma rays striking the scintillation crystal, the linear amplifier receives pulses having a wide variation in pulse height. This is fine if one is interested in detecting all the radiant energy under the probe. If, on the other hand, the operator is interested only in the activity of a specific radionuclide, additional processing is necessary. This is accomplished by the pulse height analyzer. Using this device, the

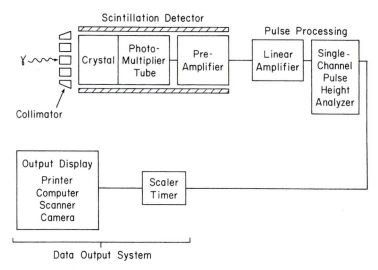

Figure 15.3   Basic scintillation detection system.

operator can discriminate against all radiant energy other than the one of interest. In this way, the activity of a specific radionuclide is analyzed by allowing only those pulses related to it to be processed. Consequently, single-channel and multichannel pulse height analyzers are commonly found in nuclear medicine laboratories. These processed data must then be displayed in such a way to provide information regarding the amount of radioactivity present and its location within the body. This information is quite important in determining the status of the organs under investigation. Studies of the amount of radioactivity present as a function of time enable physicians to ascertain whether the organ is functioning properly or not, whereas studies of the location of the radionuclide enable physicians to display the organ or abnormal tissue. Both types of studies are valuable.

To actually provide a measure of the scintillation events occurring during many nonimaging applications, it is often necessary to count these events and provide some means to display this information. The scaler is the most common type of electronic device found in nuclear medicine that accomplishes this task. A scaler is used to count the pulses produced by the detector system and processed by a variety of electronic pieces of equipment. Thus the scaler is an electronic device that accepts signal pulses representing a range of energy levels (energy of the incident radiation) and counts them. The scaler is usually designed so that the operator has a choice between accepting a certain number of counts (preset count) or a predetermined period of time over which the counts can be accumulated (preset time). With preset time the scaling device will count the number of events that occur during a set period of time and will then shut off automatically—thus the count is the variable. With preset count, the predetermined number of counts are accumulated, after which the scaler is automatically shut off. In this case, time is the variable.

Both types of data can be observed by the operator by viewing the front panel of the scaler itself. However, this information can also be supplied to other devices for analysis. Consider the case of operating the scaler in the present time mode. This information can be presented in a variety of ways. For example, it can be (1) displayed as counts per unit time (second, minute, etc.) for continual observation by the operator, (2) supplied to a digital printer to provide a running tabulation of the counts as a function of time, or (3) directed to a computer that can retain the counts in memory and perform a variety of calculations on the incoming data as they actually are being collected (if the process is slow enough) and at the same time provide the operator with a visual display (on the screen of an oscilloscope) in the form of a plot of radioactive decay. Thus, once the radiation is detected by the scintillation detector and subsequently measured by the scaler, the operator can be presented with information in a variety of formats that may be of clinical value.

Scintillation data output may also be processed through an imaging device. In general, these devices take the pulse output from a detector and the electronic processing devices already described and place the pulse in some representation spatially, according to its point of origin in the radioactive source. Obviously, there is some distortion in this representation, since the point of origin lies in a three-dimensional plane, whereas most common instruments are capable of displaying that point of origin in only two dimensions. However, in spite of this obvious shortcoming, this technique has proved to be of tremendous clinical value.

The most commonly used stationary detector system is the scintillation camera or gamma camera. This device views all parts of the radiation field continuously and is therefore capable of operating almost like a camera, building up an image quickly. Initially developed by Hal Anger during the late 1950s, the first Anger camera reflected the convergence of the disciplines of nuclear physics, electronics, optics, and data processing in a clinical setting. Its ultimate acceptance and further development has had a profound impact, not only on the practice of clinical nuclear medicine, but on the entire diagnostic process. The initial concepts introduced by Anger became basic to the art of imaging specific physiological processes.

15.3.2 The Gamma Camera

The operation of the basic gamma camera is illustrated in Figure 15.4. The detector of the gamma camera is placed over the organ to be scanned. To localize the radiation from a given point in the organ and send it to an equivalent point on the detector, a collimator is placed over the base of the scintillation crystal. Since gamma rays cannot be "bent," another technique must be used to selectively block those gamma rays that, if allowed to continue on their straight-line path, would strike the detector at sites completely unrelated to their points of origin in the subject. This process of selective interference is accomplished by the collimator. To prevent unwanted off-axis gamma rays from striking the crystal, collimators usually contain a large number of narrow parallel apertures made of heavy-metal absorbers.

Consider the multihole collimator illustrated in Figure 15.5a. In this case, the collimator consists of a flat lead plate through which narrow holes are drilled. As can be seen, only a gamma event occurring directly under each hole will penetrate the collimator, and it will be represented at only one location on the face of the crystal. If

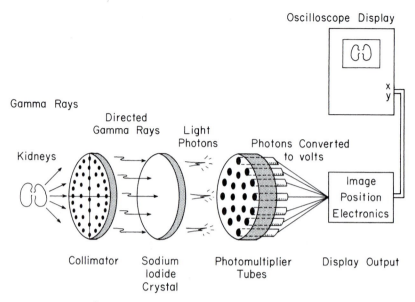

Figure 15.4 Basic elements of a gamma curve.

this gamma event occurred much further away from the collimator, such as in region Y, then it would be represented by more than one location on the face of the crystal. When this occurs, the resolution, which may be defined as the ability of the detector to distinguish between two sources at various distances from the collimator, is greatly decreased. For the multihole collimator, then, the best resolution occurs when the area of interest is close to the collimator. Thus, as the subject is moved from point Z toward the detector, the resolution is improved. Obviously, when viewing an organ that lies beneath the surface of the skin, it becomes quite important to closely approximate its distance from the probe relative to the degree of resolution required.

There are also other types of collimators, such as the pinhole collimator depicted in Figure 15.5b. The pinhole collimator permits entry of only those rays aimed at its aperture. These gamma rays enter the collimator and proceed in a straight line to the crystal where they are detected in inverted spatial correspondence to their source. When the source is located at a distance from the collimator equal to that of the pinhole to the crystal, then the source is represented on the crystal in exactly the same size as it exists. However, by proper positioning of the subject, it is possible to actually magnify or decrease the field of view of the detector. That is, as the source is moved closer to the aperture of the collimator, magnification occurs. A pinhole collimator, therefore, enlarges and inverts the image of the source located beneath it.

From this brief discussion, it can be seen that collimators in conjunction with a scintillation detector essentially "focus" the radioactivity occurring at a particular point in space within the organ to a particular point on the surface of the crystal. The

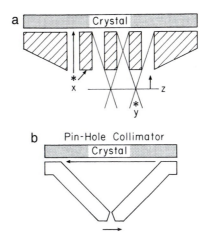

Figure 15.5   (a) Multihole collimator used in conjunction with scintillation detector. Event x is represented at only one location in the crystal, whereas event y has multiple sites associated with its occurrence. (b) The pinhole collimator allows magnification and is used for viewing small organs at short range.

radiation passing through the collimator then impinges on or interacts with the scintillation crystal.

A simple gamma camera system is currently the most important and the most basic instrument for diagnostic nuclear medicine. Interfacing with a digital computer makes the system even more versatile. Digital image-processing systems perform data collection, storage, and analyses. Data acquisition requires digitization of the image by an analog-to-digital converter (ADC), which divides a rectangular image area into small elements, or pixels, usually a 64 by 64, 128 by 128, or 256 by 256 matrix. As a result, one can select a particular region of interest to obtain certain quantitative information. Digital computation also improves dynamic studies since the regional rate of uptake and clearance pattern can easily be obtained from the serial images for any particular region, making it possible to study cardiac wall motion, for example.

15.3.3   Positron Imaging

The positrons emitted through transformation of a radionuclide can travel only a short distance in a tissue (a few millimeters), and then are annihilated. A pair of 511-keV photons that travel at 180° to one another is created. A pair of scintillation detectors can sense positron emission by measuring the two photons in coincidence. Annihilation coincidence detection provides a well-defined cylindrical path between the two detectors. Multihole collimators are unnecessary to define the position in a positron camera because electronic collimation accomplishes the task. Positron cameras have limited use; however, they have attained great importance in a highly modified form in positron-emitting transaxial tomography or positron-emission tomography (PET scanning). A large number of small NaI(TI) detectors is arranged in an annular form so that annihilation photons in coincidence at 180° permit detection of position. Tomographic images are, however, obtained by computer-assisted reconstruction techniques. Several large centers use positron emission tomography in conjunction with cyclotron-produced short-lived radiopharmaceuticals (mainly carbon 11, nitrogen 13, and fluorine 18) to provide structural as well as metabolic information.

15.4   RADIOGRAPHIC IMAGING SYSTEMS

This section discusses externally produced radiation that passes through the patient and is detected by radiation-sensitive devices behind the patient. These radiographic imaging systems rely on the differential attenuation of x-rays to produce an image. Initial systems required sizable amounts of radiation to produce distinct images of tissues with good contrast (that is, the ability to differentiate between body parts having minimally different density, such as fat and muscle). High contrast differences such as between soft tissue and air (as in the lung) or bone and muscle can be differentiated with a much smaller radiation dose. However, with the development of better film and other types of detectors, the radiation dosage has decreased for both high- and low-contrast resolution. Even more significant was the development of the computerized axial tomography (CAT) scanner in the 1970s. This device produces a cross-sectional view of a patient, instead of the traditional shadowgraph recorded by conventional x-ray systems, by using computers to reconstruct the x-ray attenuation data. In the process, it provides the clinician with high-contrast resolved images of virtually any body part and can be reconstructed in any body plane. To better understand the significance of this imaging modality, let us review some of the basic concepts underlying routine x-ray imaging.

15.4.1   Basic Concepts

All x-ray imaging systems consist of an x-ray source, a collimator, and an x-ray detector. Diagnostic medical x-ray systems utilize externally generated x-rays with energies of 20–150 keV. Since the turn of the twentieth century, conventional x-ray images have been obtained in the same way, which is by using a broad-spectrum x-ray beam and photographic film. In general, x-rays are produced by a cathode ray tube that generates a beam of x-rays when excited by a high-voltage power supply. This beam is shaped by a collimator and passes through the patient, creating a latent image in the image plane. Depending on the type of radiographic system employed, this image is detected by x-ray film, an image intensifier, or a set of x-ray detectors. Using the standard film screen technique, x rays pass through the body, projecting an image of bones, organs, air spaces, and foreign bodies onto a sheet of film (see Figure 15.6a–c). The "shadow graph" images obtained in this manner are the results of the variation in the intensity of the transmitted x-ray beam after it has passed through tissues and body fluids of different densities.

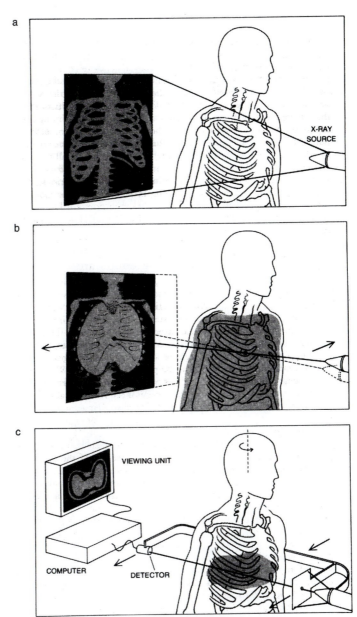

Figure 15.6   Overview of the three x-ray techniques used to obtain "medical images." (a) A conventional x-ray picture is made by having the x-rays diverge from a source, pass through the body, and then fall on a sheet of photographic film. (b) A tomogram is made by having the x-ray source move in one direction during the exposure and the film in the other direction. In the projected image, only one plane in the body remains stationary with respect to the moving film. In the picture, all other planes in the body are blurred. (c) In computerized tomography, a reconstruction from projections is made by mounting the x-ray source and an x-ray detector on a yoke and moving them past the body. The yoke is also rotated through a series of angles around the body. Data recorded by the detector are processed by a special computer algorithm or program. The computer generates a picture on a cathode-ray screen.

This technique has the advantages of offering high-resolution, high-contrast images with relatively small patient exposure and a permanent record of the image. On the other hand, its disadvantages include significant geometric distortion, inability to discern depth information and inability to provide real-time imagery. As a result, conventional radiography is the imaging method of choice for such tasks as dental, chest, and bone imagery. Since bone strongly absorbs x-rays, fractures are readily discernible by the standard radiographic technique. When this procedure is used to project three-dimensional objects into a two-dimensional plane, however, difficulties are encountered. Structures represented on the film overlap, and it becomes difficult to distinguish between tissues that are similar in density. For this reason, conventional x-ray techniques are unable to obtain distinguishable/interpretable images of the brain, which consists primarily of soft tissue. In an effort to overcome this deficiency, attempts have been made to obtain shadow graphs from a number of different angles in which the internal organs appear in different relationships to one another and to introduce a medium (such as air or iodine solutions) that is either translucent or opaque to x-rays. However, these efforts are usually time-consuming, sometimes difficult, sometimes dangerous, and often just not accurate enough.

In the early 1920s, another x-ray technique was developed for visualizing three-dimensional structures. With this technique, known as plane tomography, the imaging of specific planes or cross sections within the body became possible. In plane tomography, the x-ray source is moved in one direction while the photographic film (which is placed on the other side of the body and picks up the x-rays) is simultaneously moved in the other direction (Figure 15.6b). The result of this procedure is that, while the x-rays travel continuously changing paths through the body, each ray passes through the same point on the plane or cross section of interest throughout the exposure. Consequently, structures in the desired plane are brought sharply into focus and are displayed on film, whereas structures in all the other planes are obscured and show up only as a blur. Such an approach is clearly better than conventional methods in revealing the position and details of various structures and in providing three-dimensional information by such a two-dimensional presentation. There are, however, limitations in its use. First, it does not really localize a single plane since there is some error in the depth perception obtained. Second, large contrasts in radiodensity are usually required to obtain high-quality images that are easy to interpret. In addition, x-ray doses for tomography are higher than routine radiographs, and because the exposures are longer, patient motion may degrade the image content.

Computerized tomography (CT) represents a completely different approach. Consisting primarily of a scanning and detection system, a computer, and a display medium, it combines image-reconstruction techniques with x-ray absorption measurements in such a way as to facilitate the display of any internal organ in two-dimensional axial slices or by reconstruction in the Z axis in three dimensions. The starting point is quite similar to that used in conventional radiography. A collimated beam of x-rays is directed through the section of body being scanned to a detector that is located on the other side of the patient (Figure 15.6c). With a narrowly collimated source and detector system, it is possible to send a narrow beam of x-rays to a specific detection site. Some of the energy of the x-rays is absorbed while the remainder continues to the detector and is measured. In computerized tomography, the detector system usually consists of a crystal (such as cesium iodide or cadmium tungstate) that has the ability to scintillate or emit light photons when bombarded with x-rays. The intensity of these light photons or "bundles of energy" is in turn measured by photodetectors and provides a measure of the energy absorbed (or transmitted) by the medium that is penetrated by the x-ray beam.

Since the x-ray source and detector system are usually mounted on a frame or "scanning gantry," they can be moved together across and around the object being visualized. In early designs, for example, x-ray absorption measurements were made and recorded at each rotational position traversed by the source and detector system. The result was the generation of an absorption profile for that angular position. To obtain another absorption profile, the scanning gantry holding the x-ray source and detector was then rotated through a small angle and an additional set of absorption or transmission measurements was recorded. Each x-ray profile or projection obtained in this fashion is basically one-dimensional. It is as wide as the body but only as thick as the cross section.

The exact number of these equally spaced positions determines the dimensions to be represented by the picture elements that constitute the display. For example, to generate a 160 by 160 picture matrix, absorption measurements from 160 equally spaced positions in each translation are required. It will be recalled, however, that each one-dimensional array constitutes one x-ray profile or projection. To obtain the next profile, the scanning unit is rotated a certain number of degrees around the patient and 160 more linear readings are taken at this new position. This process is repeated again and again until the unit has been rotated a full 180°. When all the projections have been collected, 160×180, or 28,800, individual x-ray intensity measurements are available to form a reconstruction of a cross section of the patient's head or body.

At this point, the advantages of the computer become evident. Each of the measurements obtained by the previously described procedure enters the resident computer and is stored in memory. Once all the absorption data have been obtained and located in the computer's memory, the software packages developed to analyze the data by means of image-reconstruction algorithms are called into action. These image-reconstruction techniques, which are based on known mathematical constructs developed for astronomy, were not used routinely until the advent of the computer because of the number of computations required for each reconstruction. Modern computer technology made it possible to fully exploit these reconstruction techniques.

To develop an image from the stored values of x-ray absorption, the computer initially establishes a grid consisting of a number of small squares for the cross section of interest, depending on the size of the desired display. The result of this process is something like the mesh of strings in a tennis racket. Since the cross section of the body has thickness, each of these squares represents a volume of tissue, a rectangular solid whose length is determined by the slice thickness and whose width is determined by the size of the matrix. Such a three-dimensional block of tissue is referred to as a voxel (or volume element) and is on the display in two dimensions as a "pixel" (or picture element) (Fig. 15.7).

During the scanning process, each voxel is irradiated by a narrow beam of x-rays up to 180 times. Thus, the absorption caused by that voxel contributes to up to 180 absorption measurements, each measurement part of a different projection. Since each voxel affects a unique set of absorption measurements to which it has contributed, the computer calculates the total absorption due to that voxel. Using the total absorption and the dimension of the voxel, the average absorption coefficient of the tissues in that voxel is determined precisely and displayed in a corresponding pixel as a shade of grey.

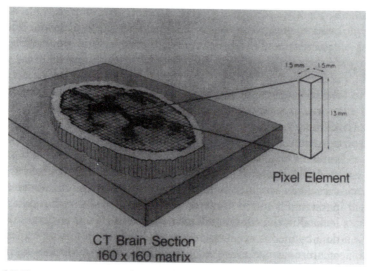

Figure 15.7   Sketch of a cross-sectional image available using computerized tomography. Each image is divided into discrete three-dimensional sections of tissue referred to as voxels (or volume elements). On the computer monitor, the slice is viewed in two dimensions as pixels, or picture elements, in this case 1.5 × 1.5 mm.

Example Problem 15.8

Demonstrate how x-rays are attenuated as they pass through a cross-sectional area of the patient's body in multiple rays.

Solution

The cross section of interest can be considered to be made up of a set of blocks of material (Fig. 15.8). Each block has an attenuating effect upon the passage of the x-ray energy or photons, absorbing some of the incident energy passing through it. In the line of blocks illustrated in Figure 15.8, the first block absorbs a fraction A_1 of the

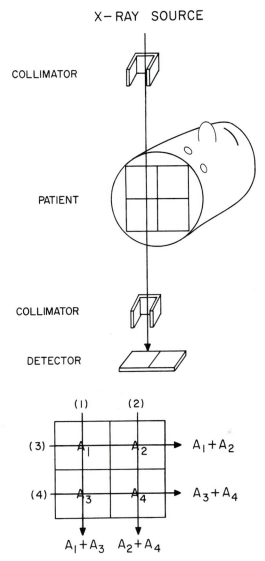

Figure 15.8   Sketch of a manner in which the absorption coefficient values for each picture element are obtained.

incident photons, the second a fraction A_2, and so on, so that the nth block absorbs a fraction A_n. The total fraction "A" absorbed through all the blocks is the product of all the fractions, and the logarithm of this total absorption fraction is defined as the measured absorption. Consider Figure 15.8 once again. The set of absorption measurements for an object that comprises only four blocks would be as follows:

- In position 1: $A(1) = A_1^* \, A_3$ (1)
- In position 2: $A(2) = A_2^* \, A_4$ (2)
- In position 3: $A(3) = A_1^* \, A_2$ (3)
- In position 4: $A(4) = A_3^* \, A_4$ (4)

In practice, only the measured absorption factors $A(1)$, $A(2)$, $A(3)$, and $A(4)$ would be known. The problem is, therefore, to compute A_1, A_2, A_3, and A_4 from the measured absorption values. The fact that the computation is possible can be seen from equations 1 to 4. There are four simultaneous equations and four unknowns, so a solution can be found. To reconstruct a cross section containing n rows of blocks and n columns, it is necessary to make at least n individual absorption measurements from at least n directions. For example, a display consisting of 320×320 picture elements requires a minimum of 320×320, or 102,400, independent absorption measurements.   ■

It was mentioned previously that these absorption measurements are taken in the form of profiles. Imagine a plane parallel to the x-ray beam as defining the required slice. An absorption profile is created if the absorption of the emergent beam along a line perpendicular to the x-ray beam is plotted. This profile represents the total absorption along each of the x-ray beams. In general, the more profiles that are obtained, the better the contrast resolution of the resulting image.

From these individual measurements, a single two-dimensional plane can be reconstructed, and by simply (i.e., for the computer) stacking the appropriate sequence of such planes, it is possible to reconstruct a full three-dimensional picture. Image reconstruction of a three-dimensional object is therefore based primarily on a process of obtaining a cross section or two-dimensional image from many one-dimensional projections. The earliest method used to accomplish this was called backprojection, which means that each of the measurement profiles was projected back over the area from which it was taken. Unfortunately, this rather simple approach was not totally successful because of blurring. To overcome this, iterative methods were introduced that successively modified the profile being backprojected until a satisfactory picture was obtained. Iteration is a good method but slow, requiring several steps to modify the original profiles into a set of profiles that can be projected back to provide an unblurred picture of the original image. To speed up the process, mathematic techniques involving convolution or filtering were introduced that permit the original profile to be modified directly into the final one.

Whichever method of reconstruction is used, the final result of the computation is the same. In each case, a file, usually known as the picture file, is created in the computer memory. The picture file contains an absorption coefficient or density reading for each element of the final picture (e.g., 25,600 for 160×160 pixels or over 100,000 for 320×320 pixels). The resultant absorption coefficients for each element of the image calculated in this manner can then be displayed as gray tones or color scales on a visual display. Most CT systems project an image onto the screen of a computer display terminal. Each element, or pixel, of the picture file has a value that represents the density (or more precisely the relative absorption coefficient) of a volume in the cross section of the body being examined.

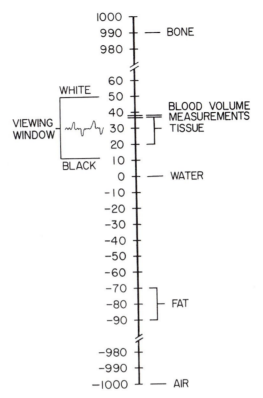

Figure 15.9   Hounsfield scale for absorption coefficients.

The scale developed by Hounsfield that is shown in Figure 15.9 demonstrates the values of absorption coefficients that range from air (−1000) at the bottom of the scale to bone at the top. This original scale covers some 1000 levels of absorption on either side of water, which is indicated as zero at the center. This is chosen for convenience, since the absorption of water is close to that of tissue due to the high percentage of water that is found in tissue. In using this scale, the number 1000 is added to the absorption coefficient of each picture element. In the process, the absorption value of air becomes zero while that of water becomes 1000. Having assigned each pixel to a particular value in this scale, the system is ready to display each element and create the reconstructed image.

Example Problem 15.9

Illustrate the process of backpropagation.

Solution

The process can be illustrated by means of a simple analogue. In the following figure a parallel beam of x-rays is directed past and through a cylinder-absorbing substance. As a result, a shadow of the cylinder is cast on the x-ray film. The density of the exposed and developed film along the line AA can be regarded as a projection of the object. If a series of such radiographs are taken at equally spaced angles around the cylinder, these radiographs then constitute the set of projections from which the cross section has to be reconstructed.

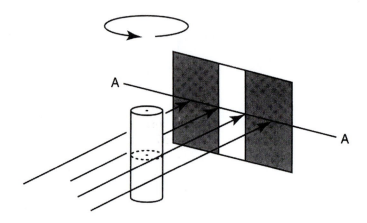

An approximate reconstruction can be reproduced by directing parallel beams of light through all the radiographs in turn from the position in which they were taken as shown in the following diagram.

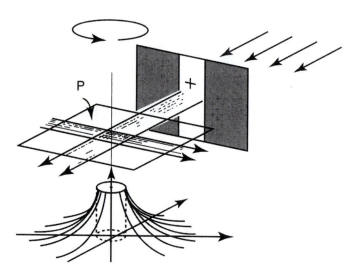

The correct cross section can be reconstructed by backprojecting the original shadow as shown in the preceding diagram and subtracting the result of backprojecting two beams placed on either side of the original shadow as shown in the following. Mathematically, this is the equivalent of taking each transmission value in the projection and subtracting from it a quantity proportional to adjacent values. This process is called convolution and is actually used to modify projections.   ■

In the display provided by any medical imaging device, factors affecting the resolution or accuracy of the image must be recognized and modified. Since the information provided by a CT scanner in essence reflects the distribution of x-ray absorption in a cross section of a patient, the smallest change in x-ray absorption that can be detected and projected in the image defines the density resolution. In addition, the smallest distance apart that two objects can be placed and still be seen as separate entities defines the spatial resolution. The density resolution reflects the sensitivity of the method to tissue change whereas spatial resolution indicates the fineness of detail possible in the image produced by the CT scanner. These two characteristics of resolution are related to each other and to the radiation dose received by the patient. The clarity of the picture, and hence the accuracy with which one can measure absorption values, are often impaired by a low signal-to-noise ratio which is caused by a reduction in the number of photons arriving at the detectors after penetrating the body. Fewer available photons, in turn, result in a reduction in the sharpness of the image since the resultant differences between the absorption coefficients of adjacent pixels are reduced. This is a situation that must be accepted as long as low radiation doses are used. Despite this limitation, the reconstructed image can be enhanced by display units that constitute another major component of the CT system.

Display units usually have the capacity to display a given number (16, 32, etc.) of colors or gray levels at one time. These different colors (or gray levels) can be used to represent a specific range of absorption coefficients. In this way, the entire range of absorption values in the picture can be represented by the entire range of available gray tones or colors (or both). To sharpen the resolution or detail of the display, the range of gray tones between black and white can be restricted to a very small part of the scale. This process of establishing a display "window" that can be raised or lowered actually enhances the imaging of specific organs. For example, if an image of the tissue of the heart is desired, this window is raised above water density, but if an image of the detail in the lung is desired, it is lowered. The overall sensitivity can be further increased by reducing the window width. This permits ever-greater differentiation between absorption coefficients of specific organs and the surrounding tissue.

Various other techniques are available to make the picture more easily interpretable. For instance, a portion of the picture can be enlarged or certain absorption values can be highlighted or made brighter. These methods do not add to the original information, but enhance the image made possible by the data available in the computer's memory. Since a human observer can see an object only if there is sufficient contrast between the object and its surroundings, enhancement modifies the subjective features of an image to increase its impact on the observer, making it easier to locate and precisely measure obscure details.

There are a number of distinct advantages of CT scanning over conventional x-ray techniques. For example, computerized tomography provides three-dimensional information concerning the internal structure of the body by presenting it in the form of a series of slices (see Fig. 15.10). Because of its contrast sensitivity it can show small differences in soft tissue clearly, which conventional radiographs cannot do. It accurately measures the x-ray absorption of various tissues, enabling the nature of these tissues to be studied, and yet the absorbed dose of x-rays given to the patient by CT scanning is often less than the absorbed dose by a conventional x-ray technique such as tomography.

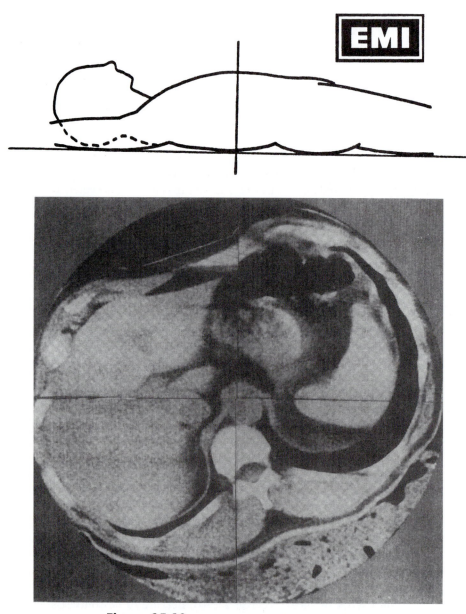

Figure 15.10   Photograph of a CT slice of the body.

In comparing the CT approach with conventional tomography, which also images a slice through the body by blurring the image from the material on either side of the slice, it becomes apparent that in the tomographic approach only a thin plane of the beam produces useful information on the slice to be viewed. The remainder of the beam, which is by far the larger component, passes through material on either side of the slice, collecting unwanted information that produces artifacts in the picture. On the other hand, the x-ray beam in computerized tomography passes along the full length of the plane of the slice, thereby permitting measurements to be taken that are 100% relevant to that slice and that slice alone. These measurements are not affected by the materials lying on either side of the section. As a result, the entire information potential of the x-ray beam is used to the fullest, and the image is more clearly defined.

The medical applications of such detailed pictures are considerable. Hard and soft tumors, cysts, blood clots, injured or dead tissue, and other abnormal morphological conditions can be observed easily. The ability to distinguish between various types of soft tissues means that many diagnostic procedures can be simplified, thus eliminating the need for a number of methods that require hospitalization, are costly, and involve a potential risk to the patient due to their invasive nature. Since computerized tomography can accurately locate the areas of the body to be radiated and can monitor the actual progress of the treatment, it is effective not only in diagnosis but in the field of radiotherapy as well.

15.4.2   CT Technology

Since its introduction, computerized tomography has undergone continuous refinement and development. In the process, CT scanning has developed into a technology of significant importance throughout the world. The manufacture of quality CT systems requires a strong base in x-ray technology, physics, crystallography, electronics, and data processing.

The evolution of CT scanners has resulted from improvements in: (1) x-ray tube performance; (2) scan times; (3) image-reconstruction time; (4) reduction and, in some cases, elimination of image artifacts; and (5) reduction in the amount of radiation exposure to patients. Achievements in these areas have focused attention on two key issues. The first is image quality and its relationship to diagnostic value. The second is scanner efficiency and its relationship to patient care and the economic impact it entails. To attain a better understanding of the operation and utility of present CT systems, it is important to understand some of the evolutionary steps.

CT scanners are usually integrated units consisting of three major elements: (1) the scanning gantry, which takes the readings in a suitable form and quantity for a picture to be reconstructed; (2) the data handling unit, which converts these readings into intelligible picture information, displays this picture information in a visual format, and provides various manipulative aides to enhance the image and thereby assist the physician in forming a diagnosis; and (3) a storage facility, which enables the information to be examined or reexamined at any time after the actual scan.

Scanning Gantry

The objective of the scanning system is to obtain enough information to reconstruct an image of the cross section of interest. All the information obtained must be relevant and accurate, and there must be enough independent readings to reconstruct a picture with sufficient spatial resolution and density discrimination to permit an accurate diagnosis. The operation of the scanning system is therefore extremely important.

CT scanners have undergone several major gantry design changes (see Figure 15.11). The earliest generation of gantries used a system known as "traverse and index." In this system, the tube and detector were mounted on a frame and a single beam of x-rays traversed the slice linearly, providing absorption measurements along one profile. At the end of the traverse, the frame indexed through 1° and the traverse was repeated. This procedure continued until 180 single traverses were made and 180 profiles were measured. Using the "traverse and index" approach, the entire scanning

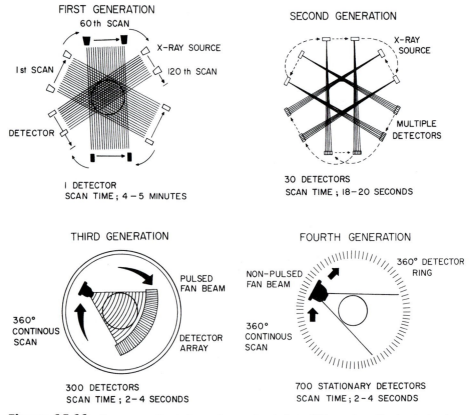

Figure 15.11   Four generations of scanning gantry designs. With modern slip ring technology, third- or fourth-generation geometry allows spiral volumetric scanning using slice widths from 1 to 10 mm and pixel matrixes to 1024^2. Typically, a 50-cm volume can be imaged with a single breath hold.

procedure took approximately 4 to 5 minutes. The images it produced had excellent picture quality and hence high diagnostic utility. However, this approach had several disadvantages. First, it was relatively slow, resulting in relatively low patient output. Second, streak artifacts were common, and although they normally were not diagnostically confusing, the image quality did suffer. Finally, and perhaps the biggest drawback, was that since the patient had to remain absolutely immobile during the scan, its use was restricted solely to brain scans.

In an attempt to solve these problems, a second generation of gantries evolved. The scanning procedures in this approach included a translation step as in the first generation, but incremented the scanning gantry 180° around the patient in 10° steps. In addition, in an effort to obtain more profiles with each traverse, a larger fan beam was used. By using a 10° fan beam, it was possible to take ten profiles—at 1° intervals—with each traverse. By indexing through 10° before taking the next set of profiles, it was possible to obtain a full set of 180 profiles by making only 18 traverses. At the rate of approximately one second for each traverse, scanning systems of this type could operate in the range of 18–20 seconds. Even in this extremely reduced scan time, the scanning and detection system was able to obtain over 300,000 precise absorption measurements during the complete (i.e., over 180°) scan. These values were then used to construct a picture of the slice under investigation. With this new generation of CT scanners, it became possible to obtain cross sections of any part of the body. However, it still was not fast enough to eliminate the streak artifacts, nor did it seem possible to further reduce scan time with this approach. In essence, the first and second generations had reached their limits. Scanning times could not be reduced further and still achieve an acceptable image quality. Nevertheless, this geometry had the potential to achieve the highest spatial and contrast resolutions of all the gantry designs.

A major redesign ensued and many of the scanning system components were improved. In the process, a third generation of gantries evolved that consisted of a pure rotational system. In these systems, the source-detector unit had a pulsing, highly collimated wide-angle (typically, 20–50°) fan beam, and a multiple detector array was rotated 360° around the patient. This single 360° smoothly rotating movement produced scan times as low as 500 ms, increased appreciably the reliability of the data because they were taken twice, and increased the quality of the reconstructed image. A by-product of this design was that the pulsing x-ray source could be synchronized with physiologic parameters, enabling rhythmic structures such as the heart to be more accurately imaged.

The main advantages of this type of system were its simplicity and its speed, but it had two major disadvantages. First, its fixed geometric system, with a fan beam usually established for the largest patient, was inefficient for smaller objects and in particular for head scanning. Second, it was particularly prone to circular artifacts. These "ring" artifacts were most obvious in early pictures from this type of machine. The magnitude of the difficulty in achieving stability can be appreciated when it is realized that an error of only one part in 10,000 can lead to an artifact that is diagnostically confusing.

Consequently, a fourth-generation gantry system consisting of a continuous ring (360°) of fixed detectors (usually numbering between 600 and 4800) was developed. In this system, the x-ray source rotates as before, and the transmitted x-rays are detected by the stationary detector ring. The fan beam is increased slightly so that the detectors on the leading and trailing edge of the fan can be continuously monitored and the data adjusted in case of shifts in detector performance. As a result, the image obtained using this approach is more reliable, and the ring artifact is usually eliminated. Fourth-generation scan times of approximately 500 ms or less with reconstruction times of 30 seconds or less are now commonplace.

The development of high-voltage slip rings has enabled third- and fourth-generation scanners to continuously rotate about the patient. By moving the patient couch at a uniform speed in or out of the scan plan, the x-ray beam will describe a helix or spiral path through the patient. Slice widths of 1–10 mm and pixel matrixes up to 1024^2 may be employed, and typically a 50-cm long volume can be imaged in a single breath hold. Because of improvements in x-ray tube technology, spatial and contrast resolution are maintained. New display techniques allow for real-time, 3-D manipulations and virtual reality fly-through of tubular structures such as the aorta or large intestine.

Another gantry design, which might be called a fifth generation, utilizes a fixed detector array arranged in a 210° arc positioned above the horizontal plane of the patient and four target rings of a 210° arc positioned below the horizontal plane and opposite to the detector array. There is no mechanical motion in the gantry except the movement of the couch into and out of the imaging plane (see Fig. 15.12).

X-rays are produced when a magnetic field causes a high-voltage focused electron beam to sweep along the path of the anode target. Up to four 1.5 mm simultaneous slices are created in 50 or 100 ms and displayed in up to a 512^2 pixel matrix. This ultra-fast CT can volumetrically examine the heart and evaluate the presence and severity of coronary calcific arthrosclerosis. Although the ultra-fast scanner can almost stop biologic motion, its contrast and spatial resolution is presently less than third- or fourth-generation mechanical scanners, therefore its general applications are somewhat less.

X-ray Tubes and Detectors

It must be emphasized that the whole purpose of the scanning procedure is to take thousands of accurate absorption measurements through the body. Taken at all angles through the cross section of interest, these measurements provide an enormous amount of information about the composition of the section of the body being scanned. Since these readings are taken by counting photons, the more photons counted, the better will be the quality of the information. With these factors in mind, it is essential that the radiation dose be sufficient to obtain good quality pictures. Since the total radiation dose is a function of the maximum power of the tube and its operating time, faster scanners have to use a significantly higher-powered tube. These aspects of dose efficiency are very important in tube and detector design.

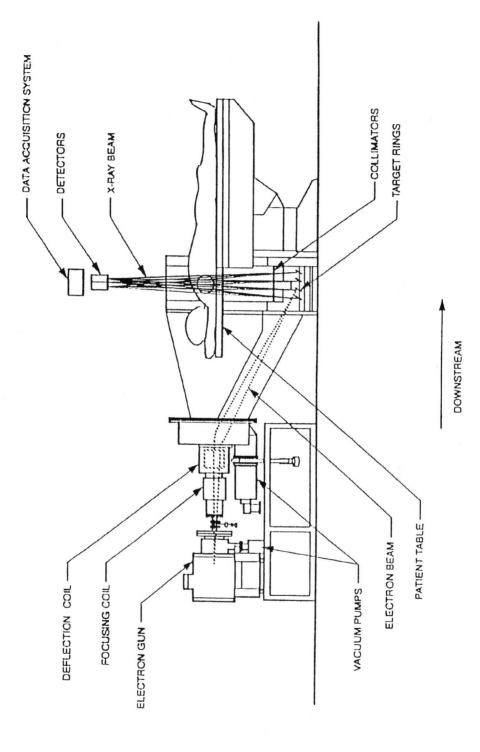

Figure 15.12 Illustration of an ultrafast CT scanner. In this device, magnetic coils focus and steer the electron beam through the evacuated drift tube where they strike the target rings. The presence of four target rings and multiple x-ray collimators allows four unique x-ray beams to pass through the patient. These beams are then interrupted by two contiguous detector arrays and form eight unique image slices (courtesy of Imatrol).

DATA ACQUISITION SYSTEM

DETECTORS

X-RAY BEAM

DEFLECTION COIL

FOCUSING COIL

ELECTRON GUN

COLLIMATORS

TARGET RINGS

DOWNSTREAM

VACUUM PUMPS

ELECTRON BEAM

PATIENT TABLE

Currently, CT systems that operate in the region of 20 s use fixed-anode, oil-cooled tubes, which run continuously during the scan time, whereas those that operate in much less than 20 s have a rotating-anode, air- or water-cooled tube, which is often used in a pulse mode. The fixed-anode tubes, being oil cooled, can be run continuously, and the sequence of measurements is obtained by electronic gating of the detectors. Faster scanners need tubes of substantially higher power, thereby requiring the rotating-anode type. There is a tendency to use these in a pulse mode largely for convenience, because it simplifies the gating of the detectors. One of the problems encountered with fast machines, however, is the difficulty in providing enough dose within the scan time. For high-quality pictures, many of these scanners run at much slower rates than their maximum possible speed.

Of equal importance in the accurate detection and measurement of x-ray absorption is the detector itself. Theoretically, detectors should have the highest possible x-ray photon capture and conversion efficiency to minimize the radiation dose to the patient. Clearly, a photon that is not detected does not contribute to the generation of an image. The detection system of a CT scanner that is able to capture and convert a larger number of photons with very little noise will produce an accurate, high-quality image when aligned with suitable reconstruction algorithms. Consequently, the selection of a detector always represents a trade-off between image quality and low radiation dose.

Three types have been used: (1) scintillation crystals (such as sodium iodide, calcium fluoride, bismuth germinate, or cadmium tungstate) combined with photomultiplier tubes, (2) gas ionization detectors containing Xenon, and (3) scintillation crystals with integral photodiodes. It will be recalled that the principal factors in judging detector quality are how well the detector captures photons and its subsequent conversion efficiency, which results in optimal dose utilization. Those systems employing scintillation crystals combined with photomultiplier tubes do not lend themselves to the dense packing necessary for maximal photon capture, even though they offer a high conversion efficiency. The less-than-50% dose utilization inherent in these detectors requires a higher dose to the patient to offset the resultant inferior quality image. Although gas detectors are relatively inexpensive and compact and lend themselves for use in large arrays, they have a low conversion efficiency and are subject to drift. Consequently, use of Xenon detectors necessitates increasing the irradiation dose to the patient.

The advantage of scintillation crystals with integral photodiodes extends beyond their 98% conversion efficiency since they lend themselves quite readily to a densely packed configuration. The excellent image quality combined with high-dose utilization (i.e., approximately 80% of the applied photon energy) and detector stability ensures that solid-state detectors will continue to gain favor in future CT systems.

Data Handling Systems

In conjunction with the operation of the scanning and detection system, the data obtained must be processed rapidly to permit viewing a scan as quickly as possible. Data handling systems incorporating the latest advances in computer technology have been developed to coincide with changes in the design of the scanning and detector components of CT scanners. For example, with the evolution of the third-generation scanners, the data handling unit had to digitize and store increasing amounts of data consisting of: (1) positional information, such as how far the scanning frame was along its scan; (2) reference information from the reference detector, which monitored the x-ray output; (3) calibration information, which was obtained at the end of each scan; and, (4) the bulk of the readings, the actual absorption information obtained from the detectors. Electronic interfaces are now available that convert this information to digital form and systematically store it in memory of the resident computer. The next step is the actual process of image reconstruction (see Fig. 15.13).

It will be recalled that the first stage of this computation process is to analyze all of the raw data and convert them into a set of profiles, normally 180 or more, and then convert these profiles into information that can be displayed as a picture and used for diagnosis. This is the heart of the operation of a CT scanner, that element that makes CT totally different from conventional x-ray techniques and most other imaging

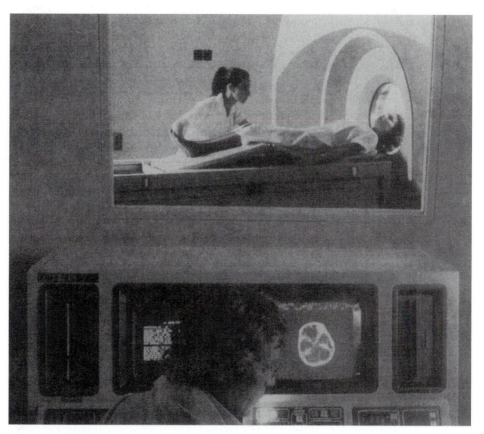

Figure 15.13 Photograph of a modern display terminal for CT scanner.

techniques. The algorithms or computer programs for reconstructing tissue x-ray absorption properties from a series of x-ray absorption profiles fall into one of the following categories: simple backprojection (sometimes called summation), Fourier transforms, integral equations, and series expansions. The choice of an algorithm for a CT scanner depends on the algorithm's speed and accuracy.

Mathematical algorithms for taking the attenuation projection data and reconstructing an image can be classified into two categories: iterative and analytic. The iterative techniques (also known as the algebraic reconstruction technique [ART]), such as the one used by Hounsfield in the first generation scanner, require an initial guess of the two-dimensional pattern of x-ray absorption. The attenuation projection data predicted by this guess is then calculated and the results compared with the measured data. The difference between the measured data and predicted values is used iteratively so that the initial guess is modified and that difference goes to zero. In general, many iterations are required for convergence, with the process usually halted when the difference between the calculated and the measured data is below a specified error limit. Several versions of the ART were developed and used with first- and second-generation CAT scanners. Later-generation scanners used analytic reconstruction techniques since the iterative methods were computationally slow and had convergence problems in the presence of noise.

Analytic techniques include the Fourier transform, backprojection, filtered backprojection, and convolution backprojection approaches. All of the analytic methods differ from the iterative methods in that the image is reconstructed directly from the attenuation projection data. Analytic techniques use the central section theorem and the two-dimensional Fourier transform, which is illustrated with the aid of Figure 15.14. Given an image $f(x, y)$, a single projection is taken along the x direction, forming a projection $g(y)$ described by

$$g(y) = \int_{-\infty}^{\infty} f(x, y) dx$$

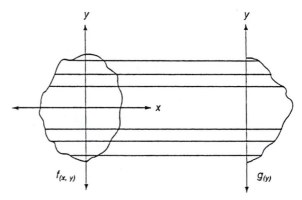

Figure 15.14   Illustration of a projection via the central-sectioning theorem.

This projection represents an array of line integrals as shown in Figure 15.14. The two-dimensional Fourier transform of $f(x, y)$ is given by

$$F(u, v) = \int\int_{-\infty}^{\infty} f(x, y) \exp[-j2\pi(ux + vy)]dxdy$$

In the Fourier domain, along the line $u = 0$, this transform becomes

$$F(0, v) = \int\int_{-\infty}^{\infty} f(x, y) \exp(-j2\pi vy)dxdy$$

which can be rewritten as

$$F(0, v) = \int\int_{-\infty}^{\infty} f(x, y)dx/\exp(-j2\pi vy)dy$$

or

$$F(0, v) = F_1[g(y)]_1$$

where $F_1[]$ represents a one-dimensional Fourier transform. It can be shown that the transform of each projection forms a radial line in $F(u, v)$, and therefore $F(u, v)$ can be determined by taking projections at many angles and taking these transforms. When $F(u, v)$ is completely described, the reconstructed image can be found by taking the inverse Fourier transform to obtain $f(x, y)$. The image reconstruction method used by most modern CAT scanners is the filtered backprojection reconstruction method. In this method, attenuation projection data for a given ray or scan angle are convolved with a spatial filter function either in the Fourier domain or by direct spatial convolution. The filtered data are then backprojected along the same line, using the measured value for each point along the line, as shown in Figure 15.15. The total backprojected image is made by summing the contributions from all scan angles. Depending on the

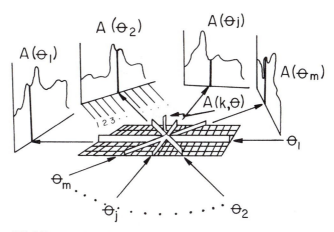

Figure 15.15   Sketch of the results obtained with simple backprojection algorithm.

filter technique, the image is obtained directly after backprojection summation or via the inverse Fourier transform of the backprojected image.

It should be obvious by now that compared to early head and body scanners, current models process information more quickly, use energy more efficiently, are more sensitive to differences in tissue density, have finer spatial resolution, and are less susceptible to artifact. CT scanners may be compared with one another by considering the following ten factors:

1. *Gantry design*, which affects scan speed, patient processing time, and cost-effectiveness.
2. *Aperture size*, which determines the maximum size of the patient along with the weight carrying capacity of the couch.
3. *The type of x-ray source*, which affects the patient radiation dose and the overall life of the scanning device.
4. *X-ray fan beam angle and scan field*, which affects resolution.
5. *The slice thickness*, as well as the number of pulses and the angular rotation of the source which are important in determining resolution.
6. *The number and types of detectors*, which are critical parameters in image quality.
7. *The type of minicomputer employed*, which is important in assessing system capability and flexibility.
8. *The type of data handling routines* available with the system, which are important user and reliability considerations.
9. *The storage capacity* of the system, which is important in ascertaining the accessibility of the stored data.
10. *Upgradeability and connectivity*. Modern CT scanners should be capable of modular upgradeability and should communicate to any available network.

In reviewing the available CT systems, it is important to compare them not only in technical terms, but against what one considers to be an ideal scanner. The ultimate objective of computerized tomography is to provide accurate diagnostic information that significantly improves patient care. In the ideal case, a CT scanner would provide an unambiguous diagnosis, thereby eliminating the necessity for further tests. Because financial considerations are also important, and CT scanners require a high capital outlay and have considerable maintenance costs, the ideal system should process a large patient load quickly, providing high "patient throughput." A good scanner, therefore, must satisfy two major criteria: it must provide good diagnostic information and it must permit high patient throughput.

EXERCISES

1. Represent the decay process discussed in Example Problem 15.2 in symbolic form.
2. Compute the energy liberated when $^{238}_{92}$U(amu = 238.050786) decays to $^{234}_{90}$Th(amu = 232.038054) via α emission.

3. Carbon 14, $^{14}_{6}$C is a radioactive isotope of carbon that has a half-life of 5730 years. If an initial sample contained $1000\,^{14}$C nuclei, how many would still be around after 22,920 years?

4. A 50-g sample of carbon is taken from the pelvis bone of a skeleton and is found to have a ^{14}C decay rate of 200 decays/min. It is known that carbon from a living organism has a decay rate of 15 decays/min × g and that ^{14}C has a half-life of 5730 years $= 3.01 \times 10^9$ min. Find the age of the skeleton.

5. The half-life of a radioactive sample is 30 min. If you start with a sample containing 3×10^{16} nuclei, how many of these nuclei remain after 10 min?

6. Find the energy liberated in the beta decay of $^{14}_{6}$C to $^{14}_{7}$N.

7. How long will it take for a sample of polonium of half-life 140 days to decay to one-tenth its original strength?

8. Suppose that you start with 10^3 g of a pure radioactive substance and 2 h later determine that only 0.25×10^3 g of the substance remains. What is the half-life of this substance?

9. The half-life of an isotope of phosphorus is 14 days. If a sample contains 3×10^{16} such nuclei, determine its activity.

10. How many radioactive atoms are present in a sample that has an activity of $0.2\,\mu$Ci and a half-life of 8.1 days?

11. A freshly prepared sample of a certain radioactive isotope has an activity of 10 mCi. After 4 h, the activity is 8 mCi.
    (a) Find the decay constant and half-life of the isotope.
    (b) How many atoms of the isotope were contained in the freshly prepared sample?
    (c) What is the sample's activity 30 h after it is prepared?

12. Tritium has a half-life of 12.33 years. What percentage of the nuclei in a tritium sample will decay in 5 years?

13. For the following process $^{23}_{10}\text{Ne}_{13} \rightarrow ^{23}_{11}\text{Ne}_{12} + ^{0}_{-1}e + v$, what is the maximum kinetic energy of the emitted electrons?

14. The half-life of ^{235}U is 7.04×10^8 years. A sample of rock that solidified with the earth 4.55×10^9 years ago contains N atoms of ^{235}U. How many ^{235}U atoms did the same rock have when it solidified?

15. Of the three basic types of radioaction—alpha, beta, and gamma—which has the greatest penetration into tissue? Explain the rationale behind your answer.

16. Discuss the principle of scintillation. How is it detected? How can it be used to generate an image?

17. One method of treating cancer of the thyroid is to insert a small radioactive source directly into the tumor. The radiation emitted by the source can destroy cancerous cells. Why do you suppose $^{131}_{53}$I is used for this treatment?

18. Provide a clinical example of the use of instrumentation employed to detect the rate of "radioactive workout."

19. In a photomultiplier tube, assume that there are seven diodes with potentials 100, 200, 300, 400, 500, 600, and 700 V. The average energy required to free an electron from the dynode surface is 10 eV. For each incident

electron released from rest, how many electrons are freed at the first dynode? At the last dynode?

20. Discuss the process of producing a tomograph. Provide figures to illustrate your answer.

21. Describe the operation of the iterative method to produce an image.

22. Provide an illustration of how the net absorption coefficient is obtained for each volume element.

Suggested Reading

Alpen, E.L. (1990). *Radiation Biophysics*. Prentice Hall, Princeton, NJ.

Bronzino, J.D. (1982). *Computer Applications in Patient Care*. Addison-Wesley, Reading, MA.

Cho, Z.H., Jones, J.P. and Singh, M. (1993). *Foundations of Medical Imaging*. John Wiley, New York.

Croft, B.F. and Tsui, B.M.N. (1995). Nuclear medicine. In *The Biomedical Engineering Handbook*. CRC, Boca Raton, FL.

Cunningham, I.A. and Juoy, P.F. (1995). Computerized tomography. In *The Biomedical Engineering Handbook*. CRC, Boca Raton, FL.

Hounsfield, G.N. (1973). Computerized transverse axial scanning (tomography) part 1, description of system. *Br. J. Radiol.* **46**, 1016–1022.

McCollough, C.H. and Morin, R.L. (1994). The technical design and performance of ultrafast computed tomography. *Cardiac Imagery* **32**, 521–536.

McCollough, C.H. (1992). Acceptance testing of a fifth-generation scanning-electron-beam computed tomography scanner. *Med. Phys.* **19**, 846.

Serway, R.A. and Faughn, J.S. (1989). *College Physics*, 2nd Ed. Saunders College, Philadelphia.

Shroy Jr., R.E., Van Lipel, M.S. and Yaffe, M.J. (1995). *X-ray In the Biomedical Engineering Handbook*. CRC, Boca Raton, FL.

16 MEDICAL IMAGING

Thomas Szabo, PhD*

Chapter Contents

*With contributions from Kirk K. Shung and Steven Wright.

At the conclusion of this chapter, the reader will be able to:

- Distinguish between the principles of pulse-echo ranging and ultrasound imaging.
- Describe how ultrasound images are formed.
- Explain the fundamentals of acoustic wave propagation, reflection, and refraction.
- Describe the operation and characteristics of a piezoelectric transducer.
- Explain the principle of an acoustic matching layer.
- Discuss the basic types of acoustic scattering.
- Calculate the effects of acoustic absorption.
- Describe the fundamentals of beam formation and focusing.
- Discuss the block diagram of an ultrasound imaging system.
- Explain basic ultrasound Doppler.
- Explain four basic principles of the interaction of magnetic fields and charges.
- Discuss spin states and precession of nuclear dipoles in a magnetic field.
- Explain the Larmor frequency and nuclear magnetic resonance.
- Explain how flip angles affect recovery and relaxation time constants.
- Distinguish between free induction decay and spin echo signals.
- Explain how detected resonance signals are spatially localized.
- List the steps involved in creating a magnetic resonance image.
- Explain what is displayed in a magnetic resonance image.
- Draw and explain the block diagram of an MRI system.
- Explain the significance of k-space in MRI.
- Discuss applications of MRI including fMRI.
- Compare major imaging modalities.

16.1 INTRODUCTION

Of the major diagnostic imaging modalities, ultrasound is the most frequently used, second only to standard plane-view x-rays. Over the years, the cumulative number of ultrasound exams completed is estimated to be in the billions. Unlike x-rays and computed tomography (CT) scanning, ultrasound imaging involves no ionizing radiation, and therefore is considered to be noninvasive. Furthermore, it is portable, easy to apply, low in cost, and provides real-time diagnostic information about the mechanical nature and motion of soft tissue and blood flow. The basic principle of ultrasound imaging is the display of pulse-echoes backscattered from tissues.

Magnetic resonance imaging (MRI) also obtains detailed anatomic information without using ionizing radiation. MRI differentiates among types of organs by sensing the spin of their atoms when a person is placed in a large, static, magnetic field. Static cross-sectional images of the body include both bone and soft tissue, and the appearance of the images can vary considerably by the selection of specific parameters. Of the major imaging modalities, MRI is the most abstract and complicated technically. In addition to its precise anatomical capability, it is often used for presurgery planning and for cancer detection. Functional MRI (fMRI) provides images of brain activity in response to various stimuli.

In this chapter, the operation and principles of both ultrasound imaging and MRI will be explained. At the end of the chapter, the main features of all major imaging modalities, including CT and x-ray, will be compared. An understanding of the basic principles of the Fourier transform, introduced in Chapter 10, will enhance comprehension of the imaging concepts introduced in this chapter. More information about the Fourier transform can be found in the following section of this chapter.

16.1.1 Review of Fourier Transforms

Because Fourier transforms simplify the understanding of the imaging principles of ultrasound imaging and MRI, their relevant properties are reviewed here. One important Fourier transform concept from Chapter 6 is the equivalence of convolution in the time domain and multiplication in the frequency domain. Another is the use of an impulse response function and a transfer function for describing system and filter responses in the time and frequency, respectively.

An impulse function is a generalized function that has the unusual property that it samples the integrand:

$$\int_{-\infty}^{\infty} \delta(t - t_0)g(t)dt = g(t_0) \tag{16.1}$$

When the transform of the impulse function is taken, the result is an exponential

$$H(\omega) = \int_{-\infty}^{\infty} \delta(t - t_0)e^{-i\omega t}dt = e^{-i\omega t_0} \tag{16.2}$$

which shows that a delay in time is equivalent to a multiplicative exponential delay factor in the frequency domain. When the impulse has no delay or $t_0 = 0$, $H(\omega) = 1.0$, a constant.

The preceding relation can be generalized to the form where $\Im$ represents the Fourier transform operation,

$$\Im[g(t - b)] = e^{-i2\pi bf}G(\omega) \tag{16.3}$$

A scaling factor can be added to this time shifting/delay theorem to make it even more useful,

$$\Im[g(a(t-b))] = \frac{e^{-i\omega b}}{|a|} G(\omega/a) \tag{16.4a}$$

Note that a similar relation exists for the inverse transform denoted by $\Im^{-1}$,

$$\Im^{-1}[G(a(\omega-b))] = \frac{e^{ibt}}{|a|} g(t/a) \tag{16.4b}$$

Finally, an important unique property of the impulse function is

$$\delta(at) = \frac{1}{|a|} \delta(t) \tag{16.5}$$

Example Problem 16.1

Find the inverse Fourier transform of $R(\omega) = \sin[3(\omega - \omega_0)t_1]$.

Solution

Recognize the basic function G in $R(\omega)$ as

$$G(\omega) = \sin \omega t_1 = \frac{e^{i\omega t_1} - e^{-i\omega t_1}}{2i}$$

Then from the Fourier transform pair expressed in Eq. 16.2,

$$\Im^{-1}[G(\omega)] = g(t) = (i/2)[\delta(t - t_1) - \delta(t + t_1)]$$

From Eq. 16.4b with $a = 3$ and $b = \omega_0$ find $r(t)$ as

$$\Im^{-1}[R(\omega)] = \left(e^{i\omega_0 t}/3\right)(i/2)\left[\delta\left(\frac{t - t_1}{3}\right) - \delta\left(\frac{t + t_1}{3}\right)\right]$$

and from scaling in the impulse function in Eq. 16.5,

$$r(t) = \left(e^{i\omega_0 t}\right)(i/2)[\delta(t - t_1) - \delta(t + t_1)] \qquad \blacksquare$$

16.2   DIAGNOSTIC ULTRASOUND IMAGING

16.2.1   Origins of Ultrasound Imaging

Though it has long been known that bats use sound for echo location, the intentional use of ultrasound (sound with frequencies above our range of hearing) for this purpose began, surprisingly, with the sinking of the Titanic in 1913. L.F. Richardson, a British scientist, filed patents within months of the Titanic disaster for echo location of icebergs (and other objects) using sound in either water or air. By the end of World War I, C. Chilowsky and P. Langevin in France invented practical implementations of echo location with high-power electronic transmitters and piezoelectric transducers for locating submarines and echo ranging. These principles of echo ranging were applied much later to electromagnetic waves to create RADAR (RAdio Detection And Ranging). The circular sweep of a RADAR echo ranging line is displayed on PPI (Plan Position Indicator) monitors, an early example of pulse-echo imaging. This type of display and technology was in turn adopted by underwater investigators to develop SONAR (SOund Navigation And Ranging).

After experiencing these technologies during World War II, several doctors wanted to apply echo location principles to the interior of the human body. Fortunately, a device—the supersonic reflectoscope—developed for finding defects in solid objects by ultrasound echo location, became available in addition to other surplus wartime equipment. This type of equipment sent out short pulses a few microseconds in length, repeated at longer intervals of a millisecond as illustrated by Figure 16.1.

Short pulses were needed to determine the location of tissue boundaries. A typical setup is shown in Figure 16.2, along with a record of echoes displayed as a function of time. Here the delay, t, to each echo is the round trip distance to the object, $2z$, divided by the speed of sound in the material, c, or $t = 2z/c$. The echo range instrument consisted of a piezoelectric transducer that converted the electrical pulses from the transmitter to acoustic pulses and reconverted received echoes from targets into electrical signals. These signals were then amplified and displayed as a time record on an oscilloscope. This type of display was known as an A-mode display, with A signifying amplitude.

In 1949, U.S. Naval doctor G. Ludwig reported his measurements of sound speed in parts of the body, and the results showed they had an average value of

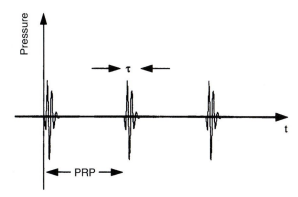

Figure 16.1 A plot of a series of transmitted acoustic pulses repeated at intervals called a pulse repetition period (PRP). Typically, the −3 dB width of a pulse is only a few microseconds in order to resolve different tissue interfaces, whereas the PRP is about a millisecond. For the purposes of illustration, the PRP has been compressed to enhance the presentation of pulses. The horizontal or time axis represents ambient pressure and positive values on the vertical axis show compressional acoustic pressure and negative values, rarefactional pressure.

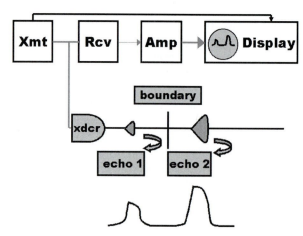

Figure 16.2   Basic echo ranging system consisting of a transmitter, transducer, receiver, amplifier, and oscilloscope display. Pulse echoes are shown below at delays corresponding to depths of reflecting objects above.

$c_0 = 1540$ m/s. More recent precise measurements showed that sound speed for most tissues varied by only a few percent as listed in Table 16.1. This important finding meant that the location of organs could be determined by the simple delay equation with a single value for sound speed, c_0, which is still used today in modern imaging systems. In other words, the geometric accuracy of the placement of organs in an ultrasound image largely depends on the uniformity of the sound speed in the field of view. Ludwig also measured the characteristic acoustic impedance of tissues, $Z = \rho c$ in which ρ and c represent the density and speed of sound of each tissue, respectively. He found that at the boundary of two different tissues, the reflection factor was related to the impedances of the individual tissues, Z_1 and Z_2:

$$RF = \frac{Z_2 - Z_1}{Z_2 + Z_1} \tag{16.6a}$$

TABLE 16.1   Acoustic Properties of Tissue

Material	C (m/s)	α(dB/MHz^y-cm)	y	ρ (kg/m³)	Z (MegaRayls)
Air	343			1.21	0.0004
Bone	3360	3.54	0.9	1789	6.00
Blood	1550	0.14	1.21	1039	1.61
Fat	1450	0.6	1.0	952	1.38
Honey	2030			1420	2.89
Liver	1570	0.45	1.05	1051	1.65
Muscle	1580	0.57	1.0	1041	1.645
Water @ 20°C	1482.3	2.17×10^{-3}	2.0	1000	1.482

Independently, Dr. D. Howry, another U.S. doctor, found that because the reflection factors were small, ultrasound penetrated through multiple layers of soft tissue with ease (except for regions with gas or bone). In 1956, he was able to make detailed anatomical maps of the body with sound and showed that they corresponded with known locations and sizes of organs and tissues. A graph of reflection factors in dB (decibels),

$$RF_{dB} = 20 \log_{10} (RF) \tag{16.6b}$$

all with reference to Z_1 for blood is shown in Figure 16.3. These factors are different enough to provide adequate discrimination, an important factor in differentiating among tissues in imaging.

Dr. J.J. Wild, an English surgeon, and J. Reid, an electrical engineer (now professor emeritus) working in Minnesota with a surplus 15 MHz radar simulator in 1951, recognized the value of ultrasound for diagnosis and attempted to use it to detect cancer in the stomach. They also made calculations on pulse-echo waveforms to infer properties of healthy and diseased tissue and began a new field now called tissue characterization. In the process, Wild and Reid developed near real-time ultrasound imaging systems and ways of placing transducers (which transmit sound and receive pulse echoes and convert them to electrical signals) directly on the skin.

In the 1950s, a number of groups around the world became interested in the diagnostic possibilities of ultrasound. S. Satomura, Y. Nimura, and T. Yoshida in Japan detected blood flow in the heart through Doppler-shifted motion, and I. Edler and C.H. Hertz in Sweden studied the motion of the heart and started echocardiography, the use of ultrasound to study the properties and dynamics of the heart.

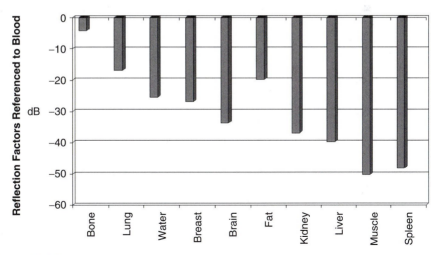

Figure 16.3 Amplitude reflection factor for tissues normalized to the impedance of blood and plotted on a decibel scale.

Investigators around the world continued to make progress with the development of ultrasound imaging. Even though internal organs and the dynamics of heart motion could be detected with ultrasound, ultrasound remained a laboratory curiosity. Until the early 1960s, mothers and fetuses were imaged by x-rays. A 1956 report by Alice Stewart, an English epidemiologist, linked deaths from cancer in children to their mothers' exposure to x-rays during pregnancy. This tragic finding gave ultrasound imaging its first commercial opportunity to provide a safe alternative for fetal imaging.

During the early 1960s, several companies developed ultrasound imaging systems suitable for imaging fetuses and other internal organs. R. Soldner of Siemens designed the first real-time mechanical ultrasound imaging scanner in 1965 in Germany. Drs. I. McDonald and T.G. Brown developed the first commercially successful diagnostic ultrasound imaging system, the Diasonograph in 1968.

These investigators eventually preferred a pulse-echo method of ultrasound imaging with equipment similar to that shown in Figure 16.4. The image is made up of a sequential arrangement of echo ranging lines where each line would correspond to each pulse in a sequence such as that depicted in Figure 16.1. The diagram shows an oval-shaped object with a sound beam piercing it. The sound is transmitted, and received echoes corresponding to the front and back boundaries of the oval at that beam location are then displayed as white dots on a vertical echo line with time increasing in the downward direction on a monitor. Next, the transducer is moved to another location and the process is repeated until all the lines have been sent. The resulting series of lines, geometrically arranged on the display to correspond to the actual positions of the transducer, result in an ultrasound image of the object. These early images were viewed either on long persistence cathode ray tubes or captured with cameras using long time exposures. The imaging presentation shown here was called B-mode, or brightness mode, in which the echo amplitudes were assigned white for maximum and black for minimum. Later, as electronics improved, a true gray scale for intermediate amplitudes could be implemented so that more subtle differences among tissues could be seen.

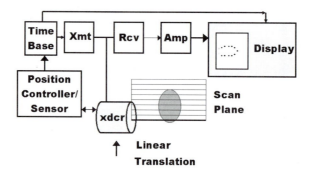

Figure 16.4   Basic elements of a pulse-echo imaging system shown with mechanical scanning. Note that a position controller/sensor is included along with a different form of information display compared to that of an echo ranging system.

Other investigators worked on arrays, an arrangement of small transducers that could be electronically switched and excited so that mechanical movement of a single transducer could be replaced by rapid real-time array imaging. Two contributors were J.C. Somer, who showed that ultrasound phased arrays were possible in 1968, and N. Bom, who developed the first linear arrays in 1971 in the Netherlands. By the 1980s almost all commercial imaging systems utilized arrays.

Many other important developments followed, some of which are highlighted in Sections 16.2.6 and 16.2.7. Some of the upcoming topics that make modern ultrasound imaging possible are scattering and absorption from tissues, transducers, focusing with arrays, imaging systems, and special signal processing methods.

16.2.2 Acoustic Wave Propagation, Reflection, and Refraction

Ultrasound imaging is based on waves that are sent into the body and received. A wave is a disturbance that moves through or along the surface of a medium. A small rock dropped on a smooth surface of water will create ripples in the form of expanding circles. In other words, the disturbances in time also have a spatial extent.

Two useful simple models for wave propagation are the plane wave and the spherical wave. A model for a spherical wave consists of a sphere expanding in radius with increasing time. A plane wave is a one-dimensional model that assumes that the lateral dimensions of the wave perpendicular to the direction of propagation are infinite in extent. A simple one-dimensional wave can be described by

$$p(z,t) = A \exp(i\omega(t - z/c_0)) + B \exp(i\omega(t + z/c_0)) \qquad (16.7)$$

where t is time, and z is the propagation coordinate.

The two arguments in the exponents represent waves traveling along the positive z axis and along the negative z axis, respectively. For a constant frequency, for example, when the argument of the first term is equal to π, the term is real and has an amplitude of -1. The next time the amplitude of the first term has this value is for an argument of 3π. For this situation, both t and z must increase; therefore these arguments can be recognized as wavefronts that propagate forward with a delay equal to the distance divided by the sound speed, $t - z/c_0$.

If k is a wavenumber defined as

$$k = 2\pi f/c_0 = \omega/c_0 \qquad (16.8)$$

and f is frequency, a sinusoidal wave can be interpreted as a function of propagation distance at a fixed time with a wavelength, $\lambda = 2\pi/k$, or as a function of time at a specific location with a period, $T = 2\pi/\omega$, as illustrated in Figure 16.5.

The ratio of a traveling pressure wave, p, to the particle velocity, v, of the fluid is called the specific acoustic or characteristic impedance,

$$Z_L = p/v_L = \rho c_L \qquad (16.9)$$

Note that Z_L is positive for forward traveling waves and negative for backward traveling waves. For fresh water at $20°C$, $c_L = 1481$ m/s, $Z_L = 1.48$ MegaRayls

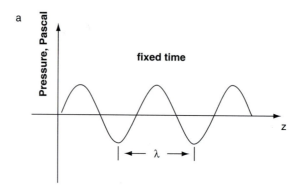

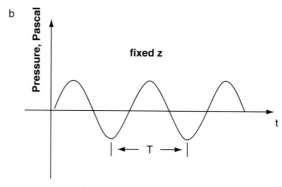

Figure 16.5   (a) Sinusoidal acoustic wave propagation as a function of space at a fixed time. (b) Sinusoidal acoustic wave propagation as a function of time at a fixed distance z; pressure is the acoustic parameter plotted here. The distance and time between two troughs or peaks are defined as the wavelength, (λ), and period, (T), of the wave, respectively.

(10^6 kg/m^2 sec), and $\rho = 998$ kg/m^3. The subscript L designates longitudinal waves, the type that are most important in imaging. In a longitudinal wave, the changes in wave amplitude are aligned along the direction of propagation with positive parts of sinusoidal waves, called compressional half cycles, and the negative ones, called rarefactional half cycles, as depicted in Figure 16.5b.

The instantaneous intensity is

$$I_L = pv^* = pp^*/Z_L = vv^*Z_L \tag{16.10}$$

To determine the amplitude of a reflected wave, a solution similar to that of Eq. 16.7 can be applied. Consider the problem of a single frequency acoustic plane wave propagating in an ideal fluid medium a distance d to a boundary with a different medium as shown in Figure 16.6. A plane wave is a sinusoidal-type wave with infinite extent in the lateral directions (x and y). For the example shown, the propagating medium has a wavenumber k_1 and an impedance Z_1 whereas the second medium has

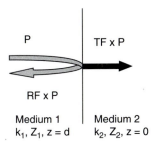

Figure 16.6 One-dimensional model of wave propagation at a boundary between media having impedances of Z_1 and Z_2.

impedance Z_2, and a wavenumber k_2. As shown in Figure 16.6, the waves to the left of this second medium can be described by a combination of forward and backward traveling waves,

$$p = p_0 \exp{(i(\omega t - k_L z))} + RFp_0 \exp{(i(\omega t + k_L z))} \tag{16.11}$$

that is a modified version of Eq. 16.7, $k_L = k_1$, and RF is a reflection factor for the amplitude of the negative going wave.

To determine the factor RF, it is useful to make analogies between acoustic variables and more well-known electrical parameters: pressure to voltage and particle velocity to electrical current. As illustrated by Figure 16.6, a source is situated at $z = d$ and the second medium is represented by a real load of impedance Z_2 located at $z = 0$, with a semi-infinite length. By analogy, the pressure at $z = 0$ is like a voltage drop across Z_2, so from Eq. 16.11, drop out common $\exp{(i\omega t)}$ terms to obtain,

$$p_2 = p_0(1 + RF) \tag{16.12}$$

The particle velocity there is like the sum of currents flowing in opposite directions corresponding to the two wave components. From Eq. 16.9,

$$v_2 = (1\text{-}RF)p_0/Z_1 \tag{16.13}$$

The impedance, Z_2, can be found from

$$Z_2 = \frac{p_2}{v_2} = \frac{(1 + RF)Z_1}{1 - RF} \tag{16.14}$$

Finally the right-hand side of Eq. 16.14 can be solved to obtain

$$RF = \frac{Z_2 - Z_1}{Z_2 + Z_1} \tag{16.15a}$$

A transmission factor, TF, can be determined from

$$TF = 1 + RF \tag{16.15b}$$

or

$$TF = \frac{2Z_2}{Z_1 + Z_2} \tag{16.15c}$$

Example Problem 16.2

Find the reflection and transmission factors for the case of a free, a perfectly matched, and a rigid boundary. What is the acoustic pressure transferred under these conditions? Assume water as medium 1.

Solution

For a free or air-type boundary or open-circuit condition, $Z_2 = 0$, so from Eq. 16.15a, there will be a 180° inversion of the incident wave, or $RF = -1$. Here the reflected wave cancels the incident, so $TF = 0$. For a matched condition, $Z_2 = Z_1$, $RF = 0$ or no reflection, and $TF = 1$ for perfect amplitude transfer. For a rigid boundary, $Z_2 = \infty$, corresponding to a short circuit condition or a stress-free boundary, and the incident wave will be reflected back, or $RF = +1$, without phase inversion. In this case, $TF = 2$.   ∎

Oblique Waves at a Liquid–Liquid Boundary

What happens when the incident wave is no longer normal to the boundary? This situation is depicted in Figure 16.7, in which a single-frequency longitudinal wave traveling in a liquid medium 1 is incident at an angle to a boundary with a different liquid medium 2 in the plane x–z. At the boundary, pressure and particle velocity are continuous. The tangential components of wavenumbers also must match, so along the boundary,

$$k_{1x} = k_1 \sin \theta_I = k_2 \sin \theta_T = k_1 \sin \theta_R \tag{16.16a}$$

where k_1 and k_2 are the wavenumbers for mediums 1 and 2, respectively. The reflected angle, θ_R, is equal to the incident angle, θ_I, and an acoustic Snell's law results:

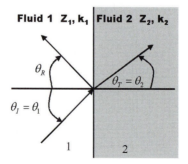

Figure 16.7   Oblique longitudinal waves at a liquid–liquid interface. The x-axis lies along the boundary and the z-axis is normal to the boundary.

$$\frac{\sin \theta_I}{\sin \theta_T} = \frac{c_1}{c_2} \tag{16.16b}$$

which can be used to find the angle θ_T. Equation 16.16a enables the determination of θ_R. From the previous relations, the wavenumber components along z are

$$\text{incident} \quad k_{Iz} = k_1 \cos \theta_I \tag{16.17a}$$

$$\text{reflected} \quad k_{Rz} = k_1 \cos \theta_R \tag{16.17b}$$

$$\text{transmitted} \quad k_{Tz} = k_2 \cos \theta_T \tag{16.17c}$$

which indicate that the effective impedances at different angles are

$$Z_{1\theta} = \frac{\rho_1 c_1}{\cos \theta_I} = \frac{Z_1}{\cos \theta_I} \tag{16.18a}$$

and

$$Z_{2\theta} = \frac{\rho_2 c_2}{\cos \theta_T} = \frac{Z_2}{\cos \theta_T} \tag{16.18b}$$

Here the impedance is a function of angle which reduces to familiar values at normal incidence, and otherwise becomes larger with angle. When the incident wave changes direction as it passes into medium 2, the bending of the wave is called refraction. For semi-infinite fluid media joined at a boundary, each medium is represented by its characteristic impedance given by Eqs. 16.18a and 16.18b. Then just before the boundary, the impedance looking toward medium 2 is given by Eq. 16.18b. The reflection coefficient there is given by Eq. 16.15a,

$$RF = \frac{Z_{2\theta} - Z_{1\theta}}{Z_{2\theta} + Z_{1\theta}} = \frac{Z_2 \cos \theta_I - Z_1 \cos \theta_T}{Z_2 \cos \theta_I + Z_1 \cos \theta_T} \tag{16.19a}$$

where the direction of the reflected wave is along θ_R, and the transmission factor along θ_T is

$$TF = \frac{2Z_{2\theta}}{Z_{1\theta} + Z_{2\theta}} = \frac{2Z_2 \cos \theta_I}{Z_2 \cos \theta_I + Z_1 \cos \theta_T} \tag{16.19b}$$

To solve these equations, θ_T is found first from Eq. 16.16b. This liquid–liquid interface is often used to model waves approximately at a tissue-to-tissue boundary.

Example Problem 16.3

For a wave incident at a water–honey interface, determine the reflection factor at $45°$ and at $50°$. The sound speeds and impedances for water (medium 1) and honey (medium 2) are 1.48 km/s and 2.05 km/s and 1.48 MegaRayls and 2.89 MegaRayls, given in Table 16.1.

Solution

From Snell's law, Eq. 16.16b, $\theta_T = \arcsin[(2.05/1.48)\sin 45] = 78.4°$. Then from Eq. 16.19a,

$$RF = \frac{2.89 \times \cos 45 - 1.48 \times \cos 78.4}{2.89 \times \cos 45 + 1.48 \times \cos 78.4} = 0.746$$

Before finding the result for 50°, it is worth finding the critical angle at which the incident wave is directed along the boundary. This angle, found by setting $\theta_T = 90°$ in Snell's law, is known as the critical angle, which is $\theta_I = 46.22°$ in this case. For an incident angle of 50° which is past the critical angle, the wave is just reflected and does not get transmitted, or $RF = 1$. ∎

16.2.3 Transducer Basics

The essential part of an ultrasound system is a means to generate and receive acoustic waves. This function is performed by the transducer, which can convert electrical signals to acoustic pressure waves and vice versa. Inside the transducer is a piezoelectric crystal or ceramic that deforms when a voltage is applied and transmits acoustic waves. A reciprocal piezoelectric effect, in which electrical charge is created by the mechanical deformation of a crystal, allows the transducer to convert returning acoustic waves back to electrical signals. The piezoelectric effect was discovered by the Curie brothers in 1880.

A simple model is presented that describes the basic acoustic and electrical characteristics of a piezoelectric transducer. Consider a transducer to consist of a rectangular piece of piezoelectric material with electrodes on the sides as shown in Figure 16.8.

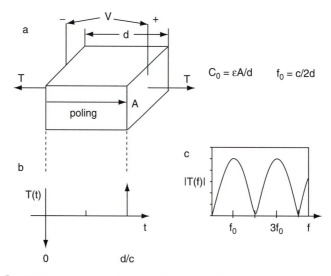

Figure 16.8   (a) Diagram for a piezoelectric crystal radiating into a medium matched to its impedance, (b) stress time response, and (c) stress frequency response.

Each side has a cross-sectional area, A, with a top and bottom that are much longer ($> 10X$) than the thickness, d. Because piezoelectric material is dielectric, it is essentially a capacitor with a clamped capacitance,

$$C_0 = \varepsilon^S A/d \qquad (16.20)$$

in which ε^S is a clamped dielectric constant under the condition of zero deformation. Because the crystal is a solid rather than a liquid, elastic waves are created internally. The elastic counterpart of pressure is stress, or force per unit area, represented by the symbol T. Another important elastic variable is strain, S, the relative change in length of the crystal divided by its original length. Both stress and strain vary with direction in a piezoelectric material, but for the single direction of wave propagation, it is possible to relate the stress to the strain as

$$T = C^D S - hD \qquad (16.21)$$

in which h is a piezoelectric constant. This equation is known as a modified Hooke's law in which the first term relates the stress to the strain and the second term relates stress to the applied electric field. The elastic stiffness constant, C^D, is obtained under a constant dielectric displacement, D, and if E is electric field,

$$D = \varepsilon^S E = \frac{\varepsilon^S AV}{dA} = C_0 V/A \qquad (16.22)$$

When a voltage impulse is applied across the electrodes, the piezoelectric effect creates impulsive forces at the sides of the transducer, given by

$$F(t) = TA = (hC_0 V/2)[-\delta(t) + \delta(t - d/c)] \qquad (16.23)$$

where the media above and below has the same acoustic impedance, Z_c, as the transducer; the speed of sound between the electrodes is given by $c = \sqrt{C^D/\rho}$, and δ represents an impulse (see Fig. 16.8b).

Since it can create acoustic waves, the crystal could be regarded as a singing capacitor with its own unique voice or spectral characteristics and resonant frequency. To obtain the spectrum of this response, take the Fourier transform of Eq. 16.23,

$$F(f) = -i(hC_0 V)e^{\frac{-i\pi fd}{c}} \sin[\pi(2n+1)f/2f_0] \qquad (16.24)$$

which has maxima at odd harmonics (note $n = 0, 1, 2, 3, \ldots$) of the fundamental resonance $f_0 = c/2d$ as shown in Figure 16.8c.

Transducer Electrical Impedance

Because of the forces generated at the sides of the transducer, the electrical impedance looking through the voltage terminals is affected. Across the wires connected to the transducer (Fig. 16.8a) a radiation impedance, Z_A, is seen in addition to the capacitive reactance so that the overall electrical impedance is

$$Z_T = Z_A - i(1/\omega C_0) = R_A(f) + i[X_A(f) - 1/\omega C_0] \qquad (16.25)$$

From the transducer electrical equivalent circuit in Figure 16.9a, Z_A is radiation impedance of which R_A and X_A are its real and imaginary parts. $R_A(f)$ can be found

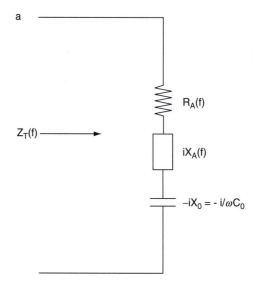

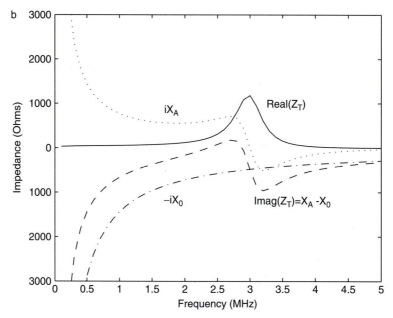

Figure 16.9   (a) Transducer equivalent circuit and (b) transducer impedance as a function of frequency.

from the total real electrical power, W_E, flowing into the transducer for an applied voltage V, and current I:

$$W_E = II^*R_A/2 = |I^2|R_A/2 \qquad (16.26a)$$

where current is $I = i\omega Q = i\omega C_0 V$ and Q is charge. The total power radiated from both sides of the transducer into a surrounding medium of a modified specific acoustic impedance (including an area factor), $Z_C = \rho c A$, equal to that of the crystal, is

$$W_A = ATT^*/(2Z_C/A) = A^2|F(f)/A|^2/2Z_C = |hC_0V \sin(\pi f/2f_0)|^2/2Z_C. \qquad (16.26b)$$

Setting the powers of Equations 16.26a and 16.26b equal leads to a solution for R_A:

$$R_A(f) = R_{AC} \sin c^2(f/2f_0) \qquad (16.27a)$$

where $\sin c(x) = \sin(\pi x)/(\pi x)$ and

$$R_{AC} = \frac{k_T^2}{4f_0C_0} = \frac{d^2k_T^2}{2A\varepsilon^S} \qquad (16.27b)$$

in which the electroacoustic coupling constant is k_T, and $k_T = h/\sqrt{C^D/\varepsilon^S}$. Note that R_{AC} is inversely proportional to the capacitance and area of the transducer and directly dependent on the square of the thickness, d. Also, note that at resonance,

$$R_A(f_0) = \frac{k_T^2}{\pi^2 f_0 C_0} \qquad (16.27c)$$

Network theory requires that the imaginary part of an impedance be related to the real part through a Hilbert transform, so the radiation reactance can be found as

$$X_A(f) = \Im_{Hi}[R_A(f)] = R_{AC}\frac{[\sin(\pi f/f_0) - \pi f/f_0]}{2(\pi f/2f_0)^2} \qquad (16.28)$$

The transducer impedance is plotted as a function of frequency in Figure 16.9b. Here R_A is maximum at the resonant frequency f_0 and there X_A is zero.

This simple model describes the essential characteristics of a piezoelectric transducer. The electrical impedance has a maximum of real radiation resistance at the resonant frequency. The force spectrum is also peaked at the resonant frequency and has a certain shape.

Example Problem 16.4

Find the transducer impedance at resonance for a square transducer 2.5 mm on a side for a resonant frequency of 3 MHz for the ceramic PZT5A with $\varepsilon^s/\varepsilon_0 = 830, c = 4.35$ km/s, $k_T = 0.49$, and $\varepsilon_0 = 8.85$ pF/m.

Solution

First find the thickness needed to achieve a resonant frequency of 3 MHz from $f_0 = c/2d$, or $d = c/2f_0 = 4.35$ mm/μs/2×3 MHz $= 0.725$ mm. Then the capacitance is

$$C_0 = \frac{830 \times 8.85 \times 10^{-12} F/m \times (2.5 \times 10^{-3} m)^2}{7.25 \times 10^{-4} m} = 63.3 \text{ pF}$$

Find R_A from Eq. 16.27c:

$$R_A = \frac{0.49^2}{\pi^2 \times 3 \times 10^6 \text{ Hz} \times 63.3 \times 10^{-12} F} = 128.2 \text{ ohms}$$

Note that the units in the denominator are $Hz - F = amps/volt = ohm^{-1}$. Then, since at resonance $X_A = 0$, from Eq. 16.25,

$$Z_T = 128 - i/(2 \times \pi \times 3 \times 10^6 \text{ Hz} \times 63.3 \times 10^{-12} F) = 128 - i838 \text{ ohms} \quad \blacksquare$$

Transducer Frequency Response

Transducer design is concerned with altering the shape of the spectrum to achieve a desired bandwidth and a short, well-behaved pulse shape, or impulse response. Another design goal is to improve the electroacoustic efficiency of the transducer. One measure of this efficiency is transducer loss, the ratio of time average acoustic power reaching the desired medium, usually tissue, W_R, divided by the maximum electrical power available from an electrical source, W_g,

$$TL(f) = \frac{W_R}{W_g} \tag{16.29a}$$

and defined in dB as

$$TL_{dB}(f) = 10 \log_{10} TL(f) \tag{16.29b}$$

The transducer loss can be broken down into an electrical loss factor, $EL(f)$, and an acoustic loss factor, $AL(f)$,

$$TL(f) = EL(f)AL(f) \tag{16.30}$$

The problem of optimizing the transfer of electrical power into the radiation resistance is that of maximizing $EL(f)$ over a desired bandwidth. Figure 16.10 shows the

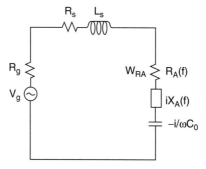

Figure 16.10 Electrical voltage source with tuning inductor connected to transducer equivalent circuit.

relevant parameters involved. The transducer electrical impedance is connected through a tuning inductor to a voltage source with impedance R_g. An expression for electrical loss for this situation is

$$EL(f) = \frac{4R_A(f)R_g}{(R_A(f) + R_g + R_s)^2 + \left(X_A(f) - \frac{1}{\omega C_0} + \omega L_s\right)^2} \qquad (16.31a)$$

If the capacitance is tuned out by a series inductor, $L_S = 1/(\omega_0^2 C_0)$ and $R_s = 0$, then at resonance,

$$EL(f_0) = \frac{4R_A(f_0)R_g}{\left[R_A(f_0) + R_g\right]^2} \qquad (16.31b)$$

Notice that if $R_A = R_g$, and $R_s \ll R_g$, then $EL(f_0) \sim 1$.

The determination of acoustic loss is a more complicated function of frequency but it is straightforward to determine its value at resonance. In the simple transducer model introduced in Section 16.2.3, a simplifying assumption was made that the acoustic loading on both sides of the piezoelectric crystal was the same acoustic impedance as that of the crystal; however, this is not the case in general, as illustrated by Figure 16.11. Note the arrangement in this figure is rotated 90° from that in Figure 16.8. On the top of this figure, the piezoelectric is loaded by a backing material and below the lens, by water or tissue. The bottom of the crystal shown here is usually regarded as the "business end" of the transducer where the forward waves propagate

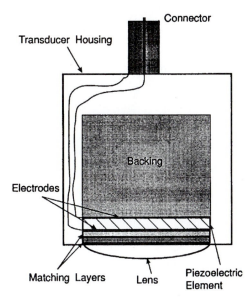

Figure 16.11 Construction of a single-crystal mechanical transducer. Forward propagation is directed downward.

along the positive z axis that is directed downward in Figure 16.11. Waves in the backward direction are suppressed or absorbed by a backing material that aids in broadening the spectrum of the transducer. By the left-right convention of Figure 16.8, if the acoustic impedance looking out of the crystal to the left is Z_L, then that to the right along the positive z axis is Z_R. In the case shown on the left of the figure, $Z_L = Z_B$ (backing impedance) and $Z_R = Z_W$ (impedance of water). The acoustic loss for this case at resonance can be found from (note all these impedances include the area factor A)

$$AL(f_0) = Z_R/(Z_L + Z_R) = Z_W/(Z_B + Z_W) \tag{16.32}$$

The physical meaning of this equation is that the electrical power reaching the radiation resistance is converted into acoustical power radiating from the two electroded faces of the piezoelectric; the proportion of power in the forward (right) direction is represented by the ratio in Eq. 16.32.

Construction of a single crystal transducer is shown in Figure 16.11. The key parts are the backing, the piezoelectric crystal coated with thin electrodes for electrical contact, matching layers, and a lens material, with an acoustic impedance close to water, for focusing (discussed in more detail in Section 16.2.6). Matching layers will be discussed shortly; in practice, more than one matching layer is used to broaden the bandwidth of the transducer.

In the first simple model, the crystal was loaded by acoustic impedances equal to that of the crystal, or $Z_C = Z_L = Z_R$ so that $AL(f_0) = 0.5$, as expected. Another example is that in which the left side is loaded by air or $Z_L = 0A$ MRayls and the right, by water, $Z_R = 1.5A$ MRayls. In this case, $AL(f_0) = 1.0$ but at the expense of extremely narrow bandwidth and a correspondingly long pulse. One case of interest is where the backing material is matched to that of the crystal so that $Z_L = Z_C = 30A$ MRayls and $Z_R = 1.5A$ MRayls, and $AL(f_0) = 0.048$, an inefficient transfer.

To improve the forward transfer efficiency, as described by the acoustic loss factor, matching layers are employed. At the resonant frequency, a matching layer is designed to be a quarter of a wavelength thick, and to be the mean value of the two impedances to be matched, or

$$Z_{ml} = \sqrt{Z_1 Z_2} \tag{16.33}$$

Example Problem 16.5

Match a crystal of impedance $Z_c = 30A$ MRayls to water, $Z_w = 1.5A$ MRayls.

Solution

From Eq. 16.33, the matching layer impedance is

$$Z_{ml} = \sqrt{Z_W Z_C} = \sqrt{1.5A \text{ MRayls} \times 30A \text{ MRayls}} = 6.7A \text{ MRayls}$$

The acoustic input impedance at resonance is

$$Z_1 = Z_{ml}^2/Z_2$$

therefore, the impedance looking to the right is

$$Z_R = Z_{ml}^2/Z_W = (6.7A)^2/1.5A = 30A \text{ MRayls}$$

leading to $A(f_0) = 30/(30 + 30) = 0.5$ for a water load with an intervening matching layer or an improvement of over 10 dB. ∎

The effect of adding a matching layer can be included in a more generalized equation for radiation resistance at resonance by adding the loading from both sides,

$$R_A(f_0) = \frac{2k_T^2}{\pi^2 f_0 C_0} \left(\frac{Z_c}{Z_L + Z_R} \right) \tag{16.34}$$

The acoustical and electrical losses can now be combined to estimate the overall transducer loss. To do this, a simplification is made for the purposes of illustration that the acoustical loss is constant over the transducer bandwidth. The actual losses can be computed accurately by a more complete equivalent circuit model. From Eqs. 16.30 and 16.31a,

$$TL(f) \approx \left[\frac{4R_A(f)R_g}{\left(R_A(f) + R_g + R_s\right)^2 + \left(X_A(f) - \frac{1}{\omega C_0} + \omega L_s\right)^2} \right] \left[\frac{Z_R}{Z_R + Z_L} \right] \tag{16.35}$$

Example Problem 16.6

Find the transducer loss at resonance for the transducer described in Example Problem 16.4 under the following two conditions for a source resistance $R_g = 50\Omega$ (ohms): (1) No tuning, $Z_L = Z_B = 6A$ MRayls and $Z_R = Z_W = 1.5A$ MRayls. (2) Add a tuning inductor and a matching layer.

Solution

(1) Here $R_A = 128$ ohms, $\frac{1}{\omega C_0} = 838$ ohms,

$$TL(f_0) = \frac{4 \times 128 \text{ ohms} \times 50 \text{ ohms}}{(128 \text{ ohms} + 50 \text{ ohms})^2 + (838 \text{ ohms})^2} \times \frac{1.5A}{6A + 1.5A}$$
$$= 6.98 \times 10^{-3} \text{ or } 10 \log_{10}(TL)$$
$$= -21.6 \text{ dB}$$

(2) R_A is determined from Eq. 16.34 and Example Problem 16.4, ohms. For tuning, find $L_S = 1/(\omega_0^2 C_0) = 838/(2\pi \times 3e6) = 44.5 \mu H$. Then,

$$TL(f_0) = \frac{4 \times 128 \text{ ohms} \times 50 \text{ ohms}}{(128 \text{ ohms} + 50 \text{ ohms})^2 + (838 \text{ ohms} - 838 \text{ ohms})^2} \times \frac{30A}{30A + 1.5A}$$
$$= .673 \text{ or } 10 \log_{10}(TL)$$
$$= -1.72 \text{ dB}$$
∎

One way of describing the difference in shapes as a figure of merit is a bandwidth is stated in dB. For example, a −6 dB bandwidth is defined relative to the peak spectral value. A center frequency is defined through the upper and lower −6 dB frequencies as

$$f_c = (f_{high} - f_{low})/2$$

An example of a −6 dB bandwidth with its associated frequencies is shown in Figure 16.12. Also included in this figure are similar widths for the pulse envelope (pulsewidths) in dB levels.

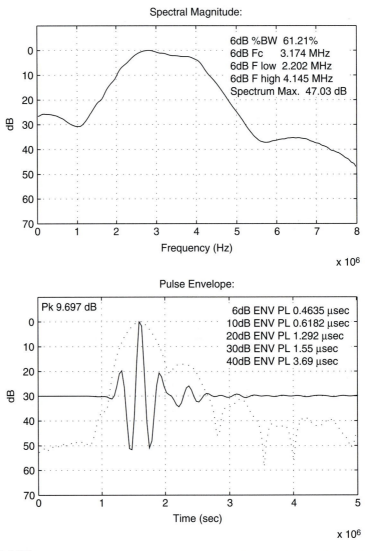

Figure 16.12 (Top) Transducer response spectral magnitude and −6 dB bandwidth dB. (Bottom) Transducer impulse response and its envelope measured in pulse-widths in dB.

Transducer Types and Arrays

Most imaging systems use arrays rather than single-crystal transducers. A one-dimensional array is an in-line arrangement of transducers called elements that are addressable individually or in small groups. The elements are spaced at regular intervals, typically one-half to two wavelengths apart in water. An array is shown in Figure 16.13. The key advantages of arrays are that they can be rapidly focused and steered electronically or electrically switched whereas a strictly mechanical single transducer has a fixed focal length and can only be steered or translated mechanically. Each element, in terms of basic design, is treated as an individual transducer. Arrays and focusing will be discussed in more detail in Section 16.2.6 on diffraction.

The two most common types of arrays are the linear array and the phased array. The linear array forms an image by translating the active aperture (a certain number of elements), one element at a time along the length of the array, as illustrated by Figure 16.14a. At each position of the active aperture, an acoustic line is created (i.e., a pulse-echo time record of a selected length or scan depth). A total number of lines, N, is formed in this way. In a phased array, the center of the active aperture is always the same and the scanning of acoustic lines is accomplished through electronic angular steering. Each line is steered by a small incremental angle from the previous one, as

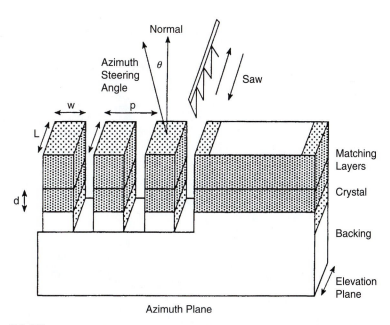

Figure 16.13 Construction of a one-dimensional array with a saw dicing up a multilayer sandwich structure into individual elements. Here the propagation direction is along the normal axis, pointing upwards.

shown in Figure 16.14b. When N lines have been received, these lines form the basis for an image frame. Examples of the image formats formed by these two array types can be seen in the B-mode images of Figures 16.15 and 16.16. A variant of the linear array is the curved linear array which operates like a linear array but on a curved convex surface rather than on a flat surface. An example of an image from a curved linear array can be seen in Figure 16.17.

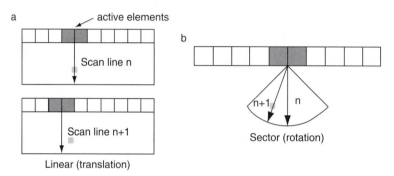

Figure 16.14 Illustration of the time sequencing used for image formation for (a) linear array format and (b) sector array format.

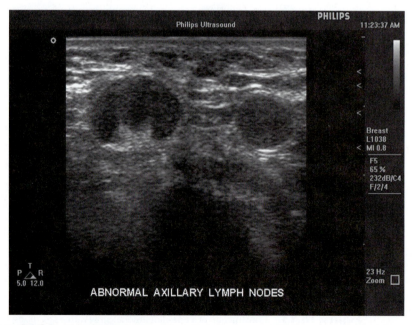

Figure 16.15 B-mode image of lymph nodes in the breast at 12 MHz, an example of a linear array format (courtesy of Philips Medical Systems).

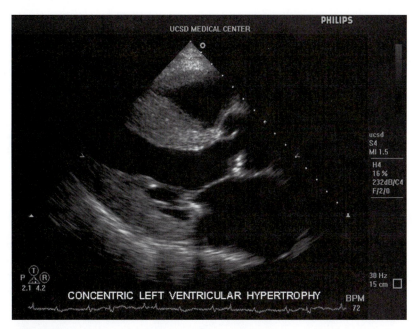

Figure 16.16 B-mode image of the heart at 4 MHz, an example of a sector array format (courtesy of Philips Medical Systems).

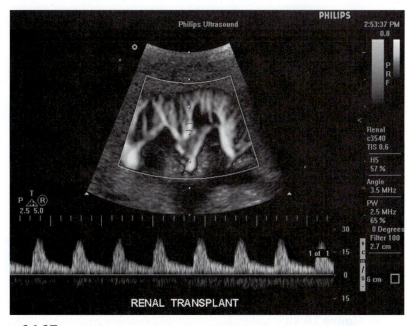

Figure 16.17 (Top) Power Doppler image of kidney with pulsed Doppler line down the center and Doppler gate in center of line. (Below) Corresponding pulsed wave Doppler spectrum. Image from a curved linear array (courtesy of Philips Medical Systems).

Example Problem 16.7

For the example shown in Figure 16.14a, determine the frame rate for a scan depth of $s_d = 150$ mm, 100 lines per frame, and $c_0 = 1.5$ mm/μs.

Solution

The round-trip time for one line is $2^*s_d/c_0 = 200$ μs. The time for a full frame is N lines/frame or, in this case 100 lines/frame $\times$ 200 μs/line $= 20$ ms/frame or 50 frames per second. ∎

Transducer arrays come in a variety of sizes, shapes, and center frequencies to suit different clinical applications, as illustrated by Figure 16.18. Access to the body is made externally through many possible "acoustic windows" where a transducer makes contact by coupling to the body with a water-based gel. Except for regions containing bones, air, or gas, which are opaque to imaging transducers, even small windows can be enough to visualize large interior regions. The limitation of accessibility to viewing certain organs clearly is offset by specialized probes such as transesophageal (down the throat) and intracardiac (inside the heart) transducers that image from within the body.

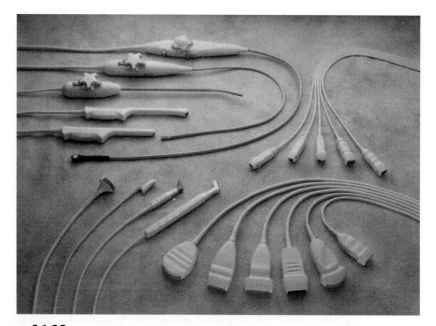

Figure 16.18   Transducers available on a modern imaging system are designed for a wide variety of clinical applications. Transducer groups: bottom right: linear and curved linear arrays; top right: phased arrays; left side: specialty probes including intraoperative, transesophageal, and transvaginal arrays (courtesy of Philips Medical Systems).

16.2.4 Scattering

Interference of waves within the transducer plays an important role in the transduction process. The crystal resonates when its thickness is half a wavelength and the matching layer is designed to be a quarter of a wavelength thick at resonance. The size of an object relative to a wavelength is also a useful way of looking at acoustic scattering from objects. Since ultrasound imaging is based on pulse echoes returning from organs, it is necessary to determine how the reflected signals are affected by the dimensions and shape of an object relative to the insonifying wavelength.

Scattering falls within three ranges of effects: specular, diffractive, and diffusive (Fig.16.19). Specular scattering is already familiar. When the dimensions of an object are much greater then a wavelength, the reflected sound returns at the reflected angle equal to the incident angle relative to the surface, as described in Section 16.2.2 on refraction. In this case, the reflected amplitude is determined by the reflection factor appropriate for the angle of incidence. At the other extreme, when an object is much smaller than a wavelength, its reflections are diffusive. Here the object is so small relative to a wavelength that its features no longer contribute to the reflected wave. If the object is a sphere, then the reflected pressure is proportional to the frequency squared, inversely with the radius cubed, and a term related to the differences in elastic constants and densities between the sphere and the surrounding medium.

The intermediate range between these two extremes, diffractive scattering, occurs when the dimensions of the object are on the order of wavelengths. If the surface of

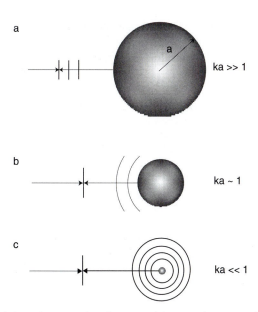

Figure 16.19 (a) Specular scattering from an object much greater than a wavelength. (b) Diffractive scattering from an object on the order of wavelengths. (c) Diffusive scattering from an object much smaller than a wavelength.

the object is divided into infinitesimal points and lines are drawn from these points to an observation point, then the differences in path lengths along these lines can be significant in terms of wavelengths. Contributions from these points could cause constructive and destructive interference effects at the observation point because of phase differences along the paths. Depending on the shape of the object and its orientation and distance to an observation point, complicated scattering patterns can result. Even for simply shaped objects such as spheres or cylinders, the scattering pattern or directivity of these objects is highly frequency dependent as the object shape and dimensions change relative to a wavelength.

Human optical vision depends mainly on specular reflections (objects much greater than wavelengths of light). An acoustic image is formed from pulse echoes along acoustic lines from an observation point to parts of the object. Unlike optics, the acoustic appearance of an object can change both with the orientation of the observer relative to the object and with the insonifying frequency.

16.2.5   Absorption

When waves propagate in real media, losses are involved. Just as forces encounter friction, pressure, and stress, waves lose energy to the medium of propagation and result in weak local heating. These small losses are called attenuation and can be described by an exponential law with distance. For a single frequency, f_c, plane wave, a multiplicative amplitude loss term can be added,

$$A(z, t) = A_0 \exp(i(\omega_c t - kz)) \exp(-\alpha z) \qquad (16.36)$$

The attenuation factor, α, is usually expressed in terms of nepers/cm in this form. Another frequently used measure of amplitude is the deciBel (dB), which is most often given as the ratio of two amplitudes, A and A_0 on a logarithmic scale,

$$\text{Ratio(dB)} = 20 \log_{10} (A/A_0) \qquad (16.37)$$

or in those cases where intensity is simply proportional to amplitude squared $(I_0 \propto A_0^2)$,

$$\text{Ratio(dB)} = 10 \log_{10} (I/I_0) = 10 \log_{10} (A/A_0)^2 \qquad (16.38)$$

Most often, α, is given in dB/cm,

$$\alpha_{dB} = 1/z\{20 \times \log_{10} [\exp(-\alpha_{nepers} z)]\} = 8.6886(\alpha_{nepers}) \qquad (16.39)$$

Data indicate that the absorption is a function of frequency (see Table 16.1). Many of these losses obey a frequency power law defined as

$$\alpha(f) = \alpha_0 |f|^y \qquad (16.40)$$

in which y is a power law exponent. The pressure amplitude can be written to first order as a function of frequency,

$$A(z, f) = A_0 \exp(-i2\pi f z/c) \exp[-\alpha_0 |f|^y z] \qquad (16.41)$$

where the first factor describes the propagation delay to z. The actual loss per wavenumber is very small, or $\alpha/k \ll 1$. Even though the loss per wavelength is small, absorption has a strong cumulative effect over many wavelengths. Absorption for a round-trip echo path usually determines the allowable tissue penetration for imaging.

Example Problem 16.8

Determine the absorption loss for 10 cm of propagation in muscle for a frequency of 5 MHz.

Solution

From Table 16.1, $\alpha_0 = 0.57$ dB/cm-MHz. Method 1 : Convert absorption coefficient to nepers from Eq. 16.39), α (nepers/MHz-cm) $= 0.57/8.6886 = 0.0656$. Use this value in Eq. 16.41 to obtain $A/A_0 = \exp(-0.0656$ nepers/MHz-cm $\times 5$ MHz $\times 10$ cm$) = 3.76 \times 10^{-2}$. In dB terms $20 \log_{10}(A/A_0) = -28.5$ dB. Method 2: If the linear answer is not needed, the answer in dB can be obtained directly by using the absorption coefficient in dB as $\alpha_0 f z = -0.57$ dB/MHz-cm $\times 5$ MHz $\times 10$ cm $= -28.5$ dB. ∎

16.2.6 Diffraction

Waves transmitted by a transducer are not plain but form a complicated pattern. The formation of these patterns, as shown in Figure 16.20, is caused by the radiation of sound waves from different locations on the aperture (transducer face) and the mutual

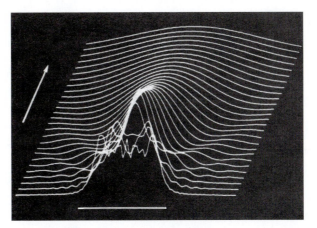

Figure 16.20 Diffracted field of a 40-wavelength-wide line aperture along the x axis. The vertical axis intensity and beam profiles are shown at intervals of about 70 wavelengths along the beam axis that is compressed relative to the lateral dimension (1920 wavelengths are shown along the z axis).

interference of these radiated waves. This phenomenon, also called diffraction, is a consequence of the aperture dimensions that are on the order of wavelengths.

The two most common aperture shapes are the circle and rectangle as shown in Figure 16.21. A slice of the three-dimensional beam in a plane is what is usually depicted in graphs. For the circle, because of symmetry, any plane through the beam axis, here the z axis, will be identical. For the rectangular aperture, the beam formation differs in all planes through the beam axis, and the most important planes for imaging are the x–z and y–z planes. The beam amplitude described by Figure 16.20 corresponds to an x–z plane from a rectangular aperture.

Beams have recognizable landmarks. A method borrowed from maps is a contour plot of the acoustic pressure magnitude, often depicted in dB relative to maximum amplitude at each depth. Of particular interest is the -6 dB contour. A cross section of the beam, perpendicular to the beam axis is called a beam plot. The width between points of this -6 dB contour on a beam plot is called the full-width-half-maximum (FWHM). A curious outcome of the radiation from these apertures is that there is a region in which the beam narrows. The depth where the last axial peak occurs is called the transition distance or natural focal length, F_N. This transition depth demarcates two regions, one with peaks and valleys, called the near field, and one with a beam with a single peak diminishing in amplitude width and broadening with distance, called the far field, as depicted in Figure 16.20. The transition depth for a circular aperture of radius a is

$$z_t = a^2/\lambda \tag{16.42a}$$

For a rectangular aperture, the transition distance for an aperture L_x in the x–z plane is

$$z_t \approx L_x^2/(\pi\lambda) \tag{16.42b}$$

The natural focal length is the distance to the last axial peak and is approximately the transition distance.

The far-field beam pattern for a rectangular aperture is the Fourier transform of the amplitude across the aperture. In the case of uniform illumination,

$$A(x_0, y_0, 0) = \prod (x_0/L_x) \prod (y_0/L_y) \tag{16.43a}$$

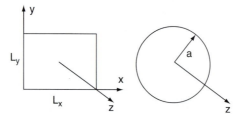

Figure 16.21   (Left) Circular aperture of radius a. (Right) Rectangular aperture with lengths of L_x parallel to the x axis and L_y parallel to the y axis. The z axis is perpendicular to the $x - y$ plane of the aperture.

where

$$\prod (x/L) = \begin{cases} 0 & |x| > L/2 \\ 1/2 & |x| = L/2 \\ 1 & |x| < L/2 \end{cases} \tag{16.43b}$$

the far-field pattern in the x–z plane is a sinc function,

$$p(x, z, \omega) = \frac{L_x\sqrt{p_0}}{\sqrt{\lambda z}} e^{i\pi/4} \frac{\sin (\pi L_x x/\lambda z)}{(\pi L_x x/\lambda z)} = \frac{L_x\sqrt{p_0}}{\sqrt{\lambda z}} e^{i\pi/4} \text{sinc}\left(\frac{L_x x}{\lambda z}\right) \tag{16.44a}$$

A plot of this pattern is given by Figure 16.22.

In the case of a uniform amplitude u_0 on a circular aperture, the far-field pattern is the two-dimensional Fourier transform of the circularly symmetric aperture function,

$$p(\bar{\rho}, z, \lambda) \approx \frac{ip_0 \pi a^2}{\lambda z} \frac{2J_1(2\pi\bar{\rho}a/(\lambda z))}{2\pi\bar{\rho}a/(\lambda z)} = ip_0 \left(\frac{\pi a^2}{\lambda z}\right) jinc\left(\frac{\bar{\rho}a}{\lambda z}\right) \tag{16.44b}$$

where J_1 is the Bessel function of the first kind, $jinc(x) = 2J_1(2\pi x)/(2\pi x)$, and $\bar{\rho}$ is the radial distance to an observation point at $(\bar{\rho}, z)$. A plot of this pattern is given by Figure 16.22. Note that the shapes of the far-field patterns are maintained with distance as their amplitudes fall and beams broaden with distance.

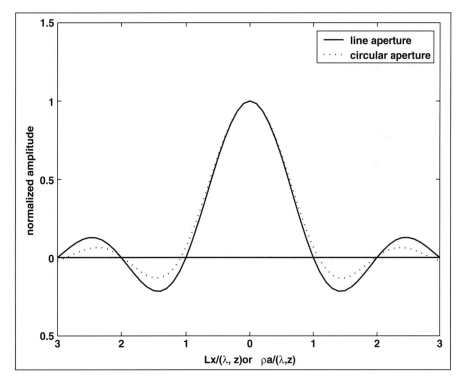

Figure 16.22 (a) Far-field jinc beam shape from a circular aperture normalized to (b) far-field sinc function from a square aperture with the same aperture area.

From these far-field patterns it is easy to determine the FWHM beam widths. For the rectangular aperture in the x–z or y–z plane,

$$\text{FWHM} = 1.206\lambda z/L \qquad (16.45a)$$

where L is the appropriate aperture for that plane. Similarly, for a circular aperture,

$$\text{FWHM} = 0.7047\lambda z/a \qquad (16.45b)$$

Example Problem 16.9

To understand the normalization in Figure 16.22, find the FWHM for a circular aperture having the same area as a square aperture and compare it to the FWHM of the square aperture.

Solution

Set $\pi a^2 = L^2$, or $a(m) = L(m)/\sqrt{\pi}$. Substitute in Eq. 16.45b, FWHM = $0.7047\lambda(m)z(m)(\sqrt{\pi}/L(m)) = 1.249\lambda(m)z(m)/L(m)$. This is close to the value for the square aperture from Eq. 16.45a, FWHM = $1.206\lambda z/L(m)$.

To narrow the beams even more and at different depths, geometric focusing is applied. Like optics, acoustic focusing can be implemented with a type of lens. Unlike optics, both concave and convex converging lenses can be made because materials exist such that their sound speeds are either greater or less than that of water (tissue). Under the principles of ray optics, the rays converge at the geometric focal point, F. From the reciprocal law of lenses, the overall total focal length is the combined effect of the natural focal length and the geometrical focal length,

$$1/F_{\text{total}} = 1/F_N + 1/F \qquad (16.46)$$

This relationship shows that the location of the axial peak for a focusing aperture is now moved in from the geometrical focal length. For example, if the natural focal length is 100 mm and the geometric focal length is 50 mm, the overall effective focal length is 33.3 mm. The shape of beams in the focal plane is the same as the far-field pattern of an unfocused beam; consequently, Eqs. 16.44–16.45 can be applied with $z = F$. ■

So far, solid apertures have been described. Arrays can also be considered to be spatially sampled apertures. To first order, the beams for both a solid and an adequately sampled aperture are similar. The one-dimensional array actually has two types of focusing as depicted in Figure 16.23. The azimuth or scan plane, here the x–z plane, is focused electronically whereas the elevation or yz plane is focused by a fixed mechanical lens.

16.2.7   Ultrasound Imaging Systems

The formation of an image can be understood through the operation of an imaging system. In Section 16.2.1 an imaging system was introduced as having a transducer

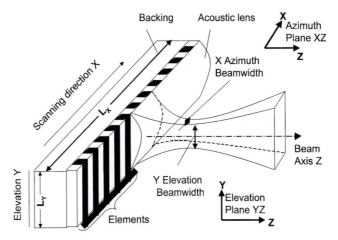

Figure 16.23 Focusing of a one-dimensional phased array in both the azimuth and elevation planes (adapted from Panda, 1998).

that could be scanned either mechanically or electronically to produce a pattern of sequential lines to form an image. With reference to Figure 16.24, the management of an array imaging system is accomplished by a computer or central processing unit (CPU). Once the scan depth and mode are selected, transmit pulses, each repeating at the time interval for a line, are sent in synchronism with a master timing clock like those in Figure 16.1. Each of these pulses initiates a set of transmit pulses from the transmit beamformer that are sent to each element, each one of which is delayed as necessary to form an electronic lens for focusing and steering the acoustic beam for the selected line direction in the azimuth plane.

Sound is scattered from tissue interfaces and inhomogeneities and is picked up as a series of pulse echoes by the array acting as a receiver. Depth-dependent time gain

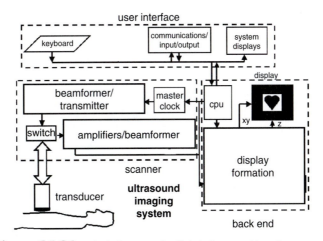

Figure 16.24 Block diagram of a digital ultrasound imaging system.

compensation (TGC) amplifiers can be controlled by the user to improve the image uniformity at different depths. The receive beamformer applies electronic dynamic focusing (nearly perfect focusing at each depth point), and signal processing (often in terms of filtering) provides the signals needed for the particular selected mode. For ordinary imaging, the envelopes of the pulse echoes are extracted and scan conversion, the process of interpolating the lines containing pulse-echo envelopes into a filled-out, gray-scale image that can be displayed on conventional PC or TV screens such as the images in Figures 16.15 through 16.17. Additional steps such as log compression are used to improve the range of pulse-echo amplitudes visible on the screen. The images in Figures 16.15 and 16.16 are of the most common type, B-mode or brightness mode.

16.2.8 Imaging and Other Modes

Other modes can supply additional information, especially about blood flow. The Doppler effect takes advantage of the apparent change in the ultrasound frequency caused by the velocity v of the blood and the angle θ the transducer makes with the vessel, as seen in Figure 16.25.

The shift in frequency from the transmitted one, f_0, or the classic Doppler shift frequency, f_D, can be expressed as

$$f_D = [2(v/c_0)\cos\theta]f_0 \qquad (16.47)$$

where c_0 is the sound speed of the intervening medium. Scattering from blood is mainly from groups of red blood cells. In a vessel there is a distribution of velocities so that what is displayed is a Doppler spectrum containing frequencies corresponding to the range of sound speeds insonified. This scattering is usually not visible by ordinary B-mode imaging but is detectable by sensitive ultrasound Doppler instrumentation. Doppler shifts are either detected along the length of an acoustic line selected to intersect a vessel of interest as continuous wave (CW) Doppler, or a small time

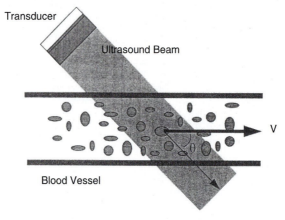

Figure 16.25 Sound beam intersecting blood moving at velocity v in a vessel tilted at angle θ.

interval is placed over the region of interest by pulsed wave (PW) Doppler as shown in the image of Figure 16.17.

A more global view of blood flow, especially for the chambers of the heart and larger blood vessels, can be obtained by color flow imaging (CFI) mode. While lacking the precision of the Doppler modes, color flow imaging provides a real time view of the approximate velocity and direction of blood flow. This information is represented by mapping of colors assigned to the velocity magnitude and direction (towards or away from the transducer). In this mode, several acoustic lines are sent in the same direction to obtain flow information over time, and then signal processing extracts the velocity information for display. A second type of Doppler image is called power Doppler in which the amplitude of the flow is presented but not the direction, as illustrated by the top of Figure 16.17.

Most ultrasound imaging is two dimensional in that a picture is created in an imaging or scan plane. A three-dimensional (3D) image can be created by scanning a volume rather than a plane. By mechanically moving the array in a direction perpendicular to the imaging planes, pausing long enough to acquire each image plane, and interpolating between planes, it is possible to fill in the overall volume with image data. The 3D images can be viewed in selected cut planes through the volumes or through surface or volume rendering. For example, a common application is 3D imaging of the fetus through surface rendering in which the boundary between the fetal skin and amniotic fluid is used to create a 3D opaque surface (Fig. 16.26).

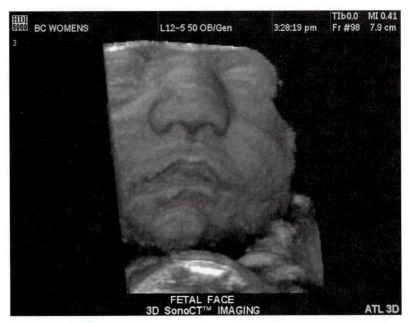

Figure 16.26   3D ultrasound image of a surface-rendered fetal head (courtesy of Philips Medical Systems).

Volume rendering allows internal detail of organs to be visualized through transparent boundaries between organs and layers. Images can also be formed through other ways of creating a set of volume data by different mechanical scanning techniques or by complete electronic scanning by a stationary two-dimensional array.

16.3   MAGNETIC RESONANCE IMAGING (MRI)

16.3.1   Introduction

Magnetic resonance has been applied successfully to medical imaging of the body because of its high water content. The hydrogen atoms in water (H_2O) and fat make up approximately 60% of the body by weight. Because there is a proton in the nucleus of each hydrogen atom, as the nucleus spins a small magnetic field or moment is created. When hydrogen is placed in a large static magnetic field, the magnetic moment of the atom spins around it like a tiny gyroscope at the Larmor frequency, which is a unique property of the material. For imaging, a radio frequency rotating field in a plane perpendicular to the static field is needed. The frequency of this field is identical to the Larmor frequency, and once the atom is excited, the applied field is shut off and the original magnetic moment decays to equilibrium and emits a signal. This voltage signal, detected by the same coils used for the applied field, and two relaxation constants are sensed. The longitudinal magnetization constant, T_1, is more sensitive to the thermal properties of tissue. The transversal magnetization relaxation constant, T_2, is affected by the local field inhomogeneities. These constants are used to discriminate among different types of tissue and for image formation. T_1 weighted images are used most often.

Today, MRI finds widespread application in the detection of disease and surgical planning. MR images are highly detailed representations of internal anatomy. These may be called parameterized images because considerable skill is involved in adjusting the instrument to obtain images that emphasize different types of tissue contrast, the discrimination among different organ types and between healthy and pathological tissues. MRI is used to examine most of the body, including the brain, abdomen, heart, large vessels, breast, bones, as well as soft tissue, joints, cartilage, muscle, and the head and neck. It is used for both children and adults and for detecting cancer pathologies, tumors, and hemorrhaging.

An early precedent to MRI was nuclear magnetic resonance (NMR), first observed as a phenomenon by Felix Bloch and Edward Purcell and their coworkers. They discovered that not only did precessing nuclei emit a radiofrequency (rf) signal but a radiofrequency could also be used to control precession at the Larmor resonant frequency and, once stopped, the nuclei would emit a detectable RF signal at the same frequency. They won a Nobel prize in 1951 for their work.

Interest shifted to determining composition of materials through unique frequency shifts associated with different chemical compounds. Eventually, biological NMR experiments were underway, and soon, detailed spectral information from phosphorus, carbon, and hydrogen nuclei were obtained. Specialized magnets were designed to accommodate parts of the body for study.

Paul Lauterbur was one of the first to realize that images could be made using NMR principles. He published an image of a heterogeneous object in 1973. Using the rf signals from NMR, he was able to localize them in space by changing the magnetic field gradient. By the mid and late 1970s, early MR images were produced of animals and the human body. At first, because the signals were so weak, these results were regarded as a laboratory curiosity. In 1971, Raymond Damadian demonstrated that the relaxation constants, T_1 and T_2, differed for malignant tumors and normal tissue. Peter Mansfield developed a mathematical model to analyze signals from within the human body in response to a strong magnetic field, as well as a very fast imaging method. Continuous research spurred the evolution of modern MRI instruments with high signal to noise and generated a considerable knowledge base of how to apply MRI to diagnostic imaging. Lauterbur and Mansfield shared the 2003 Nobel prize for medicine for their MRI discoveries.

16.3.2. Magnetic Fields and Charges

To understand how MRI works, several relevant characteristics of magnetic fields are reviewed here, in particular, the interactions between electrical charges and magnetic fields. Einstein pointed out that it is useful to consider electricity and electromagnetic fields as aspects of the same energy.

Four cases will be covered, each one useful for providing insights into aspects of MRI processes. In the first case, a magnetic field is generated when an alternating current travels along a wire. For an infinitely long wire, the Biot–Savart law reveals that a circular or circumferential magnetic field flux is generated by the current as indicated in Figure 16.27:

$$B_\phi = \frac{\mu_0 I}{2\pi r} \tag{16.48}$$

where the magnetic flux B is in units of Webers (1 Tesla(T) $= 10^4$ Webers/m^2), $\mu_0 = 4\pi 10^{-7}$ Henry/m [Weber/(amp-meter)] is the permeability of

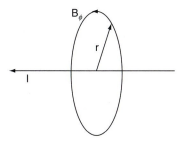

Figure 16.27 Circular magnetic field generated by an electrical current flowing down an infinitely long wire.

free space, I is current in Amperes, and r is the radial distance from the wire in meters. The right-hand rule applies: as the fingers curl about the direction of the B field, the thumb points in the direction of the current.

Example Problem 16.10

Calculate the magnetic field in T at 5 mm from a wire carrying 20 A of current.

Solution

Using Eq. 16.48,

$$B_\phi = \frac{4\pi e - 7 \times 20}{2\pi \times 5e - 3} (\text{Weber/m}^2) \times 1 \, T/10^4 (\text{Weber/m}^2) = 8 \times 10^{-6} \, T \quad \blacksquare$$

If the wire is coiled into a circular loop, a magnetic dipole with north and south poles is created. As a second case, an equivalent situation is created by a rotating charge as shown by Figure 16.28. Current I, flowing along an increment of wire dl, in a loop is equivalent to a charge q, of mass m, orbiting at a frequency, v. The magnetic dipole moment is the product of the equivalent current and area at a large distance r,

$$\hat{\mu} = (qvA)\hat{z} = qvpr^2\hat{z} = \frac{1}{2}qr^2\omega\hat{z} \qquad (16.49)$$

where the direction of $\hat{\mu}$ is along unit vector $\hat{z}$ according to the right-hand rule. A vector is a quantity that has a magnitude and a direction; in this case, a unit vector has a magnitude of one and is directed along the z axis. If the mass of the charge is m, the orbital angular momentum is

$$L = mr^2\omega \qquad (16.50a)$$

The classic gyromagnetic ratio is defined as

$$\gamma_c = \frac{\mu}{L} = \frac{q}{2m} \qquad (16.50b)$$

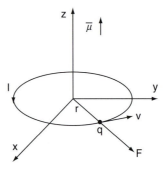

Figure 16.28   Magnetic dipole moment of a charge in a circular orbit.

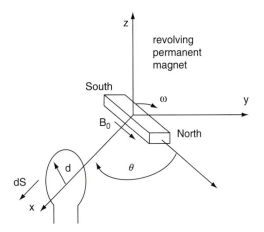

Figure 16.29 Voltage induced in wire loop of radius d by a revolving permanent magnet of strength B_0.

This overall result implies that an orbiting charge can act like a small magnet with its own north and south pole.

Can a moving magnetic field create a current or voltage in a wire? For the third case, consider the arrangement in Figure 16.29 where a wire loop of radius d is perpendicular to the x axis and where a permanent magnet of strength B_0 is whirling about the z axis at a constant angular frequency ω. The angle between the x axis and the magnet axis can be described as $\theta = \omega t$. Then, Faraday's law specifies that the voltage created in the loop by the spinning magnet can be written in terms of the electric field E around the loop. The area of the loop is $S = \pi d^2$ and its vector is perpendicular to the loop. If the field rotating relative to the x axis is $B_0 \cos \omega t$ and has a direction along vector B, then according to Faraday's law,

$$V = -\frac{d}{dt}\left(B_0 \pi d^2 \cos \omega t\right) = -\omega B_0 \pi d^2 \sin \omega t \qquad (16.51)$$

The voltage picked up from the rotating magnet is sinusoidal and is maximum when the axis of the magnet is perpendicular to the plane of the loop and is zero when the axis is parallel.

For the fourth case, a whirling charge is placed in a strong static magnetic field B_0, as depicted by Figure 16.30. Here the action of the field on the charge exerts a force on the charge described by the Lorentz force equation,

$$\hat{F} = q\hat{v} \times \hat{B}, \ or |F| = q|v||B| \sin \theta \qquad (16.52a)$$

where this vector cross product notation means the velocity vector $\hat{v}$ is tangential to the orbit at the position of the charge and the force is exerted outward perpendicular to both $\hat{v}$ and the applied field direction, $\hat{B}_0$ and the angle between $\hat{v}$ and $\hat{B}_0$, $\theta = 90°$. The magnitude of this force can be rewritten for this case as

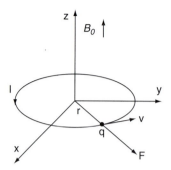

Figure 16.30   Lorentz force on a charge revolving about a static magnetic field. Vectors for velocity, *v*, force, *F*, and magnetic field, B_0, are shown at the position of the charge.

$$|F| = q|v|\,|B|\sin\theta = mv^2/r \tag{16.52b}$$

The classical angular frequency of the charge can be expressed in terms of the velocity of the charge and the radius, which can be rewritten from the previous equation, to give

$$\omega_c = v/r = v/(mv/qB_0) = (q/m)B_0 \tag{16.53a}$$

From the definition of the classic gyromagnetic ratio, Eq. 16.50, comes an important equation in MRI for the classical frequency,

$$v_c = \frac{\omega}{2\pi} = \frac{(2\gamma_c)B_0}{2\pi} = \frac{\gamma_c B_0}{\pi} \tag{16.53b}$$

which shows that the orbital frequency of the charge is proportional to the applied magnetic field. Unfortunately, this is not exactly what is needed for MRI because classical electromagnetic theory is for a charge that does not revolve on its own axis. The charge of interest in MRI is for an electron, which has its own individual spin. This situation is analogous to the revolution of the earth around the sun in combination with the revolution of the earth about its own axis. To obtain this important equation, an explanation of spin states is necessary from quantum mechanics.

16.3.3   Spin States

Based on the previous discussion, one could expect that the electron spinning on its own axis would create a miniature magnetic field; consequently, it would behave like a magnetic dipole with its own north and south pole.

Permanent bar magnets are dipoles that have a strong polarization in the form of north and south poles. If two equal permanent bar magnets are placed in the north to south setup shown in Figure 16.31a, they are strongly attracted and are said to have a strong attractive force between them. If they are placed close to each other in a north–north (or a south–south)configuration as in Figure 16.31b, the magnets are forced apart by a strong repulsive force. These two arrangements of magnets are two positions in which the strongest forces are stabilized in equilibrium.

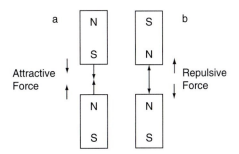

Figure 16.31   Attractive (a) and repulsive (b) magnetic forces for two arrangements of permanent magnets with north (N) and south (S) poles.

If a number of the tiny magnetic dipoles are placed in a strong static magnetic field, B_0, then they will align either with the direction of the field (parallel) or lock into a position opposing the field (antiparallel) as depicted in Figure 16.32. Most of them will line up along the applied field because less energy is required to maintain that orientation. The overall vector summation of the individual dipole moments results in an overall net magnetization,

$$\hat{M}_0 = \sum_n \hat{\mu}_n \tag{16.54}$$

as shown in the figure.

The energy required for the transition from one spin state to another is governed by

$$\Delta E = \hbar v \tag{16.55}$$

in which Plank's constant is $\hbar = 6.626 \times 10^{-34}$ J-s. The physical meaning of this relation is that a photon of frequency v is either absorbed to send an electron to a higher energy state or it is radiated for the electron to fall to a lower state. Eventually, excited electrons return to their original equilibrium state. Unlike x-ray imaging which emits ionizing radiation, only harmless photons are emitted during this mechanism which is used in MRI.

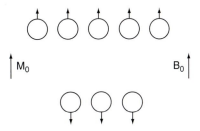

Figure 16.32   Alignment of magnetic dipoles in a static magnetic field B_0 and the net magnetization.

What is the frequency necessary for this transition? Returning to the last case of a charge revolving about the nucleus in a static magnetic field, the spin gyromagnetic ratio, γ, from quantum mechanics for a spinning electron is q/m. The orientation of the spin results in a left-hand rule (clockwise about B_0). Like Eq. 16.53b, substitution of γ a frequency equation yields

$$v_L = \frac{\omega}{2\pi} = \frac{\gamma}{2\pi}B_0 = \gamma'B_0 \qquad (16.56)$$

which is known as the Larmor frequency that can be used to calculate the transition energy. Note that the gyromagnetic constant, γ, is associated with angular frequency, and its normalized version, γ', with frequency. This is the important equation that connects frequency to applied magnetic field. For the hydrogen isotope most often used in MRI, $\gamma' = 42.58\,\text{MHz/T}$. Fortunately, the values of parameters needed for MRI fall within reasonable ranges: typical frequencies are under 100 MHz and fields are 0.1 to 4 T.

The Larmor frequency is a resonance excitation frequency used for externally producing transitions in spin states. For MRI, only isotopes that have unpaired nuclear spins that are multiples of 1/2 are of interest. For example, the spin state for ^{1}H aligned with the applied field is assigned the name "1/2," and has a dipole moment of $\mu_z = \gamma\hbar \times (+1/2)$. The antiparallel state for hydrogen has a spin designation of "–1/2," and has a moment $\mu_z = \gamma\hbar \times (-1/2)$. In Table 16.2, values of γ' are given for different isotopes. In general, $\mu_z = \gamma\hbar m$, in which m can include any of the following values $m = 0, \ldots \pm (I - 1), I$ where I is the spin-state number in Table 16.2.

How many of these spin states are there? If the number of spins in the low-energy state is called n_- and the number of those in the higher state are n_+, then Boltzmann statistics predict the ratio of spins in either of these two states at any given time,

$$n_-/n_+ = \exp(-\Delta E/KT) \qquad (16.57)$$

where the Boltzmann constant is $K = 1.3805 \times 10^{-23} J/K$, ΔE is from Eq. 16.55, and T is temperature in degrees Kelvin (K).

TABLE 16.2   Characteristics of Isotopes Relevant to MRI

Isotopes	Spin State	γ' (MHz/T)	Natural Isotropic Abundance %	Sensitivity Relative to ^{1}H
^{1}H	1/2	42.58	99.99	100
^{13}C	1/2	10.71	1.11	1.6
^{19}F	1/2	40.05	100	3.4
^{23}Na	3/2	11.26	100	9.3
^{31}P	1/2	17.24	100	6.6

Example Problem 16.11

Find the fraction of excess population in the spin-up (n_+) state for a frequency of 20 MHz and a temperature of 300 K.

Solution

Calculate the change in energy for a given frequency of 20 MHz from Eq. 16.55:

$$\Delta E = \hbar v = 6.626 \times 10^{-34} J - s \times 20 \times 10^{6}/s = 1.325 \times 10^{-26} J$$

From Eq. 16.57, the excess population can be written as

$$(n_+ - n_-)/n_+ = 1 - n_-/n_+ = 1 - \exp(-\Delta E/\underline{K}T) \qquad (16.58)$$

Substituting the necessary values,

$$(n_+ - n_-)/n_+ = 1 - \exp\left[-1.325 \times 10^{-26} J/(1.3805 \times 10^{-23} J/K)300K\right]$$
$$= 3.2 \times 10^{-6}$$
■

This is a small number based on the material under examination being entirely of one type. In reality, certain types of isotopes occur more commonly than others. The ratio of one type of isotope to the total number available in percent is called the natural isotropic abundance, and is listed in Table 16.2. A third relevant question is how many of these isotopes occur in the human body? The third factor is sensitivity relative to the hydrogen isotope, or the equivalent number of nuclei in a field, also listed in Table 16.2. Fortunately, hydrogen is plentiful in the body, especially in fat and water.

Example Problem 16.12

Compare the isotopes ^{1}H, ^{13}C, and ^{31}P for imaging. Use Table 16.2.

Solution

Note that ^{1}H is 99.98% naturally abundant. ^{13}C is not, and occurs at only 1.11%, so this isotope is not suitable. On the other hand, ^{31}P is abundant but is difficult to detect due to its low sensitivity. Because the body is 60% water by weight, it is not surprising that ^{1}H, with its high sensitivity and abundance, is usually used for MR imaging. ■

16.3.4 Precession

To excite hydrogen dipoles into a number of spin states for imaging, an external magnetic field can be applied. The natural inclination for the spin magnetic dipoles to align along the z axis makes detection by a coil difficult if the coil is placed perpendicular to the x axis as in the fourth case from Section 16.3.2. It is necessary to find a means to bring the dipoles down into the $x - y$ plane so they can be detected by a coil in a manner analogous to case 4. A force is required to push the dipoles into a precession, a downward spiraling orbit. This mechanism is similar to the action of gravity on a spinning top, which is vertical initially, and which is tilted eventually by the force of gravity into a precessing orbit and finally, into a final horizontal position.

This precession is illustrated by Figure. 16.33. At first the net magnetization vector, $\hat{M}_0$, is aligned with the static magnetic field along the z axis as depicted in Figure 16.33a. The application of a time-varying magnetic field, B_1, along the x axis at the Larmor frequency causes the magnetization to precess at an angle ϕ about the z axis at this frequency, as shown in Figure 16.33b.

To clarify what happens to $\hat{M}_0$ next, it is worth introducing a reference frame notation to simplify the description. This frame will have coordinates described by a prime notation, such as $\hat{x}'$. One reason for using this frame is that the trajectories of these spinning, precessing magnetic vectors are complicated, and simplified methods are helpful in visualizing them. Another reason is that a bundle of frequencies is usually involved in these measurements, and it is easier to track their movements relative to the reference frequency v_L.

Rotating Frame

The approach is called a rotating frame, a coordinate frame of reference that rotates with the magnetization vector whirling at a Larmor frequency, v_L, in contrast to a reference point in a fixed Cartesian coordinate system. This methodology is similar to a stroboscopic viewpoint: to make a rotating object appear still, a stroboscopic light flashes in synchronism with each revolution of the object. This light frequency serves as a reference so that if the revolving object rotates faster or slower, its deviation from this reference frequency is easily seen. Similarly, a frequency rotating clockwise relative to v_L in a rotating frame is called a positive frequency, and a frequency rotating counterclockwise is called a negative frequency.

Recall case 3 from Section 16.3.2, a description of a configuration where a signal was detected from a permanent magnet spinning around the z axis. The resulting detected signal was a sinusoidal signal with a frequency ω. Instead of a permanent magnet, consider a time-varying magnetization vector spinning at a Larmor frequency about the z axis in the $x - y$ plane at a position initially aligned with the x axis at time t,

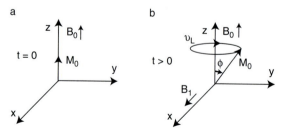

Figure 16.33   (a) Initial alignment of magnetization with the static magnetic field B_0. (b) Tipping of the net magnetization from the z axis through an angle ϕ by the application of the rf B_1 field aligned along the x axis causes M_0 to spin about the z axis at the Larmor frequency.

$$\hat{B}_x(t) = \hat{x}B_1 \cos \omega t \qquad (16.59)$$

To find out what this magnetic field vector would look like in a rotating frame, a coordinate transformation is applied to obtain the rotated components of the field in the new frame, which are part of a new vector B_1' with components,

$$B_{x'} = \cos (\omega t)B_x - \sin (\omega t)B_y \qquad (16.60a)$$

$$B_{y'} = \sin (\omega t)B_x + \cos (\omega t)B_y \qquad (16.60b)$$

which with the substitution of Eq. 16.58 becomes

$$B_{x'} = B_1 \cos^2 \omega t \qquad (16.60c)$$

$$B_{y'} = B_1 \cos \omega t \sin \omega t \qquad (16.60d)$$

From trigonometric identities for half-angles, the preceding result simplifies to

$$\hat{B}_1'(t) = \hat{x}'B_{x'} + \hat{y}'B_{y'} = \hat{x}' \left(\frac{B_1}{2} - \frac{B_1}{2} \cos 2\omega t \right) + \hat{y}' \frac{B_1}{2} \sin 2\omega t \qquad (16.60e)$$

which, with low-pass filtering to eliminate the components at twice the Larmor frequency, leaves

$$\hat{B}_1'(t) = \hat{x}' \frac{B_1}{2} \qquad (16.60f)$$

a component of B_1 that appears to be stationary (not time varying) in the rotating frame, as expected.

Flip Angles and Decay

In MRI, the ability to control the position of the magnetization vector is important. A practical configuration to accomplish this magnetic precession is shown in Figure 16.34 for rotated coordinates. First, the magnetization vector is shown aligned with the static field in Figure 16.34a. Second, a radio frequency (rf) magnetic field B_1 along the x' axis at the Larmor frequency, ν_L, is applied as a tone burst (a gated sinusoid) of

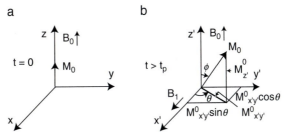

Figure 16.34 (a) Initial alignment of magnetization at $t = 0$. (b) Tipping of the net magnetization from the z axis through an angle φ by the application of rf B_1 field at the Larmor frequency aligned along x' in rotated coordinates. The projections of the net magnetization onto the axes are shown.

pulse time duration, t_p. The application of this pulse creates a force to pull the magnetization $\hat{M}_0$ (usually initially aligned in an equilibrium position along the z axis) down into a precession angle φ from the z axis, called the flip or tip angle,

$$\varphi = \gamma\left(\frac{B_1}{2}\right)t_p = \gamma B_1' t_p \tag{16.61}$$

The precessing magnetization vector can also be broken down into its Cartesian components in the rotated frame as illustrated by Figure 16.34b. In addition to the component along the z' axis, $M_{z'}^0$, there are two components in the $x' - y'$ plane with a radial magnitude, $M_{x'y'}^0$. It is convenient to look at the complicated resulting changes in time separately along the two components, $M_{z'}^0$ and $M_{x'y'}^0$, even though they occur simultaneously.

Example Problem 16.13

Find the pulse length necessary to achieve a phase rotation of π for the field calculated in Example Problem 16.10. Repeat for a coil that generates a field 100 times greater.

Solution

From Eq. 16.60, solve for t_p using Eq. 16.56 and Table 16.2.
$\gamma = 2\pi\gamma'$ from Eq. 16.56
$B_1 = B_\phi = 8 \times 10^{-6}$ T from Example Problem 16.10 and $\gamma = 42.58 \times 10^6$ MHz/T
Solve for t_p in Eq. 16.61,

$$t_p = \frac{\varphi}{2\pi\gamma'(B_1/2)} = \frac{\pi}{2\pi \times 42.58 \times 10^6 (/s - T)(4 \times 10^{-6} T)} = 2.936 \text{ msec}$$

For the second case the pulse is smaller by a factor of 100, or 29.36 microseconds. ∎

If the flip angle is chosen to be 90° or $\pi/2$ radians, the magnetization is rotated from an initial value of M_0 along the z' axis to a position lying along the y' axis as illustrated by Figure 16.35a. The vertical component is brought a value of zero in the $x' - y'$ plane and then rebounds back to its original initial value of M_0 over a period of time. This recovery can be described by the following equation and Figure 16.36:

$$M_{z'}(\theta = \pi/2) = M_0[1 - \exp(-t/T_1)] \tag{16.62}$$

where T_1 is the spin lattice recovery time and t starts after the rf pulse. It is also called the longitudinal time constant because of its association with the vertical direction. This relaxation time is affected by temperature and the viscosity of the tissue or material (it is longer as viscosity increases).

If the flip angle is chosen to be 180° or π radians, the magnetization is rotated from an initial value of M_0 along the z' axis to a position lying along the $-z'$ axis. In this antiparallel orientation, the appropriate relaxation equation is

$$M_{z'}(\theta = \pi) = M_0[1 - 2\exp(-t/T_1)] \tag{16.63}$$

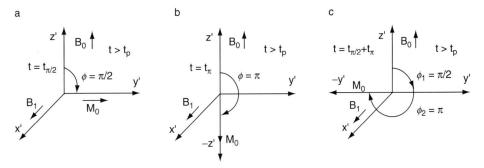

Figure 16.35 (a) Tipping through an angle of $\pi/2$ radians to the y' axis. (b) Tipping through an angle of π radians to the $-z'$ axis. (c) Tipping sequence: first tip through an angle of $\pi/2$ radians to the y' axis and then tip through an angle of π radians to the $-y'$ axis.

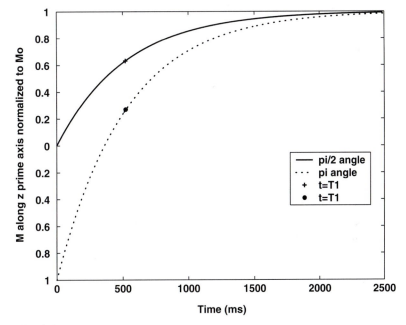

Figure 16.36 Plots of the recovery of a magnetization vector component along axis z' for a relaxation time T_1, s versus time for a $\pi/2$ rotation, and a π rotation for gray matter of the brain.

This change is illustrated by Figure 16.35b. A generalization of these equations to an arbitrary value of the $z-$ component of magnetization following the rf pulse, M_z^0, is

$$M_{z'}(t) = M_0\left[1 - \left(1 - \frac{M_z^0}{M_0}\right)\exp(-t/T_1)\right] \quad (16.64)$$

To determine the magnetization components in the $x - y$ plane, refer back to the situation of the $\pi/2$ flip angle where the magnetization is rotated to the y' axis at time $t = t_{\pi/2}$. In order for equilibrium to be restored, the initial $M^0_{x'y'}$ component at a time t beginning after the rf pulse must decay back to a zero value so that the original value of the purely vertical M_0 can be recovered,

$$M_{x'y'}(\theta = \pi/2) = M^0_{x'y'}[\exp(-t/T_2)] \tag{16.65}$$

in which T_2 is the spin–spin relaxation time as plotted in Figure 16.37. Because of the motion and orientation of this component, its interaction with the material or tissue is different than that of T_1. As Figure 16.38 shows, a dephasing process occurs as vectors sweep over an increasing sector over time so that the net sum of vectors gradually goes to zero.

T_2 is sensitive to molecular interactions and inhomogeneities in the applied field B_0. A more realistic definition of this relaxation constant includes the effects of variations ΔB_0,

$$\frac{1}{T_2^*} = \frac{1}{T_2} + \frac{\gamma \Delta B_0}{2} \tag{16.66}$$

Through a comparison of Figures 16.36 and 16.37, it is evident that the T_2 constants are much shorter than T_1 and T_2^* is the shortest of all. Typical values for relaxation times can be found in Table 16.3.

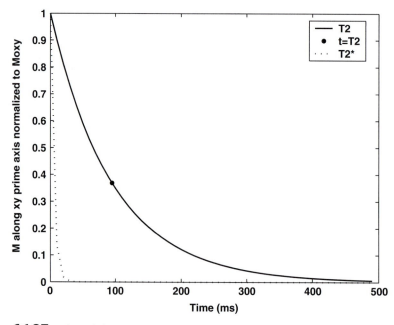

Figure 16.37   Plots of the magnetization vector component in the $x' - y'$ for a decay time T_2 versus time for a $\pi/2$ rotation for gray matter and $T_2^* = 5$ ms.

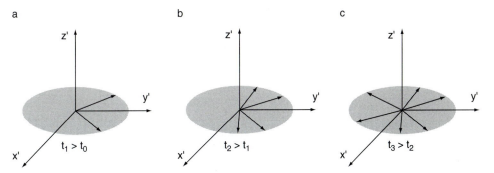

Figure 16.38 Time sequence of dephasing plots for spin–spin decay with time constant T_2 after $\pi/2$ rotation pulse occurring at $t = t_p$.

Example Problem 16.14

After a $\pi/2$ rotation, determine the percentage of recovery of the magnetization vertical component associated with T_1 in fat when the horizontal magnetization vector component has decayed to a $1/e$ value.

Solution

From Table 16.3, $T_2 = 100$ ms which would be the time required for the horizontal magnetization component to decay to a value of $1/e$ value. With the value of $T_1 = 160$ ms and Eq. 16.62, the relative recovery value is

$$1 - \exp\left(-T_2/T_1\right) = 0.465$$ ■

Detected Response Waveforms

Because sufficient background information has been given to describe the excitation and decay of a nuclear resonance, it is now possible to describe the kinds of signals detected. A setup like that of Figure 16.33a is used with a detection loop in a plane perpendicular to the x axis. A $\pi/2$ rf excitation pulse for tipping the field 90° consists

TABLE 16.3 T1 and T2 Values for Common Tissue Types

Tissue	T_1 (ms)	T_2 (ms)
Cerebro-spinal fluid	2000	1000
Fat	160	100
Gray matter	520	95
Malignant tumor	800	200
Typical edema or infarction	600	150
White matter	380	85

of a tone burst t_p long at the Larmor frequency according to Eq. 16.61. Since the duration of the pulse is much longer than any of the recovery and decay time constants, a sinusoidal excitation in an exponential form will be used to simplify analysis. The resulting transverse magnetization vector in the $x - y$ plane is

$$\hat{M}_{xy}(t) = M_{xy}^0 e^{-|t|/T_2} e^{j\omega t}(\hat{x} + \hat{y}) \qquad\qquad t < t_p \qquad\qquad (16.67)$$

The effect of this excitation is to cause a resonance and a change in the spin states of hydrogen electrons. Faraday's law can be applied to the detection of the overall net magnetization response from Section 16.3.2, case 3. The area S of a loop of diameter d can be rewritten in terms of magnetization M and an equivalent current I,

$$S = \frac{M}{I} \qquad\qquad (16.68)$$

If the direction of the area is taken to be the x' axis and the corresponding component of M is used in Faraday's law, Eq. 16.51, for the cubic volume, V_{voxel}, of an image voxel

$$v(t) = -V_{\text{voxel}}\frac{\partial}{\partial t}\left[\left|B_1'(t) \times \frac{M_{xy}}{I}\right|\right]$$

$$v(t) = -V_{\text{voxel}}\frac{\partial}{\partial t}\left[\left|\hat{B}_1'(t) \times \frac{\hat{M}_{xy}}{I}\right|\right] = -V_{\text{voxel}}\frac{\partial}{\partial t}\left[|B_1'(t)|\left|\frac{M_{xy}}{I}\right|\sin\theta\right] \qquad (16.69\text{a})$$

From Eq. 16.67 for M_{xy} and Eq. 16.60f for $\hat{B}_1'(t)$, and since the angle between $\hat{B}_1'$ and $\hat{M}_{xy}$ is $\theta = 90°$, the voltage becomes

$$v(t) = \left(\frac{B_1}{2I}\right) \cdot \left(j\omega V_{\text{oxel}}M_{xy}^0 e^{-|t|/T_2} e^{j\omega t}\right) \qquad\qquad (16.69\text{b})$$

where V_{oxel} is the cubic volume of an image voxel and the second-order terms in dM/dt are dropped. This detected voltage is called the free induction decay (FID), the real part of which is shown in Figure 16.39.

The major reason the FID waveform is unsuitable for detection is that it is severely affected by variations in the magnetic field that dominate the response over the signals

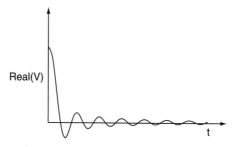

Figure 16.39   Free induction decay waveform versus time. This waveform starts at $t = 0$.

of the desired spin states. Also, the waveform weakens rapidly with the shorter time decay constants. As shown in Figure 16.38, spin states experience different magnetic fields because of their spatial location; therefore, the angular frequencies of their magnetization vectors vary, some moving faster than others, according to Eq. 16.56.

A clever alternative is called the spin echo method. If a certain time, $T_e/2$, elapses so that the spin vectors are severely but not totally dispersed as illustrated by Figure 16.35c, a second π pulse, with an amplitude twice that of the original $\pi/2$ pulse, rotates the leading vectors back to the $+y$ axis as depicted in Figure 16.40a. The individual magnetization vectors rotate in an opposite direction toward realignment. As the vectors draw into coherent phase, the detected signal increases to an alignment peak at T_e and then decays as shown in the spin echo waveform of Figure 16.40b. The beauty of this approach is that the effects of magnetic field variations are canceled out, resulting in a truer T2 tissue response. Following a derivation similar to that of the FID waveform, we can obtain an expression for the spin echo:

$$v(t) = j(B_1/2I)\omega V_{\text{oxel}}M^0_{xy}e^{-|t/T_2|}e^{j\omega t} \tag{16.70}$$

To find the spectrum of the spin echo, we use the following transform pair:

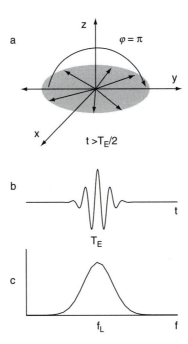

Figure 16.40   (a) A π pulse rotation causes a reversal in the rotation direction of the magnetization vectors so they move toward realignment: (b) spin echo waveform versus time, and (c) spin echo spectral magnitude versus frequency.

$$\Im[\exp(-|t|)] = \frac{2}{1 + (2\pi f)^2} \tag{16.71a}$$

along with the transform pair for the exponential (note in Eq. 16.70, $\omega = \omega_L = 2\pi f_L$, a constant, the electrical frequency corresponding to the physical Larmor frequency, ν_L),

$$\Im\left(e^{j\omega_L t}\right) = \delta(f - f_L) \tag{16.71b}$$

and the scaling theorem, Eq. 16.4a, to obtain the echo spectrum of Eq. 16.70:

$$V(f) = \Im\left(\frac{jB_1 M_{xy}^0 \omega_L V_{\text{oxel}} e^{-|t|/T_2|} e^{j\omega_L t}}{2I}\right) = \frac{jB_1 M_{xy}^0 \omega_L V_{\text{oxel}} T_2}{2I} \left\{\frac{1}{1 + [2\pi T_2(f - f_L)]^2}\right\} \tag{16.71c}$$

and the magnitude of which is shown in Figure 16.39c. If a function $S(f)$ is defined as

$$S(f) = \frac{jB_1 M_{xy}^0 \omega_L V_{\text{oxel}} T_2}{2I} \left\{\frac{1}{1 + [2\pi T_2 f]^2}\right\} \tag{16.71d}$$

then the right-hand side of Eq. 16.71c becomes

$$V(f) = S(f - f_n) \tag{16.71e}$$

in which $f_n = f_L$ for this case. In general, the index n can be associated with the excitation Larmor frequency at a location x_n. How this mapping is carried out is the subject of the next section.

16.3.5   Setup for Imaging

To create images from detected signals, a means of spatially localizing the detected waveforms must be used. Three methods are used to encode the waveform data and position, one for each Cartesian coordinate. The three-dimensional image is discretized into tiny cubes called voxels, each with a volume $V_{\text{oxel}} = \Delta x \Delta y \Delta z$.

The first step in setting up the image is to fix the desired location of the slice plane, one voxel thick, along the z axis as shown in Figure 16.41. The person to be imaged is slid into a large superconducting magnet that creates a strong static field, B_0. Recall from cases 1 and 2 from Section 16.3.2 that a current in a wire or loop can generate a magnetic field. Electrical coils are arranged within the magnet to create an electrically controlled linear magnetic field gradient along the z axis:

$$\vec{B}_z = \hat{z}(B_0 + G_z z) \tag{16.72a}$$

where G_z is a gradient constant in T/m. From Eq. 16.56, a unique Larmor frequency is associated with each spatial location z,

$$\omega = \gamma B_z = \gamma B_0 + \gamma G_z z \tag{16.72b}$$

The relative position $z = 0$ is called the isocenter. The reference frequency associated with this center is $\omega_0 = \gamma B_0$ which can be subtracted from ω by electronic mixing.

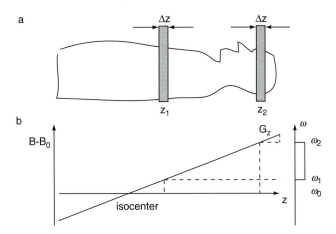

Figure 16.41 (a) Side view of selected axial image plane locations. (b) Two regions corresponding to selected locations on the z-axis magnetic field gradient. On the right vertical axis are regional gradient spectra from rf excitation pulses whose Larmor center frequencies are selected for z-axis positions. Scale exaggerated for clarity.

By applying a time excitation pulse with an appropriate center frequency as shown in Figure 16.41, a specific location is selected. The envelope of the sinusoid used for time domain rf pulse excitation is shaped to obtain a flatter field. A sinc-shaped pulse would have a flat rectangular spectral shape, a consequence of its Fourier transform relation, but, ideally, it would have to be infinitely long. This pulse and the corresponding field-spectral shape are shown in Figure 16.42. A sinc-time pulse has a Fourier transform, $\Im$, that can be expressed as a rect function of width Δf,

$$\Im[\sin c(\Delta ft)\exp(j2\pi f_1 t)] = (1/\Delta f)\prod[(f-f_1)/\Delta f] \qquad (16.73)$$

As a compromise, a truncated or tapered or modified sinc pulse of finite length is applied instead, such as the tapered shape shown in Figure 16.40b and c.

Example Problem 16.15

Find the bandwidth needed to achieve a z-slice thickness of 10 mm. Assume $G_z = 0.352\,\mathrm{T/m}$. For $B_0 = 1\,\mathrm{T}$, determine the frequencies needed to move 10 cm and 20 cm from the isocenter.

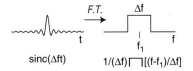

Figure 16.42 Sinc modulated excitation pulse with a selected Larmor center frequency and its Fourier transform representing its spectrum or the shape of the applied field.

Solution

From a version of Eq. 16.72b with γ', the difference in frequency is proportional to thickness, or

$$\Delta f = 42.58(MHz/T) \times 0.352(T/m) \times .010\,m = 0.15\,\text{MHz}$$

The same equation can be used with a value of the offset from the isocenter, $f_1 = 42.58\,(\text{MHz})\,[1T + 0.352\,(T/m) \times 0.1\,m] = 44.08\,\text{MHz}$. Similarly, $f_2 = 45.58\,\text{MHz}$ corresponds to the frequency needed to move 20 cm from the isocenter. ∎

In order to scan the $x - y$ slice plane located by the z-axis gradient control, as seen in Figure 16.43, the second step is to locate data points along the axis. A gradient is also applied along the x axis for spatial localization,

$$\omega = \gamma B_x = \gamma B_0 + \gamma G_x x \qquad (16.74a)$$

Again, with mixing, $\omega_{\text{mix}} = \omega_0$ for $B = B_0$, the difference frequency is proportional to the position x,

$$\omega_\Delta = \omega - \omega_{\text{mix}} = \gamma G_x x \qquad (16.74b)$$

For z slice selection, a specific frequency is applied that corresponds to a particular desired location, according to Eq. 16.72b. Although a similar principle is used for the locations along x, the implementation is somewhat different. Here, the detected signal frequency varies with distance x, a consequence of the linear magnetic field gradient, as illustrated by Figure 16.44. A number of excited resonances positioned along the

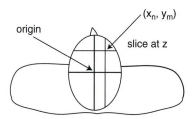

Figure 16.43   Depiction of an $x - y$ slice plane of thickness Δz at a distance z, like the positions shown in Figure 16.41.

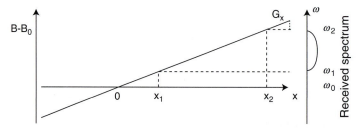

Figure 16.44   Frequency encoding method for x axis with two locations, x_1 and x_2 shown.

scan line x can contribute to the overall detected signal and are encoded in the received signal as different frequencies.

Unlike diagnostic ultrasound imaging in which specific echo delays in time are associated with reflectors at different round trip distances, MRI signals for different spatial locations are frequency encoded and added together in the same time signal position. In this case,

$$V(f) = \sum_n S(f - f_n) \tag{16.75}$$

The means of decoding these positions will be explained in a later section. Although MRI systems have employed yet another linear magnetic gradient along the y axis to provide localization for this coordinate, a different phase-encoding scheme is now more common. The mechanism employed is that of the flip angle, Eq. 16.61, with a different twist,

$$\varphi(y) = \omega t = -\left(\gamma B_0 t + \gamma G_y t_p y\right) \tag{16.76}$$

in which $y = y' + y_0$ where y_0 is the value at the center. In the previous discussion of the flip angle, an rf magnetic field with an appropriate pulse width, t_p, was applied to the spins to rotate a magnetization vector through a desired angle. In this context, Eq. 16.76 shows that either the pulse length, t_p, or the pulse gradient slope, G_y, can be used to change the phase.

The preferred method, as illustrated by Figure 16.45, is not to change pulse widths but rather to alter amplitudes and signs of the gradient slope, which can be changed by voltage amplitudes applied to the y-axis gradient coil. In order to fill out the $x - y$ plane, a means of scanning along y is needed. Typically, 128 or 256 different excitations are utilized and each one has to be uniquely linked with a specific y position. Pulse and excitation sequences for several consecutive lines are depicted in Figure 16.46.

16.3.6   MR Imaging

Almost all the necessary bits and pieces needed for the imaging process have been introduced. Now, step by step, the different controls and detection methods will be reviewed and assembled into a sequence of events for imaging. The process will be simplified slightly for clarification. In Figure 16.43 is an $x - y$ image plane that

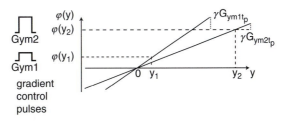

Figure 16.45   Phase encoding method for the relative y axis for two locations y_1 and y_2.

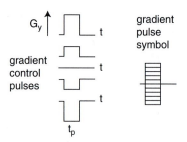

Figure 16.46   Pulse sequences for several consecutive lines. Note the amplitude of G_y pulses changes on sequential lines. T_R is the repetition time between lines. Usually the amplitudes for G_y range from a maximum value, G_{ymax}, through a negative value of the same magnitude.

contains the isocenter, or center of magnetic field coordinate system, for a head scan. How net magnetization signals from specific locations in this plane translate into magnetic fields, detected waveforms and ultimately, into an image will be demonstrated.

Once the subject is within the MRI magnet, the following sequence pulses are sent, as shown in Figure 16.47:

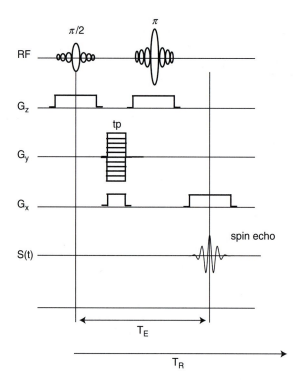

Figure 16.47   Essential rf excitation, and gradient pulse sequences, G_z, G_y, G_x, for spin-echo detection, $s(t)$.

Two rf pulses, one a $\pi/2$ pulse and another a π pulse, force the magnetization vectors to precess into the $x - y$ plane to set up spin echoes, as described in Section 16.3.4. The $G_z z$-axis gradient pulses occur over the same intervals as the rf pulses. They turn on the gradient coils which, in combination with the frequency selected for the rf pulses, select the z location of the $x - y$ image plane as detailed in Section 16.3.5.

The gradient pulses for the x axis occur during the precession phase and also over the interval during which the spin echoes are formed. The field must be on during these intervals so that the positions of the spins are translated into different frequencies that vary with position along the x axis.

Finally, a y gradient pulse excites the coils along the y axis with an amplitude appropriate to phase encode the position y.

Spin echoes are sensed at $t = T_E$. The sequence is repeated at intervals of T_R.

To form an image, the set of acquired signals and the object to be imaged (a distribution of net magnetization vectors) are assumed to be related by Fourier transforms. If a particular slice plane has been selected at z, then the object is a function of position (x,y) which implies that a double Fourier transform (for two dimensions) is involved. It will be easier to consider each dimension separately (x or y) and then combine them into a 2D transform. Signals detected from a net magnetization distribution at a position (x_n, y_m) will be considered.

Before starting with the y axis, it will be helpful to review the role of the impulse in Fourier transform theory described in Section 16.2. The impulse and its transform play an essential role in the understanding of MRI. The impulse function is a generalized function which has the unusual property that it samples the integrand as introduced in Section 16.1.1,

$$\int_{-\infty}^{\infty} \delta(f - f_0)R(f)df = R(f_0) \tag{16.77a}$$

When the inverse Fourier transform of the impulse function is taken, the result is an exponential

$$g(t) = \int_{-\infty}^{\infty} \delta(f - f_0)e^{i2\pi ft}df = e^{i2\pi f_0 t} \tag{16.77b}$$

which has the forward Fourier transform,

$$G(f) = \int_{-\infty}^{\infty} \left(e^{i2\pi ft_0}\right)e^{-i2\pi ft}dt = \int_{-\infty}^{\infty} e^{-i2\pi t(f-f_0)}dt = \delta(f - f_0) \tag{16.77c}$$

If $w(t)$ has the Fourier transform $W(f)$, then

$$\int_{-\infty}^{\infty} \left[w(t)e^{i2\pi f t_0} \right] e^{-i2\pi f t} dt = \int_{-\infty}^{\infty} w(t)e^{-i2\pi t(f - f_0)} dt = W(f - f_0) \qquad (16.77\text{d})$$

The simplest place to start is the y axis. For the y dimension, phase encoding according to Eq. 16.76 is applied and expressed with the constant phase term suppressed after mixing,

$$\varphi(y) = -\gamma G_y t_p y \qquad (16.78\text{a})$$

The phase associated with a particular location is

$$\varphi_m(y_m) = -\left(\gamma G_{y,m} t_p y_m \right) \qquad (16.78\text{b})$$

where the phase is indexed to a position y_m as well as a specific gradient slope coefficient, $G_{y,m}$. The encoded time signal can be described by

$$s_m(0, G_y) = e^{-i\gamma t_p y_m G_y} \qquad (16.78\text{c})$$

The Fourier transform pair of variables for the y dimension will be the varying amplitude slope, G_y, as indicated in Figure 16.45 and a new variable called u. Analogous to the paired variables t and f, which have reciprocal units, u will have the upside-down or inverse units of m/T. These variables, G_y and u, can be used to construct a specific phase term that later will be associated with y_m in a Fourier transform format,

$$S_m(0, u) = \int_{-\infty}^{\infty} \left(e^{i2\pi u_m G_y} \right) e^{-i2\pi u G_y} dG_y = \int_{-\infty}^{\infty} e^{-i2\pi(u - u_m)G_y} dG_y = \delta(u - u_m) \qquad (16.79)$$

Here u appears in the argument of second exponent in the integrand. The overall argument is recognized as a phase according to Eqs. 16.77c and 16.78a, so that an explicit expression for u can be obtained by equating phases,

$$2\pi u G_y = -\gamma t_p y G_y \qquad (16.80\text{a})$$

$$u = -\gamma t_p y / (2\pi) \qquad (16.80\text{b})$$

This evaluation of the Fourier transform of phase encoding follows directly from the discussion of the impulse function. As a consequence of this change in variables, the spectrum S_m in Eq. 16.79 can be restated in terms of the scaled variable from Eq. 16.80b,

$$S_m(0, u) = \delta(u - u_m) = \delta\left[\left(-\gamma t_p / (2\pi) \right)(y - y_m) \right] \qquad (16.81\text{a})$$

$$S_m(0, y) = \left[2\pi / (\gamma t_p) \right] \delta(y - y_m) \qquad (16.81\text{b})$$

where use has been made of the property $\delta(au) = \delta(u)/|a|$ from Section 16.2.

The derivation of this key equation reveals an important principle in magnetic resonance imaging. First, a local value of a gradient slope, $G_{y,m}$, encodes a position,

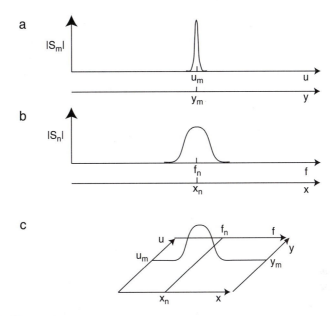

Figure 16.48 (a) Spectrally phase-encoded magnitudes plotted versus multiple scaled y and u axes. (b) Spectral frequency-encoded magnitudes plotted versus multiple scaled x and f axes. (c) Resulting pattern in xy or uf plane. Scale exaggerated for clarity.

y_m, into a phase according to Eq. 16.78b. Second, this phase, in terms of Fourier transforms, assures that the signal function has a spectrum centered on the value of u associated with the location of the corresponding net magnetization. Through simple scaling, the scaled spectra are translated into spatial locations along the y axis (Fig. 16.48a). This process maps a scaled spectral magnitude into its locations along the y axis.

To formulate the Fourier transform relation for the x axis, a single spin echo waveform is generated, for example, at the position $y = 0$ in Figure 16.40b and so that $G_y = 0$. From Eq. 16.70, the signal waveform, delayed by time T_E and decaying with a unique time constant T_{2n} associated with location x_n can be expressed as

$$s_n(t, 0) = 2S_0 e^{-|(t-T_E)/T_{2n}|} e^{j\omega_n(t-T_E)} \tag{16.82}$$

in which the s is a function of t and $G_y = 0$, S_0 is a constant, and the Larmor frequency associated with location x_n is

$$\omega_n = \gamma G_x x_n \tag{16.83}$$

after mixing removes ω_0 at the isocenter.

The one-dimensional Fourier transform of Eq. 16.82, is

$$S_n(f, 0) = \int_{-\infty}^{\infty} s(t, 0) e^{-i2\pi f t} dt = \int_{-\infty}^{\infty} \left[S_0 e^{-|t-T_E|/T_{2n}} \right] \left[e^{-i2\pi f_n T_E} e^{-i2\pi(f-f_n)t} \right] dt \tag{16.84a}$$

The form of the complex exponents is recognizable from Eq. 16.77d as the transform of an impulse function centered on f_n. This relation implies that the Fourier transform of the function in the left brackets is centered on f_n. To show this result, it is necessary to determine if the Fourier transform of the first term in brackets is of the form of Eqs. 16.71a and 16.71d with S_0 as the constant,

$$W_n(f, 0) = \int_{-\infty}^{\infty} w_n(t, 0)e^{-i2\pi ft}dt = \int_{-\infty}^{\infty} \left[S_0 e^{-|t-T_E|/T_{2n}} \right] e^{-i2\pi ft} dt = \frac{S_0}{1 + (2\pi f\, T_{2n})^2}$$

(16.84b)

then $S_n(f, 0)$ can be expressed as

$$S_n(f, 0) = e^{-i2\pi f_n T_E} W_n(f - f_n)$$

(16.84c)

From the relation for the Larmor frequency, this equation can be rewritten in terms of a scaled variable for frequency from Eq. 16.83 as

$$S_n(x, 0) = e^{-i\gamma G_x x_n T_E} W_n \left[\frac{\gamma G_x}{2\pi} (x - x_n) \right]$$

(16.84d)

The derivation of this key equation reveals a second important principle in magnetic resonance imaging. First, a locally excited resonance encodes a position into a frequency f_n as part of the phase of a time waveform. Second, this phase, in terms of Fourier transforms, assures that the signal function has a spectrum centered on the frequency associated with the location of the net magnetization density. Third, through simple scaling, the scaled spectra are translated into spatial locations along the x axis (Fig. 16.48b). This process maps scaled spectral magnitudes into their spatial locations along the x axis.

To advance to two dimensions, the one-dimensional inverse Fourier transforms are combined in a two-dimensional relation,

$$s(t,\ G_y) = \Im\Im^{-1}[I(f,\ u)] = \iint I(f,\ u)e^{i2\pi ft}e^{i2\pi u G_y}df du$$

(16.85)

To find the object distribution, I, from the measured set of signals from a location $(x_n,\ y_m)$, a forward 2D Fourier transform is

$$I(f, u) = S_{mn}(f, u) = \Im\Im[s_{mn}(t, G_y)] = \iint s_{mn}(t, G_y)e^{-i2\pi ft}e^{-i2\pi u G_y}dt dG_y$$

(16.86)

If the set of signals is taken to be simply the product of the one-dimensional signals for this simplified example,

$$s_{mn}(t, G_y) = \left[S_0 e^{-|(t-T_E)/T_{2mn}|}e^{j\omega_n(t-T_E)} \right] e^{i2\pi y_m G_y}$$

(16.87)

Then, inserting this set into the 2D transform of Eq. 16.86 yields

$$S_{mn}(x, y) = \left\{ e^{-i\gamma G_x x_n T_E} W_{mn} \left[\frac{\gamma G_x}{2\pi} (x - x_n) \right] \right\} [2\pi/(\gamma t_p)]\delta(y - y_m)$$

(16.88)

In this case, as depicted in Figure 16.48c, the intersection of these one-dimensional functions places a scaled spectrum at the location (x_n, y_m). The role of u_m is to localize the position of the spectrum along y through scaling. In an MRI image, this spectrum, S_{mn}, is displayed as amplitude mapped on a gray scale with full white as maximum amplitude in the image plane at location z. What are displayed, then, are not the detected time signals, but the amplitudes of their spectra, which through encoding methods, are mapped into their proper spatial locations from anatomy.

In reality, during actual measurements, what is sensed are the signals from all the active net magnetization vectors from all the locations being scanned. The detected voltage is the sum of all the signals picked up in a specific z plane,

$$V(t, \ G_y) = \sum_{m,n} s_{m,n}(t, G_y) \tag{16.89}$$

where the signals are not the ideal forms derived in Equations 16.87 and 16.88, but signals reflecting the characteristics and organization of tissue structures.

A convenient way of discussing the signals in terms of Fourier transforms is to describe the acquired signals as belonging to a k-space domain and the transformed signals as being in the space (x,y) domain. As shown in the following, this perspective is simply accomplished as a change in variables in the previous transform equations. The simplest place to start is in the exponential arguments of the transform equations, Equations 16.83 and 16.80a, that contain the key variables. These variables are set equal to the desired new ones,

$$2\pi f_n t = k_x x_n \tag{16.90a}$$

$$2\pi u G_y = k_y y_m \tag{16.90b}$$

From previous definitions, it is straightforward to show that

$$k_x = \gamma G_x t \tag{16.91a}$$

$$k_y = -\gamma G_y t_p \tag{16.91b}$$

or,

$$t = k_x/\gamma G_x \tag{16.92a}$$

$$G_y = -k_y/\gamma t_p \tag{16.92b}$$

Equations 16.92a and 16.92b can be substituted in Eq. 16.85 to yield

$$s(k_x/\gamma G_x, \ -k_y/\gamma t_p) = \iint \left\{ \left(\frac{\gamma G_x}{2\pi}\right) \left(\frac{-\gamma t_p}{2\pi}\right) I\left[\left(\frac{\gamma G_x}{2\pi}\right)x, \ \left(\frac{-\gamma t_p}{2\pi}\right)y\right] \right\} e^{ik_x x} e^{ik_y y} dx dy \tag{16.93}$$

or by redefining the functions in terms of the new variables, $\tilde{s}$ and $\tilde{I}$,

$$\tilde{s}(k_x, k_y) = \iint \tilde{I}(x, \ y) e^{ik_x x} e^{ik_y y} dx dy \tag{16.94}$$

where $\tilde{I}(x, y)$ is equal to the term in the braces outside of the exponentials in Eq. 16.93 and $\tilde{s}$ is redefined as well. Similarly, Eq. 16.86 becomes

$$\tilde{I}(x, y) = \iint \tilde{s}(k_x, k_y) e^{-ik_x x} e^{-ik_y y} dk_x dk_y \qquad (16.95)$$

Previous results have now been recast in xy-space and k-space variables.

16.3.7   MRI Systems

The main parts of an MRI system are illustrated by Figure 16.49. A large superconducting magnet provides a homogeneous static B_0 magnetic field within its interior. Typical clinical field values vary from 0.15 to 2 T but there are systems available in the 3 to 5 T range. Inserted within the main magnet are pairs of surface coils for producing the x, y, and z axis gradients. These typically produce variations of only a few percent of the value of the static field for spatially localizing the magnetic spin signals of the body. These signals are picked up by rf transmit/receiver coils which are sometimes in the form of surface coils placed close to the body. The rf signals are directed by a computer (CPU) through the rf pulse generator and sequencer. A switch routes the received rf signals through an amplifier and into a data acquisition unit after which they are processed and Fourier transformed to create an image. Modern

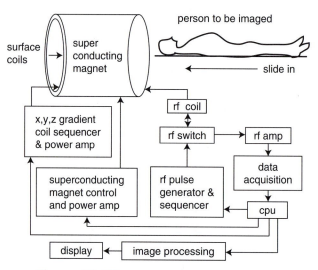

Figure 16.49   Block diagram of an MRI system.

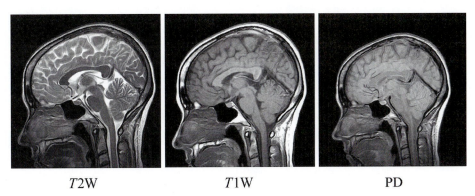

T2W   T1W   PD

Figure 16.50 MR images of sagittal brain cross sections using three types of weighting planes: (a) T2, (b) T1, (c) proton density (courtesy of Philips Medical Systems).

systems allow the operator considerable flexibility to alter the gradients and pulse sequences to achieve different image effects and weighting. Typical image acquisition times are on the order of 50 to 100 ms, though faster acquisitions are becoming available.

Different signal characteristics can be emphasized by a manipulation of the basic pulse timing sequences (see Fig. 16.47). For example, a short pulse repetition interval, TR, and a short spin echo time, TE, results in the commonly used $T1$ weighting ($T1$ W) shown in Figure 16.50b. For example, typical values are $TE = 12$ ms, $TR = 300$ ms. Prolonging TR and the delays preceding and after TE produces a T_2-weighted ($T2$ W) image as in Figure 16.50a. An example of this is $TE = 100$ ms and $TR = 3000$ ms. A combination of a long TR and a short TE creates what is called a proton density weighted (PD) image demonstrated by Figure 16.50c. For example, typical values are $TE = 30$ ms and $TR = 3000$ ms. Recall that T_1 is more sensitive to local thermal properties and T_2 depends more on local transverse magnetic field inhomogeneties. Many other alternative timing and pulse sequences are possible and are useful for different clinical applications.

An important aspect of imaging is contrast among different types of tissues. In Figure 16.51 longitudinal magnetization is shown for two tissues as a function of time, indicating their different $T1$ rise times. Note that the contrast among the tissues varies at different times as indicated by the vertical separation between the curves at different times. For example, compare the separations for times of 400 ms, 800 ms, and 1600 ms. Here it can be seen that contrast can be changed by timing intervals. Another way of enhancing contrast is to inject a paramagnetic medium or contrast agent such as gadolinium.

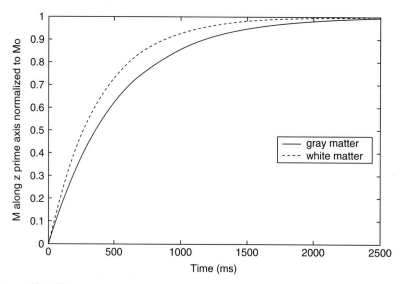

Figure 16.51   Longitudinal magnetization versus time for gray matter and white matter.

16.3.8   MRI Applications

Functional MRI (fMRI) is the use of MRI to detect localized changes in brain activity, usually in the form of changes in cerebral metabolism, blood flow, volume, or oxygenation in response to task activation. These changes are interrelated and may have opposite effects. For example, an increase in blood flow increases blood oxygenation, whereas an increase in metabolism decreases it. The most common means of detection is measuring the changes in the magnetic susceptibility of hemoglobin. Oxygenated blood is dimagnetic and deoxygenated blood is paramagnetic. These differences lead to a detection method called blood oxygen level dependent contrast (BOLD). Changes in blood oxygenation can be seen as differences in the T_2^* decay constant, so T_2 weighted images are used. Because these changes are very small, images before and after task initiation are subtracted from each other and the resultant difference image is overlaid on a standard image. Special care must be taken when obtaining these images because the effect is small and can be corrupted by several sources of noise. Typically, hundreds of images are taken for each slice plane position and statistical analysis is used to produce the final image. Sources of noise errors are thermal noise, head movement, and respiratory and cardiac cycles. An example of an fMRI image is shown in Figure 16.52.

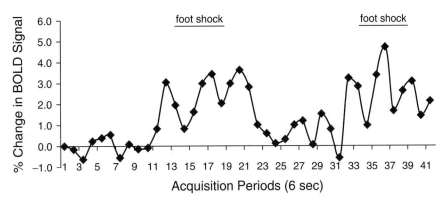

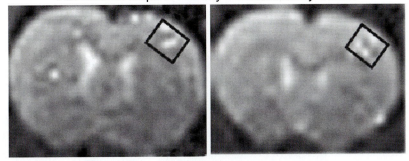

Figure 16.52   Functional imaging with fast spin echo. Images show percent change in BOLD signal in the right somatosensory cortex (boxed area in activational maps) in response to electrical stimuli to the left hind limb paw of a rat (reprinted from Ludwig et al., 2004, with permission from Elsevier).

16.4   COMPARISON OF IMAGING MODES

How can imaging modalities be compared? Each major diagnostic imaging method is examined in the following sections and the overall results are tallied in Table 16.4. Examples of three imaging modalities, CT, MRI, and ultrasound, are shown as different images of the right kidney in Figure 16.53.

TABLE 16.4 Comparison of Imaging Modalities

	Ultrasound	X-Ray	CT	MRI
What is imaged	Mechanical properties	Mean tissue absorption	Tissue absorption	Biochemistry (T1 & T2, & PD)
Access	Small windows adequate	2 sides needed	Circumferential around body	Circumferential around body
Spatial resolution	Frequency & axially dependent; 3 to 0.3 mm	~1 mm	~1 mm	~1 mm
Penetration	Frequency dependent; 3 to 25 cm	Excellent	Excellent	Excellent
Safety	Very good	Ionizing radiation	Ionizing radiation	Very good
Speed	100 frames per second	Minutes	Half-minute to minutes	Minutes
Cost	$	$	$$$$	$$$$$$$$
Portability	Excellent	Good	Poor	Poor

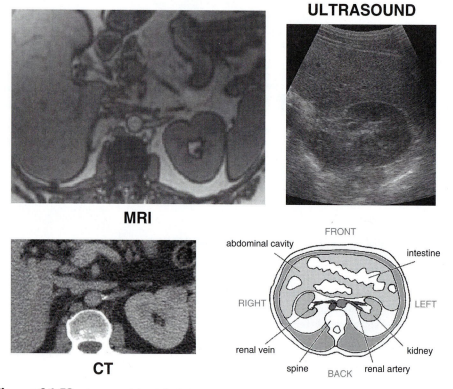

Figure 16.53 Images of the right kidney as viewed by different imaging modalities: upper left: MRI; bottom left: CT; top right: ultrasound; bottom right: Graphical depiction of abdominal cross section showing location of kidneys (courtesy of Dr. Marilyn Roubidoux).

16.4.1 Ultrasound

Unlike other imaging modalities, ultrasound resolution and penetration depend on the center frequency and type of transducer selected. The resolution is spatially variant and depends on both the size of the active aperture and the center frequency (and bandwidth) of the transducer and the selected transmit focal depth. A commonly used focal depth to aperture ratio is five, so that the half-power beam width is approximately two wavelengths at the center frequency; therefore, the transmitted lateral spatial resolution is about two wavelengths. For typical frequencies in use ranging from 1 to 15 MHz, lateral resolution ranges from 3 mm to 0.3 mm and is the smallest in the focal region and varies elsewhere in a nonuniform way because of diffraction effects caused by apertures on the order of a few to tens of wavelengths. For a short pulse, axial resolution is approximately two wavelengths.

Another major factor in determining resolution is attenuation that limits penetration. Attenuation increases with higher center frequencies and depth; therefore, penetration decreases correspondingly so that fine resolution is difficult to achieve at deeper depths.

Ultrasound images are highly detailed and geometrically correct to first-order maps of the mechanical structures of the body according to their "acoustic properties" such as differences in characteristic impedance that depend on stiffness or elasticity and density. The dynamic motion of organs such as the heart can be revealed by ultrasound operating at up to hundreds of frames per second.

Diagnostic ultrasound is noninvasive. Ultrasound is also safe and does not have any cumulative biological side effects. Two other strengths of ultrasound imaging are its relatively low cost and portability. With the widespread availability of miniature portable ultrasound systems for screening and imaging, these two factors will continue to improve.

A high skill level is needed to obtain good images with ultrasound. This expertise is necessary because of the number of access windows, differences in anatomy, the many possible planes of view, and the experience required to find relevant planes and targets of diagnostic significance and to optimize instrumentation. Furthermore, a great deal of experience is required to recognize, interpret, and measure images for diagnosis.

16.4.2 CT Imaging

Computed tomography (CT) [also known as computed axial tomography (CAT)] scanning involves x-rays and has been described in Chapter 15. As the x-rays pass through the body, they are absorbed by tissue so that an overall "mean attenuation" image results along the ray path. Spatial resolution is not determined by wavelength but by focal spot size of the x-ray tube and scatter from tissue; the state of the art is about 1 mm as of this writing. Radioactive contrast agents can be ingested or injected to improve visualization of vessels. Though exposures are short, x-rays are a form of ionizing radiation, so that dosage effects can be cumulative and extra precautions are needed for sensitive organs such as eyes and for pregnancies.

CT equipment is large and stationary so that a person can fit inside, and as a result, it is relatively expensive to operate. Consecutive pictures of a moving heart are now achievable through synchronization to ECG signals. The resolution of CT images is typically 1 mm. CT scanning creates superb images of the brain, bone, lungs, and soft tissue, making it complementary to ultrasound.

Although the taking of CT images requires training, it is not difficult. Interpretation of CT cross-sectional images demands considerable experience for definitive diagnosis.

16.4.3   Magnetic Resonance Imaging

For magnetic resonance imaging, the patient is placed in a strong static magnetic field created by a large enclosing electromagnet. The resolution is mainly determined by the gradient or shape of the magnetic field and it is typically 1 mm. Images are calculated by reconstruction algorithms based on the sensed voltages proportional to the relaxation times. Tomographic images of cross-sectional slices of the body are computed. The imaging process is fast and reasonably safe since no ionizing radiation is used. Care must be taken to keep ferromagnetic materials away from the powerful magnets used and there are limits in the strength of the applied magnetic fields and in how quickly they are switched. Because the equipment needed to make the images is expensive, exams are costly.

MRI equipment has several degrees of freedom such as the timing, orientation, and frequency of magnetic fields; therefore, a high level of skill is necessary to acquire diagnostically useful images. Diagnostic interpretation of images involves both a thorough knowledge of the settings of the system and a great deal of experience.

16.4.4   Summary

The different features and characteristics of the three imaging modalities are summarized in Table 16.4. The tissue properties being imaged are considerably different among the different modalities and can be complementary. Both CT and MRI can image the whole body with consistent resolution and contrast; however, they are the most expensive methods of imaging and are the least portable and slowest. Ultrasound, on the other hand, has variable resolution, penetration and limited access to certain portions of the body and is used to image soft tissues only but it has advantages in low cost, portability, and speed.

EXERCISES

1. Find the Fourier transform of $\prod[(t - t_0)c/L]$ given the Fourier transform pair $\Im[\prod(\tau)] = \sin(\pi\phi)/(\pi\phi) = \sin c(\phi)$.
2. Select the correct answer. Ultrasound imaging for medical applications began to grow significantly when (a) Wild and Reid demonstrated real-time imaging, (b) Edler and Hertz started the science of echocardiography,

(c) Stewart found a connection between cancer in children and prenatal x-ray exposure, (d) Satomura and his colleagues detected blood flow using Doppler techniques.

3. Compute the pressure reflection factor between blood and muscle at normal incidence in dB (use Table 16.1). Compare the reflection factor at 45° incidence to this result. Which is greater?

4. Derive the normal reflection and transmission factors in terms of intensity rather than amplitude. Determine the normal transmission intensity factor between blood and the following: liver, an ideal rigid boundary, and air. How do these compare to the corresponding amplitude results?

5. If the Q of the electrical part of a transducer equivalent circuit is defined as $Q = f_0/\Delta f = 1/[\omega_0 C_0 R_A(f_0)]$ where Δf is the -6-dB bandwidth, and f_0 is 5 MHz, compare the fractional -6-dB bandwidths (in %) for the following two transducer materials: $Ak_T = 0.7$, $\varepsilon_R = 1470$, $k_T = 0.9$, and $\varepsilon_R = 680$. Hint: see Eq. 16.27c.

6. Calculate the electrical impedances of the following two transducer designs at resonance using the constants from Example Problem 16.3: one with a direct water load and another with a matching layer between the crystal and the water load and both with $Z_B = Z_C$. Hint: use Eq. 16.34. What is the improvement in acoustic loss by including the matching layer?

7. Ultrasound is not used to image bone directly. (a) To estimate the strength of the reflection determine the normal amplitude reflection factor from muscle to bone. (b) Determine a matching layer impedance to better match the muscle to bone. (c) What are the amplitude reflection and transmission factors with the matching layer at resonance?

8. A company can only afford either a tuning inductor or a matching layer for their transducer product. Which would you recommend and why? Material details: area $A = 400\,mm^2$, thickness $d = 0.87\,mm$, crystal data $c = 4.35\,km/s$, $\varepsilon_R^S/\varepsilon_0 = 830$, $\varepsilon_0 = 8.85pF$, $k_T = 0.49$. Electrical source impedance $R_g = 50$ ohms.

9. Compare the frame rates for the following two cases: a 3-MHz cardiac sector scanner with 128 lines per frame for a depth of 10 cm and a small parts 10-MHz linear array with 300 lines per frame and a 1 cm depth.

10. If a 5-MHz transducer in an imaging system has an overall round-trip dynamic range of 100 dB. What is the greatest distance it can detect a pressure reflection between a liver–bone boundary? A liver–muscle boundary? Assume average tissue properties to the boundary are the same as that of liver.(Use Table 16.1 for relevant tissue and absorption data.)

11. A small lesion about 2 mm in diameter at a depth of 6 cm must be found. Which of the following rectangular array transducers with apertures and center frequencies would you use and why? (a) $L = 15\,mm$ and 6-MHz, (b) $L = 11\,mm$ and 5-MHz (c) $L = 12\,mm$ and 3-MHz. Assume $c = 1.5\,mm/\mu s$ and a focal length of $F = 60\,mm$ in the scan plane.

12. For a spherically focusing transducer with an aperture, $a = 5\,mm$, and a frequency of 5-MHz, show which of the following combinations places the effective focal length for a target at a depth of 40 mm: (a) $F = 50$ mm, (b) $F = 40$ mm, (c) $F = 80$ mm.

13. For the Doppler detection of blood, a certain 2-MHz ultrasound system cannot sense signals below a threshold of -35 dB. Assume a layer of muscle with negligible absorption loss above a blood vessel where blood is flowing at velocity of 1m/s. Find the range of incident angles (to the nearest degree) that will provide adequate signal strength and the corresponding Doppler frequencies. Use Table 16.1 and note the Doppler angle = $90°$ − incident angle.

14. For the arrangement of the whirling magnet in Fig. 16.29, as d is increased does the intercepted voltage increase or decrease? What is a physical explanation for this effect? (Note volt = Weber/s).

15. Determine all the nuclear magnetic dipole moments, μ_z, for ^{1}H, ^{19}F, and ^{23}Na. Use Table 16.2.

16. Calculate the net magnetization moment M_0 for water at a temperature of T = $300°$ K and a magnetic field of $B_0 = 0.5$ T and a proton density of $N = 6.7 \times 10^{19}$ protons/mm^3. Use a form of Eq. (16.54), $\hat{M}_0 = \sum_n \hat{\mu}_n = \hat{\mu}(n_+ - n_-)$ and assume $n_+ = N/2$.

17. If $\hat{B}_y(t) = \hat{y}B_1 \sin(\omega t)$, what will the field be in the rotated frame? After low-pass filtering to eliminate 2ω frequencies, what components are left?

18. (a) For a flip angle of $\pi/2$ radians and $B_1 = 10^{-5}$ T, what is t_p for ^{1}H (b) Find the total flip time for a $\pi/2$ angle followed by a π angle rotation.

19. Determine the relative decay for two individual rotations of $\pi/2$ and π in white matter and a malignant tumor at the time $t = 250$ ms. Which angle provides more contrast between the two tissue types? Repeat at $t = 500$ ms and compare the contrast to the previous case.

20. (a) Derive an expression for the spectrum of an FID pulse, $v'(t) = v(t) H(t)$ where $H(t)$ is the step function, $H(t) = 1$ for $t > 0$, and $H(t) = 0$ for $t > 0$. Utilize the Fourier transform pair, $\left[e^{-a|t|} H(t) \right] = \frac{1 - i2\pi f/a}{|a| \left[1 + (2\pi f/a)^2 \right]}$. (b) What is the complex ratio of the FID spectrum to that of the spin echo? (c) Find the value of this ratio at the Larmor frequency. (d) Can you find an explanation for this ratio based on the time waveforms involved?

21. (a) If $T_2 = T_2^* = 5$ ms for the FID pulse of Problem 20, find the spectral component at the Larmor frequency of 5 MHz for the conditions $B_1 = 20 \mu T, I = 1A, V_{oxel} = 1$ cm^3, and $M_{xy}^0 = 5 \times 10^{-12} J/(T\text{-mm}^3)$. (b) How does this value compare for a similar spectral magnitude of a spin echo with T_2 for fat at the same Larmor frequency?

22. Find the -6 dB width of the envelope of a time pulse and the two end frequencies needed to scan a 1 mm thick slice from -15 cm to $+15$ cm around an isocenter on the z axis. Assume $G_z = 0.5$ T/m.

23. (a) In order to cover a range of ± 10 cm about an isocenter along the y axis, phase encoding is applied with a pulse length $t_p = 10\,\mu s$ for ^{1}H and an overall phase shift of $\pi/2$. Find the slopes, G_{ym}, at the endpoints and the slope resolution for 256 steps.

24. Determine the relative position (x_n, y_m) from the center for ^{1}H, given $f_n = 78.5\,\text{MHz}$, $G_x = 10^{-4}\text{T/m}$, $t_p = 5\,\mu s$, and $u_m = 21.3\,\text{m/T}$.

25. Derive $s_{nm}(k_x, k_y)$ based on Eq. 16.87.

26. Answer the following:

    a. What is displayed in an MRI image?

    b. How is a location from a set of signals in the body determined?

    c. Of the three imaging axes, which are actively controlled and which are passively sensed?

27. Explain the basic physical principles underlying the differences in appearance (image content and tissue differentiation) among $T1$, $T2$, and PD weighted images.

SUGGESTED READING

Bitter, F. and Medicus, H.A. (1973). *Fields and Particles: An Introduction to Electromagnetic Wave Phenomena and Quantum Physics*. American Elsevier, New York.

Chakeres, D.W. and Schmalbrock, P. (1992). *Fundamentals of Magnetic Resonance Imaging*. Williams & Wilkins, Baltimore, MD.

Cho, Z.H., Jones, J.P. and Singh, M. (1993). *Foundations of Medical Imaging*. Wiley, New York.

Jensen, J.A. (1996). *Estimation of Blood Velocities Using Ultrasound*. Cambridge Univ. Press, Cambridge.

Kim, E.E. and Jackson, E.F. (1999). *Molecular Imaging in Oncology*. Springer-Verlag, Berlin.

Kino, G.S. (1987). *Acoustic Waves: Devices, Imaging, and Analog Signal Processing*. Prentice-Hall, Englewood Cliffs, NJ.

Kuperman, V. (2000). *Magnetic Resonance Imaging Physical Principles and Applications*. Elsevier Academic, Boston.

Lauterbur, P.C. (1973). Image formation by induced local interactions: Examples employing nuclear magnetic resonance. *Nature* **242,** 190–191.

Liang, Z.P. and Lauterbur, P.C. (2000). *Principles of Magnetic Resonance Imaging: A Signal Processing Approach*. IEEE Press, New York.

Ludwig, R., Bodgdanov, G., King, J., Allard, A. and Ferris, C.F. (2004). A dual rf resonator system for high-field functional magnetic resonance imaging of small animals. *J. Neurosci. Meth.* **132,** 125–135.

Panda, R.K. (1998). Development of novel piezoelectric composites by solid freeform fabrication techniques. Dissertation. Rutgers, New Brunswick, NJ.

Shung, K.K., Smith, M.B. and Tsui, B.M.W. (1992). *Principles of Medical Imaging*. Academic, San Diego, CA.

Szabo, T.L. (1998). Transducer arrays for medical ultrasound imaging. In *Ultrasound in Medicine, Medical Science Series, F* (A. Duck, A.C. Baker and H.C. Starritt, Eds.). Institute of Physics Pub., Bristol, U.K.

Szabo, T.L. (2004). *Diagnostic Ultrasound Imaging: Inside Out*, Elsevier Academic, Boston.

17 BIOMEDICAL OPTICS AND LASERS

Gerard L. Coté, PhD
LiHong V. Wang, PhD
Sohi Rastegar, PhD

Chapter Contents

At the conclusion of this chapter, the reader will be able to:

- Understand essential optical principles and fundamentals of light propagation in tissue as well as other biological and biochemical based media.

- Utilize basic engineering principles for developing therapeutic, diagnostic, sensing, and imaging devices with a focus on the use of lasers and optical technology.

- Describe the biochemical and biophysical interactions of optic and fiber optic systems with biological tissue.

- Understand the photothermal interactions of laser systems with biological tissue.

The use of light for therapeutic and diagnostic procedures in medicine has evolved from the use of sunlight for heat therapy and as a simple tool for the inspection of eyes, skin, and wounds to the current use of lasers and endoscopy in various medical procedures. Indeed, the introduction of photonic technology into medicine over the last decade has revolutionized many clinical procedures, and because of its low cost, relative to many other technologies, photonic technology has the potential to greatly impact health care. For example, coherent fiber optic bundles have been applied in laparoscopy cholecystectomy (minimally invasive removal of the gallbladder) to transform a once painful and expensive surgery into virtually an outpatient procedure. In the process, biomedical engineers have been instrumental in defining and demonstrating the engineering fundamentals of the interaction of light and heat with biological media, resulting in advances in dermatology, ophthalmology, cardiology, and urology.

Optical engineering has traditionally been taught as part of an electrical engineering curriculum and been primarily applied in the development of defense technology. In biomedical optics applications, however, there are many issues related to the interaction of light with participating biological tissue/media that classical optics books and curricula do not cover. Therefore, the goal of this chapter is to provide students with a better understanding of the fundamental principles associated with the growing field of biomedical optics as well as the design of optically based therapeutic, diagnostic, and monitoring devices.

17.1 INTRODUCTION TO ESSENTIAL OPTICAL PRINCIPLES

Throughout the ages great minds, including those of Galileo, Newton, Huygens, Fresnel, Maxwell, Planck, and Einstein, have studied the nature of light. Initially it was believed that light was either corpuscular or particle in nature, or it behaved as waves. After much debate and research, however, it has been concluded that there are circumstances in which light behaves like waves and those in which it behaves like particles. In this section, an overview of the behavior of light is described, beginning with the wave equation, which is commonly derived from Maxwell's equations. In addition, the optical spectrum and the fundamental governing equations for light absorption, scattering, and polarization are covered.

17.1.1 Electromagnetic Waves and the Optical Spectrum

It was in the late 1800s that J. Clerk Maxwell showed conclusively that light waves were electromagnetic in nature. This was accomplished by expressing the basic laws of electromagnetism and deriving from them the wave equation. The key to validating this derivation was that the free space solutions to the wave equation corresponded to electromagnetic waves with a velocity equal to the known experimental value of the velocity of light. Many introductory optics texts (such as the works of Hecht and Pedrotti) begin with the vectorial form for Maxwell's equations,

$$\nabla^2 \mathbf{E} = \varepsilon_0 \, \mu_0 \partial^2 \mathbf{E}/\partial t^2 \tag{17.1}$$

where $\nabla^2 \mathbf{E}$ is the partial second derivative of the electric field with respect to position $(\partial^2 \mathbf{E}/dx^2 + \partial^2 \mathbf{E}/dy^2 + \partial^2 \mathbf{E}/dz^2)$, ε is the electric permittivity of the medium, and μ is the magnetic permeability of the medium. In free space $\mu = \mu_0$ and $\varepsilon = \varepsilon_0$. From symmetry, there is a similar solution for the magnetic field, **H**. Since the wave velocity (c_0) is equal to the known value of the velocity of light, it can be expressed as follows:

$$c_0 = 1/(\varepsilon_0 \, \mu_0)^{1/2} \tag{17.2}$$

Therefore, the change in the electric or magnetic field in space is related to the velocity and the change of the field with respect to time. Given Cartesian axes O_x, O_y, O_z shown in Figure 17.1, a typical solution to this wave equation is the sinusoidal solution given by:

$$E_x = E_0 \exp\left[j(\omega t - kz)\right] \tag{17.3}$$

$$H_y = H_0 \exp\left[j(\omega t - kz)\right] \tag{17.4}$$

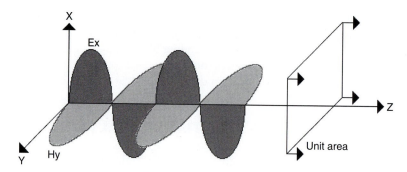

Figure 17.1 Sinusoidal electromagnetic wave.

which states that the electric field oscillates sinusoidally in the $x-z$ plane, the magnetic field oscillates in $y-z$ plane (orthogonal to the e-field and in phase), and the wave propagates in the O_z direction.

The frequency, f, and wavelength, λ, of the wave are related to the velocity, c, and given by

$$f = \omega/2p \tag{17.5}$$

$$\lambda = 2p/k \tag{17.6}$$

$$c = f\lambda = \omega/k \tag{17.7}$$

The ratio of the velocity of the wave in free space, c_0, to that in the medium through which the light propagates, c, is known as the index of refraction of the medium, $n = c_0/c$.

Example Problem 17.1

Given a red helium neon laser with a wavelength of 633 nm, determine the velocity of the light traveling through clear tissue such as the cornea that has an index of refraction of 1.33. How would this velocity change in glass that has an index of refraction of 1.5? Explain the significance of this result.

Solution

Rearranging the preceding equations,

$$c_{\text{tissue}} = c_0/n_{\text{tissue}} \simeq (2.998^*10^{\wedge}8)/1.33 = 2.25^*10^{\wedge}8 \text{m/s for clear tissue}$$

while

$$c_{\text{glass}} = c_0/n_{\text{glass}} \simeq (2.998^*10^{\wedge}8)1.5 = 2.00^*10^{\wedge}8 \text{ m/s for glass}$$

The significance of this result is that light travels faster through a material with a lower index of refraction such as clear tissue compared to glass. This has many implications, as you will see later in the chapter when it comes to determining the angle of reflection and refraction of light through differing materials and clear tissues (e.g., cornea, aqueous humor, and lens of the eye). ∎

The intensity or power per unit area of the wave can also be expressed in terms of the Poynting vector, Π, (i.e., defined in terms of the vector product of the electric and magnetic field) as

$$\Pi = \mathbf{E} \times \mathbf{H} \qquad (17.8)$$

The intensity of the wave is the average value of the Poynting vector over one period of the wave, thus if E and H are spatially orthogonal and in phase as shown in Figure 17.1 then

$$I = < |\Pi| > = \text{power unit area} = c\varepsilon E^2 \qquad (17.9)$$

where $< |\Pi| >$ represents the average value of Π. For this equation it is clear that the intensity is proportional to the square of the electric field. This is true for the propagation of a wave through an isotropic dielectric medium. From Maxwell's equations it can be shown that the intensity is also proportional to the square of the magnetic field. Therefore, if the electric and magnetic field were in phase quadrature (90° out of phase), then the intensity would be zero as shown in the following:

$$I = < |\Pi| > = <E_o \cos(\omega t) H_o \sin(\omega t)> = 0. \qquad (17.10)$$

since the average value of a sine wave times a cosine wave is zero.

In addition to intensity, light can be characterized by its wavelength (or rather electromagnetic spectrum) which ranges from low-frequency radio waves at 10^3 Hz to gamma radiation at 10^{20} Hz. As depicted in Figure 17.2, the wavelength region of light can be divided into ultraviolet which is 0.003 to 0.4 micrometers, to the visible region of 0.4 to 0.7 micrometers (7.5×10^{14} to 4.3×10^{14} Hz frequency range), to the near-infrared region 0.7 to 2.5 micrometers, to the mid-infrared region from 2.5 to 12 microns, and to the far-infrared beyond 12 microns.

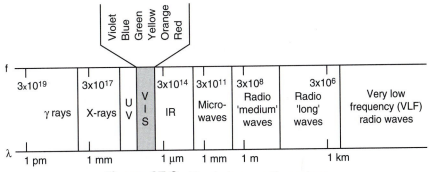

Figure 17.2　The electromagnetic spectrum.

17.1.2   Optical Polarization

In the previous section the sinusoidal solution in terms of E_x and H_y is only one of an infinite number of such sinusoidal solutions to Maxwell's equations. The general solution for a sinusoid with angular frequency ω can be written as

$$\mathbf{E}(\mathbf{r}, t) = \mathbf{E}(\mathbf{r}) \exp(j\omega t) \qquad (17.11)$$

in which $\mathbf{E}(\mathbf{r}, t)$ and $\mathbf{E}(\mathbf{r})$ are complex vectors and $\mathbf{r}$ is a real radius vector.

For simplicity, consider the plane monochromatic (single wavelength) waves propagating in free space in the O_z direction; then the general solution to the wave equation for the electric field can be written:

$$E_x = e_x \cos(\omega t - kz + \delta_x) \qquad (17.12)$$
$$E_y = e_y \cos(\omega t - kz + \delta_y) \qquad (17.13)$$

in which δ_x and δ_y are arbitrary phase angles. Thus this solution can be described completely by means of two waves (e-field in the x–z plane and e-field in the y–z plane). If these wave are observed at a particular value of z, say z_o, they take the oscillatory form:

$$E_x = e_x \cos(\omega t + \omega \delta'_x) \quad \delta'_x = \delta_x - kz_o \qquad (17.14)$$

$$E_y = e_y \cos(\omega t + \delta'_y) \quad \delta'_y = \delta_y - kz_o \qquad (17.15)$$

and the top of each vector appears to oscillate sinusoidally with time along a line. E_x is said to be linearly polarized in the direction O_x, and E_y is said to be linearly polarized in the direction O_y.

It can be seen from Equation 17.15 that if the light is fully polarized, it can be completely characterized as a 2 by 1 matrix in terms of its amplitude and phase $[e_x \exp(j\delta'_x)$ and $e_y \exp(j\delta'_y)]$. This vectoral representation is known as the Jones vector. When the optical system design includes the propagation of the light through a nonscattering, and hence nondepolarizing, medium such as a lens or clear biological sample, the medium can be completely characterized by a 2 by 2 Jones matrix. Therefore, the output of the propagation of the polarized light can be modeled as a multiplication of the Jones matrix of the optical system and the input light vector. For a system that yields both polarized and partially polarized light such as that obtained from tissue scattering, the system can be characterized using 4 by 4 matrices known as Mueller matrices and a 4 by 1 matrix known as the Stokes vector.

Returning to Equation 17.15, the tip of the polarized light vector is the vector sum of E_x and E_y which, in general, is an ellipse as depicted in Figure 17.3 whose Cartesian equation in the $x-y$ plane at $z = z_o$ is:

$$E_x^2/c_x^2 + E_y^2/c_y^2 + 2.(E_x E_y/e_x e_y)\cos\delta = \sin^2\delta \qquad (17.16)$$

in which $\delta = \delta'_y - \delta'_x$. It should be noted that the ellipse folds into a straight line producing linear light if; (1) $E_x \neq 0$ and $E_y = 0$, or (2) $E_y \neq 0$ and $E_x = 0$, or (3) $\delta = m\pi$ in which m is a positive or negative integer. The light becomes circular if $e_x = e_y$ and $\delta = (2m + 1)\pi/2$ since E_x and E_y would then have equal amplitudes and be in phase quadrature.

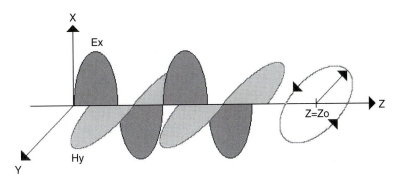

Figure 17.3   Polarization propagation of an E wave.

The polarization properties of light become particularly important for anisotropic media in which the physical properties are dependent upon direction (i.e., E_x is different from E_y and thus values e_x, e_y, and δ will vary along the propagation path as depicted with a birefringent crystal or polarization preserving optical fiber).

Example Problem 17.2

Consider a general solution to the wave equation represented as a plane electromagnetic wave (in SI units) given by the expression

$$E_x = 0, \ E_z = 0, \qquad E_y = 2\cos\left[2\pi \times 10^{14}(t - x/c) + \pi/2\right]$$

For the equations shown, determine the frequency (f), wavelength (λ), direction of motion (x, y, or z), amplitude (A), initial phase angle (ϕ), and polarization of the wave (linear, circular, elliptical).

Solution

The general form of a plane electromagnetic wave can be expressed as $E = A\cos(kx \pm \omega t + \phi)$ where $k = 2\pi/\lambda$, $\lambda = c/f$, $\omega = 2\pi f$,   and $c = 3 \times 10^8$ meters/second. The preceding equation can therefore be written in general form as $E_y = 2\cos\left[2\pi \times 10^{14}t - (2\pi \times 10^{14}x/c) + \pi/2\right]$. Thus comparing the two equations yields the following values, $f = 1 \times 10^{14}$ Hertz, $\lambda = 3 \times 10^{-6} = 3$ micrometers, motion is in the positive x direction, $A = 2$ Volts/meter, $\phi = \pi/2$, and since $E_x = E_z = 0$ the wave must be linearly polarized in the y direction.   ∎

17.1.3 Absorption, Scattering, and Luminescence

Having discussed polarization, attention will now be focused on the optical properties of light that describe the changes in the light as it passes through biological media. The optical properties of the light are defined in terms of the absorption, scattering, and anisotropy of the light. Knowing the absorption, scattering, and direction of propagation of the light has applications for both optical diagnostics and sensing in addition to optical therapeutics. For instance, many investigators are trying to quantifiably and noninvasively measure certain body chemicals such as glucose using the absorption of light. In addition, by knowing the absorption and scattering of light, investigators are modeling the amount of coagulation or "cooking" of the tissue that is expected for a given wavelength of light, such as in the case of prostate surgery.

The optical penetration depth and the volume of tissue affected by light is a function of both the light absorption and scattering. If it is assumed scattering is negligible, or rather that absorption is the dominant component, then the change in light intensity is determined by the absorption properties in the optical path and the illumination wavelength. For instance, for wavelengths above 1.5 μm the light propagation through all biological tissues, since they have large water content, is absorption dominant. The intensity relationship as a function of optical path and concentration can be described by the Beer–Lambert law as

$$I(z) = I_o \exp{(-\mu_a z)} \tag{17.17}$$

in which I is the transmitted light intensity, I_o is the incident light intensity, z is the path length of the light, and μ_a (1/cm) is the absorption coefficient. The absorption coefficient can be further broken down into two terms which include the concentration of the chemically absorbing species, C (mg/mL) times the molar absorptivity, $v(\text{cm}^2/\text{mg})$, which is a function of wavelength.

Luminescence, also known as fluorescence or phosphorescence depending on the fluorescent lifetime, is a process that occurs when photons of electromagnetic radiation are absorbed by molecules, raising them to some excited state and then, on returning to a lower energy state, the molecules emit radiation, or luminesce. Fluorescence does not involve a change in the electron spin and therefore occurs much faster. The energy absorbed in the luminescence process can be described in discrete photons by the equation

$$E = hc/\lambda \tag{17.18}$$

in which E is the energy, h is Planck's constant (6.626×10^{-34} J-s), c is the speed of light, and λ is the wavelength. This absorption and reemission process is not 100% efficient since some of the originally absorbed energy is dissipated before photon emission. Therefore, the excitation energy is greater than that emitted,

$$E_{\text{excited}} > E_{\text{emitted}} \tag{17.19}$$

and

$$\lambda_{\text{emitted}} > \lambda_{\text{excited}} \tag{17.20}$$

The energy level emitted by any fluorochrome is achieved as the molecule emits a photon and due to the discrete energy, level individual fluorochromes excited by a given wavelength typically fluoresce only at a certain emission wavelength. Originally, fluorescence was used as an extremely sensitive, relatively low-cost technique for sensing dilute solutions. Only recently have investigators used fluorescence in turbid media such as tissue for diagnostics and sensing.

It is well known that tissues scatter as well as absorb light, particularly in the visible and near-infrared wavelength regions. Scattering, unlike absorption and luminescence, need not involve a transition in energy between quantized energy levels in atoms or molecules, but rather is typically a result of random spatial variations in the dielectric constant. Therefore, the actual light distributions can be substantially different from distributions estimated using Beer's law. In fact, light scattered from a collimated beam undergoes multiple scattering events as it propagates through tissue. The transport equation that describes the transfer of energy through a turbid medium (a medium that absorbs and scatters light) is an approach that has been proven to be effective. The transport theory has been used to describe scattering, absorption, and fluorescence, but until recently polarization has not been included, in part because it takes relatively few scattering events to completely randomize the polarization of the light beam. The transport theory approach will be described in Section 17.2.2 and is a heuristic theory based on a statistical approximation of photon particle transport in a multiple scattering medium.

17.2 FUNDAMENTALS OF LIGHT PROPAGATION IN BIOLOGICAL TISSUE

In this section the propagation of light through biological media such as tissue is discussed, beginning with a simple ray optics approach for light traveling through a nonparticipating medium in which the effect of the absorption and scatter within the medium is ignored. The effects of absorption and scatter on light propagation are then discussed along with the consideration of boundary conditions and various means of measuring the optical absorption and scattering properties.

17.2.1 Light Interactions with Nonparticipating Media

Previously, we defined light as a set of electromagnetic waves traveling in the O_z direction. In this section light will be treated as a set of "rays" traveling in straight lines that, when combined, make up the plane waves described using the complex exponentials discussed previously. This approach is important in order to study the effect on the light at the boundaries between two different optical media. In the treatment of geometrical or ray optics of the light, it is first assumed that the incident, reflected, and refracted rays all lie in the same plane of incidence. The propagation at the boundary between these two interfaces as shown in Figure 17.4 is based on two basic laws: the angle of reflection equals the angle of incidence, and the sine of the angle of refraction bears a constant ratio to the sine of the angle of incidence (Snell's law). It should be noted that ray propagation can be very useful for a good majority of the applications considered with the propagation of the light through bulk optics such as in the design of lenses and prisms. It has severe limitations, however, in that it can not be used to predict the intensities of the refracted and reflected rays nor does it incorporate the effects of scatter and absorption. In addition, when apertures are used that are smaller than the bulk combined rays, a wave process known as diffraction occurs which causes the geometrical theory to break down and the beams to diverge.

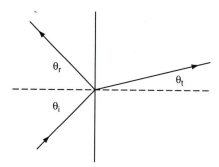

Figure 17.4   Ray propagation of light at the boundary of two interfaces. Clearly the angle of incidence equals the angle of reflection ($\theta_i = \theta_r$) and from Snell's law the sine of the angle of refraction bears a constant ratio to the sine of the angle of incidence [$n_i \sin(\theta_i) = n_t \sin(\theta_t)$].

Example Problem 17.3

Given a green light beam that hits the cornea of the eye, making an angle of $25°$ with the normal, as depicted in the following figure, (a) determine the output angle from the front surface of the cornea into the cornea given the index of refraction of air is 1.000, and that of the cornea is 1.376; and (b) what can be said about how the cornea bends the light?

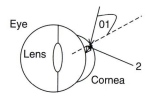

Solution

(a) Snell's law, $n1 \sin(\theta 1) = n2 \sin(\theta 2)$, can be rearranged so as to calculate $\theta 2$ in the lens.

$$\theta 2 = \sin^{-1}(1.000 \sin(25)/1.376) = 17.89 \text{ degrees}$$

(b) It has been shown and can be said that the cornea tends to bend the light toward the normal as it passes through. This makes sense because the eye is made to bend the light so that it can pass through the center iris and lens toward the retina to be imaged by the brain.   ■

Wave theory is used to describe the phenomena of reflection and refraction in an effort to determine the intensity of the light as it propagates from one medium to another. If two nonconducting media are considered, as in Figure 17.4, the boundary conditions follow from Maxwell's equations such that the tangential components of **E** and **H** are continuous across the boundary $E_i + E_p = E_t$, and the normal components of **B** (the magnetic flux density) and **D** (the electric displacement) are continuous across the boundary. These conditions can be true at all times and all places on the boundary only if the frequencies of all waves are the same. The relative amplitudes of the waves will now be considered (note the value of **E, H, D,** and **B** will depend on the direction of vibration of **E** and **H** fields of the incident wave relative to the planes of incidence). In other words, they depend on the polarization of the wave. The wave can be divided into two orthogonal polarization states and, using Maxwell's equations, Snell's law, and the law of reflection as well as imposing the boundary conditions, the following Fresnel equations can be derived. The reflected and transmitted equations in the parallel configuration are

$$r_p = \left(\frac{E_r}{E_i}\right)_p = (-n_1 \cos \theta_t + n_2 \cos \theta_i)/(n_1 \cos \theta_t + n_2 \cos \theta_i) \qquad (17.21)$$

$$t_p = \left(\frac{E_t}{E_i}\right)_p = (2n_1 \cos \theta_i)/(n_1 \cos \theta_t + n_2 \cos \theta_i) \qquad (17.22)$$

The normal or perpendicular reflected and transmitted equations are

$$r_\perp = \left(\frac{E_r}{E_i}\right)_\perp = (n_1 \cos \theta_i - n_2 \cos \theta_t)/(n_1 \cos \theta_i + n_2 \cos \theta_t) \qquad (17.23)$$

$$t_\perp = \left(\frac{E_t}{E_i}\right)_\perp = (2n_1 \cos \theta_i)/(n_1 \cos \theta_i + n_2 \cos \theta_t) \qquad (17.24)$$

Using the preceding equations, there are two noteworthy limiting cases. The first case is as follows. If $n_t > n_i$ then $\theta_i > \theta_t$, $r_\perp$ is always negative for all values of θ_i, r_p starts out positive at $\theta_i = 0$ and decreases until it equals 0 when $\theta_i + \theta_t = 90°$. This implies the refracted and reflected rays are normal to each other and from Snell's law this occurs when $\tan \theta_i = n_2/n_1$. This particular value of θ_i is known as the polarization angle or, more commonly, the Brewster angle. At this angle only the polarization with E normal to the plane of incidence is reflected and as a result this is a useful way of polarizing a wave. Now as θ_i increases beyond θ_p, r_p becomes progressively negative, reaching -1.0 at $90°$, which implies that the surface performs as a perfect mirror at this angle. On the other hand, at normal incidence $\theta_i = \theta_r = \theta_t = 0$ then $t_p = t_\perp = 2n_i/(n_i + n_t)$ and $r_p = -r_\perp = (n_2 - n_1)/(n_1 + n_2) = (n_1 - n_2)/(n_1 + n_2)$. The wave intensity, which is what can actually be measured by a detector, is defined proportional to the square of the electric field. Thus, $(I_R/I_i) = (r_p)^2 = \left(\frac{E_r}{E_i}\right)^2$ $= (n_t - n_i)^2/(n_i + n_t)^2$ and $(I_t/I_i) = (t_p)^2 = \left(\frac{E_t}{E_i}\right)^2 = 4n_i^2/(n_i + n_t)^2$. The preceding expressions are useful since they represent the amount of light lost by normal specular reflection when transmitting from one medium to another.

The second limiting case is if $n_t < n_i$, then $\theta_t > \theta_i$, $r_\perp$ is always positive for values of θ_i, and $r_\perp$ increases from its initial value at $\theta_i = 0$ until it reaches $+1.0$ at what is known as the critical angle θ_c. At this angle $\theta_i = \theta_c$, then $\theta_t = 90°$ and from Snell's law $\sin \theta_t = n_1/n_2 \sin \theta_i$. Clearly for $n_1 > n_2$ (less dense to more dense medium) $\sin \theta_t$

could be greater than one according to the previous equation. This is not possible for any real value of θ_t, and thus the angle for which $\theta_i = \theta_c$ is when $(n_1/n_2) \sin \theta_c = 1$, or rather $\theta_t = 90°$. Thus for all values of $\theta_i > \theta_c$ the light is totally reflected at the boundary. This phenomenon is used for the creation of optical fibers in which the core of the fiber (where the light is supposed to propagate) has an index of refraction slightly higher than the cladding surrounding it so that the light launched into the fiber will totally internally reflect, allowing for minimally attenuated propagation down the fiber.

17.2.2   Light Interaction with Participating Media: The Role of Absorption and Scattering

When light is incident on tissue either from a laser or from other optical devices, tissue acts as a participating medium by reflecting, absorbing, scattering, and transmitting various portions of the incident wave of radiation. Ideally, an electromagnetic analysis of the light distribution in the tissue would be performed. Unfortunately, this can be quite cumbersome and, furthermore, a reliable database of electrical properties of biological tissues would be required. An alternative practical approach to the problem is to use transport theory, which starts with the construction of the differential equation for propagation of the light intensity. In the following subsection the equation for light intensity distribution in a purely absorbing medium will be derived. With this motivation, the next subsection will introduce the general equation of transfer for a medium that both scatters and absorbs the light.

The Case of Pure Absorption

To describe the distribution and transport of laser energy in a nontransparent participating medium the medium can be viewed as having two coexisting "phases": (1) a material phase for all the masses of the system, and (2) a photon phase for the electromagnetic radiation. Figure 17.5 shows the material phase as circles and the photon phase as curved arrows which strike the material phase. The energy balance equation for the material phase is introduced and discussed in the thermodynamic descriptions in Sections 17.3 and 17.5. The energy balance equation for the photon phase is discussed next.

Consider an infinitesimal volume of the material under irradiation (Figure 17.6). The rate of change of radiative energy, $U^{(rad)}$, with time is the difference between the incoming and outgoing radiative fluxes in the element minus the rate of energy absorption by the material phase. The difference between the incoming and outgoing fluxes is, in the limit, the negative of the divergence of the radiative flux. Therefore, denoting the energy absorbed by the material phase, which is the laser heat source term, by Q_L and the radiative flux by $\vec{q}^{(rad)}$, the governing differential equation for energy rate balance in the photon phase can be written as

$$\partial U^{(rad)}/\partial t = -\vec{\nabla} \cdot \vec{q}^{(rad)} - Q_L \qquad (17.25)$$

Any possible photon emission or scattering by the material phase has been ignored.

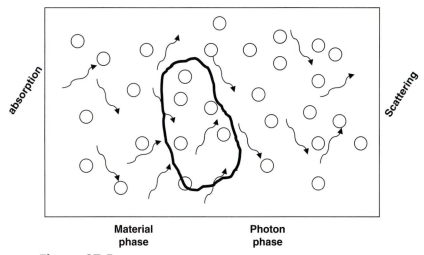

Figure 17.5   Two coexisting phases: photon phase and material phase.

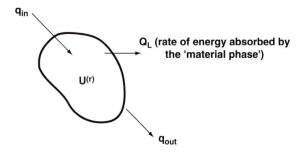

Figure 17.6   Control volume of the photon phase with radiative fluxes in and out, radiative energy, and rate of energy lost to the material phase by absorption.

For a steady-state system in which radiation is collimated and monochromatic, laser light travels only in the positive z direction, assuming only one forward flux, Eq. 17.25 reduces to

$$dq_z^{(rad)}/dz = -Q_L \qquad (17.26)$$

Considering the speed of light and the dimensions of a biological tissue, the assumption of steady-state condition is a reasonable assumption for most applications except when very fast light sources are used and/or when time-resolved analyses are considered. The important step at this point is to use the phenomenological relation

$$Q_L = \mu_a I \qquad (17.27)$$

where μ_a is defined as the absorption coefficient and I is the total light intensity. The total intensity at a point is the sum of radiative fluxes received at that point. In the case presented here, using purely absorbing tissue, a single radiative flux is used, thus

$$I = q_z^{(rad)} \tag{17.28}$$

Proceeding by replacing Equations 17.27 and 17.28 into 17.26, the following differential equation is obtained

$$\frac{dI}{dz} = -\mu_a I \tag{17.29}$$

which has the simple solution

$$I(z) = I_0 \exp\left(-\mu_a z\right) \tag{17.30}$$

where I_0 denotes the intensity at the surface, $z = 0$. This equation is the well-known Beer–Lambert law of absorption as given in Equation 17.17 for a purely absorbing medium, which was mentioned earlier in the chapter.

The laser heat source term can now be written (Eq. 17.27 and 17.30) as

$$Q_L = \mu_a I_0 \exp\left(-\mu_a z\right) \tag{17.31}$$

This equation can be generalized to an axisymmetric three-dimensional case to include the effect of radial beam profile of a laser light incident orthogonally on a slab by writing it as

$$Q_L(r, z) = \mu_a I_0 \exp\left(-\mu_a z\right) f(r) \tag{17.32}$$

where $f(r)$ is the radial profile of an axisymmetric laser beam. For a Gaussian beam profile, which is a common mode of laser irradiation,

$$f(r) = \exp\left(-\frac{2r^2}{\omega_o^2}\right) \tag{17.33}$$

where ω_o is known as the '$1/e^2$ radius' of the beam, since at $r = \omega_o$, $f(r) = 1/e^2$.

Example Problem 17.4

For a Gaussian laser beam irradiating at a wavelength of 2.1 μm the absorption coefficient is estimated to be $25\ cm^{-1}$ and scattering is negligible. The radial profile of light intensity and the rate of heat generation at the tissue surface, $z = 0$, and its axial profile along the center axis of the beam, $r = 0$, can be found using the previous equations. Graph $I(r, 0)$ for r ranging from $-2\omega_o$ to $+2\omega_o$ and $I(0, z)$ for $z = 0$ to $z = 5/\mu_a$, as well as a graph of $Q_L(r, 0)$ and $Q_L(0, z)$ are shown on the next page. For simplicity, $I_o = 1\ W/cm^2$.

Solution

Plot of the radial and axial profile of light intensity, I, and volumetric rate of absorption Q_L in a purely absorbing tissue. ∎

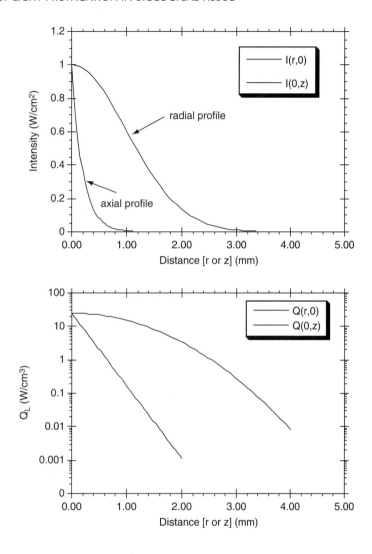

Methods for Scattering and Absorbing Media

The fundamental quantity in the transport theory approach is the total specific intensity, L, which is a function of position $\vec{r}$ for light in the direction given by a unit vector $\hat{s}$ and its units are $W.cm^{-2}.s^{-1}$. The equation of transfer satisfied by L can be written as

$$\frac{dL(\vec{r}, \hat{s})}{ds} = -(\mu_a + \mu_s)L(\vec{r}, \hat{s}) + \mu_s \int_{4\pi} p(\hat{s}, \hat{s}')L(\vec{r}, \hat{s}')d\omega' + S(\vec{r}, \hat{s}) \qquad (17.34)$$

This equation shows the decrease in L due to scattering μ_s and absorption μ_a and the increase in L due to scattering from L coming from another direction $\hat{s}'$ (see Figure

17.7). The sum of scattering and absorption, $\mu_t = \mu_a + \mu_s$, is defined as the attenuation coefficient. The function $p(\hat{s}, \hat{s}')$ is called the "phase function" and is related to the scattering amplitude of a particle, a scaled form of the probability distribution of scattering angles. Note that by ignoring scattering and assuming a collimated light source (17.34) reduces to (17.29) which was derived in the last subsection. $S(\vec{r}, \hat{s})$ is the source term which could be irradiation on the surface, fluorescence generated inside the tissue, or an internal volumetric light source.

The equation of transfer is an integrodifferential equation for which a general solution does not exist. However, several approximate solutions have been found, such as the two-flux and multiflux models, the discrete ordinate finite element method, the spherical harmonic method, the diffusion approximation, and the Monte Carlo method. Each of these is subject to certain limitations and assumptions. In this chapter the focus will be on the diffusion approximation, which is one of the more widely used solutions.

Diffusion Approximation

The diffusion approximation is a second-order differential equation that can be derived from the radiative transfer equation (Eq. 17.34) under the assumption that the scattering is "large" compared with absorption. The solution to this equation provides a useful and powerful tool for the analysis of light distribution in turbid media. The governing differential equation for the diffusion approximation is

$$\nabla^2 \phi_d - 3\mu_a \mu_{t'} \phi_d = -\frac{\mu_{s'}}{D} \phi_c \qquad (17.35)$$

where ϕ_d is the diffuse fluence rate and the parameters of the equation are: $\mu_{s'} = \mu_s(1 - g), \mu_{t'} = \mu_{a'} + \mu_{s'}$ and $D = 1/3\,\mu_{t'}$, in which g is defined as the anisotropy of the medium.

The total light fluence rate, $\phi[W/cm^2]$, is the sum of the collimated part, ϕ_c, and the diffuse part, ϕ_d. The total fluence rate as given by the following equation is a key parameter in laser–tissue interaction. It can be thought of as the total light received at a point in space, or within the medium, through a small sphere divided by the area of that sphere.

$$\phi(r, z) = \phi_c(r, z) + \phi_d(r, z). \qquad (17.36)$$

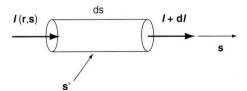

Figure 17.7   Light intensity change in the transport theory approach, including scattered light from other directions.

The collimated fluence rate is given by

$$\phi_c = I_o(r)(1 - r_{sp}) \exp(-\mu_{t'} \, z) \qquad (17.37)$$

where I_o is the surface flux density (W/cm^2) of the incident beam and r_{sp} is the specular reflectance. The optical boundary condition at the beam axis ($r = 0$) is

$$\left. \frac{\partial \phi_d}{\partial r} \right|_{r=0} = 0 \qquad (17.38)$$

The boundary condition elsewhere is

$$\phi_d - 2AD\nabla\phi_d \cdot \tilde{n} = 0 \qquad (17.39)$$

where A is the internal reflectance factor and $\bar{n}$ is the inward unit normal vector. The internal reflectance factor A can account for the effect of mismatch in the index of refraction between the boundary and the surrounding medium and is given by

$$A = \frac{1 + r_i}{1 - r_i} \qquad (17.40)$$

where r_i is evaluated by an empirical formula

$$r_i = -1.440n_{rel}^{-2} + 0.710n_{rel}^{-1} + 0.688 + 0.0636n_{rel} \qquad (17.41)$$

and n_{rel} is the ratio of the refractive indices of the tissue and the medium. The internal reflectance factor A reduces to 1 in cases where the boundary is matched (i.e., when $n_{rel} = 1$).

Equations 17.35 through 17.41 provide the governing differential equations and boundary condition for the diffusion approximation. A solution of these either by analytical or numerical methods allows the calculation and analysis of fluence rate within a scattering and absorbing medium such as biological tissues.

As an example for an isotropic point source inside an infinite medium, the solution for fluence rate as measured by a detector imbedded inside the medium at a "large" distance r from the fiber can be derived using the Green's functions solution of the previous equations as follows:

$$\phi(r) = \frac{\phi_o}{4\pi D} \frac{e^{-r/\delta}}{r} \qquad (17.42)$$

where $\delta = \sqrt{D/\mu_a}$ is the penetration depth.

Example Problem 17.5

It is desired to measure the concentration of an absorber in a scattering medium with known scattering coefficient. If the reduced scattering coefficient $\mu_{s'}$ is known and the relative intensity at a distance r_o from an isotropic source can be measured, an algebraic equation may be solved for the absorption coefficient based on the diffusion approximation, given r_o is large enough for diffusion approximation to be valid.

Solution

Given scattering coefficient $\mu_{s'}, r_o$, and $\phi(r_o)/\phi_o$, an algebraic equation for μ_a based on the diffusion approximation is required.

$$\phi(r) = \phi_o \exp(-r/\delta)/(4\pi\, D\, r)$$

$$\log[\phi(r)/\phi_o] = -r/\delta - [\log 4 + \log \pi + \log D + \log r] \qquad (ex.1)$$

$$D = 1/3\mu_{t'}$$

$\delta = \sqrt{D/\mu_a}$. The left-hand side of Equation ex.1 is a known value, say k_1

$$\therefore k_1 = -r/\delta - [\log 4 + \log \pi + \log D + \log r]$$

$$\therefore k_1 + \log 4 + \log \pi + \log r = -r/\delta - \log D \qquad (ex.2)$$

The left-hand side of Equation ex. 2 is again a known value, say k_2

$$\therefore k_2 = -r/(\sqrt{D/\mu_a}) - \log D$$

$$\therefore k_2 = -r\sqrt{\mu_a/D}) - \log D$$

$$\therefore k_2 = -r(\sqrt{\mu_a/1/(3\mu_{t'})}) - \log(1/3\mu_{t'})$$

$$\therefore k_2 = -r(\sqrt{3\mu_a\mu_{t'}}) - \log(1/3\mu_{t'})$$

$$\therefore k_2 = -r(\sqrt{3\mu_a + \mu_{s'}})) - \log(1/3(\mu_a + \mu_{s'})) \qquad (ex.3)$$

Solving Equation ex. 3, the value of the coefficient of absorption can be found. ∎

17.2.3 Measurement of Optical Properties

The measurement of optical properties, namely absorption coefficient (μ_a), scattering coefficient (μ_s), and scattering anisotropy (g) of biological tissues remains a central problem in the field of biomedical optics. Knowledge of these parameters is important in both therapeutic and diagnostic applications of light in medicine. For example, optical properties are necessary to make accurate assessments of localized fluence during irradiation procedures such as photodynamic therapy, photocoagulation, and tissue ablation. Also, in addition to having a profound impact on in vivo diagnostics such as optical imaging and fluorescence spectroscopy, the optical properties themselves can potentially be used to provide metabolic information and diagnose diseases.

To date, a number of methods have been developed to measure tissue optical properties. The collimated transmission technique can be used to measure the total interaction coefficient ($\mu_a + \mu_s$). In this technique, a collimated light beam illuminates a thin piece of tissue. Unscattered transmitted light is detected while the scattered light is rejected by use of a small aperture. The unscattered transmitted light can be calculated based on the Beer–Lambert law, which is an extension of Equation 17.30. The Beer–Lambert law including scattering media is $I(z) = I_0 \exp[-(\mu_a + \mu_s)z]$ where $I(z)$ is the unscattered transmitted light intensity after penetrating a depth of

z. In collimated transmission measurements, I_0, $I(z)$, and z are measured. Therefore, $\mu_a + \mu_s$ can be deduced.

Example Problem 17.6

A 5 mW collimated laser beam passes through a 4 cm nonabsorbing, scattering medium. The collimated transmission was measured through a small aperture to be 0.035 mW. Calculate the scattering coefficient.

Solution

Based on Beer's law, $\mu_s = \ln(5/0.035)/4 = 1.2\,\text{cm}^{-1}$. ∎

Probably the most common technique is the integrating sphere measurement. In this technique, a thin slice of tissue is sandwiched between two integrating spheres (spheres with an entrance and exit port whose inner surface is coated with a diffuse reflecting material). A collimated beam is incident upon the tissue sample. Both the diffuse reflectance and transmittance are measured by integrating the diffusely reflected and transmitted light, respectively. These two measurements are used to deduce the absorption coefficient (μ_a) and the reduced scattering coefficient ($\mu_{s'}$) with a model. The model could be based on the adding–doubling method, delta-Eddington method, Monte Carlo method, or other light transport theories.

Another technique is normal incidence video reflectometry. A collimated light beam is normally incident upon a tissue. The spatial distribution of diffuse reflectance is collected using either a CCD camera or an optical fiber bundle. Diffusion theory is used to fit the measured spatial distribution of diffuse reflectance to determine the optical properties. The measured spatial distribution of diffuse reflectance must be in absolute units unless total diffuse reflectance is measured with the diffuse reflectance profile. Calibration to absolute units is a sensitive procedure and hence this method is not ideal for a clinical setting.

It is possible to use time-resolved or frequency-domain techniques to measure optical properties. But these techniques require instrumentation that may not be cost-effective for nonresearch applications.

One recent promising approach for in vivo optical property measurements is a fiber-optic–based oblique incidence reflectometry. It is a fairly simple and accurate method for measuring the absorption and reduced scattering coefficients, μ_a and $\mu_{s'}'$, providing the sample can be regarded as a semi-infinite turbid medium, as is the case for most in vivo tissues. Therefore, this approach will be covered in more detail in this section.

Obliquely incident light produces a spatial distribution of diffuse reflectance that is not centered about the point of light entry. The amount of shift in the center of diffuse reflectance is directly related to the medium's diffusion length, D. A fiber-optic probe may be used to deliver light obliquely and sample the relative profile of diffuse reflectance. Measurement in absolute units is not necessary. From the profile, it is possible to measure D, perform a curve fit for the effective attenuation coefficient, μ_{eff}, then calculate μ_a and $\mu_{s'}$. Here μ_{eff} is defined as

$$\mu_{eff} = \sqrt{\mu_a/D} \qquad (17.43)$$

The spatial distribution of diffuse reflectance of normally incident light has previously been modeled using two isotropic point sources, one positive source located 1 mfp′ below the tissue surface and one negative image source above the tissue surface. The positive source represents a single scatter source in the tissue, and the height in z of the image source depends on the boundary condition. The transport mean free path is defined as

$$mfp' = 1/(\mu_a + \mu_{s'}) \qquad (17.44)$$

With oblique incidence, the buried source should be located at the same path length into the tissue, with this distance now measured along the new optical path as determined by Snell's law. It is assumed that (a) the angle of incidence, and (b) the indexes of refraction for both the tissue and the medium through which the light is delivered is known. The net result is a change in the positions of the point sources, particularly a shift in the x direction. These two cases are diagrammed in Figures 17.8a and b.

The diffuse reflectance profile for oblique incidence is centered about the position of the point sources, so the shift, Δx, can be measured by finding the center of diffuse reflectance relative to the light entry point.

The two-source model gives the following expression:

$$R(x) = \frac{1}{4\pi} \left[\frac{\Delta z(1 + \mu_{eff}\rho_1) \exp(-\mu_{eff}\rho_1)}{\rho_1^3} + \frac{(\Delta z + 2z_b)(1 + \mu_{eff}\rho_2) \exp(-\mu_{eff}\rho_2)}{\rho_2^3} \right]$$

$$(17.45)$$

where $\Delta z = 3D \cos \theta_{tissue}$. Equation 17.45 can be scaled arbitrarily to fit a relative reflectance profile that is not in absolute units. The effective attenuation coefficient, μ_{eff}, is defined in the preceding section, and ρ_1 and ρ_2 are the distances from the two sources to the point of interest (the point of light collection, see Figure 17.8). As can be seen in Figure 17.8b, the diffusion coefficient can be calculated from Δx:

$$D = \Delta x/(3 \sin \theta_{tissue}) \qquad (17.46)$$

Optical properties of biological tissues depend on the tissue type and optical wavelength. For instance, the liver, with its reddish color, would have a much higher absorption coefficient from a green light source such as an argon laser than would a tan piece of tissue such as chicken breast. Depending on the tissue type and exact wavelength, a typical set of optical properties for visible or near-infrared light would be: $0.1 \, \text{cm}^{-1}$ for the absorption coefficient, $100 \, \text{cm}^{-1}$ for the scattering coefficient, and 0.9 for the anisotropy. In the UV region, light absorption is dominated by proteins. In the IR region, light absorption is dominated by water. The near IR ($\sim$800 nm) is considered a diagnostic window because of the minimal absorption and relatively low scattering in this region.

An important application of optical properties is the measurement of hemoglobin oxygenation saturation, which is a critical physiological parameter. Because the oxygenation and deoxygenated hemoglobin have different absorption spectra, the

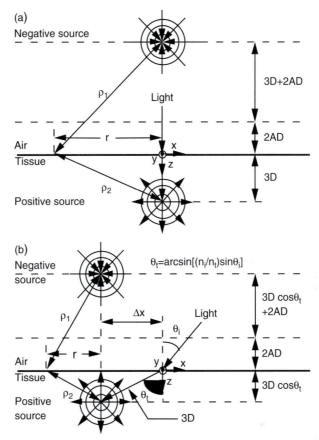

Figure 17.8   (a) Positions of point sources in diffusion theory model for normal incidence. (b) Positions of point sources in diffusion theory model for oblique incidence. The y axis points out of the page. r_1 and r_2 are the distances from the positive and negative point sources to the point of interest on the tissue surface, at a radius of r from the axis of the sources. θ_t is determined from Snell's law.

relative concentration ratio between the two forms of hemoglobin can be calculated once the optical properties of the tissue are measured.

17.3   PHYSICAL INTERACTION OF LIGHT AND PHYSICAL SENSING

After reading Section 17.2, it becomes quite apparent that most of the tissues in our body are not transparent to light (i.e., light is typically either absorbed or scattered in tissues). In this section the physical interaction of the light with the tissue is described and its use in both sensing and therapeutics. The four fundamental physical interactions that will be described are the thermal changes induced or measured using light, pressure changes induced or measured using light, velocity changes manifested as Doppler frequency shifts in the light, and path-length changes in the sample, which causes an interference pattern between two or more light beams.

17.3.1   Temperature Generation and Monitoring

All tissues in our body absorb light at various wavelengths and the absorption process can convert the light energy into heat. In addition, our body tissues, as with any object above absolute zero temperature, also generate light radiation known as blackbody radiation. Thus light can be utilized to heat tissues for therapy or be measured from tissues to determine temperature.

Temperature Monitoring

The measurement of temperature, in general, has traditionally been performed reliably and fairly inexpensively using electrically based sensors such as thermistors or thermocouples. However, these sensors are not as useful when strong electromagnetic interference is present, which is common in a hospital setting. Also, when attempting to monitor temperature rise due to laser radiation these sensors are inappropriate since they absorb the laser radiation. Therefore, to measure temperature in the body several indirect optic and fiber optic approaches have been reported. For example, liquid crystal optrodes show a dramatic change in color due to temperature differences, interferometric sensors change phase or fringes with a change in the path-length due to temperature variations, and the luminescence of many materials depends strongly on temperature.

The direct type of temperature sensor initially described is known as a thermographic or radiometer system. The primary science and engineering of this type of system has been done by the military for tasks such as detecting vehicles, personnel, and even ships in total darkness. The formula for the total emission from a blackbody at temperature T is

$$I = \varepsilon \sigma T^4 \tag{17.47}$$

where ε is the emissivity and σ is the Stefan–Boltzmann constant. At room temperature objects emit mainly in the far-infrared region of the wavelength spectrum, but as temperature rises the emission appears in the near infrared and finally in the visible spectrum. The emissivity is one ($\varepsilon = 1$) for biological tissue and hence the total emission is dependent only on the temperature T. For the military applications a thermographic picture generated from the IR emission can be used to display the temperature, and using this same technique the surface temperature of the human body can be monitored for medical applications. For example, a thermal camera may be used with a TV monitor to reveal the temperature distribution on the human chest in an effort to reveal the pathologic condition of breast cancer since the affected breast tissue will be slightly warmer than the healthy tissue. It should be mentioned that although IR imaging is less reliable than x-ray mammography for breast screening, it is totally passive and the patients are not exposed to ionizing radiation. The thermographic systems used for this IR emission measurement require a line of sight between the warm surface and the detector. When no direct line of sight exists, an infrared-transmitting optical fiber can sometimes be used to make the connection to the detector. This type of fiber optic radiometer has been proposed for treatments using microwave heating and laser tissue heating treatments.

Light-Induced Heating

As mentioned previously, in addition to monitoring the light radiation from the body, light itself can be used to heat tissues that absorb the light. Light-induced heating of tissue can be used for a variety of applications, including biostimulation, sealing or welding blood vessels, tissue necrosis, or tissue vaporization. Biostimulation has been claimed as a result of minute heating of the tissue whereby the light-induced heat stimulates nerves and/or accelerates wound healing. Higher-energy absorbed laser light can facilitate the joining of tissues, in particular blood vessels (i.e., anastomosis), and can be used to coagulate blood to stop bleeding during a surgical procedure. If the light-induced heating causes the temperature to rise above 45°C, tissue necrosis and destruction occurs, as would be desired for the treatment of cancer or enlarged prostate tissues. At even higher power densities ablation or vaporization of the tissue occurs, as is the case for the corrective eye surgery known as radial keratectomy.

The light-induced heating is typically performed with laser light. The lasers used have different wavelengths (from the ultraviolet to the infrared spectrum), power densities (i.e., ratio between beam power and irradiated area), and duration times. The amount of energy imparted to the tissue and hence temperature rise can be changed by either varying the power density or the duration of the time pulse of the laser. For high power density coagulation, necrosis, and vaporization of tissue can occur whereas at low power densities minimal heating is observed. The amount of absorption by the tissue is a function of the wavelength of the light used. For some wavelengths, for instance those of the known strong absorption bands of water, the laser beam is highly absorbed since tissue is primarily made of water. At these wavelengths the energy is then highly absorbed in a relatively thin layer near the surface where rapid heating occurs (i.e., radial keratectomy). The absorption is less for other wavelengths used and this results in slower heating of a larger volume of the tissue (i.e., for prostate coagulation).

Temperature Generation and Rate of Photon Absorption

A thermodynamically irreversible mode of interaction of light with materials is the process of absorption in which the photon energy is absorbed by the material phase. In the absence of conduction, the temperature rise in tissue is governed by a thermodynamic equation of state. The equation of state requires that the change in internal energy of the system is proportional to temperature rise. The change in internal energy of the system, in the absence of conduction and other heat transfer processes, is equal to the rate of energy deposition by the laser. Expressing this relation in terms of time derivatives gives

$$\Delta U / \Delta t = \rho C \Delta T / \Delta t \approx Q_L \tag{17.48}$$

where $\Delta T(K)$ is temperature rise, $Q_L(\text{W/m}^3)$ is volumetric rate of photon absorption by the tissue, $\rho(\text{kg/m}^3)$ is mass density, $C(\text{J/Kg.K})$ is specific heat, $\rho C(\text{J/m}^3.\text{K})$ is volumetric heat capacity, and t is the duration. A very important factor in temperature rise by photons is the rate of photon absorption Q_L which is also known as the light/laser source term and as the rate of energy deposition. For ordinary interaction processes, the rate of absorption of photons by the material is proportional to irradiance, and the constant of proportionality is the absorption coefficient.

$$Q_L = \mu_a \phi \qquad (17.49)$$

Irradiance ϕ is related to the spatial distribution of photons within the tissue.[1] For a purely absorbing materials, for instance, the Beer–Lambert law applies and if a Gaussian beam profile is additionally assumed then

$$Q_L = \mu_a \times \phi_o \exp(-\mu_a z) \exp(-2r^2/\omega^2) \qquad (17.50)$$

where ϕ_o is incident intensity, z is depth, and r is radial distance. In the presence of scattering, one of the scattering models discussed before can be used to describe ϕ_o.

Equation 17.48 assumes no other thermal interaction processes such as conduction, convection, or perfusion. If these processes can be ignored, as for example for very short laser pulses, then temperature rise can be estimated as

$$\Delta T \approx Q_L \Delta t / \rho C \qquad (17.51)$$

where Δt is exposure duration. Other thermal effects of laser–tissue interaction will be considered in more detail in Section 17.5. Equation 17.51 is valid for very short irradiations during which heat diffusion is negligible. A criteria or an estimate of upper limit for validity of Eq. 17.51 is

$$\Delta t_{\text{MAX}} = \frac{1}{\sqrt{4\mu_a^2 \alpha}} \qquad (17.52)$$

where α is the thermal diffusivity of the material. For water the value of α is about $1400 \, \text{cm}^2/\text{s}$.

17.3.2   Laser Doppler Velocimetry (LDV)

In addition to temperature, another physical interaction of light is the Doppler phenomenon which is based on a frequency shift due to the velocity of a moving object. For medical applications these objects are typically the moving red blood cells with a diameter of roughly $7 \, \mu\text{m}$. The Doppler approach is used for measuring blood flow velocity for a variety of medical applications, including heart monitoring, transluminar coronary angioplasty, coronary arterial stenosis, tissue blood flow on the surface of the body, and monitoring blood flow on the scalp of a fetus during labor.

When a light wave of frequency f and velocity c impinges on a stationary object, it is reflected at the same frequency. However, if the object moves with velocity v, the

---

[1] For multiphoton light-tissue interaction processes, the exponent of ϕ is increased. For example, for a two-photon process $Q_L = \mu_a * \phi^2$

reflected frequency f' is different from f. This difference or shift in frequency, δf, from the original light wave is known as the Doppler effect or Doppler shift, and can be described as:

$$\delta f = f - f' \qquad (17.53)$$

The Doppler shift is described in Equation 17.54 in terms of the velocity, v, of the moving red blood cells, the refractive index of the medium, n, the speed of light in the tissue, c_o, the input frequency, f, and the angle between the incident beam and the vessel, θ:

$$\delta f / f = 2v\, n \cos\theta / c_o \qquad (17.54)$$

Further, it is well known that the wavelength, λ is equal to the speed of light divided by the frequency so that

$$\delta f = 2v\, n \cos\theta / \lambda_o \qquad (17.55)$$

A comparable analogy for the Doppler effect with sound waves is a train moving toward an observer. As the train gets closer the whistle sounds like a higher pitch.

Example Problem 17.7

Using a laser Doppler system, calculate the velocity of blood given an argon laser (514 nm), an angle with the vessel of 30°, an index of refraction of 1.33, and a frequency shift of 60 KHz. Does this make sense physiologically? Does it really matter if you were off in your probe to vessel measurement angle by 10°?

Solution

Using Equation 17.55, the velocity can be calculated to be

$$v = (\delta f x \lambda)/(2n \cos(\theta)) = [(60 \times 10^3)(514 \times 10^{-9})]/[2\,(\cos(30))\,(1.33)] = 1.34\,\text{cm/s}$$

∎

This is reasonable for an average-sized vessel in the human body. If you were off by 10°—for example you thought the angle was 40° instead of 30°—this would make a pretty big difference with the newly calculated speed of 1.51 cm/s (a 13% error).

Laser Doppler is a good method for monitoring velocities, but one reason it has not gained wide use clinically is that the flow (cm^3/sec) of blood, or rather the average velocities of all the particles in the fluid, is the physical quantity of diagnostic value and flow cannot be directly measured using the Doppler approach. Calculating the flow rate for a rigid tube filled with water by knowing the velocity is a straightforward problem to solve. However, determining the flow rate of blood in the body by simply using the Doppler measured velocity is a much more difficult problem, particularly for narrow vessels. Blood is thicker than water and the flow characteristics more complex. Blood vessels also are not rigid, straight tubes, and the flow of blood is pulsatile. Lastly, a fiber optic probe is often inserted in the blood vessel and scanned to get a series of velocity measurements to determine flow, but the probe itself can alter the flow measurement.

Overall, laser Doppler velocimetry is a simple concept which can, with some effort, be used to measure relative changes in flow, but it is very difficult for this approach to be calibrated for absolute flow measurements. Thus, the standard for blood mass-flow measurements has been by thermodilation and this is widely used in clinical practice. The thermodilation method is used to determine flow by inducing a predetermined change in the heat content of the blood at one point of the circulation and detecting the resultant change in temperature at some point downstream, after the flow has caused a controlled degree of mixing across the vessel diameter.

17.3.3 Interferometry

The phenomenon of interference is another method by which physical light interaction takes place and it depends on the superposition of two or more individual waves, typically originating from the same source. Since light consists of oscillatory electric and magnetic fields which are vectors, they add vectorally. Thus, when two or more waves emanating from the same source are split and travel along different paths, they can then reunite and interfere constructively or destructively. When the constructive and destructive interference are seen to alternate in a spatial display, the interference is said to produce a series or pattern of fringes. If one of the paths in which the light travels is altered by any small perturbation such as temperature, pressure, or index of refraction changes, then, once recombined with the unaltered reference beam, the perturbation causes a shift in the fringe pattern which can be readily observed with optoelectronic techniques to about 10^{-4} of the fringe spacing. The useful information regarding the changing variable of interest can be measured quite accurately as a path-length change on the order of one-hundredth of a wavelength or 5×10^{-9} meters for visible light.

There are several variations for producing light interference, including one of the first instruments known as the Rayleigh refractometer from which came the Mach–Zehnder interferometer. For the sake of brevity, this section will focus on two more sophisticated variations of the Rayleigh idea: the dual-beam Michelson interferometer and the multiple-beam Fabry–Perot interferometer. As depicted in Figure 17.9, the Michelson interferometer begins with a light source that is split into two beams by

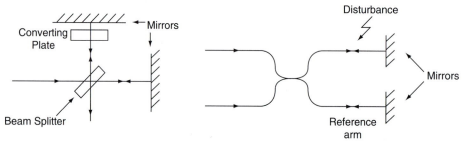

Figure 17.9   Michelson interferometer: (a) bulk optics, and (b) fiber optics.

means of a beam-splitting mirror or fiber optics which also serves to recombine the light after reflection from fully silvered mirrors. A compensating plate is sometimes needed to provide equal optical paths before introduction of any perturbation or sample to be measured. The perturbation can take the form of a pressure or temperature change causing a strain and hence path-length change in the sample arm of the fiber optic or as a change due to the addition of a tissue sample or replacement of the mirror with a tissue sample. For example, the Michelson interferometer has been investigated for measuring tissue thickness, particularly for corneal tissue as feedback for the radial keratectomy procedure (laser removal or shaving of the cornea to correct vision) and this interferometer, when used with a low coherent light source, has been researched for use in optical coherence tomographic imaging of superficial tissues. The governing equation for the irradiance of the fringe system of circles concentric with the optic axis in which the two interfering beams are of equal amplitude is given as

$$I = 4I_o \cos^2(\delta/2) \tag{17.56}$$

where I_o is the input intensity and the phase difference δ is defined as

$$\delta = 2\pi\Delta/\lambda \tag{17.57}$$

In this equation, λ is the wavelength and the net optical path difference, Δ, is defined as

$$\Delta = 2d\cos(\theta) + \lambda/2 = (m + 0.5)\lambda \tag{17.58}$$

Clearly from 17.58 the following equation can be derived

$$2d\cos(\theta) = m\lambda \tag{17.59}$$

in which $2d$ is the optical path difference or the difference in the two paths from the beam splitter, m is the number of fringes, and θ is the angle of incidence ($\theta = 0$ is a normal or on-axis beam). When a plate, gas, or a thin minimally absorbing or scattering tissue slice is assumed with constant index of refraction and is inserted in one of the paths, then $d = (n_s - n_{air})L$ in which L is the actual length of the substance, n_s is the index of refraction of the substance, and n_{air} is the index of refraction of the air.

Example Problem 17.8

A thin sheet of clear tissue, such as a section of the cornea of the eye, (n ≈ 1.33) is inserted normally into one beam of a Michelson interferometer. Using 589 nanometers of light, the fringe pattern is found to shift by 50 fringes. Determine the thickness of the tissue section.

Solution

From Equation 17.59, $2d\cos(\theta) = m\lambda$, and thus $d = (50 \times .589)/2 = 14.72\,\mu m$, which is the calculated optical path length. However, the physical length of the tissue must take into account the index of refraction of the sample and the air. Thus, as described in the paragraph after Equation 17.59, the equation for physical path length is $L = d/(n_s - n_{air}) = 14.72/(1.33 - 1.0) = 44.6$ micrometers. ∎

All the dual-beam interferometry techniques such as the Michelson and Mach–Zehnder approaches suffer from the limitation that the accuracy depends on the location of the maxima (or minima) of a sinusoidal variation as shown in Equation 17.59. For very accurate measurements, such as in precision spectroscopy, this limitation is severe. Rather than dual beam, if the interference of many beams is utilized the accuracy can be improved considerably. A Fabry–Perot interferometer uses a multiple-beam approach as depicted in Figure 17.10. As can be seen in the figure, the interferometer makes use of a plane parallel plate to produce an interference pattern by combining the multiple beams of the transmitted light. The parallel plate is typically composed of two thick glass or quartz plates which enclose a plane parallel plate of air between them. The flatness and reflectivity of the inner surfaces are important and are polished generally better than $\lambda/50$ and coated with a highly reflective layer of silver or aluminum. The silver is good for wavelengths above 400 nm in the visible but aluminum has better reflectivity below 400 nm. These film coatings must also be thin enough to be partially transmitting (~50-nm thickness for silver coatings). In many instances the outer surfaces of the glass plate are purposely formed at a small angle relative to the inner faces (several minutes of arc) to eliminate spurious fringe patterns that can arise from the glass itself acting as the parallel plate interferometer. When L is fixed, the instrument is referred to as a Fabry–Perot etalon. The nature of the superposition at point P is determined by the path difference between successive parallel beams, thus

$$\Delta = 2n_f L \cos \theta_t = m\lambda_{(minimum)} = 2L \cos \theta_t \qquad (17.60)$$

using $n_f = 1$ for air.

Other beams from different points of the source but in the same plane and making the same angle θ_t with the axis satisfy the same path difference and also arrive at P. With L fixed, the preceding equation for Δ is satisfied for certain angles θ_t and the fringe system is the familiar concentric rings due to the focusing of fringes of equal inclination, as depicted in Figure 17.11a and b. If the thickness of L varies with time, a

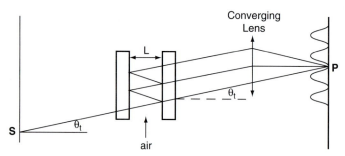

Figure 17.10   Fabry–Perot multiple-beam interferometer.

(a)

(b)

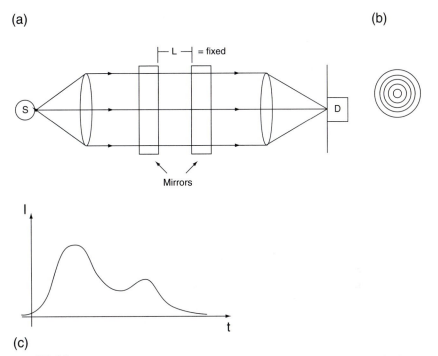

Mirrors

(c)

Figure 17.11 (a) F-P etalon with L fixed showing (b) spacial separation, and (c) temporal separation.

detector D will record an interferogram as a function of time as depicted in Figure 17.11(c).

Variation in the fringe pattern irradiance of the Fabry–Perot as a function of phase or path difference is known as the fringe profile. The sharpness of the fringes are important to the ultimate resolving power of the instrument. Using the trig identity $\cos \delta = 1 - 2 \sin^2 (\delta/2)$ the transmittance can be written

$$T = I_T/I_i = 1/(1 + [4r^2/(1 - r^2)^2] \sin^2 (\delta/2)) \text{ (Airy function)} \qquad (17.61)$$

where r is the reflectivity and the term in the square brackets in the denominator is known as the coefficient of finesse, F; thus,

$$T = 1/(1 + F \sin^2 (\delta/2)) \qquad (17.62)$$

It should be noted that as $0 < r < 1$ then $0 < F < \infty$. The coefficient of finesse, F, also represents a certain measure of fringe contrast

$$F = ((I_T)_{max} - (I_T)_{min})/I_T min = (T_{max} - T_{min})/T_{min} \qquad (17.63)$$

It should be noted here that $T_{max} = 1$ when $\sin (\delta/2) = 0$ and $T_{min} = 1/(1 + F)$ when $\sin (\delta/2) = \pm 1$. Given r, the fringe profile can be plotted as shown in Figure 17.12.

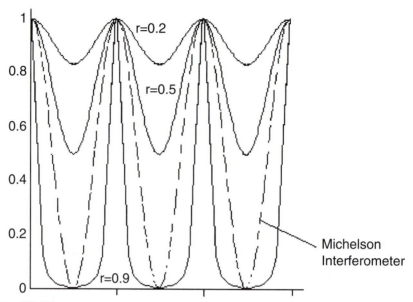

Figure 17.12   The difference in resolution between the multiple-beam Fabry–Perot interferometer with different reflectivities and the dual-beam Michelson interferometer.

As seen in the figure, $T = T_{max} = 1$ at $\delta = m2\pi$ and $T = T_{min} = 1/(1+F)$ at $\delta = (m + 1/2)2\pi$. As can be seen, $T_{max} = 1$ regardless of r, and T_{min} is never zero, but approaches zero as r approaches one. Also the transmittance peaks sharply at higher values of r, remaining near zero for most of the region between fringes. It should be noted that the Michelson interferometer produces broad fringes when it is normalized to the same max value since it has a simple $\cos^2(\delta/2)$ dependence on the phase (dashed lines in the figure). An example for a fiber optic based Fabry–Perot interferometer is the commercially available intracranial fluid pressure monitoring system for patients with severe head trauma or hydrocephalus (increased amount of cerebral spinal fluid in the ventricles and/or subarachnoid spaces of the brain).

17.4   BIOCHEMICAL MEASUREMENT TECHNIQUES USING LIGHT

In the past few years, there has been much enthusiasm and a strong effort by medical device companies and universities to perform diagnostic procedures such as cancer detection and quantifiably monitor blood chemicals such as glucose, lactic acid, albumin, and cholesterol using various optical approaches. The most well-known optical monitoring approach used clinically is the blood oxygenation sensor, which detects qualitative changes in the strong optical absorption peaks of oxygenated and deoxygenated hemoglobin. The most common optical approaches for diagnostic and sensing applications include absorption, scattering, luminescence, and polarimetry. The primary variable for each of these approaches is a change in the light intensity as it passes through the medium, which can change as a function of the wavelength or polarization of the light. For absorption and luminescence or fluorescence, the intensity will change nearly linearly for moderate analyte concentrations and nonlinearly for high concentrations.

17.4.1 Spectroscopic Measurements Using Light Absorption

Many investigators have suggested infrared absorption as a potential route to blood chemical monitoring and diagnostic sensing, in particular for glucose monitoring and cancer detection. The governing equation (17.17) for purely absorbing media as well as fluorescence was described previously using the Beer–Lambert law. Expressed logarithmically the equation becomes,

$$A = \ln(T) = \ln(I_o/I) = \mu z = z? \varepsilon_i C_i \qquad (17.64)$$

in which A is absorbance, T is transmittance, I_o is the incident light, I is the transmitted light, z is again the path length, and μ is again the absorption coefficient, or rather the sum of the multiplication of the molar absorptivity times the concentration of all the different components in the analyte. For tissue and blood this equation is valid in the mid-infrared wavelength region (wavelengths of 2.5–12 micrometers) in which the absorption peaks due to various chemicals are distinguishably sharp and the scatter is weak. Unfortunately, the absorption of the light due to water within tissue in this region is orders of magnitude stronger than any of the blood chemicals, which results in the possibility of only short path-length sample investigations (on the order of micrometers). This brings about the possibility for surface or superficial investigations on skin or, with the use of fiber optics, on internal body cavities. Unfortunately, the optical fibers available in this region tend to be toxic, hydroscopic, and/or rigid. Consequently, the combination of the water absorption and lack of good fibers makes in vivo diagnostics and sensing very difficult in the mid–IR region. The light sources used in this region include the broad-based tungsten bulb, nernst glower, nichrome wire, and globar rod, and narrowly tunable, typically liquid nitrogen–cooled, laser diodes. The detectors used include the cooled mercury cadmium telluride (MCT), thermopile, and thermistor. The optics include sodium chloride and potassium bromide, mirrors (typically gold coated), and gratings.

In the near-infrared (NIR) wavelength region the spectrum is not affected by water to the same degree as in the mid-infrared region, allowing path lengths of 1 mm to 1 cm to be used. In addition, low OH silica fibers are quite transparent across this range, and are nontoxic and nonhydroscopic. The NIR region (700–2500 nm) exhibits absorptions due to low energy electronic vibrations (700–1000 nm), as well as overtones of molecular bond stretching and combination bands (1000–2500 nm). These bands result from interactions between different bonds (–CH, – OH, NH) to the same atom. Typically, only the first, second, and third overtones of a molecular vibration are detectable and are broad in nature. Thus, only at high concentrations of the chemical species are these overtones qualitatively detectable, with the intensity dropping off rapidly as overtone order increases. The NIR absorption bands are also influenced by temperature, pressure, and hydrogen bonding effects and can overlap significantly. In the 700 to 1200 nm region they are also influenced by scattering effects. Thus, unlike mid-IR spectroscopy, NIR spectroscopy is primarily empirical and not well suited for qualitative work. However, using techniques such as multivariate statistics, quantitative analysis is possible with the NIR spectrum. The light sources used in this region include the broad-based tungsten bulb, light-emitting diodes (LEDs), laser-pumped solid-state lasers such as the tunable titanium sapphire laser, and laser diodes. The detectors used include silicon (good to 1 micrometer), germanium (good to 1.7 micrometers), and indium antiminide (good to 5.5 micrometers). The optics includes low OH glasses, quartz, glass, gratings, and mirrors.

The primary methods for varying the wavelength of the light include dispersive and nondispersive methods. The dispersive approach uses either a ruled reflective grating or transmissive prism to separate the wavelengths of light. The nondispersive systems include a series of wavelength selection filters or a Fourier transform infrared (FTIR)–based instrument. The FTIR method uses an interferometer similar to the Michelson interferometer depicted in Figure 17.9 to collect the entire spectrum and then deconvolutes it using Fourier transform techniques. Both approaches can be configured to cover the NIR and mid-IR ranges. The advantages of the dispersive systems include higher resolution and separation of closely spaced wavelength bands whereas the nondispersive systems generally have better throughput since all the light passes through the sample.

17.4.2   Monitoring Approaches Using Scattered Light

There are fundamentally two types of optical scattering for diagnostics and monitoring: elastic and inelastic. The elastic scattering can be described using Mie theory (or Rayleigh scattering for particles small compared to the wavelength) in which the intensity of the scattered radiation can be related to the concentration, size, and shape of the scattering particles. The inelastic scattering in which the polarization of the particle is not constant can be described as Raman scattering. Fluorescence is also inelastic.

When using the light scattering phenomenon for sensing, the intensity of the reflected light is usually considered. The reflection of light, however, can be divided into two forms. The first, specular reflection or a "mirror" type of reflection, occurs at the interface of a medium. The returned light yields no information about the material since it never penetrates the medium and thus the specularly reflected light is typically minimized or eliminated with the design of the optical sensor. Diffuse reflection, however, occurs when light penetrates into a medium, becomes absorbed and multiply scattered, and makes its way back to the surface of the medium. The model describing the role of diffuse scattering in tissue is based on the radiative transfer theory as was described in Section 17.2. The same theory applies for sensing.

The use of elastically scattered light has been suggested for both diagnostic procedures such as cancer detection and for monitoring analytes such as glucose noninvasively for diabetics. For use as a monitoring application of chemical changes such as glucose, researchers have used an intensity-modulated frequency-domain NIR spectrometer, capable of separating the reduced scattering and absorption coefficients to detect changes in the reduced scattering coefficient showing correlation with blood glucose in human muscle. This approach may be promising as a relative measure over time since clearly an increase in glucose concentration in the physiologic range decreases the total amount of tissue scattering. However, the drawbacks of the elastic light scattering approach for analyte monitoring are still quite daunting. The specificity of the elastic scattering approach is the biggest concern with this method since other physiologic effects unrelated to glucose concentration could produce similar variations of the reduced scattering coefficient with time and, unlike the absorption approach, elastic light scattering is nearly wavelength independent. The measurement precision of the reduced scattering coefficient and separation of scattering and absorption changes is another concern with this approach. It is difficult to measure such small changes and be insensitive to some of the larger absorption changes in the tissue, particularly hemoglobin. This approach also needs to take into account the different refractive indices of tissue. Tissue scattering is caused by a variety of substances and organelles (membranes, mitochondria, nuclei, etc.) and all of them have different refractive indices. The effect of blood glucose concentration and its distribution at the cellular level is a complex issue that needs to be investigated before this approach can be considered viable. An instrument of this type would require in vivo calibration against a gold standard, since the reduced scattering coefficient is dependent on additional factors such as cell density. Lastly, there is a need to account for factors that might change the reduced scattering coefficient such as variations in temperature, red-blood-cell concentration, electrolyte levels, and movements of extracellular and intracellular water. As a diagnostic screening tool for cancer detection, measurement of the scatter in thin tissues or cells may hold promise. Many of the changes in tissue due to cancer are morphologic rather than chemical and thus occur with changes in the size and shape of the cellular and subcellular components. Thus the changes in elastic light scatter should be larger with the morphologic tissue differences. If the wavelength of the elastic light scattering is carefully selected so as to be outside the major absorption areas due to water and hemoglobin and if the diffusely scattered light is measured as a function of angle of incidence there is potential for this approach to aid in pathologic diagnosis of disease.

Inelastic Raman spectroscopic scattering has been utilized over the past thirty years mainly by physicists and chemists. Raman spectroscopy has become a powerful tool for studying a variety of biological molecules including proteins, enzymes and immunoglobulins, nucleic acids, nucleoproteins, lipids and biological membranes, and carbohydrates, but with the advent of more powerful laser sources and more sensitive detectors it has also become useful as a diagnostic and sensing tool. The phenomenon of Raman scattering is observed when monochromatic (single-wavelength) radiation is incident upon the media. In addition to the transmitted light, a portion of the radiation is scattered. Some of the incident light of frequency w_o exhibits frequency shifts $\pm w_m$, which is associated with transitions between rotational, vibrational, and electronic levels. In general, the intensity and polarization of the scattered radiation is dependent upon the position of observation relative to the incident energy. Most studies utilize the Stokes type of scattering bands that correspond to the $w_o - w_m$ scattering. Therefore, the Raman bands of interest are shifted to longer wavelengths relative to the excitation wavelength.

As with infrared spectroscopic techniques, Raman spectra can be utilized to identify molecules since these spectra are characteristic of variations in the molecular polarizability and dipole moments. Raman spectroscopy can be considered as complementary to absorption spectroscopy as neither technique alone can resolve all of the energy states of a molecule. In fact, for certain molecules, some energy levels may not be resolved by either technique. Due to the anharmonic oscillator model for dipoles, overtone frequencies exist in addition to fundamental vibrations. It is an advantage of Raman spectroscopy that the overtones are much weaker than the fundamental tones, thus contributing to simpler spectra as compared to absorption spectroscopy. One advantage to using Raman spectroscopy in biological investigations is that the Raman spectrum of water is weaker which, unlike infrared spectroscopy, only minimally interferes with the spectrum of the solute and thus the spectrum can be obtained from aqueous solutions. However, the Raman signal is also weak and only recently, with the replacement of slow photomultiplier tubes with faster CCD arrays as well as the manufacture of higher power near-infrared laser diodes, has the technology become available to allow researchers to consider the possibility of distinguishing normal and abnormal tissue types and quantifying blood chemicals in near-real time. In addition, investigators have applied statistical methods such as partial least squares (PLS) to aid in the estimation of biochemical concentrations from Raman spectra.

As with elastic scatter, Raman spectroscopy has been used for both diagnostics and monitoring. The diagnostic approaches look for the presence of different spectral peaks and/or intensity differences in the peaks due to different chemicals present in, for instance, cancerous tissue. For quantifiable monitoring it is the intensity differences alone that are investigated. In tissue, a problem in addition to the water absorption, is the high fluorescence background signal as a result of autofluorescence incurred in heavily vascularized tissue, due to the high concentration of proteins and other fluorescent components. Instrumentation to excite in the NIR wavelength range has been proposed to overcome the autofluorescence problem since the fluorescence component falls off with increasing wavelength. Excitation in the NIR region offers the added advantage of longer wavelengths that pass through larger tissue samples with lower absorption and scatter than other spectral regions such as visible or ultraviolet. However, in addition to fluorescence falling off with wavelength, the Raman signal also falls off to the fourth power as wavelength increases. Thus, there is a tradeoff between minimizing fluorescence and maintaining the Raman signal. The eye has been suggested as a site for analyte concentration measurements using Raman spectroscopy to minimize autofluorescence; however, the disadvantage to using the eye for Raman spectroscopy is that the laser excitation powers must be kept low to prevent injury, but this significantly reduces the signal-to-noise ratio. Lastly, like infrared and NIR absorption, to quantifiably determine the inherently low concentrations of analytes in vivo the presence of different chemicals must be accounted for that yield overlapping Raman spectra.

17.4.3 Use of the Luminescence Property of Light for Measurement

As described previously, luminescence is the absorption of photons of electromagnetic radiation (light) at one wavelength and reemission of photons at another wavelength. The photons are absorbed by the molecules in the tissue or medium, raising them to some excited energy state and then, upon returning to a lower energy state, the molecules emit radiation or luminesce. The luminscent effect can be referred to as fluorescence or phosphorescence. Fluorescence is luminescence that has energy transitions that do not involve a change in electron spin and therefore the reemission occurs much faster. Consequently, fluorescence occurs only during excitation whereas phosphorescence can continue after excitation. As an example, a standard television while turned on is producing fluorescence but for a very short time after it is turned off the screen will phosphoresce.

The measurement of fluorescence has been used for both diagnostic and monitoring purposes. Obtaining diagnostic information, in particular with respect to cancer diagnosis or the total plaque in arteries, has been attempted using the extrinsic and intrinsic fluorescence of tissue. The intrinsic fluorescence is due to the naturally occurring proteins, nucleic acids, and nucleotide coenzymes whereas extrinsic fluorescence is induced by the uptake of certain "impurities" or dyes in the tissue. Extrinsic fluorescence has also been investigated, for instance, to monitor such analytes as glucose, intracellular calcium, proteins, and nucleotide coenzymes. Unlike the use of fluorescence in chemistry on dilute solutions, the intrinsic or autofluorescence of tissue as well as the scattering and absorption of the tissue act as a noise source for the extrinsic approach.

The response of a fluorescence sensor can be described in terms of the output intensity as

$$I_f = \Phi_f(I_o - I) \tag{17.65}$$

in which, I_f is the radiant intensity of fluorescence, Φ_f is the fluorescence efficiency, I_o is the radiant intensity incident on the sample, and I is the radiant intensity emerging from the sample. The fluorescence efficiency can be described as the combination of three factors. First is the quantum yield, which is the probability that an excited molecule will decay by emitting a photon rather than by losing its energy nonradiatively. This parameter varies from 1.00 to 0.05 and varies with time on the order of nanoseconds. Thus, in addition to intensity measurements, time-resolved fluorescence measurements are possible using pulsed light sources and fast detectors. The second parameter is the geometrical factor, which is the solid angle of fluorescence radiation subtended by the detector which depends on your probe design. Lastly is the efficiency of the detector itself for the emitted fluorescence wavelength.

Since fluorescence is an absorption/reemission technique it can be described in terms of Beer's law as

$$I_f = \Phi_f I_o [1 - \exp(\varepsilon C l)] \qquad (17.66)$$

in which C is the concentration of the analyte, l is the path length, and ε is the molar absorptivity. The preceding equation can be described in terms of a power series and for weakly absorbing species ($\varepsilon C l < 0.05$) only the first term in the series is significant. Therefore, under these conditions, the response of the fluorescence sensor becomes linear with analyte concentration and can be described as

$$I_f = \Phi_f I_o \varepsilon C l \qquad (17.67)$$

The primary fluorescent sensors are based on the measurement of intensity; however, lifetime measurements in the time or frequency domain are also possible. To gain the most information, particularly in a research or teaching setting, dual monochromators (grating-based wavelength separation devices) are used with either a photomultiplier tube as the detector or a CCD array detector. In a typical bench-top fluorimeter a broad, primarily ultraviolet/visible, xenon bulb is used as the light source. The light is coupled first through a monochromator, which is a wavelength separator that can be set for any excitation wavelength within the range of the source. The light then passes through the sample and is collected by a second monochromator. The light reflected from the grating within the second monochromator can be scanned so that a photomultiplier tube (PMT) receives the different wavelengths of light as a function of time. Alternatively, all the wavelengths from the grating can be collected simultaneously on a CCD detector array. The advantage of the CCD array is that it provides for real-time collection of the fluorescence spectrum. The advantage of the PMT is that it is typically a more sensitive detector. In many systems a small portion of the beam is split at the input and sent to a reference detector to allow for correction of fluctuations in the light source. Once the optimal configuration for a particular biomedical application, such as cervical cancer detection or glucose sensing, has been investigated using the bench-top machine, an intensity measurement system can be designed with a simpler, more robust, configuration. Such a system can be designed with wavelength-specific filters instead of monochromators and made to work at two or more discrete wavelengths. In addition, optical fibers can be used for delivery and collection of the light to the remote area. Since the excitation wavelength and fluorescent emission wavelengths are different, the same fiber or fibers can be used to both deliver and collect the light. In any configuration it is important to match the spectral characteristics of the source, dye (if used), sample, and detector. For instance, depending on the tissue probed there are strong absorbers, scatterers, and autofluorescence that need to be factored into any fluorescent system design.

17.4.4 Measurements Made Using Light Polarization Properties

Having discussed some of the fundamental electromagnetic theory of polarized light generation in Section 17.1.2, some of the applications of polarized light shall now be discussed. Two of the emerging applications of polarized light are for biochemical quantification such as glucose and tissue characterization, in particular, to aid in cancer identification or for use in the measurement of the nerve fiber layer of normal and glaucomatous eyes.

The rotation of linearly polarized light by optically active substances has been used for many years to quantify the amount of the substance in solution. A variety of both polarimeters, adapted to the examination of all optically active substances and saccharimeters designed solely for polarizing sugars have been developed. The concept behind these devices is that the amount of rotation of polarized light by an optically active substance depends on the thickness of the layer traversed by the light, the wavelength of the light used for the measurement, the temperature, the pH of the solvent, and the concentration of the optically active material. Historically, polarimetric measurements have been obtained under a set of standard conditions. The path length typically employed as a standard in polarimetry is 10 cm for liquids, the wavelength is usually that of the green mercury line (5461 Angstroms), and the temperature is 20°C. If the layer thickness in decimeters is L, the concentration of solute in grams per 100 ml of solution is C, α is the observed rotation in degrees, and $[\alpha]$ is the specific rotation or rotation under standard conditions which is unique for all chiral molecules, then

$$C = \frac{100\alpha}{L[\alpha]} \qquad (17.68)$$

In the previous equation the specific rotation $[\alpha]$ of a molecule is dependent upon temperature, T, wavelength, λ, and the pH of the solvent.

For polarimetry to be used as a noninvasive technique, for instance in blood glucose monitoring, the signal must be able to pass from the source, through the body, and to a detector without total depolarization of the beam. Since the skin possesses high scattering coefficients, which causes depolarization of the light, maintaining polarization information in a beam passing through a thick piece of tissue (i.e., 1 cm), which includes skin, would not be feasible. Tissue thicknesses of less than 4 mm, which include skin, may potentially be used, but the polarimetric sensing device must be able to measure millidegree rotations in the presence of greater than 95% depolarization of the light due to scattering from the tissue. As an alternative to transmitting light through the skin, the aqueous humor of the eye has been investigated as a site for detection of in vivo glucose concentrations since this sensing site is a clear biological optical media. It is also known that glucose concentration of the aqueous humor of the eye correlates well with blood glucose levels, with a minor time delay (on the order of minutes), in rabbit models. The approximate width of the average anterior chamber of a human eye is on the order of 1 cm. Therefore, an observed rotation of about 4.562 millidegrees per optical pass can be expected for a normal blood glucose level of 100 mg/100 ml, given a specific rotation of glucose at $\lambda = 633$ nm of $45.62°\text{cm}^2\text{g}^{-1}$ and a thickness of 1 centimeter. The eye as a sensing site, however, is not without its share of potential problems. For instance, potential problems with using the eye include corneal birefringence and eye motion artifact. As shown in Equation 17.68, the rotation is directly proportional to the path length and thus it is critical that this length be determined or at least kept constant for each individual subject regardless of the sensing site. If the eye is used as the sensing site, the angle of incidence on the surface of the cornea must be kept relatively constant for each patient so that not only the path length but alignment remains fixed each time a reading is taken. In most tissues, including the eye, the change in rotation due to other chiral molecules such as proteins needs to be accommodated in any final instrument. In addition, most other tissues also have a birefringence associated with them that would need to be accounted for in a final polarimetric glucose sensor.

The birefringence and retardation of the polarized light and polarized scattering of the tissue is the signal rather than the noise when using polarized light for tissue characterization. For example, a scanning laser polarimeter has been used to measure changes in retardation of the polarized light impinging on the retinal nerve fiber layer. It has been shown that scanning laser polarimetry provides statistically significant higher retardation for normal eyes in certain regions over glaucoma eyes. Images generated from the scattering of various forms of polarized light have also been shown to be able to differentiate between cancer versus normal fibroblast cells.

17.4.5   Micrometer and Nanometer Biosensing Applications

Developments in microtechnology and, in particular, nanotechnology are transforming the fields of biosensors, prosthesis and implants, and medical diagnostics. In terms of medical diagnostics, these devices are being used for external, lab-on-a-chip, high-throughput screening for analyzing blood and other samples. Many researchers and companies are developing nanotechnology applications for use inside the body for anticancer drugs, insulin pumps, and gene therapy. Others are working on prosthetic devices that include nanostructured materials.

One nanotechnology that has come to the forefront is that of quantum dots. Quantum dots are devices capable of confining electrons in three dimensions, in a space small enough that their quantum (wavelike) behavior dominates over their classical (particle-like) behavior. At room temperature, confinement spaces of 20–30 nm or smaller are typically required. Once the electrons are confined, they repel one another and no two electrons can have the same quantum state. Thus, the electrons in a quantum dot will form shells and orbitals highly reminiscent of the ones in an atom, and will exhibit many of the optical, electrical, thermal, and chemical properties of an atom. Quantum dots can be grown chemically as nanoparticles of a semiconductor surrounded by an insulating layer nearly colloidal in nature. These particles can also be deposited onto a substrate, such as a semiconductor wafer patterned with metal electrodes, or they can be crystalized into bulk solids by a variety of methods. Either substance can be stimulated with electricity or light (e.g., lasers) in order to change its properties.

Although quantum dots are nanometer-scale crystals that were developed in the mid-1980s for optoelectronic applications, one application in the biomedical area has been to probe living cells in full color over extended periods of time. Such a technique could reveal the complex processes that take place in all living organisms in unprecedented detail, such as the development of embryos. Existing imaging techniques use natural molecules that fluoresce, as discussed previously, such as organic dyes and proteins that are found in jellyfish and fireflies. However, each dye emits light over a wide range of wavelengths, which means that their spectra overlap and this makes it difficult to use more than three dyes at a time to tag and image different biological molecules simultaneously. The fluorescence of dyes also tends to fade away quickly over time, whereas semiconductor nanocrystals—quantum dots—can get around these problems. In addition to being brighter and living longer than organic fluorophores, quantum dots have a broader excitation spectrum. This means that a mixture of quantum dots of different sizes can be excited by a light source with a single wavelength, allowing simultaneous detection and imaging in color.

Although the preceding example has focused on biomedical sensing using fluorescent quantum dots, these micro- and nanoparticles can also be made of various materials and used with all of the light propagation methods. For instance, metal nanoparticles such as gold or silver can be used with Raman spectroscopy to produce an effect known as surface-enhanced Raman spectroscopy (SERS) which gives rise to signals a million times more sensitive than regular Raman signals. These same types of metal nano- or microparticles can be injected into cancerous tissue and used as absorbers that when hit with infrared light will absorb the energy, cook the cancerous tumor, and kill it. Since nanoparticle development is still in its infancy, it remains to be seen what other biomedical applications are to come from the combination of these particles with light.

17.5 FUNDAMENTALS OF PHOTOTHERMAL THERAPEUTIC EFFECTS OF LASERS

Therapeutic application of a laser is mediated by conversion of photonic energy to absorbed energy within the material phase of the tissue. The primary mode of this energy conversion manifests itself as a nonuniform temperature rise which leads to a series of thermodynamic processes. These thermodynamic processes can then be exploited as a means to affect therapeutic actions such as photothermal coagulation and ablation of tissue. Another mode of interaction is the utilization of the absorbed energy in activation of endogeneous or exogenous photosensitizing agents in a photochemical process known as photodynamic therapy.

Laser interaction with biological tissue is composed of a combination of optical and thermodynamic processes. An overview of these processes is shown in Figure 17.13. Once laser light is irradiated on tissue, the photons penetrate into the tissue and—depending on the tissue optical properties such as scattering coefficient, absorption coefficient, and refractive index—the energy is distributed within the tissue. A portion of this energy is absorbed by the tissue and is converted into thermal energy, making the laser act as a distributed heat source. This laser-induced heat source in turn initiates a nonequilibrium process of heat transfer manifesting itself by a temperature rise in tissue. The combined mechanisms of conduction, convection, and emissive radiation distribute the thermal energy in the tissue, resulting in a time- and space-dependent temperature distribution in the tissue. The temperature distribution

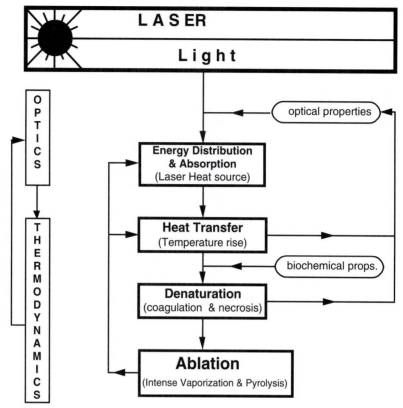

Figure 17.13   Flowchart of photothermochemical processes in thermal interaction of laser light with biological tissues.

depends on the thermal properties, conductivity, heat capacity, convective coefficients, and emissivity of the tissue.

As heat deposition and transfer continues, a certain threshold can be reached above which a process of irreversible thermal injury initiates. This process leads to coagulation of tissue caused by denaturation of enzymes and proteins and finally leads to necrosis of constituent cells. As a result of this thermchemical process of injury, properties of the tissue, especially the optical properties, start to change. The change in properties in turn influences the process of energy absorption and distribution in the tissue.

The next stage in these processes is the onset of ablation. As the temperature continues to rise, a threshold temperature is reached at which point, if the rate of heat deposition continues to exceed the rate at which the tissue can transport the energy, an intensive process of vaporization of the water content of the tissue combined with pyrolysis of tissue macromolecules initiates, which results in ablation or removal of tissue.

17.5.1 Temperature Field During Laser Coagulation

In this section the governing equations will be described for a thermodynamic analysis of laser heating of biological tissues up to the onset of ablation. First the heat conduction equation and typical boundary and initial conditions are described. Next the Arrhenius–Henriques model for quantitative analysis of irreversible thermal injury to biological tissue will be introduced.

In Section 17.2, how the laser energy is absorbed by a participating medium such as biological tissue was described. As laser energy with a rate Q_L is absorbed by a material under irradiation, there is an immediate thermal energy flux traveling in different directions. This is caused by nonuniformity of Q_L in both the radial direction, due to beam profile, and the axial direction, due to absorption. The energy rate balance equation in the material can be found as follows.

Consider an infinitesimal element of the material under laser irradiation (Figure 17.14). The rate of thermal energy storage in the element, U, depends on the difference between incoming and outgoing thermal fluxes, the rate of laser energy absorption, Q_L, and other energy rate interactions which are lumped together and labeled Q_o. The difference between influx and efflux is, in the limit, equal to the negative of the divergence of the thermal flux vector. So the energy rate balance can be written as

$$\partial U/\partial t = -\,\vec{\nabla}\cdot\vec{q}+Q_L + Q_o \tag{17.69}$$

where $\vec{\nabla}\cdot\vec{q}$ is the change in thermal flux. The equation of state for thermal energy is

$$U = \rho^{CT} \tag{17.70}$$

where T is an absolute temperature (K), ρ (kg/m^3) is the density, and C (joules/kg K) is the specific heat of the material.

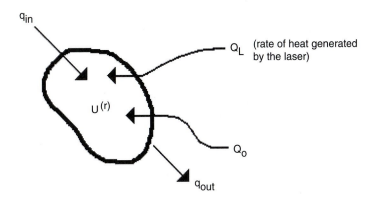

Figure 17.14 Thermal fluxes and energy in a control volume in the material phase.

At this point, as in the derivation of the photon phase, a phenomenological relation for the thermal flux is needed. The commonly used relation is known as "Fourier's law" and is written as

$$\vec{q} = -\kappa \, \vec{\nabla} T \qquad (17.71)$$

where κ (W/m K) is the thermal conductivity of the material. Proceeding now by substituting Eqs. 17.71 and 17.70 into 17.69, the heat conduction equation is obtained,

$$\partial(\rho C T)/\partial t = \vec{\nabla} \cdot \kappa \, \vec{\nabla} T + Q_L + Q_o \qquad (17.72)$$

This equation, solved with application of proper initial and boundary conditions can, in theory, predict the temperature distribution in a material under laser irradiation.

In the case of biological tissue with other energy rates involved, Q_o can be written as

$$Q_o = Q_b + Q_m \qquad (17.73)$$

where Q_b is the heat rate removed by blood perfusion, and Q_m is the rate of metabolic heat generation. These terms are not considered here. They are in general much smaller than the laser heat source term Q_L except when lower laser powers are used for longer time as in hyperthermia treatments.

The material properties κ, and ρC have been measured for many biological tissues, including human tissues. In general these properties may depend on temperature and water content, which could cause a nonlinear effect in Eq. 17.72. Another source of nonlinearity may be due to temperature dependence of the optical properties which, when present, has a more dominant role.

17.5.2   Thermal Coagulation and Necrosis: The Damage Model

The idea of quantifying the thermal denaturation process was first proposed in the 1940s. Using a single-constituent kinetic rate model this early work grossly incorporates the irreversible biochemical processes of coagulation, denaturation, and necrosis associated with thermal injury to biological tissue in terms of a single function. Defining an arbitrary nondimensional function, Ω, as an index for the severity of damage, the model assumes that the rate of change of this function follows an Arrhenius relation:

$$\frac{d\Omega}{dt} = A \exp\left(-\frac{E}{RT}\right) \qquad (17.74)$$

where R is the universal gas constant (cal/mole), T is the absolute temperature (K), and A (1/sec) and E (cal/mole) are constants to be determined experimentally. Total damage accumulated over a period t can be found by rearranging and integrating Eq. 17.74:

$$\Omega = A \int_0^t \exp\left(-\frac{E}{RT}\right) dt \qquad (17.75)$$

The experimental constants A and E for pig skin were determined such that $\Omega = 1.0$ corresponded to complete transepidermal necrosis and $\Omega = 0.53$ indicated the minimum condition to obtain irreversible epidermal injury. The reported values are $A = 3.1 \times 10^{98}$ (1/sec) and $E = 150{,}000$ (cal/mole). These values result in a threshold temperature of about $45°C$. Different values for other tissues have been reported. For instance, for human arterial vessel walls, the values are $A = 4.1 \times 10^{44}$ (1/sec) and $E = 74{,}000$ (cal/mole). For these values $\Omega = 1.0$ was defined as the threshold for histologically observed coagulation damage. It should be noted that the coagulation process (i.e., collagen denaturation) is a different damage process than that seen in skin burns.

Once the heat conduction equation (17.72) is solved for the temperature field, the temperature history at every point in the tissue can be used in Eq. 17.75 to predict the accumulated damage at that point.

17.5.3 Photothermal Ablation

The temperature rise in a biological tissue under laser irradiation cannot continue indefinitely. A threshold temperature can be reached beyond which, if the rate of heat deposition by the laser continues to be higher than the rate at which heat can be transported by heat transfer mechanisms, intense vaporization of the water content of the tissue occurs. This results in creation of vacuoles, or vacuolization, followed by pyrolysis at higher temperatures. The combined processes of intense vaporization, vacuolization, and pyrolysis of tissue macromolecule constituents is the laser ablation process. This is the primary underlying mechanism for removal of tissue by laser surgery.

The large water content of most biological tissues suggests that the dominant part of the ablation process is the intense vaporization process. Apart from a short discussion of pyrolysis, intense vaporization is the main process of concern.

In the following section, a derivation is given for the ablation interface energy balance equation which introduces the mathematical nonlinearity of the problem of ablation as a moving boundary problem. The section is concluded by a note on pyrolysis.

Ablation Interface Energy Balance Equation

The heat conduction equation alone cannot provide a mathematical model for determination of the dynamic behavior of the ablation front as a function of time. In fact, the motion of the ablation front influences the heat transfer process. An energy balance at the ablation front is needed to determine the motion of the front of ablation and its influence on temperature field.

Consider a portion of a material under laser irradiation with unit cross-sectional area and of thickness Δs as shown in Figure 17.15 which is at the ablation threshold temperature and is to be vaporized in the next interval of time Δt. Using Fourier's law for heat flux, Equation 17.71, and after mathematical manipulations, the following equation for energy balance at the ablation interface can be derived:

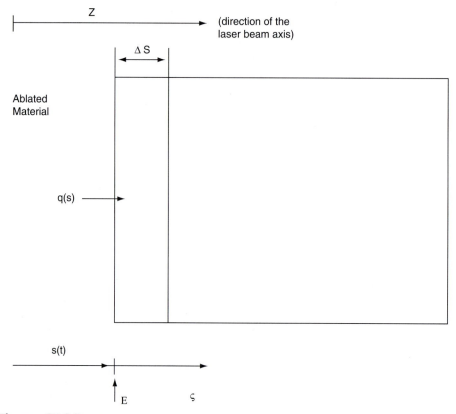

Figure 17.15   Ablation interface energy balance which is depicted as the amount of energy required to ablate a portion Δs of the tissue.

$$\kappa \frac{\partial T(s, t)}{\partial z} = \rho f_L L \frac{ds}{dt} \qquad (17.76)$$

where f_L is the water fraction parameter. Nonlinearity inherent in the problem can be realized from this equation in that it couples the temperature gradient at the ablation interface to the rate of change of the front position which is not known a priori. The solution requires an iteration procedure which requires the solution of the heat conduction equation simultaneously with the ablation front equation, with application of proper boundary and initial conditions.

Equation 17.76 is essential in providing information on the dynamics and thermodynamics of ablation. When solved simultaneously with the heat conduction equation and with the application of proper boundary and initial conditions, this equation provides the information on the position and velocity of the front of ablation.

Furthermore, an important observation can be readily made about the temperature profile within the tissue as discussed in the following paragraph.

Note that on the right-hand side of Equation 17.76 ds/dt must be positive for the ablation process to proceed. This means that on the left-hand side the temperature, $\partial T/\partial z$ must be positive. Consequently, the subsurface temperature must be higher than the surface temperature. That is, subsurface tissue must be superheated to temperatures higher than the surface ablation temperature so that thermal energy can be provided to the surface for the ablation process to proceed. This is in accordance with one of the corollaries of the second law of thermodynamics stating that heat can be transported only from hotter points to colder points. In fact, the higher than ablation threshold temperature puts the subsurface tissue in a metastable equilibrium condition which may be perturbed by internal tissue conditions and result in nucleation and vaporization initiating subsurface and manifesting itself by an explosion and mechanical tearing of the tissue surface. It should be remarked at this point that initiation of the ablation process involves nonequilibrium nucleation processes which are not considered by the present approach.

17.5.4 Analytical Solution of Laser Heating in a Purely Absorbing Medium

The governing equations for laser irradiation with ablation were described in Section 17.5.3. Here the governing equations will be repeated for a one-dimensional semi-infinite and purely absorbing medium. The governing equation for the preablation heating stage can be written as

$$\frac{\partial \rho c T}{\partial t} = \frac{\partial^2 T}{\partial z^2} + \mu_a I e^{-\mu_a z} \tag{17.77}$$

where μ_a is the absorption coefficient. This equation is valid up to the onset of ablation which is assumed to occur when the surface temperature reaches the ablation threshold temperature, T_{ab}. The details of analysis of ablation is beyond the scope of this chapter and will not be considered.

In the late 1970s a nondimensionalization of the heat conduction equation for an axisymmetric three-dimensional case of preablation laser heating of tissue by a Gaussian beam in an absorbing medium was solved. The following relations will transform the governing equations into a dimensionless moving frame:

$$\theta = \frac{T - T_0}{T_{ab} - T_0} \quad \xi = (I_0 c/kL)z \quad \tau = (I_0 c/kL)(I_0/\rho L)t \tag{17.78}$$

The variables introduced are, respectively, dimensionless temperature, θ, dimensionless coordinate, ξ, in the moving frame with origin at the ablation front, and dimensionless time, τ. A dimensionless absorption parameter, B, and a dimensionless heating parameter, λ, are also defined as

$$B = (kL/I_0 c)\alpha$$
$$\lambda = c(T_{ab} - T_0)/L \tag{17.79}$$

Analytical solutions of the nondimensional form of the governing equations can be found by Laplace transformation of the space variable ξ. The solution is as follows:

$$\theta(\xi, \tau) = \frac{1}{B\lambda} \left\{ 2B\sqrt{\tau}\, ierfc\left[\frac{\xi}{2\sqrt{\tau}}\right] - e^{-B\xi} \right.$$
$$\left. + \frac{1}{2} e^{B^2\tau} \left(e^{-B\xi} erfc\left[B\sqrt{\tau} - \frac{\xi}{2\sqrt{\tau}}\right] + e^{B\xi} erfc\left[B\sqrt{\tau} + \frac{\xi}{2\sqrt{\tau}}\right] \right) \right\}$$

(17.80)

This equation describes the temperature field as a function of the dimensionless space variable ξ and dimensionless time τ and is valid up to the onset of ablation when $\theta = 1$. The symbol *ierfc* indicates the integral of the function *erfc*, $ierfc(z) = \int_z^\infty erfc(t)dt$. This equation can be implemented with relative ease using modern user-friendly computer tools such as MATLAB™ or Mathematica™.

An example of the graph of this solution, which is the progression in time of the temperature profile as function of depth, is shown in Figure 17.16.

By letting $\theta = 1$ (i.e., $T = T_{ab}$) at $\xi = 0$ in Equation 17.78, the following transcendental algebraic equation is obtained, which can be solved numerically for the time for the onset of ablation τ_{ab}:

$$B\lambda = \frac{2}{\sqrt{\pi}} B\sqrt{\tau_{ab}} + e^{B^2 \tau_{ab}} erfc[B\sqrt{\tau_{ab}}] - 1$$

(17.81)

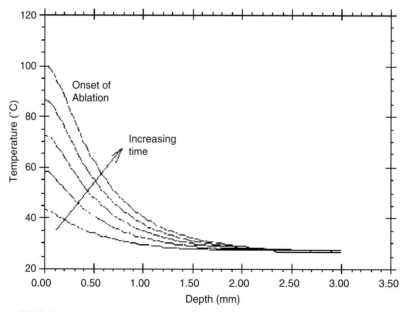

Figure 17.16 Nondimensional temperature as a function of nondimensional depth at various times prior to the onset of ablation.

Note that for large values of $B\lambda$ the behavior is almost linear with a slope of $\frac{2}{\sqrt{\pi}}$.

The present analysis did not consider the case of scattering tissue. An analytical solution for Q_L could be obtained using the diffusion approximation approach described in Section 17.2. However, implementation of that solution into Equation 17.72 to solve analytically for temperature would be highly cumbersome, at best. Therefore, a solution of temperature field would normally require a numerical approach, such as the finite difference or the finite element method.

17.5.5 Biochemical Damage Analysis

The Arrhenius model for prediction of damage was introduced in the preceding section. The damage function was defined in Section 17.5.2. For numerical computation it is more convenient (due to large values of A and E) to rewrite the integrand of this function in an equivalent form so that Equation 17.83 becomes

$$\Omega(z,\ t) = \int_0^t \exp\ (\ln A - \frac{E}{RT(z,\ t)})dt \qquad (17.82)$$

Two sets of values were used for A and E, one for pig skin and another for human vessel wall. The value of $\Omega(z,t)$ can be determined by replacing for $T(z,t)$ from the analytical solution of the last section. The preablation solution should be used up to the onset of ablation and after that the ablation stage solution should be used.

The extent of damage or the position of the damage front at a given time t can be found by setting $\Omega = 1$ in Equation 17.82 and solving for z. Alternatively, for every point z Equation 17.82 can be integrated by a controlled variable time-step procedure up to the time when $\Omega = 1$. The latter method was used to determine the position of the damage front as a function of time and the difference in the value integrand at the upper and lower bounds of integration was monitored and used to change the time step. An example of such a calculation was done for a 5-watt laser pulse of 500 ms duration. The calculation was done using both values for A and E from both the pig skin and another for human vessel wall. Results together with the position of ablation front as a function of time (details of that beyond the scope of this chapter) are shown in Figure 17.17.

The vertical line at about 0.0035 seconds indicates the onset of ablation, which helps to show that there had been biochemical damage in the tissue before the onset of ablation. After the onset of ablation, the difference between the damage front and ablation front is the actual extent of damage in the tissue intact. The important observation from this graph is the difference in prediction of the extent of damage using the two different sets of values for A and E. For the same irradiation conditions, thermal damage would propagate faster in skin than it would in the vessel wall. This is expected since the damage mechanisms for the vessel wall and skin are different.

Example Problem 17.9

Consider the skin subjected to a constant temperature heating T. For exposure durations $t = 1\,\mu s$, $t = 1\,ms$, $t = 1\,s$, $t = 60\,s$, and $t = 30$ minutes, find the critical

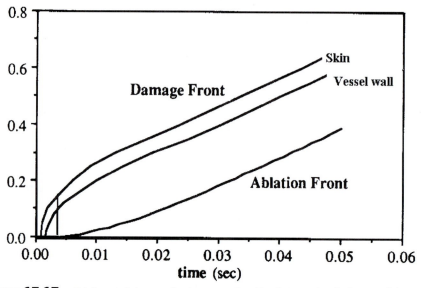

Figure 17.17   Ablation and damage front propagation. For the same irradiation conditions, there would be more thermal damage beyond the ablation front in skin than in the vessel wall.

temperature (in °C) T_c required to achieve thermal damage, $\Omega = 1$, during the exposure. Plot T_c vs exposure duration.

Solution

Given that the exposure durations are $1\,\mu s$, $1\,ms$, $1\,s$, $60\,s$, and 30 minutes and R is the universal gas constant $= 2\,cal/gm\text{-}°C$, then what is required for each case is for T_c (critical temperature at which point $\Omega = 1$) to be calculated.

Assumptions: Values of $\ln A$ and E are approximated to pig tissue values of 102.72 and 74,000, respectively.

Equation:

$$\Omega(z,\ t) = \int_0^t \exp\left(\ln A - E/(RT(z,t))\right)dt$$

$$\therefore 1 = \int_0^t \exp\left(\ln A - E/(RT_c)\right)dt$$

$$\therefore 1 = \int_0^t \exp\left(102.72 - 74{,}000/(2 \times T_c)\right)dt \quad (\text{ex.1})$$

$$\therefore 1 = \exp\left(102.72 - 74{,}000/(2 \times T\_c)\right) \int_0^t dt \quad \text{(ex.2)}$$

$$\therefore 1 = \exp\left(102.72 - 74{,}000/(2 \times T\_c)\right)t \quad \text{(ex.3)}$$

Solving Equation ex. 3 for all values of t, we can get T_c for those values. Tabulated values of t and T_c are as follows:

Exposure Time t (in s)	Critical Temperature T_c in °C
1e^{-6}	143.1548
1e^{-3}	113.153
1	87.2
60	74.56
1800	72

∎

17.5.6   Effect of Vaporization and Ablation Temperature

The coefficient f_L was introduced previously in relation to the heat of ablation as the water fraction parameter. Another nonvaporization phenomenon that can affect this coefficient is ejection of material due to subsurface nucleation. As was discussed, chunks of the material may be ejected without vaporization. Another factor that can affect the ablation rate is the excess energy required for pyrolysis which, as discussed before, manifests itself as a higher ablation threshold temperature.

Although the full solution of the ablation problem in this chapter was not considered, to understand how these phenomena affect the ablation rate the steady-state ablation velocity may be used. This equation can be directly derived to be

$$v_{ss} = \frac{I}{\rho c \Delta T_{ab} + \rho f_L L} \tag{17.83}$$

Fixing the values of I, ρ, c, and L, the effect of the parameters f_L and ΔT_{ab} on v_{ss} can be determined. In the following example the physical properties of water are used and I is chosen to be $267.8\ \text{W/cm}^2$ so that $v_{ss} = 1\ \text{mm/sec}$ for $\Delta T_{ab} = 100°\text{C}$ and $f_L = 1$. Figure 17.18 shows the effect of ablation temperature on the ablation velocity for various values of f_L from 0.01 to 1.0. The general effect of an increase either in ablation temperature or in f_L is a decrease in ablation velocity, as is also obvious from Equation 17.83. Observe, however, that the larger the value of f_L the less effect the ablation temperature has on the ablation velocity. For instance, whereas for $f_L = 0.1$ ablation velocity drops from 4 mm/sec to 1.15 mm/sec over a temperature change from 100°C to 500°C, for the same range and $f_L = 0.8$ the change is only from 1.20 mm/sec to 0.69 mm/sec. Therefore, if f_L should indicate the fraction of water in tissue, for example, a decrease in ablation rate should be of concern as a result of higher pyrolysis/ablation temperatures only for tissue with low water content (e.g.,

30% or lower). A cross section of families of curves in Figure 17.18 at $\Delta T_{ab} = 200°C$ for more values of f_L is shown in Figure 17.19.

This figure shows that a change of f_L from 0 to 1 for $\Delta T_{ab} = 200°C$ results in a change in the ablation velocity from 3.2 mm/sec to 0.86 mm/sec; that is, the ablation velocity can be over 3 times as large as its value for $f_L = 1.0$. The families of curves of Figure 17.18 are more closely packed at higher values of ablation temperature. This means that the effect of a change in fL on the ablation velocity is less significant at higher ablation temperatures.

17.6   FIBER OPTICS AND WAVEGUIDES IN MEDICINE

Rigid tubes for the examination of body cavities were used many centuries before the birth of Christ; however, it was not until the 1800s that illumination was used in the tube by means of a candle and 45° mirror. The introduction in the early 1900s of multiple lenses to transmit images led to semiflexible tubes for insertion into the body. The use of fiber optic probes based on thin and transparent threads of glass dates back to the late 1920s but lay dormant for two decades until the idea was revived in the 1950s. The first medical instrument based on the technology, a flexible fiber optic gastroscope, was developed and first used on patients in 1959. In the 1960s the first lasers were developed and in the early 1970s there was a rapid development in the

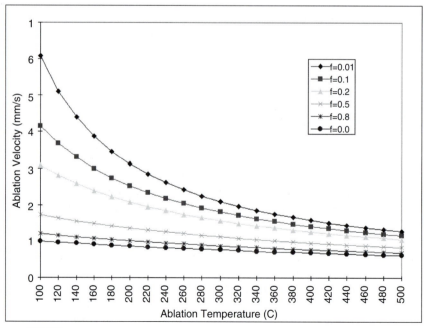

Figure 17.18   The effect of ablation temperature on ablation velocity for various water contents.

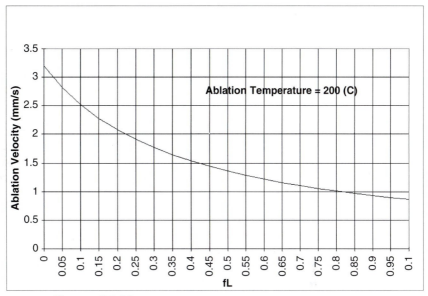

Figure 17.19   The effect of water content, f_L, on ablation velocity.

field of fiber optics for communications. All of these events have contributed to the modern fiber-optic probes and endoscopes used today.

17.6.1   Principles of Fiber Optics and Waveguides

In Section 17.2.1, the interaction of light with a nonparticipating medium was described. In that section, light was described as rays and the Fresnel equations for the interaction of the light rays at the boundaries of two media were derived. It was found that, depending on the index of refraction of two slab types of materials and the angle of incident of the light, the amount of reflection and refraction could change. Although transmission of light through an optical fiber is a complex problem, the phenomenon can be understood with a simple geometric model. As depicted in Figure 17.20, the fiber can be thought of as a long rod of transparent material in which the rod or core of the fiber has a higher index of refraction (n_1) than the surrounding cylindrical shell material or cladding (n_2). The propagation of the light occurs down the core because of the total internal reflection of the light from the core–cladding interface which, from Snell's laws, occurs when the angle of reflection at this interface is greater than the critical angle. In order for the light to be internally reflected at the core–cladding interface, the light injected into the end of the fiber must be smaller than a cone with some limiting angle θ_o defined by

$$n_0 \sin (\theta_o) = (n_1{}^2 - n_2{}^2)^{1/2} = \text{NA} \qquad (17.84)$$

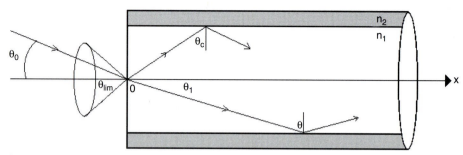

Figure 17.20   Basic model of an optical fiber with a cylindrical core with index of refraction (n_1) and cladding index (n_2) where ($n_2 < n_1$).

where NA, which stands for numerical aperture, is defined by this limiting value. In silica fibers the NA is generally between 0.2 and 0.4. As can be depicted in the previous equation, increasing the difference in the index of refraction of the core and the cladding will increase the NA and also the acceptance angle. A large NA and acceptance angle gives rise to a large number of rays with different reflection angles or different transmission modes. A different zigzag of the beam path is thus made for each mode. For a small core diameter, a single mode propagation can be generated in a fiber.

Many kinds of optical fibers have been created with different structure, geometry, and materials, depending on the ultimate application. For instance, the tips of the fibers can be changed to produce side-firing beams for therapeutic applications such as coagulation of prostate tissue. The fibers' tips can also be tapered for pinpoint application of the light beam or made as a diffuse tip for broad uniform application of the light. In general, optical fibers have been classified in terms of the refractive index profile of the core and whether there are single modes or multimodes propagating in the fiber. For instance, as depicted in Figure 17.21, if the fiber core has a uniform or constant refractive index it is called a step-index fiber; if it has a nonuniform, typically parabolic, refractive index that decreases from the center to the cladding interface it is known as a graded-index fiber; and if the fiber core is small with a low NA only a single mode will propagate. The graded-index fibers have been shown to reduce the modal dispersion by a factor of 100 times and increase the bandwidth over a comparably sized step-index fiber. These general classifications apply to most fibers, but special fiber geometries are often required, as discussed next, for endoscopic coherent fiber bundle fiber imaging, single fibers or noncoherent bundles for sensing and diagnostics, and optical fibers made for high-powered therapeutic applications.

In terms of the fiber material, three areas need to considered: the wavelength required, the power, and the biocompatibility. For therapeutic applications the amount of power, both for continuous lasers and pulsed lasers, needs to be considered. In particular, any impurities in the fiber may absorb the light and either

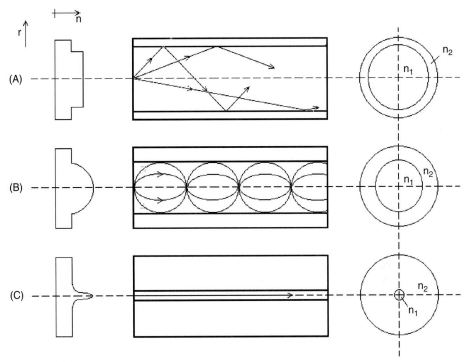

Figure 17.21 Typical geometry of the main types of optical fibers, including a (a) step-index, (b) graded-index, and (c) single-mode fiber.

decrease the amount of light reaching the probe for sensing or heat the fiber, which can then cause fiber damage. In addition to impurities, the fiber material itself can work well for one wavelength and not work at all for others. For instance, the standard silica or glass fibers transmit well in the visible wavelength range but beyond 2.0 micrometers these fibers absorb the light and thus can not be used for infrared light transmission. Other fibers such as germanium, sapphire, barium fluoride, calcogenide, or hollow waveguide (i.e., gold coated) can be used for these wavelengths but these fibers, in addition to impurities, may have problems for use in the body such as being too stiff, being hydroscopic (dissolves in water), or being toxic. Thus, the biocompatibility is a function of the material in that it needs to not be toxic to the patient and must be able to function in the body without dissolving or breaking off.

17.6.2 Coherent Bundles for Imaging

Optical fibers can be bundled together in an aligned fashion such that the orders of the fibers at both ends are identical. Such a fiber bundle is called a coherent bundle or an ordered bundle. Of course, the coherence here means the correlation between the spatial positions of the fibers at both ends and has nothing to do with the light coherence. The important property of coherent bundles is their capability of transmitting images through a flexible channel. If an image is projected onto one end of a bundle, a replicate of the image is produced at the other end. Coherent bundles of optical fibers are the key components in endoscopes.

In endoscopic applications, an internal organ is imaged and viewed outside the body in a minimally invasive fashion. An incoherent (nonordered) bundle of optical fibers is used to illuminate the portion to be imaged inside the human body. A coherent (ordered) bundle of optical fibers is used to transmit an image of the target portion. A white light source is usually used for the illumination such that a real-color image of the tissue can be obtained.

The quality of image transmission is mainly determined by two factors: light collection and image resolution. The collection power of each individual optical fiber is limited by the diameter of its core and the numerical aperture (NA). A large NA and core diameter allows good transmission of light from the illuminated object to the eye of the physician. The image resolution indicates how fine details can be seen and is limited by the core diameter, d, of the cores of the individual fibers. The resolution, in number of discernible lines per millimeter, is approximately $d/2$. The smaller the core diameter is, the better the image resolution. Due to the limited resolution, a straight line in an object may appear zigzagged in the image.

Cladding of each individual optical fiber is required to minimize or avoid crosstalk among the fibers. Unclad fibers were used in early endoscopes and had poor imaging quality. Light in unclad fibers may leak from the core and cross into other fibers. The crosstalk causes an overlay of various portions of an image during transmission and hence produces a blurred image.

The requirements of high light collection and good image resolution have conflicts. A large core diameter allows good light transmission but gives poor resolution. A thick cladding layer avoids crosstalk but limits light collection and image resolution. A trade-off has to be made. In practice, the core diameter is usually $10-20$ μm and the cladding thickness is of the order of $1.5-2.5$ μm.

Several other factors may deteriorate the image obtained with a coherent bundle. Some stray light may transmit through the cladding layers into the cores. The stray light would add an undesirable background that reduces the image contrast. Defective fibers in a coherent bundle would cause a serious problem. If a defective fiber does not transmit any light, a static dark spot appears in the image. If the ordering of the fibers is not identical at both ends, image distortion will degrade the images.

Lenses may be added at both ends of an imaging coherent bundle to adjust the magnification. Although the distance between the objective lens and the fiber bundle can be adjusted in principle to adjust the focusing of the coherent bundle, a fixed focus with a large depth of focus is often used for simplicity. A VCR may be used to record the images in real time while the images are displayed on a monitor.

Endoscopes may be specially built for imaging of various organs. The following is a list of commonly used endoscopes: angioscopes for veins and arteries, arthroscopes (or orthoscopes) for the joints, bronchoscopes for the bronchial tubes, choledoscho-scopes for the bile duct, colonoscopes for the colon, colposcopes for the vagina, cystoscopes for the bladder, esophagoscopes for the esophagus, gastroscopes for the stomach and intestines, laparoscopes for the peritoneum, laryngoscopes for the larynx, and ventriculoscopes for the ventricles in the brain.

17.6.3 Diagnostic and Sensing Fiber Probes

In terms of sensors for minimally invasive measurements into the body, fiber optic probes are relatively new to the field (1980s). To understand the potential capability of these probes, a sensor is first defined along with some of the requirements for a good sensor. A sensor is a device that transforms an input parameter or measurand into another parameter known as the signal. For instance, a displacement membrane on the tip of a fiber probe could transform a pressure signal into a light intensity change, which is then depicted as a change in voltage from a light detector. The requirements for any good sensor are specificity or the ability to pick out one parameter without interference of the other parameters, sensitivity or the capability to measure small changes in a given measurand, accuracy or closeness to the true measurement, and low cost. As with most sensors, fiber optics have to trade off these parameters within some limit; for example, you may be able to get 95% accuracy at a reasonable cost but to obtain 99% accuracy might require a huge increase in cost (i.e., military specifications of components). Fiber-optic probes also offer the potential for miniaturization, good biocompatability for the visible and near-infrared wavelengths, speed because light is used, and safety because no electrical connections to the body are required.

All fiber optic probes transmit light into the body and the light, directly or indirectly, interacts with the biological parameter of interest, be it a biological fluid, tissue, pressure, or body temperature. The interaction causes a change in the light beam or beams that travel back to the detection system. The returning signal can be physically separated from the input signal in a separate fiber or fibers, or it can be separated if the output beam is at a separate wavelength from the input beam, as is the case for fluorescence probes. Fiber optic sensors can be used for each of the chemically and physically based measurements described in Sections 17.3 and 17.4. However, rather than try to discuss each fiber optic probe for each application, the rest of this section will be focused on separating the fiber probes into two classes: indirect and direct fiber optic sensors. Direct fiber optic sensors are defined as those probes in which the light interacts directly with the sample. For instance, absorption measurements can theoretically be made to determine such things as blood oxygenation or to quantify blood analytes with either two side-firing fiber optic probes separated at a fixed distance as depicted in Figure 17.22a or an evanescent wave fiber optic that has the cladding stripped off so that some of the light traveling down the core travels out of the fiber, interacts with the sample, and then returns to the fiber core as depicted in Figure 17.22b. In Figure 17.22c a fiber optic bundle is depicted that could be used to transmit and receive light to and from tissue to distinguish between normal and cancerous or precancerous lesions using direct measurement of tissue autofluorescence or Raman spectrum. Interferometers such as the Fabry–Perot type described in Section 17.3.3 have been designed, using partially reflecting mirrors built into a fiber optic, and any physical change imparted on the fiber, such as that due to fluctuations in body

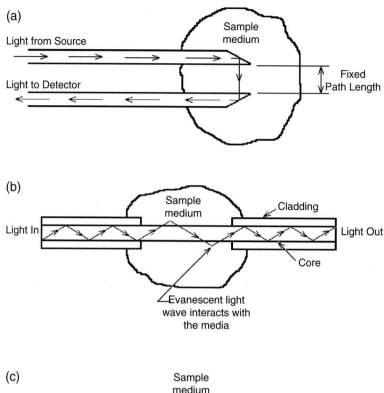

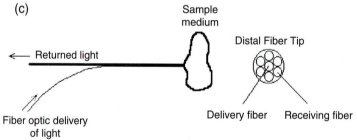

Figure 17.22   Direct fiber optic probe designs including (a) absorption probe using two side-firing fibers, (b) an evanescent wave probe in which the light transmits from the core to the sample and back into the core, and (c) a multifiber design for use in distinguishing normal from cancerous tissue by directly measuring autofluorescence or Raman spectra.

temperature or pressure, can be directly measured by the fiber in the form of a change in the interference pattern of the light. Lastly, the fiber optic probe for measurement of tissue optical properties described in Section 17.2.4 is another sample of a direct probe that measures reflections from the sample. It should be noted that each of these descriptions are somewhat oversimplified in that a great deal of design needs to be done to gain the specificity, accuracy, and sensitivity required to measure these various low level parameters in the noisy environment of the body. For indirect fiber optic measurements a miniaturized transducer (sometimes referred to as on optode) is attached to the distal end of the fiber so that the light interacts with this transducer and it is the transducer that interacts with the sample. A displacement optode such as that depicted in Figure 17.23a can be used to monitor pressure or temperature changes by simply causing less light to be specularly reflected into the fiber as the tip is moved. In addition, fluorescent chemistry can be placed within a membrane on the distal tip of the fiber, as depicted in Figure 17.23b, in which the amount of fluorescent light produced is a measure of the concentration of a particular blood analyte. These indirect methods of measurement are also simplified and the actual final design to obtain an accurate and sensitive signal can be quite complicated. Overall, the indirect fiber optic measurements typically have higher specificity over direct measurement approaches but at the expense of requiring a more complicated probe.

17.7 BIOMEDICAL OPTICAL IMAGING

Medical imaging has revolutionized the practice of medicine in the past century. Physicians are empowered to "see" through the human body for abnormalities non-invasively and to make diagnostic decisions rapidly, which has impacted the therapeutic outcomes of the detected diseases.

Medical imaging dates back to 1895 when x-rays were discovered serendipitously by German physicist W. C. Roentgen, who received the first Nobel prize in physics in 1901 (http://www.almaz.com/nobel/) for this important work. Human anatomy can be imaged easily by simple x-ray projections, such as chest x-rays, which are still being used. Contemporary medical imaging began with the invention of computerized tomography (CT) in the 1970s. G. N. Hounsfield in England produced the first computer reconstructed images experimentally, based on the theoretical foundation laid by A. M. Cormack in the United States. Both of them were awarded the Nobel prize in medicine in 1979. The essence of CT is that if an object is viewed from a number of different angles, then a cross-sectional image of it can be computed (i.e., "reconstructed").

The advent of CT has inspired other new tomographic and even 3-D imaging techniques. The application of reconstruction to conventional nuclear medicine imaging led to positron emission tomography (PET) and single photon emission computed tomography (SPECT). A similar application to the technique of nuclear magnetic resonance led to magnetic resonance imaging (MRI).

Different imaging modalities are used to detect different aspects of biological tissues through a variety of contrast mechanisms. X-ray imaging primarily detects electron density. In MRI, proton density and its associated relaxation properties are detected. Ultrasonography detects acoustic impedance, and nuclear imaging detects nuclear radiation emitted from the body after introducing a radiopharmaceutical inside the body to tag a specific biochemical function. The advantages and

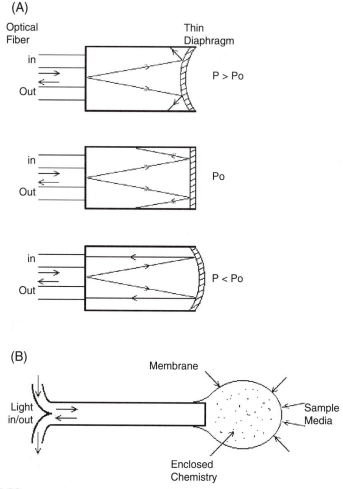

Figure 17.23 Indirect fiber optic probe designs including (a) a thin reflectance diaphragm for pressure or temperature measurements, (b) an optode with some chemistry enclosed within a membrane at the distal tip of the fiber, which can be used for instance to determine a particular analyte concentration using fluorescent chemistry specific to the analyte of interest.

disadvantages of different imaging modalities may be illustrated using breast cancer detection. Breast cancer is the most common malignant neoplasm and the leading cause of cancer deaths in women in the United States. A means for prevention of breast cancer has not been found, and early detection and treatment is the best solution to improve cure rate. At present, x-ray mammography and ultrasonography are clinically used for breast cancer detection. Mammography is currently the only reliable means of detecting nonpalpable breast cancers. As a supplementary tool, ultrasound is used to evaluate the internal matrix of circumscribed masses found using mammography or of palpable masses that are obscured by radiographically dense parenchyma using mammography. However, x-ray mammography is ionizing radiation, and imaging of radiographically dense breasts is difficult. Ultrasonography cannot detect many of the nonpalpable cancers that are not visible on mammograms of good quality. Most of the studies claiming that ultrasonography is useful in this regard compare it to poor-quality mammography or fail to compare it to mammography.

Several other techniques are under investigation for breast cancer imaging. Magnetic resonance imaging (MRI) offers great promise for imaging of the radiographically dense breast. Breast MRI is superior to mammography in differentiating solid from cystic lesions and is equivalent to mammography in providing information regarding different parenchymal patterns. Injection of intravenous contrast material with MRI increases cancer detectability in spite of the fact that breast cancer and glandular tissues have similar magnetic resonance tissue characteristics. However, breast MRI is expensive, has inferior spatial resolution to mammography, and cannot image microcalcifications. Breast CT has been investigated for the differentiation of benign from malignant solid masses. Breast CT involves the use of intravenous injection of iodinated contrast material and has limited spatial resolution and high cost; hence, it is not suited for routine breast cancer screening.

17.7.1 Optical Tomographic Imaging

Nonionizing optical tomography is a new and active research field although projection light imaging was investigated as early as 1929. The optical properties of normal and diseased tissues are usually different despite the large variation of values in optical properties of the normal tissues alone. Therefore, it is possible to detect some breast cancers based on measurements of optical properties.

The optical difference is not surprising because cancerous tissues manifest significant architectural changes at the cellular and subcellular levels, and the cellular components that cause elastic scattering have dimensions typically on the order of visible to near-IR wavelengths. Some tumors are associated with vascularization, where blood causes increased light absorption. The use of optical contrast agents may also be exploited to enhance the optical contrast between normal and abnormal tissues. Because the optical information is determined by the molecular conformations of biological tissues, optical imaging is expected to provide sensitive signatures for early cancer detection and monitoring.

Because tissues are optically turbid media that are highly scattering, light is quickly diffused inside tissues as a result of frequent scattering. The strong scattering has made optical detection of biological tissues challenging. A typical scattering coefficient for visible light in biological tissues is $100 \, \text{cm}^{-1}$ in comparison with $0.2 \, \text{cm}^{-1}$ for x-rays used in medical diagnostics. Light transmitted through tissues is classified into three categories: ballistic light, quasi-ballistic light, and diffuse light. Ballistic light experiences no scattering by tissue and thus travels straight through the tissue. Ballistic light carries direct imaging information as x-ray radiation does. Quasi-ballistic light experiences minimal scattering and carries some imaging information. Multiply-scattered diffuse light carries little direct imaging information and overshadows ballistic or quasi-ballistic light.

One of the techniques used for optical tomography is called early-photon imaging. If diffuse light is rejected, and ballistic or quasi-ballistic light is collected, buried objects can be detected much like x-ray projection. This technique uses a short-pulse laser (<1 ps pulse width) to illuminate the tissue. Only the initial portion of transmitted light is allowed to pass to a light detector, and the late-arriving light is gated off by a fast optical gate. Because the ballistic or quasi-ballistic photons travel along the shortest path length, they arrive at the detector sooner than diffuse photons. If only ballistic light is detected, the technique is called ballistic imaging. It has been shown that ballistic imaging is feasible only for tissue of thickness less than 1.4 mm or 42 mean free paths (mfp). Most ballistic imaging techniques reported in the literature have achieved approximately 30 mfp. Therefore, this approach is suitable for thin tissue samples but suffers loss of signal and resolution for thick tissues as a result of the strong scattering of light by the tissue.

Example Problem 17.10

Calculate the decay of ballistic light after penetrating a biological tissue 30 mfp thick. If the scattering coefficient of the tissue is 100 cm^{-1}, calculate the corresponding thickness in cm.

Solution

Based on Beer's law, the decay is $\exp(-30) = 9.4 \times 10^{-14}$. The thickness is $30/100 = 0.3$ cm. ∎

For tissue of clinically useful thickness (5 to 10 cm), scattered light must be used to image breast cancers. It has been shown that for a 5-cm-thick breast tissue with an assumed absorption coefficient of 0.1 cm^{-1} and reduced scattering coefficient of 10 cm^{-1}, the detector must collect transmitted light that has experienced at least 1100 scattering events in the tissue to yield enough signal. Therefore, ballistic light or even quasi-ballistic light does not exist for practical purposes. However, if a 10-mW visible or near-infrared laser is incident on one side of the 5-cm thick breast tissue, it has been estimated, using diffusion theory, that the diffuse transmittance is on the order of 10 nW/cm^2 or 10^{10} photons/(s cm^2), which is detectable using a photomultiplier tube capable of single-photon counting. Similarly, the diffuse transmittance through a 10-cm thick breast tissue would be on the order of 1 pW/cm^2 or 10^6 photons/(s cm^2). The significant transmission of light is due to the low absorption coefficient despite the high scattering coefficient.

Imaging resolution of pure laser imaging degrades linearly with increased tissue thickness. The temporal profiles of the scattered light may be detected using a streak camera. The early portion of the profiles was integrated to construct the images of buried objects in a turbid medium. This time-domain technique requires expensive short-pulse lasers and fast light detectors.

Optical-coherence tomography (OCT) has emerged as a useful clinical tool. This technique is based on Michelson interferometer (see Fig. 17.9) with a short-coherence length light source. One arm of the interferometer leads to the sample of interest, and the other leads to a reference mirror. The reflected optical beams are detected at the photodetector. The two beams coherently interfere only when the sample and the reference path lengths are equal to within the source coherence length. Heterodyne detection is performed by taking advantage of the direct Doppler frequency shift that results from the uniform high-speed scan of the reference path length. Recording the interference signal magnitude as a function of the reference mirror position profiles the reflectance of the sample, which produces an image similar to an ultrasonic A scan. OCT has achieved 10-μm resolution but a penetration depth of <1 mm.

OCT was recently extended to image blood flow in superficial vessels based on Doppler shift. The blood flow causes a Doppler shift on the frequency of the light. Frequency analysis with fast Fourier transform of the optical interference signal yields the Doppler shift. The Doppler shift is used to calculate the velocity of the blood flow.

A technique for optical imaging of thick tissue is the frequency-domain technique, which is based on photon-density waves. The governing equation of photon-density waves is the following diffusion equation:

$$\partial \Phi(r, t)/c\partial t + \mu_a \Phi(r, t) - D\nabla^2 \Phi(r, t) = S(r, t) \tag{17.85}$$

where $\Phi(r, t)$ is light fluence rate at point r and time t (W/cm^2), c is the speed of light in the medium (cm/s), μ_a is the absorption coefficient (cm^{-1}), and $S(r,t)$ is the source intensity (W/cm^{-3}). The frequency-domain imaging technique requires the use of complex inverse algorithms for image reconstruction. Amplitude-modulated (at approximately 100 MHz) laser light is used to illuminate the tissue at multiple sites. At each illumination site, an optical detector measures the amplitude and phase of diffuse light at multiple locations around the tissue. The measured diffuse light can be estimated by the diffusion equation (17.85) if the optical properties of the tissue sample are known. Conversely, the optical properties can be calculated by use of the measured diffuse light, which is the image reconstruction. A sample reconstructed image is shown in Figure 17.24, which was based on a theoretically generated data set of diffuse light.

17.7.2 Hybrid Optical Imaging

Several emerging imaging techniques are being developed by combining relatively transparent acoustic energy with strongly scattering light, which are called hybrid optical imaging. Ultrasound-modulated optical tomography, photoacoustic tomography, and sonoluminescence will be briefly discussed.

In ultrasound-modulated optical tomography, an ultrasonic wave is focused into a scattering medium to modulate the laser light passing through the medium containing buried objects. The modulated laser light collected by a photomultiplier tube is related to the local mechanical and optical properties in the zone of ultrasonic modulation. If the buried objects have optical properties that are different from those of the background scattering medium, an image can be obtained by raster scanning the device.

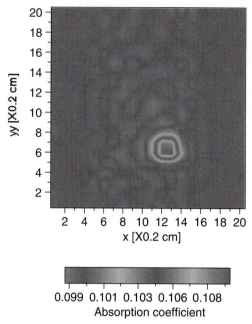

Figure 17.24   A sample reconstructed image using photon-density waves. A simulated tumor with elevated absorption coefficient is visible in the lower right quadrant.

The inverse of ultrasound-modulated optical tomography (acoustooptical tomography) is photoacoustic tomography. In photoacoustic tomography, a short-pulse light beam is injected into the scattering medium. The light is diffused in the medium and partially absorbed. If there is a strong optical absorber such as a tumor in the middle of the medium, more light will be absorbed by this optical absorber than by its neighbor. The absorbed optical energy is converted into heat. Due to thermal elastic expansion, an acoustic wave is generated. Stronger heat generation will produce a stronger acoustic wave. Therefore, a strong optical absorber emanates a strong acoustic wave. If multiple acoustic transducers are used to measure the acoustic signal around the medium, the absorber can be located based on the temporal distribution of the acoustic signals and hence produce an image of the medium. Functional images of the brain have been achieved recently in small animals.

The ultrasonic generation of light known as sonoluminescence (SL) was first reported in 1934, as multiple-bubble sonoluminescence (MBSL). SL has attracted an extraordinary amount of attention in this decade since single-bubble sonoluminescence (SBSL) was reported in 1990. Although the full explanation of SL is still in development, it is well known that light is emitted when tiny bubbles driven by ultrasound collapse. The bubbles start out with a radius of several microns and expand to ~50 microns due to a decrease in acoustic pressure in the negative half of a sinusoidal period. After the sound wave reaches the positive half of the period, the situation rapidly changes. The resulting pressure difference leads to a rapid collapse of the bubbles accompanied by emission of light. The flash time of SL has been measured to be in the tens of picoseconds. SBSL is so bright that it can be seen by the naked eye even in a lighted room, whereas MBSL is visible in a darkened room. Researchers have envisioned possible applications of SL in sonofusion, sonochemistry, and building ultrafast lasers using the ultrafast flash of light in SL.

Sonoluminescent tomography (SLT) is an application of SL that has been developed for cross-sectional imaging of strongly scattering media noninvasively. Sonoluminescence, which is generated internally in the medium by exposure to external ultrasound, is used to produce images of a scattering medium by raster scanning the medium. The spatial resolution is limited by the focal spot size of the ultrasound and can be improved by tightening the focus.

Optical imaging techniques may also measure the optical spectra of biological tissues noninvasively. Besides the absorption spectra that are usually measured for other materials, the scattering spectra may also be measured. The optical spectra may be used to quantify important physiological parameters such as the saturation of hemoglobin oxygenation.

In summary, optical imaging and biomedical optics techniques in general have the following advantages: (1) the use of nonionizing radiation, (2) the capability of measuring functional (physiological) parameters, (3) the potential high sensitivity to the pathologic state of biological tissues, and (4) low cost. However, biomedical optics is a challenging research field in its infancy and as it grows it will continue to require the participation of many diverse, talented physicians, scientists, and engineers.

EXERCISES

1. (a) Name at least three reasons why it might be desirable to use optic or fiber optic sensors for biomedical applications. (b) Name three potential drawbacks to using optic or fiber-optic biomedical sensors.

2. In one paragraph, or using bullets, explain the similarities and differences between polarization, luminescence, absorption, scatter, and Raman scattering of light.

3. Given an interface between clear tissue (i.e., water) $n_t = 1.33$ and glass ($n_g = 1.5$), compute the transmission angle for a beam incident in the water at $45°$. If the transmitted beam is reversed, so that it impinges on the interface from the glass side, calculate what the transmitted angle will be through the tissue. Does this make sense? Explain.

4. Using MATLAB, or some other software, plot ($R_\perp = r_\perp^2$) and ($R_p = r_p^2$) as a function of the input angle θ_i for a glass ($n_i = 1.5$) and air ($n_t = 1.0$) interface. Explain what it means.

5. An Nd:YAG laser operating at 1064 nm with a Gaussian beam is to be used for performing a Portwine stain treatment. Assume a heuristic model for light distribution in a homogeneous material such that $I(r, z) = I_o \exp(-\mu z) \exp(-2(r/w_b)^2)$, where $\mu = \mu_a + \mu_s$, and μ_a and μ_s

are the absorption and scattering coefficients. For the following multilayer case, at $r = 0$ give the expressions and graph intensity I and the rate of heat generation Q in the tissue by the laser, $I(r = 0, z)$ and $Q(r = 0, z)$. Assume $I_o = 1$. The absorption and scattering coefficients of each skin layer and blood vessel for this wavelength are given below. Discuss how selective photothermolysis of the blood layer is achieved.

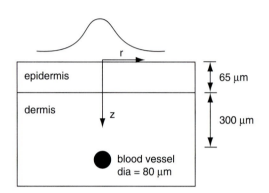

	μ_a (cm^{-1})	$\mu_{s}{}'$ (cm^{-1})
Blood	20	20
Epidermis	0.1	10
Dermis	0.5	10

6. Consider a turbid biological tissue with an absorption coefficient $\mu_a = 0.1\,\text{cm}^{-1}$. Using the diffusion approximation, graph on the same plane the logarithm of relative intensity vs radius for four different values of reduced scattering coefficient: $\mu_{s'} = 1$, 10, 50, 100 cm^{-1}, r varying from 0.1 cm to 2 cm. Describe the result in terms of the effect of scattering on the intensity profile.

7. Write a computer program in C/C++ or another computer language to implement diffusion theory calculation for oblique-incidence diffuse reflectance $R(x)$.

8. For liver tissue irradiated by the pulsed Ho:YAG laser, determine the exposure duration during which the equation for heat conduction may be ignored. Assume the absorption coefficient at the Ho:YAG laser wavelength is 30 cm^{-1} and assume the thermal diffusivity of the tissue is the same as that of water. If the threshold temperature of ablation is $150\,^{\circ}\text{C}$, determine the threshold fluence (J/cm^2) for ablation.

9. Given a Helium–Neon laser-based fiber optic Doppler probe at an angle of 60° with a blood vessel that registers a frequency shift of 63 KHz, what is the velocity of the blood? Is this a reasonable number from a physiologic point of view when compared to the average velocity in the major and minor blood vessels in the human body?

10. Assume you want to measure the thickness of a piece of tissue ($n_t = 1.33$) bounded by air ($n_t = 1.0$) using either the Michelson or Fabry–Perot interferometric approach. Your light source has a wavelength of 780 nm and your instrument can count 20 fringes. The tissue is ablated (or cut) with a high-powered pulsed laser which removes roughly 2 micrometers of tissue

per pulse. The maximum tissue thickness before it is all removed is 20 micrometers. (a) Calculate to see if each of the interferometer systems will have a long enough dynamic range, without moving the system, to be used with this piece of tissue. Explain your result. (b) Assume you want to use either of your interferometric systems as a feedback unit for the laser removal of tissue. What is the range in the number of fringes (m_{min} and m_{max}) for each of your systems in order to measure both the minimum slice thickness and the maximum tissue thickness? Explain.

11. In general, when is a Fabry–Perot interferometer preferred over a standard Michelson interferometer? Can you make a Fabry–Perot interferometer that performs worse than a Michelson interferometer? Explain.

12. For Raman spectroscopy, what are the Stokes and anti-Stokes bands? Which bands are typically used for sensing? What are the main challenges to overcome for Raman spectroscopy to be used for biomedical sensing?

13. Assume you want to measure glucose through the anterior chamber of the eye as a means of noninvasively quantifying blood glucose. Given glucose has a specific rotation of $41.89°/(dm\ g/ml)$ at a wavelength of 656 nm and the anterior chamber of the eye has a path length of roughly 0.8 cm, calculate the concentration of glucose for a rotation of 15 millidegrees. Is this a reasonable value from a physiologic point of view? Would the patient be considered normal or diabetic?

14. In the near-infrared region of the optical spectrum between 600–1100 nanometers there are well-known absorbance peaks for oxygenated and deoxygenated hemoglobin. A big assumption can be made for the moment that in this region the dominant optical signal is due to absorption (which is not generally the case). Given that you propagate through a 2-cm sample of tissue, you need to calculate three parameters: the concentration of oxyhemoglobin and deoxyhemoglobin, and a background blood absorbance. At three wavelengths (758 nm, 798 nm, 898 nm), you can measure the extinction coefficients for oxygenated (1.612, 2.072, and 2.763 $mM^{-1}\ cm^{-1}$) and deoxygenated hemoglobin (3.914, 2.072, and 2.072 $mM^{-1}cm^{-1}$), respectively. The total absorption coefficients at these three wavelengths in the tissue can be measured using a time-resolved system as (0.293 cm^{-1}, 0.1704 cm^{-1}, 0.1903 cm^{-1}), respectively. Calculate the oxygenated and deoxygenated hemoglobin levels given these parameters. Is this reasonable from a physiologic point of view? Explain.

15. Using the parameters of the equation for problem 10 give a graph of temperature vs time at $z = 0$, for t varying from 0 to 1 ms, using the analytical solution of the heat conduction equation given in the chapter. On the same graph plot temperature, ignoring the effect of conduction, which assumes linear variation with time. Comment on your findings in relation to diffusion time found in problem 10.

16. Consider skin tissue that is exposed to linear heating, for example as in pulsed laser coagulation, such that $T = T_o + mt$ when T_o is initial temperature (e.g., 37°C) and m is the rate of heating (e.g., 10^3°C/s). Describe how

you would find the critical temperature for thermal damage. Do you expect that to be lower or higher than a constant temperature case of the same exposure duration?

17. For the study of ablation onset time, plot a graph of $B\sqrt{\tau_{ab}}$ vs $B\lambda$. Measure the slope for "large" values of $B\lambda$; it should be $\frac{2}{\sqrt{\pi}}$. For absorption coefficient 100/cm and a laser intensity 1000 W/cm^2 find the time for the onset of ablation using water thermal properties, assuming ablation initiates at 150°C.

18. In the visible optical region, tissue scattering dominates absorption. Assume the scattering coefficient of a 5-cm thick tissue is 100 cm^{-1} and the wavelength of light is 0.5 μm. In order to detect on average a single ballistic photon transmitted through the tissue, what is the energy in Joules that is required for the incident light? Compare the energy of the incident light with the rest energy of the earth using Einstein's mass–energy equivalence equation $E = mc^2$, where the mass of the earth is 6×10^{24} kg.

19. Use diffusion theory to estimate light penetration in biological tissues. A 10-mW isotropic point source is buried in an infinite turbid medium. Assume that the absorption coefficient is 0.1 cm^{-1}, the scattering coefficient of a 5-cm thick tissue is 100 cm^{-1}, the scattering anisotropy is 0.9, and the wavelength of light is 0.5 μm. Calculate the light fluence rate 5 cm from the source.

20. Use diffusion theory to estimate sonoluminescence light transmission in biological tissues. A unit-power isotropic point source is buried in an infinite turbid medium. Assume that the absorption coefficient is 0.1 cm^{-1}, the scattering coefficient of a 5-cm thick tissue is 100 cm^{-1}, and the scattering anisotropy is 0.9. Calculate the light fluence rate integrated over a sphere of a 5-cm radius centered at the source.

21. What additional information does the phase measurement in a frequency-domain imaging technique provide compared with continuous wave techniques that measure only the amplitude of the diffuse light?

SUGGESTED READING

Boisdé, G. and Harmer, A. (1996). *Chemical and Biochemical Sensing With Optical Fibers and Waveguides.* Artech, Boston.

Born, M. and Wolf, E. (1999). *Principles of Optics,* 7th Expanded Ed. Cambridge Univ. Press, Cambridge, UK.

Culshaw, B. and Dakin, J. (1989). *Optical Fiber Sensors: Vol. I &II.* Artech, Boston.

Duck, F.A. (1990). *Physical Properties of Tissue.* Academic, London.

Hecht, E. (1987). *Optics.* Addison-Wesley, Reading, MA.

Katzir, A. (1993). *Lasers and Optical Fibers in Medicine.* Academic, San Diego, CA.

Lakowicz, J. (1999). *Principles of Fluorescence Spectroscopy,* 2nd Ed. Kluwer Academic/Plenum Publishers, New York, NY.

Pedrotti, F.L. and Pedrotti, L.S. (1993). *Introduction to Optics,* 2nd Ed. Prentice Hall, Upper Saddle River, NJ.

Tuchin, V. (2000). *Tissue Optics: Light Scattering Methods and Instruments for Medical Diagnosis*. Society of Photo-Optical Instrumentation Engineers Press, Bellingham, WA.

Vo-Dinh, T. (2003). *Biomedical Photonics Handbook*. CRC, Boca Raton, FL.

Welch, A.J. and van Gemert, M.J.C. (1995). *Optical-Thermal Response of Laser Irradiated Tissue*. Plenum, New York & London.

Willard, H.H., Merritt, L.L., Jr., Dean, J.A. and Settle, F.A., Jr. (1988). *Instrumental Methods of Analysis,* 7th Ed. Wadsworth, Belmont, CA.

APPENDIX

John Enderle, PhD

A.1 MATLAB

MATLAB is a high-level computer program that performs technical computing and plotting based on matrices. It is a programming language that has many of the same capabilities as FORTRAN, C and other programming languages, but is much easier to use. This section introduces the reader to MATLAB in sufficient detail so that the program can be used to solve relevant problems in this book. Other MATLAB features are introduced in later chapters as needed.

A.1.1 Matrix Basics

MATLAB was originally developed as MATrix LABoratory, with most commands stated in terms of matrices. A matrix is a rectangular array consisting of n rows and m columns of elements. The array $\mathbf{A}$, denoted with a bold uppercase letter, is denoted as

$$\mathbf{A} = \begin{bmatrix} a_{11} & a_{12} & \cdots & a_{1m} \\ a_{21} & a_{22} & \cdots & a_{2m} \\ \vdots & \vdots & & \vdots \\ a_{n1} & a_{n2} & \cdots & a_{nm} \end{bmatrix} \tag{A.1}$$

where each element of the array, a_{ij}, is a constant or a function. The order of array $\mathbf{A}$ is referred to as $n \times m$. If $n = m$, $\mathbf{A}$ is called a square matrix of order n. For our purposes, the matrix $\mathbf{A}$ usually stores the parameters for a system, such as values for resistors, capacitors and inductors.

An array of one column is referred to as a column vector, that is, an array of order $n \times 1$. We typically denote a column vector with a bold lower-case letter, such as

$$x = \begin{bmatrix} x_1 \\ x_2 \\ \vdots \\ x_n \end{bmatrix} \tag{A.2}$$

where each element x_i is a variable, constant or a function of time. The column vector is useful for representing variables and inputs for a system. An array of one row is referred to as a row vector, that is, an array of order $1 \times n$. The row vector is useful for representing a polynomial of the characteristic function (defined later) for a system. A matrix of order 1×1 is a scalar.

Addition and subtraction of matrices are valid only for matrices of the same order. Either operation is applied to corresponding elements term by term, that is,

$$\begin{bmatrix} a_{11} & a_{12} & \cdots & a_{1m} \\ a_{21} & a_{22} & \cdots & a_{2m} \\ \vdots & \vdots & & \vdots \\ a_{n1} & a_{n2} & \cdots & a_{nm} \end{bmatrix} + \begin{bmatrix} b_{11} & b_{12} & \cdots & b_{1m} \\ b_{21} & b_{22} & \cdots & b_{2m} \\ \vdots & \vdots & & \vdots \\ b_{n1} & b_{n2} & \cdots & b_{nm} \end{bmatrix}$$

$$= \begin{bmatrix} c_{11} = a_{11} + b_{11} & c_{12} = a_{12} + b_{12} & \cdots & c_{1m} = a_{1m} + b_{1m} \\ c_{21} = a_{21} + b_{21} & c_{22} = a_{22} + b_{22} & \cdots & c_{2m} = a_{2m} + b_{2m} \\ \vdots & \vdots & & \vdots \\ c_{n1} = a_{n1} + b_{n1} & c_{n2} = a_{n2} + b_{n2} & \cdots & c_{nm} = a_{nm} + b_{nm} \end{bmatrix} \tag{A.3}$$

where, in general $c_{ij} = a_{ij} + b_{ij}$. Subtraction follows similarly with $c_{ij} = a_{ij} - b_{ij}$.

Matrix multiplication is valid only when the number of columns in the first matrix is equal to the number of rows in the second matrix. If $\mathbf{A}$, order $n \times m$, is multiplied by $\mathbf{B}$, order $m \times n$, then the order of the resulting matrix $\mathbf{C}$ is $n \times n$. Each element of $\mathbf{C}$, c_{ij}, is equal to the sum of products of the ith row of $\mathbf{A}$ with the jth column of $\mathbf{B}$, that is

$$c_{ij} = \sum_{k=1}^{n} a_{ik} b_{kj} \tag{A.4}$$

for $i = 1, 2, \ldots, n$ and $j = 1, 2, \ldots, m$. In general, matrix multiplication for matrices of the same order is not commutative, that is, $\mathbf{AB} \neq \mathbf{BA}$. The multiplication of a matrix by a scalar α equals the product of each element of the matrix by α.

The identity matrix $\mathbf{I}$ is a square matrix whose nondiagonal elements are zero and whose diagonal elements are one, that is

$$I = \begin{bmatrix} 1 & 0 & 0 & \cdots & 0 \\ 0 & 1 & 0 & \cdots & 0 \\ 0 & 0 & 1 & \cdots & 0 \\ \vdots & \vdots & \vdots & & \vdots \\ 0 & 0 & 0 & \cdots & 1 \end{bmatrix} \tag{A.5}$$

The null matrix, $\mathbf{0}$, has 0 for all elements in the matrix.

Matrix operations of addition, subtraction, multiplication and division follow much the same processes that these operations do with real numbers. For arbitrary and appropriately defined matrices **A**, **B**, and **C**, we have

Commutative Property: $A + B = B + A$
Associative Property: $A + (B + C) = (A + B) + C$
Distributive Property: $A(B + C) = AB + AC$
Identities Involving I and 0:

$$AI = A$$
$$0A = 0$$
$$A0 = 0$$
$$A + 0 = A$$

The inverse of a square matrix **A**, denoted, A^{-1}, is defined by

$$A^{-1}A = AA^{-1} = I \tag{A.6}$$

The use of the matrix inverse is important in solving a set of simultaneous equations that describe a system.

The transpose of matrix **A** is denoted as either A' or A^T, where we interchange rows and columns of **A** to form A^T as follows from Equation A.1.

$$A^T = \begin{bmatrix} a_{11} & a_{21} & \cdots & a_{n1} \\ a_{12} & a_{22} & \cdots & a_{n2} \\ \vdots & \vdots & & \vdots \\ a_{1m} & a_{2m} & \cdots & a_{nm} \end{bmatrix}$$

Simultaneous Equations and Matrices

A set of equations with unknown variables often results in the process of analyzing a system, after applying interconnection laws or other techniques. At other times, a set of equations with unknown parameters results in solving the coefficients of a differential equation with initial conditions.

Suppose the following set of equations describes the currents in an electric circuit, i_i, i_2, and i_3, after applying Kirchhoff's voltage law.

$$2i_1 - i_2 + 0i_3 = 6$$
$$-2i_1 + 3i_2 + i_3 = 0 \tag{A.7}$$
$$-i_1 + 5i_2 - 4i_3 = 0$$

The constants multiplying the currents involve the resistors in the circuit, with the value of 6 due to the input to the circuit. In writing each equation in A.7 we have included all variables, written in order, even if a variable is multiplied by 0.

Equation A.7 is written in matrix form as

$$\begin{bmatrix} 2 & -1 & 0 \\ -2 & 3 & 1 \\ -1 & 5 & -4 \end{bmatrix} \begin{bmatrix} i_1 \\ i_2 \\ i_3 \end{bmatrix} = \begin{bmatrix} 6 \\ 0 \\ 0 \end{bmatrix} \tag{A.8}$$

or

$$\mathbf{Ai} = \mathbf{F} \tag{A.9}$$

with appropriately defined matrices

$$\mathbf{A} = \begin{bmatrix} 2 & -1 & 0 \\ -2 & 3 & 1 \\ -1 & 5 & -4 \end{bmatrix}, \mathbf{i} = \begin{bmatrix} i_1 \\ i_2 \\ i_3 \end{bmatrix}, \text{and } \mathbf{F} = \begin{bmatrix} 6 \\ 0 \\ 0 \end{bmatrix}$$

To solve Equation A.8 for the current vector $\mathbf{i}$, we premultiply both sides of the equation by the inverse $\mathbf{A}^{-1}$, that is

$$\mathbf{A}^{-1}\mathbf{Ai} = \mathbf{A}^{-1}\mathbf{F} \tag{A.10}$$

or

$$\mathbf{i} = \mathbf{A}^{-1}\mathbf{F} \tag{A.11}$$

A.1.2 Using MATLAB

MATLAB is a program that evaluates mathematical expressions, plots results, and comes with a wide assortment of toolboxes (applications) for various engineering fields. The toolboxes include neural networks, imaging, digital signal processing, control theory, linear algebra, signals and systems, numerical methods, symbolic math, and SIMULINK. SIMULINK provides an iterative solution (i.e., an approximation) to very complex differential-integral equations. The plotting capabilities of MATLAB are sufficient for most work. For more sophisticated or professional plotting, output from MATLAB can be exported into Microsoft® EXCEL.

The material presented here is a very brief introduction to MATLAB. Readers interested in learning more about the program should look in the library and bookstores to find many books on this powerful program.

Getting Started

To begin a MATLAB session in Microsoft Windows, click "Start", "Programs", the folder "MATLAB", and then the program "MATLAB". A command window opens with three parts as shown in Figure A.1.

Our work at this stage is done in the MATLAB Command Window on the right side of Figure A.1. To begin, we start typing in the Command Window after the ">>". Text can be typed interactively with MATLAB executing the command after a carriage return (called CR hereafter) by clicking the "Enter" key. Keep in mind that MATLAB differentiates between lower and uppercase in executing commands (so "a" is different from "A"), and it does not use italic or bold characters as defined in Section A.1.

Figure A.1   MATLAB program with three windows. (1) Launch Pad Window, (2) Command Window, and (3) Command History.

MATLAB evaluates scalars as well as matrices. For example, if we enter

>> A = 10/5

MATLAB returns with

A =

    2

To enter a matrix, we start with the matrix name, then an equals sign followed by a left bracket, the elements of the array, and then a right bracket. Each element in a row is entered, beginning with the first row, separated by a blank space or a comma. To move from one row to the next row, a semicolon or a CR is entered. For example, matrix **A** is entered in MATLAB by writing

>> A = [5 9 6; 3 2 3; 5 9 1]

MATLAB returns with

A =

    5   9   6
    3   2   3
    5   9   1

Matrix elements can also be entered as a function, such as

>> A = [− 5/2, sqrt(4), 3∗(6 − 4)]

MATLAB returns with

A =

    −2.5000   2.0000   6.0000

Note that a ";" at the end of any line suppresses printing of that line. A line can also be extended by typing "..." and a CR at the end of any line so that the next line is just a continuation of the previous line.

For example,

```
>> C = [1 ...
2 3
4 5 6
7 8 9]
C =
     1   2   3
     4   5   6
     7   8   9
```

Here the first row is written on two lines with "..." at the end of the first line and a CR rather than ";"separating the rows.

To invert matrix **A**, given by

```
>> A = [5 9 6; 3 2 3; 5 9 1]
```

type

```
>> B = inv(A)
```

MATLAB returns with

```
B =
    −0.2941      0.5294      0.1765
     0.1412     −0.2941      0.0353
     0.2000      0          −0.2000
```

There are many special functions within MATLAB like the exponential and π. For example, to use π in a calculation, we use the word "pi"; enter the following and note the response.

```
>> pi
ans =
     3.1416
```

MATLAB returns a numerical result for each command entered. The exponential function uses the syntax "exp (− 0.5)", which, when entered, gives 0.6065. Most standard elementary functions in a good scientific calculator including built-in functions, are included in MATLAB like the sine and cosine functions. The entire list of MATLAB functions is given from the "HELP" menu: select "MATLAB HELP", then "Functions – Alphabetical List".

1.1.3 Arithmetic Expressions

Arithmetic operations are carried out in MATLAB with the following symbols:

$$\wedge \quad * \quad / \quad \backslash \quad + \quad -$$

for power operator, multiplication, right division, left division, addition and subtraction, respectively. Note that 1/4 and 4\1 both equal 0.25, and represent 1 divided by 4 and 4 divided into 1, respectively. We often use the "\" operation when solving simultaneous equations as in Ex. A.6.

MATLAB performs operations in the order previously listed, that is, $\wedge$ first, then * or / or \ calculated in order from left to right, and then + or − calculated in order from left to right. For example,

$$>> 5 + 6\wedge 2*3/7*2$$

gives

ans =
    35.8571

This operation is done by first calculating $6 \wedge 2 = 36$, then multiplying 36 by 3, followed by dividing this result by 7, then multiplying the entire result by 2, and then finally adding 5. The sequence of like math operations works from left to right, with parentheses used to change the order of calculation. For example,

$$>> 3*(6 - 4)$$

gives

ans =
    6

The first calculation is 6–4 and then this result is multiplied by 3.

Trigonometric functions are calculated in MATLAB with the angle in radians. For example, to evaluate $\sin(45°)$, we enter

$$>> \sin(pi/4)$$

which gives

ans =
    0.7071

Example A.1

Solve the following set of simultaneous equations for $i_1, \ldots, i_5$.

$$4i_1 - i_2 - i_3 - i_4 - i_5 = 240$$
$$-i_1 + 8i_2 - i_3 + 0 \times i_4 + 0 \times i_5 = 0$$
$$-i_1 - i_2 + 5i_3 - i_4 + 0 \times i_5 = 0$$
$$-i_1 + 0 \times i_2 - i_3 + 5i_4 - i_5 = 0$$
$$-i_1 + 0 \times i_2 + 0 \times i_3 - i_4 + 8i_5 = 0$$

Solution

The simultaneous equations are written in matrix format as

$$\mathbf{Ai} = \mathbf{F}$$

with

$$\mathbf{A} = \begin{bmatrix} 4 & -1 & -1 & -1 & -1 \\ -1 & 8 & -1 & 0 & 0 \\ -1 & -1 & 5 & -1 & 0 \\ -1 & 0 & -1 & 5 & -1 \\ -1 & 0 & 0 & -1 & 8 \end{bmatrix} ; \quad \mathbf{i} = \begin{bmatrix} i_1 \\ i_2 \\ i_3 \\ i_4 \\ i_5 \end{bmatrix} ; \quad \mathbf{F} = \begin{bmatrix} 240 \\ 0 \\ 0 \\ 0 \\ 0 \end{bmatrix}$$

Column vector *i* is easily solved by using the reverse division operation on the previous equation, that is,

$$\mathbf{A} \backslash \mathbf{Ai} = \mathbf{i} = \mathbf{A} \backslash \mathbf{F}$$

The MATLAB commands to solve for *i* are

```
>> A = [4 −1 −1 −1 −1; −1 8 −1 0 0; −1 −1 5 −1 0; −1 0 −1 5 −1; −1 0 0 −1 8];
>> F = [240;0;0;0;0];
>> i = A \ F
```

```
i =
   77.5000
   12.5000
   22.5000
   22.5000
   12.5000
```

■

A.1.4 Vectors

MATLAB stores vectors as a row vector with the first element as the **i** component, the second the **j** component and the third the **k** component, all written within brackets. For example, the vector $\mathbf{F} = 3\mathbf{i} - 4\mathbf{j} + 2\mathbf{k}$ is entered in MATLAB as

```
>> F = [3 −4 2]
```

which gives

```
F =
    3 −4  2
```

To calculate the magnitude of the vector, the MATLAB command "norm" is used. For example, the magnitude of the vector F is calculated as

>> Fmagnitude = norm(F)

Fmagnitude =
    5.3852

MATLAB also computes the dot product of two vectors with the command "dot(A,B)". For example, if $\mathbf{A} = 5\mathbf{i} - 8\mathbf{j} - 2\mathbf{k}$ and $\mathbf{B} = 6\mathbf{i} + 4\mathbf{j} - 3\mathbf{k}$, the dot product is computed in MATLAB as

>> A = [5 −8 −2];
>> B = [6 4 −3];
>> ABdot = dot(A, B)
ABdot =
    4

The cross product is also a built-in function in MATLAB using the command "cross(A,B)". For example, $\mathbf{A} \times \mathbf{B}$ is written as

>> A = [5 −8 −2];
>> B = [6 4 −3];
>> ABcross = cross(A, B)
ABcross =
   32   3   68

A.1.5 Complex Numbers

MATLAB stores all numbers as complex numbers and uses either i or j to represent the imaginary component of complex numbers. In MATLAB $z = 3 + i4$ and $z = 3 + j4$ are equivalent, that is

>> z = 3 + 4j

gives
z =
    3.0000 + 4.0000i

and

>> z = 3 + 4i

gives

z =
    3.0000 + 4.0000i

As mentioned previously, here we use the symbol i for current (as in Ex. A.1) and the symbol j for the complex numbers. MATLAB, however, defaults to the symbol i. The polar form for complex numbers is also valid in MATLAB; that is, for any arbitrary real value x and y,

$$z = x + jy = |z|e^{j\theta}$$

where

$$|z| = \sqrt{x^2 + y^2}$$

$$\theta = \tan^{-1}\frac{y}{x}$$

MATLAB has many built-in functions for working with complex numbers, with a few essential ones given in Table A.1.

TABLE A.1   MATLAB Commands for Complex Numbers with $z = x + jy$

Value	Syntax
Re(z)	real(z)
Im(z)	imag(z)
$|z| = \sqrt{x^2 + y^2}$	abs(z)
$\theta = \tan^{-1}\frac{y}{x}$	angle(z)

Example A.2

Use MATLAB to determine $z_3 = \frac{z_1}{z_2}$ if $z_1 = 3 + j2$ and $z_2 = 2 - j$.

Solution

In MATLAB we write

```
>> z1 = 3 + j*2
z1 =
   3.0000 + 2.0000i
>> z2 = 2-j
z2 =
   2.0000 − 1.0000i
>> z3 = z1/z2
z3 =
   0.8000 + 1.4000i
```

∎

Example A.3

Evaluate

$$z = (60 + j30) + 100e^{-j0.4887}$$

Solution

To evaluate z, we use the following MATLAB commands, mixing the rectangular and polar forms of a complex number.

$>> z = 60 + j*30 + 100*exp(-j*0.4887)$
$z =$
   $1.4829e + 002 - 1.6948e + 001i$ ∎

Example A.4

Evaluate the magnitude and angle of the following complex number:

$$z = (5 - j4)^3$$

Solution

$>> z = (5 - j*4)^{\wedge}3$
$z =$
   $-1.1500e + 002 - 2.3600e + 002i$

$>> r = abs(z)$
$r =$
  262.5281

$>> theta = angle(z)$
$theta =$
    -2.0242 ∎

A.1.6 Polynomials and Roots

MATLAB works with polynomials using a row vector and can calculate the roots of the polynomial using a built-in function. A polynomial is written as a row vector of the polynomial coefficients in descending order starting with the highest order. Consider the polynomial

$$x^4 - 12x^3 + 0x^2 + 25x + 116 = 0 \qquad (A.12)$$

which is entered in MATLAB as

$>> p = [1 - 12 \ \ 0 \ \ 25 \ \ 116]$

returning

$p =$
  $1 \ \ -12 \ 0 \ 25 \ 116$

It is important to enter every term, even terms with a 0 coefficient as in Equation A.12. To find the roots of the polynomial, we use the built-in function "roots". To find the roots of Equation A.12, we write

>> r = roots(p)

and MATLAB returns with two real roots and a pair of complex conjugate roots:

r =
  11.7473
  2.7028
  −1.2251 + 1.4672i
  −1.2251 − 1.4672i

The convention for MATLAB is that polynomials are row vectors and roots are column vectors. One can work in the opposite direction and find the polynomial given as a set of roots via the MATLAB "poly" command. Continuing from the last MATLAB calculations to find the roots of a polynomial, use the "poly" command to restore the polynomial

>> pp = poly(r)

which returns

pp =
    1.0000 −12.0000 −0.0000 25.0000 116.0000

Polynomial multiplication is carried out using the MATLAB function "conv", which performs the convolution of the two polynomials. Consider multiplying the following two polynomials using MATLAB.

$$A(x) = x^3 + 2x^2 + 3x + 4 \tag{A.13}$$

$$B(x) = x^3 + 4x^2 + 9x + 16 \tag{A.14}$$

In MATLAB we type

>> A = [1 2 3 4];
>> B = [1 4 9 16];
>> AB = conv(A, B)

which returns

AB =
    1   6   20   50   75   84   64

Note that the semicolon after the matrix A and B suppresses the echo of the command. Symbolically, the polynomial product of Equation A.13 and Equation A.14 is given by

$$AB(x) = x^6 + 6x^5 + 20x^4 + 50x^3 + 75x^2 + 84x + 64 \tag{A.15}$$

To add polynomials together, the polynomials need to be of the same order. For example, consider adding the following two polynomials:

$$S(x) = x^5 + 4x^4 + 5x^3 + 6x^2 + 7x + 2 \tag{A.16}$$

and

$$T(x) = 2x^4 + 9x^3 + 8x^2 - 4x + 5 \qquad \text{(A.17)}$$

Since the polynomial in Equation A.17 is 4th order, we pad the row vector with a zero in place of a x^5 coefficient. Thus the polynomials are entered in MATLAB as

```
>> S = [1 4 5 6 7 2];
>> T = [0 2 9 8 −4 5];
>> U = S + T
```

which returns

```
U =
   1   6   14   14   3   7
```

or symbolically

$$x^5 + 6x^4 + 14x^3 + 14x^2 + 3x + 7 \qquad \text{(A.18)}$$

To divide one polynomial by another, the MATLAB command "deconv(A,B)" is used, which represents B divided into A, or equivalently A divided by B. As there is usually a remainder from polynomial division, the syntax used is [Q,V] = deconv(A,B), where Q is the quotient polynomial and V is the remainder. Using MATLAB, the ratio of polynomials given by Equations A.16 and A.17, that is, $x^5 + 4x^4 + 5x^3 + 6x^2 + 7x + 2$ divided by $2x^4 + 9x^3 + 8x^2 - 4x + 5$ is written

```
>> S = [1 4 5 6 7 2];
>> T = [2 9 8 −4 5];
>> [Q, R] = deconv(S, T)
Q =
   0.5000 − 0.2500
R =
   0   0   3.2500   10.0000   3.5000   3.2500
```

Symbolically, this result is written as

$$\frac{1}{2}x - \frac{1}{4} + \frac{3.25x^3 + 10x^2 + 3.5x + 3.25}{x^5 + 2x^4 + 9x^3 + 8x^2 - 4x + 5}$$

To illustrate a simpler example with no remainder, divide $x^2 - 2x + 1$ by $x - 1$, giving

```
>> S = [1, −2, 1];
>> T = [1, −1];
>> [Q, V] = deconv(S, T)
Q =
   1 −1
V =
   0   0   0
```

The result is $x - 1$.

Example A.5

Consider the following equation:

$$\frac{(1600 - 3x)^3}{27x^3} = \frac{x}{500 - x}$$

Find x given that it must be between 0 and 500.

Solution

We use "format long" in this solution so that enough significant digits are carried through. This is necessary because ordinarily MATLAB evaluates expressions with "format short". First evaluate $(1600 - 3x)^3$ as follows:

```
>> format long
>> y = [- 3 1600];
>> r = conv(y, y);
>> w = conv(y, r)
```

which returns

```
w =
   1.0e + 009*
 -0.00000002700000   0.00004320000000   - 0.02304000000000   4.09600000000000
```

where "w" is the expression $(1600 - 3x)^3$. Notice that "1.0e + 009" multiplies all of the coefficients in the previous polynomial.

We cross-multiply to eliminate the denominator term on the right side as follows:

```
>> z = [- 1 500];
```

and multiplying w*z is done by another "conv"

```
>> s = conv(z, w)
```

resulting in

```
s =
   1.0e + 012*
```

Columns 1 through 4
0.00000000002700   -0.00000005670000   0.00004464000000   - 0.01561600000000
Column 5
2.04800000000000

The left-side denominator term $27x^3$ multiplied by the right side numerator term x is $(27x^4)$, written in MATLAB as

```
>> q = [27 0 0 0 0]
```

giving

```
q =
   27  0  0  0  0
```

Next, we subtract the two polynomials, using the MATLAB command

```
>> t = s - q
```

resulting in

t =

1.0e + 012 ∗

Columns 1 through 4

$\quad$ 0 − 0.00000005670000 $\quad$ 0.00004464000000 $\quad$ −0.01561600000000

Column 5

2.04800000000000

The polynomial evaluated by the previous steps can be written as

$$-5.67 \times 10^4 x^3 + 4.464 \times 10^7 x^2 + 1.5616 \times 10^{10} x + 2.048 \times 10^{12} = 0$$

The final step in the problem is to evaluate the roots of the resulting polynomial:

>> v = roots(t)

which yields

v =

1.0e + 002 ∗

2.62504409150663 + 2.62297313283571i

2.62504409150663 − 2.62297313283571i

2.62292769000261

Two of the roots are complex and are eliminated because they fall outside the problem constraints, giving a solution of

$$x = 262.3$$

$\blacksquare$

Table A.2 summarizes some MATLAB commands used for evaluating polynomials.

TABLE A.2 MATLAB Commands for Polynomials

Operation	Symbolic Expression	Syntax
Polynomial	$a_n x^n + a_{n-1} x^{n-1} + \cdots + a_1 x + a_0$	$P = [a_n a_{n-1} \cdots a_1 a_0]$
Roots of a polynomial	$(x - r_1)(x - r_2) \cdots (x - r_n)$	$r = \text{roots}(P)$
Write a polynomial from the roots	$a_n x^n + a_{n-1} x^{n-1} + \cdots + a_1 x + a_0$	$\text{poly}(r)$
Multiply two polynomials $S(x) = A(x) \times B(x)$	$S(x) = (a_n x^n + a_{n-1} x^{n-1} + \cdots + a_1 x + a_0) \times$ $\quad (b_m x^m + b_{m-1} x^{m-1} + \cdots + b_1 x + b_0)$	$S = \text{conv}(A, B)$
Add two polynomials $S(x) = A(x) + B(x)$; note that the order of the polynomials must be the same	$S(x) = (a_n x^n + a_{n-1} x^{n-1} + \cdots + a_1 x + a_0) +$ $\quad (0 x^n + \cdots + b_m x^m + b_{m-1} x^{m-1} + \cdots + b_1 x + b_0)$	$S = A + B$
Divide two polynomials $S(x) = \frac{A(x)}{B(x)} = Q(x) + V(x)$;	$S(x) = \dfrac{a_n x^n + a_{n-1} x^{n-1} + \cdots + a_1 x + a_0}{b_m x^m + b_{m-1} x^{m-1} + \cdots + b_1 x + b_0} = q_p x^p + \cdots q_1 x + q_0$ $\quad + \dfrac{v_l x^l + \cdots + v_1 x + v_0}{b_m x^m + b_{m-1} x^{m-1} + \cdots + b_1 x + b_0}$	$[Q, V] = \text{deconv}(A, B)$

A.1.7 Plotting with MATLAB

MATLAB has extensive plotting capabilities for two and three-dimensional graphing of vectors and matrices for data visualization and presentation graphics. The appearance of the graphs, including line widths, color, annotations and labeling can be customized to meet the needs of the user.

fplot

The basic plotting command in MATLAB is "fplot('f',[T_1, T_2])", which plots the function f over the interval T_1 to T_2. Note that the expression "f" is placed within apostrophes so that MATLAB associates this as a string expression rather than a function to be evaluated at time "t". Consider the plotting $f = 2e^{-t}$ in the interval 0 to 5 using MATLAB. We enter the following command:

$>>$ fplot('2∗exp (− t)', [0, 5])

The command "fplot" graphs the function f in the time interval from 0 to 5 in a separate window as shown in Figure A.2. To edit the properties of the plot shown in Figure A.2, select the arrow icon (next to the print icon) and then double-click on the graph. Another command window opens whereby one can change the color of the line, line styles, fonts, axes labels, axes, etc.

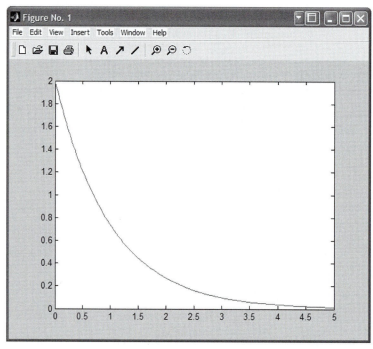

Figure A.2   MATLAB plot of the function $f = 2e^{-t}$ plotted using the command fplot.

Graphs are easily moved from MATLAB to other applications by copying the graph to the clipboard (Edit, Copy Fig.) and then pasting it into another application such as Microsoft® Word. One has control of the graph copied to clipboard by editing the Copy Options from the Edit menu. The "Metafile" format copies the graph without background color. The "Bitmap" option copies an exact replica of the graph including the background color.

plot

Another plotting command is "plot(t,y)", which plots y vs. t with points connected by straight lines. The independent variable t is easily defined using the "linspace" command with the syntax "t = linspace(a,b,n)" that generates a row vector of n linearly spaced elements over the timespan a to b.

For example, we can plot the function $y = 210\sqrt{2}e^{-2t}\sin(\sqrt{2}t)$ using the following commands:

```
>> t = linspace(0, 8, 1000);
>> y = 297*sin (1.414*t).*exp ( −2*t);
>> plot(t, y)
```

The first command creates a 1000-element row vector t, with values ranging from 0 to 8. The second command creates a 1000-element row vector y by multiplying $\sin(1.414*t)$ by $\exp(-2*t)$ with the ".*" operation. The ".*" operation is the dot multiplication function that performs an element-by-element multiplication. MATLAB generates an error if one uses "*" instead of ".*" in the previous calculation, because it associates t as a 1×1000 row vector, which forces $\sin(1.414*t)$ as a 1×1000 row vector and $\exp(-2*t)$ as a 1×1000 row vector. In this case, the matrix multiplication is not valid because the order of the matrices does not agree. The command "plot" graphs the function y in the time interval from 0 to 8 with 1000 connected points as shown in Figure A.3.

It is also possible to plot two functions on a graph using the "plot" command. To plot $f = 100e^{-t}$ and $y = 210\sqrt{2}e^{-2t}\sin(\sqrt{2}t)$ on the same graph, enter the following:

```
>> t = linspace(0, 8, 1000);
>> y = 297*sin (1.414*t).*exp ( −2*t);
>> f = 100*exp ( − t);
>> plot(t, y, t, f)
```

which generates the graph in Figure A.4. As before, the graph is fully customizable, with the line color, line styles, fonts, axes labels, axes, etc., changed to the user's needs.

There are many more plotting functions available within MATLAB, including three-dimensional graphs, histograms, pie charts, etc. The reader is directed to the Help function in MATLAB to discover all of the plotting capabilities.

Plotting with Microsoft Excel

While MATLAB provides very good plotting capabilities, exporting the data to other applications such as Microsoft® Excel gives graphics a more professional appearance.

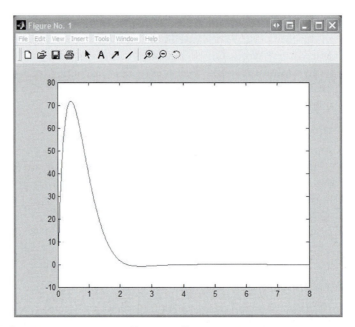

Figure A.3 The function $y = 210\sqrt{2}e^{-2t}\sin(\sqrt{2}t)$ graphed using the MATLAB "plot" command.

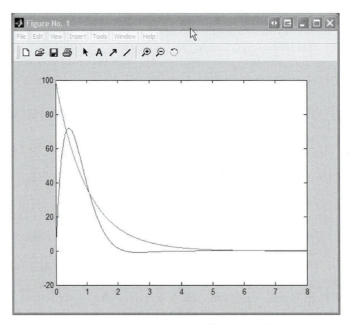

Figure A.4 The functions $f = 100e^{-t}$ and $y = 210\sqrt{2}e^{-2t}\sin(\sqrt{2}t)$ graphed using the MATLAB "plot" command.

Before plotting the data in Excel, we need to first export the data from MATLAB and then import it into Excel. All data in MATLAB is stored under variable names in a common workspace accessible to all toolboxes, including SIMULINK. All variables used in the MATLAB workspace are given by the "who" or "whos" command. The "who" command simply lists the variables and the "whos" command lists the variables along with their sizes, number of bytes used and class (i.e., logical character, integer array, floating point number array).

Before exporting the data, it is easiest to collect the data in a single matrix with the independent variable listed first, along with the dependent variables—each evaluated at the points listed for the independent variable. Consider exporting the function shown in Figure A.3, where t and y are given by

$$>> t = linspace(0, 8, 1000);$$
$$>> y = 297*sin(1.414*t).*exp(-2*t);$$

Keep in mind that t and y are stored as 1×1000 row vectors, so we must take the transpose of each row vector when concatenating to form the new matrix *data* by

$$>> data = [t', y'];$$

Data is stored in the matrix *data* with order 1000×2. If there are additional vectors to be exported as in Figure A.4, we concatenate all the data with the command "data = $[t', y', f']$;".

To export the data to an arbitrary ASCII file "output.txt", with tabs separating each variable, we use the following command:

$$>> save\ output.txt\ data - ascii - tabs$$

This file is stored in the MATLAB subfolder "work" by default. Next, open the file "output.txt" in Excel. At this time, the data are stored in the first two columns in the workspace. From here one can use the step-by-step "chart wizard" to create a "scatter chart" like that shown in Figure A.5.

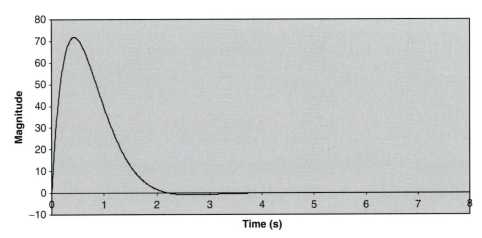

Figure A.5   The function $y = 210\sqrt{2}e^{-2t}\sin(\sqrt{2}t)$ drawn using Microsoft Excel.

A.1.8 Loading, Quitting and Saving the Workspace

Most problems can be solved interactively using MATLAB by entering information on the command line, pressing Enter, and having MATLAB evaluate the expression. For more challenging and lengthy projects, one can use m-files to store work to be retrieved later for further analysis. As a user becomes more proficient in MATLAB, m-files are used more frequently to store MATLAB commands, data and input-output. The default extension for MATLAB is .mat. To create an m-file, one can save the workspace, or write the commands in Windows Notepad and load the commands in MATLAB with the load command. All of the expressions in the m-file are computed when loaded.

To open an m-file, type

$>>$ load "filename.mat"

The command "save" is used to save the workspace to an m-file. If no filename is given, MATLAB defaults to "matlab.mat."

A.2 SOLVING DIFFERENTIAL EQUATIONS USING MATLAB

MATLAB provides the Symbolic Math Toolbox for symbolically solving many calculus and algebraic problems. The Symbolic Math Toolbox is a collection of commands that operate on functions based on a program called Maple® developed at the University of Waterloo in Ontario, Canada. Unlike previous exercises in which the answers MATLAB provided were numbers, using this toolbox results in answers that are symbolic functions. Examples of symbolic functions are

$$\tan(t^2), \quad \frac{d}{dt}(4y^2 + 2y + 4), \quad \int \sin(y)e^{-2y}dy, \text{ and } 4\ddot{y} + 2\dot{y} + 6 = e^{-2t}.$$

MATLAB has commands among its features in this toolbox that determine the exact symbolic solution for a derivative or an integral of a function and for differential equations. For example, if we want to compute the derivative of $\tan(t^2)$, MATLAB provides the answer $2t(1 + \tan(t^2)^2)$. In this section, we introduce a few of these features, which are appropriate for solving problems in this chapter.

Before performing symbolic computations, we need to declare all of the symbolic variables. To write a symbolic variable we enclose it between single quotation marks using the "sym" MATLAB command or use the "syms" command. For example, the command

$>>$ y = sym('y')

creates a symbolic variable "y". Instead of defining a symbolic variable using the "sym" command, we can write it directly by writing the function within single quotes as follows:

$>>$ f = 'tan (t^2)'
  f =
    tan (t^2)

which creates a symbolic variable "f" that is stored in memory as "tan (t^2)". To declare more than one symbolic variable, use the "syms" command. For example,

>> syms x y w z

or as

>> syms('w', 'x', 'y', 'z')

declares two four symbolic variables. This command is equivalent to the "sym" command for each variable separately. Whenever a symbolic variable is used after it is declared as a symbolic variable, any function using it is also a symbolic function. For example,

>> f = tan (y^2)
    f =
        tan (y^2)

Notice that in the previous MATLAB command, we did not need to write f = 'tan (t^2)' (with apostrophes) because "y" is already declared as a symbolic variable. These variables are now available to be used with any MATLAB command to create symbolic results. With "w", "x", "y" and "z" declared as symbolic variables, we can create a symbolic determinant by using a symbolic matrix using the MATLAB "det" command as follows:

>> m = [w, x; y, z]
    m =
        [w, x]
        [y, z]
>> det(m)
    ans =
        w∗z − x∗y

The MATLAB commands "diff" and "int" are the symbolic derivative and the indefinite integral functions, respectively. The following illustrates these two symbolic functions with some common expressions.

>> x = diff(' cos (y)')
    x =
        − sin (y)

and

>> x = int(' sin (y)')
    x =
        − cos (y)

Sometimes the result of a MATLAB operation gives an answer that is not easily recognizable. The MATLAB command "simplify" algebraically simplifies the result to a form that is more readily recognized. For example,

>> simplify((x^2 + 7∗x + 12)/(x + 4))
    =
        x + 3

As we will see, this command is very useful during integration and differentiating operations.

MATLAB also calculates definite integrals by including the limits of integration as arguments in the "int" command. For example, to calculate $\int_0^\pi \cos{(y)}dy$, we use the following MATLAB command

```
>> x = int(' sin (y)', '0', 'pi')
        x =
        2
```

The limits of integration do not have to be numbers, but can be symbolic variables. For example,

```
>> x = int(' sin (y)', 'a', 'b')
        x =
        - cos (b) + cos (a)
```

The "solve" command is convenient for solving symbolic algebraic expressions. For example, to solve for the roots of the polynomial $x^4 + 14x^3 + 71x^2 + 154x + 120 = 0$, we use

```
>> solve('x^4 + 14*x^3 + 71*x^2 + 154*x + 120 = 0')
ans =
     [ - 5]
     [ - 4]
     [ - 3]
     [ - 2]
```

Recall from the previous chapter that we could calculate the roots of a polynomial using the "roots" command, that is

```
>> p = [1 14 71 154 120];
>> r = roots(p)
        r =
        -5.0000
        -4.0000
        -3.0000
        -2.0000
```

The difference between the two commands is that "solve" allows us to symbolically solve for the roots without specifying the coefficients of the polynomial such as the well known quadratic equation:

```
>> syms a b c y x
>> solve('a*x^2 + b*x + c = 0')
ans =
     [1/2/a*( - b + (b^2 - 4*a*c)^(1/2) )]
     [1/2/a*( - b - (b^2 - 4*a*c)^(1/2) )]
```

In solving for the response in many dynamic systems we need to solve a system of equations for several unknown variables such as the coefficients in the natural response. During the solution of Ex. 2.8, we needed to solve $K_1 + K_2 = 3$ and $-5K_1 - 3K_2 = 299$, which we did using a matrix approach. We can also solve a system of equations using the "solve" command by typing

```
>> syms k1 k2
>> [k1, k2] = solve('k1 + k2 = 3', '−5*k1 − 3*k2 = 299')
    k1 =
        −154
    k2 =
        157
```

There are many other syntaxes available using the "solve" command; use the MATLAB help command to learn about them or consult other references.

The MATLAB command for solving ordinary differential equations is "dsolve". The syntax involves using the capital letter "D" to denote a derivative, "D2" to denote the second derivative, "D3" to denote the third derivative, and so on. The syntax for the argument of this command involves writing the given differential equation and the initial conditions, each separated by a comma and enclosed in a single quote. For example, to solve the following differential equation for $t \geq 0$

$$\ddot{y} + 4\dot{y} + 3y = 0$$

with initial conditions $y(0) = 1$ and $\dot{y}(0) = 0$ using MATLAB's "dsolve", we have

```
>> dsolve('D2y + 4*Dy + 3*y = 0', 'Dy(0) = 0', 'y(0) = 1')
    ans =
        −1/2*exp (− 3*t) + 3/2*exp (− t)
```

To plot these results, we use the MATLAB plotting function called "ezplot", an easy way to plot functions defined by symbolic functions. To plot the previous result, we execute

```
>> ezplot(y, [0, 5])
```

which gives the results shown in Figure A.6.

The argument enclosed in the brackets of the "ezplot" command is the initial and final points.

If the initial conditions are not entered in the "dsolve" command, the solution is calculated in terms of unknown coefficients. To illustrate, consider entering the command "dsolve" without initial conditions,

```
>> dsolve('D2y + 4*Dy + 3*y = 0')
```

which gives us

$$C1*exp (− t) + C2*exp (− 3*t)$$

The values for C1 and C2 are determined from the initial conditions.

To solve

$$\ddot{y} + 16,000\dot{y} + 10^8 y = 5 \times 10^8 t$$

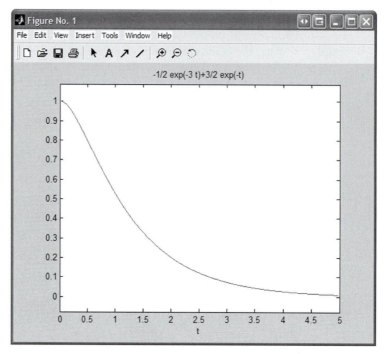

Figure A.6 An illustration of the MATLAB plotting function, "ezplot."

with $\dot{y}(0) = 300$ and $y(0) = -45 \times 10^{-3}$ using MATLAB, we have

$$\gg \mathrm{dsolve('D2y} + 16000*\mathrm{Dy} + 10^{\wedge}8*\mathrm{y} = 5*10^{\wedge}8*\mathrm{t'},$$
$$\mathrm{'Dy(0)} = 300', \mathrm{'y(0)} = -45*10^{\wedge}(-3)')$$

ans $=$

$$-1/1250 + 5*\mathrm{t} - 293/30000*\exp(-8000*\mathrm{t})*$$
$$\sin(6000*t) - 221/5000*\exp(-8000*t)*\cos(6000*t)$$

To solve a set of simultaneous differential equations, we use the command "dsolve" with the argument syntax consisting of the given differential equations and initial conditions, each separated by a comma and enclosed in a single quote. For example, suppose we wish to solve:

$$5\dot{y}_2 + 10\dot{y}_1 + 60y_1 = 300u(t)$$
$$5\dot{y}_2 + 40y_2 + 5\dot{y}_1 = 0$$

With zero initial conditions at $t = 0$. Using MATLAB, we write

$$>> [x, y] = dsolve(`5*Dy + 10*Dx + 60*y = 300',$$
$$`5*Dy + 40*y + 5*Dx = 0', `x(0) = 0', `y(0) = 0')$$

$$x =$$
$$-45 + 45*\exp{(12/17*t)}$$
$$y =$$
$$-5*\exp{(12/17*t)} + 5$$

On occasion, MATLAB produces incorrect or misleading results to the solution of differential equations, so care should be taken in verifying the solution by plotting, checking the initial conditions and final conditions, or simulating the solution with SIMULINK.

A.3 BLOCK DIAGRAMS AND SIMULINK

As discussed earlier, solving a differential equation or a set of simultaneous differential equations involves a considerable amount of work. In the previous section, MATLAB's symbolic math toolbox provided an easier solution method when it worked, but it sometimes failed to produce the correct solution. In this section, we use another MATLAB toolbox called SIMULINK to numerically calculate the solution for linear and nonlinear differential equations. For nonlinear differential equations, SIMULINK is usually the only solution option. To implement a numerical solution in SIMULINK, we draw the system of equations in a block diagram.

A.3.1 Block Diagrams

A block diagram is a graphical representation of the system's differential equations using basic mathematical operations such as constants, gains, summing junctions or summers, and integrators. To deal with nonlinear elements, we can create special blocks that contain mathematical functions or standard functions such as the square root, exponential and logarithmic functions. The basic drawing elements for the block diagram are shown in Figure A.7.

To create a block diagram, we first take the highest derivative term output variable and put it on the left-hand side of the equal sign and the other terms on the right-hand side. Next, the block diagram is formed by connecting elements on the right-hand side to a summing junction representing the equation side, with the output being the left-hand side of the equation. From the summer, we include as many integrator blocks as necessary to reduce the highest order derivative to the output variable (i.e., $\frac{d^3y}{dt^3}$ requires three integrators). Finally, we use a gain block and connect the output variable and lower-order derivatives to appropriate gain blocks back to the summer. If there are any inputs, these are added to the summer using a constant block for a unit step function, or an appropriate function block such as t, or e^{-t}.

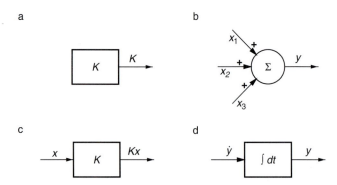

a

b

c

d

Figure A.7   Basic drawing elements in a block diagram: a. constant, b. summer, c. gain, and d. integrator.

Example A.6

Draw a block diagram for $\ddot{y} + 4\dot{y} + 3y = 0$.

Solution

To begin the process of drawing a block diagram, the differential equation is re-arranged as

$$\ddot{y} = -4\dot{y} - 3y$$

and the block diagram is started by drawing the summer as shown in the following figure.

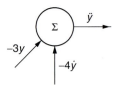

Next, we insert two integrator blocks to the output of the summer to obtain $\dot{y}$ and y as illustrated in the following diagram.

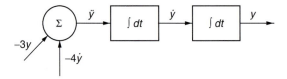

Finally, the block diagram is completed by connecting $\dot{y}$ and y to the summer via a gain block as shown in the following figure.

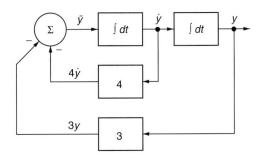

A block diagram can also be drawn for a set of simultaneous differential equations using the same approach as in Ex. A.6, drawing a block diagram for each equation and connecting the output variables and its derivatives to the appropriate summer. ∎

Example A.7

Draw a block diagram for the following set of simultaneous differential equations.

$$3\ddot{y}_1 + 10\dot{y}_1 + 60y_1 + 5\dot{y}_2 = f(t)$$
$$5y_1 + 4\ddot{y}_2 + 5\dot{y}_2 + 40y_2 = 0$$

where $f(t)$ is an unspecified input function.

Solution

We begin by solving for the highest order derivative in each equation.

$$\ddot{y}_1 = \tfrac{1}{3}(-10\dot{y}_1 - 60y_1 - 5\dot{y}_2 + f(t))$$
$$\ddot{y}_2 = \tfrac{1}{4}(-5y_1 - 5\dot{y}_2 - 40y_2)$$

The block diagram is drawn by first using two summers, with $\ddot{y}_1$ and $\ddot{y}_2$ as the output of each summer, followed by two integrators for each output variable. Next, we complete the diagram by including the terms on the right-hand side of the previous equations as the input to each summer with appropriate gain elements, giving us the following figure. ∎

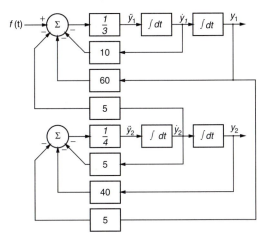

A.4 SIMULINK

SIMULINK is a toolbox in MATLAB that simulates linear and nonlinear, continuous and discrete dynamic systems. We program a system from the block diagram as described in the previous section using SIMULINK by dragging and dropping blocks and connecting the blocks together according to the system equation or equations. After building the model we run the program to plot results or export the results to MATLAB.

To start SIMULINK, start MATLAB as before, then type "simulink" in lower case in the command window. This brings us to another window containing the icons for implementing a SIMULINK simulation as shown in Figure A.8. Next, select "File" on the menu bar, then "New" and "Model". This brings up another window, a workspace, that is used to create the SIMULINK model as shown in Figure A.9. The icons shown make up the basic subsystem of blocks used to create a simulation, with each providing a set of blocks arranged into libraries. The libraries are a collection of blocks grouped into similar functions.

Double-clicking on any of the icons opens up the library of blocks. For example double-clicking on the "Math Operations" library opens the window shown in Figure A.10 with common operations such as absolute value (Abs), gain, math function, etc. The library "Continuous" contains the blocks for the derivative and integrator among other blocks. The library "Sources" contains the blocks for a constant, clock, step, etc., all blocks that produce an output without an input. The block "clock" outputs the current simulation time as a variable, useful in decision blocks and outputting variables to MATLAB. The library "Sinks" contains the blocks associated that have an input but no output, such as the scope that displays a variable as a function of time as if on an oscilloscope. Another important block in this library is the block "To Workspace", which allows exporting a variable to MATAB for further analysis and plotting.

Blocks are copied from the "SIMULINK Library Browser" window to the workspace window in Figure A.9 by simply dragging them across from one to the other. Each block is given a unique name below the block once it is placed in the workspace. If a block is repeated, its name is the generic block name followed by an integer. There is also a block identifier placed inside the block. The block name can be edited by placing the mouse pointer on the name, clicking the left mouse button, and then editing the name. Once in the workspace, the block's behavior is set by double-clicking the block, which opens another window where parameters are specified. The block size can be changed by selecting the block with the mouse pointer, and then, while the left mouse button is held down on the handle, drag the handle to its new size. To remove a block, simply select and delete it. For convenience, one can use the copy and paste commands to reproduce a block.

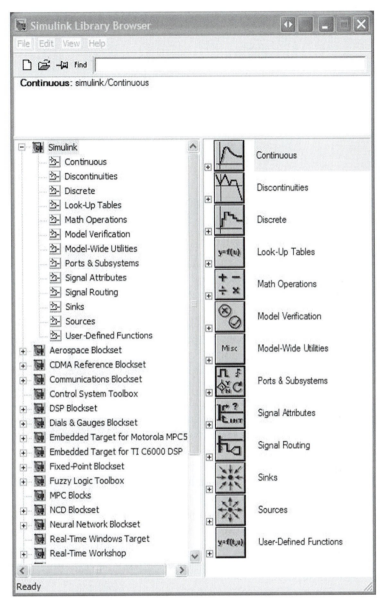

Figure A.8 SIMULINK Library of blocks on the right. Other MATLAB toolboxes are shown on the left.

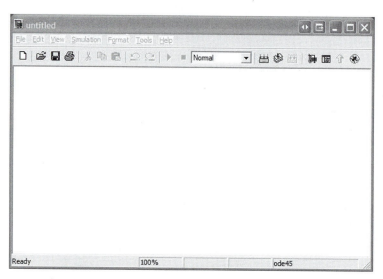

Figure A.9   SIMULINK workspace used to create the model.

Blocks are connected together with line segments. To connect two blocks, move the mouse pointer to the angle bracket (>) on the output of one block, hold the left mouse button down, and then drag the pointer to the input < of another block. If the line segment is drawn incorrectly, use the mouse to select, and then delete it. The line segment can be manipulated by selecting it and while the left mouse button is held down, drag the line segment to its desired shape. One can also break into a line segment by holding the mouse pointer over the line segment with the "CTRL" key pressed down, click the left mouse button and drag the pointer to the desired location.

Shown in Figure A.11 (Top) is the workspace that consists of a sine wave, integrator and scope, connected together as the output of the sine wave to the input of the integrator to the output of the integrator to the input of the scope. Double-clicking on the sine wave reveals its properties as shown in Figure A.11 (Middle). Double-clicking on the scope block results in a graph shown in Figure A.11 (Bottom). The scope is an excellent debugging tool to verify all simulation outputs are correct, but other graphing capabilities in MATLAB and EXCEL provide far superior plots for professional reports.

Once a SIMULINK model is defined, a simulation of the model is run whereby all outputs and inputs are computed from the simulation start time to the end time for each time period. Before running the simulation, the simulation parameters need to be specified by clicking "Simulation" on the menu and then selecting the pull down "Simulation Parameters". Shown in Figure A.12 is the "Simulation Parameter" window with default parameters. The default values are usually acceptable for many simple applications. Obviously, the "Stop time" should be set to be at least 5 time constants (dominant time constant). The start time can be set to a value appropriate for the simulation. The default simulation algorithm (ode45, the Dormand-Prince

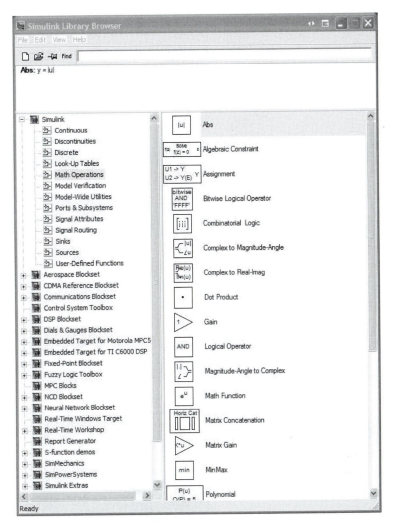

Figure A.10   The "Math Operations" library.

formula) is usually adequate for most applications. This algorithm is the best first choice as it involves calculating the output value at the next state based only on the previous output value.

Simulations can be run with either a fixed-step solver or a variable-step solver. The fixed-step solver calculates the model at equally spaced time steps, beginning at the start time. The step size must be small enough so that the approximations are valid, but not too small so that the program takes too long to run. A good rule of thumb is to start at

$$\text{step size} = \frac{\text{Total simulation time}}{500 \text{ to } 5000}$$

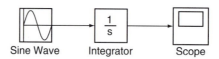

Sine Wave   Integrator   Scope

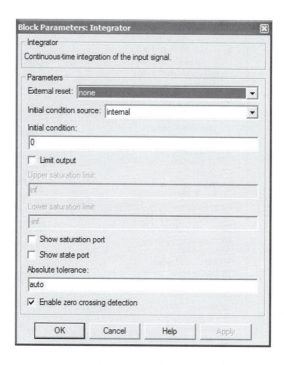

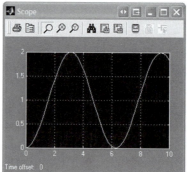

Figure A.11 (Top) SIMULINK blocks. (Middle) Block parameters for the integrator block. (Bottom) The output of the scope block.

and then to reduce the step size by 10. If the simulation output does not change, retain the step size, otherwise continue to reduce the step size by 10 until the output stabilizes. The variable-step size adjusts the step size, decreasing the size when the

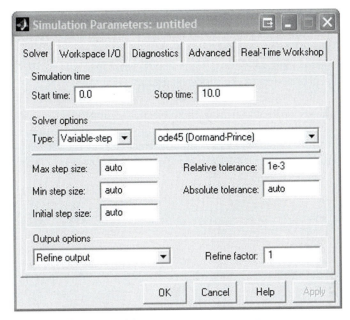

Figure A.12   Simulation Parameters Window.

outputs are changing rapidly and increasing the step size when the outputs are changing slowly. The variable-step size simulation often saves simulation calculation time resulting in a faster simulation. To run the simulation, select the command "Start" from the "Simulation" menu or the run command icon.

After creating a SIMULINK model, use "Save" or "Save As" from the "File" menu to save the model. Many simulation models have a number of parameters, some of which are changed from simulation to simulation. For convenience, it makes sense to define the parameters in MATLAB (i.e., K=5, M=3, etc.), with the parameter values saved in an m-file. The m-file can then be opened in MATLAB and these parameters can then be used in the SIMULINK model. To print the SIMULINK model, select the "Print" command from the "File" menu, or select the "Copy Model to Clipboard" command from the "Edit" menu and paste it into another application like Microsoft Word.

Example A.8

Simulate the response of the system described in Ex. A.7 with zero initial conditions if the input $f(t)$ is a pulse with magnitude 120 and duration 10 applied at $t = 0$.

Solution

Whenever we create a SIMULINK simulation, it is best to begin by drawing the block diagram for the system. In Ex. A.7, we already created the block diagram where the highest derivative was first solved in each equation and all of the other terms were placed on the right-hand side of the equation. The block diagram was then constructed using a summer for the equal sign, the integrator to remove the derivative terms, and gains for the feedback elements. The SIMULINK model is essentially a reproduction of the block diagram.

To create the SIMULINK model, we open a SIMULINK session, open a SIMULINK workspace, and then open the appropriate library to begin dragging and dropping the icons to create the model. The final SIMULINK model is shown in the following figure. Details on how to create the model are provided in the remainder of the example.

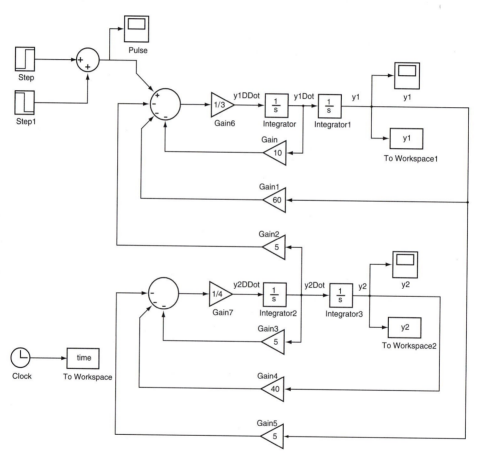

To begin drawing the model, first drag the summer, which is located in the "Math Operations" library. The summer icon looks like ⊕. This block can be changed to a rectangle from the default circle by double-clicking the block, which opens the "Block Parameters Sum" window. Also, the number of inputs can be increased by adding additional "pluses" or "minuses" in the list of signs. For the first equation, we have four elements connected to the summer (a positive for the input and three negative feedback elements), so the summer is modified by removing one positive sign and adding three negative signs, giving ⟳.

The input is a pulse with magnitude 120 and duration of 10. To create a pulse, we use two step inputs from the Library "Sources". The step input block looks like ⬛ to enter the magnitude and duration, double-click on the block to open the "Block Parameters Step" and enter for the first block 0 for "Step Time", 0 for the "Initial value" and 120 for the "Final value". The second step function uses the values 10 for the "Step Time", 0 for the "Initial value" and −120 for the "Final value". Another summer with line segments from the output of the step to the input of the summer is used to connect the two inputs. The output of the input summer is connected to the "+" input of the first summer.

After the summer, a gain of 1/3 is needed. The gain block looks like ▷, which is found in the "Math Operations" library. After dragging it from the library, the magnitude is set by double-clicking and entering a value of 1/3. The number inside the gain block is the value of the gain if it can fit, otherwise "-k-" is used. A line segment is used to connect the summer to the gain.

Next, two integrators are dragged across to the workspace from the "Continuous" library. The integrator has the symbol $\frac{1}{s}$ in it that comes from the Laplace transform representation of integration. While not necessary in this example, integrators sometimes have nonzero initial conditions. To enter the initial condition for an integrator, double-click on the integrator block, which opens the "Block Parameters Integrator" window, where the initial condition can be entered. By default, the initial condition for the integrator are zero. The output of the summer is connected with a line segment to the input of the first integrator, the output of the first integrator is connected to the input of the second integrator, and the output of the second integrator is connected to the input of the third integrator. It is generally good practice to label the output of each integrator. A label can be entered by clicking anywhere in the open workspace. A convenient label for $\ddot{y}_1$ is "y1DDot" (borrowed from the D-operator), $\dot{y}_1$ is "y1Dot" and so on.

Three gain blocks are then dragged across to the workspace from the "Math Operations" library. To rotate the gain block so that it is in the correct orientation, select it and press the "CTRL and R" keys together twice for a 180° counterclockwise rotation (each click rotates 90°). The value for each gain block is entered by double-clicking and entering 10, 60, and 5. Line segments are used to connect the output of the integrators to the gain block by moving the mouse pointer over the integrator output line segment, pressing "Ctrl" and left clicking the mouse (to break into the line), and then moving the mouse pointer to the input of the gain block. The output of the gain blocks are then connected to the summer. Note that the term $\dot{y}_2$ is not connected to the gain block until the second equation is drawn.

The same steps are used to create the SIMULINK model for the second equation. Notice that the input for Gain2 block is $\dot{y}_2$, and the input for Gain5 is y_1.

To view the output, two scopes are used from the "Sinks" Library. For clarity the labels on each scope are renamed to "y1" and "y2" by double-clicking on the labels below the scope and making the change. Double-clicking on the scope icon opens a new window that contains a graph for the output variable. A scope is also drawn for the pulse input as good practice to insure that the input is correctly specified. The binocular icon on the menu of the scope graph auto scales the graph.

The "Clock" block from the "Sources" library is used to output the simulation time at each simulation step. To move the simulation data from SIMULINK to the MATLAB workspace for further analysis and plotting, the "To Workspace" block is used from the "Sinks" library. Three "To Workspace" blocks are used in the example to output data for the variables y_1, y_2 and *Time*. The "To Workspace" writes the data to an array specified by the block's "Variable name". By default, the "To Workspace" uses the "Variable name" "simout" for the first, "simout2" for the second and so on. These names are easily renamed by double-clinking on the block and entering meaningful names, such as y1, y2 and Time in this example. While the block is open, the "Save format" should be changed to "Array" and the "Sample Time" to a number that provides enough data points. In this example, the simulation time is 20, so the sample time is set at .2 to provide 100 data points for each variable in the MATLAB workspace. Once in the workspace, one can manipulate or plot the variables, i.e., using the MATLAB command "plot(Time,y1)".

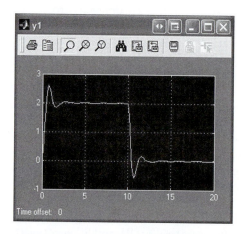

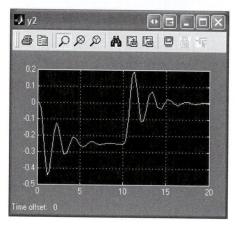

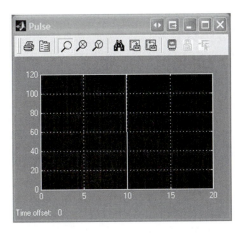

The final step is to open the "Simulation Parameters" to change the simulation "Stop time" to 20 and click the run button. The preceding figure displays the output of the three scopes.

In some simulations, a model's performance is evaluated by changing parameter values. For the model in this example, the block parameters would need to be changed individually. For a model with hundreds of parameters, this would entail a considerable effort. It is far easier to enter the parameter values in MATLAB and use the parameter names in the blocks rather than values. For instance, enter the following in MATLAB

```
>> PulseGain = 120;
>> PulseDuration = 10;
>> Gain = 10;
>> Gain1 = 60;
>> Gain2 = 5;
>> Gain3 = 5;
>> Gain4 = 40;
>> Gain5 = 5;
>> Gain6 = 1/3;
>> Gain7 = 1/4;
```

Then in each gain block, use the parameter name rather than the value. Any changes in the MATLAB values are automatically updated in SIMULINK. Defining initial conditions for integrator blocks can also be done in MATLAB. To avoid entering the parameter values each time MATLAB is opened, store the values in an m-file. ∎

It is often convenient to use subsystems to collect blocks together when drawing a SIMULINK model to make a model more readable. In the previous example, for instance, the pulse input is easily drawn as a subsystem by selecting the blocks by collecting them in a bounding box with the mouse pointer (i.e., use the mouse to draw a box around the elements), right-clicking the mouse and then selecting "Create subsystem". The inputs and the outputs to the subsystem are those of the elements in the selected blocks. Shown in Figure A.13 is part of the model from the previous example using a subsystem for the pulse. Double-clicking the subsystem block opens a new window with the blocks in the subsystem with the output from the subsystem indicated by the Out1 block.

Two blocks that are useful in simulations are the "Math Function" and "Fcn". The "Math Function" from the "Math Operations" library is convenient for entering mathematical functions such as the exponential, logarithm and power functions from a drop down list. For example, with the input variable name is "u" and the function "exp", the output is "exp(u)". The label within the block is the math function. The "Fcn" from the "User-Defined Functions" library is convenient for entering general expressions. Once again, the input variable name is "u". Any arithmetic operator $(+ - */)$ and mathematical function (cos, sin, tan, exp, power, ...) can be used in the block, as well as any parameter from MATLAB. For instance, in writing the potassium conductance for a membrane, the expression $\bar{g}_k n^4$ is needed, where $\bar{g}_k$ is a constant and n is a voltage and time dependent rate constant. In SIMULINK, this expression is written using the "Fnc" block with the parameter "Gkbar*u^4" where "Gkbar" is defined in MATLAB and "u" the input to the block representing n.

In modeling systems, some parameters may change based on threshold conditions. For example, in the saccadic eye movement system, the input to the agonist muscle, F_{ag}, is a low-pass filtered plus-step waveform given by the following equation.

$$\dot{E}_{ag} = \frac{N_{ag} - F_{ag}}{\tau_{ag}}$$

where N_{ag} is the neural control input (pulse-step waveform), and

$$\tau_{ag} = \tau_{ac}(u(t) - u(t - t_1)) + \tau_{de}u(t - t_1)$$

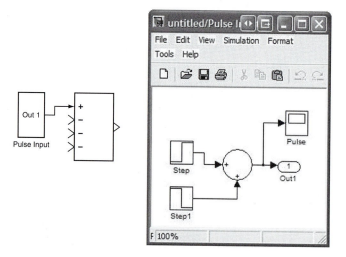

Figure A.13 (Left) A subsystem for the pulse input from the previous example. (Right) Double-clicking the subsystem block opens a new window that displays the blocks within the subsystem.

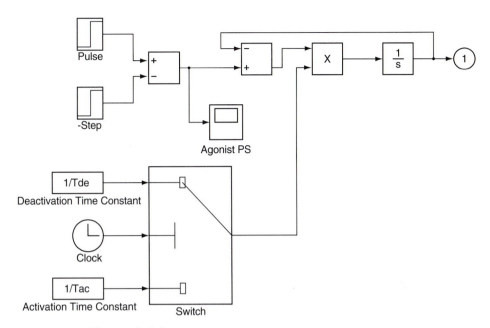

Figure A.14   A low-pass filter with a time varying time constant.

where τ_{ag} is the time-varying time constant made up of τ_{ac} and τ_{de}, the activation and deactivation time constants. To simulate this subsystem we use the "Switch" block in the "Signal Routing" library as shown in Figure A.14. The "clock" block is used to monitor the simulation time. At the start of the simulation, the time constant τ_{ac} is used. When the simulation time is greater than t_1, threshold is reached and the switch changes to the time constant τ_{de}. The remainder of the blocks in Figure A.13 solves for F_{ag}.

For information on the other operations in SIMULINK, the reader is encouraged to consult the help menu.

INDEX